en or b

HUMAN GROWTH

2
Postnatal Growth

HUMAN GROWTH

HUMAN GROWTH

2
Postnatal Growth

Edited by

Frank Falkner
The Fels Research Institute
Wright State University School of Medicine
Yellow Springs, Ohio

and

J. M. Tanner
Institute of Child Health
London, England

BAILLIÈRE TINDALL · London

Library of Congress Cataloging in Publication Data

Main entry under title:

Human growth.

Includes bibliographies and index.
CONTENTS: v. 1. Principles and prenatal growth. – v. 2. Postnatal growth. – v. 3. Neurobiology and nutrition.
1. Human growth–Collected work. I. Falkner, Frank Tardrew, 1918-   II. Tanner, James Mourilyan. [DNLM: 1. Growth. 2. Gestational age. WS103 H918]
QP84.H76   612'5   78-1440
ISBN 0-306-34462-9 (v. 2)

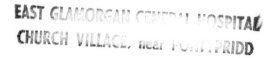
A BAILLIÈRE TINDALL book published by Cassell Ltd.
35 Red Lion Square, London WC1R 4SG
and at Sydney, Auckland, Toronto, Johannesburg
an affiliate of Macmillan Publishing Co. Inc., New York

HUMAN GROWTH

Volume 1   Principles and Prenatal Growth   ISBN 0 7020 0731 5
Volume 2   Postnatal Growth   ISBN 0 7020 0732 3
Volume 3   Neurobiology and Nutrition   ISBN 0 7020 0733 1

©1978 Plenum Press, New York
A Division of Plenum Publishing Corporation
227 West 17th Street, New York, N.Y. 10011

First published 1978

Printed in the United States of America

Contributors

DONALD A. BAILEY
Professor
College of Physical Education
University of Saskatchewan
Saskatoon, Canada

FREDERICK C. BATTAGLIA
Professor and Chairman
Department of Pediatrics
University of Colorado School of
   Medicine
Denver, Colorado

INGEBORG BRANDT
University Children's Hospital
University of Bonn
Bonn, Germany

JO ANNE BRASEL
Associate Professor of Pediatrics
Director, Division of Growth and
   Development
Institute of Human Nutrition and
   Department of Pediatrics
Columbia University College of
   Physicians and Surgeons
New York, New York

C. G. D. BROOK
Consultant Pediatrician
The Middlesex Hospital
London, England

NOËL CAMERON
Lecturer in Human Auxology
Department of Growth and
   Development
Institute of Child Health
London, England

ARTO DEMIRJIAN
Director, Growth Center
Université de Montréal
Montréal, Quebec, Canada

GILBERT B. FORBES
Professor of Pediatrics and of
   Radiation Biology and Biophysics
The University of Rochester, School
   of Medicine and Dentistry
Rochester, New York

RHODA K. GRUEN
Institute of Human Nutrition and
   Department of Pediatrics
Columbia University College of
   Physicians and Surgeons
New York, New York

MELVIN M. GRUMBACH
Professor of Pediatrics
University of California San Francisco
San Francisco, California

MALCOLM A. HOLLIDAY
Professor of Pediatrics
The Children's Renal Center,
   Department of Pediatrics
University of California San Francisco
San Francisco, California

HARRY ISRAEL, III
Chief, Dental Research Section
Children's Medical Center
Dayton, Ohio

FRANCIS E. JOHNSTON
Professor of Anthropology
University of Pennsylvania
University Museum
Philadelphia, Pennsylvania

JEROME L. KNITTLE
Professor of Pediatrics
Director, Division of Nutrition and
   Metabolism
Mt. Sinai School of Medicine
New York, New York

ROBERT M. MALINA
Professor
Department of Anthropology
University of Texas
Austin, Texas

W. A. MARSHALL
Professor of Human Biology
Department of Human Sciences
Loughborough University
Loughborough, Leicestershire,
   England

ROY L. RASMUSSEN
Assistant Professor
Department of Physical Education
St. Francis Xavier University
Antigonish, Nova Scotia, Canada

ALEX F. ROCHE
Senior Scientist
The Fels Research Institute
   and Department of Pediatrics
Wright State University School of
   Medicine
Yellow Springs, Ohio

MICHAEL A. SIMMONS
Departments of Pediatrics and
   Obstetrics
The Johns Hopkins University School
   of Medicine
Baltimore, Maryland

PATRICK G. SULLIVAN
Reader in Orthodontics
London Hospital Medical College
Dental School
London, England

J. C. VAN WIERINGEN
University Children's Hospital
Het Wilhelmina Kinderziekenhuis
Utrecht, The Netherlands

JEREMY S. D. WINTER
University of Manitoba
Endocrine–Metabolism Section
Health Sciences Center
Winnipeg, Manitoba, Canada

Preface

Growth, as we conceive it, is the study of change in an organism not yet mature. Differential growth creates form: external form through growth rates which vary from one part of the body to another and one tissue to another; and internal form through the series of time-entrained events which build up in each cell the specialized complexity of its particular function. We make no distinction, then, between growth and development, and if we have not included accounts of differentiation it is simply because we had to draw a quite arbitrary line somewhere.

It is only rather recently that those involved in pediatrics and child health have come to realize that growth is the basic science peculiar to their art. It is a science which uses and incorporates the traditional disciplines of anatomy, physiology, biophysics, biochemistry, and biology. It is indeed a part of biology, and the study of human growth is a part of the curriculum of the rejuvenated science of Human Biology. What growth is not is a series of charts of height and weight. Growth standards are useful and necessary, and their construction is by no means void of intellectual challenge. They are a basic instrument in pediatric epidemiology. But they do not appear in this book, any more than clinical accounts of growth disorders.

This appears to be the first large handbook—in three volumes—devoted to Human Growth. Smaller textbooks on the subject began to appear in the late nineteenth century, some written by pediatricians and some by anthropologists. There have been magnificent mavericks like D'Arcy Thompson's *Growth and Form*. In the last five years, indeed, more texts on growth and its disorders have appeared than in all the preceeding fifty (or five hundred). But our treatise sets out to cover the subject with greater breadth than earlier works.

We have refrained from dictating too closely the form of the contributions; some contributors have discussed important general issues in relatively short chapters (for example, Richard Goss, our opener, and Michael Healy); others have provided comprehensive and authoritative surveys of the current state of their fields of work (for example, Robert Balázs and his co-authors). Most contributions deal with the human, but where important advances are being made although data from the human are still lacking, we have included some basic experimental work on animals.

Inevitably, there are gaps in our coverage, reflecting our private scotomata, doubtless, and sometimes our judgment that no suitable contributor in a particular field existed, or could be persuaded to write for us (the latter only in a couple of instances, however). Two chapters died on the hoof, as it were. Every reader will notice the lack of a chapter on ultrasonic studies of the growth of the fetus; the manuscript, repeatedly promised, simply failed to arrive. We had hoped, also, to include a chapter on the very rapidly evolving field of the development of the visual processes, but here also events conspired against us. We hope to repair these omissions in a second edition if one should be called for; and we solicit correspondence, too, on suggestions for other subjects.

We hope the book will be useful to pediatricians, human biologists, and all concerned with child health, and to biometrists, physiologists, and biochemists working in the field of growth. We thank heartily the contributors for their labors and their collective, and remarkable, good temper in the face of often bluntish editorial comment. No words of praise suffice for our secretaries, on whom very much of the burden has fallen. Karen Phelps, at Fels, handled all the administrative arrangements regarding what increasingly seemed like innumerable manuscripts and rumors of manuscripts, retyped huge chunks of text, and maintained an unruffled and humorous calm through the whole three years. Jan Baines, at the Institute of Child Health, somehow found time to keep track of the interactions of editors and manuscripts, and applied a gentle but insistent persuasion when any pair seemed inclined to go their separate ways. We wish to thank also the publishers for being so uniformly helpful, and above all the contributors for the time and care they have given to making this book.

Frank Falkner
James Tanner

Yellow Springs and London

Contents

Chapter 4

Somatic Growth of the Infant and Preschool Child

Francis E. Johnston

Chapter 5

Body Composition and Energy Needs during Growth

Malcolm A. Holliday

Chapter 6

Puberty

W. A. Marshall

Chapter 7

Prepubertal and Pubertal Endocrinology

Jeremy S. D. Winter

Chapter 8

The Central Nervous System and the Onset of Puberty

Melvin M. Grumbach

Chapter 9

Body Composition in Adolescence

Gilbert B. Forbes

Chapter 14

Skull, Jaw, and Teeth Growth Patterns

    Patrick G. Sullivan

Chapter 15

Dentition

    Arto Demirjian

Chapter 16

Secular Growth Changes

    J. C. van Wieringen

Contents of Volumes 1 and 3

V
POSTNATAL GROWTH

1

Cellular Growth: Brain, Liver, Muscle, and Lung

JO ANNE BRASEL and RHODA K. GRUEN

1. Introduction

The growth of tissues which are not self-renewing occurs by a combination of increase in the number and increase in the size of the component cells. Work in animals (Enesco and Leblond, 1962; Winick and Noble, 1965) in the last 15 years has shown that initially cell division and increase in number predominate; later the rate of cell division slows and increase in size begins. When cell division ceases, cells continue to grow in size causing further tissue enlargement until maturity is reached, when no further net increase in number or size occurs. These conclusions were based on the measurement of total tissue DNA and protein content. Since the DNA per diploid nucleus is constant within any one species and since essentially all cellular DNA is chromosomal (Boivin *et al.,* 1948; Mirsky and Ris, 1949), the tissue DNA content is a reflection of growth in cell number. Such a biochemical measure of growth does not define which type of cell is increasing in number, and in tissues of multiple cell types its usefulness is limited in the absence of accompanying histologic information. In tissues where polyploidy or multinucleated cells develop during the growth process, total DNA content no longer reflects growth in cell number if a cell is to be strictly defined by membrane boundaries. It is, nonetheless, a reflection of the extent of DNA synthesis during growth and is an accurate reflection of the formation of "diploid cell units," which, as discussed later, has some real biological significance.

Growth in cell mass can be assessed by calculation of the ratio of weight to DNA or of protein to DNA. Since the former may be affected by changes in hydration unrelated to changes in cell size, the latter is more commonly used. By this criterion any increase in protein content out of proportion to the increase in DNA content over time indicates growth in the average cell size or more precisely, growth per "diploid cell unit." Again, in tissues of mixed cell types the ratio should be combined with histologic data for a more complete picture of cellular growth.

JO ANNE BRASEL and RHODA K. GRUEN • Institute of Human Nutrition, Department of Pediatrics, Columbia University College of Physicians and Surgeons, New York, New York.

*JO ANNE BRASEL AND
RHODA K. GRUEN*

Although the growth phases of increase in cell number alone (hyperplasia), combined increase in number and size (hyperplasia and hypertrophy), and cell size increase alone (hypertrophy) are common to all the non-self-renewing tissues, the duration of the phases varies considerably among the tissues. Some tissues, such as brain, reach a full adult complement of DNA early in postnatal life; others, like skeletal muscle, do not reach adult values of DNA until after adolescence. In later discussion the importance of these differences in timing will be amplified, and data will be presented to support the hypothesis that the cellular response of a tissue or organ to a growth stimulus varies depending on the phase of cell growth it is in when the stimulus is applied.

In studies of animal growth it is a simple enough matter to remove and analyze an entire tissue for DNA and protein content. This is not possible for the living human, yet this report will deal almost exclusively with human data. These data are relatively small in number and come from the following sources: aborted fetuses, autopsy material from normally grown subjects who experienced sudden, fatal injury or disease, and biopsy material in which the tissue was felt to be homogeneous in composition and whose total mass can be indirectly measured. Animal studies will be cited only to amplify or further strengthen concepts developed from the human work. For each tissue selected for description, certain recent papers, especially of a review nature, will be cited. Specific individual citations will be made only in relation to quoted data, usually numerical in form. This chapter is not meant to be an extensive review of the subject of cellular growth in the human, but rather an overview to provide a broad understanding.

2. Brain

Data on cellular growth of the human brain have come largely from three laboratories: those of Dobbing, Howard, and Winick. The first data to appear were those of Winick (1968) in which 31 whole brains were analyzed. The 31 subjects, ranged in age from 13 weeks gestation to 13 postnatal months and were felt to be normal or without central nervous system defects. These data suggested that growth in DNA ceased at 5–6 months of age at a value of 900–1000 mg, but only nine specimens older than 6 months were analyzed. Increases in weight and protein remained linear up to 13 months of age with no indication that growth was about to cease.

In 1969 Howard *et al.* (1969) published data on 28 fetal specimens aged 10–31 weeks gestation. Cerebrum and cerebellum were studied separately; the description of the dissection suggests that the cerebrum consisted of the cerebral hemispheres plus the brain stem. Therefore the two values have been summed to obtain total brain DNA. These data are in excellent agreement with the prenatal specimens studied by Winick. The DNA increase from the 10th through the 13th week in the cerebrum was extraordinary; DNA content rose nearly tenfold, from 3.55 to 28.6 mg. The equation for the line was determined to be: log cerebral $DNA_{10-13} = -2.1057 + (0.25978 \times age)$. Thereafter the rate of DNA accumulation declined and the equation, becoming linear, was: cerebral $DNA_{14-31} = -175.31 + (14.727 \times age)$. The DNA rose from 33.0 mg at 14 weeks to 264 mg at 31 weeks. The cerebellar DNA increased at the same logarithmic rate from 10 to 31 weeks; the equation was: log cerebellar $DNA = -1.4638 + (0.090936 \times age)$. The DNA rose from 0.383 to 22.2 mg over the time period studied.

In a 1969 study of normal and malnourished subjects, Winick and Rosso (1969) reported data on whole brain DNA from ten additional control subjects, ranging in age from 15 weeks gestation to 11½ postnatal months, which completely overlap the data previously reported (Winick, 1968). In 1970 Winick *et al.* (1970) reported studies of postnatal regional growth in brains from control and malnourished children. The first discrepancy in the brain cell growth story arises with these data. Contrary to the earlier report, total brain DNA continued to increase to at least a year of age and perhaps beyond to 18 months, especially in cerebellum. At two years of age 1000 mg of DNA were present in cerebrum, 500 mg in cerebellum, and 25 mg in brain stem. These values were 50% greater than those of the earlier report (Winick, 1968); even the one value for a 6-month control subject (1187 mg) was approximately 30% above the previous 6-month value of 900 mg. The level of 1500 mg was also approximately 50% greater than other 1- to 2-year values to be cited later. The reason(s) for these discrepancies are unknown, but since some of the analyses were carried out on lyophilized samples rather than on fresh or frozen specimens, methodological differences may play a role.

In 1970 the first paper from Dobbing and Sands (1970) appeared; it reported data from 76 specimens from 10 weeks of gestation to one year of age. The forebrain, cerebellum, and brain stem were analyzed separately but only the summed results were presented in this report; additionally the data were reported as micromoles of DNA-P, rather than total mg of DNA, which means the results, unfortunately, cannot be compared to the data of Howard *et al.* (1969) or Winick (1968). In 1973 Dobbing and Sands (1973) extended the initial series to a total of 139, consisting of 29 fetal brains of 10–22 weeks gestational age, 36 fetal brains from 25 weeks to term, 65 brains from birth to 7 years and nine brains from 10 to 78 years. The data for the three brain regions and whole brain were again reported as DNA-P. These investigators concluded that there were two periods of rapid growth in total brain DNA (Figure 1) (Dobbing, 1970). One occurred at approximately 18 weeks of gestation and was felt to represent neuronal multiplication; the second occurred approximately three months after birth and was felt to be due to glial replication. The biphasic nature of the DNA velocity curve is in distinct contrast to the single growth peak of velocity seen in brain weight or in cholesterol accumulation, an index of myelination. It is interesting to note that the data of Kapeller-Adler and Hammad (1972) show a fall in DNA concentration and a rise in the protein-to-DNA ratio in whole brain at the 17- to 20-week data point, which may reflect initial growth in mean cell size once the spurt in neuronal division is over. When one examines the patterns of growth, increases in DNA in whole brain plateau at approximately 18 months of age; the cessation is a little earlier in cerebellum and a little later in forebrain. The cerebellum grows particularly rapidly postnatally, and although it is much smaller in weight than the forebrain, it contains more than half as many cells as the forebrain. Thus Dobbing and Sands (1973) agree with Winick *et al.* (1970) that increase in DNA continues into the second year of life and that cerebellum contains half as many cells as cerebrum; but they disagree with the patterns of regional growth since cerebellar DNA reached plateau levels before forebrain in their study. Since no numerical data are given and since the figures are plotted as DNA-P, no direct comparisons of the absolute levels can be made between the two studies.

There are some further data which add to the confusion. Chase *et al.* (1974) have reported DNA values for five control brains from subjects ranging in age from 11 to 22 months. The mean total DNA was 936 mg, a figure close to the first reported plateau level of Winick (1968). However, the cerebellar DNA content (431 mg)

JO ANNE BRASEL AND
RHODA K. GRUEN

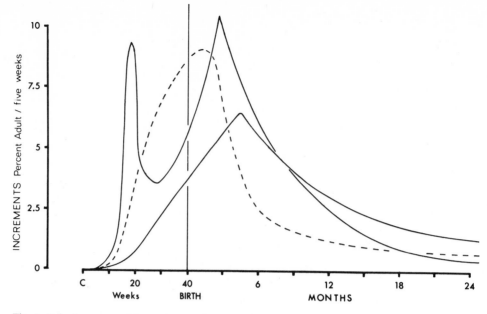

Fig. 1. Velocity curves of the various components of human brain growth. The incremental growth in DNA (two-peaked solid line) shows both a prenatal and postnatal peak in contrast to brain weight (dashed line) and cholesterol (one-peaked solid line) which show a single postnatal peak. (Reprinted with permission from Dobbing, 1970.)

agreed more closely with the later report (Winick *et al.*, 1970). Additionally nearly half of the total DNA content was located in the cerebellum, rather than one third as noted by the other workers.

Cheek (1973, 1975) has wrestled with these contradictions and has utilized a computer to reevaluate the Dobbing and Sands data. He has converted the micro-moles of DNA-P to milligrams of DNA, but the values for the molecular weight of human DNA and the percent P of human DNA needed to make this conversion are not cited. The computer derived a linear regression with an r of 0.95 for the data. Unfortunately there is no statement as to the age range which the regression equation describes, but below 13 weeks of gestation the equation predicts a negative DNA content and in the published figure (Cheek, 1973) the line stops at 20 weeks postnatally. The prenatal data of Howard *et al.* (1969), Winick (1968), and Winick and Rosso (1969) agree well with results expected from the Cheek regression line. Postnatally, the values of Winick and co-workers are higher, especially those of Winick *et al.* (1970) where lyophilized brains were analyzed. Cheek (1975) has concluded that the data of Dobbing and Sands are correct. We have also replotted the data (Figure 2) and are unable to draw as firm a conclusion. The Howard *et al.* (1969) data were calculated from the equations cited in the publication, summed to obtain a value for whole brain, and plotted for 10, 11, 12, 13, 14, 16, 18, 20, 25, and 31 gestational weeks. The data of Dobbing and Sands (1973) were calculated from the Cheek (1973) equation and plotted at 13, 14, 16, 18, 20, 23, 28, 30, 35, and 40 gestational weeks plus 4, 10, 16, and 20 weeks postnatally. Individual data points from the early Winick reports (Winick, 1968; Winick and Rosso, 1969) are also plotted. The later very high values (Winick *et al.*, 1970) have been omitted. The mean value of Chase *et al.* (1974) for five infants between 11 and 22 months of age have also been recorded. Thus the trends in the data from the various groups can be examined, but each data point does not carry equal weight in terms of the number of analyses it represents. Unfortunately the values used to convert micromoles of DNA-P to milligrams of DNA cannot be critically evaluated, and it would not require much of a change to shift the Dobbing and Sands data slightly to the left. Therefore we feel that the only statements which can be made are: (1) Most neuronal cell division, at least in the cerebrum, occurs prenatally, largely during the second trimester. (2) Glial cell division is largely a postnatal event, with peak replication at three months of age. (3) Growth in DNA in whole brain and its two major regions continues into the second year of postnatal life. The time of plateauing of this growth is not precisely known, but probably is approximately 18 months of age. (4) In the nearly mature brain the cerebellum is densely packed with cells

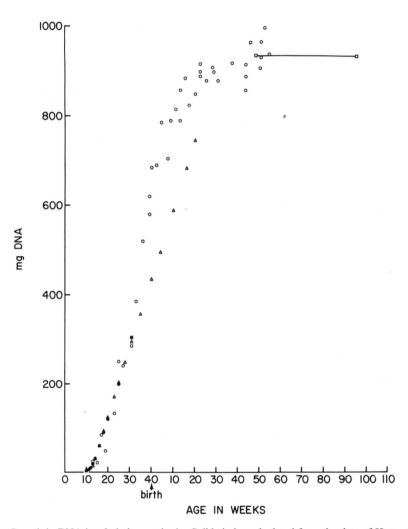

Fig. 2. Growth in DNA in whole human brain. Solid circles calculated from the data of Howard *et al.* (1969); triangles calculated from the data of Dobbing and Sands (1973); using the equation derived by Cheek (1973); squares connected by the horizontal line represent the mean value reported by Chase *et al.* (1974); open circles are individual points of data from Winick (1968) and Winick and Rosso (1969). See text for explanation of the figure.

JO ANNE BRASEL AND
RHODA K. GRUEN

and contains about 50% as much DNA as the cerebrum. (5) The final amount of total DNA attained is unknown but may be between 1000 and 1500 mg.

However, even if precise and accurate information on the patterns of DNA growth in the brain are obtained, its usefulness is limited. Because brain consists of many regions, each containing numerous different cell types and undergoing varying rates of cell division, migration, and maturation, a decrease in the total DNA content, even of a specific region, provides no insight into what type of cellular abnormality has occurred. Therefore, it is particularly important in the study of brain pathology to combine careful histological studies with biochemical analyses.

3. Liver

Mammalian liver grows by increase in the number and size of its cells and by forming multinucleate cells and other cells with single nuclei containing tetraploid and octaploid amounts of DNA. Human liver growth is no exception and several studies of the ploidy development in humans have been reported (Leuchtenberger et al., 1954; Swartz, 1956; Ranek et al., 1975). The data of Swartz (1956) reveal that until 6 years of age the liver contains essentially only diploid (2N) nuclei; between 6 and 10 years mitotic figures are noted in liver tissue and a rare tetraploid (4N) cell may be encountered. During the pubertal years a definite tetraploid class of cells is established and until 20 years both 2N and 4N cells are present. With cessation of body growth, beginning after 20 years, a definite octaploid (8N) nuclear class is noted. From then until death all three classes exist in the liver with the number of 4N and 8N cells increasing slowly with advancing age. Even in adulthood, however, over 80% of the nuclei are of the 2N class. The frequency of polyploid nuclei is independent of sex. There are differing reports as to whether the frequency of binucleate cells is proportionate to the extent of nuclear ploidy or not and whether the male displays more binucleate cells than the female.

Since polyploidy is an established, age-dependent feature of liver growth, the use of the protein/DNA ratio as an index of cell size must be reexamined. Epstein (1967) determined cell volume in mammalian liver samples and found that the volume was directly proportional to cell ploidy regardless of whether the cell contained one or more nuclei of variable ploidy. This is consistent with the operation of gene dosage in the control of rates of synthesis of cell constituents in polyploid liver cells. Therefore, even in mature liver, the protein/DNA ratio is a valid index of the average cell mass per "diploid cell unit."

Data on total liver DNA and protein content during normal human development are scanty. Widdowson et al. (1972) have published data from a number of fetuses plus values from five adults. At 13–14 weeks of gestation total liver DNA was approximately 13 mg, at 20–22 weeks 130 mg, at 30 weeks 320 mg, and at birth approximately 600 mg. The adult values ranged from 2770 to 4600 mg, indicating that considerable DNA synthesis accompanies postnatal liver growth. A certain, but unknown portion, of the DNA in liver is due to hematopoetic tissue, especially before and shortly after birth. The protein/DNA ratios in fetal liver were stable at approximately 13 between 15 and 30 weeks of gestation and then more than doubled to nearly 30 at birth; the mean adult value was 56 with a range of 52–64. Thus postnatal growth in liver is accomplished by proportionately greater increases in "diploid cell units" than in mean cell size.

One unique aspect of liver growth is its ability to regenerate; this property is not seen in other tissues which do not continually renew their cells through life. More than half the liver can be removed surgically, even in adult animals, and the deficit will be replaced within days. Extensive synthesis of RNA, protein, and DNA are required since new cells are formed to replace those which are lost. The fact that DNA replication is possible in liver throughout life, as evidenced by the continued formation of 4N and 8N cells into old age, probably explains why liver can regenerate, while other tissues, which cease DNA replication at some finite point in development, cannot.

4. Skeletal Muscle

Various aspects of skeletal muscle growth have been reviewed by Widdowson (1974) and Mastaglia (1974). Quantitatively skeletal muscle is the most important soft tissue of the body; at birth it accounts for 25% of body weight and by adulthood approximately 40%. The muscle fibers arise from the fusion of the long, narrow, multinucleated myoblasts. Contrary to previous belief, the number of muscle fibers does increase after birth; the age at cessation of fiber formation is not known but is said to be between birth and 4 months of age by Montgomery (1962). However, Adams and De Rueck (1973) have reported that fiber number increases up to the middle of the fifth decade.

Growth in muscle mass has been assessed by creatinine excretion. In normal children on a low-creatinine diet, Graystone (1968) has determined that each gram of urinary creatinine excreted per 24 hr is the equivalent of 20 kg of muscle mass. Figure 3, from the work of Cheek (1974), demonstrates that from 5 to 18 years muscle mass in males increases more than fivefold and in females approximately

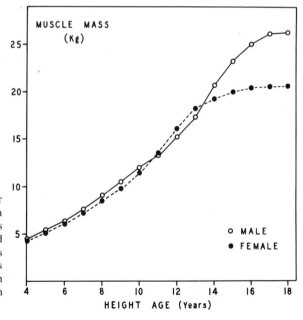

Fig. 3. Growth in muscle mass for both sexes during childhood. Growth in muscle mass is similar in both sexes until puberty when the greater and more prolonged growth in males leads to the well-known sexual differences in the amount of muscle tissue in adults. (Reprinted with permission from Cheek, 1974.)

fourfold. The amount of muscle per unit height is similar in the two sexes until 13 years of age when the male adolescent growth spurt begins and proportionately more muscle tissue is deposited in the male per unit height gained. From 11 to 17 years the male doubles his muscle mass; the female doubles her muscle mass from 9 to 15 years. The difference in ultimate muscle mass achieved is importantly related to the continued growth in the male between 13 and 18 years, a time when only small increments in muscle mass are noted in the female (see also Chapter 10).

Widdowson *et al.* (1972) have directly measured the DNA content of gastrocnemius muscle from human fetuses; levels rose from 10–15 mg at 15 weeks of gestation, to 160 mg at 25 weeks, to approximately 300 mg at birth. Determination of muscle nuclear number during postnatal growth has been made largely from extrapolation of information from biopsy samples. Earlier it had been held that muscle fiber nuclei did not undergo mitosis; yet both histologic and biochemical studies demonstrated an increase in muscle nuclear number. It is now felt that the additional subsarcolemmal nuclei arise from the mitosis and incorporation of undifferentiated mononucleated satellite cells into the muscle fiber (Moss and Leblond, 1970). Since muscle tissue contains both fiber and nonfiber nuclei and since skeletal muscle may not be homogeneous in composition throughout the body, it is important to deal with these issues before citing the data derived from biopsy. In one study of four different limb muscles in male rats at three different ages, Enesco and Puddy (1964) found that the number of nuclei which lay outside the fiber was consistently 35%. Under these circumstances, therefore, an increase in DNA content, even if uncorrected for nonfiber nuclei, would reflect proportionate growth in muscle fiber nuclei. The constancy of DNA concentration in muscles throughout the body has also been studied. Cheek *et al.* (1971) report that DNA concentrations among four muscles of the Sprague-Dawley rat and among five muscles of the young macaque monkey were in agreement. Yet Enesco and Puddy (1964) found agreement in only certain muscles of the Sherman rat; additionally, the extent of agreement varied with age and disease state. No systematic study of DNA concentration in various muscles at different ages has been carried out in the human. Therefore although the importance of the human data derived from biopsy material should not be underestimated, it must be recalled that the validity of the extrapolation has not been unequivocally established in man.

The work of Cheek (1968, 1974) and Cheek and Hill (1970) utilized biopsy extrapolation in which the DNA concentration in gluteal or abdominal rectus muscle samples was multiplied by the amount of muscle mass derived from creatinine excretion. Forty normal male subjects from infancy through 16 years and 23 normal female subjects from 3 to 17 years were studied; there are practically no data, however, for either sex between 18 months and 5 years. The data reveal a 14-fold increase in muscle nuclear number in males from 2 months to 16 years; this increase is not achieved by a steady accumulation of new nuclei with time. Three separate linear equations were derived for males 0.18–1.5 years, 5–10.5 years, and 10.5–16 years of age. The slopes of these three separate lines are, in general, similar to those for postnatal growth in height and weight; that is, the slope or rate of accumulation of new nuclei is greater in infancy than in the middle childhood years, while at adolescence the slope again increases. From 5 to 10 years the increase is from 0.90×10^{12} to 1.22×10^{12} nuclei, from 10 to 16 years the increase is from 1.22 to 3.10×10^{12} nuclei. The data over this entire age-span can also be expressed as a cubic equation, as shown in Figure 4 (Cheek and Hill, 1970). The pattern of muscle

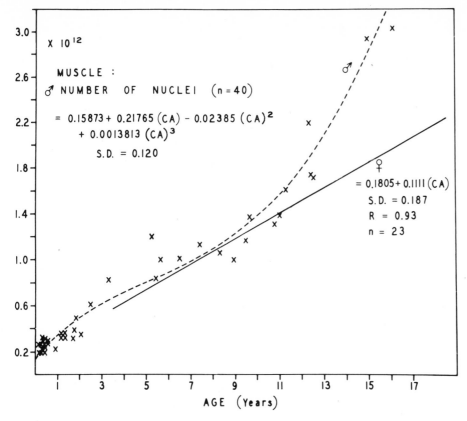

Fig. 4. Growth in muscle nuclear number in humans during childhood. The x's represent individual points for males; individual data points for females are not shown. The greater growth in muscle nuclei in males at the time of adolescence is clearly demonstrated; prior to that time there is little difference between the sexes. (Reprinted with permission from Cheek and Hill, 1970.)

nuclear growth in female subjects over the age of 5 years is linear through adolescence (Figure 4); in contrast to males no acceleration in nuclear accumulation occurred at adolescence. At 16 years of age the equation would predict a nuclear number of 1.96×10^{12} in females, a value 63% of the male figure. Since muscle mass of the female is approximately 80% of the male value at this age (Figure 3), the nuclear density must be less in the female. This is, in fact, confirmed by protein-to-DNA ratios which are slightly greater in females during the childhood years (Cheek, 1968). With further growth of muscle mass in males beyond 16 years of age, it is likely that nuclear density and protein/DNA ratio will equal or exceed the values in females.

The protein/DNA ratio has been used to assess average cell size or cell mass in skeletal muscle, even though the muscle fibers are multinucleated structures in which the individual nuclei are believed to be only diploid in nature. In this context the protein/DNA ratio can be considered an index of the average mass or protein content per "diploid cell unit" even though this unit has no physical membrane boundaries. Widdowson *et al.* (1972) determined protein/DNA ratios in fetal muscle samples. The increases before 30 weeks of gestation were modest in comparison to the sharp increases noted in the last 10 weeks of pregnancy. The values for the gastrocnemius at 40 weeks were in accord with values in abdominal rectus muscles obtained by Cheek (1968) in older infants (Figure 5), which suggests that the rapid growth phase initiated *in utero* persists for some period after birth. In five adult gastrocnemii the protein/DNA ratio ranged from 121 to 280 (Widdowson *et al.*, 1972); the value reported by Cheek (1968) for adolescents is approximately 300 for both sexes, and one might wonder if the adult values reflect some degree of involutional atrophy.

The studies cited thus far pertain to normal growth and development of muscle. Similar techniques have been utilized to study muscle growth following nutritional and hormonal alterations. One of the greatest difficulties in interpreting these data lies in the fact that it has been assumed, *a priori* in some instances, that the ratio of fiber to nonfiber nuclei, the constancy of DNA concentration among the various muscle groups, and the ratio of 1 g creatinine/20 kg muscle mass pertain in disease as well as in health. This may not be so; indeed there is some evidence to the contrary (Enesco and Puddy, 1964; Beach and Kostyo, 1968; Alleyne, 1968). With these caveats in mind, the following generalizations can be made. Since the increase in nuclear number occurs throughout prenatal and postnatal growth through adolescence, this process is vulnerable to insults occurring at any time during this period. Growth-hormone deficiency (Brasel and Cheek, 1968) and undernutrition (Graham *et al.*, 1969; Cheek *et al.*, 1970*a*) in the immature human subject reduce muscle nuclear number, as well as total muscle mass, when values are compared to age

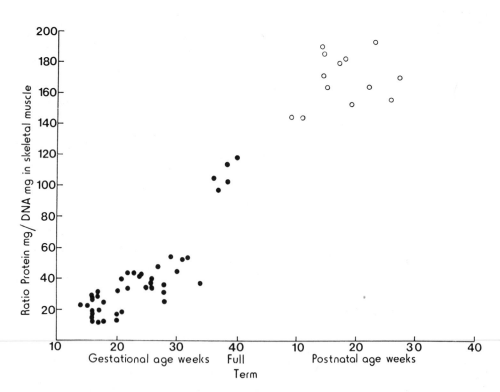

Fig. 5. Protein-to-DNA ratios in skeletal muscle of human fetuses and infants. The prenatal data of Widdowson *et al.* (1972) (solid circles) are in good agreement with the postnatal data of Cheek (1968) (open circles) which suggests that the rapid growth phase initiated *in utero* persists for some period after birth. (Reprinted with permission from Widdowson *et al.*, 1972.)

mates; replacement hormone therapy and nutritional rehabilitation will restore these deficits toward normal. Overnutrition in early childhood increases total muscle mass (Forbes, 1964) and muscle nuclear number (Cheek *et al.*, 1970*b*); reduction in body weight associated with some loss of muscle mass would not be expected to alter muscle nuclear number but might well decrease the average protein-to-DNA ratio. In animal studies (Winick and Noble, 1966) undernutrition in immature animals reduces the DNA content of skeletal muscle, and recovery with rehabilitation is incomplete. A similar period of undernutrition in adult animals reduces muscle mass by decreasing the protein/DNA ratio without affecting DNA content; with rehabilitation recovery is complete. Thus the timing of an insult to muscle growth may be more important in determining its effects on muscle composition than the nature of the insult itself.

5. Heart

The various component structures of the human heart develop very early in gestation (Dische, 1972). By 37 days, when the embryo measures only 1.5 cm, this differentiation has occurred, and thereafter the shape and form of the heart develop in response to changes in intraluminal pressure and volume of intrauterine blood flow. In the first two trimesters of pregnancy the heart makes a proportionately greater contribution to body weight than later; the heart-to-body-weight ratio is 1:68. In the last trimester the ratio falls to reach 1:138 at birth. In the first six postnatal months the more rapid growth of the body as a whole further decreases the ratio to 1:200 which changes little thereafter throughout the rest of life under normal conditions.

Cardiac growth is accomplished by increases in the number of myocardial fibers and by nuclear division which continues for a short, but ill-defined, period after birth. The extent of myocardial DNA synthesis after birth is controversial; one review (Dische, 1972) states that shortly after birth each nucleus undergoes one further amitotic division to reach the finite nuclear number of adulthood. Adler and Hueck (1971), on the other hand, found that the DNA content of the heart rose from 10 mg in infancy to 80–90 mg at 22 years. However, Widdowson *et al.* (1972) also measured DNA content in developing heart; total DNA rose from about 4 mg at 20 weeks of gestation to slightly above 16 mg at 30 weeks and reached 35–40 mg at birth; values in five adult hearts ranged from 193 to 355 mg. The data of Winick *et al.* (1972) from a small number of fetal and infant human hearts are in good general agreement with the data of Widdowson *et al.* (1972). There are no ready explanations for these widely discrepant results. Since solely biochemical studies will not differentiate between myocardial muscle and nonmuscle nuclei, the several-fold increase in DNA from birth to adulthood cannot be assumed to represent solely increases in muscle nuclei. In fact, a significant portion of the total DNA growth undoubtedly represents growth in nonmuscle cells (Sandritter and Adler, 1971; Rabinowitz and Zak, 1972). Like liver nuclei, myocardial muscle nuclei undergo significant polyploid formation (Kompmann *et al.*, 1966; Adler and Hueck, 1971; Sandritter and Adler, 1971). Not only do the number of binucleate myocardial fibers increase with age, but also the extent of polyploid nuclei increases markedly. Between birth and seven years of age, 80–90% of the nuclei are diploid (2N). A significant tetraploid (4N) class is established during the next few years; by 16–20 years approximately 70–80% of the nuclei are tetraploid, the rest are about half

*JO ANNE BRASEL AND
RHODA K. GRUEN*

diploid and half octaploid (8N). Some of the measurable increase in total heart DNA must reflect this polyploidy occurring between 7 and 20 years of age.

Protein-to-DNA ratios have been determined in heart muscle (Widdowson *et al.*, 1972; Kapeller-Adler and Hammad, 1972). From 6 to 20 weeks of gestation the ratio rises from approximately 10 to 20; between 20 and 30 weeks gestation the ratio is between 20 and 40; it rises sharply in the last 10 weeks to nearly double the value by birth. In five adult hearts the values ranged from 141 to 234.

Since no studies of the relationship between ploidy and cell mass have been performed on the heart, there is no information as to the appropriateness of the protein-to-DNA ratio as an index of average protein mass per "diploid cell unit." In this instance microscopic measures of fiber diameters are perhaps a more accurate representation of hypertrophic growth. Roberts and Wearn (1941) reported that the diameter of left ventricular muscle cells increased from 6 μm in newborns to 14 μm in normal adults. When the workload of the heart is increased, the heart enlarges (Zak, 1973). In the adult, muscle fibers increase in diameter and length; when the enlargement is extreme, the number of muscle cells may also be increased (Kompmann *et al.*, 1966; Zak, 1974). In adult rats a measurable increase in DNA content, radiothymidine uptake, and mitotic indices is seen, but this activity is almost entirely limited to connective tissue cells (Morkin and Ashford, 1968; Grove *et al.*, 1969*a,b*; Neffgen and Korecky, 1972). In the young animal the growth response is primarily a hyperplastic one, and the muscle cells are involved in the synthetic activity along with connective tissue cells (Sasaki *et al.*, 1970; Bishop, 1971; Neffgen and Korecky, 1972; Levin *et al.*, 1975). An increased number of muscle cells and little change in diameter have been noted in animal studies. In congenitally malformed human hearts the onset of polyploidy occurs much earlier than normal and is significantly more extensive (Adler and Hueck, 1971). In several other studies of the enlarged human heart where age of onset or type of cardiac abnormality is not defined, a cellular hyperplastic response or an increase in ploidy has been described (Kompmann *et al.*, 1966; Morishita *et al.*, 1970; Astorri *et al.*, 1971; Sandritter and Adler, 1971; Rabinowitz and Zak, 1972; Zak, 1973, 1974). Therefore, although the extent of enlargement may be proportionately the same in the young and adult heart, the cellular makeup of the heart may be considerably different, and it is probably not accurate to speak of cardiac hypertrophy in many cases of cardiomegaly, especially in the immature subject.

6. Lung

Lung growth and development have been reviewed by Charnock and Doershuk (1973) and Thurlbeck (1975). In human lung 65–75% of bronchial branching occurs between the 10th and 14th weeks of gestation, and bronchial development is complete by the 16th week. Cilia are noted at the 13th week. Between the 16th and 26th weeks proliferation of the mesenchyme occurs, a rich blood supply develops within the mesenchyme, the epithelial lining of the airway becomes flattened, and toward the end of the period type I and type II cells can be distinguished in the epithelial lining. From 24–26 weeks of gestation to 40 weeks "terminal sacs" differentiate further and at birth a considerable number of alveoli are present. However, the time of onset of alveolar development is uncertain, and there are varying reports as to the number present at birth. Most investigators agree that the majority of alveoli develop postnatally, but there is disagreement as to the age of cessation of alveolar multiplication, with estimates ranging from one year of age to puberty. Differences in the methodology used to assess alveolar number and large variations in the total number normally reached at different ages may explain some of the discrepancy among the reports on the age of complete alveolar formation. For example, adult alveolar number may vary from 200 to 600 million, according to one study (Angus and Thurlbeck, 1972). Other parts ot the lung, especially the nonalveolated airways, grow in parallel to changes in body size. Tracheal diameter increases directly with the increase in chest circumference. The size of major bronchi is related to stature while the bronchioles are constant in number, regardless of body size. Alveolar number may also be related to body length. Thus a tall adult is likely to have larger bronchi and bronchioles than a short adult and more alveoli of probably the same size.

In the past two decades considerable attention has been paid to the origin and function of the type I and type II epithelial cells (Campiche *et al.,* 1963; Weibel, 1974; McDougall and Smith, 1975). Both cells are of endodermal origin and radiothymidine uptake studies have conclusively shown that the type II cell divides to give rise to the type I cell. Using the electron microscope, the cells are easily differentiated since type I cells show a small perinuclear body and extensive cytoplasmic projections which line a considerable part of the alveolar surface; its cytoplasm is poor in subcellular organelles. Type II cells appear larger, more rounded or cuboidal on section; early in development glycogen is prominent but later glycogen disappears and subcellular organelles become extensive. The presence of mitochondria, rough endoplasmic reticulum, and an extensive Golgi system suggests active synthetic activities. The type II cells also contain a highly specific granule in the form of an osmiophilic lamellar body; these are the secretory granules for the phospholipid component of pulmonary surfactant, the important surface-lowering compound.

There are very few reports of growth in DNA and protein in the human lung. Winick *et al.* (1972) reported DNA values from a few fetal specimens between the 13th and 40th weeks of gestation. Total DNA content increased more than 50-fold during this period; a particularly marked increase in total DNA was seen in the last 10 weeks of pregnancy. Kapeller-Adler and Hammad (1972) reported DNA and protein concentrations and protein-to-DNA ratios in fetal specimens from 6 to 20 weeks of gestation. The DNA and protein concentrations rose in a similar fashion so that the ratio changed little during this period. Since tissue weight is not recorded, no estimation of total DNA content can be made. Bradley *et al.* (1975) reported that the DNA concentration per milligram dry lung weight decreased threefold between the 17th week of gestation and adulthood, which reflects a threefold increase in mean cell size over the period, but again total DNA content cannot be calculated from the data presented.

Functional removal of lung tissue causes changes in the remaining tissue. The decrease in functional mass can occur through surgical removal, interruption of blood supply, airway obstruction, or external compression. The remaining lung tissue undergoes varying degrees of expansion to fill the void left in the chest cavity and increases its functional capacity. These adjustments may or may not completely make up for the loss. The growth processes accompanying the expansion and functional changes have not been much studied in the human. Studies in animals have revealed that increases in wet weight, dry weight, protein, collagen, RNA, and

JO ANNE BRASEL AND
RHODA K. GRUEN

DNA may all occur (Addis, 1928; Romanova *et al.*, 1967; Buhain and Brody, 1973; Fisher and Simnett, 1973; Nattie *et al.*, 1974; Inselman *et al.*, 1976). However, the extent of the increase, if it occurs at all, is dependent upon the technique used to remove functional lung tissue as well as on the age of the animal. In general pneumonectomy produces a greater growth response than various types of atelectasis, and young animals undergo a more prompt and greater response than older animals, especially in regard to growth in DNA. Controversy exists as to whether any new alveoli are formed at any age or under any circumstances following functional removal of lung tissue. Uptake of radiothymidine into newly synthesized DNA occurs to a small extent, even in adult lung tissue; this likely reflects a renewal of cells lining the airways and not a net increase in the number of cells present. When such lining cells are injured or destroyed, there is considerable capacity for regeneration which undoubtedly involves cell division. Thus although there is controversy over the ability of the lung under stress to form new functional units, i.e., alveoli, especially in adulthood, the cells lining the respiratory tract can be replaced to some extent after injury even in adulthood.

7. Summary and Conclusions

This chapter has summarized the available data on the cellular growth of nonrenewing tissues in the human. These data are scanty but, not surprisingly, are most complete for fetal tissues as a result of availability of specimens. Although the information is incomplete, the evidence suggests that the patterns of cellular growth in the human follow those previously described in animal tissues. Early growth is hyperplastic in nature, then as cell division nears completion, hypertrophic growth commences and continues after hyperplasia has ceased. In human brain the hyperplastic phase ends sometime during the second year of life. In skeletal muscle it continues through adolescence. In liver and heart polyploidy occurs well into adulthood and in the former DNA replication appears to be possible under the appropriate stimulation even in the aged animal. The data for lung are too sparse to make a judgment as to when cell division ceases.

In animal work the hyperplastic growth phase has been shown to be particularly vulnerable to growth-affecting stimuli (Winick and Noble, 1966). Both increases and decreases in cell number brought about during early proliferative growth are apt to be permanent even when treatment is begun before the proliferative phase is over. The available data suggest the same occurs in humans. Thus brain DNA content is reduced in malnourished infants (Winick and Rosso, 1969; Winick *et al.*, 1970; Chase *et al.*, 1974) and polyploidy in cardiac muscle is increased in congenital heart disease. Similar growth-affecting stimuli applied once proliferative growth is complete change only cell size, and these abnormalities are readily corrected with therapy. Perhaps one of the most important concepts to arise from the cell growth studies is that the timing of an insult may be more important in determining the nature of the growth response, the long-term consequences, and the recoverability than the nature of the insult itself. This information has characterized the nature of some types of growth failure at a cellular level. Furthermore it has provided a mechanistic explanation for the particular vulnerability of the rapidly growing organism to stimuli which affect growth.

Adams, R. D., and De Rueck, J., 1973, Metrics of muscle, in: *Basic Research in Myology,* Proceedings of the II International Congress on Muscle Diseases, Part I, pp. 3–11, Excerpta Medica, ICN Series #294, Amsterdam.

Addis, T., 1928, Compensatory hypertrophy of the lung after unilateral pneumonectomy, *J. Exp. Med.* **47:**51.

Adler, C. P., and Hueck, C., 1971, DNA content in growing human hearts, *Verh. Dtsch. Ges. Pathol.* **55:**464.

Alleyne, G. A. O., 1968, Studies on total body potassium in infantile malnutrition: The relation to body fluid spaces and urinary creatinine, *Clin. Sci. (London)* **34:**199.

Angus, G. E., and Thurlbeck, W. M., 1972, Number of alveoli in the human lung, *J. Appl. Physiol.* **32:**483.

Astorri, K., Chizzola, A., Visioli, O., Anversa, P., Olivetti, G., and Vitali-Mazza, L., 1971, Right ventricular hypertrophy—a cytometric study on 55 human hearts, *J. Mol. Cell. Cardiol.* **2:**99.

Beach, R. K., and Kostyo, J. L., 1968, Effect of growth hormone on the DNA content of muscles of young hypophysectomized rats, *Endocrinology* **82:**882.

Bishop, S. P., 1971, Myocardial cell growth in the normal and hypertrophying neonatal heart, *Circulation* **44**(Suppl. 2):142 (abstract).

Boivin, A., Vendrely, R., and Vendrely, C., 1948, L'acide désoxyribonucléique du noyau cellulaire, dépositaire des caractères héréditaires; arguments d'ordre analytique, *C.R. Acad. Sci.* **226:**1061.

Bradley, K., McConnell-Bruel, S., and Crystal, R. G., 1975, Collagen in the human lung, *J. Clin. Invest.* **55:**543.

Brasel, J. A., and Cheek, D. B., 1968, The effect of growth hormone on the cellular mass of hypopituitary dwarfs, in: *Growth Hormone* (A. Pecile and E. Muller, eds.), pp. 433–440, Excerpta Medica, Amsterdam.

Buhain, W. J., and Brody, J. S., 1973, Compensatory growth of the lung following pneumonectomy, *J. Appl. Physiol.* **33:**898.

Campiche, M. A., Gautier, A., Hernandez, E. I., and Reymond, A., 1963, An electron microscopic study of the fetal development of human lung, *Pediatrics* **32:**976.

Charnock, E. L., and Doershuk, C. F., 1973, Developmental aspects of the human lung, *Pediatr. Clin. North Am.* **20:**275.

Chase, H. P., Canosa, C. A., Dabiere, C. S., Welch, N. N., and O'Brien, D., 1974, Postnatal undernutrition and human brain development, *J. Ment. Defic. Res.* **18:**355.

Cheek, D. B., 1968, Muscle cell growth in normal children, in: *Human Growth* (D. B. Cheek, ed.), pp. 337–351, Lea and Febiger, Philadelphia.

Cheek, D. B., 1973, Brain growth and nucleic acids: The effect of nutritional deprivation, in: *Brain and Intelligence: Ecology of Child Human Development* (F. Richardson, ed.), pp. 237–256, National Educational Press, Hyattsville, Maryland.

Cheek, D. B., 1974, Body composition, hormones, nutrition and adolescent growth, in: *Control of Onset of Puberty* (M. M. Grumbach, G. D. Grave, and F. E. Mayer, eds.), pp. 426–447, John Wiley and Sons, New York.

Cheek, D. B., 1975, The fetus, in: *Fetal and Postnatal Cellular Growth* (D. B. Cheek, ed.), pp. 3–22, John Wiley and Sons, New York.

Cheek, D. B., and Hill, D. E., 1970, Muscle and liver cell growth: Role of hormones and nutritional factors, *Fed. Proc.* **29:**1503.

Cheek, D. B., Hill, D. E., Cordano, A., and Graham, G., 1970*a,* Malnutrition in infancy: Changes in muscle and adipose tissue before and after rehabilitation, *Pediatr. Res.* **4:**135.

Cheek, D. B., Schultz, R. B., Parra, A., and Reba, R. C., 1970*b,* Overgrowth of lean and adipose tissues in adolescent obesity, *Pediatr. Res.* **4:**268.

Cheek, D. B., Holt, A. B., Hill, D. E. and Talbert, J. L., 1971, Skeletal muscle cell mass and growth: The concept of the deoxyribonucleic acid unit, *Pediatr. Res.* **5:**312.

Dische, M. R., 1972, Observations on the morphological changes of the developing heart, *Cardiovasc. Clin.* **4:**175.

Dobbing, J., 1970, Undernutrition and the developing brain: The relevance of animal models to the human problem, *Am. J. Dis. Child.* **120:**411.

Dobbing, J., and Sands, J., 1970, Timing of neuroblast multiplication in developing human brain, *Nature* **226:**639.

Dobbing, J., and Sands, J., 1973, Quantitative growth and development of human brain, *Arch. Dis. Child.* **48:**757.

Enesco, M., and Leblond, C. P., 1962, Increase in cell number as a factor in the growth of the young male rat, *J. Embryol. Exp. Morphol.* **10:**530.

Enesco, M., and Puddy, D., 1964, Increase in the number of nuclei and weights in skeletal muscle of rats of various ages, *Am. J. Anat.* **114:**235.

Epstein, C. J., 1967, Cell size, nuclear content, and the development of polyploidy in the mammalian liver, *Proc. Natl. Acad. Sci. U.S.A.* **57:**327.

Fisher, J. M., and Simnett, J. D., 1973, Morphogenetic and proliferative changes in the regenerating lung of the rat, *Anat. Rec.* **176:**389.

Forbes, G. B., 1964, Lean body mass and fat in obese children, *Pediatrics* **34:**308.

Graham, G. D. C., Cordano, A., Blizzard, R., and Cheek, D. B., 1969, Infantile malnutrition: Changes in body composition during rehabilitation, *Pediatr. Res.* **3:**579.

Graystone, J. E., 1968, Creatinine excretion during growth, in: *Human Growth* (D. B. Cheek, ed.), pp. 182–197, Lea and Febiger, Philadelphia.

Grove, D., Nair, K. G., and Zak, R., 1969*a,* Biochemical correlates of cardiac hypertrophy III. Changes in DNA content; the relative contributions of polyploidy and mitotic activity, *Circ. Res.* **25:**463.

Grove, D., Zak, R., Nair, K. G., and Aschenbrenner, V., 1969*b,* Biochemical correlates of cardiac hypertrophy IV. Observations on the cellular organization of growth during myocardial hypertrophy in the rat, *Circ. Res.* **25:**473.

Howard, E., Granoff, D. M., and Bujnovszky, P., 1969, DNA, RNA and cholesterol increases in cerebrum and cerebellum during development of human fetus, *Brain Res.* **14:**697.

Inselman, L. S., Mellins, R. B., and Brasel, J. A., 1976, Effects of atelectasis on compensatory lung growth, Presented at the annual meeting of the American Thoracic Society, *Am. Rev. Resp. Dis.* **113:**41.

Kapeller-Adler, R., and Hammad, W. A., 1972, A biochemical study on nucleic acids and protein synthesis in the human fetus and its correlation with relevant embryological data, *J. Obstet. Gynaecol. Br. Commonw.* **79:**924.

Kompmann, M., Paddags, I., and Sandritter, W., 1966, Feulgen cytophotometric DNA determinations on human hearts, *Arch. Pathol.* **82:**303.

Leuchtenberger, C., Leuchtenberger, R., and Davis, A., 1954, A microspectrophotometric study of the desoxyribose nucleic acid (DNA) content in cells of normal and malignant human tissues, *Am. J. Pathol.* **30:**65.

Levin, A. R., Brasel, J. A., Deely, W., Redo, S. F., Ogata, K., and Winick, M., 1975, The response of the right ventricle to experimentally induced pulmonary artery obstruction, *Growth* **39:**107.

Mastaglia, F. L., 1974, The growth and development of skeletal muscles, in: *Scientific Foundations of Paediatrics* (J. A. Davis and J. Dobbing, eds.), pp. 348–375, W. B. Saunders, Philadelphia.

McDougall, J., and Smith, J. F., 1975, The development of the human type II pneumocyte, *J. Pathol.* **115:**245.

Mirsky, A. E., and Ris, H., 1949, Variable and constant components of chromosomes, *Nature* **163:**666.

Montgomery, R. D., 1962, Growth of human striated muscle, *Nature* **195:**194.

Morishita, T., Sasaki, R., and Yamagata, S., 1970, Studies on deoxyribonucleic acid content and cell count of human heart muscle, *Jpn. Heart J.* **11:**36.

Morkin, E., and Ashford, T. P., 1968, Myocardial DNA synthesis in experimental cardiac hypertrophy, *Am. J. Physiol.* **215:**1409.

Moss, F. P., and Leblond, C. P., 1970, Nature of dividing nuclei in skeletal muscle of growing rats, *J. Cell Biol.* **44:**459.

Nattie, E. E., Wiley, C. W., and Bartlett, D., Jr., 1974, Adaptive growth of the lung following pneumonectomy in rats, *J. Appl. Physiol.* **37:**491.

Neffgen, J. F., and Korecky, B., 1972, Cellular hyperplasia and hypertrophy in cardiomegalies induced by anemia in young and adult rats, *Circ. Res.* **30:**104.

Rabinowitz, M., and Zak, R., 1972, Biochemical and cellular changes in cardiac hypertrophy, *Annu. Rev. Med.* **23:**245.

Ranek, L., Keiding, N., and Jensen, S. T., 1975, A morphometric study of normal human liver cell nuclei, *Acta Pathol. Microbiol. Scand. Sect. A* **83:**467.

Roberts, J. T., and Wearn, J. T., 1941, Quantitative changes in the capillary-muscle relationship in human hearts during normal growth and hypertrophy, *Am. Heart J.* **21:**617.

Romanova, L. K., Leikina, E. M., and Antipova, K. K., 1967, Nucleic acid synthesis and mitotic activity during development of compensatory hypertrophy of the lung in rats, *Bull. Exp. Biol. Med. (U.S.S.R.)* **63**:303.

Sandritter, W., and Adler, C. P., 1971, Numerical hyperplasia in human heart hypertrophy, *Experientia* **27**:1435.

Sasaki, R., Morishita, T., Ichikawa, S., and Yamagata, S., 1970, Autoradiographic studies and mitosis of heart muscle cells in experimental cardiac hypertrophy, *Tohoku J. Exp. Med.* **102**:159.

Swartz, F. J., 1956, The development in the human liver of multiple desoxyribose nucleic acid (DNA) classes and their relationship to the age of the individual, *Chromosoma* **8**:53.

Thurlbeck, W. M., 1975, Postnatal growth and development of the lung, *Am. Rev. Resp. Dis.* **111**:803.

Weibel, E. R., 1974, A note on differentiation and divisibility of alveolar epithelial cells, *Chest* **65**:19S (suppl.).

Widdowson, E. M., 1974, Changes in body proportions and composition during growth, in: *Scientific Foundations of Paediatrics* (J. A. Davis and J. Dobbing, eds.), pp. 153–163, W. B. Saunders, Philadelphia.

Widdowson, E. M., Crabb, D. E., and Milner, R. D. G., 1972, Cellular development of some human organs before birth, *Arch. Dis. Child.* **47**:652.

Winick, M., 1968, Changes in nucleic acid and protein content of the human brain during growth, *Pediatr. Res.* **2**:352.

Winick, M., and Noble, A., 1965, Quantitative changes in DNA, RNA, and protein during prenatal and postnatal growth in the rat, *Dev. Biol.* **12**:451.

Winick, M., and Noble, A., 1966, Cellular response in rats during malnutrition at various ages, *J. Nutr.* **89**:300.

Winick, M., and Rosso, P., 1969, The effect of severe early malnutrition on cellular growth of human brain, *Pediatr. Res.* **3**:181.

Winick, M., Rosso, P., and Waterlow, J., 1970, Cellular growth of cerebrum, cerebellum, and brain stem in normal and marasmic children, *Exp. Neurol.* **26**:393.

Winick, M., Brasel, J. A., and Rosso, P., 1972, Nutrition and cell growth, in: *Nutrition and Development* (M. Winick, ed.), pp. 49–97, John Wiley and Sons, New York.

Zak, R., 1973, Cell proliferation during cardiac growth, *Am. J. Cardiol.* **31**:211.

Zak, R., 1974, Development and proliferative capacity of cardiac muscle cells, *Circ. Res.* Suppl. II:17.

2

Cellular Growth: Adipose Tissue

C. G. D. Brook

1. Introduction

There are two very distinct types of fat in the human, which have been named brown fat and white fat because of their gross macroscopic appearance. Brown fat has a primarily thermogenetic role and is active in humans in the early neonatal period. Although in other mammals it retains its function into later life and is used in times of hibernation, it is probable that after early infancy it has no further function in the human. White adipose tissue has a multiplicity of functions; as an organ containing triglyceride it is the major store of energy for the body, but it also has functions of insulation, of protection, and, in the different sixes, of considerable attraction. The biochemistry and nervous control of lipogenesis and lipolysis are similar in white and brown fat, the differences being more in terms of quantity than quality. Later discussion will emphasize a few of the structural differences which enable these differences to be achieved.

It is perhaps worth prefacing a chapter on the growth of body fat by a reminder to the reader that the human organism cannot live without fat, such is the importance of its many roles. In Western society it has come to be much undervalued in cultural terms because of the continual emphasis on the hazards of obesity. This has led to a state of chronic ignorance among doctors about the adipose organ, since it has largely been regarded as an object of ridicule. Study of its growth highlights biological principles in a way that the study of growth of other less accessible organs cannot do.

2. Brown Adipose Tissue

Brown fat is primarily concerned with the provision of heat, mainly in the newborn period. For this reason its morphology and development are sufficiently different from white fat to merit brief consideration separately.

C. G. D. BROOK ● The Middlesex Hospital, London, England.

Brown fat consists of cells containing many fat vacuoles around which are arranged many large and complex mitochondria. The object of this arrangement is presumably to facilitate rapid lipolysis and heat production, which can be detected by a change of skin temperature in babies (Grausz, 1970). Increased blood flow from the fat, which is situated around the great vessels and kidneys, provides the necessary transport medium for this central heating system (Hull, 1974).

The main role of brown adipose tissue being neonatal thermogenesis, it develops in advance of white adipose tissue and its vacuoles are filled with fat before birth. After birth, at least in rabbits, the amount of brown adipose tissue increases with growth, but with the establishment of feeding the cells gradually become, like those in white adipose tissue, unilocular and the brown fat disappears at a rate which is dependent on the ambient temperature. What happens in the human is not yet known, but the idea that there may be more or less metabolically active depots of fat is an old one (Carlson and Hallberg 1968); brown fat would be a good candidate for producing fatty acids rapidly in response to stress.

3. White Adipose Tissue

The adipose organ in man is unique in being easily sampled and measured by simple anthropometric techniques. For these reasons more is known about its growth at a cellular level than about any other organ. Nevertheless, because of the complicated roles it plays as an organ of storage, of insulation, of thermogenesis, and of energy provision, there remain many points of doubt and ignorance which will become clear in the following description of its growth.

3.1. Growth of the Adipose Organ

Methods for studying body composition in man fall broadly into two groups. Physiological methods include underwater weighing or other densitometric techniques, such as gas displacement, to estimate the specific gravity of the body, and measurement of the amount of ^{40}K, the naturally occurring isotope of potassium, in the body, which leads to the calculation of lean body mass. The amount of adipose tissue can be obtained from these estimations by relatively easy calculations, but the methods themselves suffer from being extremely difficult to perform and from requiring very expensive apparatus for their performance.

Consequently many workers have used varieties of isotopic or chemical dilution methods for measuring body water or body fat directly, and still others have used techniques such as radiography or ultrasound. Plainly none of these techniques is appropriate for every day use, and many of them are impossible to use in children for technical or ethical reasons. Thus many workers have sought to relate objective measurements of body fat to measurements of skinfold thickness.

These measurements are extremely simple to perform, rapid, painless, and repeatable to an accuracy of about 5% (Edwards *et al.*, 1955). They have been shown to correlate well with other measurements of body fat, both in adults and in children (Durnin and Rahaman, 1967; Brook, 1971*a*). Pařisková and Roth (1972) have shown that the accuracy of the prediction of body fat from skinfold thicknesses is not necessarily increased by measuring a large number of skinfolds and that a combination of the measurements of triceps and subscapular skinfold thicknesses, which are representative of limb and body fat, respectively, give a good indication of total body fat.

Body fat appears in the fetus rather late in gestation, at about the seventh month, and its period of most rapid growth appears to last from that time to around the end of the first year of postnatal life. Because of the difficulties in estimating body fat accurately in babies, since body composition changes rapidly in the early weeks of life, we have to rely almost entirely on measurements of skinfold thickness to estimate this growth, and relatively few accurate data are available for the early months of life. The best are incorporated into the currently available standards for triceps and subscapular skinfolds in British children (Tanner and Whitehouse, 1975). There are differences in the exact patterns of growth of body fat between different races (Eveleth and Tanner, 1976).

Plotting the sum of the 50th centile values of measurements of triceps and subscapular skinfold thicknesses in British children on a logarithmic scale in millimeters against chronological age gives an indication of how body fat accumulates in man after birth since total body fat is related to the log sum of these two skinfolds (Figure 1). Such evidence as exists about the increase in body fat before birth suggests a continuation backwards of the initial increase shown here. After a very rapid increase in body fat over the first 9 months of life, an increase which occurs slightly earlier and is slightly greater in males than females, fat is lost in both sexes during the early years of childhood. This loss is less in girls than in boys. From the age of eight, when the earliest-maturing girls will be starting puberty, there is a gradual increase in body fat which continues throughout puberty. In males there is the interesting phenomenon of a preadolescent fat spurt, which we do not understand at all, and then a further considerable increase during puberty proper. Since body weight shows a fairly steady increase during childhood, with a marked change in tempo at puberty (Tanner *et al.*, 1966), the fluctuations in skinfold thicknesses suggest cyclical accumulations of fat and lean tissue during childhood with fat accumulation predominating in early childhood and at puberty. Mathematical examination of figures for body weight and body composition in children confirm this suggestion (Dugdale and Payne, 1975).

In adult women skinfold thicknesses show a continuing steady increase with advancing years up to the sixth decade, when they decline again (Wessel *et al.*, 1963). In men, measurements of body fat remain fairly constant (Norris *et al.*, 1963), but common observation indicates a redistribution of fat with a relative gain in trunk fat. Thus the span of growth of the adipose organ is longer than that of any other organ, and it remains more adaptable than most in accommodating environmental changes.

There are other curiosities of adipose tissue which are, as yet, very incompletely understood. The first is the obvious fact that, throughout life, it can be made to alter in size in response to changes in environmental circumstances. These changes enter into the clinical problems of obesity and malnutrition. No other organ can adjust itself like this. Secondly, adipose tissue changes in configuration as a consequence of various stimuli; one has only to think of the shape differences between men and women, the patterns of accumulation of fat with age, and the distribution of fat in Cushing's disease to realize that we remain ignorant of many facts concerning the growth of adipose tissue.

3.2. Methods of Examining Adipose Tissue Cellularity

Biopsy of the adipose organ is extremely simple (Hirsch and Goldrick, 1964) and can safely be performed even in small premature babies. Since body fat is dependent upon the number of adipose cells and their size, many methods have

C. G. D. BROOK

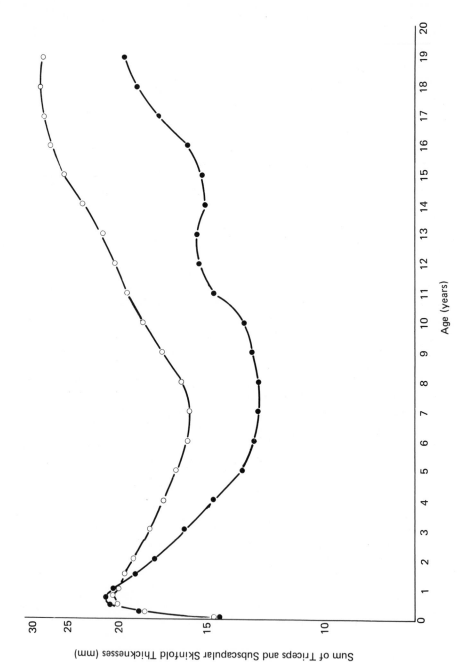

Sum of Triceps and Subscapular Skinfold Thicknesses (mm)

Age (years)

Fig. 1. Change in total body fat with age in children. ●, Boys; ○, girls.

been devised for handling biopsy specimens. Methods of sizing broadly comprise microscopic examination of appropriate histological preparations (Bjurulf, 1959; Sjöström *et al.,* 1971) or separation of the adipose cells and the application of various sizing techniques to the free cells (Rodbell, 1964; Hirsch and Gallian, 1968). While much print has been used in promoting one or other technique of sizing adipose cells, the remarkable finding is the uniformity of results achieved by different groups of workers.

Nevertheless examination of adipose cells freed from connective tissue by collagenase digestion or by fixation of the triglyceride within the cell with osmium tetroxide suffers from two serious disadvantages. First, no information is given about the distribution of cell size *in situ* and, secondly, there is a danger that very small fat cells may be missed. The importance of this will become clear in discussion of departures from normal growth, but there is little doubt that examination of fixed sections is an essential complement to any other method of adipose cell sizing.

Estimation of the number of adipose cells is much more difficult. It has been done by measuring the weight of lipid in the "average" fat cell and by determining how many cells are required to contain the whole weight of body fat. Apart from the methodological difficulties of measuring body fat, there are major problems in the definition and measurement of the average fat cell. At every site of adipose tissue deposition there is a range of adipose cell size which varies somewhat from one subcutaneous site to another (Salans *et al.,* 1973) and very considerably from subcutaneous sites to visceral ones (Brook, 1971*b*). Accordingly, estimates of fat cell number based on data derived from single-site biopsy specimens must be treated with caution. These methodological problems would properly vitiate all further discussion of effects on adipose tissue cellularity were not the magnitudes of the difference from normal caused by environmental effects so great and reported so consistently by different groups of workers using different techniques. Nevertheless, only broad conclusions and biological principles can be enunciated from studies of adipose cell number. If values in individual subjects are needed, these must be based on averages of determinations of adipose cell size from several sites.

3.3. Cellular Development of Adipose Tissue

In 1969 Hirsch and Han reported studies on the size, number, and rate of formation of fat cells in rats. Early growth was accompanied by a progressive enlargement as well as by an increase in the number of fat cells. Beyond the 15th week of extrauterine life, however, when a rat is skeletally more than 90% mature (Hughes and Tanner, 1970), the adipose organ changed only through the size of fat cells and not through cell number.

As far as humans are concerned, the data are rather few because of the difficulties of obtaining samples from normal children. Thus knowledge of the relative contributions which increases of cell size or number make to the growth of the adipose organ is far from complete. The data of Hirsch and Knittle (1970) on 34 infants and children from birth to the age of 13, using the osmium tetroxide fixation method, indicated that there was roughly a threefold increase in cell size from birth to age 6, with little change between ages 6 and 13. However, at 13 the adult cell size had not been reached, so a further increase in size must have taken place during puberty. Cell number showed a rapid threefold increase in the first year of life and then a more gradual but continuous increase throughout childhood. As with cell size, the number of adipose cells had not reached an adult value by age 13.

Dauncey and Gairdner (1975), using the microscopic technique, focused their attention on the period around birth and demonstrated in 59 infants a steady increase in adipose cell size from 25 weeks of gestation up to 40 weeks of postnatal life. A very simple calculation indicates that this increase would be adequate to accommodate all the fat laid down during that period, which leaves little room for an increase in the number of cells during that time (Gairdner and Dauncey, 1974). The size of the cells at 40 weeks of age was similar to the size found by other workers in older children so it may be inferred that adipose cell size remained fairly constant during childhood, which confirms the work of Hirsch and Knittle (1970). As far as numbers of cells are concerned, especially with the large site-to-site differences in children (Duckerts and Bonnet, 1973), it is perhaps better only to write in generalities, but it would appear that most of the earliest very rapid enlargement of the adipose organ takes place by enlargement of cells and that an increase in number takes over later. As very large cells have not been found, even in the most obese subjects, it may be that cell size is the trigger for cell replication in adipose tissue.

In older children my own data, using the method of Hirsch and Gallian (1968) (Figure 2), confirm that cell size changes rather little during childhood. The increase in body fat which takes place must be, therefore, due to a fairly gradual increase in the number of adipose cells, and much the best sense would be made if this increase roughly followed the pattern of the accumulation of body fat shown in Figure 1.

It seems then that the earliest increases probably occur in cell size with presumably some increase in cell number, while increases in cell number are more important for growth later. Variations of fatness between individual subjects are probably mostly accommodated by variations of cell size: thus, for example, women, who have more body fat than men, have about the same number of cells as men but each cell is larger.

From the over-all viewpoint the question of whether increases of cell size predominate over increases of cell number at a particular point in time is probably not of fundamental importance, since external events can almost certainly modify the process (see below). What matters is that the adipose organ increases by a combination of both processes, although increases in the number of cells appear not to occur after the end of puberty.

Even this may well prove incorrect since two groups of workers have recently shown independently, working with guinea pigs and mice, respectively, that pockets of very small cells can be found among other "normal" adipose cells which could presumably enlarge if external events required more cells (Kirtland *et al.*, 1975; Ashwell *et al.*, 1975*a*). How widespread this phenomenon is or whether it occurs in man is not known at present, but it indicates how fluid our ideas about adipose cell growth must remain, since these cells would certainly not be "counted" by many of the techniques already discussed, which may, therefore, seriously have underestimated total numbers of adipose cells. This finding indicates clearly the importance of inspecting fixed sections of adipose tissue as well as counting cells in any future developments on this front.

3.4. Influences on the Growth of Adipose Tissue

3.4.1. Genetic Factors

There are very few data which pertain to the genetic determination of body fat, even though it is a common enough observation that children resemble their parents in fatness. Majority opinion favored an environmental explanation for this similar-

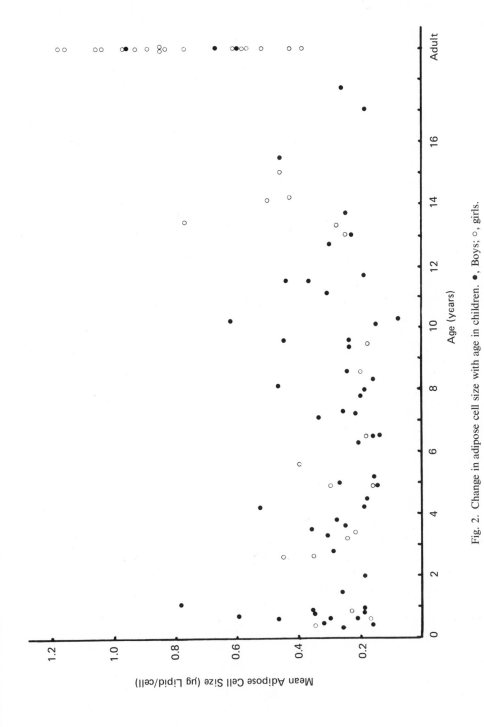

Fig. 2. Change in adipose cell size with age in children. ●, Boys; ○, girls.

ity, but quantification of the heritability of skinfold thickness reveals a very strong genetic effect in the determination of the variation of fatness (Brook *et al.,* 1975). Work in progress indicates that obese mothers tend to have obese babies (A. G. L. Whitelaw, personal communication) which accords with a strong genetic component to body fatness.

It is important to reiterate here that any genetic influence can, of course, be overridden by a strong enough environmental stimulus. An extreme example to illustrate this would be the reduction of adult height, which is strongly genetically determined, by surgical means. A similarly extreme environmental stimulus appears to be exerted with regard to the fatness of obese young children. In older children values of heritability of skinfold thickness are high but, like adipose cell size, there seem to be considerable site differences (Osborn and de George, 1959), and all data on the genetic determination of fatness must be looked at in this light.

Nothing is yet known of the genetic determination of the number of adipose cells.

3.4.2. Environmental Factors

Overnutrition. The classic experiments of Widdowson and McCance (1960) indicated that early overfeeding had long-term effects on the weight and body composition of experimental animals. Knittle and Hirsch (1968) showed that it also affected the size and number of adipose cells in the weanling rat, although Hollenberg and Vost (1968) showed that during periods of rapid weight gain in adult rats accretion of fat was due to the deposition of fat in existing fat cells rather than by addition of new cells. This was confirmed in a short-term study of experimental obesity in man (Sims *et al.,* 1968).

From these experiments a voluminous literature about adipose tissue cellularity in human obesity has emerged and has led to the establishment of a few fairly simple facts. In human obesity, both in adults and in children, the size of the fat cells is increased. This is hardly surprising, since in normal subjects fat cell size can be predicted from estimations of body fat even in infants (Figure 3) and obese subjects emphasize the trend. Cell size, however, varies considerably among the fat depots of obese subjects as well as normal ones (Salans *et al.,* 1973), and this makes statements about numbers of fat cells in obesity suspect. Nevertheless large differences have emerged between the numbers of adipose cells in normal and obese subjects, allowing a division of obesity into hyperplastic (an excessive number of cells of moderate size) and hypertrophic (a normal number of very large cells) types.

Many variables have been examined to determine what makes an individual fall into one of the two groups. Of the likely factors, that is, degree, duration, and age of onset of obesity, the last seems to be the most highly favored (Hirsch and Knittle, 1970; Brook *et al.,* 1972; Salans *et al.,* 1973; Sjöström and Björntorp, 1974). Nevertheless, duration of obesity alone or degree of obesity by itself may well also be determinants of adipose tissue cellularity in adults (Ashwell *et al.,* 1975*b*).

With regard to age of onset of obesity, Salans *et al.* (1973) conclude that there are two distinct periods early in life during which overfeeding may lead to hypercellularity of adipose tissue. These periods are within the first years of life and between ages 9 and 13, periods which accord well with those in which rapid accumulation of body fat occurs (Figure 1). This suggests that the rapidly increasing organ is sensitive at these times to the environmental stimulus of overnutrition and responds to it by cell division or, perhaps, by cell recruitment. Such a hypothesis is not at

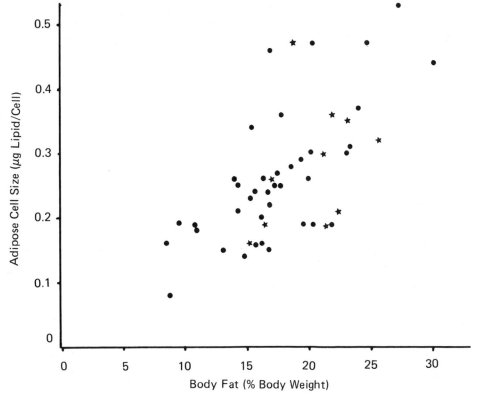

Fig. 3. Relation of the size of adipose cells to body fat in children. •, Boys aged 1–17 years ($N = 40$); ⋆, boys aged 0–1 years ($N = 11$).

odds with present information on the cellular growth of adipose tissue and tallies well with other parameters of growth advanced by overnutrition in a sensitive period, such as the height and skeletal maturity of obese children (Brook, 1973a).

What remains unclear is the natural history of body fatness and whether it matters at all how many cells an individual has. These problems will only be resolved by more data and, best of all, by longitudinal data.

Undernutrition. Experiments of undernutrition are epidemiologically more common in the world as a whole but less well researched from the point of view of adipose cell growth than those of overnutrition. There is evidence that rat adipose tissue is sensitive to undernutrition in a way analogous to its sensitivity to overnurition (Knittle and Hirsch, 1967), but a comparable experimental situation does not (or should not) occur in the human in postnatal life. It is generally agreed that once adipose cells have been formed they cannot be lost through malnutrition which causes only a change in adipose cell size.

In children of short stature who had been small-for-dates babies, Brook (1972a) showed that the number of adipose cells was low and Bonnet and Daune (1976) indirectly confirmed this by demonstrating that the incidence of hyperplastic obesity was relatively less in obese children of birth weight under 3000 g. This suggests that the later months of pregnancy are at least associated with the programing of changes in numbers of adipose cells and that the malnutrition due to placental insufficiency, which low birth weight implies, affects this number.

Berglund *et al.* (1974) were, however, unable to detect a difference between the number of fat cells in men who had suffered from malnutrition for 3 months due to congenital hypertrophic pyloric stenosis in infancy compared with controls. This is, perhaps, not very surprising not only because the period of malnutrition was short but also because, if most of the enlargement of the adipose organ was occurring through changes in cell size at that time, it would not be expected much to affect cell number. We may take this as further indirect evidence of the different contribution of cell size and number increase to the growth of the adipose organ at different times.

Summary. We remain ignorant of the quantitative genetic influence on adipose cell growth, but there is little doubt that environmental stimuli, such as overnutrition or undernutrition, if applied at critical periods, do exert an influence on the number of adipose cells in man. At present we can only guess how severe the stimulus has to be, for how long it has to be applied, and when exactly the critical periods for the adipose organ occur. Nor do we know whether the number of adipose cells an individual has makes any difference to his future fatness. There is a clear need for systematic longitudinal studies of adipose cell growth to provide answers to all these questions.

3.4.3. Hormones

We are extremely ignorant of the endocrine processes which influence cell division. There is evidence that injection of growth hormone (GH) in rats produces an increase in muscle nuclei and some indication that insulin may be responsible for increases in cell size (Cheek and Graystone, 1969). It appears that obese children with advanced skeletal maturity have smaller muscle cells and more of them compared to controls, while children without advanced bone age fall within the normal range (Cheek *et al.*, 1970). Similar findings apply to obese children and their adipose cells (Brook, 1972*a*), and it is tempting to attribute such changes to common hormonal mechanisms.

Overfeeding is likely to result in increased insulin secretion and, of course, the hyperinsulinism of the obese subject is a well-established fact. In children a glucose load causes suppression and then a secondary release of growth hormone (Theodoridis *et al.*, 1969), but there is paradoxical secretion of growth hormone by infants in response to a glucose load (Milner *et al.*, 1972). Consequently, if changes in cell size and number are dependent on hormonal influences, there is plenty of scope for the effects of overfeeding in infancy.

Two natural models exist for the study of hormonal influences on adipose cellularity. The first is the obesity of overweight infants born to mothers with diabetes mellitus, which occurs both in alloxan-treated diabetic rats (Picon, 1967) and man (Osler, 1960). In the former an increase in the number of fat cells was found in one fat depot but not in another (Björntorp *et al.*, 1974). These authors could not, however, identify any differences in adipose cell size or number between adult men who had been the infants of inadequately treated insulin-dependent diabetic mothers and control subjects. They concluded, therefore, that excessive glucose *in utero* might influence the number of fat cells at birth but did not lead to permanent changes.

In fact the obese infants of diabetic mothers do seem to stay fatter than normal children (Farquhar, 1969), so there must be some permanent effect on adipose tissue. Work presently in progress suggests that the size of adipose cells of infants of diabetic mothers correlates well with the degree of hyperglycemia in pregnancy but that the number is not so related (A. G. L. Whitelaw, personal communication). This would tally well with Cheek and Graystone's (1969) data on rat muscle cells and with the demonstration that administration of insulin to rats at birth mainly affects the size of fat cells rather than the number (Salans *et al.,* 1972).

The second model concerns hypopituitary (GH-deficient) children. These have a depleted complement of adipose cells (Brook, 1972*b*) which increases under the influence of exogenously administered human growth hormone (Brook, 1973*b;* Bonnet *et al.,* 1974). The increase in the number of cells is greater than the increase seen in normal children of the same age, but whether it exceeds the catch-up in the rest of growth (i.e., restores the complement of adipose cells to normal) is not certain.

These two models suggest that insulin and growth hormone are both necessary for adipose cell growth and that they cause increase in size and number of cells, respectively.

4. Conclusion

The adipose organ in man appears late in gestation and grows throughout childhood through a combination of increases in cell size and cell number. There is some indirect evidence that the former may be mediated by insulin and the latter by growth hormone. Of the genetic determination of adipose cellularity, nothing is known.

Environmental circumstances can alter both the size and the number of adipose cells. Short-term changes in nutritional status are entirely accommodated by changes in adipose cell size, but longer-term stresses, especially in early childhood or at puberty, can alter cell number. Whether there is a potentiality for recruitment of cells to form new adipocytes in response to overnutrition is not certain, but there seems no doubt that adipose cells, once formed, cannot be lost even in response to marked malnutrition.

Many facts remain obscure about the growth of body fat; only long-term longitudinal studies will reveal them.

5. References

Ashwell, M., Priest, P., and Sowter, C., 1975*a*, Importance of sections in the study of adipose tissue cellularity, *Nature* **256**:724.

Ashwell, M., Priest, P., and Boudoux, M., 1975*b*, Adipose tissue cellularity in obese women, *Rec. Adv. Ob. Res.* **1**:74.

Berglund, G., Björntorp, P., Sjöström, L., and Smith, U., 1974, The effects of early malnutrition in man on body composition and adipose cellularity at adult age, *Acta Med. Scand.* **195**:213.

Björntorp, P., Enzi, G., Karlsson, K., Krotiewski, M., Sjöström, L., and Smith, U., 1974, The effect of maternal diabetes on adipose tissue cellularity in man and rat, *Diabetologia* **10**:205.

Bjurulf, P., 1959, Atherosclerosis and body build with special reference to size and number of subcutaneous fat cells, *Acta Med. Scand.* **166**(Suppl.):349.

Bonnet, F. P., and Daune, M. R., 1976, Factors affecting the subcutaneous adipose cell total number of the obese child, in: Proceedings of the 1st Beilinson Symposium on the Adipose Child (Z. Zaron, ed.), *J. Paediatr. Adol. Endocrinol.* **1**:112.

Bonnet, F. P., Vanderschueren-Lodeweyckx, M., Eeckels, R., and Malvaux, P., 1974, Subcutaneous adipose tissue and lipids in blood in growth hormone deficiency before and after treatment with human growth hormone, *Pediatr. Res.* **8**:800.

Brook, C. G. D., 1971a, Determination of body composition of children from skinfold measurements, Arch. Dis. Child. 46:182.

Brook, C. G. D., 1971b, Composition of human adipose tissue from deep and subcutaneous sites, Br. J. Nutr. 25:377.

Brook, C. G. D., 1972a, Obesity in childhood, M. D. thesis, University of Cambridge.

Brook, C. G. D., 1972b, Evidence for a sensitive period in adipose cell replication in man, Lancet 22:624.

Brook, C. G. D., 1973a, Fat children, Br. J. Hosp. Med. 10:30.

Brook, C. G. D., 1973b, Effect of human growth hormone treatment on adipose tissue in children, Arch. Dis. Child. 48:725.

Brook, C. G. D., Huntley, R. M., and Slack, J., 1975, Influence of heredity and environment in determination of skinfold thickness in children, Br. Med. J. 2:719.

Brook, C. G. D., Lloyd, J. K., and Wolff, O. H., 1972, Relation between age of onset of obesity and size and number of adipose cells, Br. Med. J. 2:25.

Carlson, L. A., and Hallberg, D., 1968, Basal lipolysis and effects of norepinephrine and prostaglandin E_1 on lipolysis in human subcutaneous and omental adipose tissue, J. Lab. Clin. Med. 71:368.

Cheek, D. B., and Graystone, J. E., 1969, The action of insulin, growth hormone and epinephrine on cell growth in liver, muscle and brain of the hypophysectomised rat, Pediatr. Res. 3:77.

Cheek, D. B., Schultz, R. B., Parra, A., and Reba, R. C., 1970, Overgrowth of lean and adipose tissues in adolescent obesity, Pediatr. Res. 4:268.

Dauncey, M. J., and Gairdner, D., 1975, Size of adipose cells in infancy, Arch. Dis. Child. 50:286.

Duckerts, M., and Bonnet, F., 1973, Adipose cell size differences related to the site of adipose tissue biopsy in children, Biomedicine 19:214.

Dugdale, A. E., and Payne, P. R., 1975, Pattern of fat and lean tissue deposition in children, Nature 256:725.

Durnin, J. V. G. A., and Rahaman, M. M., 1967, The assessment of the amount of fat in the human body from measurements of skinfold thickness, Br. J. Nutr. 21:681.

Edwards, D. A. W., Hammond, W. H., Healy, M. J. R., Tanner, J. M., and Whitehouse, R. H., 1955, Design and accuracy of calipers for measuring subcutaneous tissue thickness, Br. J. Nutr. 9:133.

Eveleth, P., and Tanner, J. M., 1976, World-wide Variation in Human Growth, Cambridge University Press, London.

Farquhar, J. W., 1969, Prognosis of babies born to diabetic mothers in Edinburgh, Arch. Dis. Child. 44:36.

Gairdner, D., and Dauncey, J., 1974, The effect of diet on the development of the adipose organ, Proc. Nutr. Soc. 33:119.

Grausz, J. P., 1970, Interscapular skin temperatures in the newborn infant, J. Pediatr. 76:752.

Hirsch, J., and Gallian, E., 1968, Methods for the determination of adipose cell size in man and animals, J. Lipid Res. 9:110.

Hirsch, J., and Goldrick, A. B., 1964, Serial studies on the metabolism of human adipose tissue. I. Lipogenesis and free fatty acid uptake and release in small aspirated samples of subcutaneous fat, J. Clin. Invest. 43:1776.

Hirsch, J., and Han, P. W., 1969, Cellularity of rat adipose tissue: Effect of growth, starvation and obesity, J. Lipid Res. 10:77.

Hirsch, J., and Knittle, J. L., 1970, Cellularity of obese and nonobese human adipose tissue, Fed. Proc. 29:1516.

Hollenberg, C. H., and Vost, A., 1968, Regulation of DNA synthesis in fat cells and stromal elements from rat adipose tissue, J. Clin. Invest. 47:2485.

Hughes, P. T. R., and Tanner, J. M., 1970, The assessment of skeletal maturity in the growing rat, J. Anat. 106:371.

Hull, D., 1974, The function and development of adipose tissue, in: Scientific Foundation of Paediatrics (J. A. Davis and J. Dobbing, eds.), pp. 440–455, Heineman, London.

Kirtland, J., Gurr, M. I., Saville, G., and Widdowson, E. M., 1975, Occurrence of pockets of very small cells in adipose tissue of the guinea pig, Nature 256:723.

Knittle, J. L., and Hirsch, J., 1968, Effect of early nutrition on the development of rat epididymal fat pads: Cellularity and metabolism, J. Clin. Invest. 47:2091.

Milner, R. D. G., Fekete, M., and Assan, R., 1972, Glucagon insulin and growth hormone response to exchange transfusion in premature term infants, Arch. Dis. Child. 47:186.

Norris, A. H., Lundy, T., and Shock, N. W., 1963, Trends in selected indices of body composition in men between the ages 30 and 80 years, Ann. N.Y. Acad. Sci. 110:623.

Osborn, R. H., and de George, F. V., 1959, Genetic Basis of Morphological Variation, Harvard University Press, Cambridge, Massachusetts.

Osler, M., 1960, Body fat of newborn infants of diabetic mothers, *Acta Endocrinol.* **34**:277.

Pařisková, J., and Roth, Z., 1972, The assessment of depot fat in children from skinfold thickness measurements by Holtain caliper, *Hum. Biol.* **44**:613.

Picon, L., 1967, Effect of insulin on growth and biochemical composition on the rat fetus, *Endocrinology* **81**:1419.

Rodbell, M., 1964, Localisation of lipoprotein lipase in fat cells of rat adipose tissue, *J. Biol. Chem.* **238**:753.

Salans, L. B., Zarnowski, M. J., and Segel, R., 1972, Effect of insulin upon the cellular character of rat adipose tissue, *J. Lipid Res.* **13**:616.

Salans, L. B., Cushman, S. W., and Weismann, R. E.,1973, Adipose cell size and number in nonobese and obese patients, *J. Clin. Invest.* **52**:929.

Sims, E. A. H., Goldman, R. F., Gluck, C. M., Horton, E. S., Kelleher, P. C., and Row, D. W., 1968, Experimental obesity in man, *Trans. Assoc. Am. Physicians* **81**:153.

Sjöström, L., and Björntorp, P., 1974, Body composition and adipose tissue cellularity in human obesity, *Acta Med. Scand.* **195**:201.

Sjöström, P., Björntrop, P., and Vrána, J., 1971, Microscopic fat cell size measurements on frozen-cut adipose tissue in comparison with automatic determinations of osmium fixed fat cells, *J. Lipid Res.* **12**:521.

Tanner, J. M., and Whitehouse, R. H., 1975, Revised standards for triceps and subscapular skinfolds in British children, *Arch. Dis. Child.* **50**:142.

Tanner, J. M., Whitehouse, R. H., and Takaishi, M., 1966, Standards from birth to maturity for height, weight, height velocity and weight velocity, British children 1965, *Arch. Dis. Child.* **41**:454, 613.

Theodoridis, C. G., Brown, P. A., Chance, G. A., and Rayner, P. H. W., 1969, Growth hormone response to oral glucose in children with simple obesity, *Lancet* **1**:1068.

Wessel, J. A., Ufer, A., Van Huss, W. D., and Cederquist, D., 1963, Age trends of various components of body composition and functional characteristics in women aged 20–69 years, *Ann. N.Y. Acad. Sci.* **110**:608.

Widdowson, E. M., and McCance, R. A., 1960, Some effects of accelerating growth, *Proc. R. Soc. London Ser.* B **152**:188.

3

The Methods of Auxological Anthropometry

NOËL CAMERON

1. Introduction

Anthropometry is the technique of expressing quantitatively the form of the body. Hrdlicka (1947) defines it as a system of techniques, the systematized art of measuring and taking observations on man, his skeleton, his brain, and other organs, by the most reliable means and methods for scientific purposes. It is limited only by the problems to which it is applied and no treatise, however large, can account for all the factors present in an original study. Time and again the anthropometrist will be presented with problems for which there are no guidelines, and he, alone or in discussion with his colleagues, must find satisfactory solutions.

The object of this chapter is to lay down some ground rules for the anthropometry of somatic growth, or as I have termed it, auxological anthropometry. This area is one that has been largely neglected in most standard anthropometric writings because the science of anthropometry developed mostly in an anthropological context and was concerned with studies of differences between adults.

Techniques used by Lawrence, Broca, Haddon, Keith, and Martin, some of the most eminent anthropologists of the last two centuries, have been borrowed and refined to cope with the very special problem of measuring changing parameters of an individual from birth to maturity and senescence. The individual cannot be remeasured after six months to check old measurements; he has grown or withered, for the body is in a constantly dynamic state and will not produce the same answer after one week has elapsed, let alone a number of months. Therefore the anthropometry of somatic growth has this problem: how to take a measurement, record it, and be certain, within the limits of known error, that the measurement represents that parameter at that particular moment in time, so that if the measurement is

NOËL CAMERON • Department of Growth and Development, Institute of Child Health, London, England.

repeated at a future time, then any positive or negative increment is a real change and not the result of error.

At first sight this problem appears quite simple. The sophisticated instrumentation that is now available, a good knowledge of the landmarks which must be palpated, and the correct positioning of instrument and subject should result in the correct measurement. But this is not the case. It is not only the subject that is a dynamic, constantly changing individual; so too is the observer. His accuracy is susceptible to changes of mood, fatigue, health, and environment, and therefore he requires to control himself as much as the instrumentation and subject.

The anthropometry of somatic growth thus intimately involves the subject, instrument, and observer in an interactive situation that the observer is constantly attempting to standardize in order to measure the only important change—that of the subject. It is enough at this juncture to make the reader aware of this problem and prevent his dismissing anthropometry as a rather simple, but necessary, part of research or diagnosis. The research director or the hospital physician must know the extent to which his subjects are growing and that this information has been accurately acquired.

Longitudinal growth studies are the means by which we learn of individual growth patterns and the differences in tempo of growth between individuals. Count Philibert Gueneau de Montbeillard is usually credited with the first documented longitudinal growth study surviving to the present day (Scammon, 1927). During the years 1759–1777, he measured the growth of his son and published the resulting data in a supplement to Buffon's *Histoire Naturelle*. Unfortunately no record of the techniques he used to measure stature is available, but we must assume from the classical form of the unsmoothed distance curve that he was extremely accurate and used a standard technique. The study of growth was to lie fallow for the next century until Franz Boas' writings at the end of the 19th century. During this dormant period considerable interest was being aroused among anthropologists as to the origins of man. One of the earliest prime movers in this arousal was Louis-Jean-Marie Daubenton (1715–1799), a comparative anatomist who provided much of the basic data for Buffon's contribution to natural history. Following Daubenton, Blumenbach, J. C. Prichard, and W. Lawrence contributed much to physical anthropology. All this was before Darwin's *Origin of Species* in 1859. After this, anthropologists emerged in relative abundance, but nearly all were interested principally in the origins of man and used anthropometry to this end. They were interested not in how man was growing and developing in their own times but on how he had come to be there at all. "How has mature man come to be this shape and size?" was a far more pertinent question than "What processes of growth do we go through to become mature men?" To answer their questions anthropologists used the data from "cross-sectional" studies on primitive man and, with few exceptions, their instruments changed little from 1850 to 1950.

However, anthropologists did acutely realize that standardization of anthropometric techiques and procedures was a necessity. Until 1870 the Broca techniques of anthropological anthropometry were universal. Growing individualism and the isolationism of Germany resulted in three conferences dealing with German anthropological anthropometry during the 19th century: the Congress of the German Anthropological Society in 1874, the Craniometric Conference in Munich in 1877, and the Berlin Conference in 1880. The result of these conferences was the "Frankfurt Agreement" adopted at the 13th General Congress of the German Anthropological Society held at Frankfurt-on-Main. Thus the French and German schools of anthropometry were established. Further attempts at unification occurred during the 1890s in Paris under the individual initiative of R. Collignon (1892) and at the 12th International Congress of Prehistoric Anthropology and Archaeology of 1892 in Moscow. According to Hrdlicka (1947), neither of these accomplished anything substantial, but standardization became at least a recognized necessity. At the International Congress of 1912 in Geneva an "International Agreement for the Unification of Anthropometric Measurements to be made on the Living Subject" was finally drawn up. This agreement gave general principles of measurement and detailed definitions of 49 measurements, some of which are extremely interesting when viewed in the light of the current state of the science. It was agreed, for instance, to measure the left side of the body except for the measurements of acromial and trochanteric height, yet still today many studies, especially in the field of nutrition, are undertaken using the right side of the body. Observers were requested to indicate in every case, with precision, their method of measurement and the instruments employed. Wilton Krogman in the preface to Garrett and Kennedy's *Collation of Anthropometry* (1971) cites the case in which he had to dismiss nearly 300 of 600 studies on child growth while compiling the volume *Growth of Man* for Dr. T. Wingate Todd. The studies were dismissed because of inadequacies of sample size, poor statistical analysis, and "sloppy" definition of measurements.

So, although unification has been agreed to in principle, the adherence to internationally recognized methods of measurement is a far different thing and will probably only come about when scientific periodicals insist on precise indications of methods used and instruments employed in every study involving anthropometry.

The anthropometry peculiar to somatic growth was advanced in no small degree by the longitudinal growth studies made in the United States in the first half of this century. From 1904 to 1948, 17 such studies were started and 11 completed. At present only one or two are still in operation. Their complexity varied from the relatively simple elucidation of the development of height and weight to data yielding intercorrelations of behavior, personality, social background, and physical development. One factor common to all was the desire to maintain accuracy of measurement. Administrative problems and staff changes meant that this was not always possible but runs of ten years with the same measurement team are to be found in the Berkeley Growth Study of 1927 (Bayley, 1940), and the Yale study of the same year (Gessell and Thompson, 1938). Research workers became aware of the need for comparability of measurements and published precise accounts of their methods and techniques, with suitable adaptations for the measurement of growth. Three of the best accounts of techniques used for these early studies are those of Frank Shuttleworth (1937) for the Harvard Growth Study of 1922 (made in the School of Education), Harold Stuart (1939) for the Center for Research in Child Health and Development Study of 1930, and Katherine Simmons in her reports of the Brush Foundation Studies of 1931 (Simmons, 1944). It is indicative of the problems involved in anthropometry that Shuttleworth's and Simmons' subjects were measured while still clothed. Stuart (1939) provides the most complete account of measurements used for growth study using the techniques laid down in the International Congress of 1912, "with diversions from these techniques at appropriate times." H. V. Meredith (1936) stands alone in his perception of the problems involved in the measurement of human growth at the time of the early U.S.A. studies. The multitude of papers he published on the growth of children from Iowa City, Iowa; Massachusetts; Alabama; Toronto, Canada; and Minnesota contain excellent examples of reliability control (see Section 3). He was convinced that long-term studies of physical growth would only be facilitated by preliminary detailed investigations into the accuracy with which body parameters could be measured during growth. If the growth increment from one age to the next was less than the 90th centile of the differences between repeated measurements of a chosen parameter, he thought it unwise to take the measurement. Detailed descriptions of the measurements his team used are included in his reports, and although they are not always the best techniques available, he and his colleague Virginia Knott (1941) did criticize the sparsity of modern techniques and the lack of information pertaining to their reliabilities.

These studies did much to influence more recent researchers in the field of growth and development on both sides of the Atlantic. In Britain few worthwhile longitudinal growth studies were undertaken before 1949. Fleming's study of stature and head measurements (1933) was the only one to be published for use as standards of reference by 1950 (Tanner, 1952). However, the Harpenden Growth Study initiated in 1949 by Professor J. M. Tanner became something of a turning point for British auxological anthropometry. R. H. Whitehouse, the anthropometrist who measured the subjects from the beginning, was to stay with this study for its duration, and at the time of writing he is still measuring some of the subjects who are now almost 30 years old. The record he created, of every measurement on every subject on every occasion for 25 years having been made by him personally, is a model for longitudinal studies for all time. The team of Tanner and Whitehouse radically altered the approach to anthropometric measurement of growth. Whitehouse found that the instruments used worldwide for anthropometry were cumbersome and tiring to use for an 8-hr day. Using the best principles of ergonomic design, and with due regard to the accuracy needed for growth measurements, Tanner and Whitehouse designed the Harpenden range of anthropometric instruments, discussed in detail below, which have become accepted internationally for their accuracy, consistency, and ease of use. Alongside developments in instrumentation came the development of techniques, and much research was undertaken, both in England and abroad, to test their accuracy and reliability compared to the older methods (Damar and Goldman, 1964; Sloane and Shapiro, 1972; Parizkova and Roth, 1972). More attention was paid to the meaning of statistically significant errors of measurement, and modern laboratories were designed to provide privacy for the subject and efficiency for the anthropometrist.

Today the standard of anthropometry can be assessed from the way in which it is used. In Great Britain anthropometry is relied upon to provide accurate records of the growth velocity of abnormally growing children. Some 1500 children, with suspected growth disorders, are measured each year by the London team of a nationwide network of anthropometrists and physicians who work in close cooperation to evaluate their growth patterns. Should the parameters of height, weight, limb circumferences, skinfolds, limb lengths, or their relative ratios, when compared to normal standards, cause concern to the physician, they are followed up to check whether their growth velocities are of an acceptable rate for their age and sex. In this way growth disorders, difficult to diagnose accurately by any other means, are evaluated, and a treatment or management regime is initiated to promote either a normal or supernormal growth velocity as measured by the clinical auxologist.

Growth studies are still of paramount importance to developing countries. The Cuban Growth Study started in 1972 (Jordan *et al.*, 1975) illustrates a textbook approach to growth analysis using anthropometry with a full awareness of its capabilities—and of its errors in unskilled hands. Nine measurement teams of seven individuals were trained by one expert over a 2-week period and then attended quality control meetings every two months for the duration of the study to test their reliability and validity. The resulting low standard errors reflect the care taken in the training and monitoring of these teams.

The sensitivity of certain anthropometric measurements to dietary changes means that this science can be confidently used to check the nutritional status of different communities (Buzina and Uemura, 1974). Commonly height, weight, arm circumference, and skinfolds provide the best indices of malnutrition, and this includes overnutrition as well as undernutrition. There is, however, some disagreement as to which of these parameters provides the single most useful measurement. Arnhold (1969) and Loewenstein and Phillips (1973) maintain that the comparison of upper arm circumference to height, as utilized in the QUAC stick, constitutes a quick and accurate means of grading malnutrition in children. Sastry and Vijayava-ghaven (1973) and latterly the World Health Organization (Waterlow *et al.,* 1977) favor the comparison of height to age and weight to height and restrict the use of arm circumference to "emergency situations." It must be emphasized that the determination of nutritional status may often be achieved using very simple means (Morley, 1974), and as long as the standards on which the means are based are relevant to the population under investigation, then there is frequently no need to jeopardize one's assessment by using techniques requiring greater skill than is available.

2. The Auxological Anthropometrist

Accurate instrumentation has overcome many of the inherent errors in anthropometry, but the greatest source of error comes from the measurer himself. Hrdlicka (1947), in his treatise on practical anthropometry, described the qualities needed by an anthropometrist; good eyesight for distance and color; "freedom from halitosis and other unpleasant odors"; sympathy, perseverence, orderliness, thorough honesty and carefulness; he should be "careful of the sensibilities of his subjects; careful in technique, careful in reading the scale of his subject, careful in recording and capable of concentration on his work." The lack of these qualities has produced more errors than all other causes together. It is interesting to note that he thought men to be superior to women as anthropometrists; although women excelled in carefulness and devotion to work, they lacked the stamina for field work and marriage would seriously hamper their careers. The outcry against such male chauvinism would be strong today—and all the Cuban anthropometrists were female—but the qualities outlined by Hrdlicka are still very relevant. To these must be added basic qualifications in anatomy and physiology, so numerous and important are the landmarks that must be palpated for accurate measurement. The anthropometrist's relationship with both the instruments and his subjects should be such that he is relaxed and confident with both and almost obsessional about acquiring and maintaining accuracy. Stamina and fitness are pertinent qualities. Design of the anthropometric laboratory so that the subject is made to move rather than the measurer will mean efficient and less tiring work may be done. Most of the techniques described below require two positions of the subject for each single position of the measurer.

Anthropometry is such a varied science that no amount of instruction will cover all possible difficulties that will arise during the course of measurement. It is at this time that sound judgment, common sense, and a real knowledge of the science will be the anthropometrist's most worthwhile qualities. Translations of some of the early anthropometric writings in German and French are available, but for speed and the purists' approach the works of Hrdlicka (1947), Ashley Montagu (1960), and Garrett and Kennedy (1971) are almost a necessity. Writings specific to the field of growth and development are few, and very often the diversity of techniques is misleading. Publications from the Fels Research Institute in America or the Institute of Child Health in England are probably the best and most carefully controlled in the anthropometric methods used for their studies. The accuracy achieved by these workers, necessary because of the small linear changes that they often have to measure, is at times quite remarkable and can seldom be bettered by purely automated schemes of measurement. Physical anthropologists involved only with cross-sectional measurements, are not in a position to see their repeated measurements on the same subject over a period of years and thus cannot gauge their accuracy by looking at the smooth growth curve obtained by the most accurate of auxological anthropometrists when measuring the growth of a subject. Referral of an intending measurer to a growth center is often advisable for this specific reason: that his accuracy must be better than the physical anthropologist if growth is to be properly recorded. With spiraling costs of research it is important that investigators do their "homework" before embarking on their chosen study and doubly important that the anthropometrist collecting the basic data realizes the scope of his science and the errors inherent in his measurements.

Once the basic reading and learning has been done, practical instruction from a trained anthropometrist must be obtained. If subjects are available this may last a matter of days, but it is useless unless the learner then practices his newfound art on all suitable occasions. One of the best schemes for practice is to contact a local school and carry out a study of one's accuracy and reliability on all age groups and both sexes. If a maternity hospital is at hand then practice on neonates is advisable. This testing of reliability ought to be repeated every year to check departures from accepted techniques and enforce first principles of measurement. If possible, the anthropometrist should refer back to his original teacher for comments on any unnoticed change in style and points that need correction. As with all practical skills, the objective view is usually more critical and therefore more helpful.

Familiarity with the instruments is of paramount importance to any anthropometrist. He should be able to service and mend his instruments in the same way, and with the same regard to perfection, as a racing car driver is acquainted with and able to service and repair his vehicle. To the latter it is a question of life or death, to the former a question of his professional standing and scientific integrity. Not that the instruments recommended below are subject to inaccuracy or frequent breakdown. But, as with all precision instruments that are in continual use, they will at times require calibration and servicing.

Finally, the anthropometrist must always be aware that his techniques are not unchangeable. New techniques and procedures ought to be looked for in each situation that presents itself. For too long anthropological anthropometry did not advance and instrumentation did not change because the men who controlled the advancement of the science were satisfied that their techniques were right. The auxological anthropometrist is at the same time a reactionary and an innovator, keeping what is best but trying to improve his techniques to cope with the vagaries of the growing body and the multitude of differences between individuals. He must communicate any advancement to his fellows. To reemphasize the recommendations of the 1912 International Congress: precise and complete accounts of techniques and instruments should be included in all research reports.

3. Reliability

Definitive works of the reliability of auxological anthropometry are few. Knott (1941), Tanner and Weiner (1949), and Gavan (1950) each expressed their concern at this state of affairs. Tanner and Weiner (1949) thought that this was perhaps due to the profusion and variety of measurements made on living subjects. The indecision among anthropometrists in determining which measurements and techniques are best to use could well be due to this sparseness of published reliability data. Certainly very few authors carry out reliability studies, or if they do they seldom feel them important enough to report their results in their papers. The reasons for this are obscure but a powerful one is that there is (or was) a lack of standardized terminology to describe the reliability of measurement in a clearly understandable statistical format.

It is as well, therefore, to begin this section by defining what we mean by reliability and describing the ways in which it may be assessed and reported.

Knott (1941), in her authoritative paper on anthropometric reliability, uses the term "reliability" to refer to the comparison of independent measurements repeated on the same subject within a short time interval. One is concerned, of course, with the fluctuations in measurements, commonly called errors, and whether they are systematically positive or negative. Knott (1941), by way of explanation, divides errors into two sorts: "accidental" and "technical." The former covers such things as misreading the instrument and the misrecording of a correctly read dimension, the latter such things as definitions of landmarks, bad subject positioning, errors in locating properly described landmarks, variations in slanting the instrument, application of incorrect pressure, and finally—and this is very important in auxological anthropometry where children are one's main subject sample—positional changes of the subject after initially correct positioning. Healy (1958), who must be regarded as a pioneer in this field, simply subdivides the term "errors" into actual errors of measurement and real fluctuations in the dimension being measured. By real fluctuations he does not mean the growth of a parameter from one occasion to the next, but implies from the terminology the situation in which a subject has a stable value about which random variations occur.

The assessment of errors is basically done in two ways. Spielman *et al.* (1972) suggested that these be called "repeats" and "duplicates," but perhaps a more descriptive terminology would be "within-observer replicates" and "between-observer replicates." The first, within-observer replicates, involves one observer carrying out repeated measurements on a number of subjects, while the second involves a number of observers each measuring each subject. Where large-scale surveys are involved the second system is obviously preferable, but the first is effective and convenient for small surveys and individual observer assessment. In the simplest case with p subjects and q observers, Healy (1958) recommends an analysis of variance of the form shown in Table I.

An example of how this is done is furnished by Edwards *et al.* (1955). They were concerned with the introduction of a new skinfold caliper and therefore wished to test its consistency compared with others in current use. These latter calipers had different spring pressures and face areas. Five experiments were involved of which four used the statistic of the standard deviation of the differences between duplicate measurements taken by one or all of a number of observers. These four experiments were to test: (1) the effect of different spring pressures; (2) all combinations of pressures and face areas; (3) the reliability of the new calipers; and (4) the errors of measurement of the new calipers at various spring pressures. The fifth experiment was designed to provide a more comprehensive trial of the new caliper, and involved three models of the new caliper each producing a slightly different pressure at the jaw faces. To test the average difference between calipers, whether these differences vary with the subject or observer, the presence of observer bias, or the observer–subject interaction, an analysis of variance for six observers, three calipers, and 180 duplicate readings was set up as in Table II. The interpretation of this analysis follows the format suggested by Healy (1958).

The following model for the observations is implied from the analysis of variance in Table I:

$$y = y_s + E_o + E_{so} + E_D$$

where y is the observed reading, y_s is the stable value for the subject concerned, E_o, E_{so}, and E_D relate to the error of the particular observer, subject–observer interaction, and the particular observation, respectively. They are assumed to be random and independently distributed variables with variances s_o^2, s_{so}^2, and s_D^2, respectively, which are thus used to calculate the expected mean squares in the last column of Table I. E_D represents the actual error of measurement, E_o allows for an average personal bias (and is therefore constant for each observer), and E_{so}, the interaction term, allows for the fact that an observer's bias may in practice vary from one subject to another. The assumption of independence implies that the subject–observer interaction is not systematic, i.e., it is not true that some observers tend to overestimate large measurements and underestimate small measurements relative to others. The variance components may be readily obtained from Table I. That for the duplicates is equal to the mean square (V_D); for the subject–observer interaction it is half the differences between the interaction mean square and the duplicate mean square ($V_{so} - V_D/2$); for the observer it is the difference between the observer mean square and the interaction mean square divided by $2p$; ($V_o - V_{so}/2p$). In the case of within-observer replicates with a single observer, the differences between the measurements is $E_{D1} - E_{D2}$ and the SD $= \sqrt{2s_D}$. In the case of between-observer replicates with more than one observer, the difference involves the interaction errors (E_o and E_{so}) and the $SD = \sqrt{2(s_o^2 + s_{so}^2 + s_D^2)}$.

One further statistic in common use is the standard error of measurement (s_{meas}), calculated from the formulas:

$$s_{meas} = s_d/\sqrt{2} \qquad \text{or} \qquad s_{meas} = s_1\sqrt{1 - r_{12}}$$

where s_d is the standard deviation of the differences, s_1 is the pooled SD of the total of all measurements, and r_{12} the coefficient of reliability or objectivity.

The design of reliability studies by the method of within- or between-observer replicates yields data on the absolute values and the magnitude and direction of the

Table I. Analysis of Variance Model for the Assessment of Reliability[a]

	DF	Mean squares	Expected mean squares
Subjects	$p - 1$	V_s	
Observers	$q - 1$	V_o	$2ps_o^2 + 2s_{so}^2 + s_D^2$
Subjects × observers	$(p - 1)(q - 1)$	V_{so}	$2s_{so}^2 + s_D^2$
Duplicates	pq	V_D	s_D^2

[a]From Healy (1958). Reproduced by kind permission of the Wayne State University Press and M. J. R. Healy, *Hum. Biol.* **30**(3):210–218.

Table II. Analysis of Variance of Measurements with Three Similar Dial-Type Calipers (in Logarithmic Scale)[a]

Source of variation	DF	Mean square			
		Biceps	Triceps	Subscapular	Suprailiac
Calipers	2	72.70	69.17[b]	115.30[b]	201.74[b]
Subjects × calipers	58	41.65	3.46	2.01	8.60
Observers	5	3560.46[b]	109.64[b]	125.90[b]	699.66[b]
Observers × calipers	10	57.52	5.61	1.66	3.73
Observers × subjects	25	170.60[b]	27.75[b]	8.95[b]	66.03[b]
Observers × subjects × calipers	50	37.22	2.61	1.96	9.33
Duplicates	180	42.09	4.38	1.81	7.69

[a]From Edwards (1955). Reproduced by kind permission of The Cambridge University Press.
[b]Significant at $P < 0.01$.

differences between repeats by one observer or duplicates by a number of observers. The statistical treatment and interpretation of these differences, either in isolation or related to the absolute values for the dimension, constitute the knowledge to be gained from a reliability study.

Meredith (1936) and his co-worker Knott (1941) studied the distribution of the absolute differences calculated without regard to sign. They reported the median, 90th centile, and maximum differences for each dimension. The median difference and the calculated 90th centile difference were also reported as percentages of the mean size of the dimension. Thus a median and (effectively) maximum percentage error were reported. Meredith related these to the expected normal growth increment to find what age intervals would in general produce increments greater than this measurement error. Thus stature, for instance, had a 90th centile difference of 6.5 mm and an annual increment of 65.2 mm at age 8–9 in boys and could be measured bimonthly, whereas ''arm girth'' with a 90th centile difference of 4.2 mm and an annual increment of 5.2 mm could better be measured annually. These suggestions are certainly important if one is looking for significance between two periods of measurement, although if curve fitting is involved in a longitudinal study there is no reason why measurements should not be taken more often. Meredith (1936) emphasized the important fact that the error in relation to the absolute measurement may be small but in relation to the growth increment may be large. To illustrate this point, an error of 1.0 cm in the height of the average English boy at age 10 is only 0.7% of his actual height but 20% of his annual growth increment of 5.0 cm/yr.

Although Meredith (1936) and Knott (1941) used the distribution of absolute differences without regard to sign, leading growth statisticians advocate the inclusion of the sign in calculations of reliability. One is, after all, seeking to determine whether there is systematic bias, and using the absolute value only introduces confusion on this issue.

Gavan (1950) used the results of his ''consistency study'' so that the differences between replicated measurements were not reported at all. Instead he used the mean values for each dimension and the size of their standard deviations to divide the dimensions into high, medium, and low consistency groups. The reasons for doing this are apparent from his experimental design. Six measurement teams were trained and the work so divided that any one observer took only one fourth of the total number of measurements. In all there were eight series of measurements taken by six teams on five subjects. One subject was measured by three teams, but no subject was measured by all. The material he presents covers 62 measurements, each taken at least 10 times by one team on one subject; most were taken 30 times, with a maximum of 49. Every team did not take the full complement of measurements. Gavan calculated the mean value, its standard deviation, and the coefficient of variation (CV) for each dimension by averaging the results of all the teams that had taken that dimension. He then plotted mean against standard deviation to obtain a scattergram representing all 62 dimensions. From visually appraising the vertical distribution, i.e., the standard deviation distribution, he was able to insert arbitrary boundaries where there was a break in this distribution. The validation of this procedure was apparent when he listed the measurements in each of the three groups so determined. From high-consistency measurements, i.e., those with small SDs, to low-consistency measurements, i.e., those with large SDs, he found an increasing coefficient of variation.

In Table III are listed some of Gavan's 62 measurements with their means, SDs, and CVs.

Gavan included an important discussion on why consistency should change and came to the conclusion that consistency increases as: (1) the number of technicians decreases, (2) the amount of subcutaneous tissue decreases, (3) the experience of the technician increases, and (4) the landmarks are more clearly defined. Therefore in any growth study the smallest number of the most experienced anthropometrists should be used and the landmarks upon which they depend should be well defined and/or easily located.

Kemper and Pieters (1974) set out to test the objectivity of several anthropometric measurements taken by two institutes on the same subjects. Most of their interpretation is based on the coefficient of objectivity (the correlation coefficient between repeated sets of measurements), but they did indicate the direction of the

Table III. Consistency Groups for a Selection of Parameters[a]

	N	Mean (mm)	SD (mm)	CV (%)
High consistency				
Stature	196	1755.1	5.1	0.3
Sitting height	195	912.6	5.9	0.6
Head circumference	197	524.2	3.4	0.6
Forearm/hand length	150	474.2	3.8	0.8
Foot length	194	264.7	2.7	1.0
Arm length	196	786.8	6.7	0.8
Medium consistency				
Chest circumference	195	931.6	13.1	1.4
Calf circumference	70	364.6	4.7	1.3
Upper-arm length	149	343.2	6.8	2.0
Biiliac diameter	147	294.0	7.7	2.6
Mid-upper-arm circumference	70	288.4	6.7	2.4
Chest breadth	151	282.7	5.4	1.9
Low consistency				
Lower-leg length	148	465.1	11.6	2.5
Upper-leg circumference	69	349.7	9.2	2.6
Waist circumference	195	816.2	17.7	2.2

[a]From Gavan (1950).

*Table IV.   Standard Deviation (cm) of Differences between Duplicate Measurements, Averaged
for the Two Measurers of Each Team; Stature in Adolescents (Each Based on 10 Duplicates)*[a]

Team	Control No.							Average
	1	2	3	4	5	6	7	
Pinar del Rio	0.212	0.205	0.363	0.240	—[b]	0.234	—[c]	0.251
Havana (metropolitan)	—[b]	—[b]	0.264	0.254	0.183	0.149	0.184	0.207
Interior	0.234	0.240	0.166	0.186	0.242	0.190	—[c]	0.210
Mantanzas	0.240	0.122	0.084	0.180	0.166	0.151	—[c]	0.157
Las Villas	0.290	0.172	0.198	0.230	0.209	0.279	0.236	0.231
Camaguey	0.252	0.230	0.213	0.225	0.198	0.267	—[c]	0.231
N. Oriente	0.264	0.161	0.196	0.200	0.154	0.303	0.206	0.212
S. Oriente	0.262	0.148	0.206	0.185	0.156	0.176	0.186	0.188
All teams	0.251	0.183	0.211	0.213	0.187	0.219	0.203	0.211

[a]From Jordan (1975).
[b]The team did not participate due to illness of one measurer.
[c]The study in these provinces had ended.

differences between the repeated sets of measurements by correlating the actual
difference and the mean absolute measurement for the two institutes. Suprailiac
skinfold, for instance, had a correlation coefficient of -0.306 which was significant
at the 5% level and indicated that differences for high values were smaller than
those for low values.

The 1972 Cuban National Child Growth Study reported by Jordan *et al.* (1975)
contains data for the reliability of eight measurement teams, all trained by one
expert, mostly in terms of the standard deviation of duplicates. The results for
stature may be seen in Table IV. Seven quality-control meetings were organized
and, as can be seen, the general trend is for an increase in reliability shown by the
decreasing standard deviations from left to right. No significant differences
occurred between individual measurers within teams, nor between teams, nor
between different quality-control occasions, even though the first appear worse
than the others. However, if there had been a loss of reliability, it would have
shown up quite clearly and suitable steps taken to correct the measurement team.

Martorell *et al.* (1975) investigated and evaluated the measurement variability
in the anthropometry of preschool children. They presented a simple technique of
evaluating measurement variability for cross-sectional and longitudinal measurements by expressing the total measurement variance as a percentage of the appropriate intersubject variance. The statistic produced by this procedure estimates the
portion of the intersubject variance accounted for by measurement variance and
serves as an index of the relative reproducibility of anthropometric measurements.
Eighteen variables were used to illustrate the approach of these authors to reliability
testing. Their results indicate that chest circumference has poor reliability when
compared to the measurements of supine length, calf and upper-arm circumference,
and subscapular skinfold.

We have looked then at reliability, its assessment, and the practical ways in
which it has been employed and reported on. The methods of Meredith (1936) and
Knott (1941) are excellent and very pertinent for growth studies. The design of
Edwards *et al.* (1955) is elegant and allows a great deal of interpretation from the
one analysis of variance. Jordan *et al.* (1975) illustrate the extreme usefulness and
importance of continuous reliability checks during the course of the study. Marto-

rell *et al.* (1975) provide a statistically sound index of anthropometric reproducibility which should serve as a model for further studies.

To summarize: reliability may be tested by experiments of a test–retest design, resulting in data pertaining to the absolute values and the differences between two occasions. These differences may be analyzed in terms of the standard deviation, coefficient of variation, and mean. The analysis of variance , as proposed by Healy (1958) and used by Edwards *et al.* (1955), is perhaps the best statistical design producing most information from the data.

4. Recommended Instruments

Advances in instrumentation in recent years have meant that there are currently available a large number of anthropometric instruments which are purported to give accurate results. Most have been in use, in one form or another, since the last century. Their development from physical anthropology has endowed them with characteristics which are, at times, unsuitable for auxological anthropometry. The problem posed by the latter science is that the subject is usually young, sometimes uncooperative, perpetually in motion, and always growing. Therefore, the instruments common to auxology have to be rather different from their anthropological counterparts.

Biacromial diameter may be measured using the Martin anthropometer described in detail by Hrdlicka (1947), or the Harpenden anthropometer described below (Section 4.6). Martin's anthropometer is characteristically difficult to use because of the high levels of friction involved owing to its design. Soft tissue, which must be compressed at the lateral borders of the acromial processes, accentuates the need for an instrument with very low frictional forces and an easy-to-read digital display. Such an instrument is the Harpenden anthropometer. It was designed with the knowledge that soft tissue often masks palpable bony landmarks and the measuring instrument must therefore be able to compress this tissue and allow an accurate reading of the desired parameter. To measure the same parameter with Martin's instrument would require the anthropometrist to locate the bony landmarks, apply the blades of the anthropometer to these landmarks, and then to remove the instrument before being able to read the measurement. The Harpenden anthropometer simply requires palpation and application of the blades to the landmarks and then direct reading of the measurement while the blades are still in position.

Digital counter displays are one of the refinements that epitomize the approach of Tanner and Whitehouse to anthropometric instruments. The counter system has little to do with the actual measurement technique, it concerns the relationship between the observer and what he is observing—the measurement to be recorded. Previously anthropological instruments were designed with much less emphasis on the observer–instrument relationship. The instruments recommended below, therefore, have taken the sophistication of anthropological measurement one step further and incorporated devices to negate inaccuracy in reading the instruments. Also because the counters display measurements in whole units of the smallest unit required, i.e., millimeters, the often precarious procedure of the observer including a decimal point in his measurements is eliminated. (In the case of anthropometry in Great Britain the placing of decimals is left to the computer.) Instruments will be described later in this chapter that automate the whole process of reading and recording measurements. They do, however, have various disadvantages which make their inclusion in this section inadvisable.

The following instruments are manufactured by Holtain Ltd., Crymmych, Pembrokeshire, Wales, Great Britain, and are available from them and also from Siber Hegner, Talstrasse, Zürich, Switzerland. Technical data relating to their construction applies at the time of writing.

4.1. Harpenden Stadiometer

The Harpenden stadiometer (Figure 3) is used to measure stature. It is composed of a rigid vertical backboard, faced in formica for easy cleaning, and a rigid horizontal wooden headboard moving up and down the backboard by virtue of miniature ball-bearing rollers in order to ensure a movement which is free yet without cross-play. The headboard is 600 mm from ground level at its lowest point and may be raised to 2100 mm for maximum height readings. The engineering is precise enough to ensure that the headboard moves downward at a slow speed under the influence of gravity, and need not, therefore, be pushed downward, thereby risking damage to the counter display unit fixed to the right of the headboard. Damage to these counters is unlikely as they are tested up to 5000 rpm but, especially where children are concerned, they must be protected from undue strain. The counter displays the measurement in millimeters and, as with all counter mechanisms, should be read to the last completed unit. Thus, if the last wheel is between four and five, four is the unit recorded.

Fixed and portable forms of the stadiometer are available. The former is recommended for hospital, laboratory, school, and health clinic use while the latter is particularly convenient for field work where transportation is a problem.

The main frame of this instrument is made of light alloy angle and the fixed form is provided with adjustable wall brackets for mounting purposes. Its total weight is approximately 12.7 kg (28 lb), and the portable form, including a base into which the stadiometer may be packed, weighs only 20.5 kg (45 lb).

4.2. The Harpenden Supine Measuring Table

This instrument is similar in construction to the Harpenden stadiometer, but on a horizontal rather than vertical plane. In the horizontal position the headboard is fixed and the footboard is free-moving and mounted on ball bearings. It has the same ease of movement and potential for speedy and error-free direct readings, to the nearest millimeter as the stadiometer, but a larger range of 300 mm to 2100 mm. The supine measuring table weighs 29.5 kg (65 lb) and ought to be mounted on permanent wall brackets, although adjustable legs may be supplied.

4.3. The Harpenden Infantometer

The infantometer is designed for postneonatal growth studies, and its use is recommended in conjunction with, and as a follow-up instrument to, the Harpenden neonatometer described below. Its construction is similar to that of the supine measuring table but on a smaller scale with a range of 300–940 mm. The movable footboard is operated via a constant pressure lever which locks the footboard when a pressure of 0.5 kg (1 lb) is applied to it. This refinement is particularly important when very young children are being measured who invariably bend their legs in protest at being manipulated on the apparatus (see techniques for measuring supine or recumbent length, Section 6.4.).*

4.4. The Neonatometer

The neonatometer was specifically designed for neonatal growth studies, in preterm as well as full-term infants, as a high-accuracy, counter-reading instrument (Davies and Holding, 1972). The Welsh National School of Medicine collaborated in its development, and it may be used in incubators. This author also strongly recommends its use in neonatal health clinics and ''well-baby'' clinics where the standard of measurement is, at times, appalling due to a naive approach to measurement and clinical assessment based upon measurement.

Its construction is of a light alloy frame within which are a curved fixed headboard and ball-bearing-mounted, free-moving footboard operated via a constant-pressure lever. The whole apparatus, weighing only 2.0 kg (4.4 lb), may be placed over the neonate who may then lie undisturbed while the observer prepares for measurement.

4.5. The Harpenden Sitting-Height Table

Sitting height may be measured with the Harpenden anthropometer but, where space and money allow, the facility of a sitting-height table is recommended (Section 6.3.1., Figure 4). Its construction is of a table approximately 60 cm high along the length of which runs a carriage, fitted with an antireverse carriage lock. This carriage has a vertical, rigid backboard mounted on it with a ball-bearing counterbalanced head block giving accurate and direct readings from 320 to 1090 mm. The antireverse carriage lock allows for variation in upper-leg length and an adjustable foot rest compensates for lower-leg variations.

4.6. The Harpenden Anthropometer

The Harpenden anthropometer is the most versatile instrument available to the anthropometrist (Figure 1). This instrument is derived from, but has significant advantages over, the more conventional Martin anthropometer beloved of physical anthropologists for more than half a century. These advantages arise from the design of the instrument. As with other Harpenden instruments, the movable parts run on miniature ball-bearing rollers allowing a movement that is free, without cross-play, and is effortless to operate. It can be operated from the tips of the branches or blades that fit into the counter-housing and the end of the main beam. Its user can, therefore, by means of his free fingertips, palpate bony landmarks while holding the instrument. Degrees of accuracy can thus be obtained which are impossible with other anthropometers.

---

*The Fels length measurer is available from Waters Instruments, Inc., Rochester, Minnesota, U.S.A. This is a less-sophisticated portable or fixed combined length and stature anthropometer that allows acceptable accuracy and techniques, particularly for field studies.

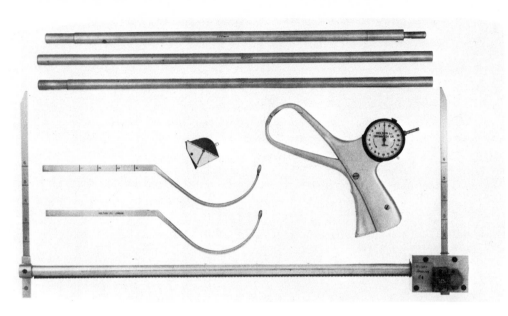

Fig. 1. The Harpenden anthropometer, Holtain (Tanner/Whitehouse) skinfold caliper, and meter Micromatic tape measure.

The counter gives a direct reading over a range of 50–570 mm. To allow measurement of parameters, e.g., sitting height, beam extensions may be added to the main beam, thereby extending the anthropometer to a maximum length of 2 meters. When using these extensions a constant must be added to the counter reading. Recurved branches may be inserted instead of straight ones to measure such parameters as chest depth, cranial width, etc.

4.7. The Holtain (Tanner/Whitehouse) Skinfold Calipers

Skinfold calipers (Figure 1) have undergone dramatic changes since their initial usage in soft-tissue anthropometry. Early designs of this instrument, current in the 1930s and 1940s, had the disadvantage of increasing spring pressure at greater openings of the jaws, lack of standardized spring pressures, poor visual displays of the measurement, and a lack of ergonomic design. These disadvantages meant that there were few data which were comparable between different growth studies. The Holtain caliper was designed by a team led by J. M. Tanner and R. H. Whitehouse and has now been generally accepted as the most consistently reliable and convenient caliper of its kind (Damar and Goldman, 1964; Sloane and Shapiro, 1972; Parizkova and Roth, 1972; Edwards *et al.,* 1955). It was accepted as the recommended caliper for use in the International Biological Program for a number of studies around the world (Weiner and Lourie, 1969).

The jaws exert a constant pressure of 10 g/mm^2 over a range of 0–48 mm. The clock-face counter is calibrated to 0.2 mm allowing a high degree of accuracy in trained hands. One word of warning on the use of skinfold calipers: skinfolds are notoriously difficult to measure accurately and should only be measured after a good deal of training and practice. Highly trained observers can maintain an accuracy of 0.3–0.6 mm at jaw openings of 30 mm (Edwards *et al.*, 1955).

4.8. Tape Measures

Tape measures (Figure 8) are used to measure circumferences when quantitative assessments of this parameter are needed. I make this point because unnecessary errors can be introduced to a study involving, say, states of under- and overnutrition, by the use of quantitative measurements when qualitative measurements, i.e., "too fat," may do just as well.

Having decided to use a tape measure, however, certain criteria must be satisfied. These are suppleness, strength, accuracy, consistency, and ease of use. Linen tapes favored by many workers do stretch and therefore are inconsistent for continual use. Ideally a plastic, perhaps disposable, nonstretch fabric could be used, but to date none have proved completely satisfactory for general use.

Suppleness and strength are incorporated in metallic tape measures; these also have the advantage that they may be thin, roughly 0.5 cm, to allow measurement of the smallest arm with the greatest accuracy. Metallic tapes are consistent in their measurements as they will not stretch with use. Ease of use is a quality that depends on over-all design. Here I recommend the meter Micromatic tape manufactured by Mabo-Stanley of France. This tape satisfies all criteria and is self-retracting.

4.9. The Holtain Bicondylar Vernier

Bicondylar measurements may be made using a number of instruments. The Holtain bicondylar vernier is one such instrument, but there are others that will do the job with equal accuracy; spreading calipers could, for instance, be used.

4.10. Weighing Scales

The measurement of weight is perhaps the easiest of anthropometric techniques and yet instruments of inadequate design are often used.

Accuracy to 0.1 kg is quite acceptable as long as regular calibration ensures the minimum chance of error. Greater tolerance must be allowed where babies are concerned as 0.1 kg forms a higher percentage of total body weight than it does in adolescents and adults. The instrument best suited to repeatedly accurate weight measurement is that designed on the balance-arm principle with two balance arms: one major arm measuring to 150 kg in 10-kg steps, and the other minor arm to 10 kg in 0.1-kg steps. The graduations should be facing the observer for ease of operation. CMS Weighing Equipment Ltd., 18 Camden High Street, London, manufacture a number of such instruments, incorporating all these recommendations.

Most recently CMS have developed a portable-type field-survey scale in collaboration with the Tropical Child Health Unit and the Department of Growth and Development of the Institute of Child Health, London University. This scale, model number MPS 110, incorporates the principles outlined above, with a smaller capacity of 110 kg and a carrying weight of 13.5 kg. It is fast gaining acceptance as an extremely robust machine ideal for the rigors of field surveys and is currently being used in England, Italy, Saudi Arabia, and Zaire.

The following surface landmarks are described in some detail because of their importance to the accuracy of measurement. Student anthropometrists should take time to become fully acquainted with the technique of palpation and learn some surface anatomy before attempting to measure.

The vast majority of anatomical expressions used below are from *Gray's Anatomy,* but where greater understanding can be achieved from colloquial expressions, these have been used. The reader is advised to familiarize himself with the accompanying illustration indicating the positioning of most of the landmarks and orient himself with the directions of medial/lateral/distal/proximal for each limb segment or body part (Figure 2).

Frankfurt Plane. This is at the lower border of the orbit of the eye in the same horizontal plane as the upper margin of the external auditory meatus. For the supinated Frankfurt plane used in recumbent and crown–rump length, substitute ''vertical plane'' for ''horizontal plane.''

Mastoid Process. This is the conical projection below the mastoid portion of the temporal bone. It may be palpated immediately behind the lobule of the ear.

Lateral Border of the Acromion. The acromion can be traced forwards along its medial margin from the acromioclavicular joint to its tip, and then backwards along its lateral margin across the top of the shoulder until it meets the crest of the scapular spine at the prominent acromial angle. This angle is usually but not always the most lateral point of the acromion, because of great diversity in positioning different subjects. Palpation of the most lateral portion may best be conducted by running the anthropometer blades laterally along the shoulders until they drop below the acromia. By pushing medially against each acromion, its most lateral portion must be closer to the anthropometer blades and may then be felt below the surface marks left by the blades.

Iliac Crest and Anterior Superior Iliac Spine. The iliac crest may be palpated as the most superior edge of the ilium and is posterior to the anterior superior spine which lies at the lateral end of the fold of the groin and can be felt without difficulty in the living subject.

Medial Border and Inferior Angle of the Scapula. The medial border is hidden in its upper point by the trapezius muscle, but below the scapular spine it can be traced downwards and slightly laterally to the inferior angle, which overlies the seventh intercostal space at the level of the spine of the seventh thoracic vertebra. In the obese subject palpation is difficult and must be effected by pressing the fingertips medially below the scapular spine until the resistance of the medial border is felt.

Lateral and Medial Epicondyles of the Femur. The lateral epicondyle is the most prominent point on the lateral aspect of the lateral condyle and may be palpated as the bony protuberance superior and lateral to the femurotibial joint space. The medial epicondyle is the most prominent point on the medial surface of the medial condyle and lies below and a little in front of the adductor tubercle and may be palpated directly opposite the lateral epicondyle.

Lateral and Medial Epicondyles of the Humerus. The lateral epicondyle occupies the lateral part of the nonarticular portion of the lower end of the humerus and may be palpated opposite and a little above the medial epicondyle which forms a conspicuous blunt projection on the medial side of the lower end of the humerus.

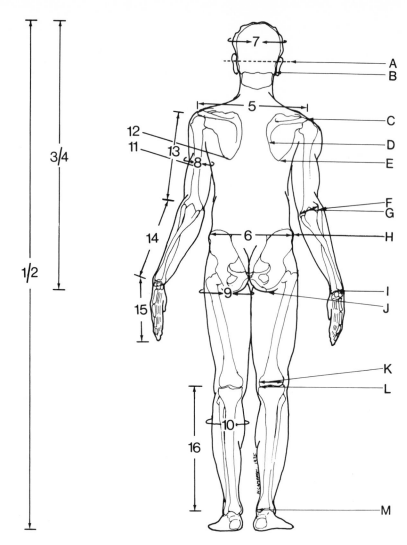

Fig. 2. The major surface measurements and landmarks. Measurements: 1/2. stature—recumbent length; 3/4. sitting Height (crown–rump length); 5. biacromial diameter; 6. biiliac diameter; 7. head circumference; 8. extended upper-arm circumference; 9. upper-thigh circumference; 10. maximum calf circumference; 11. triceps skinfold; 12. subscapular skinfold; 13. upper-arm length; 14. forearm length; 15. hand length; 16. tibial length. Landmarks: A. Frankfurt plane; B. mastoid process; C. lateral border of the acromial; D. medial border of the scapula; E. inferior angle of the scapula; F. head of the radius; G. medial and lateral humeral epicondyles; H. iliac crest; I. distal end of the radius; J. gluteal fold; K. medial and lateral femoral epicondyles; L. medial aspect of the femerotibial joint space; M. medial malleolus.

Head of the Radius. This may be palpated as the inverted, U-shaped bony protuberance immediately distal to the lateral epicondyle of the humerus when the arm is relaxed with the palm of the hand facing forwards.

Distal End of the Radius. This is at the border of the radius proximal to the dorsal–superior borders of the lunate and scaphoid and medial to the radial styloid. It is palpated by moving the fingers medially and proximally from the radial styloid.

Superior (Dorsal) Border of the Distal End of the Radius. This may be palpated by moving the fingers proximally and medially along the distal ends of the radius from the sytoid process.

Distal (Inferior) Border of the Medial Malleolus. This is the inferior border of the bony protuberance on the medial side of the ankle which is the medial malleolus.

Femurotibial Joint Space. This may be palpated by moving the fingers medially from the inferior border of the patella to the bony protuberance of the inferior border of the medial epicondyle of the femur. Immediately inferior to this and superior to the lateral condyle of the tibia is the joint space.

Gluteal Fold. This is viewed as the fold formed by the crossing of the gluteus maximus and the long head of the biceps femoris and the semitendinosus. This is usually present in most subjects but some, perhaps because of a lack of gluteal development, will not have an actual fold of skin. In this case the crossing of the muscle groups must be judged from the lateral profile of the buttocks and posterior thigh.

6. Measurement Techniques

The following section describes the techniques required for accurate measurement of the following parameters: (1) weight; (2) stature; (3) sitting height; (4) supine or recumbent length; (5) crown–rump length; (6) biacromial diameter; (7) biiliac diameter, biiliocristale, bicristal or pelvic breadth; (8) head circumference; (9) extended upper arm circumference; (10) upper thigh circumference; (11) maximum calf circumference, relaxed; (12) triceps skinfold; (13) biceps skinfold; (14) subscapular skinfold; (15) suprailiac skinfold; (16) hand length; (17) forearm length; (18) upper arm length; (19) cubit length; (20) tibial length; (21) foot length; (22) bicondylar femur; and (23) bicondylar humerus.

In some cases, e.g., stature, it has been necessary to describe alternative techniques which are widely used in anthropometric research. This is not through choice but through necessity—ideally, standardization of one technique to be used, universally, for one measurement is the final aim of auxologists. Until this situation is apparent, I feel it necessary to point out to the reader that other methods are available and may have been used in comparative research. It must be emphasized that each measurer or measuring team must investigate their reliability and variability before seriously using the techniques described.

6.1. Weight (Beam Scales)

The subject should preferably be weighed in the nude or else with the minimum of standard clothing supplied by the investigator and corrected for by adjusting the machine to read zero when the clothing is placed on it. If the machine is accurately calibrated, errors in this measurement ought to be negligible and only arise from reading and recording the subject's weight. It is advisable, as with all techniques, not to read out a decimal point but leave the insertion of this to the computer or statistician. Weiner and Lourie (1969) recommend that the presence of visible edema should be recorded. When neonates or very young children are to be weighed, it is often a good point of technique to weigh the mother and child together, then transfer the child to an assistant and weigh the mother alone. The baby's weight can thus be calculated by difference [(weight of mother + baby) − (weight of mother)], and the child is left undisturbed by the whole procedure.

6.2. Stature (Stadiometer)

6.2.1. Laboratory Technique

The subject's shoes and socks must be removed prior to measurement, not only because they will affect the height of the subject but they may conceal slight raising of the heels. The subject stands straight so that his heels, buttocks, and shoulders are in contact with the vertical backboard of the stadiometer (Figure 3).

It may be necessary, especially when young children are being measured, for an assistant to press down on the feet so that the undersides of the heels are in contact with the ground. The heels are placed either together, such that the medial malleoli are touching or, occasionally, if the subject has "knock-knees," such that the knees are together and feet slightly spread. The shoulders should be relaxed and sloping forward in a natural position to minimize lordosis. In order to ensure this relaxed position, the observer may find it helpful to run his hands along the shoulders and feel that there is no tension. The hands and arms are loose and relaxed with the palms facing medially. The head is positioned in the Frankfurt plane and the headboard is slid down the backboard until it touches the subject's head. To ensure that the head stays in contact with the headboard and to minimize the effect of hair thickness producing a false measurement, it is necessary to place a weight of about 1.0 kg on the headboard.

With the subject in this position he is then instructed to "take a deep breath and stand tall." This is done to aid the straightening of the spine. The air so inhaled should be exhaled by the subject when he is then told to "relax." At times the subject may tend to "hunch-up" the thoracic region in an attempt to inhale the greatest possible amount of air. This should be discouraged as it will tend to reduce the actual measured parameter. Stretching of the spine is assisted by gentle upward pressure applied beneath the mastoid processes by the measurer. This stretching minimizes diurnal variation which may be as much as 20 mm without it (Strickland and Shearin, 1972) but is normally less than 4.6 mm over the whole day using this method of measurement (Whitehouse *et al.,* 1974). The subject is then told to "relax the shoulders" and "stretch up as much as you can—keep your heels on the ground though," and the stature is read from the counter to the nearest completed unit. If the last digit on the counter is between two values, the lower is always recorded.

If the subject has obvious signs of asymmetry in the legs, it is advisable to record two heights: one standing on the longer leg with the shorter supported by a solid wedge or suitable block and the other measurement on the shorter leg with the longer partially flexed to provide stability.

6.2.2. Field Technique

In the field situation, it is not always possible to use the Harpenden stadiometer, and in this case a number of alternatives are open to the anthropometrist. These alternatives fall into four categories (see footnote, p. 48): (1) the Harpenden portable stadiometer, (2) the Harpenden portable length-measuring instrument, (3) the Harpenden anthropometer, and (4) an instrument of one's own design (Falkner, 1961).

The techniques for the portable stadiometer and portable length-measuring instrument (which doubles as the supine-length measurer) are the same as that described above so long as the instruments are properly set up and rigid in their vertical and horizontal dimensions.

Using the anthropometer requires a good deal of skill and fine attention to the details of positioning. The addition of extensions to the basic anthropometer makes

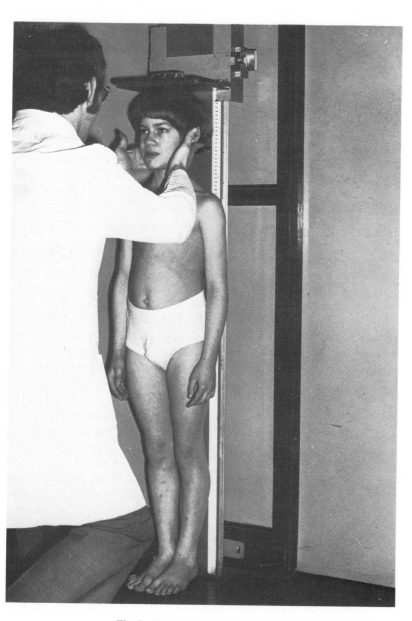

Fig. 3. The measurement of stature.

it possible to measure up to two meters, but a predetermined constant must be added to the counter reading to give the correct measurement. The anthropometer must be held vertically throughout the procedure. The following technique is proposed for the measurement of stature.

The subject stands perfectly straight, with his arms relaxed by his sides and his ankles or knees together. He should be encouraged to stand as tall as possible before measurement starts (if the measurer is working alone, this last point is a necessity because only one hand may be used to apply the gentle pressure to the spine, while the other hand holds the anthropometer). The subject's head is positioned in the Frankfurt plane, and the head is held in this position by the left hand of the measurer, the fingers to one side and the thumb to the other, applying pressure to the angle of the mandible with the chin below the level of the hand. The anthropometer is positioned behind the subject so that its lower end stands between the heels and the beam passes vertically between the buttocks, touching the back of the head. The anthropometer blade is brought down on top of the head with the right hand ensuring that the blade passes through the sagittal plane of the head, and the measurement is taken, to the nearest completed unit, after instructing the subject to "take a deep breath, stand tall, and relax."

Really, two people are needed to take this measurement in the most reliable way. The first positions and instructs the subject while the second holds the instrument, checks its position, and calls the measurement.

The measurement of stature has, of course, received the most attention from anthropometrists. It is normally the baseline measurement common to every growth study, and generally the techniques used for its measurement are of a high order. Of late the most sophisticated, and the most expensive, instrument for the measurement of stature has been developed by Hamill *et al.* (1973). Their "stadiometer" consists of a level platform with a vertical bar and steel tape attached. Positioning of the subject was as described above. The important addition to their technique, however, was the attachment of a Polaroid camera to the horizontal bar of the apparatus. This camera recorded the exact reading, with no parallax, and an identification number of the subject. The advantages are, of course, that one has a permanent recording of the measurement, there is the minimum observer error, and parallax is totally eliminated. "Instruments of one's own design" may cover (or expose) a multitude of sins, but practically as long as a rigid, vertical backboard and horizontal headboard are used and properly calibrated any inaccuracy will be due to recording error or bad positioning.

The general principles of measuring stature are the same for all current authors, but differences do exist that should be pointed out. Krogman (1950) recommends the "standing erect position" of the subject for all measurements requiring a standing posture. In principle this position is equivalent to the one described above. Unlike us, Krogman always measured stature free of any vertical backboard, using the Martin anthropometer, because of the ease with which other measurements could then be taken without disturbing the stance of the subject. He does, however, stress that when stature alone is to be measured a wall scale is advisable. Montagu (1960) favors the military position for the subject: at attention with the head erect, looking straight ahead with the visual axis parallel to the floor surface. Why Montagu chose this latter approximation to the Frankfurt plane and not the plane itself is a mystery. Martin (1914) and Martin and Saller (1957) go to great lengths to describe the Frankfurt plane but do not actually name it as such, neither does Hrdlicka (1947); there is no doubt, however, that the Frankfurt plane is the correct head position. In adults it is very unusual that the back of the head is posterior enough to touch the backboard at the same time as the buttocks and heels, but in children, who of course have not acquired their adult kyphosis, the spine may often be straight enough to achieve this position.

The correctness of applying gentle pressure to the subject in order to decrease postural and diurnal effects must ultimately be decided by the investigator. Certainly many studies have not used this technique, and in recent years only Weiner and Lourie (1969) have recommended that it should be used. Hertzberg (1968) and his colleagues rejected the pressure method in favor of a free-standing technique. When measuring children, however, the application of upward pressure gives more reliable readings from day to day than other techniques.

6.3. Sitting Height (Sitting-Height Table, Anthropometer)

6.3.1. Laboratory Technique

The subject is positioned in the sitting-height table so that the head is in the Frankfurt plane, the back is straight, the thighs horizontal or comfortably positioned, so that the tendons of the biceps femoris are approximately one inch clear of the table. The feet are supported on the footboard or metal bar incorporated in the apparatus, and the hands rest comfortably in the subject's lap (Figure 4). To ensure that the subject's back is fully extended the observer may run his index finger up the spine, applying pressure to the lumbar and sacral regions, thus causing the subject to sit up as a reflex action. Having lowered the headboard, which, as in the measurement of stature, may have a small weight on it, gentle upward pressure is applied to the mastoid processes and the subject is instructed to "sit tall" or "sit up, take a deep breath, and relax."

The measurement value is read by the observer who is standing in front of the table looking horizontally at the counter while applying traction. Once again the technique of stretching the subject reduces the effect of diurnal variation in this parameter and aids reproducibility of measurement (Whitehouse *et al.,* 1974).

6.3.2. Field Technique (Anthropometer)

In the field situation the observer stands to the left of the subject, who is sitting on a table or other convenient surface adopting the same position as for the laboratory measurement, i.e., head in the Frankfurt plane, back straight, sitting tall (Figure 5). The anthropometer with only the counter housing blade in position lies behind the subject so that it may be manipulated by the observer's right hand. The observer's left hand holds the subject's head in the Frankfurt plane while his right hand grasps the anthropometer just below the counter housing and stands it vertically behind the subject such that the base is between the buttocks and the beam touches the sacral and interscapular regions of the spine. The blade is then applied to the sagittal plane of the head and, leaving the third, fourth, and fifth fingers to hold the beam, the index finger and thumb are applied either side of the occiput to provide stability to the head position and help in the upward pressure of the left hand during traction. At the usual command the subject sits tall, takes a deep breath, and relaxes while the observer records the measurement from the counter facing him.

Close agreement as to the techniques for obtaining sitting height is apparent in the literature. Hrdlicka (1947) and Montagu (1960) recommend that the subject sit with his back against a wall on which there is a scale. Martin (1914) favors the field technique described above, and interestingly he recommends that the subject be encouraged to achieve a maximal stretch of the spine and that the spine be stretched

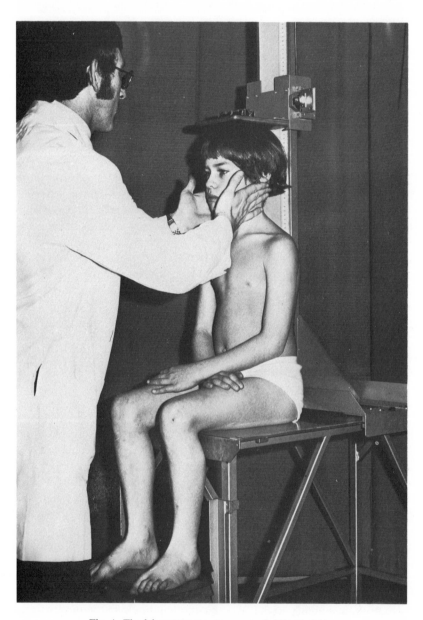

Fig. 4. The laboratory measurement of sitting height.

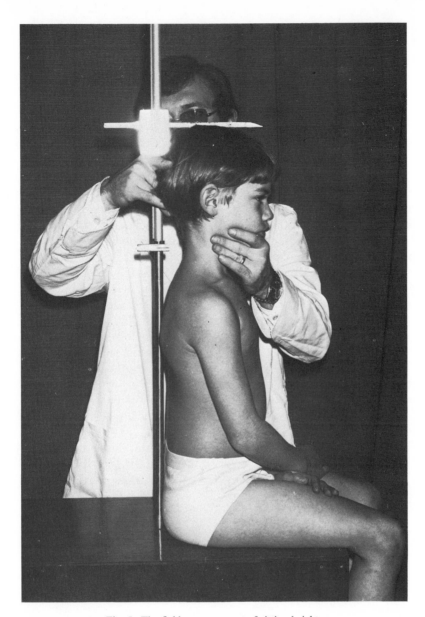

Fig. 5. The field measurement of sitting height.

as far as possible. Martin would thus seem to advocate the upward pressure method. One is reminded of the corollary of Hrdlicka's advice that it would be incongruous to take the total height in one standard position and the sitting height in another. The sitting-height table, used for the laboratory method above, effectively acts as a movable "wall," allowing for differences in femur length of the subjects.

The technique of sitting on the floor with the legs outstretched is not recommended. This would tend to put the hamstrings in a state of tension and may involve compensatory attempts by the subject to lean back and reduce this tension.

6.4. Supine or Recumbent Length (Supine-Length Table, Neonatometer, Infantometer)

6.4.1. Laboratory Technique

The subject lies on the supine-length table, or if younger than nine months in the neonatometer, with the head held in the supinated Frankfurt plane, i.e., the eyes looking directly upwards and the lower orbit of the eyes in the same vertical plane as the upper margin of the external auditory meatus. The head is held in this position by an assistant standing directly behind the fixed headboard of the table or neonatometer. When measuring particularly young children and always when using the neonatometer, the assistant must apply downward pressure on the shoulders to prevent arching of the back. This is best done with the index fingers but is difficult and proficiency in the measurement of babies takes considerable time and patience. Having secured the head and shoulders of the subject, the assistant applies gentle pressure to bring the head into contact with the fixed headboard while the observer, ensuring the legs are straight, ankles at right angles, and toes pointing directly upwards, brings the footboard into firm contact against the subject's heels. Only one heel is positioned when using the neonatometer. Such is the lack of cooperation from most babies that the observer will often find it necessary to apply downward pressure to the knee while maintaining a firm grip on the subject's foot. Once the feet are in position it is usual, when measuring infants, for the head holder to remark that the head is no longer against the board. This is good, for the head is now gently but firmly brought back to the headboard, the foot holder not giving way too much. This means that the infant is stretched very gently, which is exactly what is required (Falkner, 1966).

With cooperative subjects this is a very convenient measurement as it can be done from birth to maturity without changing the apparatus. Indeed, Roche and Davila (1974) have used recumbent length extensively in the Fels growth study and make a simple adjustment of 1.3 cm to compare their results to stature measurements. However, they emphasize that because of systematic differences among studies as to the actual adjustment to be made, there must be sufficient internal evidence from the same study as to the differences between recumbent length and stature within individuals.

6.4.2. Field Technique

To date, supine length seems not to have been much measured in a field situation. The measurement of infants up to nine months would obviously require exactly the same technique as for laboratory measurement, using the neonatometer which is eminently portable.

To attempt the measurement of adults in the field would require specially constructed instruments or the anthropometer could conceivably be clamped into a horizontal position to substitute for a supine-length table. Stuart (1939) did use this design, with a Martin anthropometer, to measure recumbent length of subjects of the Harvard University growth study, in the laboratory situation. Should the anthropometrist decide that this is a viable measurement for a field study, then he must develop instrumentation and test its validity, reliability, and accuracy before using it in the study proper.

6.5.1. Laboratory Technique

Crown–rump length is the equivalent of sitting height and is used when measuring children up to two years of age or later if they are unable to sit.

Once again the child is lying down with the head in a supinated Frankfurt plane. The head is held in this position by an assistant standing directly behind the fixed headboard of the supine-length table. Gentle pressure is applied by the assistant to bring the head into contact with the headboard while the observer ensures that the knees are bent at a right angle, the thighs vertical, and the movable footboard is brought into firm contact with the buttocks.

The point of measurement can at times be difficult to ascertain due to gluteal development and gluteal fat pads. Infantometers are now fitted with a pressure lever to lock the footboard at 0.45 kg (1 lb) pressure. This is a quite acceptable standard which ought to become universally applied. If a pressure lever is not available, then the point of measurement must be when the footboard is in firm contact with the buttocks.

6.5.2. Field Technique

Once again no record is available for the measurement of crown–rump in the field situation. The reader is referred to the previous remarks for the field techniques of recumbent length.

6.6. Biacromial Diameter (Anthropometer)

6.6.1. Laboratory and Field Technique

It is important when measuring this parameter to ensure that maximum shoulder width is being obtained (Figure 6). Errors of 2–3 cm may be encountered by the subject trying to broaden his shoulders by standing too stiffly (Whitehouse, 1974). The action of the rhomboids, levator scapuli, and trapezius will actually be to narrow the shoulders by bringing the scapuli into closer proximity to one another and elevating them.

The best way of obtaining the maximal biacromial diameter is to tell the subject to relax the shoulders; then, sitting or standing directly behind the subject, the observer runs the palms of his hands along the lines of the shoulders from the base of the neck outwards, feeling their natural forward and downward slope and relaxing any tension. Next, the most lateral borders of the acromial processes are identified by palpation and, holding the anthropometer so that the blades rest medially to the index fingers and over the angle formed by the thumb and index finger, the blades are applied to the most lateral borders of the acromia and each pressed firmly against these protuberances. Sufficient pressure is applied to minimize the thickness of the overlying soft tissue and the diameter is read to the last completed unit of the counter.

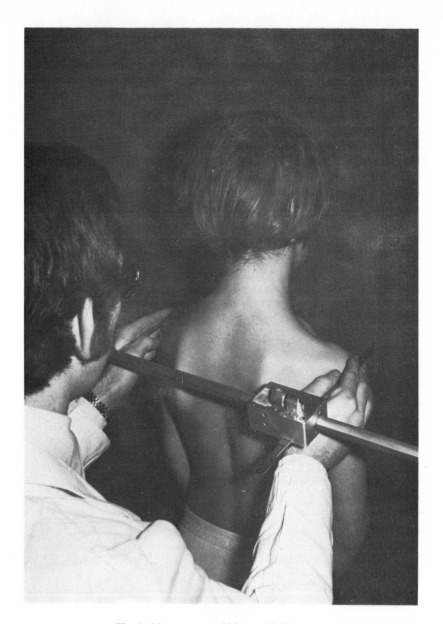

Fig. 6. Measurement of biacromial diameter.

Few current references mention that pressure must be applied when taking this measurement. Randell *et al.* (1946), Oberman (1965), and Ashe and Ashe (1943) are exceptions, all of whom mention "firm contact." The shoulder position has stimulated comment from Martin (1914) and Randell *et al.* (1946), the former requiring there to be no slouch to shorten the measurement and the latter poetically describing the shoulders as " . . . relaxed but not collapsed."

6.7. Biiliac Diameter, Biiliocristale, Bicristal, or Pelvic Breadth (Anthropometer)

Laboratory and Field Technique. The subject stands with the heels together, back to the observer and arms folded or out from his side (Figure 7). The anthropometer blades, held in the same manner as when measuring biacromial diameter, are brought into contact with the iliac crest from behind and at a slight downward angle to ensure clearance of subcutaneous fat deposits immediately behind the crest and buttocks. By "rolling" the blades over the crests, it will be seen on the digital counter that at a particular point the distance between them is greatest; this is the point of measurement. Considerable difficulty will be found in taking this measure-

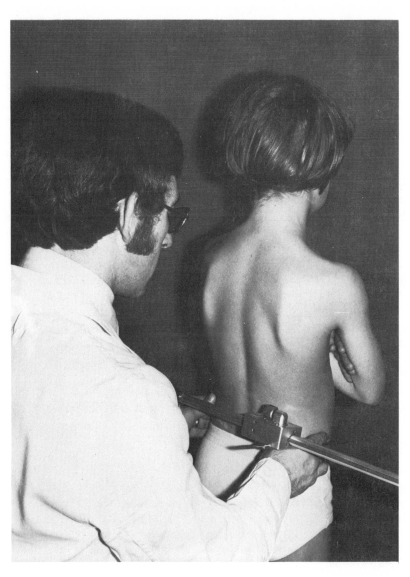

Fig. 7. Measurement of biiliac diameter.

ment on the obese subject and at times it must be avoided if measurement causes distress to the subject. In other cases, however, the iliac crests may be easily palpated, but pressure must again be applied to minimize the effect of overlying soft tissue.

This variously named measurement differs between authors only in the extent to which pressure must be applied on the soft tissue. Hrdlicka (1947) requires the pressure to be sufficient to feel bone resistance, Martin (1914) requires slight pressure, and Randell *et al.* (1946) and Ashe and Ashe (1943) require firm pressure. Comas (1960), Duckworth (1912), and Montagu (1960) omit any reference to pressure. It would seem reasonable to apply pressure as described, otherwise the contribution of subcutaneous soft tissue to the measurement will be to decrease repeatability. Bitrochanteric diameter, however, is a different measurement, taken over the most lateral part of the greater trochanters.

6.8. Circumferences (Tape Measure)

The measurement of circumferences is necessary for nutritional as well as growth surveys, but the degree of accuracy required must be fully investigated. For the general nutritional screening process it would seem rather a waste of time to subject one's results to the errors of numerical data for, let us say, upper-arm circumference, when a piece of colored string around the arm would indicate under- or overnutrition (Morley, 1974). Similarly, colored string would be no good for the accuracy required to compare circumferences with radiographically determined muscle or fat widths.

The next step is to learn to manipulate the instrument, string, or tape with a high degree of proficiency. This, once accomplished, is perfectly easy but, to begin with, causes problems to the most kinesthetically oriented. If one is right-handed the body of the tape is held in the left hand, with some 6–8 cm of tape extending from it. This part of the tape is then placed behind the limb and grasped with the index finger and thumb of the right hand; keeping the left hand still, the tape is pulled out from the housing and brought around the outside of the limb; the left hand then comes around from the opposite direction to meet it; the tape housing and end are exchanged—housing to right hand, end to left hand—and the tape pulled into contact with the limb. Enough tension is applied so that the tape rests against the skin but does not compress the tissue. This whole operation should be practiced using any tubular frame before attempting it on the subject.

For circumference measurements of the head, upper arm, and thigh, the subject stands sideways to the measurer with the head in the Frankfurt plane, arms relaxed, and legs apart.

6.8.1. Head Circumference

Laboratory and Field Technique. The subject stands sideways to the measurer with the head in the Frankfurt plane and arms relaxed (Figure 8).

The tape is passed around the head as previously described and, using the index or middle fingers of the left hand, the tape is held on the most anterior protuberance of the forehead. The right hand, maintaining tension on the tape, is used to move the tape over the most posterior protuberance of the occiput so that measurement is that of the maximum head circumference. This circumference may not necessarily be in a plane parallel to the Frankfurt plane, but the measurement of a maximum circumference cancels any error encountered in judging the position of the tape relative to the true Frankfurt plane. Having located the maximum circumference, the tape is pulled tight to compress the hair and the measurement read to the last completed unit.

There is general agreement that head circumference should be the maximum, but the actual positioning of the tape is variously described. Weiner and Lourie (1969) stipulate that it should be above (but not including) the brow ridges, Hrdlicka (1947) would appear to agree with this by stipulating that the tape passes above the supraorbital ridges, and both Martin (1914) and Montagu (1960) favor the glabella–opistocranion plane. Numerous other authors, e.g., Kay (1961), Oshima (1962), Snow and Snyder (1965), and Randell *et al.* (1946), all state that the tape should be

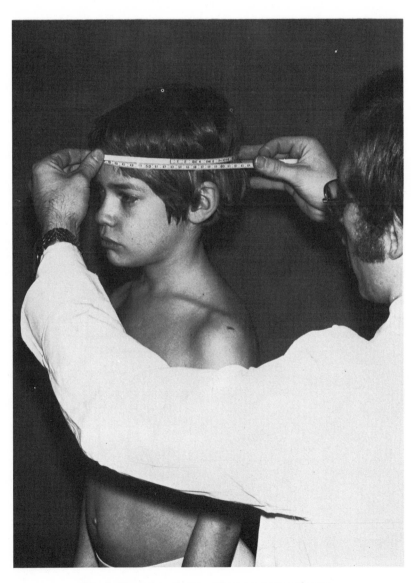

Fig. 8. Measurement of maximum head circumference.

above the brow ridges. No workers mention that pressure must be applied but I feel this to be an important, if not obvious, point of technique. Head circumference is used extensively on very young children where the problem of determining a level above the brow ridges does not arise as they are not prominent until adolescence.

6.8.2. Extended Upper-Arm Circumference

Laboratory and Field Technique. The subject is positioned as for the measurement of head circumference with the left arm completely relaxed and extended by his side (Figure 9). Any contraction of the biceps will increase the circumference and decrease repeatability and must therefore be discouraged.

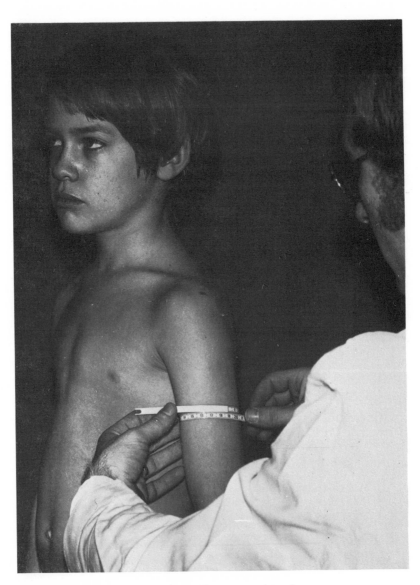

Fig. 9. Measurement of upper-arm circumference.

A mark is drawn on the lateral side of the upper arm midway between the acromion and olecranon. The tape is then passed around the arm as previously described so that it is touching the skin but not compressing the tissue, and the measurement read to the last completed unit. It is important to ensure that the tape is in a plane perpendicular to the long axis of the limb.

Upper-arm or biceps circumference has been measured in both the flexed and extended positions because of its application in nutritional assessment and clothing design. The techniques for these measurements fall into two schools. The first believes the reference point to be over the most prominent part of the biceps and the second that the reference point is midway between the acromial process and the olecranon. The differences incurred by siding with either of the schools are probably negligible, but it is advisable to consult the literature relevant to one's investigation before deciding on the landmark. The actual technique of measurement is the same in all cases. It is also important to check which arm has been used; in growth it is normal to use the left.

6.8.3. Upper-Thigh Circumference

Laboratory and Field Technique. The subject stands sideways as before, with his weight equally distributed on both feet and with his legs sufficiently apart to ensure free movement of the tape between the thighs (Figure 10). The tape is passed around the limb in a plane perpendicular to the long axis of the limb, with its upper edge in the gluteal fold. Making sure that the tape is touching the skin but not compressing the tissue, the measurement is taken to the last complete unit.

Descriptions of thigh circumference technique suffer from a lack of standard terminology. These circumferences have been termed "crotch thigh," "maximum thigh," and "upper thigh" (Randell *et al.*, 1946; Hansen and Cornag, 1958; Montagu, 1960; Hertzberg *et al.*, 1963), and these are different from thigh circumference as described by Martin (1914) and Daniels *et al.* (1953). The basic difference is in the vertical positioning of the tape and whether this should be as high in the crotch as possible (Gifford *et al.*, 1965; Hertzberg *et al.*, 1963), halfway between crotch and knee (Randell *et al.*, 1946), at the greatest bulging of the muscles (Garrett and Kennedy, 1971) or, as Martin (1914) describes it, "at the greatest medial protrusion of the muscles . . . but not in the gluteal furrow" (Garrett and Kennedy, 1971). The solution to such confusion is obviously to use standard terminology such that the title of the technique describes the site of measurement, as suggested by Hertzberg (1968). I have thus described the technique as one for "upper-thigh circumference"; Randell and Baer (1951) correctly entitle their measurement "middle-thigh circumference," and one assumes "lower-thigh circumference" would be at or just above the level of the femoral epicondyles. Such titles as "maximum thigh circumference" give no indication of whether this is at the gluteal fold level, as in Montagu's technique (1960), or not, as in Martin's technique (1914).

6.8.4. Maximum Calf Circumference, Relaxed

Laboratory and Field Techniques. For this measurement the subject must be sitting on a table such that the knees extend over the edge and the calves are relaxed. Any flexion or extension of the foot will increase calf circumference and decrease repeatability. The tape is passed around the calf in a plane perpendicular to the long axis of the limb, at the maximum circumference as judged by the profile of the gastrocnemius. This position may be checked by moving the tape above and below the observed maximum, smaller readings at these places will guarantee a correct measurement point.

Most authors favor measurement of calf circumference with the subject in a standing position and the weight evenly spread on both feet. Hrdlicka (1947) requires the left foot to be placed on a bench so that the leg forms an angle of about 90° with the thigh. It would seem better, especially where children are concerned, to take this measurement in a non-weight-bearing position with the leg completely relaxed so that errors in determining whether weight is evenly spread are avoided. Once again, it is important to consult relevant literature before finally deciding on the technique to be employed.

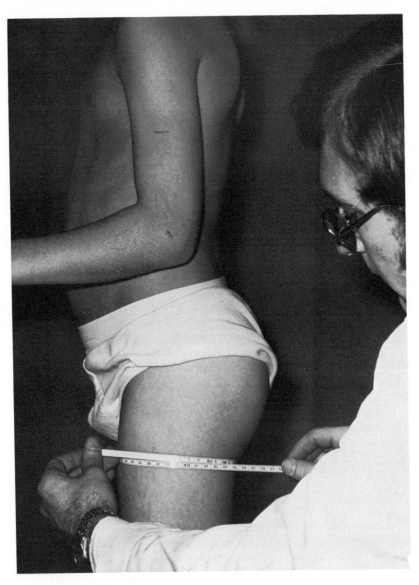

Fig. 10. Measurement of upper-thigh circumference.

Naturally there are a great many circumferences which may be taken on the living subject. The most commonly used of these are chest (maximal expiration and inspiration), ankle, maximum forearm, wrist, and waist.

The same principles of measurement as previously described apply to all these circumferences. There must be no compression of any tissue, the tape must be always perpendicular to the long axis of the limb or body segment, and the reading is always to the last completed unit.

The phrase "chest circumference" can imply eight different levels and three different states of respiration, these being maximal expiration, maximal inspiration, and normal breathing. The level and state of respiration depends on the exact requirements of the study. Martin (1914) and Martin and Saller (1957) stipulate that the level is under the bottom edge of the scapulae. Montagu (1960) is almost alone in correctly prefixing the phrase "chest circumference" with the pertinent landmark, e.g., "axillary chest girth," which is taken well up in the axillary fossae, and "mesosternal chest girth," which is taken at the level of the 4th rib. It is therefore important to check the literature.

I shall describe the general technique for use on men, boys, and prepubertal girls: the subject stands erect, his arms initially raised, then lowered after the tape is in position. Holding the tape in a horizontal plane at the level of the nipples, measure the maximum circumference of the chest during normal breathing, at full inspiration, or at full expiration.

Postpubertal girls and women obviously present a special problem which has been overcome by Randell and Baer (1951) and Hansen and Cornag (1958) by describing a breast circumference, passing over both nipples and a chest circumference passing below the breasts. Daniels *et al.* (1953) describes a bust circumference at maximal chest girth and two chest circumferences, the first above the fullness of the breasts and the second below this fullness. Difficulties must arise when these criteria are applied to postmenopausal females in whom the characteristic redistribution of fat deposits causes a shift downwards of nipple level and fullness. It would appear prudent to state in one's description the exact level, respiratory state of the subject, and degree of development in the case of adolescent girls and women at or near menopause.

Minimal ankle circumference is generally measured at a level immediately proximal to the malleoli, but O'Brien and Shelton (1941) and Randell and Baer (1951) described an ankle circumference passing through the medial point of the medial malleolus or in a plane passing through the malleoli. Accurate descriptive terminology will determine the exact reference point.

Maximum forearm circumference is usually measured at a point immediately distal to the elbow joint. Montagu (1960) requires the limb to be relaxed, and Randell *et al.* 1946; Randell and Baer (1951) make no mention of the state of tension; other authors tend to state whether the limb is flexed or relaxed. When flexed the arm is bent, fist clenched, and arm muscles flexed. Measurement is at the greatest bulge of the muscles.

Wrist circumference is taken at a point proximal to the styloid process of the ulna (Martin, 1914; Montagu, 1960). Garrett (1969*a,b*) requires the measurement to be taken at the level of the wrist crease, and other authors, e.g., Hertzberg *et al.* (1963), used the most distal point of the radius. The most frequently used landmark is the former, proximal to the styloid process.

White (1966) described waist circumference as the maximal horizontal circumference at the level of the navel. The same title used by Randell *et al.* (1946) requires the minimum circumference of the waist to be measured. Generally the multitude of descriptions tend to suggest a measurement at the minimum circumference of the abdomen between the iliac crests and the lowest ribs.

6.9. Skinfolds (Skinfold Caliper)

6.9.1. Harpenden Technique for All Skinfolds

The skinfold is often described as a "pinch" (Tanner and Whitehouse, 1975; Durnin and Rahaman, 1967; Garn and Gorman, 1956; Clarke *et al.,* 1956), but the action to obtain it is to sweep the index or middle finger and thumb together over the surface of the skin from about 6–8 cm apart and collect the subcutaneous tissue pushed away from the underlying muscle fascia by this action. To "pinch" the subject suggests a very small and painful pincer movement of the fingers, and this is not the movement made. Firstly, the measurement of skinfolds should not cause undue pain to the subject, who may be apprehensive anyway from the appearance of the calipers and will tend to pull away from the measurer. Secondly, a pincer or pinching action does not collect the quantity of subcutaneous tissue normally measured. The experienced anthropometrist knows that this measurement is prone to many errors. Location of the correct sites is critical (Ruiz *et al.,* 1971), but the variation in the consistency of subcutaneous tissue and the individual way in which each measurer collects the fold of tissue seem to be the main sources of error. A search of the relevant literature shows that the technique of skinfold measurement is hardly ever described, even by eminent workers in the field, and it is therefore not surprising to encounter very poor reliabilities and standard errors in written work, if indeed they are mentioned at all. Edwards *et al.* (1955) describe their standard deviation of the differences between duplicate measurements taken by a single observer at the triceps, subscapular, and suprailiac sites as being 0.3–0.6 mm at a jaw opening of 7 mm. The equivalent figures for different observers were roughly twice these values even when the sites of measurement were marked. This accuracy is described as being "sufficient for any presently conceivable purpose." At greater openings, the absolute error increases but the percentage error between duplicate readings stays constant at about ±5% for two thirds of repeated readings.

Only proper training, experience, a good deal of practice, and repeated consistency tests will provide the efficiency needed for this measurement, and expert advice should always be sought before embarking on skinfold measurements.

6.9.2. Corrections to Be Applied Before Statistical Analysis

If a large number of skinfold measurements are taken from different subjects, they do not follow a normal distribution (except in the very young or in malnourished communities) and the error of measurement is proportional to the absolute measurement recorded in each case. In order to apply simple statistical techniques, it is necessary that the data be converted to a normal (Gaussian) distribution. For this purpose they should be transformed according to the formula:

$$\text{Skinfold transform (2)} = 100 \log_{10} (\text{reading } 1/10 \text{ mm} - 18)$$

Multiplying by 100 eliminates negative values. The constant 18 was first evaluated on purely statistical grounds, being the amount necessary to produce the most nearly Gaussian distribution in a series of measurements of young men; however, it has subsequently turned out to be approximately double the skin thickness. A table of transforms is given by Edwards *et al.* (1955). In comparing a reading with the standard chart published by Tanner and Whitehouse (1975), for triceps and subscapular skinfolds, it is not necessary to apply this transform as it is done automatically on the scale against which the standards are plotted.

6.9.3. Triceps Skinfold

Laboratory and Field Technique. The subject stands with his back to the measurer and his arm relaxed with the palm facing the lateral thigh. The tips of the acromial process and olecranon are palpated, and a point halfway between is marked on the skin. The skinfold is picked up over the posterior surface of the tricep muscle 1 cm above the mark, on a vertical line passing upwards from the olecranon to the acromion, and the caliper jaws are applied at the marked level (Tanner and Whitehouse, 1975). The caliper is held as illustrated in Figure 11, and the fingers of the right hand relax their grip to ensure that the caliper exerts its full pressure of 10 g/mm². The left hand maintains the pinch throughout the measurement. This usually results in a stable reading, up to 20 mm, but above this, the measurement registered sometimes decreases as the measurer watches the dial. A firmer grip by the left hand may prevent this. If not, then the reading should be taken two seconds after the caliper is applied. The dial of the Holtain caliper is calibrated to 0.2 mm, but the measurement can be conveniently estimated to the last completed 0.1 mm. It is advantageous to request that the subject bend the arm before taking the skinfold and straighten it before applying the caliper. This causes any muscle which may have been picked up by the overzealous measurer to be pulled out from the skinfold by the contracting action of the triceps at arm extension.

6.9.4. Biceps Skinfold

Laboratory and Field Technique. The subject faces the measurer with the arm held relaxed at his side and the palm facing forwards. The skinfold is picked up over the belly of the biceps and 1 cm above the line marked for the upper-arm circumference and triceps skinfold on a vertical line, joining the center of the antecubital fossa to the head of the humerus. The caliper jaws are applied to the marked level.

6.9.5. Subscapular Skinfold

Laboratory and Field Technique. The subject stands as for triceps skinfold with the shoulders and arms relaxed (Figure 12). It is quite easy on the average subject to determine the inferior angle of the scapula below which the skinfold should be taken, but it is not so simple on the obese subject. To locate this point, palpate the medial border of the scapula and run the fingers of the left hand downwards along its full length until the inferior angle is located. The skinfold is picked up immediately below the inferior angle of the scapula with the fold either in

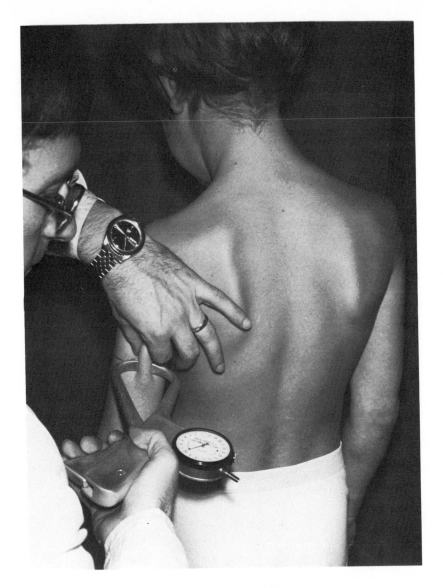

Fig. 11. Triceps skinfold.

the vertical line or slightly inclined, downwards and laterally, in the natural cleavage of the skin (Tanner and Whitehouse, 1975).

6.9.6. Suprailiac Skinfold

Laboratory and Field Technique. With the subject standing sideways with his arms folded, the skinfold is picked up vertically about 1 cm above and 2 cm medial to the anterior suprailiac spine. The caliper is applied just below the fingers. This site varies, depending on the position of the superior anterior iliac spine and may be in the mid-axillary line or anterior to it.

Edwards (1950) mentions the existence of 93 possible skinfold sites of which he used 53 to represent the distribution of subcutaneous fat. This startling fact gives an indication of the criteria for choosing a particular site; as long as there is measurable subcutaneous tissue at a locatable site, and the attachment to the deep fascia is loose and the texture not too fibrous, then it can be used. Many skinfolds are not representative of the total quantity of subcutaneous or body fat, however, and research workers have tended to concentrate on the four described above because they are the smallest number representative of body fat. The use of multiple

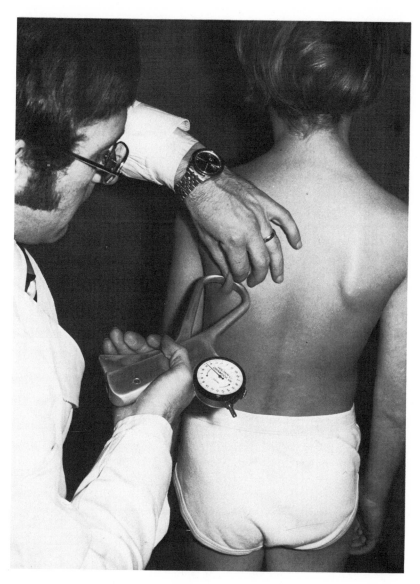

Fig. 12. Subscapular skinfold.

skinfolds also reduces the effect of measurement error which becomes critical when only one or two skinfolds are being used to estimate total body fat (Durnin and Rahaman, 1967).

6.10. Limb Lengths (Anthropometer)

6.10.1. Hand Length

Laboratory and Field Technique. The hand and forearm are placed supine on a horizontal surface, palm downwards and fingers extended, such that the third phalanx is in a direct line with the central axis of the forearm. The anthropometer blades are applied from the lateral side. The tip of the right blade is placed on the marked distal–superior border of the radius and forced distally and downwards to rest closely against the distal end of the radius, while the tip of the left blade is brought into contact with the most distal fingertip, usually that of the third phalanx.

This measurement may also be taken by applying the anthropometer from the medial side. In this case the inside of the tip of the left blade is applied to the distal–superior border of the radius and the right blade is brought into contact with the most distal fingertip.

A difference of about 1 cm is apparent when comparing these two techniques because in the former one is applying the anthropometer blade to a supposed gap between the radius and lunate and in the latter case to the superior surface of the radius. Using this knowledge a correction may be applied to standardize these measurements before statistical analysis, but the magnitude of the difference between the two techniques will depend upon the observer. Therefore, as with recumbent length, sufficient internal evidence from the same study as to the differences between techniques within subjects must be present before correction.

Hrdlicka (1947), Montagu (1960), and Martin (1914) take this measurement from the midpoint of a line connecting the styli of the radius and ulna which is equivalent to the landmark described above. The alternative is to use the method put forward by Randell *et al.* (1946), White (1966), and Ashe and Ashe (1943), in which they use the proximal end of the navicular as their landmark. This parameter is subject to error caused by the underlying tendons of the wrist preventing good palpation of the distal end of the radius and proximal border of the lunate and scaphoid (Cameron *et al.*, 1976).

6.10.2. Forearm Length

Laboratory and Field Technique. The subject stands with his back to the observer and his arms relaxed. The fixed blade of the anthropometer is applied to the distal end of the radius and the movable end to the marked head of the radius (Figure 13). Measurement is taken to the last completed unit. Arm positioning for this measurement is standard in all references with the landmarks described as radial to stylon (Montagu, 1960; Weiner and Lourie, 1969; Martin, 1914).

6.10.3. Upper-Arm Length

Laboratory and Field Technique. The subject stands as for the measurement of forearm length. The laterosuperior margin of the head of the radius and the lateral border of the acromion are marked, the fixed blade of the anthropometer is placed

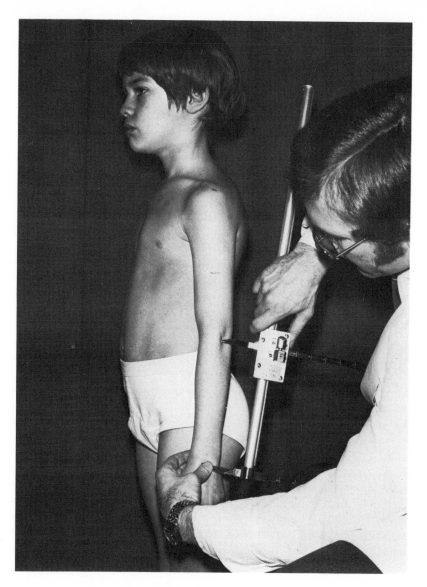

Fig. 13. Forearm length.

on the head of the radius and held there with the thumb of the left hand. The upper, movable blade is then placed immediately inferior to the acromial angle and the measurement read to the last completed unit. Upper-arm length is one of the most standard measurements in anthropometry, and all authors agree with the landmarks above as being the reference points for measurement.

6.10.4. Cubit Length

Laboratory and Field Technique. Positioning the forearm as for measurement of hand length, the anthropometer blades are applied laterally to the marked head of the radius and the most distal fingertip. Cubit length may be substituted for the controversial measurement of hand length, and this parameter estimated by subtracting forearm length from cubit length. The standard error, however, is greater for hand length using this method than for direct measurement, and its use will therefore depend upon the accuracy required in a particular survey (Cameron *et al.*, 1976).

This measurement should not be confused with forearm–hand length which is measured from the most posterior point of the olecranon and not the head of the radius.

6.10.5. Tibial Length

Laboratory and Field Technique. The subject sits facing the observer, with his left ankle or calf resting on his right knee so that the medial aspect of the tibia faces upwards. The anthropometer blades are applied to the marked proximal–medial border of the tibia and the marked distal border of the medial malleolus (Figure 14). The measurement is read to the last completed unit.

Udjus (1964) describes an equivalent measurement to this but all other authors favor a "tibiale" measurement from the floor with the subject in an erect posture (Garrett and Kennedy, 1971). For the analysis of growth, however, it is recommended that the former measurement be used to negate the confounding effect of including the foot in a parameter of the lower leg.

6.10.6. Foot Length

Laboratory and Field Technique. The subject stands with his weight on his right foot and his left foot resting sole downwards on a horizontal surface 25–30 cm from the ground. The calf should be perpendicular to the surface and the toes relaxed. It is often necessary for the observer to extend the subject's big toe as this is normally (quite unconsciously) flexed. The anthropometer blades are applied to the most posterior part of the heel and the most anterior of the toes, normally the big toe or second toe. If there is no distinct posterior protuberance of the heel then the blades of the anthropometer should be applied about 2 cm up from the supporting horizontal surface.

As with the measurement of calf length, two schools of technique are apparent for this measurement. The first, of which we are a part, favors a non-weight-bearing method; the second, composed primarily of Martin (1914) and Montagu (1960), favors a weight-bearing method. In defense of the former technique I must emphasize that flexion of the big toe, when in a weight-bearing position, will affect the measurement; thus it would seem reasonable to cancel out this potential error by using a non-weight-bearing technique.

6.10.7. Biepicondylar Femur (Bicondylar Vernier)

Laboratory and Field Technique. The subject sits on a table with the knees relaxed and extending over the edge. The blades of the bicondylar vernier are applied to the palpated distal epicondyles of the femur and the tissue over the epicondyles is compressed before taking the reading to the last completed unit.

6.10.8. Biepicondylar Humerus (Bicondylar Vernier)

Laboratory and Field Technique. Sitting as before, the subject's elbow is flexed at a right angle and the distal epicondyles of the humerus palpated. The

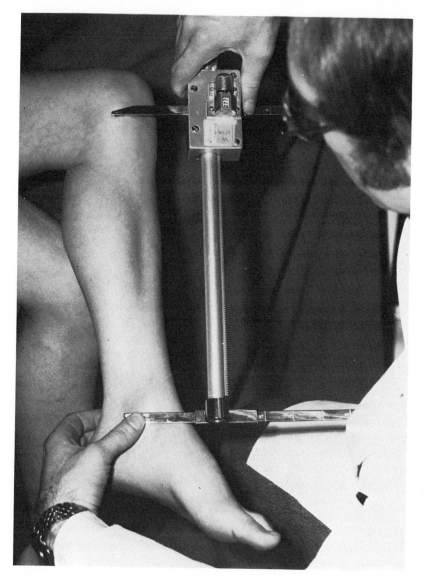

Fig. 14. Sitting tibial length.

vernier blades are placed on the epicondyles, and pressure is exerted to compress the tissues while the distance is recorded as before. This is usually an oblique measurement since the medial condyle is lower relative to the horizontal when the humerus is in this position.

6.11. Other Measurements in Current Use

The above measurements are those recommended for the analysis of human growth. I have said previously that the scope of anthropometry is vast and each investigation requires different measurements. This being the case, I make no excuses for including only those measurements which describe over-all individual growth.

Weiner and Lourie (1969) list 38 measurements for use in studies on human growth and physique; of these I have described 20 which may be used by the general physician or pediatrician to evaluate the normality of his patients and by anthropologists to set up standards of reference for different ethnic groups.

For further information on anthropometric parameters not described in this chapter, the reader is advised to consult Weiner and Lourie (1969), Garrett and Kennedy (1971), and Montagu (1960).

7. Photogrammetric Anthropometry

The technique of taking a photograph of a subject posed in a particular position to facilitate measurement and visual appraisal of the body has its roots in the works of Sheldon (1940, 1954), Tanner and Weiner (1949), and Dupertuis and Tanner (1950).

Sheldon categorized body shapes according to three basic classifications: endomorphy, mesomorphy, and ectomorphy, based on visual appraisal. Tanner, working with Sheldon's associate, Dupertuis, and subsequently with Weiner, was interested in measuring the parameters of an unmovable subject whose shape and size at the time of examination had been permanently recorded. This immobility and permanence is one of the great advantages of photogrammetry, added to which is the fact that many measurements can be carried out on a subject who has been seen just long enough to pose him and take three photographs. Later, Tanner used this technique to classify the body shape of Olympic athletes (Tanner, 1964), and currently it is being used for a variety of purposes, not the least of which is the visual assessment of children suffering from abnormalities of growth which often manifest themselves in disproportionate body shape. The asymmetry frequently evident in Silver-Russell syndrome (Tanner *et al.*, 1975*b*) is sometimes difficult to see when the child stands in front of the physician but becomes obvious when the child has been correctly posed in front of a photogrammetric camera. Marshall and Ahmed (1976) have used photogrammetry to construct charts for upper-arm and forearm lengths of normal British girls. These charts provide an effective means of defining the exact nature of abnormalities of growth in the arms or of differences between populations.

These examples, however, are relatively new. Photogrammetry was not recognized as an important and accurate anthropometric technique until Sheldon's work in the 1940s. Gavan *et al.* (1952) emphasize the fact that anthropologists—pre-Sheldon—doubted the accuracy of a photographic record. The problems they saw in taking accurate measurements were those of enlargement, distortion due to paper shrinkage, nonstandard lens–subject distances, lighting difficulties and, quite naturally, the extra time and effort that the solution of these problems entailed. Gavan *et al.* (1952) make the point that all these problems can be eliminated provided the total photographic procedure is planned for the specific purpose of producing a standard-scale picture suitable for taking measurements and observations. Without doubt the techniques outlined below have done a great deal to establish norms of practice for photogrammetry. The anthropometrist who studies these source references in detail will gain a comprehensive understanding of modern photogrammetry.

The problem with taking a two-dimensional photograph is that one encounters "parallax." To summarize the problem of parallax: the subject stands on a small turntable to be posed for the photographs. The center of this turntable is known as the center of rotation (CR) and is, at all times, a fixed distance from the lens. Say the CR–lens distance is 10 meters, all points on this center of rotation would thus be exactly 10 meters from the lens and in the same plane as the vertical meter marker, thereby making their accurate measurement feasible. Objects or points in front of or behind the CR are closer to or further from the lens and therefore cannot be measured to the same degree of accuracy as points on the CR. Tanner and Weiner (1949) calculated the error thus involved to be 1% for every 10 cm the dimension measured lies in front of the CR. If the CR–lens distance is halved to Sheldon's recommended 5 meters, the error is 1% for every 5 cm distance. Parallax error may be calculated and therefore compensated for by knowing the exact dimensions used to place the subject at the center of rotation. Knowing how far back his heels are and how far forward his toes are allows the addition and subtraction of percentage errors due to parallax from the measurements taken (see Tanner and Weiner, 1949; Gavan *et al.,* 1952; for detailed discussion of parallax and distortion).

The equipment with which photogrammetric photographs may be taken depends on the requirements of the study, the amount of money available for the purchase of equipment, and the back-up facilities in terms of developing and enlarging the photographs. Running costs for such procedures can be quite high when one considers the cost of new film and precise, reliable developing.

Sheldon (1954) used a portrait camera to obtain 5-in. × 7-in. contact prints of his subjects. To obtain the three views on one frame of the film he employed a technique that has been frequently copied and modified; he "masked" the lens with a three-panel sliding back so that the exposure was always of the central third of the frame, i.e., that part in direct line with the subject. The subject was positioned 14 ft 9 in. from the lens, allowing a 6-ft 8-in. man to be photographed with a 9.5-in. focal-length lens. This relatively long distance was to avoid parallax errors, the light rays being nearly parallel.

Lighting of the subject and a light-colored background was produced by the use of electronic flash. Trials with a dark background showed it to be less preferable than a cream or white one. One interesting feature of Sheldon's method is that at the same time as taking the photogrammetric pictures, he took 35-mm transparencies for anthroposcopic use. This single-lens-reflex camera was equipped with a 50-mm focal-length lens.

Posing of the subject was critical, and with little alteration the same methods are used today. As can be seen in Figure 15, the subject stands erect in the frontal picture. The observer ensures that the shoulders are centered on the mid-frontal plane and that the chest is relaxed, the arms forcibly extended, and the elbows locked. The subject's hands are positioned so that the fingers are extended and together, the thumbs being along the forefingers and the arms, in relation to the body, are sloping outward such that a gap of about 12–13 cm remains between the palms of the hands and the thighs. Notice that the fingers are pointing vertically downwards and the wrists are therefore bent. The subject's face should be directly facing the camera with the head in the eye–ear plane. Tanner and Weiner (1949) slightly altered this position by making the subject hold his hands in the "anatomical" or supinated position, i.e., with the palms facing forward. The former method is recommended by Dupertuis and Tanner (1950) as the sole one to be used when somatotyping, the latter being suitable for other work. These authors include a detailed methodology for the posing of the photogrammetric subject, and their paper should be consulted if serious work is to be undertaken in this field. The posture required for the side and back photographs is the same as for the front view.

Tanner and Weiner (1949) used a 35-mm camera technique for their work because of the considerable savings in cost and its extreme portability. Using a 135-mm focal-length lens, they stood the subject 10 m from the lens to reduce parallax and posed the subjects as detailed above. An important addition to their technique is the inclusion of meter and half-meter markers on the photograph to instantly give one a magnification factor. The main disadvantage of this 35-mm technique was the small size of the negatives. Stanford and Tanner (1949) overcame this by using an English F24 or American K24 aerial mapping camera to obtain their photographs. Such was the success of this equipment that it is still in use today. A 20-in. focal-length lens was used 10 m from the subject. Negatives were $5 \times 2\frac{1}{2}$ in., thereby canceling the disadvantage of the miniature technique, and up to 500 exposures could be obtained without changing the film. In the last 20 years advances in photographic equipment now allow a much greater choice of equipment to the investigator. Somatotypes of the Olympic athletes in Munich were taken using a Hasselblad system which lends itself particularly well to the speed and precision of modern anthropometry laboratories. Currently we use the Hasselblad 500 EL/M camera system with a 150-mm lens at a distance of 6.72 meters from the center of rotation.

One final point remains to be discussed in photogrammetric anthropometry: the relationship of photographically determined parameters with those determined on

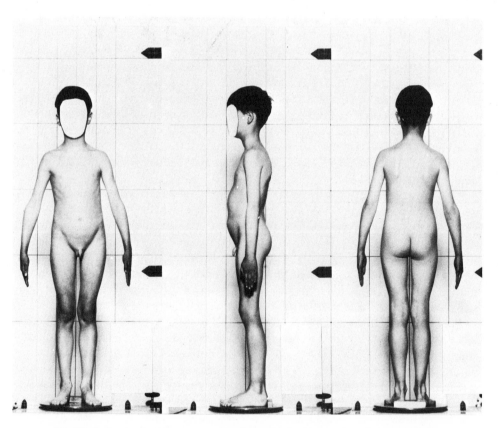

Fig. 15. Posing in photogrammetry after the method of Dupertuis and Tanner (1950).

the living subject by direct anthropometric means. Tanner and Weiner (1949) maintain that the correspondence between the two methods is good, but differences and errors in posing constitute two thirds of the error in repeat procedures of photogrammetric anthropometry. Measuring error constitutes one fifth and observer error the rest. It would be unreasonable to suggest that there is a direct carry-over from the photograph to living anthropometry, since the correlations between photographically determined dimensions and anthropometrically determined parameters are in the range 0.57–0.93, the vast majority being in the lower half of this range. Photogrammetric anthropometry is, however, an extremely useful and powerful anthropometric tool when used in the correct circumstances and for the purposes for which it was developed.

The problem of parallax, however, may be completely overcome and the third dimension included by using stereophotogrammetry. This is an anthropometric tool that owes much of its development to the pioneer work of Hertzberg *et al.* (1957), Miskin (1960), Beard and Burke (1967), and Burke (1971). The technique is to accurately map the dimensions and curvatures of the body and, by establishing where and how much of the surface of the body protrudes from a known plane of reference, a three-dimensional model is built up with obvious advantages over two-dimensional photogrammetry. The principle used to obtain these body maps is that of binocular vision; the two eyes see slightly different views of an object and cerebral fusion produces an impression of depth, the third dimension, as well as length and breadth. Similarly, if two stereophotographs are juxtaposed so that the left eye sees the left photograph and the right eye the right photograph in proper relation, the perception of depth may be clearly established.

Hertzberg *et al.* (1957) accurately contoured the entire human body. Their technical setup involved two K17 aerial cameras 12 ft from the subject plane, 3.5 ft from the floor, and 5 ft apart. A third K17 camera was positioned 17 feet from the subject for the purpose of taking a somatotype photograph at the same time. Subject positioning was important; he stood on a turntable with the feet slightly apart and the arms raised laterally to the horizontal. It was necessary to support the arms as two photographs, front and back, were taken.

Diapositives of the stereo pair of photographs were used on a Kelsh-type stereoscopic plotter. This plotter works on the same principle as the Multiplex system described by Beard and Burke (1967). The diapositives are projected onto a tracing table using a red light source for one and a blue light source for the other.

The observer wears spectacles fitted with red and blue filters so that the appropriate images are received by the left and right eyes, thus enabling the formation of a spatial model by cerebral fusion. A small mobile stand, which is adjustable for height, in the upper surface of which is set an illuminated point, is placed on the table. The stand is moved over the table with the point of light adjusted so that it appears to maintain continuous contact with the surface of the model at any desired level. A lead pencil located directly beneath the point of light marks out a linear contour on dimensionally stable drawing paper fixed to the table top.

Miskin (1960) recommends the Zeiss stereotype plotter as a cheaper substitute if one can be satisfied with spatial coordinates of points and restricted plotting.

The contour map similar to that obtained by Burke (1971) of the human face is illustrated in Figure 16. The contour interval on this particular example is 2.0 mm, but it may be greater or smaller; Hertzberg *et al.* (1957) used an interval of 0.5 in.

for whole-body contours. The shaded areas on the contour map in Figure 16 are "altitude" or "*z* coordinate" lines which serve to guide the plotter in recombining the stereoimages. For Burke's purpose these altitude lines were painted on the subject in white photographic ink close to the anatomical landmarks he wished to measure.

Contourographs are yet another development from stereophotogrammetry, the results of which are photographs of the subject with contour lines already projected onto the skin by shining a light through a grating positioned between subject and camera. The Moiré contourograph system was developed by Meadows *et al.* (1970) and advanced by Takasaki (1970) before Terada's (1974) definitive work, explaining the Moiré system in geometric detail. Its advantage over stereophotogrammetry is cheapness, but so far no data on its reliability have been published.

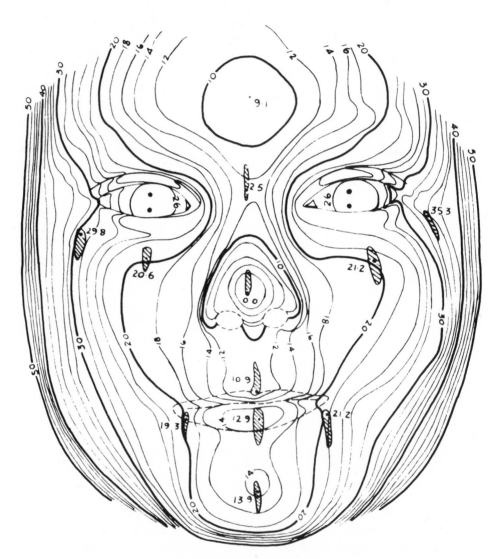

Fig. 16. Stereophotogrammetric contour map of the face of a child. The contour interval is 2 mm, and the shaded areas on the chin, lips, cheeks, nose and cheek bone correspond to marks made on the subject.

In other sections of this chapter we have discussed anthropometry in terms of measuring the external dimensions of the body and thereby describing the changes which take place during growth. The external characteristics of growth and development are, of course, the result of changes to the skeleton, muscular system, and main organ systems of the body. In order to study and quantify these changes, anthropometry utilizes radiography. Radiography allows us to study these three areas of change in detail due to the fact that by altering the nature of the X rays being transmitted to the body we can cast shadows of soft tissue and/or the bony and cartilaginous skeleton.

8.1. Hand–Wrist Radiographs

Elsewhere in this treatise the current systems for assessing the skeletal maturity of a child are discussed in detail, and the reader should consult the relevant sections for an understanding of the importance of determining skeletal age and predicting adult height. The techniques required to obtain a good radiograph of the hand and wrist for the purpose of skeletal aging are very important. To assess skeletal age one is dealing with changes in shape of the bones; any inconsistency in radiographic techniques or posing of the subject will result in nonstandard radiographs which then make comparison of bones from one age to the next unnecessarily difficult. Indeed, faulty posing may result in some bones having different appearances from those described in the bone-aging system (Tanner *et al.*, 1975*a*).

The left hand is used throughout, and placed palm downwards in contact with the film cassette, such that the axis of the middle finger is in direct line with the axis of the forearm. The upper arm and forearm should be in the same horizontal plane as the hand. The fingers are slightly spread so that they are not touching, and the thumb is placed at an approximate angle of 30° with the first finger. The palm is pressed lightly downwards on the cassette by the subject or, if the child is too young to comply with these instructions, the observer applies this pressure. In order to obtain a correctly exposed radiograph the position of the tube must be exact. It is centered above the head of the third metacarpal, at a tube–film distance of 76 cm. The skin dose is 8–10 millirads, and adequate precautions against X-ray scatter should be taken to protect the child's gonads. Development of the film should be light.

8.2. Limb Tissue Radiographs

The soft tissues of the growing body are more or less in a constant state of change. The results of this changing tissue are most easily appraised by observing the changing shape of the individual. Detailed quantitative analysis of body composition has been carried out using densitometry, dilution, and radiographic techniques. The former two techniques estimate the "size" of the various tissue components of fat, muscle, and bone, while the latter technique measures "shape" by allowing us to see where the soft tissue is situated.

In auxological anthropometry radiographic workers have been most concerned with the changing tissue patterns of the arm and calf because of the relative simplicity of X-ray technique using these areas and the fact that direct exposure of the gonads to X rays is almost totally eliminated by concentrating on the arm and leg. Appropriately exposed radiographs of the limbs (60 kV, 20 milliamp-sec for the arm and 65 kV, 20 milliamp-sec for the calf) make it a relatively simple job to distinguish bone, muscle, and subcutaneous fat (Tanner, 1965). The technique requires only that the subject hold still for one tenth of a second and that the

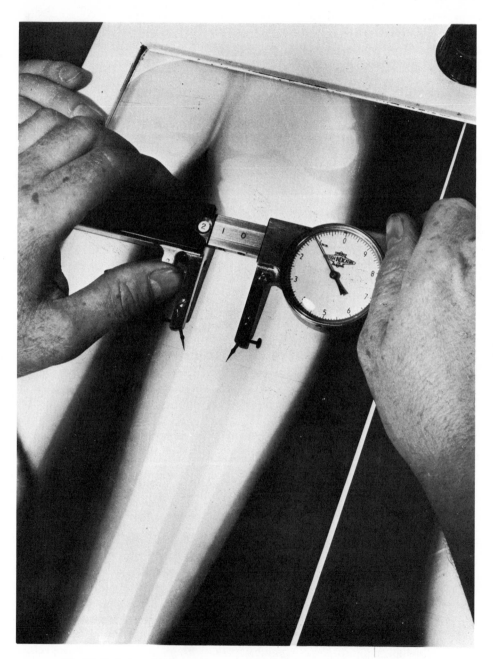

Fig. 17. Technique for the measurement of soft-tissue layers from a radiograph.

observer ensure correct positioning, correct exposure, and adequate protection for the subject.

Garn (1961) and Tanner (1965) give comprehensive notes on the historical development of radiographic techniques. Suffice it to say that the techniques are American in origin, dating from the Harvard School of Public Health study in the 1930s, and have been incorporated in the Fels' longitudinal study (Reynolds, 1944) and later in the Harpenden growth study (Tanner, 1962). This latter study involved modifications to ensure a constant object–film distance throughout growth, thereby keeping the magnification factor constant, and an anode–film distance of 2.5 m so that the beam of X rays hitting the limb is, as nearly as possible, parallel, and errors of posing the limb are minimized.

In any radiographic procedure, protection of the subject from excess X rays is of great importance. On the whole, skin doses are low; 15 mR for the calf and 10 mR for the arm, but the gonads must be protected by suitable lead screening.

Although the measurements of the soft tissue are by relatively simple procedures, Garn (1961) emphasizes that training and experience are necessary to accurately distinguish fat from muscle at many sites. In addition to the arm and calf, as used by Tanner, Garn suggests the scalp (lateral), lower arm (AP), deltoid, lower thoracic, iliac, and trochanteric sites for estimation of soft tissue. He includes appropriate data on exposure levels.

The selection of sites for measurement depends upon four criteria: the technical or radiographic difficulty of the site, the positioning error, the presence or lack of suitable underlying landmarks, and the size of the measuring error relative to the median fat thickness. In the absence of previous detailed discussion of this point, the reader is referred to Garn (1961).

Measurements of width are taken using needlepoint calipers as illustrated in Figure 17, which shows the measurement of the calf at maximal width of the tissue. This point is marked previous to measurement. The site for the upper arm is halfway between the acromion and the head of the radius on a line perpendicular to the long axis of the limb and, as nearly as possible, parallel to the two skin borders (Tanner, 1962, 1965). Muscle width in the upper arm is obtained by simple subtraction of bone and fat widths from total arm width, but some authors include bone in the "muscle" width of the maximum calf, therefore checking of comparable data for conformity is necessary.

Reliabilities for this type of measurement are quite high. Tanner (1965) quotes a standard error of measurement of 1–2% of the mean for calf muscle and bone and 3–4% for fat. In the arm all three components gave figures of 3–4%. Garn (1961) gives a reliability for fat thickness measurements of ±5% for most measurements.

9. Measuring the Disabled

I include this brief note on the measurement of the disabled because it is an area of anthropometry that is seldom, if ever, mentioned, and yet some guidelines for intending investigators might prove useful.

Obviously this type of measurement constitutes an extremely specialized area in anthropometry, and yet the problem of accurate and repeatable measurement is not insurmountable. The point of the utmost importance to the subject–observer relationship is that the subject must at no time be required to adopt positions that may appear to him undignified or stress his disablement. Thus a great deal of thought by the observer must accompany his techniques.

Alternative techniques to those recommended above must be used if the subject cannot adopt standard positions. Thus, recumbent length must be measured on subjects who are unable to stand and continued as the main length measurement even if the subject, at some later date, acquires the necessary ability to stand unsupported. In cases of hypoextensibility of arms, upper-arm circumference must be taken in the flexed position.

Secondly, standard techniques must be adapted to a nonstandard situation. For instance, it may be necessary to construct supporting apparatus to help the subject maintain a standard, or near-standard, position for an accepted technique of measurement. Silver-Russell patients often suffer from asymmetry of the legs; one can compensate for this by constructing a supporting board for the shorter leg so that an erect standing position can be achieved for measuring height. This board is of known thickness and vertically adjustable to allow for growth of the short leg.

Thirdly, the anthropometrist must attempt to find subject positions which are repeatable. For instance, the growth of a patient's deformed arm could be measured by positioning the arm flexed at a particular angle.

Measurement of the disabled must be approached in an humane manner. The greatest barrier to collecting good data on such subjects will be the breakdown of subject–observer relationships due to unconscious callousness on the part of the observer.

10. Automated Anthropometry

The automation of anthropometric instruments has been a constant subject for discussion among anthropometrists since Garn (1962) suggested the linking of two-dimensional measurement devices to electronic analyzers and recorders. Five years later he published an account of such a procedure using a Helios pinpoint micrometer caliper connected to a GADRS-4 (Gerber analog data reduction system) (Garn and Helmrich, 1967). By coupling a 10-turn potentiometer to the micrometer, adding a foot switch, and then connecting to the GADRS-4 console, direct readings, to 0.01 mm, were possible and directly transferable to an IBM typewriter, punch-cards, paper, or magnetic tape. These modifications negated the errors of visual readout, transcription, precoding, and keypunching which are inherent in the conventional anthropometric procedures. However, the severe disadvantages of cost and the necessity of an electronics expert being on virtually constant call could not be overcome.

Prahl-Andersen *et al.* (1972) took the concept of automated anthropometry one step further and, to date, their automated laboratory in Nymegen is the only one of its kind in Europe and therefore worth describing in detail. The measurements possible with their equipment are distance, diameter, circumference, stature, weight, skinfold, and grip strength. These are accomplished by the automation of the Harpenden anthropometer, Holtain skinfold caliper, Molenschott TBM balance, Bettendorff hand-grip dynamometer, and instruments of their own design to measure stature and circumferences.

The anthropometer, stadiometer, and circumference instrument are coupled to 10-turn potentiometers rotated via gears. The stadiometer is designed with a slide allowing a measurement range of 500 mm, and it may be set at three positions to measure 500–1000, 1000–1500, and 1500–2000 mm with an accuracy of ±1 mm/1000. The circumference instrument has a range of 0–1000 mm with an accuracy similar to the stadiometer. A flexible band constitutes the tape and is connected via a drum and gears to the potentiometer. Automatic tightening of the band is achieved by pushbutton control, and a light tension is maintained by a spring mechanism.

The skinfold caliper has a reduced range of 20 mm and an accuracy of ±1%. The spindle of the indicator is coupled to a magnetic displacement transducer.

The balance and grip dynamometers have single-turn potentiometers, giving the former a range of 100 kg and the latter a range of 70 kg. Interestingly the dynamometer is also connected to a slot machine which supplies a prize of marbles. If the second and third attempts at gripping exceed the first, more marbles are "won." The highest value for the three attempts is remembered and recorded.

All the instruments are operated by pushbutton control, and the data are displayed, printed, and punched onto paper tape.

Dr. Prahl-Andersen and her colleagues found an accuracy of ±2% on all the measurements except for the skinfolds, and these were accurate at a ±5% level. Strength testing, of course, is not subject to measurement error.

In common with Garn and Helmrich (1967), Prahl-Andersen *et al.* (1972) recognize the need for an electronics expert to be constantly available. This very fact has stopped many interested investigators from developing their own automated laboratories. Anthropometry is, however, moving into an automated age, and the workers in the science will have to become acquainted with electronic procedures just as they have learned to program and operate computers.

11. References

Arnhold, R., 1969, The Quac Stick: A field measure used by the Quaker service team, Nigeria, *J. Trop. Pediatr.* **15**:243.

Ashe, W. R., and Ashe, H. F., 1943, Anthropometric Measurements, Project 9, No. 741-3, Armored Force Medical Research Laboratory, U.S. Army Medical Research Laboratory, Fort Knox, Kentucky.

Bayley, N., 1940, *Studies in the Development of Young Children,* University of California Press, Berkeley, California.

Bayley, N., and Pinneau, S. R., 1952, Tables for predicting adult height from skeletal age: Revised for use with the Greulich–Pyle hand standards, *J. Pediatr.* **40**:423.

Beard, L. F. H., and Burke, P. H., 1967, Evolution of a system of stereophotogrammetry for the study of facial morphology, *Med. Biol. Illus.* **17**:20.

Burke, P. H., 1971, Stereophotogrammetric measurement of normal facial asymmetry in children, *Hum. Biol.* **43**:536.

Buzina, R., and Uemura, K., 1974, Selection of the minimal anthropometric characteristics to assess nutritional status, *Adv. Exp. Med. Biol.* **49**:271.

Cameron, N., Hughes, P. C. R., and Whitehouse, R. H., 1976, The reliability of certain limb measurements in adolescents (unpublished data).

Clarke, H. H., Geser, L. R., and Hundson, S. B., 1956, Comparison of upper arm measurements by use of roentgenogram and anthropometric techniques, *Res. Q. Am. Assoc. Hlth. Phys. Educ.* **27**:379.

Collignon, R., 1892, Projet d'entente internationale au sujet des researches anthropometriques dans les conseils de revision, *Bull. Soc. Anthropol. Paris,* Ser. 4 **3**:186.

Comas, J., 1960, *Manual of Physical Anthropology,* English edition, Charles C Thomas, Springfield, Illinois.

Damar, A., and Goldman, R. F., 1964, Predicting fat from body measurements: Densitometric evaluation of ten anthropometric equations, *Hum. Biol.* **36**:32.

Daniels, G. S., Meyers, H. C., and Worrall, S. H., 1953, Anthropometry of WAF Basic Trainees, Technical Report 53-12, Wright Air Development Center, Wright-Patterson Air Force Base, Ohio.

Davies, D. P., and Holding, R. E., 1972, Neonatometer: A new infant length measurer. *Arch. Dis. Child.* **47**:938.

Duckworth, W. L. H., 1912, *The International Agreement for the Unification of Anthropometric Measurements to Be Made on the Living Subject* (English translation of the official version), Anthropology Laboratory, University New Museum, Cambridge.

Dupertuis, G. W., and Tanner, J. M., 1950, The pose of the subject for photogrammetric anthropometry, with special reference to somatotyping, *Am. J. Phys. Anthropol.* **8**:27.

Durnin, J. V. G. A., and Rahaman, M. M., 1967, The assessment of the amount of fat in the human body from measurements of skinfold thickness, *Br. J. Nutr.* **21**:681.

Durnin, J. V. G. A., and Womersley, J., 1974, Body fat assessed from total body density and its estimation from skinfold thickness: Measurements on 481 males and females aged from 16 to 27 years, *Br. J. Nutr.* **32**:77.

Edwards, D. A. W., 1950, Observations on the distribution of subcutaneous fat, *Clin. Sci.* **9**:259.

Edwards, D. A. W., 1955, The estimation of the proportion of fat in the body by the measurement of skinfold thicknesses, *Voeding* **16**:57.

Edwards, D. A. W., Hammond, W. H., Healy, M. J. R., Tanner, J. M., and Whitehouse, R. H., 1955, Design and accuracy of calipers for measuring subcutaneous tissue thickness, *Br. J. Nutr.* **9**:133.

Falkner, F., 1961, Office measurement of physical growth, *Pediatr. Clin. North Am.* **8**:13.

Falkner, F., 1966, *Human Development,* W. B. Saunders, Philadelphia.

Fleming, R. M., 1933, Special Report Senior Medical Research Council No. 190, in: J. M. Tanner, 1952, The assessment of growth and development in children, *Arch. Dis. Child.* **27**:10.

Garn, S. M., 1961, Radiographic analysis of body composition, in: *Techniques for Measuring Body Composition* (J. Brozek and A. Henschel, eds.), National Academy of Science, Washington, D.C.

Garn, S. M., 1962, Automation in anthropometry, *Am. J. Phys. Anthropol.* **20**:387.

Garn, S. M., and Gorman, E. L., 1956, Comparison of pinch-caliper and teleroentgenogrammetric measurements of subcutaneous fat, *Hum. Biol.* **28**:4.

Garn, S. M., and Helmrich, R. H., 1967, Next step in automated anthropometry, *Am. J. Phys. Anthropol.* **26**:97.

Garrett, J. W., 1969*a*, Anthropometry of the Air Force Female Hand, AMRL-TR-69-26. Aerospace Medical Research Laboratory, Wright-Patterson Air Force Base, Ohio.

Garrett, J. W., 1969*b*, Anthropometry of the Hands of Male Air Force Flight Personnel, AMRL-TR-69-42. Aerospace Medical Research Laboratory, Wright-Patterson Air Force Base, Ohio.

Garrett, J. W., and Kennedy, K. W., 1971, A Collation of Anthropometry, Vols. I and II, Aerospace Medical Research Laboratory, Aerospace Medical Division, Air Force Systems Commands Wright-Patterson Air Force Base, Ohio, Library of Congress Catalog Card No. 74-607818.

Gavan, J. A., 1950, The consistency of anthropometric measurements, *Am. J. Phys. Anthropol.* **8**:417.

Gavan, J. A., Washburn, S. L., and Lewis, P. H., 1952, Photography: An anthropometric tool, *Am. J. Phys. Anthropol.* **10**:331.

Gessell, A., and Thompson, H., 1938, *The Psychology of Early Growth,* Macmillan, New York.

Gifford, E. C., Provost, J. R., and Lazo, J., 1965, Anthropometry of Naval Aviators—1964. NAEZ-ACEL-533, Aerospace Crew Equipment Laboratory, U.S. Naval Air Engineering Center, Philadelphia.

Hamill, P. V. V., Johnston, F. E., and Lemeshow, S., 1973, Height and Weight of Youths 12–17 Years: United States, U.S. Dept. HEW Publ. (HSM)73-1606, Rockville, Md.

Hansen, R., and Cornag, D. Y., 1958, Annotated Bibliography of Applied Physical Anthropology in Human Engineering, Technical Report 56-30, Aero Medical Laboratory, Wright Air Development Center, Wright-Patterson Air Force Base, Ohio.

Healy, M. J. R., 1958, Variations within individuals in human biology, *Hum. Biol.* **30**:210.

Hertzberg, H. T. E., 1968, Report: The conference on standardisation of anthropometric techniques and terminology, *Am. J. Phys. Anthropol.* **28**:1.

Hertzberg, H. T. E., Dupertuis, C. W., and Emanuel, I., 1957, Stereophotogrammetry as an anthropometric tool, *Photogramm. Eng.* **1957**:942.

Hertzberg, H. T. E., Churchill, E., Dupertuis, C. W., White, R. M., and Damon, A., 1963, *Anthropometric Survey of Turkey, Greece and Italy,* Pergamon Press, Oxford.

Hoerr, N. L., Pyle, S. I., and Francis, C. C., 1962, *Radiographic Atlas of Skeletal Development of the Foot and Ankle,* Charles C Thomas, Springfield, Illinois.

Hrdlicka, A., 1947, *Hrdlicka's Practical Anthropometry,* 3rd ed. (T. D. Stewart, ed.), The Wistar Institute, Philadelphia.

Jordan, J., Ruben, M., Hernandez, J., Bebelagua, A., Tanner, J. M., and Goldstein, H., 1975, The 1972

Cuban National Child Growth Study as an example of population health monitoring: Design and methods, *Ann. Hum. Biol.* **2**:153.

Kay, W. C., 1961, Anthropometry of the ROKAF Pilots, *Repub. Korea Air Force J. Aviat. Med.* **9**:61.

Kemper, H. C. G., and Pieters, J. J. L., 1974, Comparative study of anthropometric measurements of the same subjects in two different institutes, *Am. J. Phys. Anthropol.* **40**:341.

Knott, V. B., 1941, Physical measurements of young children, *Univ. Iowa Stud. Child Welfare* **18**:3.

Krogman, W. M., 1950, A handbook of the measurement and interpretation of height and weight in the growing child, *Child Dev. Monogr.* **13**:3.

Loewenstein, M. S., and Phillips, J. F., 1973, Evaluation of arm circumference measurement for determining nutritional status of children and its use in acute epidemics of malnutrition: Omeri, Nigeria, following the Nigerian Civil War, *Am. J. Clin. Nutr.* **26**:226.

Marshall, W. A., 1966, Basic anthropometric measurements, in: *Somatic Growth of the Child* (J. J. van der Werff ten Bosch and A. Haak, eds.), Leiden, Stenfert Kroese, N.V.

Marshall, W. A., and Ahmed, L., 1976, Variation in upper arm length and forearm length in normal British girls: Photogrammetric standards, *Ann. Hum. Biol.* **3**:61.

Martin, R., 1914, *Lehrbuch der Anthropologie,* Verlag von Gustav Fischer, Jena.

Martin, R., and Saller, K., 1957, *Lehrbuch der Anthropologie,* Band I, Gustav Fischer Verlag, Stuttgart.

Martorell, R., Habicht, J.-P., Yarbrough, C., Guzman, G., and Klein, R. E., 1975, The identification and evaluation of measurement variability in the anthropometry of pre-school children, *Am. J. Phys. Anthropol.* **43**:352.

Meadows, D. M., Johnson, W. O., and Allen, J. B., 1940, Generation of surface contours by moiré pattern, *Appl. Opt.* **9**:942.

Meredith, H. V., 1936, The reliability of anthropometric measurements taken on eight- and nine-year-old white males, *Child Dev.* **7**:262.

Miskin, E. A., 1960, Simple photogrammetric methods in medicine, *Med. Biol. Illus.* **10**:230.

Montagu, M. F. A., 1960, *A Handbook of Anthropometry,* Charles C. Thomas, Springfield, Illinois.

Morley, D., 1974, Measuring malnutrition, *Lancet* **1**:758.

Oberman, A., 1965, The thousand aviator study: Distributions and intercorrelations of selected variables, Monogr. 12, U.S. Naval Aerospace Medical Institute, U.S. Naval Aviation Center, Pensacola, Florida.

O'Brien, R., and Shelton, W. C., 1941, Women's measurements for garment and pattern construction, Miscellaneous Publ. No. 454, U.S. Department of Agriculture and Work Projects Administration, Bureau of Home Economics, Textiles and Clothing Division, Washington, D.C.

Oshima, M., 1962, quoted in Garrett and Kennedy, 1971.

Parizkova, J., and Goldstein, H., 1970, A comparison of skinfold measurements using the Best and Harpenden calipers, *Hum. Biol.* **42**:436.

Parizkova, J., and Roth, Z, 1972, The assessment of depot fat in children from skinfold thickness measurements by Holtain (Tanner/Whitehouse) calipers, *Hum. Biol.* **44**:613.

Peters, C. C., and Van Voorhis, W. R., 1939, *Statistical Procedures and Their Mathematical Bases,* McGraw-Hill, New York.

Prahl-Andersen, B., Pollmann, A. J., Raaken, D. J., and Peters, K. A., 1972, Automated anthropometry, *Am. J. Phys. Anthropol.* **37**:151.

Pyle, S. I., and Hoerr, N. L., 1955, *Radiographic Atlas of Skeletal Development of the Knee,* Charles C. Thomas, Springfield, Illinois.

Randell, F. E., and Baer, M. J., 1951, Survey of body size of army personnel, male and female methodology. Report 122, Office of the Quartermaster General, Research and Development Division, Quartermaster Climatic Research Laboratory, Lawrence, Massachusetts.

Randell, F. E., Damon, A., Benton, R. S., and Patt, D. I., 1946, Human body size in military aircraft and personnel equipment, AAF Technical Report 5501, Air Material Command, Wright Field, Dayton, Ohio.

Reynolds, E. L., 1944, Differential tissue growth in the leg during childhood, *Child Dev.* **15**:181.

Roche, A. F., and Davila, G. H., 1974, Differences between recumbent length and stature within individuals, *Growth* **38**:3.

Ruiz, L., Colley, J. R. T., and Hamilton, P. J. S., 1971, Measurement of triceps skinfold thickness, *Br. J. Prev. Soc. Med.* **25**:165.

Sastry, J. G., and Vijayavaghaven, K., 1973, Use of anthropometry in grading malnutrition in children, *Indian J. Med. Res.* **61**:1225.

Scammon, R. E., 1927, The first seriatum study of human growth, *Am. J. Phys. Anthropol.* **10**:329.

Sheldon, W. H., 1940, *The Varieties of Human Physique,* Harpers, New York.

Sheldon, W. H., 1954, *Atlas of Men,* Harpers, New York.

Shuttleworth, F. K., 1937, Sexual maturation and the physical growth of girls aged six to sixteen, *Child Dev. Monogr.* **2**:1.

Simmons, K., 1944, The Brush Foundation study of child growth and development, II. Physical growth and development, *Child Dev. Monogr.* **9**:1.

Sloane, A. N., and Shapiro, M., 1972, A comparison of skinfold measurements with three standard calipers, *Hum. Biol.* **44**:29.

Snow, C. C., and Snyder, R. G., 1965, Anthropometry of air traffic control trainees. AM 65-26, Federal Aviation Agency, Office of Aviation Medicine, Oklahoma City, Oklahoma.

Spielman, R. S., Da Rocha, F. J., Weitkamp. L. R., Ward, R. H., Neel, J. V., and Chagnon, N. A., 1972, The genetic structure of a tribal population, the Yanomama Indians. VII. Anthropometric differences among Yanomama villages, *Am. J. Phys. Anthropol.* **37**:345.

Stanford, B., and Tanner, J. M., 1949, An aircraft camera technique with flash or tungsten lighting, addendum to Tanner, J. M., and Weiner, J. S., 1949, *Am. J. Phys. Anthropol.* **7**:145.

Steggerda, M., 1942, Anthropometry of the living: A study on checking of techniques, *Anthropol. Briefs* **2**:7.

Stewart, T. D., 1952, *Hrdlicka's Practical Anthropometry,* 4th ed., The Wistar Institute of Anatomy and Biology, Philadelphia.

Strickland, A. L., and Shearin, R. B., 1972, Dirunal height variation in children, *Pediatrics* **80**:1023.

Stuart, H. C., 1939, Studies from the Center for Research in Child Health and Development, School of Public Health, Harvard University. I. The Center, the group under observation, sources of information and studies in progress, *Child Dev. Monogr.* **4**:1.

Takasaki, H., 1970, Moiré topography, *Appl. Opt.* **9**:1457.

Tanner, J. M., 1952, The assessment of growth and development in children, *Arch. Dis. Child.* **27**:10.

Tanner, J. M., 1962, *Growth at Adolescence,* 2nd ed., Blackwell Scientific Publications, Oxford.

Tanner, J. M., 1964, *The Physique of the Olympic Athlete,* George Allen and Unwin Ltd., London.

Tanner, J. M., 1965, Radiographic studies of body composition in children and adults, in: Body Composition, *Symp. Soc. Study Hum. Biol.* **7**:211.

Tanner, J. M., and Weiner, J. S., 1949, The reliability of the photogrammetric method of anthropometry, with a description of a miniature camera technique, *Am. J. Phys. Anthropol.* **7**:145.

Tanner, J. M., and Whitehouse, R. H., 1959, *Standards for Skeletal Maturity,* Part I. International Children's Centre, Paris.

Tanner, J. M., and Whitehouse, R. H., 1975, Revised standards for triceps and subscapular skinfolds in British Children, *Arch. Dis. Child.* **50**:142.

Tanner, J. M., Whitehouse, R. H., and Powell, J. H., 1958, Armadillo: A protective clothing as a shield from X-radiation, *Lancet* **2**:779.

Tanner, J. M., Whitehouse, R. H., Marshall, W. A., Healy, M. J. R., and Goldstein, H., 1975*a,* *Assessment of Skeletal Maturity and Prediction of Adult Height (TW2 Method),* Academic Press, London.

Tanner, J. M., Lejarraga, H., and Cameron, N., 1975*b,* The natural history of the Silver-Russell syndrome: A longitudinal study of 39 cases, *Pediatr. Res.* **9**:611.

Terada, H., 1974, A new apparatus for stereometry: Moiré contourograph, in: *Nutrition and Malnutrition* (A. F. Roche and F. Falkner, eds.), pp. 27–46, Plenum Press, New York.

Udjus, L. G., 1964, Anthropometrical changes in Norwegian men in the twentieth century, *Norwegian Monograph on Medical Science,* Universitets-forlaget, Oslo.

Waterlow, J. C., Buzina, R. Keller, W., Lane, J. M., Nichaman, M. Z. and Tanner, J. M., 1977, The presentation and use of height and weight data for comparing the nutritional status of groups of children under the age of 10 years, *Bull WHO* **55**(4):489–498.

Weiner, J. S., and Lourie, S. A., 1969, *Human Biology: A Guide to Field Methods,* IBP Handbook No. 9, Blackwell Scientific Publications, Oxford.

White, R. M., 1966, U.S. Army Anthropometry—1966, Pioneering Research Laboratory, U.S. Army Natick Laboratories, Natick, Massachusetts.

Whitehouse, R. H., 1974, personal communication.

Whitehouse, R. H., Tanner, J. M., and Healy, M. J. R., 1974, Diurnal variation in stature and sitting height in 12–14 year-old boys, *Ann. Hum. Biol.* **1**:103.

4

Somatic Growth of the Infant and Preschool Child

FRANCIS E. JOHNSTON

1. Infant and Preschool Years

This period comprises the first five years of life, that is, from birth through age four. In general, this age range spans that period during which the individual makes the adjustment from a sharply limited intrauterine environment, in which all stimuli are mediated through the maternal biological system, to the rich, varied, and often hostile, extrauterine environment, where the regulatory responses are direct functions of interactions between this environment and the infant's or child's biological system.

The first five postnatal years are, therefore, years of change, compensation, and initiation. There is adjustment to the striking environmental changes experienced; there is compensation for the new levels of stress which accompany the changes; and there is initiation of developmental and physiological regulation based upon mechanisms internal to the individual. It is no wonder that the preschool years are those which, apart from senescence, display the highest mortality rates; as many as 50% of all babies born will die by five years of age (Jelliffe, 1968). When viewed in the light of the very high growth rates that characterize this period, one accepts as logical the findings of many investigators that a significant percentage of the biological and behavioral variability encountered in later years has its genesis in the preschool period (see, e.g., Heald and Hollander, 1965; Huenemann, 1974a; Mack and Johnston, 1976).

1.1. Transition from Prenatal to Postnatal Life

Timiras (1972) has summarized the major contrasts between the prenatal and postnatal environments (Table I). The contrasts involve the physical environment,

FRANCIS E. JOHNSTON • University of Pennsylvania, University Museum, Philadelphia, Pennsylvania.

Table I. Contrast between Prenatal and Postnatal Environments[a]

	Prenatal	Postnatal
Physical environment	Fluid	Gaseous
External temperature	Generally constant	Fluctuating
Sensory stimulation	Primarily kinesthetic or vibratory	Variety of stimuli
Nutrition	Dependent upon nutrients in mother's blood	Dependent upon food availability and digestive sufficiency
Oxygen supply	Passed from mother to fetus at placenta	Passed from lung surfaces to pulmonary vessels
Elimination of metabolic products	Discharged into mother's bloodstream	Eliminated by lungs, skin, kidneys, and gastrointestinal tract

[a]Adapted from Timiras (1972).

physiological process, nutrient sources, and communication of sensory stimuli. During the prenatal period, when the transition from one environment to the other is effected, the infant undergoes systematic physiological changes which are partly due to the stresses of labor and delivery. The O_2 pressure in the umbilical artery may drop to zero, associated with the profound metabolic acidosis (James, 1959) of a pH of 6.8 or lower. Changes in other systems and organs are also apparent.

The newborn infant must successfully make the transition from a regulatory system which was largely dependent upon characteristics of the maternal organism to one which is based upon his or her own genetic and homeostatic mechanisms. The continuing process of development, which has gone on since the initial cell divisions, becomes, during the preschool years, canalized, i.e., regularized and predictive of future events. Body size during the first year of life is poorly correlated with the adult size of the individual, but by two years of age self-correlations of the size of an individual with his or her adult height have stabilized at a value of about 0.8 for well-nourished populations (Figure 1).

Despite the transition to a pattern of self-regulation, the infant and young child is still heavily dependent upon older persons. The high degree of immaturity in the young human, with respect to other mammals, is by now well known. This is particularly significant when viewed relative to the primates, our closest relatives, and has been described by a number of authors (e.g., LaBarre, 1954).

In the midst of the above, the process of growth as an intrinsic property of the organism, continues. Changes in the body's cell mass, and its supporting components, are evident from the earliest embryonic stages and constitute the broader set of dimensional changes which are called "growth." Of the postnatal years, rates of growth are most rapid during infancy and childhood and therefore are especially sensitive to the surrounding environment. The growth status of an individual during these years is a widely used measure of the quality of this environment and, by extrapolation, of the environment of the total population.

2. Description of Somatic Growth

Despite the fact that there is no single definition of growth, virtually everyone defines it in quantitative terms; i.e., growth is not elaboration of function, differentiation of tissues, nor the laying down of metabolic pathways. Rather it is the increase or decrease of some measurable quantity. To the cellular biologist, growth may be thought of as either increase in the size of cells (hypertrophy) or increase in their number (hyperplasia) and is seen as an ongoing process throughout the life of an organism (Goss, 1964). To the person more concerned with the organizational level of the living individual, growth may be thought of as an increase in some measurable dimension. This "organismic" growth results, of course, from the sum of many changes at the cellular level.

Since cellular dynamics involve not only hypertrophy and hyperplasia, but also cellular destruction, growth may be negative, where the decrease exceeds the increase, or at equilibrium, where the two are equal within a population of cells (Goss, 1964).

Somatic growth is "described," i.e., measured, by the technique of anthropometry, a methodology originally developed by physical anthropologists, but now utilized, and refined as well, by a number of disciplines (see Cameron, Chapter 3). Anthropometric measurements are carefully done using specialized instruments, standardized techniques, and, ideally, operate within a matrix of known error of measurement.

The number of measurements which can be made upon the body are almost limitless. The selection of measurements must be dictated by the purposes of the study. The design engineer, interested in an adequate workspace, a safe piece of equipment, or clothing that fits, will be interested in one class of dimensions (e.g., Hertzberg *et al.,* 1963; Clauser *et al.,* 1972) which will establish certain parameters. The human biologist, interested in cause and effect and in relating physiological and ecological factors, will be interested in another class (Weiner and Lourie, 1969), and the pediatrician, evaluating health as a function of growth (Falkner, 1966; Jelliffe, 1966), yet another. Some measurements will be common to virtually any study, others unique to only one; however, each investigator is interested in a set of dimensions which change with age and which interact with the individual's environment.

As a convenience, somatic growth, measured anthropometrically, may be visualized in terms of a few "clusters" of related measurements. These clusters relate biologically similar aspects of growth and permit a more rational selection of dimensions.

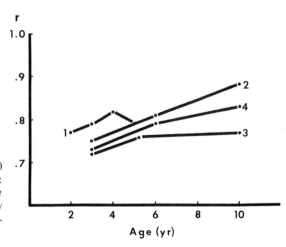

Fig. 1. Self-correlations of stature at 2–10 years with mature stature in four samples: (1) Aberdeen, sexes combined (Tanner *et al.,* 1956); (2) Berkeley boys; (3) Berkeley girls; (4) Berkeley, sexes combined (Tuddenham and Snyder, 1954).

2.1. Measures of Size

FRANCIS E. JOHNSTON   Various measurements are included within this category, all of which measure growth along some linear axis. Most common are dimensions along the longitudinal axis of the body, i.e., along the trunk and lower extremity, along the shafts of the long bones, and so forth. These dimensions are statistically related (Lindegaard, 1953) and, for practical purposes, may often be reduced to relatively few. In general, they reflect skeletal growth at the metaphyses of endochondrally formed bones. The most common is, of course, stature. However, a number of other dimensions may also be recorded, for example, the sitting height which, when subtracted from stature, yields an estimate of leg length. Other longitudinal measures include the length of individual bones, taken either from the living or from radiographs.

Size may also be measured in terms of body breadths, that is, in general measurements made in a plane perpendicular to its long axis. Most frequent are the biacromial breadth of the shoulder and the bicristal breadth of the hip; they reflect complex internal changes in the architecture of the shoulder and hip regions, including associated musculature and adipose tissue. Included here are measurements of the breadths of the extremities of certain bones, the biepicondylar breadth of the humerus and the bicondylar breadth of the femur, and also the breadths of the wrist and ankle joints, taken across the distal radius/ulna and tibia/fibula. These are utilized in estimating the breadth factor of the skeleton or of the specific bones involved.

One set of size measurements that is somewhat different from the above is that dealing with the head and face. These measurements are usually taken for specific purposes and may involve detailed measurements of the face, the cranial base, or the brain case. Head circumference is one such measurement widely used in the evaluation of the health and nutritional status of infants and young children.

2.2. Measures of Mass and Composition

The basic measurement of body mass is, of course, weight. This composite dimension, together with stature, comprises the most widely taken pair of growth measurements. However, since it is such a composite, body weight is frequently divided into its components or into smaller units. This may be done in a laboratory, where the measurement of the weights of body segments (e.g., the arm) can be accomplished with greater precision and accuracy. The measurement of body composition, e.g., the determination of fat or lean body mass, may also be done by the techniques of densitometry, whole-body counting (minerals), or fluid-space measurement.

However, in many cases, it is both necessary and desirable to utilize somatic measurements in an attempt to estimate body composition. This is true in studying large samples away from a laboratory, as well as when working with young age groups that are not suited for many of these techniques.

Somatic measurements of body composition include the measurement of skinfolds, using appropriate techniques and instruments (Brŏzek, 1963; Weiner and Lourie, 1969), and body circumferences. For many age groups, there are formulas for converting these data into whole-body estimates but for others one must utilize only the raw measurements.

By and large, humans grow proportionately, and the relationships of measurements to each other provide additional information of considerable value. The ratio of sitting height to stature can assist in diagnosing particular types of growth failure as can the ratio of fat to lean body mass, arm-span to stature, biacromial to bicristal breadth, and the like.

Measurements may be related to each other by means of an index, or by using appropriate regression techniques. One of the most widely utilized is weight-for-height, of particular value in the study of obesity. This may be done by one of several indices (weight/height^2; weight/height$^{1.75}$; $\sqrt[3]{\text{weight/height}}$, by the calculation of relative weight [$100 \times$ (weight/standard weight)], or by referring to various percentile tables. (See Newens and Goldstein, 1972, and Billewicz *et al.*, 1962, for a discussion).

These constitute the main approaches to somatic growth in infants and preschool children. Others, involving the body as a whole, in terms of "constitution" (Carter, 1972), or utilizing sophisticated noncartesian methods (Herron, 1972), while available, are almost always applied to older age groups.

3. Growth during Preschool Years

3.1. Size at Birth

Body size and composition at birth reflect, of course, fetal growth and need not concern us here. However, it is of considerable importance that, even at this time, significant variability within and between populations exists. Birth weights, for example, vary widely and systematically. On a worldwide basis (Meredith, 1970), there is an impressive range, and one may identify between-population factors (Adams and Niswander, 1973) which probably reflect both hereditary and ecological mechanisms.

The amount of within-sample variation at birth is likewise impressive. Table II presents the coefficients of variation (100 $\bar{x}/s$) for weight and length at selected ages, from the mixed longitudinal data of the Denver Child Research Council (Hansman, 1970). The relative variability at birth is similar to that seen at 5, 10, and 15 years.

Table II.   Coefficients of Variation of Body Length and Weight at Selected Ages Among Denver Children[a]

Length (%)		Age (yr)	Weight (%)	
M	F		M	F
4.5	3.5	Birth	15.3	13.7
3.0	3.8	5	9.1	10.7
3.3	4.6	10	13.7	17.5
4.0	3.6	15	15.2	14.4

[a]Calculated from means and standard deviations presented by Hansman (1970).

Body composition at birth differs by sample as well. Anthropometric measurements of newborns of black, white, and Puerto Rican ancestry (Johnston and Beller, 1976) indicate differences in skinfold thickness and estimated arm-muscle mass by sex and socioeconomic status.

Thus, postnatal growth begins with a significant component of variation in almost any measurement we may make. Since growth is cumulative through time, any attempt at explaining differences in the first five years of life must begin from an initial realization of the considerable variation arising during the prenatal period.

3.2. Infancy

The first year of life is characterized by extremely rapid growth. Among middle-class Finnish children the differences between the birth and 1-year lengths of individuals are given in Table III. This represents an increase of 50–55% in length and 180–200% in weight.

Postnatal growth velocities are highest, by far, immediately after birth, falling off sharply by 12 months. In the Finnish study the length velocities fell from 45.8 cm/yr between birth and 3 months to 14.4 cm/yr in the 9- to 12-month interval in boys; their weight velocities dropped from 10.3 to 3.4 kg/yr. The velocities in females were 14% and 13% less for length and weight from birth to 3 months, but, in the final 9- to 12-month interval, no sex difference was apparent. These are similar to the data presented by Tanner *et al.* (1966*a,b*), who have reported instantaneous velocities at 0.16 years of 40.0 cm/yr in boys and 36.0 in girls, dropping to 14.5 cm/yr and 15.9, respectively, at 0.87 years.

Information on the growth of other body dimensions, studied longitudinally between birth and 1 year is much less common. Some of the most complete data come from the Denver Child Research Council (Hansman, 1970). From the means at birth and 12 months of their mixed longitudinal sample, we may calculate the percent increase of various anthropometric dimensions over the first year of life. The increase in length, hip (biiliac) breadth, and shoulder (biacromial) breadth averaged about 55% of the birth value in boys and about 54% in girls. Crown–rump length (the equivalent of sitting height) showed a 45% increase in boys and 42% in girls, while for head circumference, the values were 39% and 35%. Growth is high in infancy and, insofar as linear measurements are concerned, differential increases among various dimensions are already demonstrable.

Additional data on other measurements during infancy are available, although they are cross-sectioned, from the measurements of 4027 infants and children by Snyder *et al.* (1975). Although the sample numbers vary by measurement, 344 of the subjects were under 12 months of age. The sampling was conducted so as to provide a representative cross-section, ethnically and socioeconomically, from 76

Table III.   Mean Increments of Length and Weight in Finnish Children, Birth to 1 Year, Mixed Longitudinal Data [a]

	Length (cm)	Weight (kg)
Boys	28.3	7.22
Girls	25.9	6.34

[a]From Kantero and Tiisala (1971*a*).

Table IV. *Skinfold Thickness in Newborn Infants*

	Number	Triceps		Subscapular	
		$\bar{x}$	SD	$\bar{x}$	SD
United States					
Whites[a]					
Males	32	4.0	1.1	3.8	1.2
Females	32	4.2	1.1	3.8	1.2
Blacks[a]					
Males	33	3.9	1.1	3.6	1.0
Females	34	4.3	1.1	4.0	1.1
Puerto Rican[a]					
Males	31	3.4	1.1	3.3	1.2
Females	34	3.9	1.3	3.5	1.3
England[b]					
Males	187	4.7	0.9	4.9	1.1
Females	144	4.7	1.0	5.2	1.2
Guatemala[c]					
Males	29	4.5	1.0	4.4	1.2
Females	16	4.7	0.9	4.3	0.8

[a]Johnston and Beller (1976).
[b]Gampel (1965).
[c]Malina *et al.* (1974).

locations in 8 states of the U.S.A. The authors present descriptive statistics for 30 linear dimensions and 9 circumferences grouped, during infancy, into four 3-month periods. Many of the dimensions are of design-engineering significance, and this study is one of the most comprehensive for the infant period, in terms of range of measurements and the attention given to techniques of measurement and design of equipment.

There is but little information on age changes in skinfolds at birth, and certainly not enough to furnish comprehensive data on percentile distributions. Table IV presents the mean thicknesses of the triceps and subscapular folds in newborns, as reported in various studies; virtually all infants were measured within the first 3 days after birth. The range of variation among the means is quite striking: the differences between the low and high values range from 20 to 50% of the low. The roles of prenatal growth, gestation, and other factors are obviously of great significance.

After birth the skinfold thicknesses at the triceps and subscapular sites rise markedly to a peak, before commencing to decline during the subsequent years. The revised standards of Tanner and Whitehouse (1975) place this peak at 6 months for the subscapular, at 6 months for the triceps in males, and 1–1½ years for the triceps in females. These peaks for the triceps are, however, valid only for the "50th percentile child," since they occur at other ages, earlier and later, for children on other percentiles. On the other hand, among mildly to moderately malnourished Guatemalan infants (Malina *et al.*, 1974), the peaks occurred in boys and girls, for both sites, at 3 months of age.

Finally, the difficulty in characterizing infant growth curves of skinfolds is indicated by the data of Fry *et al.* (1975). In contrast to the British or Guatemalan children, with a clear peak in skinfold thickness, there is only a slight dip in the means (sexes combined) of triceps and subscapular folds of Black, White, and Mexican–American infants. This occurred in the 4- to 6-month group among Blacks and in the 6- to 8-month group among the other two samples. While a rapid rise to a peak, followed by a subsequent decline, seems to be the case in most studies, the age of the peak and the duration and the intensity of the decline are not at all clear.

Small sample size may contribute significantly to this variability in pattern, and its role cannot be adequately assessed at present. We need distributions of skinfolds at specific ages during infancy in large enough samples to reduce sampling error to a minimum. For example, in a recent study, Huenemann (1974a) has reported the means of four skinfolds (and their sum) from 228 male and 220 female 6-month-old infants from Berkeley, California. These and other means are presented in Table V. For the skinfolds, samples of this size have reduced the standard error of the mean to about 1–2%.

Despite the probable decline in mean skinfold thickness after the infant peak (whenever it occurs), the arm continues to increase in size. The 3-month means of Snyder et al. (1975) for the circumference of the upper arm show a steady increase throughout this period, even at the 5th percentile. This suggests two possibilities: (1) a decrease in fatness in the upper arm concomitant with an increase in the underlying lean tissue, or (2) a thinning of the fat "ring" around the arm as it is carried outward by the expansion of the underlying tissue. In the latter case, the actual amount of fat, as estimated by the cross-sectional area, may increase despite a decrease in skinfold thickness (Tanner, 1962a; Johnston et al., 1974).

There are innumerable studies on the growth of height and weight during the first year of life. The time-honored data of Stuart's Boston study (Vaughan, 1964), although based on a small sample, have been used over and over throughout the world to describe and compare, and they need not be reprinted here. In addition, for developed nations, there are other data sets from a variety of samples: American Whites (Hansman, 1970), American Blacks and Mexican–Americans (Fry et al., 1975), British (Falkner, 1958; Tanner et al., 1966a,b), Finnish (Bäckström and Kantero, 1971), Japanese (Terada and Hoshi, 1965a), and Polish (Chrzastek-Spruch and Wolanski, 1969), to name but a few.

For the developing countries of the world, there are also many reports, for example, of infants from Guatemala (Yarbrough et al., 1975), Mexico (Faulhaber,

Table V. Means of Anthropometric Dimensions of 228 Male and 220 Female Infants, 6 Months of Age, from Berkeley, California[a]

	Male		Female	
	Mean	SD	Mean	SD
Length (cm)	68.2	2.3	66.3	2.1
Weight (kg)	8.23	0.83	7.60	0.75
Skinfold (mm)				
Triceps	8.0	1.6	8.0	1.8
Subscapular	6.2	1.4	6.5	1.6
Subprailiac	4.9	1.7	5.4	1.9
Chest	3.7	0.9	3.8	1.2
(Sum)	22.9	4.5	23.8	5.2
Circumference (cm)				
Head	44.4	1.1	43.2	1.1
Upper arm (biceps)	15.4	0.9	15.0	0.9

[a]From Huenemann (1974a).

1961; Scholl, 1975), Iran (Amirhakimi, 1974), Nigeria (Rea, 1971), and among technologically simple societies such as the Bushnegroes of Guyana (Doornbos and Jonxis, 1968), again to indicate but a few.

99

SOMATIC GROWTH OF THE INFANT AND PRESCHOOL CHILD

Despite the wide disparity in ecologies, ethnic backgrounds, and socioeconomic conditions, the variation in patterns of growth in infancy is not as large as might be expected. Habicht *et al.* (1974) have demonstrated, based upon several studies, that up until 6 months of age, the children from a wide range of groups ''. . . generally grow rather uniformly in length and weight.'' After 6 months, they note that infants from developed countries and from high socioeconomic strata in developing countries cluster together. Those from lower strata in poorer countries are arrayed below this cluster.

Regardless of the group analyzed, somatic growth during infancy (excluding skinfolds) follows a simple curvilinear path. Various regression models have successfully been used to represent growth during this period, each involving a linear plus a curvilinear function of age (Vandenberg and Falkner, 1965; Bialik *et al.*, 1973; Manwani and Agarwal, 1973; Scholl, 1975). The fit has been quite good and has permitted the estimation of the various curve parameters as formal tools for analyzing growth. Deviations from this mathematical regularity among individuals (Manwani and Agarwal, 1973; Scholl, 1975) are associated with nutritional and disease-related disturbances.

3.3. Early Childhood

The remaining preschool years, 1–4, may be called the years of early childhood. From the standpoint of somatic growth, they are the years of developmental canalization, the achievement of regularity in the rate of increase of most dimensions. Many of the measurements which were previously increasing at a rapid and changing rate now show a striking reduction in growth rate and achieve a steady, almost linear, increase until the onset of the adolescent spurt. This is particularly true of those structures which display Scammon's "neural" (or brain and head) type of growth (see Tanner, 1962*a*, p. 11). For example, the circumference of the head, which has increased by some 35% from birth to 1 year, increases only by about 10% through the 5th year, with most of that increase coming between the ages of 1 and 2.

Table VI presents the means of this dimension as reported in four mixed longitudinal growth studies. The values are very close, at similar ages, and clearly demonstrate the deceleration in growth rate mentioned above. This deceleration may be seen in Figure 2, which presents the annual velocities of growth of head circumference, in centimeters per year, for the Finnish children (Kantero and Tiisala, 1971*b*) taken at 6-month measurements. Data from other investigations are not presented since they cannot be converted to strictly comparable growth rates (i.e., based upon longitudinal data and calculated semiannually). However, the shape of the curves of Figure 2 is characteristic of essentially all studies; by 2 years of age, the increases are only in the range of 2 cm/yr.

These data suggest a general similarity in the pattern of growth in head circumference during the preschool years. This is supported by research from other human ecosystems as well. Thus among urban Jamaican infants (Grantham-McGregor and Desai, 1973), longitudinal measurements indicate that the circumference at 1 year is 33% greater than that at birth. A similar difference is seen in

Table VI. Head Circumference (cm) in Four Mixed Longitudinal Samples

Age (mo)	Finland[a] Mean	SD	Japan[b] Mean	SD	England[c] Mean	SD	USA[d] Mean	SD
Male								
3	41.0	1.2	41.1	0.9	45.7[e]	1.3	40.3	1.2
12	47.4	1.5	47.0	1.0	46.8	1.3	46.6	1.2
24	49.8	1.5	49.0	1.1	49.3	1.4	49.2	1.2
36	50.7	1.3	50.2	1.1	50.4	1.3	50.3	1.2
48	51.6	1.3					51.0	1.2
60	52.3	1.3					51.6	1.2
Female								
3	39.9	1.2	39.8	1.0	44.6[e]	1.2	39.5	1.1
12	46.3	1.3	45.6	1.1	45.7	1.2	45.6	1.3
24	48.8	1.4	47.9	1.3	47.9	1.1	48.1	1.4
36	50.0	1.4	49.0	1.4	49.5	1.1	49.2	1.4
48	50.3	1.4					50.2	1.5
60	51.3	1.5					50.7	1.5

[a]Kantero and Tiisala (1971b). [c]Falkner (1958). [e]Age = 39 weeks.
[b]Terada and Hoshi (1965b). [d]Hansman (1970).

Guatemalan Ladinos (Malina *et al.*, 1975); again, between 1 and 5 years, the means of these children had increased but 10%, consistent with the data from Europeans cited above.

Despite a general similarity in the shape of the curve of growth of head circumference, there are significant differences among groups. Meredith (1946, 1971) has effectively synthesized the literature on growth of this dimension in the first two years of life in his earlier paper and into the adult years in the latter. In particular, he has demonstrated the existence of racial differences even at birth and

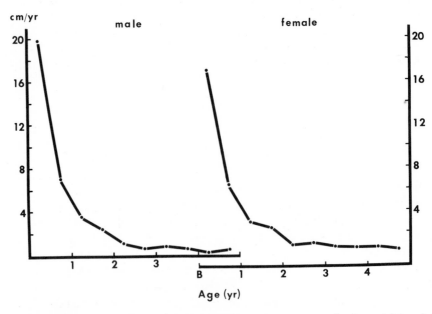

Fig. 2. Annual velocities of head circumference in Finnish children, measured at 6-month intervals (data from Kantero and Tiisala, 1971b).

questioned the notion of universal applicability of reference standards (Meredith, 1971).

The analysis of somatic growth during early childhood is rendered more difficult by problems of measurement at these ages. In particular, it is important to determine whether length was measured in the erect or the supine position, since the two techniques yield different results. Roche and Davila (1974) found that, between 2 and 5 years of age, recumbent length exceeded stature by median values ranging from 2.3 to 1.3 cm, decreasing sharply over the period. Different investigators have utilized various approaches. Kantero and Tiisala (1971a) measured stature after 21 months; Cruise (1973), in studying the growth of children who were of low birth weight, measured supine length under 3 years, as did Faulhaber (1961) in her study of Mexican children. On the other hand, Falkner (1958) and Terada and Hoshi (1965a) took all lengths in the supine position for the 3 years of their studies, while in the study of Guatemalan Ladino children sponsored by the Nutrition Institute of Central America and Panama (INCAP), all measurements are made supine, even through age 7 (Yarbrough *et al.*, 1975). Differences in technique create systematic measuring error and spuriously inflate the variance between samples.

3.4. Length and Weight

From 1 through 4 years of age, growth in length and weight approach a nearly linear rate of increase, at least as indicated by distance curves of size against age. Figure 3 presents, as an example, the curves of length and weight against age in males for two disparate samples of children: the first is taken from the survey of 3199 Slovenian children from Yugoslavia (Pogačnik, 1970), birth through 6 years of age, and the second from a study of 5684 London children from 2 weeks through 4 years of age (Gore and Palmer, 1949). Both samples are cross-sectioned, and the

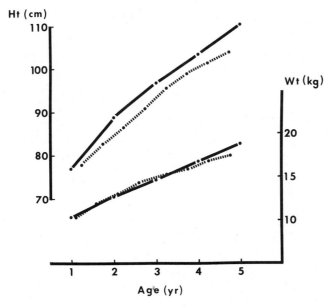

Fig. 3. Distance curves of length and weight against age in children from Yugoslavia (Pogačnik, 1970, solid line) and England (Gore and Palmer, 1949, broken line).

*Table VII.   Annual Increments of Length and Weight in Females
from 1 through 5 Years from Three Studies*

Age (yr)	U.S.A.[a]	U.K.[b]	Finland[c]
Length (cm)			
1–2	12.8	11.4	11.2
2–3	8.7	8.4	8.5
3–4	8.0	7.5	8.4
4–5	7.1	6.7	8.5
Weight (kg)			
1–2	2.5	2.7	2.8
2–3	2.0	2.0	1.9
3–4	2.1	2.0	2.1
4–5	2.2	2.2	2.3

[a]Differences between means of mixed longitudinal sample (Hansman, 1970).
[b]Average smoothed velocities, longitudinal data (Tanner *et al.,* 1966*b*).
[c]Summed increments (3-monthly between 1 and 2 years, semiannual thereafter)
(Kantero and Tiisala, 1971*a*).

curves indicate a similarity of patterns (no smoothing has been done). While a slight deceleration is indicated, these means indicate a regular difference from age group to age group.

Pure longitudinal data over this age range are simply not available for samples of any size. However, Table VII presents incremental data from three studies in which a high level of sample stability is to be found from year to year. Growth increments, over 12-month periods, have been calculated, as indicated in the table, for weight and length among females as an example of the change observed in

*Table VIII.   Percentage Differences between Length and Weight at 1 and Height
and Weight at 4 Years of Age from Various Studies*

	Length		Weight	
	Male	Female	Male	Female
Cross-sectional				
U.S.A. (white)[a]	26.6	31.4	47.5	42.4
U.S.A. (black)[a]	29.5	29.5	61.6	53.1
U.S.A. (black)[b]	31.9	30.1	63.5	63.1
U.S.A. (mixed)[c]	28.4	30.4	54.6	62.2
U.K. (mixed)[d]	27.7	30.3	49.0	57.2
Finland[e]	25.5	24.2	39.7	36.8
Czechoslovakia[f]	29.0	30.1	50.4	56.4
Yugoslavia[g]	29.3	30.2	52.9	56.0
Hong Kong[h]	28.9	29.3	52.4	53.7
Indian[i]	29.9	30.3	60.7	65.4
Mixed longitudinal				
U.S.A. (Boston)[j]	30.4	32.0	52.4	57.2
U.S.A. (Denver)[k]	29.2	31.2	54.4	61.3
U.K.[l]	27.9	28.9	50.9	55.0
Guatemala[m]	30.9	32.8	65.8	72.2

[a]Abraham *et al.* (1975).
[b]Verghese *et al.* (1969).
[c]Snyder *et al.* (1975).
[d]Gore and Palmer (1949).
[e]Bäckström and Kantero (1971).
[f]Fetter and Suchy (1967).
[g]Pogačnik (1970).
[h]Chang *et al.* (1965).
[i]Indian Council of Medical Research (1972).
[j]Stuart and Meredith (1946).
[k]Hansman (1970).
[l]Tanner *et al.* (1966*b*).
[m]Yarbrough *et al.* (1975) (supine length at 4).

individuals. A decreasing growth rate seems indicated but, after 2 years, this decrease is not consistent in all samples. An increase in the weight increment from 4 to 5 years, over that between 3 and 4, is indicated by all three studies.

The estimated percentage increase between 1 and 4 years may be seen, as calculated for 14 samples, in Table VIII. The data represent the differences between means, when adjusted to 1½ or 4½ years. There is a general consistency within each of the four columns, although the variability is greater for weight than for length. Even for weight, the only unusual values are for Finnish children with increases of less than 40%. The percentage increase in body weight is on the order of two times that for length, although it may be as little as 1.5 (Finnish females) or as great as 2.3 (Guatemalan females). In general, the increase is greater in females than males, although exceptions do occur.

3.5. Skinfold Thickness

Figures 4–7 present skinfold thicknesses from five samples of children of various nationalities and ethnic groups. The sample sizes are all quite adequate and some, such as the Yugoslavian sample (Pogačnik, 1970) are based upon a total size of over 3000. This study reported means, which are plotted on the graph; for the others, the medians are used as the statistic of preference, although skewness at this age is small. Values for the British children are only estimated, taken from the graphs of Tanner and Whitehouse (1975).

In each case the age trends are the same; there is a gradual decrease in the thickness of the adipose layer, with, however, a number of exceptions. Between-sample variation is greatest at the younger ages, and the decrease in the medians seems to be greater where the skinfold thickness at 1 year is greater. In fact, among the Ladino sample from Guatemala (Malina et al., 1974), where the medians are lowest, markedly so for the triceps, there is no real decrease with age. There are, in the literature, other reports of skinfolds among children of this age range (see, e.g., Schell and Johnston, 1976, for a review of native Americans). However, sample sizes are generally small and any patterns that emerge confirm the conclusion of a continuation of the loss of fat, which began in infancy, throughout the preschool years. The response of body fat to nutritional factors, however, may result in considerable individual and group variation that may cause reversals in this trend.

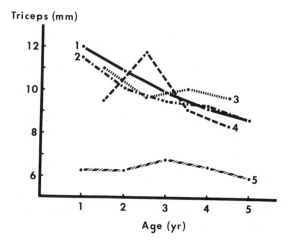

Fig. 4. Mean thickness of triceps skinfolds for males, 1 through 5 years, in five samples: (1) United Kingdom (Tanner and Whitehouse, 1975); (2) Yugoslavia (Pogačnik, 1970); (3) United States, White (Abraham et al., 1975); (4) United States, Black (Abraham et al., 1975); (5) Guatemalan Ladino (Malina et al., 1974).

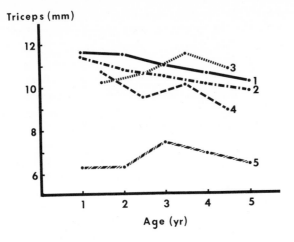

Fig. 5. Mean thickness of triceps skin-folds for females, 1 through 5 years, in five samples: (1) United Kingdom (Tanner and Whitehouse, 1975); (2) Yugoslavia (Pogačnik, 1970); (3) United States, White (Abraham *et al.*, 1975); (4) United States, Black (Abraham *et al.*, 1975); (5) Guatemalan Ladino (Malina *et al.*, 1974).

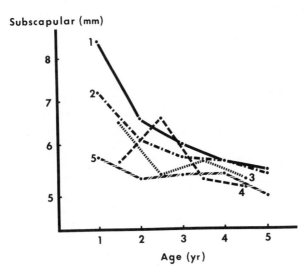

Fig. 6. Mean thickness of subscapular skinfolds for males, 1 through 5 years, in five samples: (1) United Kingdom (Tanner and Whitehouse, 1975); (2) Yugoslavia (Pogaňik, 1970); (3) United States, White (Abraham *et al.*, 1975); (4) United States, Black (Abraham *et al.*, 1975); (5) Guatemalan Ladino (Malina *et al.*, 1974).

Other dimensions grow in a more or less linear fashion, similar to height and weight. These include crown–rump length (or sitting height) (Falkner, 1958; Pogač-nik, 1970; Abraham *et al.*, 1975; Snyder *et al.*, 1975), extremity lengths (Terada and Hoshi, 1966; Anderson *et al.*, 1956, 1964; Snyder *et al.*, 1975), and trunk and extremity circumferences (Wolanski, 1961; Fetter and Suchy, 1967; Pogačnik, 1970; Indian Council of Medical Research, 1972; Malina *et al.*, 1975).

In general, these dimensions all tend to grow similarly to the body or the general curve of Scammon (see Tanner, 1962*a*). That is, during the preschool years there is, as with stature, a decreasing rate of growth until a linear rate is established. Growth in height, in fact, serves as a general model for most of the other somatic dimensions, save those of the head and those which reflect adipose fat.

Even though most dimensions grow in the manner of stature, this does not necessarily mean that their rates of change are the same. Rather they are related in some exponential function, such that $y = x^a$, where y and x are dimensions and a is the exponent which reflects the systematic relationship between them.

This constant relationship between dimensions of growth is called *allometry* (Huxley, 1932) and is characteristic of many living forms. The exponent *a* is called the coefficient of allometry. An example of this is seen in Figure 8, which shows the means of stature (crown–heel length in infants) and sitting height (crown–rump length) for two samples of boys from 3 months through 4 years of age, the mixed longitudinal American sample from Denver (Hansman, 1970) and the large cross-sectional survey of Indian children and youth (Indian Council of Medical Research, 1972). The relationship, in either sample, between the two dimensions is constant, but the increase in sitting height is less than that for stature.

The result of allometric growth is a steady change in various proportions of the body. For example, the legs become longer relative to the trunk while the trunk lengthens relative to its breadth as well as to the size of the head. In other words, the preschool years are those years in which the "chunky" physique of the newborn becomes transformed into the more elongated physique of the child. This trend continues throughout the years of childhood until the circumpubertal period.

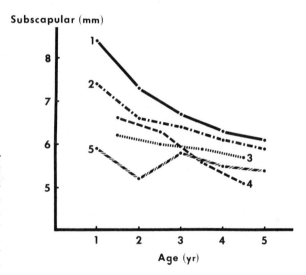

Fig. 7. Mean thickness of subscapular skinfolds for females, 1 through 5 years, in five samples: (1) United Kingdom (Tanner and Whitehouse, 1975); (2) Yugoslavia (Pogačnik, 1970); (3) United States, White (Abraham *et al.,* 1975); (4) United States, Black (Abraham *et al.,* 1975); (5) Guatemalan Ladino (Malina *et al.,* 1974).

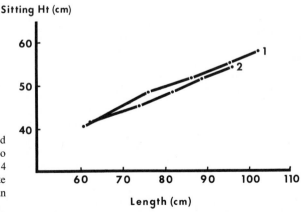

Fig. 8. Mean sitting height graphed against mean body length in two samples of boys, 3 months through 4 years of age: (1) United States, White (Hansman, 1970); (2) Indian (Indian Council of Medical Research, 1972).

Along with this series of shape changes, a considerable degree of independence exists in the growth of individuals from year to year. Increments of growth show relatively little correlation within individuals between various age periods. For example, Falkner (1958) computed the *r*'s for the recumbent length increment over the period 1–3 months and the increment over the period 12–18 months. In 54 boys the correlation was 0.07, and in 50 girls it was 0.23; among all 104 the value was 0.08, clearly not predictive. Increments over adjacent age periods will frequently show a negative correlation if the periods are shorter than 1 year. This may reflect canalization of development, seasonal effects, or simply measuring error (Tanner, 1962*b*; Billewicz, 1967).

Increments of different dimensions do not seem to be highly correlated. Meredith (1962) calculated 240 correlations between pairs of increments in 67 white males over the period of 5–10 years. While this is beyond the age range being considered, the findings are relevant. Of the total of 240, 103 were not significant, and another 71 showed only a "low" relationship (*r* = 0.31 to 0.45).

Table IX presents the correlations between either length or weight at various ages, as determined in three samples of children from the British Isles. The values are almost all significant but, except for the 1- to 2-year value, are all low to moderate. Of interest is the general similarity of the *r*'s for birth and either 3 or 5 years to the *r*'s for birth and adult weight. The progressive canalization is seen to be emerging and to have reached a more or less stable level by this time, although considerable independence still exists. Reed and Stuart (1959) have commented on the dependence as well as the independence in their description of individual variation in growth patterns.

Even with the independence indicated by the above, a basic regularity is seen to exist in growth between 1 and 4 years. This regularity permits the utilization of regression models for formal analysis, usually combining this age period with the subsequent years of childhood and, in some cases, utilizing the first year of life as well. These models may be a function of a simple parabola of the form $y = a + bt + c \log t$, where t = age (Tanner *et al.*, 1956; Scholl, 1975), or a logistic curve which, when combined with a second logistic, permits analysis from 1 year to maturity (Bock *et al.*, 1973). Other models include those discussed by Wingerd (1970) and

Table IX. *Coefficients of Correlation of Size at Various Ages from Three Samples of Children from Great Britain*

	Aberdeen[a]		Newcastle upon Tyne[b]	London[c]	
	Male	Female	Male and female	Male	Female
Weight					
Birth–3 yr	0.49	0.17	0.36		
Birth–5 yr	0.37	0.16	0.30		
Birth–adult	0.38	0.42	0.25		
Length					
Birth–3 yr	0.52	0.36			
Birth–5 yr	0.42	0.34			
Birth–adult	0.25	0.29			
1 mo–1 yr				0.19	0.54
1 yr–2 yr	0.85	0.76		0.84	0.80

[a]42 males, 38 females (Tanner *et al.*, 1956).
[b]201 males, 241 females (Miller *et al.*, 1972).
[c]60 males, 69 females (maximum) (Falkner, 1958).

the Jenss curve (Jenss and Bayley, 1937; Deming and Washburn, 1963) which approaches as an asymptote a straight line; its form is $y = (c + dt) - e^a + bt$ where t, again, is time (see discussion by Marubini, Volume 1, Chapter 7).

These models provide specific analytic frameworks for analyzing components of growth, e.g., the linear component and the intensity of deceleration. In particular they are valuable in studying individual variation across growth time, but they may be used in comparing samples as well.

4. Determinants of Growth and Its Variability in Infants and Preschool Children

Within the individual child, growth is a steady and regular process. As Tanner (1965) has noted, it does not proceed in "fits and starts" but rather shows a basic, underlying constancy (except, as noted above, in terms of certain increments). All indications are that growth is a self-regulated process, to the point that a conceptual model of the operation of the internal agents involved may be developed (Tanner, 1963).

At the same time, the amount of growth variability which may be catalogued is quite striking. This variation may be separated into significant between-individual and between-sample components. Any interpretation of growth up to 5 years of age must deal with the operation of those factors which promote growth variation. While a detailed coverage of the subject is not the major purpose of this chapter, nevertheless, it is necessary to view physical growth during these years in the context of factors which create, enhance, and maintain the variation which we may so easily observe.

4.1. Factors of Heredity

Basic to any discussion of growth variation is the influence of inheritance upon variability. In this respect there are factors associated with family lines, which promote variability among individuals, and factors associated with populations, which promote variability among groups.

The contrast between monozygotic and dizygotic twins provides a classic design for determining heritability, even though the degree to which data from twins may be generalized is still unclear. Nevertheless, Vandenberg and Falkner (1965) have analyzed, by regression, the growth of 29 MZ and 31 DZ twin pairs from birth to 2 years. They found, using F ratios of within-pair variances, a significant hereditary effect upon the growth rate and, particularly, upon its deceleration. On the other hand, they found that the "initial status" (i.e., at age zero), as estimated from the regression analysis showed no genetic component. Similar findings have been reported by Wilson (1976), who analyzed twins from the same study. MZ twins showed a steadily increasing concordance of size, relative to DZ twins, until 3 years of age. This is in agreement with data on the genetics of birth weight (see the review by Hunt, 1966) which show only a minor contribution of the fetus' genotype to status at birth (see Robson, Vol. 1, Chapter 10).

Starting, then, from the position at birth, in which genetic factors have played only a small part, one may examine the expression of the infant's and child's genotype as a function of age. This has been done most commonly by means of the correlation of parental size with the size of the child at specific ages; a related strategy is to contrast the size of children grouped on the basis of parental size. There have been many such analyses and, for example, Mueller (1975) has presented the results of 20 of the studies. However, his work did not include any data on children under 5 years of age.

Table X presents the r's between parental height and, as an example, the height of their daughters from birth to 4 years of age, as taken from several studies. The correlation coefficients are seen to vary considerably among the studies, by as much as tenfold. Even with the same-sex parent, by 4 years, when some stabilization has been reached, the values, e.g., in fathers, range from 0.11 to 0.35. This may reflect the influence of between- and within-family components of environmental variance and demonstrates once again that heritability, far from being a fixed notion, is sample specific and will vary significantly with the nature of the environment. On the other hand, sample sizes are small and the confidence limits of the r's are quite wide; sampling error cannot be overlooked.

Parental size has also been shown to affect the height of children by Garn (1961, 1966) and Tanner et al. (1970). In his earlier paper, Garn, reviewing genetic factors in growth, has shown that the statures of children differ from 1 year of age onward by parental mating type: tall × tall matings have taller preschoolers than short × short. The other two cited publications have presented reference standards for height which take into account the mid-parental height. Garn indicates that a difference of 12 cm in mid-parental height accounts for a difference in the child, at 1.0 years, of 4 cm in boys and 1.6 cm in girls. At 4.0 years, this difference had increased to 6.8 and 7.0 cm. Tanner et al. (1970) found regressions of child on mid-parental height ranging from 0.27 cm/cm to 0.38 between 2 and 4 years; their values fall within those indicated by Garn.

Now, such correlations indicate the operation of hereditary factors. However, because of environmental similarities within families, the r's may be substantially

Table X. Correlations between Parental Height and Height of Daughters from Several Studies

| | | Age of daughter (years) | | | | |
	Birth	1	2	3	4	F/M[a]
Tanner and Israelsohn (1963)						0.13
Fathers		0.14	0.17	0.17	0.11	
Mothers		0.50	0.58	0.56	0.61	
Bouchalová and Gerylová (1970)[b]						0.43
Fathers	0.10	0.23	0.33	0.35	0.33	
Mothers	0.20	0.30	0.31	0.34	0.32	
Chrzastek-Spruch and Wolanski (1969)[c]						0.30
Fathers	−0.02					
Mothers	0.25					
Livson et al. (1962)[d]						
Fathers	0.26	0.36	0.36	0.36	0.35	
Mothers	0.22	0.46	0.42	0.43	0.44	
Scholl (1975)						
Fathers	0.11	0.32	0.42	0.46		
Mothers	0.21	0.43	0.28	0.30		

[a]Correlation between stature of father and mother.
[b]Partial r, corrected for father/mother correlation.
[c]Firstborn.
[d]Pooled data from studies in California, Ohio, and Oxford; r's read from graphs.

inflated or decreased. Mueller (1975) has summarized the problem and pointed to the role of environmental covariance. Among older children and youths, he has found a combined r (parent/offspring) of 0.38 for children of European origin and 0.21 for those of non-European origin. He suggests that the differences in correlation indicate the harsher environment of most of the latter group.

The only controlled study of hereditary factors in skinfold thickness among preschool children is by Brook *et al.* (1975), who calculated heritability estimates in MZ and DZ twins 3–15 years of age. Among those below 10, the following heritabilities were obtained (based upon intraclass correlations):

	Boys	Girls
Triceps	-0.25 ± 0.43	0.56 ± 0.41
Subscapular	1.23 ± 0.35	0.47 ± 0.37

The large standard errors obviously account for the aberrant values in males and prevent firm conclusions. The authors did suggest that environmental effects are strong below 10 years of age but are less important thereafter.

4.2. Nutritional Factors

Nutritional factors are important sources of growth variation during the preschool years. The quantities of nutrients and their proportions within the diet interact with other variables such as physical activity and health status to regulate, to a significant extent, the rate of growth. The exact nature of the nutritional ecosystem differs markedly from one society, and one socioeconomic level, to another. The relative abundance of calories, as well as their source, varies as does the amount and quality of available protein. Specific nutrients, such as vitamins and trace elements, complete the range of dietary considerations.

Among populations not characterized by either undernutrition or overnutrition, the relationship of diet to growth is less clear and, perhaps, more determined by other factors. For example, Roche and Cahn (1962) could not explain differences in skinfold thickness between British and Australian children on the basis of caloric intakes of the two groups. Comparisons of other samples, where the diet may be estimated, are generally not available for this age range. In one study, Sims and Morris (1974) found virtually no relationship between triceps skinfold thickness and biochemical indicators of nutritional status in 163 well-nourished preschool children. Body weight showed more significant correlations, but the results were still not completely clear. Significant "clusters" emerged only when certain socioeconomic and behavioral characteristics of the family were considered.

Within-sample relationships between growth and nutrient intake are likewise generally lacking for infants and preschool children. In one study, Mack and Kleinhenz (1974) followed a small group ($N = 5$) of infants from 8 through 56 days; total milk consumption was measured as were activity levels. The correlations were numerically high, but frequently, not significant because of sample size. In general, the least active infants consumed more calories and gained weight more rapidly.

Early nutritional excess has been suggested to contribute significantly to later obesity, as indicated by a relationship between relative weight in infancy up to age 7 and measures of relative weight up to 22 years of age (Miller *et al.*, 1972; Mack and Ipsen, 1974; Fisch *et al.*, 1975; Sveger *et al.*, 1975). The role of maternal feeding behavior, while possibly contributing to this potential for obesity (Huenemann, 1974*b*), is not clear, but bottle-feeding in place of breast-feeding, and maternal overfeeding as a result, have been implicated (Malcolm, 1973; Ounsted and Sleigh,

1975). The further statistical relationships among relative weight at 1 year, relative weight during adolescence, and the pregravid weight of the mother (Mack and Johnston, 1976) suggests a significant relationship among maternal characteristics, feeding practices, infant nutrition, and subsequent growth.

Where there is undernutrition, the situation is much clearer and the relationship between protein–calorie malnutrition (PCM) and growth is, by now, firmly established. It is common practice to attribute the small body size of children and infants from developing countries and poorer socioeconomic strata to nutritional deficiencies, and the direct relationship between nutritional status and growth status is accepted. Many publications and reports have documented this. For example, Dean (1960) has, in an early paper, documented the effects of malnutrition from birth; he notes reductions of from 4% (head circumference) to 14% (bicondylar breadth of the femur) for anthropometric dimensions except for weight. Body weight reduction, of course, is used in the diagnosis of PCM itself (Gomez *et al.,* 1956; *Lancet,* 1970).

Other descriptions of the effects of undernutrition on growth during infancy and early childhood come from Central America (Cravioto, 1968; Cravioto and de Licardie, 1971; Yarbrough *et al.,* 1975), North Africa (Wishaky and Khattab, 1962; Bouterline *et al.,* 1973), Turkey (Rharon, 1962), Africa (McFie and Welbrown, 1962), and Australia (Kirke, 1969), to cite but a few.

The response of the infant and preschool child, previously malnourished, to nutritional supplementation and therapy is still unclear. The general potential for "catch-up growth" has been described (Prader *et al.,* 1963), and some researchers have presented data to show striking catch-up, with no permanent physical deficit, after nutritional rehabilitation (Čabak and Najdanvic, 1965; Garrow and Pike, 1967). On the other hand, Graham (1968) has documented the persistent effect of early infant malnutrition, while in a more recent analysis (Graham and Adrianzen, 1972) has furthered this discussion with a note on the need to consider the many complexities of the nutritional ecosystem in assessing any potential for catch-up after nutritional rehabilitation.

4.3. Illness and Health Status

This rather heterogeneous collection of factors involves the effects upon growth of disease and, as included in this section, children whose birth weights were abnormally low. Low birth weight, of course, may be the result of a shortened gestation, i.e., prematurity, or of intrauterine growth retardation, i.e., small-for-dates. The etiology of these two causes of low birth weight are, of course, quite different and must be considered separately.

The role of illness in reducing growth among infants and preschoolers is not completely clear. More often than not, groups characterized by high frequencies of illness also display other growth-deviating mechanisms, e.g., poor nutrition. In many cases, it is difficult to separate the causes and, when done, the effects of the other factors may outweigh those of illness.

In an early study, Evans (1944) contrasted increments of growth in several dimensions among 93 middle-class Iowa children, 2–5 years of age, grouped by frequency of respiratory and digestive-tract illness. There was no relationship between illness, as measured by nursery school absences, and growth.

Somewhat later, Hewitt *et al.* (1955) analyzed annual height increments by illness severity, using five categories of illness, in the 650 children of the Oxford

Child Health Survey, studied from 1 through 5 years. They found a "small but definite diminution" in growth related to severity of illness, especially in boys.

These two studies failed to agree in their findings; possibly there were associated socioeconomic differences which interacted with illness and biased the results. Both samples were from English-speaking populations, although the Iowa children were drawn from a higher social class.

One would expect clearer results among the developing nations, where episodes of illness are more frequent and acute. Among Guatemalan Ladinos, diarrheal disease has been associated with reduced increments in both length and weight, amounting to about a 10% deficit (Martorell *et al.*, 1975). However, among rural children from southern Mexico (Condon-Paoloni *et al.*, 1977), the effects were not so severe. There, annual increments of length and weight from birth through 2 years were contrasted in children with high and low frequencies of illness in three categories: upper respiratory, lower respiratory, and diarrhea. Significant reductions were found only for weight increment and diarrhea, where there was a reduction of some 6% from the mean weight by 3 years of age.

4.4. Socioeconomic and Microenvironmental Factors

Interacting with the above are various factors which are socioeconomic, sociocultural, and, in general, a part of the microenvironmental surroundings of the child. Some of these affect the individual directly and others exert their influence through intermediate mechanisms, e.g., nutrition. In sum, they form the complex developmental ecosystem which determines the development of the child during this crucial period. Many writers have discussed the variety of factors which, in fact, impinge upon growth, but the ecosystem has been represented the most elegantly by Cravioto *et al.* (1967; Cravioto, 1970), who have indicated the complexity of the situation in a developing country (see Chapter 16, Volume 3).

In their study of Lebanese children, Kanawati and McLaren (1973) have identified 20 factors differentiating, although not necessarily causing, growth differences between thriving and nonthriving children. These factors ranged from disease prevalence, presence or absence of vaccination, and caloric intake, to the existence of an indoor toilet, father's education, and the size of the house. The authors conclude that, in low socioeconomic conditions, "factors other than food and health are involved in adequate growth of the preschool child." A similar conclusion was drawn by Scholl (1975), who identified demographic and parental biological characteristics associated with growth failure in Mexican children, and by Christiansen *et al.* (1975), who identified social factors in Bogotá.

Socioeconomic conditions, themselves, are significantly associated with infant and child growth and, as an example, we may cite the data of Bouchalová and Gerylová (1970), Rea (1971), and Amirhakimi (1974) as typical for most locales. Such factors may differentiate rural and urban settings, and Wolanski and Lasota (1964) have reported quite striking differences between the two in Poland.

Even in the more affluent countries, socioeconomic effects are detectable in this age range. Garn and Clark (1975), reporting on the U.S. Ten-State Nutrition Survey, have compared children of low to higher income groups. From 2 through 4, the difference in height reaches a maximum of 2%; weight, 5%; and triceps skinfold, 12%.

Birth weight is influenced by these factors as well, and Tanner (1974) has reviewed their role. However, the magnitude of their effects may vary in different settings (Bjerre and Värendh, 1975). That is, the results of the interaction of the various factors may be relatively specific for the ecosystem.

Other factors identified include smoking (Russell *et al.*, 1968; Hardy and Mellits, 1972) and changing educational levels (de Licardie *et al.*, 1972).

5. Summary and Conclusions

The preschool years, from birth through age 4, are crucial years in determining future developmental events and health status even into adulthood. The transition from prenatal to postnatal existence, with an exponential increase in stimuli and stress, is accentuated by the dependence of the infant and small child upon the surrounding world.

The high growth rates and sharp inflections in various curves characteristic of infancy change to the regularity of childhood, associated with canalization, directionality, and self-regulation. However, the sensitivity to environmental circumstances remains such that the preschool years provide us with an excellent age range in which to evaluate developmental adequacy and, by extension, the general health and nutritional status of the population to which these children belong.

ACKNOWLEDGMENTS

The assistance of Virginia Lathbury and Jean MacFadyen in preparing this manuscript is gratefully acknowledged.

6. References

Abraham, S., Lowenstein, F. W., and O'Connell, D. E., 1975, Preliminary Findings of the First Health and Nutrition Examination Survey, United States, 1971–72, Anthropometric and Clinical Findings, Department of Health, Education and Welfare Publ. (HRA) 75-1229, U.S. Government Printing Office, Washington, D.C.

Adams, M. S., and Niswander, J. D., 1973, Birth weight of North American Indians: A correction and amplification, *Hum. Biol.* **45**:351.

Amirhakimi, G. H., 1974, Growth from birth to two years of rich urban and poor rural Iranian children compared with Western norms, *Ann. Hum. Biol.* **1**:427.

Anderson, M., Blais, M., and Green, W. T., 1956, Growth of the normal foot during childhood and adolescence, Length of the foot and interrelations of foot, stature, and lower extremity as seen in serial records of children between 1 and 18 years of age, *Am. J. Phys. Anthropol.* **14**:287.

Anderson, M., Messner, M. B., and Green, W. T., 1964, Distributions of lengths of the normal femur and tibia in children from 1 to 18 years of age, *J. Bone Joint Surg.* **46A**:1197.

Bäckström, L., and Kantero, R.-L., 1971, II. Cross-sectional studies of height and weight in Finnish children aged from birth to 20 years, *Acta Paediatr. Scand., Suppl.* **220**:9.

Bialik, O., Peritz, E., and Arnon, A., 1973, Weight growth in infancy: A regression analysis, *Hum. Biol.* **45**:81.

Billewicz, W. Z., 1967, A note on body weight measurements and seasonal variation, *Hum. Biol.* **39**:241.

Billewicz, W. Z., Kemsley, W. F. F., and Thomson, A. M., 1962, Indices of adiposity, *Br. J. Prev. Soc. Med.* **16**:183.

Bjerre, I., and Värendh, G., 1975, A study of biological and socioeconomic factors in low birthweight, *Acta Paediatr. Scand.* **64**:605.

Bock, R. D., Wainer, H., Petersen, A., Thissen, D., Murray, J., and Roche, A., 1973, A parameterization for individual human growth curves, *Hum. Biol.* **45**:63.

Bouchalová, M., and Gerylová, A., 1970, The influence of some social and biological factors upon the growth of children from birth to 5 years in Brno, 11th Reun. Coord. des Res. sur la Developpement de l'Enfant Normal, **2**:43, Centre Internationale de l'Enfance, Paris.

Bouterline, E., Tesi, G., Kerr, G. R., Stare, F. J., Kallal, Z., Turki, M., and Hemaidan, N., 1973, Nutritional correlates of child development in southern Tunisia. II. Mass measurements, *Growth* **37**:91.

Brook, C. G. D., Huntley, R. M. C., and Slack, J., 1975, Influence of heredity and environment in determination of skinfold thickness in children, *Br. Med. J.* **2**:719.

Brŏzek, J., 1963, Quantitative description of body composition: Physical anthropology's "fourth" dimension, *Curr. Anthropol.* **4**:3.

Čabak, V., and Najdanvic, R., 1965, Effect of undernutrition in early life on physical and mental development, *Arch. Dis. Child.* **40**:532.

Carter, J. E. L., 1972, *The Heath-Carter Somatotype Method,* Department of Physical Education, San Diego State College, California.

Chang, K. S. F., Lee, M. M. C., Low, W. D., Chui, S., and Chow, M., 1965, Standards of height and weight of southern Chinese children, *Far East Med. J.* **1**:101.

Christiansen, N., Mora, J. O., and Herrera, M. G., 1975, Family social characteristics related to physical growth of young children, *Br. J. Prev. Soc. Med.* **29**:121.

Chrzastek-Spruch, H., and Wolanski, N., 1969, Body length and weight in newborns, and infant growth connected with parents' stature and age, *Genet. Polon.* **10**:257.

Clauser, C. E., Tucker, P. E., McConville, J. T., Churchill, E., and Laubach, L. L., 1972, Anthropometry of Air Force Women, AMRL-TR-70-5, Aerospace Medical Research Laboratory, Wright-Patterson Air Force Base, Ohio.

Condon-Paoloni, D., Johnston, F. E., Cravioto, J., de Licardie, E. R., and Scholl, T. O., 1977, Morbidity and growth of infants and young children in a rural Mexican village, *Am. J. Public Health* **67**:651–656.

Cravioto, J., 1968, Modificacion postnatal del fenotipo causada por la desnutricion, *Sobret. Gaceta Med. Mex.* **98**:523.

Cravioto, J., 1970, Complexity of factors involved in protein–calorie malnutrition, *Bib. Nutr. Dieta* **14**:7.

Cravioto. J.. and de Licardie. E. R., 1971, The long-term consequences of protein–calorie malnutrition, *Nutr. Rev.* **29**:107.

Cravioto. J., Birch, H. G., de Licardie, E. R., and Rosales, L., 1967, The ecology of weight gain in a preindustrial society, *Acta Paediatr. Scand.* **56**:71.

Cruise, M. O., 1973, A longitudinal study of the growth of low birth weight infants: I. Velocity and distance growth, birth to 3 years, *Pediatrics* **51**:620.

Dean, R. F. A., 1960, The effects of malnutrition on the growth of young children, in: *Modern Problems in Pediatrics* (F. Falkner, ed.), pp. 111–122, Karger, New York.

de Licardie, E., Cravioto, J., and Zaldivar, S., 1972, Cambio educativo intergeneracional en la madre y crecimiento fisico del niño rural en el primer año de la vida, *Bol. Med. Hosp. Infant.* **29**:575.

Deming, J., and Washburn, A. H., 1963, Application of the Jenss curve to the observed pattern of growth during the first 8 years of life in 40 boys and 40 girls, *Hum. Biol.* **35**:484.

Doornbos, L., and Jonxis, J. H. P., 1968, Growth of Bushnegro children on the Tapanahony River in Dutch Guyana, *Hum. Biol.* **40**:396.

Evans, M. E., 1944, Illness history and physical growth, *Am. J. Dis. Child.* **58**:390.

Falkner, F., 1958, Some physical measurements in the first three years of life, *Arch. Dis. Child.* **33**:1.

Falkner, F. (ed.), 1966, *Human Development,* W. B. Saunders, Philadelphia.

Faulhaber, J., 1961, El crecimiento en un grupo de niños Mexicanos, Instituto Nacional de Antropologia e Historia, Mexico.

Fetter, V., and Suchy, J., 1967, Anthropologic parameters of Czechoslovak young generation, current state, *Cesk. Pediatr.* **22**:97.

Fisch, R. D., Bilek, M. K., and Ulstrom, R., 1975, Obesity and leanness at birth and their relationship to body habitus in later childhood, *Pediatrics* **45**:521.

Fry, P. C., Howard, J. E., and Logan, B. C., 1975, Body weight and skinfold thickness in black, Mexican–American, and white infants, *Nutr. Rep. Int.* **11**:155.

Gampel, B., 1965, The relation of skinfold thickness in the neonate to sex, length of gestation, size at birth, and maternal skinfold, *Hum. Biol.* **37**:29.

Garn, S. M., 1961, The genetics of normal human growth, *Genet. Med.* **2**:413.

Garn, S. M., 1966, Body size and its implications, in: *Review of Child Development Research* (L. W. Hoffman and M. L. Hoffman, eds.), pp. 529–561, Russell Sage Foundation, New York.

Garn, S. M., and Clark, D. C., 1975, Nutrition, growth, development, and maturation: Findings from the Ten-State Nutrition Survey of 1968–1970, *Pediatrics* **56**:306.

Garrow, J. S., and Pike, M. D., 1967, The long term prognosis of severe infantile malnutrition, *Lancet* **1**:4.

Gomez, F., Ramos-Galvan, R., Cravioto, J., Chavez, R., and Vasques, J., 1956, Mortality in second and third degree malnutrition, *J. Trop. Pediatr.* **2**:77.

Gore, A. T., and Palmer, W. T., 1949, Growth of the preschool child in London, *Lancet* **1**:385.

Goss, R. J., 1964, *Adaptive Growth,* Academic Press, New York.

Graham, G. G., 1968, The later growth of malnourished infants; effects of age, severity, and subsequent diet, in: *Calorie Deficiencies and Protein Deficiencies* (R. A. McCance and E. M. Widdowson, eds.), pp. 301–314, Little, Brown, Boston.

Graham, G. G., and Adrianzen, B., 1972, Late "catch-up" growth after severe infantile malnutrition, *Johns Hopkins Med. J.* **131**:204.

Grantham-McGregor, S. M., and Desai, P., 1973, Head circumferences of Jamaican infants, *Dev. Med. Child. Neurol.* **15**:441.

Habicht, J. P., Martorell, R., Yarbrough, C., Malina, R. M., and Klein, R. E., 1974, Height and weight standards for preschool children. How relevant are ethnic differences in growth potential? *Lancet* **1**:611.

Hansman, C., 1970, Anthropometry and related data, anthropometry skinfold thickness measurements, in: *Human Growth and Development* (R. W. McCammon, ed.), pp. 103–154, Charles C Thomas, Springfield, Illinois.

Hardy, J. B., and Mellits, E. D., 1972, Does maternal smoking during pregnancy have a long-term effect on the child? *Lancet* **2**:1332.

Heald, F. P., and Hollander, R. J., 1965, The relationship between obesity in adolescence and early growth, *Pediatrics* **67**:35.

Herron, R. E., 1972, Biostereometric measurement of body form, *Yearb. Phys. Anthropol.* **16**:80.

Hertzberg, H. T. E., Churchill, E., Dupertuis, C. W., White, R. M., and Damon, A., 1963, *Anthropometric Survey of Turkey, Greece, and Italy,* Macmillan, New York.

Hewitt, D., Westropp, C. K., and Acheson, R. M., 1955, Oxford Child Health Survey, effect of childish ailments on skeletal development, *Br. J. Prev. Soc. Med.* **9**:179.

Huenemann, R. L., 1974*a*, Environmental factors associated with preschool obesity. I. Obesity in six-month old children, *J. Am. Diet. Assoc.* **64**:480.

Huenemann, R. L., 1974*b*, Environmental factors associated with preschool obesity. II. Obesity and food practices of children at successive age levels, *J. Am. Diet. Assoc.* **64**:488.

Hunt, E. E., 1966, The developmental genetics of man, in: *Human Development* (F. Falkner, ed.), pp. 76–122, W. B. Saunders, Philadelphia.

Huxley, J., 1932, *Problems of Relative Growth,* Methuen, London.

Indian Council of Medical Research, 1972, *Growth and Physical Development of Indian Infants and Children,* Medical Enclave, New Delhi.

James, L. S., 1959, Biochemical aspects of asphyxia at birth, in: *Adaptation to Extrauterine Life* (T. K. Oliver, ed.), pp. 66–71, 31st Ross Conference on Pediatric Research, Ross Laboratories, Columbus, Ohio.

Jelliffe, D. B., 1966, *The Assessment of the Nutritional Status of the Community,* WHO, Geneva.

Jelliffe, D. B., 1968, Child Nutrition in Developing Countries, PHS Publication 1822, U.S. Government Printing Office, Washington, D.C.

Jenss, R. M., and Bayley, N., 1937, A mathematical method for studying the growth of a child, *Hum. Biol.* **9**:556.

Johnston, F. E., and Beller, A., 1976, Anthropometric evaluation of the body composition of black, white and Puerto Rican newborns, *Am. J. Clin. Nutr.* **29**:61.

Johnston, F. E., Hamill, P. V. V., and Lemeshow, S., 1974, Skinfold thickness of youths 12–17 years, United States. National Center for Health Statistics, Ser. 11, No. 132, U.S. Government Printing Office, Washington, D.C.

Kanawati, A. A., and McLaren, D. S., 1973, Failure to thrive, in Lebanon. II. An investigation of the causes, *Acta Paediatr. Scand.* **62**:571.

Kantero, R.-L. and Tiisala, R., 1971*a*, IV. Height, weight, and sitting height increments for children from birth to 10 years, *Acta Paediatr. Scand., Suppl.* **220**:18.

Kantero, R.-L. and Tiisala, R., 1971*b*, V. Growth of head circumference from birth to 10 years, *Acta Paediatr. Scand., Suppl.* **220**:27.

Kirke, D. K., 1969, Growth rates of aboriginal children in central Australia, *Med. J. Aust.* **2**:1005.

LaBarre, W., 1954, *The Human Animal,* University of Chicago Press, Chicago.

Lancet, 1970, Classification of infantile malnutrition, **2**:302.

Lindegaard, B., 1953, Variations in human body build. A somatometric and X-ray cephalometric investigation on Scandinavian adults, *Acta Psychiatr. Suppl.* **86**.

Livson, N., McNeill, D., and Thomas, K., 1962, Pooled estimates of parent–child correlations in stature from birth to maturity, *Science* **138**:818.

Mack, R. W., and Ipsen, J., 1974, The height-weight relationship in early childhood. Birth to 48 month correlations in an urban low-income Negro population, *Hum. Biol.* **46**:21.

Mack, R. W., and Johnston, F. E., 1976, The relationship between growth in infancy and growth in adolescence, *Hum. Biol.* **48**:693.

Mack, R. W., and Kleinhenz, M. E., 1974, Growth, caloric intake, and activity levels in early infancy: A preliminary report, *Hum. Biol.* **46**:354.

Malcolm, L. A., 1973, Ecological factors relating to child growth and nutritional status, in: *Nutrition and Malnutrition, Identification, and Measurement* (A. F. Roche and F. Falkner, eds.), pp. 329–352, Plenum Press, New York.

Malina, R. M., Habicht, J.-P., Yarbrough, C., Martorell, R., and Klein, R. E., 1974, Skinfold thickness at seven sites in rural Guatemalan Ladino children birth through 7 years of age, *Hum. Biol.* **46**:453.

Malina, R. M., Habicht, J.-P., Martorell, R., Lechtig, A., Yarbrough, C., and Klein, R. E., 1975, Head and chest circumferences in rural Guatemalan Ladino children, birth to 7 years of age, *Am. J. Clin. Nutr.* **28**:1061.

Manwani, A. H., and Agarwal, K. N., 1973, The growth pattern of Indian infants during the first year of life, *Hum. Biol.* **45**:341.

Martorell, R., Yarbrough, C., Lechtig, A., Habicht, J. -P., and Klein, R. E., 1975, Diarrheal diseases and growth retardation in preschool Guatemalan children, *Am. J. Phys. Anthropol.* **43**:341.

McFie, J., and Welbrown, H. F., 1962, Effect of malnutrition in infancy on the development of bone, muscle, and fat, *J. Nutr.* **76**:97.

Meredith, H. V., 1946, Physical growth from birth to two years: II. Head circumference, *Child. Dev.* **17**:1.

Meredith, H. V., 1962, Childhood interrelations of anatomic growth rates, *Growth* **26**:23.

Meredith, H. V., 1970, Body weight at birth of viable human infants: A world wide comparative treatise, *Hum. Biol.* **42**: 217.

Meredith, H. V., 1971, Human head circumference from birth to early adulthood: Racial, regional, and sex comparisons, *Growth* **35**:233.

Miller, F. J. W., Billewicz, W. Z., and Thomson, A. M., 1972, Growth from birth to adult life of 442 Newcastle Upon Tyne schoolchildren, *Br. J. Prev. Soc. Med.* **26**:224.

Mueller, W. H., 1975, Parent–Child and Sibling Correlations and Heritability of Body Measurements in a Rural Colombian Population, Ph.D. thesis, University of Texas, Austin.

Newens, E. M., and Goldstein, H., 1972, Height, weight and the assessment of obesity in children, *Br. J. Prev. Soc. Med.* **26**:33.

Ounsted, M., and Sleigh, G., 1975, The infant's self-regulation of food intake and weight gain, *Lancet* **1**:1393.

Pogačnik, T., 1970, Antropometrijski standardi otrok v. Sloveniji, *Glas. Antropol. Drus. Jugosl.* **7**:91.

Prader, A., Tanner, J. M., and von Harnack, G. A., 1963, Catch-up growth following illness or starvation, *J. Pediatr.* **62**:646.

Rea, J. N., 1971, Social and economic influences on the growth of pre-school children in Lagos, *Hum. Biol.* **43**:46.

Reed, R. B., and Stuart, H. C., 1959, Patterns of growth in height and weight from birth to 18 years of age, *Pediatrics* **24**:904.

Rharon, H., 1962, Some observations on head circumferences in children with nutritional deficiencies, *Turk. J. Pediatr.* **4**:169.

Roche, A. F., and Cahn, A., 1962, Subcutaneous fat thickness and caloric intake in Melbourne children, *Med. J. Aust.* **1**:595.

Roche, A. F., and Davila, G. H., 1974, Differences between recumbent length and stature within individuals, *Growth* **38**:313.

Russell, C. S., Taylor, R., and Low, C. E., 1968, Smoking in pregnancy, maternal blood pressure, pregnancy outcome, baby weight and growth, and other related factors, *Br. J. Prev. Soc. Med.* **22**:119.

Schell, L. M., and Johnston, F. E., 1978, Physical growth and development of North American Indian and Eskimo children and youth, in: *Handbook of North American Indians* (F. S. Hulse, ed.), U.S. Govt. Print. Off., Washington.

Scholl, T. O., 1975, Body Size in Developing Nations: Is Bigger Really Better? Ph.D. thesis, Temple University, Philadelphia.

Sims, L. S., and Morris, P. M., 1974, Nutritional status of preschoolers, *J. Am. Diet. Assoc.* **64**:492.

Snyder, R. G., Spencer, M. L., Owings, C. L., and Schneider, L. W., 1975, Anthropometry of U.S. infants and children, SP-394, Society of Automotive Engineers, Warrendale, Pennsylvania.

Stuart, H. C., and Meredith, H. V., 1946, Use of body measurements in the school health program, *Am. J. Public Health* **36:**1365.

Sveger, T., Lindberg, T., Weibul, B., and Olsson, U. L., 1975, Nutrition, overnutrition, and obesity in the first year of life in Malmö, Sweden, *Acta Paediatr. Scand.* **64:**635.

Tanner, J. M., 1962*a*, *Growth at Adolescence,* 2nd ed., Blackwell, Oxford.

Tanner, J. M., 1962*b*, The evaluation of growth and maturity in children, in: *Protein Metabolism* (F. Gross, ed.), pp. 361–382, Springer-Verlag, Berlin.

Tanner, J. M., 1963, The regulation of human growth, *Child Dev.* **34:**817.

Tanner, J. M., 1965, Growth, in: *Physical Medicine in Pediatrics* (B. Kiernander, ed.), pp. 5–31, Butterworths, London.

Tanner, J. M., 1974, Variability of growth and maturity in newborn infants, in: *The Effect of the Infant on Its Caregiver* (M. Lewis and L. A. Rosenblum, eds.), pp. 77–103, John Wiley and Sons, New York.

Tanner, J. M., and Israelsohn, W. J., 1963, Parent–child correlations for body measurements of children between the ages of 1 month and 7 years, *Ann. Hum. Genet.* **26:**245.

Tanner, J. M., and Whitehouse, R. H., 1975, Revised standards for triceps and subscapular skinfolds in British children, *Arch. Dis. Child.* **50:**142.

Tanner, J. M., Healy, M. J. R., Lockhart, R. D., MacKenzie, J. D., and Whitehouse, R. H., 1956, Aberdeen growth study, I. The prediction of adult body measurements from measurements taken each year from birth to 5 years, *Arch. Dis. Child.* **31:**372.

Tanner, J. M., Whitehouse, R. H., and Takaishi, M., 1966*a*, Standards from birth to maturity for height, weight, height velocity, and weight velocity: British children, 1965–I, *Arch. Dis. Child.* **41:**454.

Tanner, J. M., Whitehouse, R. H., and Takaishi, M., 1966*b*, Standards from birth to maturity for height, weight, height velocity, and weight velocity: British children, 1965–II, *Arch. Dis. Child.* **41:**613.

Tanner, J. M., Goldstein, H., and Whitehouse, R. H., 1970, Standards for children's height at ages 2–9 years allowing for height of parents, *Arch. Dis. Child.* **45:**755.

Terada, H., and Hoshi, H., 1965*a*, Longitudinal study on the physical growth in Japanese. (2) Growth in stature and body weight during the first three years of life, *Acta Anat. Nippon.* **40:**166.

Terada, H., and Hoshi, H., 1965*b*, Longitudinal study on the physical growth in Japanese. (3) Growth in chest and head circumferences during the first three years of life, *Acta Anat. Nippon.* **40:**368.

Terada, H., and Hoshi, H., 1966, Longitudinal study on the physical growth in Japanese. (4) Growth in length of extremities during the first three years of life, *Acta Anat. Nippon.* **41:**313.

Timiras, P. S., 1972, *Developmental Physiology and Aging,* Macmillan, New York.

Tuddenham, R. D., and Snyder, M. M., 1954, Physical growth of California boys and girls from birth to eighteen years, *Univ. Calif. Publ. Child Dev.* **1:**183.

Vandenberg, S. G., and Falkner, F., 1965, Hereditary factors in human growth, *Hum. Biol.* **37:**357.

Vaughan, V. C., 1964, Growth and development, in: *Textbook of Pediatrics* (W. E. Nelson, ed.), pp. 15–71, W. B. Saunders, Philadelphia.

Verghese, K. P., Scott, R. B., Teixeira, G., and Ferguson, A. D., 1969, Studies in growth and development, XII. Physical growth of North American Negro children, *Pediatrics* **44:**243.

Weiner, J. S., and Lourie, J. A., 1969, *Human Biology, A Guide to Field Methods,* IBP Handbook No. 9, F. A. Davis, Philadelphia.

Wilson, R. S., 1976, Concordance in physical growth for monozygotic and dizygotic twins, *Ann. Hum. Biol.* **3:**1.

Wingerd, J., 1970, The relation of growth from birth to 2 years to sex, parental size and other factors, using Rao's method of the transformed time scale, *Hum. Biol.* **42:**105.

Wishaky, A., and Khattab, A., 1962, The effect of failure of growth due to undernutrition on the different parts of the body in infancy, *J. Egypt. Med. Assoc.* **45:**1029.

Wolanski, N., 1961, The new graphic method of the evaluation of the tempo and harmony of child physical growth, in: *Proceedings 7th Meeting Deutsch Gesellschaft für Anthropologie* (W. Gieseler and I. Tillner, eds.), pp. 131–139, Musterschmidt-Verlag, Göttingen.

Wolanski, N., and Lasota, A., 1964, Physical development of countryside children and youth aged 2 to 20 years compared with the development of town youth of the same age, *Z. Morphol. Anthropol.* **54:**272.

Yarbrough, C., Habicht, J.-P., Malina, R. M., Lechtig, A., and Klein, R. E., 1975, Length and weight in rural Guatemalan Ladino children: Birth to seven years of age, *Am. J. Phys. Anthropol.* **42:**439.

5

Body Composition and Energy Needs during Growth

MALCOLM A. HOLLIDAY

1. Introduction

The study of body composition, dividing the body into component parts, is a logical step in the process of trying to correlate structure with function on a "whole-body" scale. Characterizing changes in body composition is a means for understanding the process of growth and change in function that affect nutritional needs as growth proceeds. Anatomic divisions are rather obvious, e.g., organ mass and muscle mass. Fluid–mineral divisions are less readily visualized. The body is divided into total body water and solids; total body water is further divided into an extracellular and intracellular phase. Extracellular fluid (ECF) includes plasma, interstitial fluid, and connective tissue fluids; intracellular fluid (ICF)—the fluid phase of cells—is mostly in organs and muscle (Widdowson and Dickerson, 1964). The functional aspects of body composition can be viewed in terms of organ function (brain, liver, etc.), locomotion (muscle mass), energy reserve (fat mass), environment for cells (extracellular fluid), and supporting structures (connective tissue and bone) (Figure 1). These functions, like their structures, are rather self-evident. The major organs and muscle constitute the bulk of cell proteins in the body. Of these, muscle protein is the principal reservoir for amino acids when diet is deficient or when there is a need for gluconeogenesis from amino acids. While glycogen and protein provide some reserve for energy, fat is the real reservoir for energy when the diet is deficient. Plasma and interstitial fluid are the environment and transportation system for the cells. Supporting structures—connective tissue and bone, etc., although containing protein, are not sources for protein during diet deficiency. The ECF of connective tissue is a reserve for interstitial fluid and plasma when dehydration occurs. Bone contains a reserve for calcium, phosphorus, and some other minerals.

MALCOLM A. HOLLIDAY • The Children's Renal Center, Department of Pediatrics, University of California San Francisco, San Francisco, California.

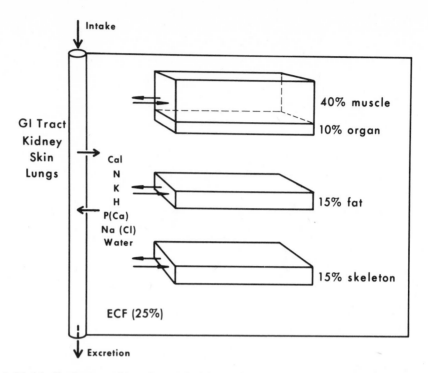

Fig. 1. Model of body composition of an adult male, in which size of specific components is indicated. Major nutrients exchanged between environment and body system through extracellular fluid (ECF) are indicated (Holliday, 1976).

Just as different components of the body serve different functions and provide for specific reserves, each has need for substrates to sustain normal functions. We will focus discussion in this chapter on components of body composition and the specific substrates they require to meet energy needs. Changes in body composition during growth are related to changes in energy requirement. The effects of under- and overnutrition upon body composition and the energy requirements are reviewed here.

2. Body Composition

2.1. Brain and Organs

The brain and other major organs are a small fraction of total cell mass. They are considered separately from muscle mass because the major organs—brain, liver, heart, and kidney (ΣOW)—have metabolic rates 20 or more times that of resting muscle (Brozek and Grande, 1955). These organs account for a large fraction of basal metabolic rate (BMR) (Table I). Furthermore, brain under normal conditions has a unique requirement for glucose as an energy source, so that brain must be considered as a special case. The growth rate of these organs is quite different from muscle, and hence their influence upon energy requirements has to be separated from that of cell mass as an entity (Holliday *et al.*, 1967; Holliday, 1971).

Table I. *Organ Metabolic Rate (OMR) and Organ Size in Adults: Relation to BMR[a]*

	Weight (kg)	V_{O_2} (ml/100 g/m)	OMR	OMR/BMR[b] (%)
Brain	1.40	4.2	414	23.2
Liver	1.60	4.1	464	26.1
Heart	0.30	8.2	182	10.2
Kidney	0.30	5.5	116	7.1
Total	3.60		1177	66.7
Skeletal muscle	28.3	0.26	500	28.1
Total	31.9		1677	94.2

[a]Table adapted from Holliday (1971).
[b]BMR for 70-kg man from Talbot (1938).

Brain size can be predicted in an individual when age and sex are known. Liver and splanchic mass, heart and kidney weight can be estimated from height, age, and sex, unless there is a pathologic and usually obvious exception from the norm (Coppoletta and Wolbach, 1933).

2.2. Muscle

Muscle mass is the largest and most variable component of cell mass. It varies because of genetic differences expressed as body types and because of differing levels of physical training. It is decreased with loss of nerve supply or immobility. The most common variables affecting muscle mass, however, are undernutrition and depressed physical activity. Calorie deficiency or inadequate food intake directly limit energy for muscle protein synthesis and indirectly depress physical activity. Both cause a loss of muscle mass (Keys *et al.,* 1950).

Muscle mass may be estimated by measuring total cell mass and subtracting a constant for organ mass (ΣOW). Because organ mass is small, variations in cell mass are due almost entirely to changes in muscle mass (Cheek, 1968).

Cell mass can only be measured indirectly.* One method derives cell mass from intracellular fluid (ICF). ICF is the difference between total body water (TBW) and ECF volume (Cheek, 1968). TBW is equal to the volume of distribution of deuterium or tritium oxide, or the volume of antipyrene or urea. It is measured with an accuracy of 2% and a precision of ±4%. ECF volume, however, is not so accurately measured (see p. 121). Hence the estimate of ICF is relatively imprecise.

Cell mass also has been estimated from the measurement of total body potassium (K) (Flynn *et al.,* 1972) or from exchangeable K (Corsa *et al.,* 1950). Total body K can be measured with a precision of ±5% using liquid scintillation wholebody counting, and where the method is available it is a useful measure. Measuring exchangeable K entails technical problems owing to the short half-life of ^{42}K, and radiation exposure is too high to use for children. Another problem that arises in using total body K as an index of cell mass is that the calculation of muscle mass from total body K assumes K concentration in cells to be constant. While this is a

*Cell mass differs from lean body mass (LBM). Lean body mass is normal fat free body weight plus a standard minimum of body fat. Because variation in body fat has little effect on cell mass, ECF volume or supporting structures, LBM is useful as a reference standard in describing variations in body fat among normal individuals (see p. 123).

safe assumption for healthy subjects, this is not the case in malnutrition (Metcoff *et al.*, 1960) or potassium deficiency (Patrick, 1977).

Muscle mass has been estimated from the quantity of creatinine excreted in a 24-hr period—the creatinine index. This method is simple and noninvasive where accurately timed and complete urine samples can be collected. Several studies (Cheek, 1968; Forbes and Bruining, 1976) have concluded from available evidence that the creatinine index is a satisfactory predictor of muscle mass in children.

Muscle mass estimated from measuring arm circumference is useful only for detecting gross deficiencies related to malnutrition or other conditions associated with severe muscle wasting.

The measurement of muscle protein and protein–DNA (Cheek, 1968), protein–K, and protein–P ratios (Delaporte *et al.*, 1976) also are useful in detecting major deviations of muscle composition from normal—principally those associated with muscle wasting.

2.3. Adipose Tissue

In the average individual 15–30% of body weight is adipose tissue. Adipose tissue is made up of adipocytes and their supporting structures (Hirsch and Knittle, 1970) and is mostly located in skin or subcutaneous tissues. Most body fat is in adipose tissue. In normal individuals 65–70% of adipose tissue is fat, and 10–20% of body weight is fat. Adipose tissue either stores or releases fat as fatty acids and glycerol, depending on whether dietary energy exceeds expenditure or the converse.

People may have less than normal amounts of body fat—as low as 5%—and be healthy so long as they are not faced with sustained starvation or stress. Individuals with severe calorie deficiency have virtually no body fat (Garrow *et al.*, 1965). Excess body fat may range from slight overweight to gross obesity. The degree of obesity correlates with early death related to heart disease, hypertension, and late-onset diabetes. Consequently, excess body fat has become a major medical concern.

Undernutrition and overnutrition are characterized by changes in body fat which usually are reflected in changes in body weight. The most commonly used method for estimating nutritional status relative to body fat is the weight-for-height index where the weight of a patient is expressed as a percent of the average weight for an individual of the same height. Undernutrition is suspected where the index is less than 90% and is considered to exist when the index is less than 80% (Waterlow and Allayne, 1971). Among a population of severely undernourished children, the range was from 50 to 90% and the average was 70% (Kerr *et al.*, 1973). Individuals are classified as overweight when their index is 110–120% and obese when it exceeds 120% (Davidson and Passmore, 1969). While a good correlation exists between body fat and weight/height index, there are obvious potential errors, e.g., changes in muscle mass (see above) or ECF volume also affect weight/height index. Variations in ECF volume are common in undernutrition; dehydration occurs with acute gastroenteritis, and edema is a common feature of chronic calorie undernutrition—both clinical (Waterlow and Allayne, 1971) and experimental (Keys *et al.*, 1950). Obese individuals may not simply be fat: obese young people often have a larger muscle mass (Forbes, 1964).

More direct measurements of body fat are difficult to make in humans (Keys and Brozek, 1953). Estimates of body fat are derived from measurement of body water; the derivation is based upon the assumption that fat-free body mass has a water content of 72%. While this assumption is appropriate in normal and obese people, it is very uncertain in undernourished or edematous people. Estimates of body fat also can be derived from measurements of specific gravity or relative density obtained by weighing individuals under water, or by estimations of body volume by air displacement. This derivation depends on fat-free body mass having a fixed and high specific gravity and adipose tissue having a lower and constant specific gravity. The limitations of this method are similar to those deriving body fat estimations from measurements of TBW.

Body fat can be estimated from measurements of skinfold thickness when the latter is measured in a systematic and consistent manner (Tanner and Whitehouse, 1962). This is the preferred method in most instances. Body fat estimated by skinfold thickness agrees well with simultaneous estimates by densitometry in adults (Durnin, 1967) or body water in children (Brook, 1971).

2.4. Extracellular Fluid

ECF volume is a major component of body weight. It consists of plasma, interstitial fluid, connective tissue fluid, and fluids such as gastrointestinal and cerebrospinal fluid or so-called transcellular fluids (Edelman and Liebman, 1959). ECF is the environment for cells (Gamble, 1947) and also is the interface between cells and the outside environment. Oxygen and nutrients from the environment reach cells through plasma and interstitial fluid, and products of cell metabolism and of diet in excess leave cells via ECF.

However, ECF is not an energy-using or -generating system—aside from its need to be circulated by heart. ECF is important because variations in its size affect interpretations of body weight as an index of body size.

ECF consists of several components: plasma, lymph, interstitial fluid, and connective tissue fluid. Changes in either concentration or volume of plasma and interstitial fluid are reflected in like changes in connective tissue fluid.

Measurement of ECF is imprecise. Its estimation is derived from the volume of distribution of substances that are distributed between plasma and interstitial fluid but are excluded from cells. One set of compounds, the saccharides (inulin, mannitol, and sucrose), penetrate connective tissue fluids slowly and underestimate ECF volume (Nichols *et al.*, 1953). They also are rapidly cleared so that conditions for measuring volume of distribution are poor. Isotopes of sodium, chloride, or bromide, or stable bromide readily penetrate connective tissue fluid but also are distributed to a degree in cell fluid, leading to an overestimate of ECF volume. Cheek (1961), in an extensive review, compared the various volumes of distributions measured by different compounds in relation to the various phases of ECF (Figure 2). He concluded that a corrected bromide space was the preferred method for estimating ECF in humans. Barratt and Walser (1968, 1969) subsequently compared sulfate spaces *in vivo* and in tissue with bromide spaces and concluded that sulfate rapidly penetrated connective tissue and did not enter cells. Kaufman and Wilson (1973) have described a method for measuring bromide using fluorescent excitation analysis that makes measurement of bromide space (using stable

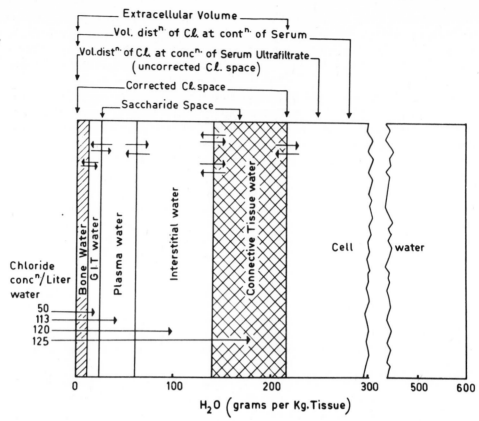

Fig. 2. Diagram of the anatomy of ECF and how various compounds used in measuring ECF volume are distributed (Cheek, 1961).

bromide) safe, rapid, and precise. Friis-Hansen (1957), who has reported ECF volumes of children, used thiosulfate which appears to measure a space similar to sulfate space.

When the technique for fluorescent excitation analysis is available, measurement of bromide space is most suited for estimating ECF volume in patients. Sulfate space both for *in vivo* and *in vitro* analyses probably is the most accurate measure of ECF volume.

2.5. Summary

The value of knowing the size of the brain and other major organs is in appreciating the larger demands for energy that these organs impose. Except in unusual circumstances, their size can be estimated from normal data and their demand for energy (OMR) is a function of their mass. Muscle mass is a large and variable cell mass among normal individuals, and it is very sensitive to nutritional deficiency. Measurement of cell mass, or muscle mass, is a useful means for evaluating nutritional status, and serial measures are useful in observing changes in nutrition status. Muscle mass is the protein reserve of the body, and its size reflects the state of that reserve as well as the state of physical activity. Fat mass is the energy balance between energy input and output. A normal reserve of fat is between 10 and 20% of body weight or 900–1800 kcal/kg. Fat composes two thirds of adipose tissue, so that 15–30% of body weight is adipose tissue. Body fat content less than 10% of body weight is suggestive of calorie deficiency, and more than 20% is suggestive of obesity. Because fat is the major variable in body composition in normal individuals, variations in weight-to-height ratio are most often due to changes in fat (see p. 120). However, changes in muscle mass and ECF volume affect the accuracy of deriving fat from the weight-for-height index. Skinfold thickness, on the other hand, is an accurate method. ECF volume in health is a derivative of plasma, interstitial fluid, lymph, and connective tissue. Normal variations in body size affect ECF volume proportionally. Connective tissue, including bone, is a relatively consistent 15% of lean body mass (see footnote on p. 119) among healthy individuals. Interstitial fluid and plasma also are functions of lean body weight. Hence ECF volume may be related either to height or lean body mass. ECF volume may be decreased (dehydration) or increased (edema) with malnutrition, and its estimation, under these conditions, can greatly increase the value of body weight in following nutritional status.

3. The Changing Pattern of Body Composition as Growth Proceeds

Growth from birth to maturity encompasses an increase in weight from 3.5 to 50–70 kg. If low-birth-weight infants are included, the lower range is 1 kg. Weight is the common reference base for describing body composition changes during growth. Usually this is done by expressing the weight or volume of a particular component as a percent of body weight (Table II). This method is descriptive of body composition at a point, but it does not convey the dynamics of change with growth. Height is a preferred standard of reference in evaluating energy expenditure (Talbot, 1938) or ECF volume (Cheek, 1968) of individuals of abnormal body

Table II. Body Composition at Different Stages of Growth

Subject and age (years)	Height (cm)	Body weight (kg)	Body weight (%)			
			Σ organ weight[a]	Muscle mass[b]	Body fat[c]	ECF volume[d]
Low birth weight		1.1	21	<10	3	50
Newborn	50	3.5	18	20	12	40
Child 0.25	60	5.5	15	22	11	32
Child 1.5	80	11.0	14	23	20	26
Child 5	110	19.0	10	35	15	24
Child 10	140	31.0	8.4	37	15	25
Male 14	160	50	5.7	42	12	21
Male adult	180	70	5.2	40	11	19
Female 13	160	45	4.8	39	18	19
Female adult	162	55	4.4	35	20	16

[a]Sum of brain, liver, heart, and kidney (Holliday, 1971).
[b]Derived from creatinine excretion (Holliday, 1971).
[c]Interpolated values from literature (Widdowson and Dickerson, 1964; Friis-Hansen, 1971).
[d]Average of bromide space (Gamble, 1946) and thiosulfate space (Cheek, 1961; Friis-Hansen, 1957).

composition. A plot of data comparing weights of any component to body weight over time expresses the relation between the two more accurately than using percent body weights (Miller and Weil, 1963). Growth rates of a particular component relative to body weight can be derived from an allometric plot, i.e., log–log plot, of component weight to body weight over time (Kleiber, 1947).

3.1. Brain and Organs

The brain, the principal user of glucose under normal conditions (Kety, 1956), is a large fraction of an infant's weight: 17% of body weight and 75% of the four major organs (Holliday, 1971). It grows rapidly in the first year of life, and then its growth rate decreases. Growth is nearly complete by the fifth year when body weight is less than a third of its mature size (Table III). The weight of brain, liver, heart, and kidney combined (ΣOW) increases just as body weight increases in the first year. From 3 to 18 months the ΣOW as percent of body weight is relatively stable—around 15%. Thereafter its growth rate is slower, and its percent body weight decreases to 5.2% at maturity (Table I).

3.2. Muscle

By contrast, muscle mass increases from a value of 22% of body weight at 3 months of age to 35% at 5 years of age and in males to more than 40% at maturity (Table I) (Cheek, 1968). Females show a similar increase until puberty when the percentage tends to decrease. The sum of organ mass and muscle mass (cell mass) averages 37% during infancy. By 5 years cell mass averages 45% of body weight and remains so. Cell mass as a fraction of body weight decreases in women after puberty.

Table III. Brain Size and Energy Requirement Relative to Body Weight and BMR at Different Stages of Growth; Glucose Needed for Brain Metabolic Rate (BrMR)

| Body weight (kg) | Brain weight (g) | Brain wt./ body wt. (%) | kcal/day | | BrMR/ BMR (%) | Glucose needed[c] (g/day) |
			Brain MR[a]	BMR[b]		
1.1	190	17		41		10[d]
3.5	475	14	(140)	161	(87)	38
5.5	650	12	(192)	300	(64)	52
11	1045	10	(311)	590	(53)	84
19 (5 yr)	1235	6.5	367	830	44	99
31	1350	4.4	400	1160	34	108
50	1360	2.7	403	1480	27	109
70	1400	2.0	414	1800	23	112

[a]Derived from data of Kennedy and Sokoloff (1957), who presented data in children from 5 years of age and up. BrMR/100 g brain weight was constant. BrMR for children under 5 years is assumed to be proportional to brain weight (Figure 5). This assumption is not extended to the low-birth-weight infant (Figure 9).
[b]Derived from data presented in Holliday (1971).
[c]Glucose needed is calculated by dividing BrMR by 3.7 kcal/g.
[d]Glucose needed to meet BMR of low-birth-weight infant.

Fat content of low-birth-weight infants is small, less than 3%. Normally during the last trimester fat is added so that infants born at term have 12% of body weight as fat. Fat continues to increase during the first year (Foman, 1966). During the second year of life, as infants turn their attention from eating and growing to walking and playing, the percentage of fat decreases. Thereafter the child's fat content remains a stable fraction of body weight, or increases. Often fat content decreases during the pubertal growth spurt. In Western societies there is a tendency for body fat content to increase after puberty. This is particularly true in the case of girls as they reach sexual maturity (Frisch *et al.*, 1973). This change may be related to the endocrine changes that affect sexual maturation and create the potential for child-bearing. Energy reserves as fat may have a biological advantage during pregnancy and lactation.

There is much greater variation in body fat at any age or stage of growth than in other components of body composition. In a simplistic view this reflects variations in calorie balance. When appetite is trimmed to activity, body fat will remain a relatively constant fraction of body weight. The controls that regulate body fat are those that sense activity and set appetite. The nature of these controls is poorly understood, and their degree of efficiency among individuals varies greatly. The precision with which energy balance is controlled probably has a strong genetic component (Garrow, 1974). Extreme forms of obesity and anorexia nervosa may be related to disorders in learning in the first years of feeding. Lesser forms may be influenced by inappropriate use of food that disassociates feeding from physiologically induced hunger (Bruch, 1973).

3.4. Extracellular Fluid

Extracellular fluid is a large fraction of body weight at birth and an even larger one in infants of low birth weight. In the first postnatal weeks there is a contraction of ECF volume (Cheek, 1961). During infancy ECF volume does not increase as rapidly as body weight so that it decreases as a percent of body weight (Friis-Hansen, 1957). This trend continues, but to a lesser degree, after 1 year of age. The changes that occur during growth from infancy are influenced by the method of measurement. Cheek (1961) reports normal values for bromide space in infants in the first week of life of 37% and for adults 22%. Friis-Hansen (1957), using thiosulfate, reported analogous values of 44% and 19%. The reason for the greater decline in values where thiosulfate was used is not clear. One possibility is that connective tissue in newborn infants is more readily penetrated by thiosulfate and becomes less so as growth proceeds (Table II).

3.5. Summary

Changes in body composition during growth reflect differential growth rates of the components. Organs grow proportionally with body weight in the first year, but they grow slower thereafter. Muscle mass, by contrast, grows at a more rapid rate than weight after the first year but, thereafter, decreases or becomes stable in most individuals. In women, body fat increases at puberty. Expressing these changes as fractions of body weight at different ages provides a statement of composition, but it does not accurately portray the dynamics of a changing relationship.

Organ mass cannot be measured during growth, but its mass can be predicted from height and sex using normal data. Muscle mass can be estimated from creatinine excretion and body fat from measures of skinfold thickness. ECF volume cannot be estimated accurately except by direct measurements of bromide or sulfate space. Changes in ECF volume following hydration (Cheek, 1956) or diuresis can be estimated from changes in body weight.

4. Body Composition and Energy Balance during Growth

The energy required as intake to balance energy expended is termed "maintenance energy requirement." Some have restricted the term to mean a minimum requirement for an individual who is sedentary and must exert little physical effort beyond feeding (Payne and Waterlow, 1971; Joint FAO/WHO, 1973). Using this definition, maintenance energy requirement is approximately $1.5 \times$ BMR. The more inclusive definition encompasses energy needed to balance all expenditures. It is preferred, because the more restricted meaning requires a rigid definition of minimum activity and an artificial set of conditions to fulfill the definition.

A state of energy balance (i.e., intake equals expenditure) implies stable weight and body composition. In particular, body potential energy as fat and protein do not change. Daily or weekly oscillations occur affecting glycogen mass most directly, and fat and muscle to a lesser degree. Body fluid, electrolyte, and mineral balances usually remain stable under conditions of energy balance except for normal oscillations of ECF volume.

The energy required to meet expenditure depends upon body composition, metabolic state, and the level of physical activity. The components of expenditure are BMR, physical activity, specific dynamic action or thermic effect of food,* and stress-induced increases in expenditure.

4.1. Substrate Systems—Brain and BMR

The nutrients needed to provide substrate for energy are carbohydrate, fat, and protein. The substrates derived from these are glucose, pyruvate, fatty acids, glycerol, ketones, and amino acids. Where nutrients are sufficient to provide substrate, energy balance exists. In most cases substrates can be used interchangeably and the mix of dietary carbohydrate, fat, and protein as precursors for substrates is not commonly a problem. Two exceptions are important. Glucose is a preferred substrate under certain conditions and amino acids, especially essential amino acids, are required to sustain body protein. Glucose can only be derived from carbohydrate, glycerol, lactate, pyruvate, and some but not all amino acids.

Under normal conditions the requirements for glucose are set by the energy expenditure of brain (Kety, 1956). A small additional quantity is needed to provide substrate to red cells (Cahill, 1966). For an adult man 25% of substrate for BMR (1600 kcal/day) should be glucose. Where brain metabolism is a larger fraction of BMR, as it is in childhood, the percent of substrate as glucose should be greater (Holliday, 1971; Cornblath and Schwartz, 1976). Stress incurs an increase in total

---

*Thermic effect of food is a term used by Garrow (1974) as a more suitable definition of the variable increase in energy expenditure which follows ingestion of food and is more appropriate than the commonly used term specific dynamic action.

energy expended, and the preferred substrate is glucose (Cuthbertson, 1964). When carbohydrate needs are not met from the diet, body protein will be catabolized to provide substrate for glucose synthesis. The normal male deprived of food but given 100 g glucose reduces protein losses to half (Gamble, 1946).

4.2. BMR and Organ Mass

The changes in body composition that accompany growth have important effects on energy requirements. Both BMR and total energy intake increase with growth; during the first year they increase in proportion with weight gain, but thereafter they increase at a slower rate. Consequently, energy metabolism and requirements per kilogram decline after the first year of life. The major share of basal energy is used by the organs—brain, liver, heart, and kidney (Table I). These organs are relatively larger at birth (Table II). The pattern of growth of these organs in relation to change in body weight parallels the pattern of increase in BMR (Holliday, 1971) (Figure 3). The parallelism suggests an important relation between organ mass [as reflected by the sum of brain, liver, heart, and kidneys (ΣOW)] and BMR during growth.

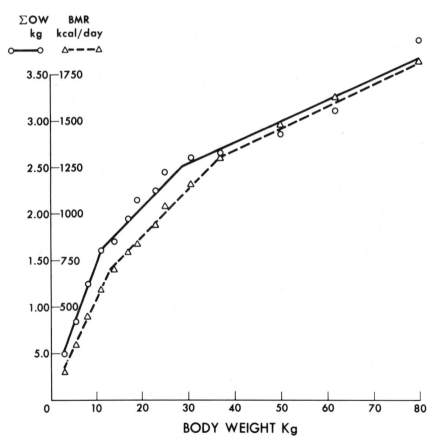

Fig. 3. The parallel relation between BMR vs. body weight and ΣOW vs. body weight from infancy to maturity (males) (Holliday, 1971).

This relation is illustrated in Figure 4, where BMR is plotted against ΣOW. The slope of increase of BMR upon organ weight is 555 kcal/kg, and the intercept is −232 (Figure 4, line A). However, if BMR is corrected for the fraction that is from muscle metabolism, the balance is a composite of organ metabolism or organ metabolic rate (OMR). The slope of OMR upon ΣOW (381 kcal/kg) is close to the weighted average OMR of adult organs (328 kcal/kg) (Table I) and the intercept (−53) is near zero (Figure 4, line B). This analysis indicates OMR is proportional to organ weight during growth. The slope of regression of log OMR against log ΣOW is 1.08 ± 0.04, (Figure 5) and defines the relative rate of increase in OMR to ΣOW. OMR increases at the same rate as ΣOW. This relation exists because the OMR of the major organs is similar, and the sum of their weights represents a relatively homogeneous mass with respect to energy metabolism. The growth rates of individual organs vary considerably from the growth rate of their sum (Figure 6). In the first year of life brain metabolic rate accounts for more than half of BMR (Figure 7).

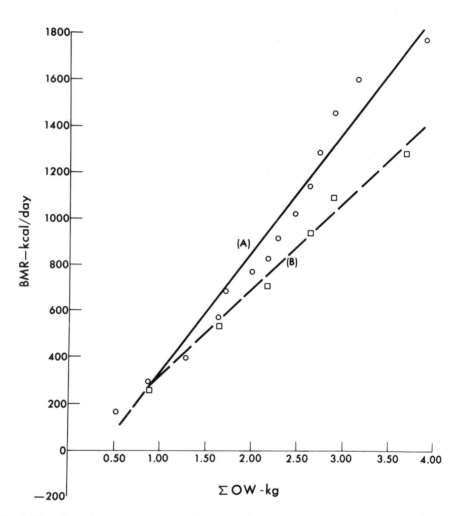

Fig. 4. The relation between (A) BMR and sum of organ weight (ΣOW) and (B) BMR minus muscle metabolic rate vs. ΣOW from infancy to maturity (males) (Holliday, 1971). (A), BMR (kcal/day) = 555 ΣOW (kg) − 232, $r = 0.98$; (B), OMR (kcal/day) = 381 ΣOW (kg) − 53, $r = 0.99$.

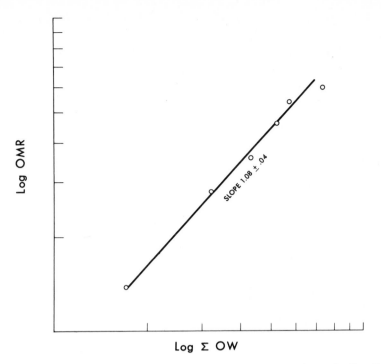

Fig. 5. Log organ metabolic rate (OMR) vs. log ΣOW. Slope of 1.08 ± 0.04 is not significantly different from 1.0. Increases in OMR and ΣOW occur at the same rate (adapted from Holliday, 1971).

Brain metabolism does not change under many conditions (Kety, 1956), and brain, among the major organs, is least affected in weight by somatic growth rate and nutrition. The magnitude of brain metabolic rate determines the carbohydrate needed as substrate (Kety, 1956; Owen *et al.*, 1967); for example, an 11-kg child may need 78 g glucose (7 g/kg/day) to prevent gluconeogenesis whereas an adult can accomplish this with 100 g (1.5 g/kg/day) (Table III) (Gamble, 1946). The size of visceral organs is closely related to height (Coppoletta and Wolbach, 1933). The energy requirement of organs probably can be predicted in healthy individuals when height is known with a satisfactory degree of accuracy, and this may account for the use of height to predict BMR (Talbot, 1938).

The relationship of weight to OMR and organ size is different in low-birth-weight infants (Sinclair *et al.*, 1967; Holliday, 1971). Organ mass grows at a rate slightly less than body weight as infants grow from 1.1 to 3.2 kg. (7–10 months gestation); however, BMR/kg increases. The slope of log BMR upon log ΣOW is 1.37 (Figure 8). Because muscle mass is so small, BMR is very nearly equal to OMR in low-birth-weight infants. OMR (kcal/100 g) is increasing more than ΣOW over this period, corresponding to the third trimester of gestation. This reflects a maturation of organs during that period.

4.3. BMR and Standard BMR

This description of BMR and OMR in relation to body size is useful in seeking to understand the relationship of body size to energy metabolism. The search for a

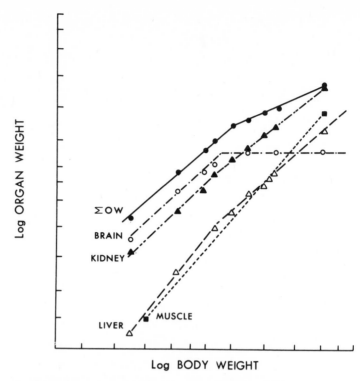

Fig. 6. Log of weights of brain, liver, kidney, muscle and ΣOW at different stages of growth. Log of individual weights adjusted so that all data might fit on a single log scale. The log of ΣOW is also provided.

universal reference standard has led to adoption of surface area and weight$^{3/4}$. Enough has been said about surface area (Kleiber, 1947; Brozek and Grande, 1955). Weight$^{3/4}$ has proved useful in comparing energy metabolism among species (Kleiber, 1947). It has been adopted for use in children of different sizes in search of a "metabolic constant" (Miller and Payne, 1963; Payne and Waterlow, 1971; Joint FAO/WHO, 1973). However, the observed relationship of BMR to weight does not parallel either the surface area or weight$^{3/4}$ relation to weight (Figure 9). For these reasons, it seems desirable to accept the empirical relation between energy metabolism and weight over the weight span encompassed by children during growth (3.5–70 kg or more). An entirely different relation applies in low-birth-weight infants (1–3.5 kg) (Sinclair *et al.*, 1967; Holliday, 1971).

4.4. Energy and Muscle

Muscle mass during growth increases relative to weight. There is an increase between the ages of 1.5–5 yr and a second increase during adolescence (Table II). These increases correspond to times when physical activity becomes a more important aspect of living. At any age and body height, muscle mass will vary, as it does in maturity, reflecting genetic factors determining body habitus and varying levels of physical activity and training. A relationship between physical activity and energy intake is intuitively perceived. It is logical to infer a relation between energy intake and muscle mass. However, certain types of muscular activity increase mass more than others. A shift in use of energy from growth to activity occurs at the end

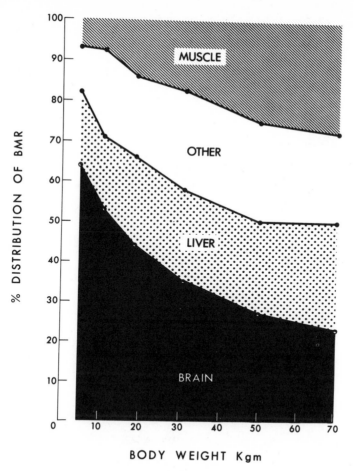

Fig. 7. Distribution of brain, liver, and muscle metabolic rates as percent of total BMR at different body weights (Holliday, 1971).

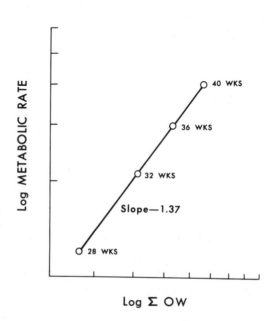

Fig. 8. Log metabolic rate vs. ΣOW of low-birth-weight infants. Metabolic rate derived from V_{O_2} measured in 92 infants appropriate for gestational age under standard cónditions (Sinclair *et al.,* 1967) who weighed between 730 and 3840 g. Points interpolated at 28–40 weeks gestational age based on data divided into weight groups appropriate to that gestational age (adapted from Holliday, 1971).

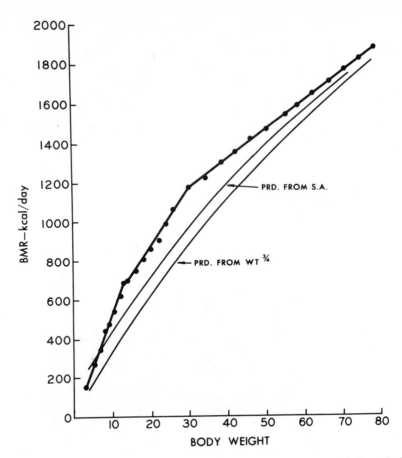

Fig. 9. Comparing observed BMR vs. weight of children from 3.5 to 70 kg with theoretical relation predicted from surface area and from weight[3/4]. The largest discrepancies are noted in the weight range 10–30 kg.

of the first year of age, and muscle mass begins its increased rate of growth shortly thereafter. The voracious appetite and impressive food consumption of adolescent boys corresponds with a high level of physical activity and a second increase in muscle mass. While these relationships are reasonable, a direct relationship between muscle mass, physical activity, and energy intake is not defined. The average intake of children in Western societies at different ages parallels BMR not muscle mass (Table IV). Energy expenditure greatest among individuals who sustain a high level of activity is not consistently proportional to muscle mass.

4.5. Energy and Fat Balance

Too few data exist to describe detailed changes in fat as related to growth. Infants increase fat in the first year of life (Foman, 1966), children may decrease fat content at adolescence (Friis-Hansen, 1971), and women may increase fat content as they reach sexual maturity (Frisch *et al.,* 1973). When children are obese throughout growth, they also are taller and have a larger cell mass, although the adipose tissue mass is most affected. This macrosomia of obese children (Forbes, 1964) is the converse of the slower growth rates and smaller bodies of children who have limited energy intakes.

Table IV.   *Total Energy Intake, BMR, and Muscle Mass at Different Stages of Growth*

Weight	Total energy intake (kcal/day)[a]	BMR[b] (kcal/day)	Muscle mass[b] (MM) (g/kg)	Intake/ BMR	(Intake − BMR)/MM
5.5	600	300	1.21	(2.00)[c]	(248)[c]
11.0	1100	590	2.57	1.86	198
19.0	1790	830	6.65	2.15	144
31.0	2380	1160	11.6	2.05	105
50.0	3020	1480	21.2	2.04	72
70.0	3100	1600	28.3	1.94	53

[a]Total daily energy intakes from Joint FAO/WHO (1973).
[b]From Holliday (1971).
[c]A significant percentage of energy intake is used for growth which is included in the term "Intake −BMR." Were the term Intake −BMR adjusted by subtracting energy used for growth, the ratios derived for the 5.5-kg child would be less.

4.6. Energy and ECF

There is no physiological relationship between energy intake and ECF volume. In one study, where BMR of healthy adult males was compared with several indices of body composition, the best correlation was found with ECF volume (Wedgewood *et al.*, 1953). This represents the type of statistical observation that comes from random variable pairing, particularly where the range of values is small. Over the entire range of growth, BMR will correlate with ECF as it will with foot length or any other index of growth. The correlation is without meaning (Brozek and Grande, 1955).

4.7. Energy and Kidney Function

There is a correlation between BMR and kidney function as measured by glomerular filtration rate (GFR) during growth after 1 year of age (Table V). Since kidney size and GFR increase at the same rate during growth, BMR parallels kidney function and weight from 1 year of age on (Holliday, 1972). This also might be fortuitous. On the other hand, this association may represent a selection process whereby kidney function that paralleled energy need presented a biological advantage. The capacity to survive and grow at an earlier time was dependent on the capacity of the individual to meet energy needs. The foods used in meeting these

Table V.   *Relation of Energy Intake and BMR to Glomerular Filtration Rate (GFR) at Different Stages of Growth*

Age	Weight	Total energy intake (kcal/day)[a]	BMR[b]	GFR (ml/min)[c]	GFR (ml/min/100 kcal) Total intake	GFR (ml/min/100 kcal) BMR
0.25	5.5	600	300	(11)	1.8	3.7
1.5	11	1100	590	29	2.6	4.9
5.5	19	1790	830	60	3.4	7.2
9.5	31	2380	1160	83	3.5	7.1
14.0	50	3020	1480	112	3.7	7.6
Adult	70	3100	1600	130	4.2	8.1

[a]From Joint FAO/WHO (1973).
[b]From Holliday (1971).
[c]From Holliday (1972).

needs were varied in respect to composition and the timing of feeding was irregular, so that a capacity for tolerating a varied diet was needed. This tolerance is provided by the kidney. The parallelism between energy requirement and renal function (food tolerance) during growth is, from this perspective, an advantage. The modern counterpoint for this relationship is that patients who have impaired renal function have a diminished tolerance to dietary variation (Holliday, 1976).

The fat of adipose tissue is a large reservoir that buffers the effects of disparities between energy intake and expenditure. Glycogen (about 250 g or 1000 kcal in man) is a smaller, more finely tuned buffer. Both body fat and glycogen are used in the normal overnight fast when intake is zero. The adipose tissue of normal individuals is 15–30% of body weight and provides 60,000–120,000 kcal in man or 900–1800/kg at any age. This assumes that adipose tissue is 65% fat, 5% protein, and 30% fluid and has an energy potential of 6 kcal/g. Synthesis of 1 g of fat requires 11–13 kcal, including 9 stored as potential energy and 2–4 used in synthesis. Addition of 1 g of adipose tissue including its protein and fluid is 8–10 kcal. Adipose tissue is metabolically active but less so than muscle. Its contribution to BMR is small (Garrow, 1974). During the period of growth where adipose tissue is a major fraction of weight gained, the energy cost of weight gain is correspondingly high (Foman, 1966; Foman *et al.*, 1971).

4.8. Energy and Growth Rate

Growth in children is most often measured either by weight gain or by increase in height. Energy intake during growth is related to weight gain. This is most easily examined in normal children during the first 6 months of life when growth rates are highest: 3–6 g/kg/day. The level of energy intake is an important factor affecting rate of weight gain. Foman *et al.* (1971) demonstrated a correlation of energy intake with weight gain in infants given several different types of formulas. Protein intake, so long as the amount given exceeded that provided from human milk, had little effect. However, the slope of regression of weight gain upon energy intake indicates the calorie cost of an increment in weight to be 15–20 kcal/g. The composition of the weight gain is estimated to be 40% fat and 11% protein (Foman, 1966).

4.9. Summary

The composition of the body and the pattern of change that occurs with growth, are genetically determined as are the metabolic characteristics of the various cells. The changes that occur with growth influence the relation of energy need to body size. The change in basal energy requirement that occurs with growth is largely a function of the change in weight of brain, liver, heart, and kidney, because these organs share a high and relatively constant metabolic rate throughout growth. Gross deviations of BMR from normal reflect either gross abnormalities in brain and/or visceral metabolic rate, e.g., hypothyroidism, or gross changes in size, e.g., anencephaly.

Muscle mass increases during growth as activity increases, and with this development energy requirement increases. However, energy needed for physical activity varies considerably, more in relation to individuals' habits than their muscle size. The average energy intake of children in England (when there was free access to food) was proportional to BMR, not muscle mass. Except for the first year of life, energy intake is mostly determined by energy expenditure. The energy cost of growth is appreciable only in the first year of life.

The assessment of an individual for undernutrition or overnutrition is done in part by assessing body composition. The assessment is concerned with organ mass, muscle mass, and fat mass. Organ mass determines BMR; muscle mass is an index of potential for activity and of protein stores; fat is a measure of energy resources. Determining ECF volume is useful, in this context, only as a means of adjusting body weight—for changes due to dehydration or edema. This may be a very important adjustment because body weight is the most often used measurement in evaluating nutritional status.

5.1. Effects of Undernutrition

The effects of undernutrition or energy deficiency depend on its severity and length and the stage in growth and development when it occurs. These effects are also influenced by illness or stress.

Starvation imposed upon a healthy individual is the most dramatic form of energy deficiency. It may occur accidentally among the adventuresome; it is used as therapy for massively obese patients; and it is a common complication of major disease and injury. The response of obese or normal individuals to starvation is well studied. For the first week there is a rapid weight loss (Forbes, 1970) associated with a catabolism of body protein (Cahill *et al.*, 1966) which is largely muscle protein (Young *et al.*, 1973). The catabolism of protein is partly a response to the need for glucose during fasting to sustain brain metabolism (Kety, 1956). The amino acids released in catabolism of muscle protein are precursors for glucose via gluconeogenic pathways.

The oxidation of 250 g of glycogen in the first week leads to a release of 750 g water and a 1-kg weight loss (Olsson and Salten, 1970). The cumulative catabolism of protein (300–400 g in the first week) and the release of its associated water results in a 1.5-kg weight loss. The use of fat and the associated catabolism of adipose tissue leads to loss of another 1.5 kg. These and the loss of ECF result in weight losses of 5–7 kg in 1 week with fasting; BMR also declines (Keys *et al.*, 1950). Since brain metabolism is not reduced (Kety, 1956), the decrease in BMR is most likely related to lower splanchic activity.

Following a week of starvation, rate of weight loss slows and nitrogen losses decrease to one third. Body fat is the predominant source for energy (Owens *et al.*, 1967). Catabolism of amino acids accounts for less than 5% of the energy used (Garrow, 1974).

Fasting is not a condition where substrate levels are low, but it is a condition where fat and muscle protein rather than food provide substrate. These same sources for substrate provide some other essential nutrients, e.g., essential fatty acids, limiting amino acids, and some minerals.

The effects of undernutrition in adult humans, in whom energy expenditure exceeds intake, are similar to those seen in fasting except the adjustment is slower. Both fat and muscle mass are lost. However, ECF volume usually stays the same or increases (Keys *et al.*, 1950).

Where individuals have adapted to a chronic state of limited energy input, they have less fat and protein and they spend energy more parsimoniously. In the eyes of the well fed, such individuals may seem indolent. In fact, they husband their energy supply carefully (Ashworth, 1968).

Children cannot survive a fast as long as adults because they have smaller stores of energy relative to their rate of energy expenditure. Children whose energy intake is limited may achieve balance and grow in length and organ mass. If the limitation in energy intake is small and can be offset by a reduction in activity, then fat and muscle mass may increase as well. However, growth will be slow.

5.2. Characteristics of Recovery from Undernutrition

Infants who are severely undernourished, i.e., have a low weight-for-height index and a low body-fat content and muscle mass, may regain weight at an accelerated rate. The rate of weight gain is closely related to calorie intake (Ashworth, 1968; Kerr *et al.,* 1973). The energy cost of weight gain in the latter study was 5–6 kcal/g. The composition of weight is estimated to be 20–30% fat and 15–20% protein (Ashworth, 1968). The effect of undernutrition upon BMR is variously interpreted, depending on how the values are reported. BMR of malnourished children when related to weight is normal and when related to height is below normal since weight-for-height ratios are low (Ashworth, 1969). As the children gain weight, BMR/kg remains the same although BMR related to height increases. The composition of weight gained is mostly adipose tissue and muscle.

Organ mass, when corrected for fat, probably changes very little (Waterlow and Allayne, 1971). The change in BMR may reflect an increase in organ metabolic rate and an increase in muscle and adipose tissue mass with a corresponding increase in their metabolic activity. The changes in body composition and energy metabolism that occur with weight gain during recovery from malnutrition have similarities with those that occur during normal growth. However, there are important differences—both qualitative and quantitative. The energy cost of weight gain (kcal/g) is greater in normal infants than in undernourished infants during recovery. The composition of weight gain also differs: weight gain in normal infants has more fat. This will not be true in infants at later stages of normal growth. These infants are also adding collagen and organ mass. Infants recovering from malnutrition, on the other hand, must replenish muscle and fat to a much greater extent.

5.3. Overnutrition and Obesity

Overnutrition and obesity are attended by characteristic changes in body composition. The massively obese individual has adipose tissue which is 80% fat, 15% water, and 2–3% protein (Garrow, 1974).* The obese individual who is or becomes inactive may have a decrease in muscle mass. By contrast, an active but overweight individual may have an increase in cell mass. This is true for some whose obesity dates to childhood. These individuals also tend to be taller (Forbes, 1964). BMR for height may be slightly higher because cell mass is greater, but BMR/kg is lower.

5.4. Summary

Body compositional changes in under- and overnutrition reflect the organism's capacity to cope with energy imbalances. It is evident that we abide by the first law of thermodynamics. We use mass—our own—to provide substrate for energy when

---

*Adipose tissue in normal individuals is 65% fat, 30% water, and 5% protein.

we do not obtain substrate from our diet; we gain mass when we obtain more potential energy than we expend. Recent work has improved the definition of the specific nature of these changes in mass with energy imbalances and has developed some tools for the clinician to use.

The relationship between maintenance energy requirement and body size changes as growth proceeds because body composition changes with growth. Energy imbalances experienced by children affect growth patterns as well as existing body composition. In the case of undernutrition growth rate is reduced; restoration of weight depends on energy intake. Weight for height is first restored and the infants voluntarily ingest extra energy. Restoration of height to appropriate height for age occurs at a slower rate and is not associated with high energy intakes. Catch-up growth often is incomplete.

Infants who are overfed become obese, i.e., adipose tissue increases. Some evidence suggests that early obesity may increase the number of fat cells, adipocytes, as well as the fat content of each cell. Children may become obese because they learn to overeat. They tend to have a larger cell mass and to be taller.

Energy imbalances, both under and over, are associated with measurable changes in body composition. When they are severe or prolonged and occur during childhood, they permanently affect growth and body composition.

6. References

Ashworth, A., 1968, An investigation of very low calorie intakes reported in Jamaica, *Br. J. Nutr.* **22**:341.

Ashworth, A., 1969, Metabolic rates during recovery from protein–calorie malnutrition: The need for a new concept of specific dynamic action, *Nature* **223**:407.

Barratt, T. M., and Walser, M., 1968, Extracellular volume in skeletal muscle: A comparison of radiosulfated radiobromide spaces, *Clin. Sci.* **35**:525.

Barratt, M., and Walser, M., 1969, Extracellular fluid in individual tissues and in whole animals: The distribution of radiosulfate and radiobromide, *J. Clin. Invest.* **48**:56.

Brook, C. G. D., 1971, Determination of body composition of children from skin fold measurements, *Arch. Dis. Child.* **46**:182.

Brozek, J., and Grande, 1955, Body composition and basal metabolism in man: Correlation analysis versus physiological approach, *Hum. Biol.* **27**:22.

Bruch, H., 1973, *Eating Disorders: Obesity, Anorexia and the Person Within,* Basic Books, New York.

Cahill, G. F., Herrera, M. G., Morgan, A. P., Soeldner, J. S., Steinke, J., Levy, P. L., Reichard, G. A., and Kipnis, D. M., 1966, Hormones—fuel interrelationships during fasting, *J. Clin. Invest.* **45**:1751.

Cheek, D. B., 1956, Change in total chloride and acid base balance with gastroenteritis following treatment with large and small loads of NaCl, *Pediatrics* **17**:839.

Cheek, D. B., 1961, ECF volume: Its structure and measurement and the influences of age and disease, *J. Pediatr.* **58**:103.

Cheek, D. B., 1968, *Human Growth; Body Composition, Cell Growth, Energy and Intelligence,* Lea & Febiger, Philadelphia.

Coppoletta, J. M., and Wolbach, S. B., 1933, Body length and organ weights of infants and children, *Am. J. Pathol.* **9**:55.

Cornblath, M., and Schwartz, R., 1976, *Disorders of Carbohydrate Metabolism in Infancy,* 2nd ed., W. B. Saunders, Philadelphia.

Corsa, L., Olney, J. M., Steinberg, R. W., Ball, M. R., and Moore, F. D., 1950, The measurement of exchangeable potassium in man by isotope dilution, *J. Clin. Invest.* **29**:1280.

Cuthbertson, D. P., 1964, Physical injury and its effects on protein metabolism, in: *Mammalian Protein Metabolism,* Vol. II (H. N. Munro and J. B. Allison, eds.), p. 373, Academic Press, New York.

Davidson, S., and Passmore, R., 1969, in: *Human Nutrition and Dietetics,* 4th ed., p. 385, Livingstone, Edinburgh.

Delaporte, C., Bergstrom, J., and Broyer, M., 1976, Variations in muscle cell protein of severely uremic children, *Kidney Int.* **10**:239.

Durnin, J. V. G. A., and Rahaman, M. M., 1967, The assessment of the amount of fat in the human body from measurement of skin fold thickness, *Br. J. Nutr.* **21**:681.

Edelman, I. S., and Leibman, J., 1959, Anatomy of the body fluids, *Am. J. Med.* **27**:256.

Flynn, M. A., Woodruff, C., Clark, J., and Chase, G., 1972, Total body potassium in normal children, *Pediatr. Res.* **6**:239.

Foman, S. J., 1966, Body composition of the male reference infant during the first year of life, *Pediatrics* **40**:863.

Foman, S. J., Thomas, L. N., Filer, L. J., Ziegler, E. E., and Leanard, M. T., 1971, Food composition and growth of normal infants fed milk-based formulas, *Acta Pediatr. Scand., Suppl.* **223**:1.

Forbes, G. B., 1964, Lean body mass and fat in obese children, *Pediatrics* **34**:308.

Forbes, G. B., 1970, Weight loss during fasting: Implications for the obese, *Am. J. Clin. Nutr.* **23**:1212.

Forbes, G. B., and Bruining, G. J., 1976, Urinary creatinine excretion and lean body mass, *Am. J. Clin. Nutr.* **29**:1359.

Friis-Hansen, B., 1957, Changes in body water compartments during growth, *Acta Pediatr. Scand.* **46**(Suppl. 110):1.

Friis-Hansen, B., 1971, Body composition during growth, *Pediatrics* **47**:264.

Frisch, R. E., Revelle, R., and Cook, S., 1973, Components of the critical weight at menarche and at initiation of the adolescent spurt: Estimated total water, lean body mass and fat, *Hum. Biol.* **45**:469.

Gamble, J. L., 1946, Physiological information gained from studies on the life raft ration, *Harvey Lect. Ser.* **42**:247.

Gamble, J. L., 1947, *Syllabus, Chemical Anatomy, Physiology and Pathology of ECF,* Harvard University Press, Cambridge.

Garrow, J. S., Fletcher, K., and Halliday, D., 1965, Body composition in severe infantile malnutrition, *J. Clin. Invest.* **44**:417.

Garrow, J. S., 1974, *Energy Balance and Obesity in Man,* pp. 5–29, American Elsevier, New York.

Hirsch, J., and Knittle, J. L., 1970, Cellularity of obese and non-obese human adipose tissue, *Fed. Proc.* **29**:1516.

Holliday, M., Potter, D., Jarrah, A., and Bearg, S., 1967, Relation of metabolic rate to body weight and organ size, a review, *Pediatr. Res.* **1**:185.

Holliday, M. A., 1971, Metabolic rate and organ size during growth from infancy to maturity, *Pediatrics* **47**:169.

Holliday, M. A., 1972, Body fluid physiology during growth, in: *Clinical Disorders of Fluid & Electrolyte Metabolism,* Chapter 12 (M. H. Maxwell and C. R. Kleeman, eds.), McGraw-Hill, New York.

Holliday, M. A., 1976, Management of the child with renal insufficiency, in: *Clinical Pediatric Nephrology,* (E. Leiberman, ed.), Chapter 20, Part 1, p. 397, J. B. Lippincott, Philadelphia.

Joint FAO/WHO ad hoc expert committee, 1973, Energy and protein requirements, *WHO Technical Report Series,* No. 522.

Kennedy, C., and Sokoloff, L., 1957, An adaptation of the nitrous oxide method to the study of the cerebral circulation in children; normal values for cerebral blood flow and cerebral metabolic rate in childhood, *J. Clin. Invest.* **36**:1130.

Kaufman, L., and Wilson, C. J., 1973, Determination of extracellular fluid volume by fluorescent excitation analysis of bromide, *J. Nucl. Med.* **14**:812.

Kerr, D., Ashworth, A., Picou, D., Poulter, N., Seakin, A., Spady, D., and Wheeler, E., 1973, Accelerated recovery from infant malnutrition with high calorie feeding, in: *Endocrine Aspects of Malnutrition,* (L. L. Gardner and P. Amacher, eds.), pp. 467–479, Kroc Foundation, Santa Ynez, California.

Kety, S. S., 1956, The general metabolism of the brain *in vivo,* in: *International Neurochem Symposium,* 2nd ed. (D. Richter, ed.), Pergamon Press, London.

Keys, A., Brozek, J., Hanschel, A., Mickelson, O., and Taylor, H. L., 1950, *The Biology of Human Starvation,* University of Minnesota Press, Minneapolis, Minnesota.

Keys, A., and Brozek, J., 1953, Body fat in adult man, *Physiol. Rev.* **33**:245.

Kleiber, M., 1947, Body size and metabolic rate, *Physiol. Rev.* **27**:511.

Metcoff, J., Silvestre, F., Antonwiez, I., Gordillo, G., and Lopez, E., 1960, Relations of intracellular ions to metabolic sequences, *Pediatrics* **26**:960.

Miller, D. S., and Payne, P. R., 1963, A theory of protein metabolism, *J. Theor. Biol.* **5**:398.

Miller, I., and Weil, W. B., Jr., 1963, Some problems in expressing and comparing body composition determined by direct analysis, *Ann. N.Y. Acad. Sci.* **110**:153.

Nichols, G., Nichols, N., Weil, W., and Wallace, W. M., 1953, The direct measurement of the extracellular phase of tissues, *J. Clin. Invest.* **32**:1299.

Olsson, K. E., and Salten, B., 1970, Variations in total body water with muscle glycogen in man, *Acta Physiol. Scand.* **80**:11–18.

Owen, O. E., Morgan, A. P., Kemp, H. G., Sullivan, J. M., Herrera, M. G., and Cahill, G. F., 1967, Brain metabolism during fasting, *J. Clin. Invest.* **46:**1519.

Patrick, J., 1977, Assessment of body potassium stones, *Kidney Int.* **11:**476.

Payne, P. R., and Waterlow, J. C., 1971, Relative requirements for maintenance growth and physical activity, *Lancet* **2:**210.

Sinclair, J. C., Scopes, J. W., and Silverman, W. A., 1967, Metabolic reference standards for the neonate, *Pediatrics* **39:**724.

Talbot, F. B., 1938, Basal metabolism standards for children, *Am. J. Dis. Child.* **55:**455.

Tanner, J. M., and Whitehouse, R. H., 1962, Standards for subcutaneous fat in British children, *Br. Med. J.* **1:**446.

Waterlow, J. C., and Allayne, G. A. O., 1971, Protein malnutrition in children: Advances in knowledge in the last ten years, *Adv. Protein Chem.* **25:**117.

Wedgewood, R. J., Bass, D. E., Klimas, J. A., Kleeman, C. R., and Quinn, M., 1953, Metabolic rate in normal man, *J. Appl. Physiol.* **6:**317.

Widdowson, E. M., and Dickerson, J. W. T., 1964, Chemical composition of the body, in: *Minimal Metabolism,* Vol. IIA (C. L. Comar and F. Bronner, eds.), p. 1, Academic Press, New York.

Young, V. R., Haverberg, L. N., Bilmazes, C., and Munro, H. N., 1973, Potential use of 3-methyl histidine excretion as an index of progressive reduction in muscle protein catabolism during starvation, *Metabolism,* **22:**1429.

6

Puberty

W. A. MARSHALL

1. Definition

In this chapter the word ''puberty'' will be used as a collective term embracing all those morphological and physiological changes which occur in the growing boy or girl as the gonads change from their infantile to their adult state. These changes involve nearly all the organs and structures of the body. They do not begin at the same age or take the same length of time to reach completion in all individuals. Also, the order in which different structures begin to undergo their adolescent change varies from one subject to another. Puberty is not complete until the individual has the physical capacity to conceive and successfully rear children.

The principal manifestations of puberty are:

1. The adolescent growth spurt. This is an acceleration, followed by a deceleration, of growth in most skeletal dimensions and in several internal organs.
2. The development of the gonads.
3. The development of the secondary reproductive organs and the secondary sex characters.
4. Changes in body composition resulting mainly from changes in the quantity and distribution of fat in association with growth of the skeleton and musculature.
5. Development of the circulatory and respiratory systems leading, particularly in boys, to an increase in strength and endurance.

These and other minor changes which accompany them will be the subject of this chapter. Puberty is, of course, the result of complex developmental processes in the central nervous system and endocrine systems, but these are discussed elsewhere.

W. A. MARSHALL • Department of Human Sciences, Loughborough University, Loughborough, Leicestershire, England.

2. The Adolescent Growth Spurt

If the statures of a typical boy and girl are measured repeatedly through childhood, and the measurements are plotted against age, curves similar to those shown in Figure 1 are obtained. The girl's curve begins to rise more steeply at about the age of 11 and the boy's at about 13. This inflection represents the adolescent spurt in stature which occurs, on the average, about two years earlier in girls than in boys. The illustrated curves represent individuals who were of average stature and who experienced the adolescent spurt at the average age for their sex. As we shall see later, the actual age at which the spurt occurs varies greatly from one individual to another in either sex.

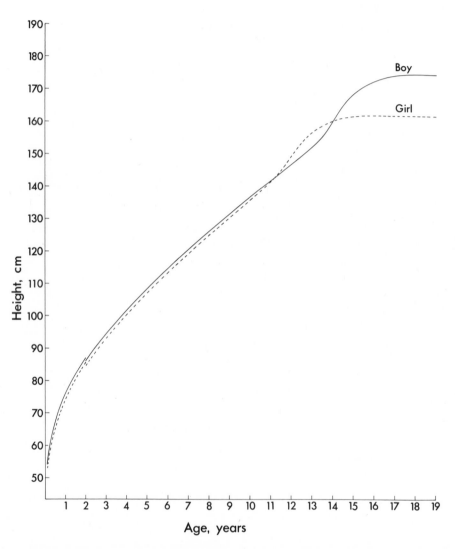

Fig. 1. Height at different ages of a hypothetical boy and girl of mean birth length, who grew at the mean rate and experienced the adolescent growth spurt at the mean age for their sex. Each finally reached the mean adult stature. (Reproduced from Tanner *et al.,* 1966, by kind permission of the authors and the editor of *Archives of Disease in Childhood.*)

Velocity curves, in which the speed of growth (in centimeters per year) is plotted against age, show the adolescent spurt more clearly. It appears as a sharp increase in velocity, which rises to a maximum and then immediately begins to decrease again (Figure 2). The maximum value is referred to as peak height velocity (PHV). When we have repeated measurements of an individual, so that we can construct a growth velocity curve, we can clearly recognize both the value, in centimeters per year, of peak height velocity and the age at which it occurs. PHV is an important landmark in the growth process and can be usefully related to other changes which occur at puberty.

The absolute value of peak height velocity varies from one child to another. Marshall and Tanner (1969, 1970) found a mean value of 10.3 cm/yr with a standard deviation of 1.54 cm/yr in 49 healthy boys who were measured every three months by a single skilled observer. The average peak velocity for 41 girls in the same study was 9.0 cm/yr with a standard deviation of 1.03 cm/yr. The value for each subject was estimated by drawing a smooth curve through a plot of the velocities. It is only by fitting a curve in some way that the moment of peak height velocity can be identified with reasonable confidence. The velocity measured over 3 months, 6 months, or a year does not represent the actual peak velocity because the growth rate passes through the final stages of its acceleration and begins to decelerate within a very short period of time. When the velocity is calculated over a whole year centered on the peak, that is, including 6 months before and 6 months after it, the average is 9.5 cm/yr for boys and 8.4 cm/yr for girls. Thus, for about a year during adolescence the velocity of growth is nearly twice the velocity in either sex just before the adolescent spurt begins (about 5 cm/yr). During the year in which a boy attains his peak height velocity he usually gains between 7 and 12 cm in stature, while a girl in the corresponding year gains between 6 and 11 cm. The maximum growth velocity is usually greater in children who reach their peak at an early age than those who do so later, but this is not a very precise relationship as the correlation between peak height velocity and the age at which it occurs is only about -0.45.

In girls studied by Marshall and Tanner (1969), the mean age at peak height velocity was 12.14 ± 0.14 years with a standard deviation of 0.88 years. The mean for boys was 14.06 ± 0.14 years with a standard deviation of 0.92 years.

So far we have referred only to stature, but the growth of nearly all skeletal dimensions accelerates at adolescence. However, the increase in growth rate is not uniform throughout the skeleton, and during the adolescent spurt in stature there is a greater increase in the length of the trunk than in that of the legs. Also, the spurt does not begin simultaneously in all parts of the body. The differences in the time at which the spurt occurs in different regions reflect the maturity gradients which are present *in utero*. At birth the head is much nearer to its adult size than the trunk or legs and represents about one fifth of the child's total length, as compared with about one eighth in the adult. The legs are relatively small at birth because their maturation is delayed in comparison with that of the cranium. Within the limbs there is apparently a reverse gradient so that the hands and feet are more mature than the lower leg and forearm which, in turn, are more advanced than the thigh or the upper arm. The hands and feet are therefore larger in comparison to the remainder of the limbs than they will be in later life. The distal segment of the limbs is correspondingly big in relation to the proximal segment. Leg length probably reaches its peak growth velocity before the trunk. The foot probably attains its maximum size before any other region with the possible exception of the head.

Certainly some adolescents have disproportionately large feet, and girls are occasionally worried by this. However, the situation is usually self-limiting as the earlier cessation of growth in the feet allows the remainder of the body to grow to a size more in keeping with that of the foot, which is not disproportionately big by the time the child is fully grown.

In the arm there appears to be a similar gradient to that in the leg, so that the forearm reaches its peak velocity about 6 months before the upper arm but some time after the hand (Maresh, 1955).

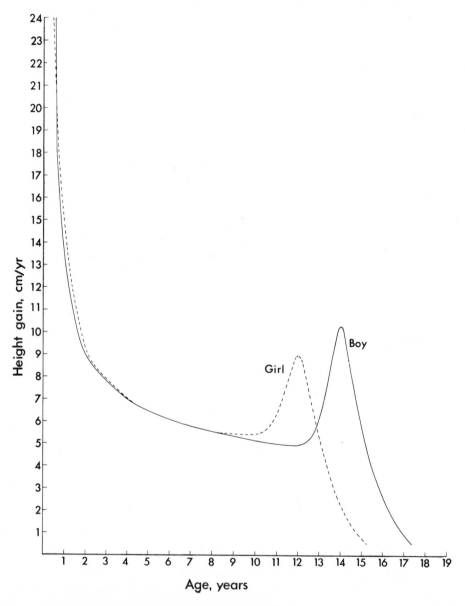

Fig. 2. Growth velocity, in centimeters per year, at different ages of the boy and girl whose statures are shown in Figure 1. (Reproduced from Tanner *et al.,* 1966, by kind permission of the authors and editor of *Archives of Disease in Childhood.*)

Leg length, measured as stature minus sitting height, reaches its peak growth velocity before the trunk. However, because the spurt in sitting height is greater than that in the lower limb, the proportion of the total stature which is due to the trunk increases during adolescence. Because the spurt in stature is due to the acceleration of both trunk length and leg length, peak height velocity is reached when the sum of the velocities of these two segments is maximal. This is usually at about the same time as sitting height reaches its peak velocity, but in those children who experience a substantial and earlier acceleration of growth in the lower limbs, stature will reach its peak velocity after the legs do so but before the trunk.

The rate of increase of the width of the shoulders and hips becomes maximal at about the same time as sitting height reaches its peak velocity. The cranium grows very slowly after early childhood but the growth of both head length and head breadth is accelerated slightly at adolescence (Shuttleworth, 1939). There is some disagreement as to whether growth in the exterior diameter of the head at adolescence is due mainly to expansion of the cranial cavity, increase in the thickness of bone, or greater thickness of the soft tissue. One radiographic longitudinal study showed that the lengths of frontal and parietal arcs increased only by 1–2% after age 12, but the thickness of the bones and the scalp tissue increased by some 15% at adolescence (Roche, 1953; Young, 1957, 1959). However, Singh *et al.* (1967), who carried out a radiographic longitudinal study of head width from age 9 to 14 years, found that growth in both sexes was due mainly to increase in the thickness of soft tissue and in the width of the cranial cavity, while the thickness of bone increased very little. The width of the cranial cavity increased more in boys than in the girls, who had a greater increase in soft tissue width.

The forward growth of the forehead at adolescence is mainly due to the development of the brow ridges and frontal sinuses, according to Björk (1955), who carried out a radiographic study of 243 boys at age 12 and again at age 20. However, the middle and posterior cranial fossae do enlarge, and the cranial base posterior to the sella turcica becomes lower (Zuckermann, 1955). The cranial base also increases in length (Ford, 1958).

In the face, the growth of most dimensions accelerates to reach a maximum velocity a few months after peak velocity in stature (Nanda, 1955). The greatest change is in the mandible where 25% of the total growth in height of the ramus is completed between the ages of 12 and 20 years, in contrast to the cranial base where only 6–7% of the total growth occurs during this period. According to Savara and Tracey (1967), the increase during adolescence is greatest in mandibular length and least for bigonial width. The growth in height of the ramus follows a pattern similar to growth in length. The result of this is that the jaw becomes longer and thicker and projects further forwards in relation to the face. These changes are much more marked in boys than in girls. The forward growth of the chin, however, is not completed until after puberty, as further apposition of bone at the mandibular symphysis usually occurs between 16 and 23 years of age.

The forward growth of the maxilla is also accelerated. This is most obvious directly above the upper incisor teeth so that the prognathism of both jaws is increased, although this increase is less marked in the upper jaw than the lower. This description of maxillary growth has been disputed by Savara and Singh (1968), who found that some measures of height and width showed greater acceleration than length. The literature on maxillary growth is somewhat unsatisfactory and difficult to interpret as different authors have employed different techniques.

In boys, the nose usually grows forward and downward relative to the rest of the face, but this change is not detectable in all children.

The adolescent spurt may be described mathematically by fitting curves to series of measurements taken on the same children at different ages. Repeated measurements of this kind constitute "longitudinal" data as distinct from the "cross-sectional" data which are obtained by measuring a number of different individuals at each age. The Gompertz and logistic functions, both of which give sigmoid curves, are the models which have been most widely used.

The Gompertz curve is defined by the equation

$$Y = P + Ke^{-e^{a-bt}}$$

where Y is a dependent variable, e.g., stature; t is an independent variable, e.g., age; P is the lower asymptote, i.e., the value of stature or other parameter at the beginning of the adolescent spurt; K is the total increase in Y, i.e., the actual gain in height during the adolescent spurt; a is a constant of integration which depends on the position of the origin; b is a constant representing the rate at which the proportion $(Y - P)/K$ is increasing to equal powers of itself. The product bKe^{-1} represents the maximum rate of growth attained, i.e., peak height velocity.

The use of the Gompertz curves in describing growth during adolescence is discussed in detail by Deming (1957) and by Marubini et al. (1972). Marubini et al. found that the logistic curve fitted their data slightly better than the Gompertz function, and they recommend it as the best equation for fitting growth data on skeletal dimensions at adolescence.

They fitted both Gompertz and logistic functions to measurements of height, sitting height, leg length, and biacromial diameter taken at three-monthly intervals in 23 boys and 18 girls. The fit of the curves was equally good for both sexes and was slightly better for sitting height, leg length, and biacromial diameter than for stature. The residuals, that is the discrepancies between the actual measurements and the fitted curves, were not significantly greater than the errors of measurement.

The logistic curve has the equation

$$Y = P + \frac{K}{1 + e^{a-bt}}$$

where the symbols have the same meaning as in the case of the Gompertz curve.

4. Growth of the Heart, Lungs, and Viscera

Several authors have described an adolescent spurt in growth of the transverse diameter of the heart, as measured radiographically (Bliss and Young, 1950; Lincoln and Spillman, 1928; Maresh, 1948). According to Simon et al. (1972), who carried out a longitudinal study of chest radiographs, this spurt is of about equal magnitude in both sexes, and the age at peak velocity for heart diameter and lung coincides with the age at peak height velocity. Growth in length of the lung apparently occurs somewhat later, and the peak velocity is not reached until about 6 months later than the peak for lung width. The magnitude of the spurt in the lungs is similar in both sexes, but the actual mean width of the lung in girls does not exceed that of boys at any stage. This contrasts with other measurements, such as stature, which become greater in girls during adolescence than in boys of the same age and remain so until the boys enter their adolescent spurt somewhat later.

There is probably an adolescent spurt in all the abdominal viscera including the liver, kidneys, and nonlymphatic portion of the spleen. Evidence on this point, however, is limited. The growth in length of the pharynx is accelerated with the result that the hyoid bone becomes much lower in relation to the mandible. Before puberty the two bones are at approximately the same level. There is less marked adolescent growth in the anteroposterior diameter of the pharynx (King, 1952).

5. Sex Differences in Size and Shape Arising at Adolescence

Apart from the development of the secondary sex characters, the most noticeable sex differences which arise at puberty are the greater stature of the male and his broader shoulders as compared with the wider hips of the female. Until girls enter

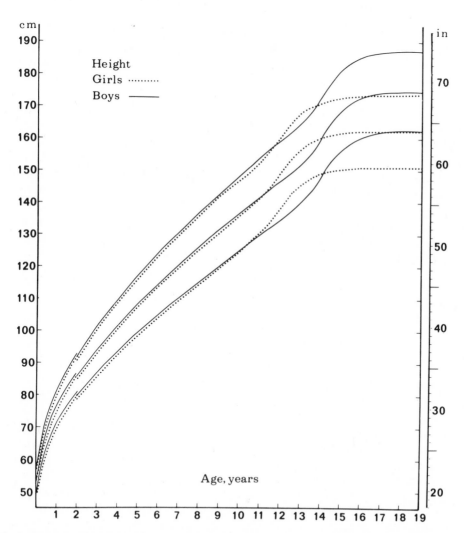

Fig. 3. Growth curves of three girls and three boys in whom the adolescent spurt occurred at the average age for their sex and produced the average increase in height. The statures of the tallest boy and girl are consistently near the 97th centile for their sex. The lower pairs are near the 50th and 3rd centiles, respectively. (Reproduced from Marshall, 1970, by kind permission of the publishers.)

their adolescent growth spurt there is little difference in stature between the two sexes. At this point, the girls tend to become taller than boys; however, this difference is not so great as some authors suggest.

Figure 3 shows the growth curves of three boys and three girls, each of whom experienced the adolescent spurt at the average age for their sex. One child of each sex is of average stature while one is at the 97th centile, i.e., tall. The other is at the third centile, i.e., short. The sex difference in stature, which arises from the earlier adolescent spurt in girls is of sufficient magnitude to make the short girl taller than the short boy, but no taller than the average or tall boy. The average girl becomes taller than the average boy but only the tall girl is taller than the 97th centile boy. A calculation based on the data of Tanner *et al.* (1966) shows that, at 13.9 years, 7% of girls who reach PHV at the average age are taller than 13-year-old boys who are at the 97th centile and also reach PHV at the average age for their sex. In practical terms the difference is most important to the tall girl and the small boy. The former may realize one day that she is taller than all her male peers and be concerned lest she should become an abnormally tall woman. Similarly, the small boy finds that he is shorter than all the girls of his age, and this too may cause concern although, in both cases, the situation is rectified when the boys experience their adolescent spurts. Boys have, on average, about two more years of preadolescent growth than girls and are therefore some 10 cm taller when they begin their adolescent spurts than girls were at the corresponding stage. The longer period of preadolescent growth in boys is also largely responsible for the fact that men's legs are longer than women's in relation to the length of the trunk because the legs grow relatively faster than the trunk immediately before adolescence.

The sex difference in the relative widths of shoulders and hips is due to the fact that girls have a larger adolescent spurt in hip width than boys when this parameter is related to stature, although, in absolute terms, the increase is no greater than in boys. The girls' spurt in other dimensions is considerably less than that in the hips. These comments apply to the over-all distance between the iliac crests and not to the pelvic outlet which is already bigger at birth in girls than in boys. The adolescent changes affect chiefly the pelvic inlet.

Even before puberty boys have wider shoulders than girls, and this difference increases at adolescence. Figure 4 shows the relative growth of the shoulders and hips in the two sexes.

6. Lymphatic Tissues

Apart from the subcutaneous fat, which is discussed below, the lymphatic tissues and thymus are the only major structures which do not show an adolescent spurt in growth. The weight of the thymus increases from birth to a maximum within the age range 11–15 and then decreases (Boyd, 1936). This observation has been confirmed by other authors but, of course, is based only on cross-sectional data from material obtained at autopsy. If it were possible to carry out a longitudinal study of growth of lymphoid tissue, the relationship between its growth and decline, in comparison with the remainder of the body, could be observed more accurately. Turpin *et al.* (1939) have carried out a study of the thymus shadow in tomographic X rays which indicate that the involution takes place at the time of the adolescent spurt. The thymus was found to be at its maximum size at about age 12 in girls and

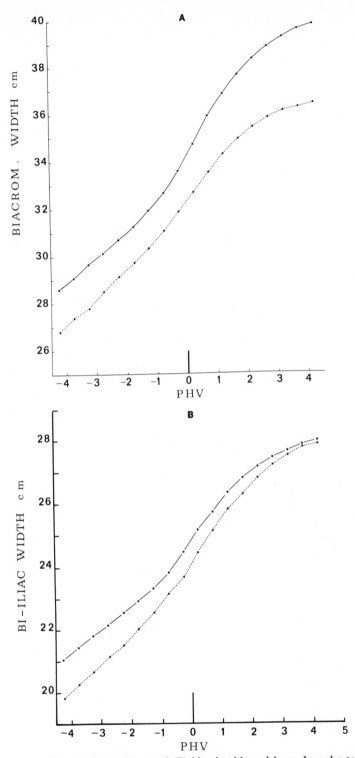

Fig. 4. Growth in width of (A) shoulders and (B) hips in girls and boys. In order to eliminate the difference in age at which the adolescent spurt occurs in the two sexes, the measurements have been plotted against a scale of years before and after peak height velocity (PHV). Girls, · · · ·; boys, ·———·. (Reproduced from Marshall, 1970, by kind permission of Harper and Row Publishers Inc.)

14 in boys with a decrease occurring the following year, but these data too were cross-sectional and the number of subjects at each age was relatively small. The lymphatic parenchyma of the thymus may reach its maximum size slightly before the gland as a whole, as a small proportion of the organ is composed of connective tissue whose absolute weight continues to increase throughout puberty. The lymphatic tissue in other organs such as the spleen, intestine, appendix, and mesenteric lymph nodes regresses at about the same time as the thymus (Scammon, 1930). In a series of 300 cases of death within 10 hr from an injury, Hwang and Krumbhaar (1940) showed that the amount of lymphatic tissue in the appendix was less in the 11–20 age group than it was in the 1–10 age group, although the remainder of the appendix showed an increase with age and some indication of an adolescent spurt. In the spleen, the absolute amount of lymphatic tissue increased slightly between these two 10-year periods, but this increase was not nearly as great as in the nonlymphatic tissue, so that the percentage of total weight accounted for by lymphatic structures dropped from 10.8 to 7.7 (Hwang *et al.*, 1938). Scammon's data on the spleen showed an actual decrease in lymphatic tissue at adolescence.

7. Development of the Reproductive Organs and Secondary Sex Characters—Male

7.1. The Testes

For many years the study of testis growth in the living subject was hampered by lack of satisfactory and acceptable methods of measuring their size. Nowadays, however, measurements which are sufficiently accurate for clinical purposes can be taken with the "Prader orchidometer" (Figure 5), which is a series of models of known volume with the shape of ellipsoids. These are mounted on a string in order of increasing size and each model is numbered according to its volume in milliliters. The instrument is held in the observer's hand and the testis is palpated gently with the the other hand. The models are similarly palpated until the one that appears to be the same size as the testis is identified. The volume of this model is then read.

Testes with volumes of 1, 2, and sometimes 3 ml on this scale are found in prepubertal boys, but a greater volume almost always indicates that puberty has begun. The volume of the adult testis varies between 12 and 25 ml when measured by this technique. According to Zachmann *et al.* (1974), the volume of the adult testis shows a negative correlation ($r = -0.47$) with age at peak height velocity. In other words, there is a tendency for the testes of early maturers to become larger than those of boys who reach puberty at a later age. There is very little increase in size of the testes before puberty; at puberty rapid growth occurs. Figure 6 shows the 10th, 50th, and 90th centiles of testis size in boys at different ages (Zachmann *et al.*, 1974). These diagrams are based on measurements taken with the Prader orchidometer. At peak height velocity the mean testicular volume was 10.5 ml.

Growth in size of the testes at puberty is associated with the development of their reproductive and endocrine functions. In the prepubertal testis, the interstitial tissue has a loose appearance and contains no Leydig cells. The seminiferous tubules are cordlike structures ranging between 50 and 80 μm in diameter and having no lumen. The lumen does not become distinct until puberty, although the

Fig. 5. The Prader orchidometer.

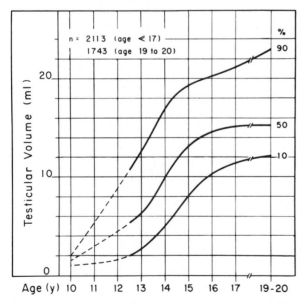

Fig. 6. The 10th, 50th, and 90th centiles of testicular volume in relation to chronological age. (Reproduced from Zachman *et al.,* 1974, by kind permission of the authors and the editors of *Helvetica Paediatrica Acta.*)

first signs of it usually appear round the age of 6 years. During childhood the number of spermatogonia in the tubules increases, although the Sertoli cells remain undifferentiated. At puberty, there is a considerable increase in the size and tortuosity of the tubules. A thin tunica propria, containing elastic fibers, develops. Differentiation of Sertoli cells and a division of the basally situated spermatogonia represent the beginning of the changes in the germinal epithelium which lead gradually to spermatogenesis. The exact chronological relationship between the beginning of maturation in the seminiferous tubules and the differentiation of the Leydig cells is uncertain, but it is probable that the Leydig cells appear at the same time as mitotic activity begins in the germinal epithelium. By the time the Leydig cells are fully differentiated histologically, the meiotic divisions which lead eventually to the development of spermatids and spermatozoa have already begun.

7.2. Accessory Sex Organs

The epididymis, seminal vesicles, and prostate increase very little in weight before puberty and then show a rapid increase, beginning at about the same time as enlargement of the testes (Figure 7).

7.3. Penis and Scrotum

There is no precise scale for measuring the development of the external genitalia, but for descriptive purposes the process may be divided into five stages as shown in Figure 8. The criteria for each stage (modified from Tanner, 1962) are as follows:

Stage 1. The infantile state which persists from birth until puberty begins. During this time the genitalia increase slightly in over-all size, but there is little change in general appearance.

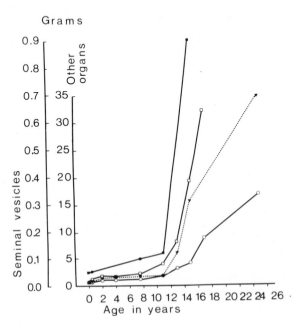

Fig. 7. Mean weights of male reproductive organs at various ages. Seminal vesicles, ■——■; testes, ▼ · · · ▼; testes plus epididymides, □——□; prostate, ○——○. (Reproduced from Marshall and Tanner, 1974, by kind permission of William Heinemann, Medical Books Ltd.)

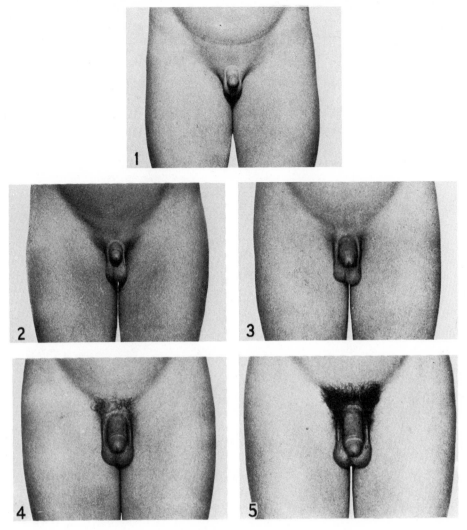

Fig. 8. Stages of development of the penis and scrotum. (Reproduced from Tanner, 1962, by kind permission of the author and Blackwell Scientific Publications Ltd.)

Stage 2. The scrotum has begun to enlarge, and there is some reddening and change in texture of the scrotal skin.

Stage 3. The penis has increased in length and there is a smaller increase in breadth. There has been further growth of the scrotum.

Stage 4. The length and breadth of the penis have increased further and the glans has developed. The scrotum is further enlarged, and the scrotal skin has become darker.

Stage 5. The genitalia are adult in size and shape.

The appearance of the genitalia may satisfy the criteria for one of these stages for a considerable time before the penis and scrotum are sufficiently developed to be classified as belonging to the next stage. It is sometimes important to distinguish between the moment at which a child's genitalia first fulfill the conditions of a particular stage, e.g., stage 3, and the remainder of the time until it becomes

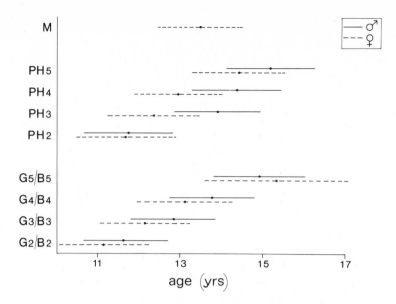

Fig. 9. Variation in age on reaching stages of puberty in boys and girls. PH2, PH3, etc., = pubic-hair stages; G2, G3, etc., = stages for penis and scrotum; B2, B3, etc., = female breast stages; M = menarche. Each horizontal line extends for one standard deviation on either side of the mean.

legitimate to classify them as stage 4. The moment of arrival at a given stage is usually indicated by the abbreviation G2, G3, etc., while at other times the genitalia are described, for example, as being "in stage 3."

The variation in ages at which different boys arrive at each stage of genital development is shown in Figure 9, which is based on data reported by Marshall and Tanner (1970) and by van Wieringen *et al.* (1968). These studies were concerned with boys of British and Dutch origin, respectively. The range indicated in the figure is one standard deviation on either side of the mean, i.e., it includes about 66% of the population. Double this range would include 95% of the population. The genitalia may attain stage 2 at any time after the 9th birthday but need not do so before the 14th, i.e., a 14-year-old, or occasionally even a 15-year-old, boy whose genitalia remain infantile may not be abnormal. Complete maturity of the genitalia (stage 5) may be reached before the 13th birthday, but some boys are not fully developed until they are 18 or occasionally even older.

7.4. Pubic Hair

The increase in distribution of the pubic hair, like the development of the genitalia, may be described in five stages. After the first recognizable pubic hair has appeared (stage 2), the stages are based entirely on the distribution of the hair and not its density. The observer must not classify sparse hair as being in a lower stage than its over-all distribution allows.

Stage 1. There is no true pubic hair, although there may be a fine velus over the pubes similar to that over other parts of the abdomen.

Stage 2. Sparse growth of lightly pigmented hair which is usually straight or only slightly curled. This usually begins at either side of the base of the penis.

Stage 3. The hair spreads over the pubic symphysis and is considerably darker and coarser and usually more curled.

Stage 4. The hair is now adult in character but covers an area considerably smaller than in most adults. There is no spread to the medial surface of the thighs.

Stage 5. The hair is distributed in an inverse triangle as in the female. It has spread to the medial surface of the thighs but not up the linea alba or elsewhere above the base of the triangle.

Stage 5 may conveniently be taken as the endpoint of adolescent growth in the pubic hair, although in early manhood the hair usually spreads beyond this triangular pattern. Some authors have used "stage 6" to indicate the spread of hair onto the abdominal wall. However, the full adult hair distribution is seldom reached before the mid-twenties. As in the case of the genitalia, we must distinguish between the moment at which a child's pubic hair reaches a given stage, e.g., PH3, and the period of time for which he is "in stage 3." Figure 9 shows the variation of age at which the various stages of pubic hair growth are reached in European boys (Marshall and Tanner, 1970; van Wieringen *et al.*, 1968). Neyzi *et al.* (1975*b*) have reported the first appearance of pubic hair at the mean age of 12.28 years in their sample of 1530 Turkish boys, although the mean was 11.8 years in the most privileged part of their sample.

7.5. Axillary and Facial Hair

Neyzi *et al.* (1975*b*) observed the first appearance of axillary hair at an average age of 13.5 in Turkish boys, i.e., about 1.3 years later than the onset of pubic hair growth. The beginning of facial hair growth at the corners of the upper lip was observed at a mean age of 14.7 years. The next stage of facial hair growth, i.e., with hair on the whole of the upper lip and some on the cheeks and chin, was not seen before the age of 16.5 years, even in the most privileged part of the sample, and they did not observe it at all in the poorer members. These are the most precise data currently available. In general it appears that axillary hair is not usually seen until the development of the genitalia is well advanced and one or two years after the first pubic hair has developed. However, this interval varies greatly from one child to another, and sometimes the axillary hair may appear before the pubic hair. The relationship between facial and axillary hair is also variable, and hair may appear at the two sites almost simultaneously. The spread of facial hair is, however, reasonably constant. Usually the first sign is an increase in length and pigmentation of the hairs at the corner of the upper lip, and this hair then spreads medially until the whole of the upper lip is covered. The upper part of the cheek and the region just below the lower lip in the midline are usually the next sites at which hair appears, followed later by the sides and lower part of the chin. Hair seldom grows on the chin before the development of both genitalia and pubic hair is complete (Tanner, 1962).

7.6. The Breast

During puberty the horizontal diameter of the male areola increases on average from about 14 mm to about 20 mm. The vertical diameter increases from approximately 11.3 to 17.0 mm (Roche *et al.*, 1971). In many boys there is also enlargement of the underlying breast tissue, sometimes to an extent which may cause discomfort or embarrassment. This, however, is usually a temporary situation, and the enlargement regresses in a year or so, although occasionally it may persist or even increase. Rarely, the enlargement of the breasts may be sufficient to cause real psychological or social difficulties and may not regress. Although this condition may warrant surgical treatment on cosmetic grounds, it is probably an extreme variation of normal rather than a pathological state (Roche *et al.*, 1971).

7.7. Variation in Rate of Progress through Pubertal Changes

Boys vary greatly, not only in the age at which they reach any given stage of development of the genitalia or the pubic hair, but also in the time they take to pass from one stage to another or through the whole sequence of changes leading to sexual maturity. If we know at what age a boy's genitalia reached stage 2, this gives us no reliable information about when he will reach stage 5 or even stage 3.

Marshall and Tanner (1970) studied the variation in the time which normal boys take to pass through the various stages of genital and pubic-hair development. The duration of any one stage can be expressed as the difference between the age at which the boy reached it (e.g., G4) and the age at which he reached the next one (G5). The duration of the whole process of genital development is the time interval between G2 and G5. The duration of pubic hair stages is similarly defined. Table I gives centiles for the variations in length of the different stages in the genitalia and pubic hair. In this table the duration of pubic-hair stage 2 (PH2–PH3) and the total duration of pubic-hair growth (PH2–PH5) is probably underestimated because the observations were made on serial photographs which probably did not reveal the first appearance of pubic hair. It should be noted that the 97.5 centile for G2–G3 is less than the 2.5 centile for G2–G5. This implies that some boys pass through the whole process of their genital development in less time than others take to progress from G2 to G3. Some boys will complete their genital development in less than two years, while others may take five or more years to complete the process. There is no clear relationship between the rate of progress through the stages of genital or pubic-hair development and the age at which they begin.

7.8. Interrelationships of the Changes at Puberty in Boys

At any one time, a boy's genitalia and pubic hair are not necessarily at the same stage of development. Also, different boys reach peak height velocity when they are

Table I. Centiles of Variation in Length of Different Stages[a]

Interval	Centiles		
	2.5	50	97.5
G2–G3	0.41	1.12	2.18
G3–G4	0.24	0.81	1.64
G4–G5	0.38	1.01	1.92
G2–G5	1.86	3.05	4.72
PH2–PH3	0.11	0.44	0.87
PH3–PH4	0.31	0.42	0.54
PH4–PH5	0.20	0.72	1.45
PH2–PH5	0.82	1.59	2.67

[a]Data from Marshall and Tanner (1970).

Table II. Percentage of Boys in Each Stage of Genital Development When First Seen in Each Pubic-Hair Stage[a]

Pubic-hair stage	Percent in genital stage				
	1	2	3	4	5
2	1	13	45	41	0
3	0	4	17	75	4
4	0	0	6	65	29
5	0	0	0	10	90

[a]Data from Marshall and Tanner (1970).

in different stages of either pubic-hair or genital development. Table II shows the percentage of boys studied by Marshall and Tanner (1970) who were in each stage of genital development when they were observed for the first time (in three monthly examinations) to be in each successive pubic-hair stage. The great majority of subjects were in genital stages 3 or 4 when the first growth of pubic hair was seen, and only 1% had any growth of pubic hair before there were visible changes in the external genitalia. Thus pubic-hair growth without genital development is very unusual, while it is entirely normal for the genitalia to reach stage 4 before there is any growth of hair. Van Wieringen *et al.* (1968) have shown that testis volume also varies greatly in boys who are in any given stage of either pubic-hair growth or development of the penis and scrotum.

8. The Reproductive Organs and Secondary Sex Characters—Female

8.1. The Ovaries

At birth, the cortex of the ovary is made of groups of primordial follicles which are separated by connective tissue. Each primordial follicle contains one primordial oocyte ringed by a layer of small undifferentiated cells, and those follicles which are furthest from the surface epithelium are the largest. At this time the medulla consists of loose fibrous connective tissue, blood vessels, and nerves, while the epithelium consists mainly of cuboidal cells. As puberty approaches the number of large follicles increases and some may grow to a considerable size before they regress. As menarche approaches an even greater number grow to a large size, but most of these will still regress and there are usually a number of anovulatory menstrual cycles before the first ovulation occurs.

As the pelvic cavity enlarges, the ovaries, uterine tubes, and uterus come to lie relatively lower in the pelvis and by the time of menarche they are in their adult position (the ovaries having been at the pelvic rim at birth). The uterine tubes increase in diameter, and more complex folds develop in the tubal mucosa where the epithelium becomes ciliated.

During the two years preceding menarche the ovaries increase considerably in size and are said to weigh between 3 and 4 g one year before menarche as compared with about 6 g at the time of the first period. The oocytes which mature during adult life are probably drawn from those already present at birth rather than from new ones formed by continuous oogenesis.

W. A. MARSHALL

8.2. The Vagina

The vagina, which is approximately 4 cm long at birth, increases only 0.5–1 cm during childhood. Further growth usually begins before the secondary sex characters appear and continues until menarche or a little later. By the time the effects of estrogen on the vaginal mucosa can be detected, the length is in the region of 7–8.5 cm and at menarche is usually between 10.5 and 11.5 cm. In childhood, the lower third of the vagina is fixed and inelastic owing to the rigidity of the musculofascial components of the pelvic floor.

At birth, the mucosa is hypertrophied and folded into high ridges as the result of the action of maternal hormones. The thickness is due mainly to the superficial layer which consists of 30–40 layers of cells which are large and irregularly shaped. There is little cornification and large quantities of glycogen are present. The mucosal hypertrophy in the newborn is more marked than that seen at any other time of life except during pregnancy. Smears taken from the vaginas of newborn infants are cytologically very similar to those of postmenarcheal girls. Later, when the effects of maternal estrogen have disappeared, the immature vaginal epithelium is similar to that over the external cervix, with a well-defined basal layer and intermediate zone but little or no superficial layer. There is no cornification. Smears taken at this stage are made up of small round cells with darkly staining cytoplasm and large nuclei. Cytological changes in the vaginal epithelium begin before there is any development of the breast or the pubic hair and are usually the first clear indication that puberty is beginning. In late childhood vaginal smears show fewer of the small cells typical of the basal layer than are found in smears taken at an earlier age. The intermediate cells are more frequent and larger, with medium-sized nuclei, and their cytoplasm stains less darkly. As the amount of estrogen in the circulation rises, the number of superficial cells in the smear becomes greater until an adult smear is obtained. Usually 5–15% of the cells are from the superficial layer before there are other signs of sexual development and before the mucosa of the vulva or distal half of the vagina shows any visible sign of estrogen action.

Immediately after birth, the pH of the vaginal contents is in the range 5.5–7.0. Within the first day or so, lactic acid, produced by lactobacillae appearing in the vaginal flora, lowers the pH to between 5 and 4 and then, as estrogen stimulation is withdrawn over the next few days, the pH becomes neutral and then alkaline. In early childhood the pH is about the same as it is at birth, although little vaginal fluid is present. A year or so before menarche the amount of fluid begins to increase and its reaction again becomes acid.

8.3. The Vulva

At birth, the genital or labioscrotal swellings which develop in the fetus are still present. They are large at first but flatten out after a few weeks. These folds are the precursors of the labia majora which do not develop as distinct structures until late childhood.

The labia minora and the clitoris are also much larger at birth in relation to the other vulval structures than they will be later, and the hymen is an inverted cone protruding outwards into the vestibule. It is quite thick but becomes thinner very soon as the stimulus of maternal estrogen is withdrawn. Its central opening at this time is usually only about 0.5 cm in diameter.

About a week after birth the effects of maternal hormone wear off, and the vulva attains the appearance which persists for about 7 years. The genital swelling and labia minora are flatter, while the clitoris does not grow as much as other structures and therefore appears smaller. During early childhood the vulval mucosa is thin, and the hymen is also thin and less protruding. Gradually, deposition of fat thickens the mons pubis and enlarges the labia majora, the surfaces of which begin to develop fine wrinkles which become more marked during the immediate premenarcheal period. The clitoris increases slightly in size while the urethral hillock becomes more prominent. The hymen thickens, and its central orifice enlarges to about 1 cm. The vestibular (Bartholin's) glands become active.

8.4. The Uterus

The uterus of the newborn is usually between 2.5 and 3.5 cm long. The cervix comprises two thirds of the whole organ, but it shrinks rapidly in the postnatal period. The external os is not completely formed.

The myometrium is thick, but the endometrium measures only 0.2–0.4 mm in depth and consists of sparse stromal cells with a surface layer of low cuboidal epithelium. The uterine glands of some infants are quite well developed. According to Ober and Bernstein (1955), 68% of 169 newborn infants had uteri with endometria in the proliferative phase, while 27% showed some degree of secretory activity and progestational changes were found in 5%. However, this activity subsides very quickly, and the endometrium remains in a quiescent state from a short time after birth until shortly before menarche.

At the age of 6 months the uterus is only about 80% of its size at birth, most of this regression having occurred at the cervix while the size of the corpus is reduced only slightly. By the fifth year the uterus has again returned to approximately its neonatal size and growth continues slowly after this, but it is not until the premenarcheal period that the uterus gains a size and shape similar to that of the adult organ. During childhood the axis of the uterus is in the craniocaudal plane, and there is no uterine flexion. There is very little growth of the corpus in early childhood, and it is not until about the tenth year that its length is approximately equal to that of the cervix. At this stage, the growth of the corpus is due to myometrial proliferation, and there is little development of the endometrium until shortly before menarche. Before this, the uterine cavity is lined by a single layer of low cuboidal cells, and there is no evidence of secretory activity.

Signs of stimulation of the cervix by maternal hormones usually disappear within the first three weeks after birth, and the endocervical canal becomes narrower. The mucosal epithelium consists of a single layer of low cuboidal cells. The squamous epithelium which covers the external surface of the cervix also becomes thinner in the first few weeks of life. The endocervical mucosa changes very little during childhood. Shortly before menarche the cervix develops its adult shape and increases in size. The cervical canal becomes larger as the cervical glands become active, although the cervix is still rather long as compared with the corpus.

Copious secretions are produced by the cervical epithelium shortly before menarche. These secretions are clear and tend to form threads. When thin preparations are dried, fernlike crystals are deposited. This type of secretion is an index of estrogen stimulation and is found in the mid-portion of the ovarian cycle in postmenarcheal girls and women.

W. A. MARSHALL

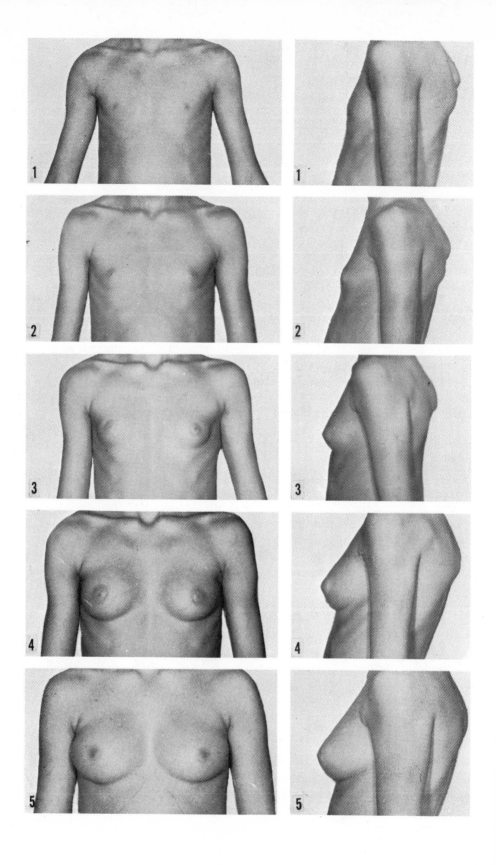

The adolescent growth of the uterus is greater in the corpus than in the cervix so that, by the time of menarche the two portions are of approximately equal length.

Little is known of the motility or secretions of the uterine tubes in premenarcheal children. Tubal movements increase in amplitude and frequency during the preovulatory phase of the ovarian cycle. After ovulation the movements decrease but secretory activity is greatest in the ovulatory and postovulatory phases of the cycle.

8.5. The Breasts

The breasts are frequently engorged at birth in both sexes and sometimes produce a whitish secretion known in folklore as "witch's milk." The engorgement and secretion disappear very quickly, and the further development of the breasts may conveniently be divided for descriptive purposes into five stages (Figure 10).

Stage 1. The infantile stage which persists from the immediate postnatal period until the onset of puberty.

Stage 2. The "bud" stage. The breast and papilla are elevated as a small mound, and the diameter of the areola is increased. The development of this appearance is the first indication of pubertal development of the breast.

Stage 3. The breast and areola are further enlarged and present an appearance rather like that of a small adult mammary gland with a continuous rounded contour.

Stage 4. The areola and papilla are further enlarged and form a secondary mound projecting above the corpus of the breast.

Stage 5. The typical adult stage with a smooth rounded contour, the secondary mound present in stage 4 having disappeared.

Some girls apparently never attain stage 5 and stage 4 persists until the first pregnancy or even beyond, while a few girls never exhibit stage 4 and pass directly from stage 3 to stage 5. The range of ages at which each stage of breast development is reached in British girls (Marshall and Tanner, 1969) is shown in Figure 9. These data are in good agreement with those of other authors who have studied similar populations. As in the case of the boy's genitalia, we must distinguish in the breast between the moment of attaining a given stage (B2, B3, etc.) and the time which is spent in that stage.

The breasts may begin to develop at any age from about 8 years onwards, and they do so in most normal girls before the 13th birthday. In the more advanced girls the breasts may be fully mature by the 12th birthday, but in others the process of development is not completed before the age of 19 and occasionally even later.

8.6. Pubic Hair

The five stages used to describe pubic-hair growth in girls are essentially the same as those for boys, but the site at which hair is first seen is usually the labia although sometimes it is the mons pubis. Figure 9 shows the variation of age at reaching each pubic-hair stage in the girls studied by Marshall and Tanner (1969), but their values for PH2 are probably too high as their photographic technique does

Fig. 10. Stages of breast development. (Reproduced from Tanner, 1962, by kind permission of the author and Blackwell Scientific Publications Ltd.)

not reveal minimal hair growth on the labia. Van Wieringen *et al.* (1968) gave a 50th centile age of 11.3 for PH2 with the 10th and 90th centiles at 9.5 and 13.0 years, respectively. These results were based on direct inspection of the children and are probably more accurate than those given by Marshall and Tanner. For the later stages of pubic hair, there is no significant discrepancy between the two studies. The range of ages within which the pubic hair may begin to grow is very similar to that in which the breasts may begin to develop but, in any individual girl, pubic-hair growth and breast development do not necessarily begin at the same time. The adult distribution of hair is usually attained between the ages of 12 and 17.

8.7. Cutaneous Glands

The apocrine glands of the axilla and vulva usually begin to function early in puberty and at about the same time. The sebaceous glands and sweat glands of the general body skin also become more active at about this time.

8.8. Interrelationship between Events of Puberty in Girls

The breasts and pubic hair do not necessarily begin to develop at the same time. Most girls' breasts have begun to develop, and some even reach stage 4, before there is any growth of pubic hair. However, the reverse situation, with the pubic hair appearing before the breasts begin to develop is entirely normal. Axillary hair usually appears when the breasts are in stage 3 or 4, but in some normal girls there may be growth of axillary hair before the breasts have begun to develop. According to Neyzi *et al.* (1975*a*), the onset of axillary hair development in Turkish girls occurred on the average about 1.5 years before menarche.

Menarche is usually a rather late event in puberty and occurs most commonly when the breasts are in stage 4, although about 25% of girls experience their first menses while they are at breast stage 3 and about 10% do not do so until their breasts are fully mature.

In girls the adolescent growth spurt begins early in relation to other changes of puberty, and it is extremely rare for a girl to menstruate before she has passed peak height velocity. Some 30–40% of girls attain PHV while their breasts are in stage 2, and their growth is slowing down throughout the remainder of puberty. It is therefore not justifiable to assume that a girl whose breasts are just beginning to develop has still the greater part of the adolescent spurt before her. She may have already experienced it. Boys are quite different from girls in this respect, in that the growth spurt occurs relatively late in relation to the development of the genitalia, and when the genitalia are beginning to develop one can assume that the whole of the adolescent spurt is yet to come. Most girls reach peak height velocity when their breasts are in stage 3, although about 10% do not do so until their breasts are in stage 4.

9. Sex Differences in Timing of Events

As we have seen above, the adolescent growth spurt occurs on the average about two years earlier in girls than in boys, so that the typical 12-year-old girl is near her maximum growth rate (PHV) whereas most boys at this age are still growing at a preadolescent rate or are in the early stages of the spurt. There are, of course, many early-maturing boys who will experience the adolescent growth spurt before late-maturing girls.

Despite the marked difference in the timing of the adolescent growth spurt, there is only a slight difference between girls and boys in the age at which the first secondary sex characteristics appear. On the average, the boys' genitalia begin to develop only a few months later than the girls' breasts and complete sexual maturity is reached at a very similar age in the two sexes. Nevertheless, from the social point of view, the girls appear to have an early puberty because their growth spurt and breast development both affect their appearance, even when they are fully clothed, while corresponding changes in boys, i.e., the development of the genitalia, are not obvious when they are clothed. Those changes which do affect the external appearance of boys, i.e., the growth spurt, the development of facial hair, and the breaking of the voice, do not occur until the genitalia are approaching maturity and hence occur at a later age than the visible changes in girls.

10. Variation among Population and Social Groups

There is a considerable amount of information about the age of menarche in different populations, but there is very little recent information about other aspects of sexual development.

The importance of genetic factors in determining the age at menarche was demonstrated by Tisserand-Perrier (1953), who found that the mean difference in menarcheal age between identical twins, in a French sample, was 2 months, while that between nonidentical twins was 8 months. In clinical practice, one often finds that children with late puberty have a history of similarly late maturity in one or both parents. However, attempts to document this relationship between age at puberty in parents and children have been largely unsatisfactory, owing to the difficulties created by the tendency for puberty to occur earlier in succeeding generations together with the fact that children are often brought up in environmental conditions greatly different from those experienced by their parents during childhood. However, when the environment is reasonably uniform, and provides all the basic requirements of good nutrition and health, the variability of menarcheal age within the population is presumably due mainly to genetic differences.

Differences between populations in age at menarche are due to the interaction of both genetic and environmental factors and may sometimes be magnified or obscured by the use of unreliable methods to estimate the age of menarche. It is therefore important that the advantages and disadvantages of the different methods be recognized.

10.1. The Cross-Sectional Retrospective Method

In this method, each subject is questioned on only one occasion and is asked the age at which she began to menstruate. This is unsatisfactory because, in many cases, the subject's recollection will be inaccurate, while some girls, particularly those who were much earlier or later than their peers, may deliberately give false answers. In a recent study (Bergsten-Brucefors, 1976) the known ages at menarche in 339 girls were compared with the ages recalled some 4 years after the event. The correlation between the actual and recalled age was $r = 0.81 \pm 0.05$. Only 63% of girls recalled the date to within ± 3 months. Damon *et al.* (1969) found a correlation of 0.78 between the actual date in 60 women and the date recalled 19 years later. The result will be further biased if the sample includes many girls who have not yet begun to menstruate. This point was illustrated by Wilson and Sutherland (1949), who showed that the difference in the mean age, calculated from recalled data, in two samples, was due to the inclusion of more premenarcheal girls in the sample having the lower mean. Further bias results from the common practice of stating one's age as that at the preceding birthday. Thus, for example, a girl who experienced menarche at the age of 12.75 years will possibly say that the event occurred when she was 12. In a large sample this kind of error would lead to the mean being underestimated by 0.5 years but, in a smaller sample, the bias is less consistent. This unsatisfactory method of data collection was used in nearly all the older surveys.

10.2. The Longitudinal Method

A group of subjects is seen repeatedly and each is asked on every occasion if she has yet begun to menstruate. If the interval between visits is short, the exact date can usually be determined with reasonable confidence. This is the most accurate method of determining the age at menarche in individuals, but a very large sample would be required to extend the process to give a reliable estimate of the population mean. In view of the financial and administrative difficulties involved, it has not yet been possible to study populations in this way.

10.3. The Status Quo Method

In order to carry out a study by this method, a suitably representative sample of the population must be selected, but each subject need only be asked her exact age at the time of the questioning and whether or not she has yet begun to menstruate. This leads to a record of the percentage of "yes" answers at each age. To this percentage distribution, a probit or logit transformation is applied, and the mean and variance can be estimated. This method does not reveal the age at menarche in any individual, but it is the best method for estimating the median and variance for the population. Figure 11 illustrates this method. The percentage of postmenarcheal girls at each age is shown, and the probits corresponding to these percentages are plotted. A line is fitted to these probits, and the age corresponding to 50% as indicated by this line is the median.

10.4. Age at Menarche in Different Populations

Table III lists the ages at menarche in various populations which have been studied in recent years. Only mean values based on the status quo method are given. The variation between populations is considerable. The relative importance of genetic and environmental factors in determining these differences is not entirely clear. Many authors have demonstrated the influence of the environment on age at menarche. Differences between socioeconomic groups are clearly important. In the Netherlands, de Wijn (1966) demonstrated a mean age at menarche of 13.8 years among girls whose fathers were in the lower and middle socioeconomic groups, while the daughters of those in the higher classes gave a value of 13.5. In the United Kingdom there is a variation of up to 6 months between different parts of the

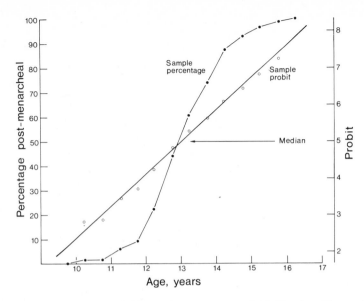

Fig. 11. Determination of age at menarche by the status quo method. The sigmoid curve indicates the percentage of girls in the sample (made at London, 1966) who had experienced menarche by each age. From this, the probits were determined and a straight line fitted through them. (Reproduced by kind permission of Professor J. M. Tanner.)

country, and this seems to be largely associated with socioeconomic conditions. In the London area, social-class differences are largely associated with differences in the number of children in the family and, once allowance is made for the number of siblings, a further social-class difference is no longer obvious. In some other countries there are wider differences in ages at menarche between social groups and these may reflect greater variation in socioeconomic conditions. For example, in northeast Slovenia, mean menarcheal ages of 13.3, 13.7, and 14.2 years were reported for girls of good, medium, and poor social standing (Kralj-Cercek, 1956). In Turkish girls, Neyzi *et al.* (1975*a*) observed that not only menarche, but all stages of breast development and pubic and axillary hair growth, occurred later in the lower socioeconomic classes as compared with the higher.

Both the number of siblings and the birth order are related to age at menarche. For a family of given size, the girls who were born later had menarche ages which were earlier by about 0.19 years per birth rank, while different family size was expressed by a delay of 0.17 years per sibling (Roberts and Dann, 1967).

The effect of social status and family size on age at menarche is probably due to a combination of several factors, of which nutrition is generally thought to be the most important. However, there is no conclusive evidence on this point because, even in children suffering from very severe undernutrition, the effects of inadequate diet are difficult to separate from the other adverse environmental conditions with which it is associated.

Modern European populations differ only slightly in their average ages at menarche, as Table III shows. In the western part of the continent most population means lie between about 12.9 years for Sweden (Ljung *et al.*, 1974) and 13.4 years for the Netherlands (van Wieringen *et al.*, 1968). Italian girls, however, reach menarche appreciably earlier. Mean values of 12.5 have been reported in a poor

Table III.   Age at Menarche in Various Populations

Country	Year	Mean ± SE	Source
Australia (Sydney)	1970	13.0	Jones *et al.* (1973)
Canada (Montreal)	1969	13.1 ± 0.04	Jenicek and Demirjian (1974)
Chile (Santiago—middle class)	1970	12.6 ± 0.12	Rona and Pereira (1973)
Cuba (Havana)	1973	12.8	Jordan *et al.* (1974)
(Rural)	1973	13.3	Jordan *et al.* (1974)
Egypt (Nubians)	1966	15.2 ± 0.3	Valsik *et al.* (1970)
England (London)	1966	13.0 ± 0.03	Tanner (1973)
Finland	1969	13.2 ± 0.02	Kantero and Widholm (1971)
Holland	1965	13.4 ± 0.03	van Wieringen *et al.* (1968)
Hungary (West)	1965	13.1 ± 0.01	Eiben (1972)
India (Lucknow)	1967	14.5 ± 0.17	Koshi *et al.* (1971)
Iraq (Baghdad—well off)	1969	13.6 ± 0.06	Shakir (1971)
(Baghdad—poorly off)	1969	14.0 ± 0.05	Shakir (1971)
Italy (Cararra)	1968	12.6 ± 0.04	Marubini and Barghini (1969)
(Rural—near Naples)	1969	12.5 ± 0.02	Carfagna *et al.* (1972)
Japan	1966–1967	12.9 ± 0.01	Yanagisawa and Kondo (1973)
Mexico (Xochimilco)	1966	12.8 ± 0.18	Diaz de Mathman *et al.* (1968)
New Guinea (Bundi—highlands)	1967	18.0 ± 0.19	Malcolm (1970)
(Kaipit—lowlands)	1967	15.6 ± 0.25	Malcolm (1969)
New Zealand (Maori)	1969	12.7 ± 0.07	New Zealand Dept. of Health (1971)
(non-Maori)	1969	13.0 ± 0.02	New Zealand Dept. of Health (1971)
Norway (Oslo)	1970	13.2 ± 0.01	Bruntland and Walløe (1973)
Poland (Warsaw)	1965	13.0 ± 0.04	Milicer and Szczotka (1966)
(Rural)	1967	14.0 ± 0.02	Laska-Mierzejewska (1970)
Senegal (Dakar)	1970	14.6 ± 0.08	Bouthreuil *et al.* (1972)
Singapore (rich)	1968	12.4 ± 0.09	Aw and Tye (1970)
(average)	1968	12.7 ± 0.09	Aw and Tye (1970)
(poor)	1968	13.0 ± 0.04	Aw and Tye (1970)
Turkey (Istanbul)	1967–1969	12.36 ± 0.01	Neyzi *et al.* (1975*a*)
U.S.A. (European origin)	1968	12.5 ± 0.11	Macmahon (1973)
(African origin)	1968	12.8 ± 0.04	Macmahon (1973)
U.S.S.R. (Moscow)	1970	13.0 ± 0.08	Miklashevskaya *et al.* (1972)

rural population near Naples (Carfagna *et al.*, 1972) and 12.6 in an urban population (Carrara) in northern Italy (Marubini and Barghini, 1969). In North America, girls of European ancestry reach menarche slightly earlier than those living in Europe. In a national sample of the United States, the mean value was 12.8 years (Macmahon, 1973) and 12.9 for French Canadians living in Montreal (Jenicek and Demirjian, 1974). In Australia and New Zealand the medians are similar to those in London, i.e., about 13.0 years (Jones *et al.*, 1973).

There are some racial differences which appear to override the effects of social conditions. Well-off Chinese girls living in Hong Kong have menarche at a median age of 12.5 (Chang, 1969), but even poor Chinese girls in the same population begin to menstruate as early as Europeans in much better economic circumstances (Lee *et al.*, 1963). Japanese living in Brazil are also early, i.e., 12.8 (Eveleth and Souza Freitas, 1969), and Japanese living in Japan are similar, a median of 12.9 ± 0.01 being calculated from the data of Yanagisawa and Kondo (1973). In New Zealand, the Maori at 12.7 are earlier than New Zealanders of European ancestry (New Zealand Department of Health, 1971).

Among Africans living in Africa, menarche is generally late, e.g., 13.4 in Uganda and 14.1 in the Nigerian Ibo (Burgess and Burgess, 1964; Tanner and O'Keefe, 1962). However, earlier means may yet be demonstrated in well-off

Africans. Afro-Americans are usually early with an average age of 12.5 (Macmahon, 1973). The Cuban population, largely of European–African mixture, averages 13.2 over the whole country, but in Havana the mean is about 12.4 years as compared with about 13.3 in rural areas (Jordan *et al.*, 1974). This implies that the late menarche among Africans in Africa may be due entirely to environmental factors, even if these are not always readily detectable. In favorable conditions, Africans may mature as early as Asiatics and southern Europeans. They may be earlier than the populations of northwest Europe.

In most Middle Eastern populations the median age of menarche is rather late, but this may be the result of economic circumstances, as well-off girls living in Istanbul have one of the earliest menarches so far recorded, 12.4 ± 0.1 (Neyzi *et al.*, 1975*a*). On the other hand, in Baghdad, even well-off girls are late at 13.6 ± 0.06 (Shakir, 1971). In India, there is a difference of nearly 2 years between the rich, who average about 12.8 (Madhavan, 1965), and the poor, who average 14.4. The latest recorded median ages of menarche are in the Melanesians of New Guinea where different studies have given values ranging from 15.5 to 18.4 years (Malcolm, 1970; Wark and Malcolm, 1969). There is a clear difference in age at menarche between town and country girls, the urban girls being earlier in every comparison so far reported (Eveleth and Tanner, 1976).

The effects of climate on age at menarche are difficult to distinguish from those of other associated factors such as race. Roberts (1969) approached this problem by using data from 39 samples studied by the status quo method. His sample included 21 from Europe, 5 from eastern Asia, 6 from India and neighboring territories in south Asia, and 7 from Africa. Multiple covariance analyses showed that, within races, the regression of menarcheal age on mean temperature was significant ($r = -0.40$). A significant relationship was also shown between menarcheal age on mean temperature and year of investigation combined ($r = +0.52$). After these effects had been taken into account, the differences between the adjusted means for racial groups remained highly significant.

10.5. Population Differences in Boys

There is very little recent information about differences in age at puberty among boys of various racial and social groups. In general, there is good agreement among western European studies. Neyzi *et al.* (1975*b*) has reported a social-class difference among boys living in Istanbul, and a very early mean onset of pubic-hair growth in Israeli boys of European ancestry was reported by Laron (1963).

11. The Trend toward Earlier Menarche

In the last century the age at menarche has become progressively earlier in Europe, North America, and several other parts of the world. The data collected over the past 100 years are naturally variable in their reliability, but when the results from several studies are combined as in Figure 12 they are remarkably consistent in indicating that menarche has been getting earlier at an average rate of about three or four months per decade. Data on children's height and weight are in keeping with the hypothesis that the adolescent spurt has also become progressively earlier. In many countries the trend toward earlier menarche is apparently continuing, but there is evidence that it may be stopping in London and Oslo (Tanner, 1973; Bruntland and Walløe, 1973).

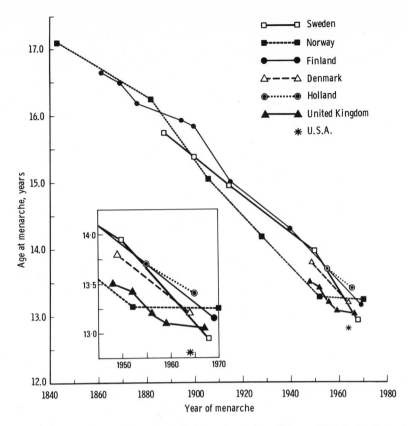

Fig. 12. Change in age at menarche 1840–1970. (Reproduced from Tanner, 1975, by kind permission of the author and W. B. Saunders Company Limited.)

12. Adolescent Sterility

Menarche does not necessarily imply fertility and neither does amenorrhea necessarily denote sterility. There are several recorded instances of conception occurring before menarche but, in most girls, ovulation does not occur until there have been a number of menstrual periods.

There are no reliable data on the relationship between menarche and the attainment of fertility in modern advanced populations and, in view of current social practices and widespread contraception, it is unlikely that such data will be forthcoming. In populations where early sexual intercourse, without contraception, is common, it is apparently unusual for pregnancy to occur until sometime after menarche. A striking example of this was reported by Gorer (1938). The Himalayan Lepchas believed that puberty in girls was the result of sexual intercourse. Girls were therefore betrothed and occasionally married from about the age of 8 years and boys from the age of 12. From this time onwards intercourse might take place regularly and yet pregnancy rarely occured before the 22nd year. Menarche in this population probably occurred about the age of 13 and the first seminal emission in boys at the age of about 15. However, there appeared to be a high over-all rate of adult sterility in this population, and it would be unwise to suggest that the results were typical of other groups. For further discussion the reader is referred to the review by Ashley Montagu (1957).

There is no clear relationship between skeletal maturation, represented as bone age, and the development of the secondary sex characters in normal children. Marshall (1974) showed that the chronological age of 74 girls, whose breasts had just begun to develop (B2), had a mean value of 11.0 with a standard deviation of 1.07, whereas the RUS skeletal age measured by the TW2 method (Tanner *et al.*, 1975) had a mean value of 11.1 with a standard deviation of 1.05. In other words, the skeletal age was just as variable as the chronological age. This was true also of a sample of 97 boys where the mean ages at G2 were 11.5 for chronological age and 11.7 for RUS age with standard deviation of 1.07 and 1.13, respectively. Figure 13 shows the distribution of skeletal and chronological age at B2 in girls. When maturity is reached, the situation is similar. In 35 girls at B5 the SD (standard deviation) of chronological age was 0.95 and that of the RUS age was 0.91. In 70

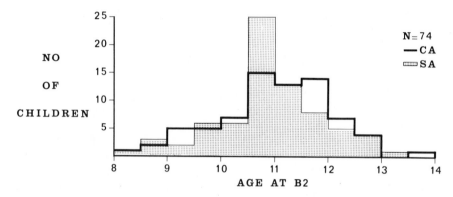

Fig. 13. Distribution of chronological age and of skeletal age in girls on reaching breast stage 2 (B2). (Reproduced from Marshall, 1974, by kind permission of the editors of *Annals of Human Biology*.)

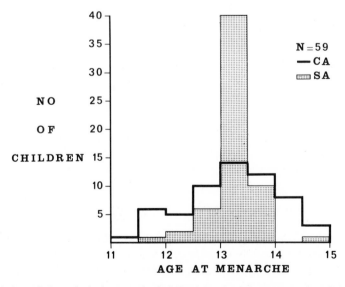

Fig. 14. Variation of chronological age and of skeletal age in girls at menarche. (Reproduced from Marshall, 1974, by kind permission of the editors of *Annals of Human Biology*.)

boys at G5 the SD of chronological age was 0.99 and that of the RUS age was 0.81. None of these small differences was statistically significant. At peak height velocity the SD of RUS age was slightly less than that of chronological age, but the difference was not statistically significant, although when children of either sex attained 95% of their mature height the SDs of their RUS ages (0.27 and 0.22 for girls and boys, respectively) were considerably less than those of chronological age (0.77 and 0.74, $P < 0.001$). At menarche (see Figure 14) the SD of chronological age in 59 girls was 0.84, whereas that of their RUS age was 0.48 ($P < 0.001$).

Clearly the widely held view that puberty begins at a more or less constant bone age is untenable as far as children who reach puberty within the normal range of variation are concerned. However, when puberty is unusually advanced or delayed, some relationship to skeletal maturation is apparent. In precocious puberty the bone age is always advanced, although this might be interpreted as the inevitable result of the circulating sex hormones which caused the early puberty rather than a demonstration of a direct relationship between puberty and bone age. When puberty is unusually delayed, as it is in many children with growth-hormone deficiency, the secondary sex characters do not usually begin to develop until the bone age is in the range within which these changes take place in normal children. Table IV shows the correlation between skeletal and chronological age in boys and girls when they reach various stages of development. All these are positive and statistically significant, indicating a tendency for those who reach puberty at an early bone age to do so at an early chronological age and *vice versa*. This is much as we would expect, although this relationship is not demonstrated in the case of 95% mature height where the samples were rather small.

The small variation of bone age, as compared with chronological age, at the attainment of 95% mature height no doubt reflects an over-all relationship between the skeletal age and the proportion of the child's total growth which has been completed. Such a relationship is implied by the usefulness of bone age in predicting adult height (Tanner *et al.*, 1975). However, by the same argument, we should not expect skeletal age to be of any help in predicting when any of the changes of puberty, other than menarche, will occur.

14. Prediction of Age at Menarche

If we know the mean age at menarche in a population and also the standard deviation, we can predict when menarche is most likely to occur in any premenarcheal girl. We can also give an upper age limit, before which she will almost certainly have begun to menstruate, if she is normal. The range of error in these predictions can be reduced if we make proper use of additional information such as the bone age. It is, however, essential to recognize that the statistical basis of the prediction changes as the age of the premenarcheal girl increases.

14.1. Prediction on the Basis of Chronological Age Alone

Let us consider a hypothetical population in which the mean age at menarche is 13.0 years and the standard deviation is 1 year. Figure 15A shows this distribution of ages at menarche. A girl in the population whose age is less than about 10.5 years may expect menarche to occur when she is at any age within the distribution, and there is a 95% probability that it will do so between the ages of 11.0 and 15.0. In the

Table IV.   *Correlation between RUS Skeletal Age and Chronological Age on Reaching Stages of Development*[a]

Stage	N	r	P
Girls			
B2	74	0.64	<0.001
B5	35	0.46	<0.01
Menarche	59	0.35	<0.01
PHV	33	0.72	<0.001
95% mature height	27	0.24	>0.10
Boys			
G2	97	0.52	<0.001
G5	70	0.45	<0.001
PHV	48	0.36	<0.01
95% mature height	33	−0.31	>0.05; <0.10

[a]Data from Marshall (1974).

absence of any additional information, such as bone age, it must be concluded that she is most likely to menstruate at the modal age for the population, i.e., 13.0 years.

When a girl has already reached an age within the distribution, i.e., over 10.5 years, but has not menstruated, the properties of a truncated normal distribution must be taken into account. In Figure 15B the shading indicates the distribution of age at menarche in girls who have reached the age of 12.0 years without experiencing menarche. Such a girl might menstruate almost immediately, but she is most likely to do so at the modal age of this distribution which is again 13.0. The 95% upper limit of the age range will, however, be slightly higher than that of the distribution in Figure 15A. Those girls who are still premenarcheal when they reach an age beyond the population mode, e.g., 14.0 years, belong to a population whose ages at menarche are distributed according to the shaded areas in Figure 15C. Statistically speaking, each of these girls is more likely to menstruate immediately than at any other later time. The age by which 95% of them will have experienced menarche will be higher than that of the population described in Figure 15B.

The method of calculating the 95% upper limit in each of the above distributions is as follows: in distributions similar to those of Figure 15A we consult a statistical table, e.g., Table IX of Fisher and Yates (1963), to find the normal deviate, i.e., the number of standard deviations from the mean corresponding to 95% of the population. This had the value 1.64. Therefore where the mean is 13.0 years and the SD 1.0 year, 95% of girls whose age is 10 or less can be expected to reach menarche by the age of 14.64.

To meet the case of premenarcheal 12-year-old girls (Figure 15), we have to estimate the percentage of the total population who are still premenarcheal at this age: 12.0 years is one standard deviation below the mean, i.e., has a normal deviate of −1 which corresponds to 16% of the population. Therefore the premenarcheal girls represent 84% of all 12-year-olds. The age by which 95% of them will have menstruated corresponds to the normal deviate representing the 16% who have already menstruated, plus 95% of the remainder, i.e.,

$$16 + \frac{(95 \times 84)}{100} = 95.8\% \text{ of the total population}$$

The normal deviate is 1.73. Therefore 95% of girls, premenarcheal at age 12, will have menstruated by the age of $13.0 + 1.73 = 14.73$ years. However, any individual at the age of 12 remains most likely to menstruate at 13.0.

W. A. MARSHALL

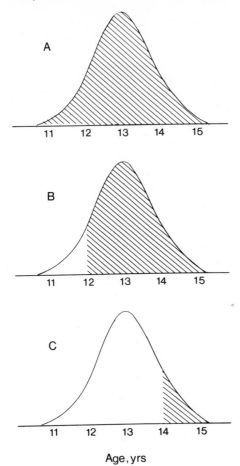

Fig. 15. Changes in distribution of possible age at menarche in premenarcheal girls of different chronological ages. (For discussion see text.) (Reproduced from Marshall and de Limogni, 1976, by kind permission of the editors of *Annals of Human Biology*.)

In those premenarcheal girls who are above the modal age (Figure 15C) and therefore most likely to menstruate immediately, the upper limit is calculated in the same way as for 12-year-olds. This method can be applied to premenarcheal girls of any age up to three standard deviations above the mean for the population.

Marshall and de Limogni (1976) used the above method to calculate the age by which menarche will have occurred in 95% of girls who are premenarcheal at given chronological ages. Figure 16 is based on these calculations and also shows the ages at which the girls in different age groups are most likely to menstruate.

14.2. Using Chronological and Skeletal Age

Information about bone age may be incorporated into the calculations by computing a multiple regression of age at menarche with bone age and chronological age. Alternatively, a simple regression of age at menarche on the difference between bone age and chronological age (BA − CA) may be calculated. The latter approach was adopted by Marshall and de Limogni (1976). It was assumed that age at menarche in girls at any given value of (BA − CA) had a gaussian distribution about the regression line with a mean represented by the corresponding age on the

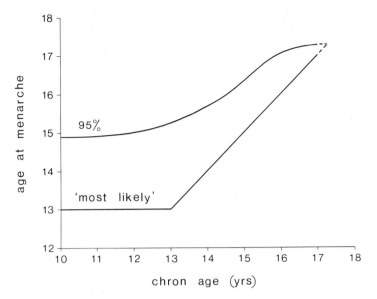

Fig. 16. Lower line: most likely age at menarche in premenarcheal girls of given chronological ages. Upper line: age by which menarche will have occurred in 95% of girls who are premenarcheal at given chronological ages. (Reproduced from Marshall and de Limogni, 1976, by kind permission of the editors of *Annals of Human Biology*.)

vertical axis, as indicated by the regression line and a standard deviation equal to the residual standard deviation (RSD) about the regression.

At any given value of (BA − CA), the properties of the truncated normal distribution must be taken into account as above. Thus a girl whose chronological age is more than 2 RSD below the regression line (A in Figure 17) may experience menarche at any age defined by the limits ±2 RSD but is most likely to at the age indicated by the regression line A^1. A girl whose age is less than 2 RSD below the line (B in Figure 17) may menstruate immediately and is most likely to do so at the age indicated by the line B^1, while the 95% upper limit is slightly higher than in the

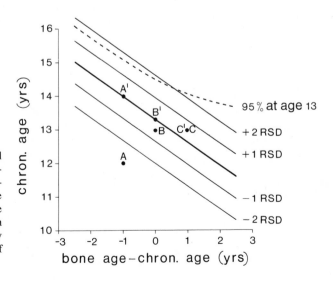

Fig. 17. Regression (with residual standard deviations) of age at menarche on (BA − CA). For explanation of examples A, B, C, and the curved line marked "95% at age 13," see text. (Reproduced from Marshall and de Limogni, 1976, by kind permission of the editors of *Annals of Human Biology*.)

case of the previous girl. A girl whose age is already greater than that indicated by the regression line C is most likely to experience menarche immediately (C[1]), while the 95% upper limit is raised and may be more than 2 RSD above the line.

The 95% upper limit for any given chronological age can be calculated at given values of BA − CA in exactly the same way as they were calculated when the bone age was not considered (above). The values at each (BA − CA) may then be joined to form a line as illustrated in Figure 17, which also shows the regression line and RSD. Figure 18 shows the age at which menarche will have occurred in 95% of girls who are premenarcheal at given values of BA − CA and given chronological ages.

Marshall and de Limogni (1976) demonstrated that, as bone age advances in relation to chronological age, the advantage of including bone age in the prediction increases. When bone age is advanced by 2 years, the error of prediction is reduced by 50% or more, and the prediction is improved in all cases except 13- and 14-year-olds with a very retarded bone age.

For most practical purposes it is less important to know the most probable age at menarche than it is to know the age by which 95% of premenarcheal girls with a given chronological age or (BA − CA) will have menstruated. This is the age by which a premenarcheal girl should have begun to menstruate and at which we should question her normality if she has not done so.

Skeletal age is not of much help in predicting age at menarche in girls below the age of 10 years. In an earlier unpublished study, Marshall found a correlation of 0.49 between bone age at age 8 and age at menarche. Simmons and Greulich (1943) found a similar correlation at age 7. However, this level of correlation is not high enough to be useful in predicting age at menarche for an individual child.

Frisancho *et al.* (1969) used the radiographic appearance of the adductor sesamoid and the fusion of the epiphyses of the distal phalanx of the second finger to estimate age at menarche. However, as fusion does not occur in this phalanx until after menarche, their technique cannot be used to predict when menarche will occur in an individual, although it may be of some value in retrospective studies of populations.

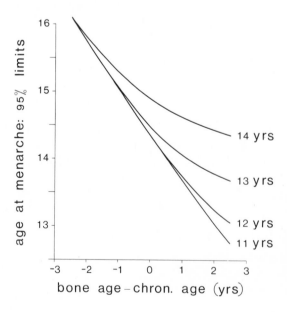

Fig. 18. Age at which menarche will have occurred in 95% of girls who are premenarcheal at given values of (BA − CA). Separate lines apply to girls of different chronological ages as indicated. (Reproduced from Marshall and de Limongi, 1976, by permission of the editors of *Annals of Human Biology*.)

The secondary sex characters are of less value than one may have hoped in predicting age at menarche. As we have seen above, Marshall and Tanner (1969) demonstrated that the interval from the beginning of breast development to menarche varies in different girls from 6 months to more than 4 years. Their observation that 26% of girls begin to menstruate in breast stage 3 and 62% do so in breast stage 4 might seem to provide a basis for saying that any girl in breast stage 4 is likely to experience menarche in the next year or so. However, this deduction would prove wrong in an appreciable number of cases because some girls remain in breast stage 4 for more than 2 years and may just have entered that stage at the time the deduction was made. Also, there remain some 10% of girls who do not menstruate until after their breasts have reached stage 5.

Frisch (1974) reported a technique for predicting age at menarche on the basis of height and weight which she used to estimate the water content of the body. The range of error in these predictions appears to be similar to that found when chronological age alone was used, as above, with due allowance for the properties of the truncated Normal distribution. There is as yet no entirely satisfactory method for predicting age at menarche but, at most ages, the most accurate method currently available is that which uses bone age, estimated by the TW2 method, in conjunction with the chronological age (Marshall and de Limogni, 1976).

15. Body Composition at Puberty

The lean body mass, i.e., the weight of the body without its fat content may be estimated by the potassium-40 method if certain assumptions are accepted (Forbes and Hursh, 1963; Forbes *et al.* 1961). The method involves counting the gamma rays which emanate from naturally occurring ^{40}K in the body.

When the lean body mass is known, the total body mass may be divided into lean and fat components. The lean body mass increases quickly during adolescence to reach a maximum of about 63 kg at age 20 in males and about 42 kg in females at age 17 (Forbes, 1972). The sex difference in lean body mass is greater than that in either height or weight, and the spurt at adolescence is much more pronounced in males. Females have a higher fat content, which increases throughout adolescence, whereas values for total body fat in males decline between the ages of 15 and 17 and then increase slightly.

Tanner (1965) measured the widths of bone, muscle, and fat in the limbs, as shown on radiographs taken by a special technique to eliminate parallax errors. Figure 19 shows the results of these measurements in 28 boys and 21 girls during puberty. In each limb, the widths of bone, muscle, and fat were measured at the same level in all subjects, and the cross-sectional area of each tissue was calculated on the assumption that the limb was circular. The values plotted in the figure were obtained by adding the measurements for calf, thigh, and upper arm to half the value for the thigh. In order to avoid the difficulties created by the adolescent changes occurring at different ages in different subjects, the average measurements for the group were calculated according to a time scale in which the zero point was the moment of peak height velocity for each child. The axis therefore shows time in years before and after PHV. This allows us to relate the changes in body composition to the adolescent growth spurt in the whole group.

In boys the rate of growth of muscle in the limbs becomes maximal at approximately the same time as growth in stature. The rate of gain in fat changes in

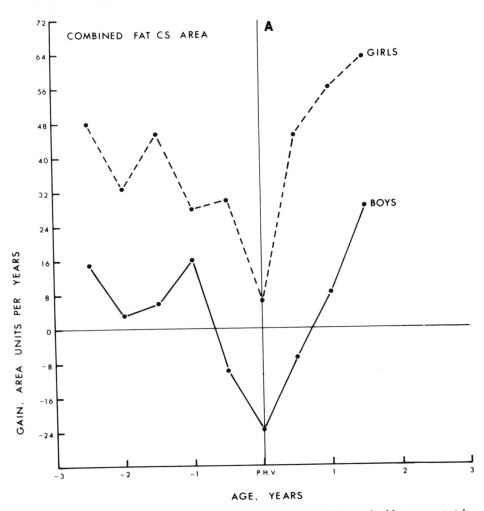

Fig. 19. Mean rate of increase in cross-sectional area of (A) fat and (B) muscle. Measurements taken from calf, arm, and thigh combined. (Reproduced from Tanner, 1965, by kind permission of the author and Pergamon Press.)

the opposite direction. Most children gain fat steadily from the age of about 8 years until adolescence, but the rate of increase in thickness of the subcutaneous fat layer becomes slower as the growth of the skeleton and muscle begins to accelerate. The minimal rate of fat gain is almost coincident with the maximum gain in bone and muscle. It should be noted in Figure 19 that the maximum rate of gain in fat in boys has a negative value, i.e., the fat is being lost. The typical boy therefore becomes thinner during adolescence, while the average girl continues to get fatter but at a slower rate during the adolescent spurt than either before or after. The decrease in rate of fat accumulation is more marked in the limbs than in the trunk.

Changes in body density have also been measured during puberty. On the basis of Archimedes' principle, the density can be calculated if the subjects are weighed under water (Heald *et al.*, 1963; Parizkova, 1961). The fact that boys gain more muscle and tend to lose fat is reflected by a greater increase in body density among boys than girls.

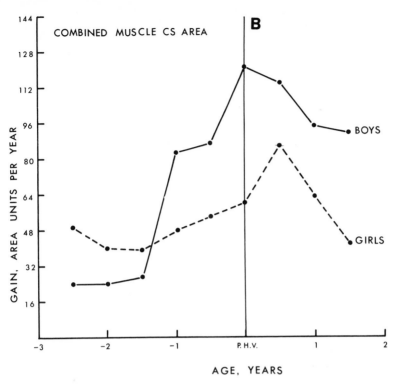

Fig. 19. (*continued*)

16. Strength and Work Capacity

As the muscles increase in size they also increase in strength. The benefit of increasing the sheer amount of muscle may be further reinforced by biochemical changes which increase the force exerted by contraction of a given amount of muscle (Morris, 1948). During the adolescent spurt, the heart and lungs become bigger not only in absolute terms but also in relation to total body size. The systolic blood pressure rises and the heart rate becomes slower, while a considerable rise in the hemoglobin concentration of the blood gives it a greater capacity for carrying oxygen. The chemical products of exercise such as lactic acid are also neutralized more efficiently. As a result of these changes the individual becomes not only stronger but able to endure hard physical work over a longer period of time. There is no justification for the popular belief that boys "outgrow their strength" in puberty. They become progressively stronger but their maximum strength may not be reached until their growth in height and sexual development are more or less complete. There may therefore be a period during which a boy looks like a mature man but has not yet developed the strength and stamina suggested by his appearance. Nevertheless he is considerably stronger than he was the previous year.

Those boys who experience the adolescent growth spurt earlier than their fellows have an advantage not only in size but also in strength and endurance. They therefore have an advantage in sports and other activities which may influence their social status in the peer group. These advantages, however, will be lost as the others approach maturity in size and strength. It is sometimes advisable to warn the early-maturing athletic boy that his physical prowess may be only temporary and that he will not always be bigger and stronger than his peers. On the other hand, late-maturing boys sometimes need reassurance to the effect that they will not always be smaller and weaker than the majority of their peers.

17. References

Aw, E., and Tye, C. Y., 1970, Age of menarche of a group of Singapore girls, *Hum. Biol.* **42**:329.

Bergsten-Brucefors, A., 1976, A note on the accuracy of recalled age at menarche, *Ann. Hum. Biol.* **3**:71.

Björk, A., 1955, Cranial base development. A follow-up X-ray study of the individual variation in growth occurring between the ages of 12 and 20 years and its relation to brain case and face development, *Am. J. Orthod.* **41**:198

Bliss, C. I., and Young, M. S., 1950, An analysis of heart measurements of growing boys, *Hum. Biol.* **22**:271.

Bouthreuil, E., Niang, I., Michaut, E., and Darr, V., 1972, Étude préliminaire de la puberté de la fille à Dakar, *Ann. Pediatr.* **19**:685.

Boyd, E., 1936, Weight of the thymus and its component parts and number of Hassall corpuscles in health and disease, *Am. J. Dis. Child.* **51**:313.

Bruntland, G. H., and Walløe, L., 1973, Menarcheal age in Norway, *Nature,* **241**:478.

Burgess, A. P., and Burgess, H. J. L., 1964, The growth pattern of East African schoolgirls, *Hum. Biol.* **36**:177.

Carfagna, M., Figurelli, E., Matarese, G., and Matarese, S., 1972, Menarcheal age of schoolgirls in the district of Naples, Italy, in 1969–70, *Hum. Biol.* **44**:117.

Chang, K. S. F., 1969, *Growth and Development of Chinese Children and Youth in Hong Kong,* University of Hong Kong, 3 volumes.

Damon, A., Damon, S. T., Reed, R. B., and Valadian, I., 1969, Age at menarche of mothers and daughters, with a note on accuracy of recall, *Hum. Biol.* **41**:161.

Deming, J., 1957, Application of the Gompertz curve to the observed pattern of growth in length of 48 individual boys and girls during the adolescent cycle of growth, *Hum. Biol.* **29**:83.

de Wijn, J. F., 1966, Estimation of age at menarche in a population, in: *Somatic Growth of the Child* (J. J. van der Werff ten Bosch and A. Haak, eds.), pp. 16–24, Stenfert-Kroese, Leiden.

Diaz de Mathman, C., Rico, V. M. L., and Galvan, R. R., 1968, Crecimiento y desarrallo en adolescentes femeninos, 2. edad de la menarquia, *Bol. Med. Hosp. Infant. (Mex.)* **25**:787.

Eiben, O. G., 1972, Genetische und demographische Faktoren und Menarchealter, *Anthropol. Anz.* **33**:205.

Eveleth, P. B., and Souza Freitas, J. A., 1969, Tooth eruption and menarche in Brazilian-born children of Japanese ancestry, *Hum. Biol.* **41**:176.

Eveleth, P. B., and Tanner, J. M., 1976, *World-wide Variation in Human Growth,* International Biological Programme Series, Cambridge University Press, London.

Fisher, R. A., and Yates, F., 1963, *Statistical Tables for Biological, Agricultural and Medical Research,* Oliver and Boyd, Edinburgh.

Forbes, G. B., 1972, Growth of the lean body mass in man, *Growth* **36**:325.

Forbes, G. B. and Hursh, J. B., 1963, Age and sex trends in lean body mass calculated from K40 measurements with a note on the theoretical basis for the procedure, *Ann. N.Y. Acad. Sci.* **110**:255.

Forbes, G. B., Gallup, J., and Hursh, J. B., 1961, Estimation of total body fat from potassium content, *Science* **133**:101.

Ford, E. H. R., 1958, Growth of the human cranial base, *Am. J. Orthod.* **44**:498.

Frisch, R. E., 1974, A method of prediction of age of menarche from height and weight at ages 9 through 13 years, *Pediatrics* **53**:384.

Frisancho, A. R., Garn, S. M., and Rohmann, C. G., 1969, Age at menarche. A new method of predicting and retrospective assessment based on hand X-rays, *Hum. Biol.* **41**:42.

Gorer, G., 1938, *Himmalayan Village,* Michael Joseph, London.

Heald, S. P., Hunt, E. E., Schwartz, R., Cook, C. D., Elliott, O., and Vajdan, B., 1963, Measures of body fat and hydration in adolescent boys, *Paediatrics* **31**:226

Hwang, J. M. S., and Krumbhaar, E. B., 1940, The amount of lymphoid tissue of the human appendix and its weight at different age periods, *Am. J. Med. Sci.* **199**:75.

Hwang, J. M. S., Lipincott, S. W., and Krumbhaar, E. B., 1938, The amount of splenic lymphatic tissue at different ages, *Am. J. Pathol.* **14**:809.

Jenicek, M., and Demirjian, A., 1974, Age at menarche in urban French Canadian girls, *Ann. Hum. Biol.* **1**:339.

Jones, E. L., Hemphill, W., and Mayers, E. S. A., 1973, Height, Weight, and Other Physical Characteristics of New South Wales Children. Part I, Children Aged 5 Years and Over, New South Wales Department of Health, G96543A, K5705.

Jordan, A., Bebelagna, A., Ruben, M., and Hernandez, J., 1974, Pubertal development and the age of menarche, Cuba, 1973, in: *Paediatrica XIV: Proceedings of the 14th International Conference of Paediatrics,* pp. 46–48, Editorial Medica Panamerica S.A., Buenos Aires.

Kantero, R. L., and Widholm, O., 1971, II. The age of menarche in Finnish girls in 1969, *Acta Obstet. Gynecol. Scand., Suppl.* **14**:7.

King, E. W., 1952, A roentgenographic study of pharyngeal growth, *Angle Ortho.* **22**:23.

Koshi, E. P., Brasad, B. G., and Bhushan, V., 1971, A study of the menstrual pattern of school girls in an urban area, *Indian J. Med. Res.* **58**:1647.

Kralj-Cercek, L., 1956, The influence of food, body build and social origin on the age at menarche, *Hum. Biol.* **28**:398.

Laron, Z., 1963, Sexual maturation in Israeli boys of European and Middle Eastern origin, preliminary study, *Acta Paediatr.* **52**:132.

Laska-Mierzejewska, T., 1970, Effect of ecological and socio-economic factors on the age at menarche, body height and weight of rural girls in Poland, *Hum. Biol.* **42**:284.

Lee, M. C., Chang, K. S. F., and Chan, M. M. C., 1963, Sexual maturation of Chinese girls in Hong Kong, *Paediatrics* **42**:389.

Lincoln, E. M., and Spillman, R., 1928, Studies on the hearts of children. II Roentgen-ray studies, *Am. J. Dis. Child.* **35**:791.

Ljung, B. O., Bergsten-Brucefors, A., and Lindgren, G., 1974, Menarche in Swedish girls, *Ann. Hum. Biol.* **1**:245.

Macmahon, B., 1973, Age at Menarche: United States, DHEW Publication No. (HRA) 74-1615, NHS Series 11, No. 133, National Center of Health Statistics, Rockville, Maryland.

Madhavan, S., 1965, Age at menarche of South Indian girls, *Indian J. Med. Res.* **53**:669.

Malcolm, L. A., 1969, Growth and development of the Kaipit children of the Markham Valley, New Guinea, *Am. J. Phys. Anthropol.* **31**:39.

Malcolm, L. A., 1970, Growth and development in the Bundi child of the New Guinea highlands, *Hum. Biol.* **42**:293.

Maresh, M. M., 1948, Growth of the heart related to bodily growth during childhood and adolescence, *Paediatrics* **2**:382.

Maresh, M. M., 1955, Linear growth of long bones of extremities from infancy through adolescence, *Am. J. Dis. Child.* **89**:725.

Marshall, W. A., 1974, Interrelationships of skeletal maturation, sexual development and somatic growth in man, *Ann. Hum. Biol.* **1**:29.

Marshall, W. A., and de Limogni, Y., 1976, Skeletal maturity and the prediction of age at menarche, *Ann. Hum. Biol.* **3**:235.

Marshall, W. A., and Tanner, J. M., 1969, Variation in the pattern of pubertal changes in girls, *Arch. Dis. Child.* **44**:291.

Marshall, W. A., and Tanner, J. M., 1970, Variation in the pattern of pubertal changes in boys, *Arch. Dis. Child.* **45**:13.

Marshall, W. A., and Tanner, J. M., 1974, Puberty, in: *Scientific Foundations of Paediatrics* (J. A. Davis and J. Dobbing, eds.), William Heinemann, London.

Marubini, E., and Barghini, G., 1969, Richerche sull'eta media di comparsa della puberta nella populazione scolare feminile di Cararra, *Minerva Pediatr.* **21**:281.

Marubini, E., Resele, L. F., Tanner, J. M., and Whitehouse, R. H., 1972, The fit of the Gompertz and logistic curves to longitudinal data during adolescence on height, sitting height and biacromial diameter in boys and girls of the Harpenden Growth Study, *Hum. Biol.* **44**:511.

Miklashevskaya, N., Solovyeva, J. S., Godina, E. Z., and Kondik, V. M., 1972, Growth processes in man under conditions of the high mountains, *Trans. Moscow Soc. Nat.* **43**:181; cited by Eveleth, P. B., and Tanner, J. M., 1976, in: *World-wide Variations in Human Growth,* Cambridge University Press, London.

Milicer, H., and Szczotka, F., 1966, The age at menarche in Warsaw girls in 1965, *Hum. Biol.* **38**:199.

Montagu, M. F. A., 1957, *The Reproductive Development of the Female,* Julian Press, New York.

Morris, C. B., 1948, The measurement of the strength of muscle relative to the cross-section, *Res. Q. Am. Health* **19**:295.

Nanda, R. S., 1955, The rates of growth of several facial components measured from serial cephalometric roentgenograms, *Am. J. Orthod.* **41**:658.

New Zealand Department of Health, 1971, New Zealand Department of Health: Physical Development of New Zealand School Children, 1969. Special Report No. 38, Health Services Research Unit, Department of Health, Wellington, New Zealand.

Neyzi, O., Alp, H., and Orhon, A., 1975a, Sexual maturation in Turkish girls, *Ann. Hum. Biol.* **2**:49.

Neyzi, O., Alp, H., Yalcindag, A., Yakacikli, S., and Orhon, A. 1975b, Sexual maturation in Turkish boys, *Ann. Hum. Biol.* **2**:251.

Ober, W., and Bernstein, J., 1955, Observations on the endometrium and ovary in the newborn, *Pediatrics* **16**:445.

Parizkova, J., 1961, Age trends in fat in normal and obese subjects, *J. Appl. Physiol.* **16**:173.

Roberts, D. F., 1969, Race, genetics and growth, *J. Biosoc. Sci. Suppl.* **1**:43.

Roberts, D. F., and Dann, T. C., 1967, Influences on menarcheal age in girls in a Welsh college, *Br. J. Prev. Soc. Med.* **21**:170.

Roche, A. F., 1953, Increase in cranial thickness during growth, *Hum. Biol.* **25**:81.

Roche, A. F., French, N.Y., and Devilla, D. H., 1971, Areola size during pubescence, *Hum. Biol.* **43**:210.

Rona, R., and Pereira, G., 1973, Genetic factors that influence age of menarche in girls in Santiago, Chile, *Hum. Biol.* **46**:33.

Savara, B. S., and Singh, I. J., 1968, Norms of size and annual increments of seven anatomical measures of the maxillae in boys from 3 to 16 years of age, *Angle Orthod.* **38**:104.

Savara, B. S., and Tracy, W. E., 1967, Norms of size and annual increments for five anatomical measures of the mandible from 3 to 16 years of age, *Arch. Oral Biol.* **12**:469.

Scammon, R. E., 1930, The measurement of the body in childhood, in: *The Measurement of Man* (J. A. Harris, C. M. Jackson, D. G. Paterson, and R. E. Scammon, eds.), pp. 173–215, University of Minnesota Press, Minneapolis.

Shakir, A., 1971, The age of menarche in girls attending schools in Baghdad, *Hum. Biol.* **43**:265.

Shuttleworth, F. K., 1939, The physical and mental growth of girls and boys aged 6 to 19 in relation to age at maximum growth, *Monogr. Soc. Res. Child. Dev.* **4** (No. 3).

Simmons, K. E., and Greulich, W. W., 1943, Menarcheal age and the height, weight and skeletal age of girls aged 7 to 17 years, *J. Paediatr.* **22**:518.

Simon, G., Reid, L., Tanner, J. M., Goldstein, H., and Benjamin, B., 1972, Growth of radiologically determined heart diameter, lung width, lung length from 5 to 19 years with standards for clinical use, *Arch. Dis. Child.* **47**:373.

Singh, I. J., Savara, B. S., and Newman, M. T., 1967, Growth of the skeletal and non-skeletal components of head width from 9–14 years of age, *Hum. Biol.* **39**:182.

Tanner, J. M., 1962, *Growth at Adolescence,* 2nd ed., p. 325, Blackwell, Oxford.

Tanner, J. M., 1965, Radiographic studies of body composition, in: *Body Composition, Symposia of the Society for the Study of Human Biology,* Vol. 7 (J. Brozek, ed.), pp. 211–236, Pergamon, Oxford.

Tanner, J. M., 1973, Trend towards earlier menarche in London, Oslo, Copenhagen, the Netherlands and Hungary, *Nature* **243**:95.

Tanner, J. M., 1975, Growth and endocrinology of the adolescent, in: *Endocrine and Genetic Diseases of Childhood and Adolescence* (L. I. Gardner, ed.), pp. 14–64, Sanders, Philadelphia.

Tanner, J. M., and O'Keefe, B., 1962, Age at menarche in Nigerian schoolgirls with a note on their heights and weights from age twelve to nineteen, *Hum. Biol.* **34**:187.

Tanner, J. M., Whitehouse, R. H., and Takaishi, M., 1966, Standards from birth to maturity for height, weight, height velocity and weight velocity; British children, 1965, *Arch. Dis. Child.* **41**:454, 613.

Tanner, J. M., Whitehouse, R. H., Marshall, W. A., Healy, M. J. R., and Goldstein, H., 1975, *Assessment of Skeletal Maturity and Prediction of Adult Height (TW2 Method),* Academic Press, London.

Tisserand-Perrier, M., 1953, Études de certains processus de croissance chez les jumeaux, *J. Genet. Hum.* **2**:87.

Turpin, R., Chassagne, P., and Lefebvre, J., 1939, La mégalothymie prépubertaire. Etude plainigraphique du thymus au cours de la croissance, *Ann. Endocrinol.* **1**:358.

Valsik, J. A., Strouhal, E., Hussein, F. H., and El-Nofely, A., 1970, Biology of man in Egyptian Nubia, *Mater. Pr. Antropol.* **78**:93.

van Wieringen, J. C., Wafelbakker, F., Vebrugge, H. P., and de Haas, J. H., 1968, *Groediagrammen nederland 1965,* Wolterns-Noordhoff, Groningen.

Wark, M. L., and Malcolm, M. A., 1969, Growth and development of the Lumi children of the Sepik district of New Guinea, *Med. J. Aust.* **2**:129.

Wilson, D. C., and Sutherland, I., 1949, The age at menarche, *Br. Med. J.* **2**:130.

Yanagisawa, S., and Kondo, S., 1973, Modernization and physical features of the Japanese with special reference to leg length and head form, *J. Hum. Ergol.* **2**:97.

Young, R. W., 1957, Post-natal growth of the frontal and parietal bones in white males, *Am. J. Phys. Anthropol.* **15:**367.

Young, R. W., 1959, Age changes in the thickness of the scalp in white males, *Hum. Biol.* **31:**74.

Zachmann, M., Prader, A., Kind, H. P., Häflinger, H., and Budlinger, H., 1974, Testicular volume during adolescence, *Helv. Paediatr. Acta* **29:**61.

Zuckerman, S., 1955, Age changes in the basicranial axis of the human skull, *Am. J. Phys. Anthropol.* **13:**521.

7

Prepubertal and Pubertal Endocrinology

JEREMY S. D. WINTER

1. Introduction

Growth is intrinsically linked with both an increment in cell mass and an acceleration of cellular proliferation. Growing mammalian cells have in common a characteristic set of biochemical phenomena, including an increase in protein synthesis, increased incorporation of nucleic acid precursors into RNA and DNA, and polysome aggregation (Hershko *et al.,* 1971). These growth-mediating processes are limited by the availability of various nutrients, metabolites, and ions in the extracellular environment, but are also susceptible to regulation by a variety of agents, including hormones. Control points for this hormonal influence upon growth include: membrane-associated processes involved with the transport of amino acids, energy substrates, and ions; transcriptional processes involved in the expression of genetic information through the synthesis of new RNA; and translational processes as reflected in the numbers and efficiency of ribosomal protein-synthesizing units.

In recent years the availability of much-improved endocrine assay techniques has focused attention upon hormones as major determinants of human growth and development. It is now possible to identify important growth-influencing peptide and steroid hormones and to correlate changes in their rate of production with clinical syndromes of disordered growth. We have gained some insight into variables which control secretion of these hormones or influence responsiveness of their target tissues, but there are still major gaps in our knowledge concerning how endocrine processes interact with each other and with genetic and nutritional factors to control the growth process. For this reason it must be stated at the outset that presently known hormonal effects do not explain most common variations or

JEREMY S. D. WINTER • University of Manitoba, Endocrine–Metabolism Section, Health Sciences Center, Winnipeg, Manitoba, Canada.

secular trends in size of healthy individuals, nor can it be shown that the majority of clinically significant disturbances of growth have their bases in deranged endocrine function.

In this review an effort has been made to provide a brief summary of the metabolic effects and patterns of secretion of an arbitrarily selected group of hormones which appear to influence human growth in a direct and significant fashion. These hormones include growth hormone, insulin, thyroxine, cortisol, and the sex steroids. In addition mention is made of prolactin, a hormone whose *raison d'être* remains somewhat obscure. A decision has been made to exclude from discussion hormones such as glucagon and the catecholamines, which play important roles in the provision of energy substrates for growth, and parathormone, vasopressin, and aldosterone, which act to maintain a stable extracellular ionic environment as a prerequisite to normal growth.

2. Growth Hormone and Somatomedin(s)

Since the first observations that pituitary extracts could correct posthypophysectomy growth failure and could even produce gigantism (Evans and Long, 1921; Smith, 1930), the importance of pituitary growth hormone in the regulation of growth has been confirmed. The first bioassay for human growth hormone was based on the growth response of tibial cartilage to pituitary extracts (Evans *et al.*, 1943). More recently radioimmunoassay techniques for the determination of growth hormone in serum have made possible the precise diagnosis of clinical states of excessive or inadequate secretion (Utiger *et al.*, 1962) and therefore appropriate therapy. Unfortunately human subjects respond only to primate growth hormones (Knobil and Greep, 1959); thus for the replacement therapy of hypopituitary children it is necessary to utilize growth hormone extracted from human pituitaries (Li and Papkoff, 1956; Aceto *et al.*, 1972; Guyda *et al.*, 1975). The typical effect of treatment with human growth hormone upon the height velocity of a child with growth-hormone deficiency is shown in Figure 1.

Human growth hormone is a single-chain polypeptide of 191 amino acids with two intrachain disulfide bridges (Niall *et al.*, 1973). Although the structure of human growth hormone is unique, it does show several areas of structural homology with other mammalian growth hormones, and there is continuing hope that a common biologically active component might be identified which would permit the use of growth hormone from domestic animals in human therapy. Endogenous growth hormone exists in plasma in two and possibly more discrete forms which differ in molecular size and possibly in biologic properties (Gorden *et al.*, 1973; Goodman *et al.*, 1974). At times of high growth-hormone secretion at least a fraction of this circulating material that is capable of reacting in a radioimmunoassay appears to be lacking in biologic activity (Sneid *et al.*, 1975).

The biochemical changes which follow administration of human growth hormone have been extensively reviewed (Daughaday and Parker, 1965; Trygstad, 1969; Root, 1972; Merimee and Rabin, 1973; Daughaday *et al.*, 1975; Talwar *et al.*, 1975). The earliest actions of growth hormone are insulinlike, involving enhanced uptake of amino acids and glucose by muscle and adipose tissue. It is likely that these effects reflect an alteration in membrane permeability, presumably a result of direct interaction between the hormone and its membrane receptors. The observation that these early direct effects of growth hormone are seen only at supraphysiologic concentrations raises serious questions concerning their significance for nor-

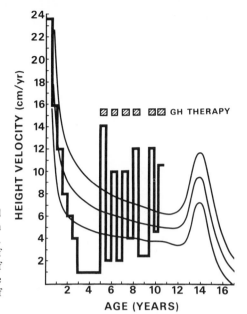

Fig. 1. Height velocity hcart of a boy with isolated growth-hormone deficiency, compared to the mean ±2 standard deviations for the normal population. Growth was maintained for the first 18 months of life. The shaded areas indicate 6-month courses of human-growth-hormone injections. Note the "catch-up" growth velocity during periods of growth-hormone replacement.

mal metabolism and growth (Adamson and Cerasi, 1975). Administration of physiologic amounts of growth hormone elicits effects which are delayed for several hours and which can be blocked by inhibitors of RNA synthesis. These delayed effects include inhibition of glucose uptake and oxidation, enhanced incorporation of amino acids into proteins, increased mobilization of fatty acids, and a secondary rise in insulin secretion. These delayed anabolic effects are not direct actions of growth hormone but appear to be mediated through the synthesis of one or more secondary growth factors, collectively termed somatomedin.

In skeletal tissue growth occurs at the point where epiphyseal chondrocytes proliferate and synthesize a matrix of collagen and sulfonated polysaccharide. These changes are stimulated by the administration of growth hormone. In their classical studies Salmon and Daughaday (1957) demonstrated that the stimulatory effect of growth hormone on tibial cartilage could be reproduced *in vitro* by normal serum but not by growth hormone itself nor by serum from hypophysectomized animals. Treatment of the hypophysectomized animal with growth hormone restored the growth-promoting ability of its serum. More recently the term somatomedin has been used to designate a family of growth-hormone-induced peptides which act collectively to promote synthesis of DNA, RNA, and protein in cartilage and to stimulate the incorporation of sulfur into cartilage glucosaminoglycans (Daughaday *et al.*, 1972). These somatomedins produce effects very similar to those of insulin, acting to stimulate glucose and amino acid transport (Kostyo *et al.*, 1973) and to promote glucose oxidation and lipid synthesis (Clemmons *et al.*, 1974). Indeed somatomedin appears to be capable of binding to insulin receptors on adipocyte, chondrocyte, and hepatocyte membranes (Hintz *et al.*, 1972); however, it seems likely that another set of "growth" receptors exist on many cells which are responsive to smaller amounts of somatomedin and are primarily responsible for its growth-promoting effects (Van Wyk *et al.*, 1975).

There is an obvious paradox between these insulinlike actions of somatomedin *in vitro* and the delayed contrainsulin effects of growth hormone *in vivo* which were described previously. It remains to be clarified whether these delayed metabolic effects are in fact mediated through somatomedin generation, or whether some other tissue growth factor is involved, or whether these effects are only the result of some contaminant in the available growth-hormone preparations. As a final caveat it should be remembered that, although somatomedin is a potent stimulator of cell growth in tissue culture systems, insufficient amounts have as yet been purified to confirm its growth-promoting activity *in vivo*.

To date three peptides with full somatomedin activity have been isolated from human plasma and designated somatomedin A, somatomedin C, and NSILA-S (nonsuppressible insulinlike activity soluble in acid–ethanol) (Van Wyk *et al.*, 1973; Hall and Luft, 1974; Froesch *et al.*, 1967). In addition another peptide, termed somatomedin B, has been shown to be growth-hormone-dependent and to stimulate [^{3}H]thymidine uptake into human glial cells and fibroblasts (Westermak and Wasteson, 1975), but it has no insulinlike activity. However, none of these peptides can substitute entirely for serum in the maintenance of *in vitro* cell growth (Chochinov and Daughaday, 1976).

These somatomedin peptides are probably produced in the liver (McConaghey, 1972; Wu *et al.*, 1974; Girard *et al.*, 1975), although muscle and kidney may contribute in a limited fashion. They all circulate in plasma bound to a carrier protein, the synthesis of which also appears to be growth-hormone-dependent (Zapf *et al.*, 1975; Cohen and Nissley, 1976).

The availability of a sensitive radioimmunoassay for human growth hormone in serum has permitted detailed study of the factors which act to regulate its secretion (Glick *et al.*, 1965). Many and varied stimuli, such as sleep, exercise, stress, ingestion of protein, a fall in circulating levels of glucose or free fatty acids, or administration of arginine, glucagon, vasopressin, or L-dopa all cause a rise in growth-hormone secretion in healthy individuals and are in general use for the diagnosis of growth-hormone deficiency (Frasier, 1974; Eddy *et al.*, 1974). All of these stimuli appear to operate through neural mechanisms, with final control of pituitary synthesis and release of growth hormone being mediated by an as yet unidentified hypothalamic growth-hormone-releasing factor (Guillemin, 1973). In addition growth-hormone release is influenced by a hypothalamic release-inhibiting hormone called somatostatin, which also acts to inhibit secretion of TSH, glucagon, and insulin (Vale *et al.*, 1973). Norepinephrine, dopamine, and possibly serotonin have been implicated as putative transmitters in the neural pathways underlying control of growth-hormone release (Imura *et al.*, 1971; Muller, 1973; Martin, 1973). Circulating growth hormone probably acts through a "short" feedback loop at a pituitary or hypothalamic level to inhibit its own secretion (Abrams *et al.*, 1971), but the existence of an additional "long" feedback loop mediated through somatomedin is suggested by the observation that children with a defect in somatomedin generation show high circulating growth-hormone levels (Laron, 1974; Van den Brande *et al.*, 1974).

Basal concentrations of growth hormone in serum from children or adults range from less than 1 to 5 ng/ml (2–10 mIU/liter). When sufficiently frequent samples are obtained, it can be shown that growth hormone is secreted in intermittent bursts, the result of intrinsic rhythmicity in the neural control mechanisms and the superimposed effects of sleep, stress, exercise, and changes in circulating levels of glucose, amino acids, and free fatty acids. Most of these secretory episodes occur during early sleep and mean serum growth hormone concentrations are highest at this time (Finkelstein *et al.*, 1972*b*; Alford *et al.*, 1973; Plotnick *et al.*, 1975).

Very high serum concentrations of growth hormone (30–180 ng/ml or 60–360 mIU/liter) are found in neonates, but these decline to basal levels within a few weeks (Shaywitz *et al.*, 1971; Finkelstein *et al.*, 1971). Plasma somatomedin activity is low at birth, possibly as a result of the suppressive effect of placental estrogens on somatomedin generation (Wiedemann and Schwartz, 1972) or possibly as a reflection of low levels of somatomedin-binding protein (D'Ercole *et al.*, 1976). During the first week of life somatomedin levels rise, but thereafter they fall together with the decline in growth-hormone levels (Tato *et al.*, 1975). The temporal organization of growth-hormone release with increased secretion during sleep becomes apparent after the third month of life (Vigneri and D'Agata, 1971).

During infancy growth hormone plays an important role in the maintenance of normoglycemia during fasting, but it may not become essential for normal growth until some time during the first year of life. Thus, as can be seen in the example in Figure 1, some patients with congenital growth-hormone deficiency may not show an obvious impairment of growth velocity until as late as 1–2 years of age. It is not clear just what endocrine factors stimulate growth during this postnatal period, but the possibility exists that somatomedin generation during early infancy is to some extent independent of growth-hormone control.

In healthy individuals plasma somatomedin activity remains fairly constant throughout childhood and adolescence (Almqvist and Rune, 1961; Kogut *et al.*, 1963; Van den Brande and Du Caju, 1973). It is likely that growth-hormone production increases gradually during this time, since the clearance rate of growth hormone is a function of body size (MacGillivray *et al.*, 1970). There is some evidence that growth-hormone secretion increases slightly during early puberty (Parlow, 1973; Finkelstein *et al.*, 1972b; Kantero *et al.*, 1975), but other investigators have not shown a corresponding pubertal rise in mean serum growth-hormone concentrations (Thompson *et al.*, 1972b; Plotnick *et al.*, 1974). It does appear that the magnitude of the growth-hormone response to stimuli such as hypoglycemia or arginine infusion is increased during puberty or after sex-steroid administration (Frasier *et al.*, 1970; Sperling *et al.*, 1970; Lippe *et al.*, 1971; Laron *et al.*, 1972a).

In children with growth-hormone deficiency, not unexpectedly, somatomedin levels are low; with growth-hormone therapy plasma somatomedin activity increases as growth accelerates. But to what extent can we explain individual size differences or growth impairment due to nonendocrine disease in terms of changes in growth-hormone secretion, somatomedin generation, or tissue response to these growth factors? In children with growth failure secondary to congenital heart disease or renal failure, growth-hormone secretion is normal but somatomedin generation may be impaired (Fahrer *et al.*, 1974; Saenger *et al.*, 1974). In both healthy children and children with nonendocrine growth disturbances there appears to be some correlation between growth velocity and plasma somatomedin activity (Hall and Filipson, 1975). Similarly, when human growth hormone is administered to healthy short children, the growth response is quite variable, being proportional to the degree of retardation of skeletal maturation (Rudman *et al.*, 1974; Lenko and Perheentupa, 1975). Thus it seems possible that changes in the efficiency of somatomedin generation or in somatomedin receptor function, rather than changes in growth-hormone secretion, may underlie many cases of nonendocrine growth retardation and may be a mechanism through which both genetic and environmental factors operate to influence body size (Beaton and Singh, 1975).

In addition to these growth-related activities, growth hormone also plays an important role in the maintenance of energy metabolism and homeostasis during a prolonged fast (Merimee and Fineberg, 1974; Haymond *et al.*, 1976). Thus chronic malnutrition results in high basal levels of growth hormone, although the response to stimulation may be impaired (Pimstone *et al.*, 1968; Samuel and Deshpande,

1972). In spite of these high growth-hormone levels, plasma somatomedin activity is low in chronic malnutrition, a finding which suggests an induced defect in somatomedin generation (Grant *et al.*, 1973). In obesity, by way of contrast, growth-hormone secretion is often impaired, but somatomedin activity remains normal (Van den Brande and Du Caju, 1973; Chaussain *et al.*, 1975). Additional evidence that in some situations somatomedin generation and growth itself may occur in the absence of growth hormone is provided by the several observations of sustained normal growth and somatomedin activity in hyperphagic, obese, growth-hormone-deficient children (Finkelstein *et al.*, 1972a; Kenny *et al.*, 1973; Costin *et al.*, 1976).

3. Insulin and Insulinlike Peptides

Awareness of the insulinlike actions of somatomedin has refocused attention upon insulin itself as a hormone necessary for human growth. It is well known that insulin stimulates protein synthesis (Manchester, 1972) and can stimulate growth of cartilage or other cells *in vitro* (Salmon and DuVall, 1970; Schwartz and Amos, 1968). While insulin is not an obligate mediator of the actions of growth hormone, full expression of the latter's anabolic effects requires the presence of insulin (Milman *et al.*, 1951). In turn growth hormone stimulates the production of insulin as well as that of somatomedin (Curry and Bennett, 1973). Both insulin and somatomedin decrease adenyl cyclase activity (Tell *et al.*, 1973). It is possible that these hormones act synergistically to regulate cell growth not only by their effects upon nutrient supply but also through changes in intracellular concentrations of cyclic nucleotides (Otten *et al.*, 1972; Kram *et al.*, 1973; Kram and Tomkins, 1973).

Neither glucose tolerance nor the insulin response to glucose administration change significantly through childhood and puberty (Loeb, 1966; Pickens *et al.*, 1967; Cole, 1973). A progressive rise in mean fasting serum insulin concentrations from 2 mIU/liter in the first year of life to 16 mIU/liter during puberty was found in one study (Grant, 1967) but was not confirmed by others (Slone *et al.*, 1966). There is no information available as yet concerning possible changes in insulin-receptor concentrations or target-cell sensitivity during growth (Roth *et al.*, 1975).

Several clinical situations exist in which insulinopenia is associated with poor growth or hyperinsulinism with increased growth (Laron *et al.*, 1972b). In children with growth-hormone deficiency insulin levels and responses to stimulation are depressed (Sizonenko *et al.*, 1975). Uncontrolled diabetes mellitus is accompanied by growth retardation with low serum somatomedin activity even though growth hormone levels may be increased (Birkbeck, 1972; Hansen and Johansen, 1970; Phillips and Young, 1976). In severe malnutrition with growth retardation insulin secretion is significantly reduced (Rao and Raghuramulu, 1972; Lunn *et al.*, 1973). Conversely obesity results in both hyperinsulinism and enhanced growth (Cacciari *et al.*, 1972; Sims and Danforth, 1974; Garn and Haskell, 1960). In Figure 2 is shown the height velocity chart of a girl with isolated growth-hormone deficiency; it is possible that her above-average growth during the first 3 years of life was a result of her hyperphagia, extreme obesity, and resultant hyperinsulinism, since caloric restriction produced an immediate cessation of growth that could only be corrected by the administration of human growth hormone. Recent evidence supports the suggestion that hyperinsulinism may stimulate somatomedin generation in the absence of growth hormone (Daughaday *et al.*, 1976).

True insulin accounts for only about 7% of the total insulinlike activity in serum that can be measured by bioassay (Froesch *et al.*, 1963). The remainder, which

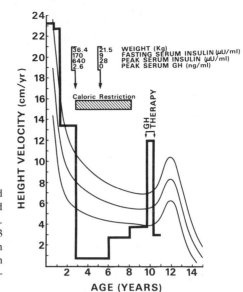

Fig. 2. Height velocity chart of a girl with isolated growth-hormone deficiency, who also showed hyperphagia, extreme obesity, and hyperinsulinism. Above-average height growth was maintained until 3 years of age; moderate caloric restriction resulted in almost complete cessation of growth coincident with return of insulin levels to normal. Note the "catch-up" growth response to growth-hormone therapy.

cannot be blocked by the addition of antiinsulin antibodies, is termed the nonsuppressible insulinlike activity (NSILA) and represents the effects of a family of peptides, some of which are growth-hormone-dependent. Obviously the two somatomedins (A and C) are components of this total activity. A third peptide, which has been designated NSILA-S, has similar properties and probably should be considered another somatomedin. This peptide inhibits lipolysis, stimulates glucose uptake and glycogen synthesis by muscle (Froesch *et al.,* 1967), inhibits adenylcyclase activity (Hepp, 1972), stimulates uptake of sulfate and thymidine by cartilage (Zingg and Froesch, 1973; Raben *et al.,* 1972), promotes growth of fibroblasts in tissue culture (Morell and Froesch, 1973), and is increased acutely after administration of growth hormone (Megyesi *et al.,* 1975).

The list of biological growth factors isolated from serum or other tissues is growing yearly. Some of these are growth-hormone-dependent and some show insulin-like activity; all will promote growth of particular cells in certain circumstances. Such a list would include MSA or multiplication-stimulating factor (Pierson and Temin, 1972; Smith and Temin, 1974), epidermal growth factor (Cohen and Savage, 1974), nerve growth factor (Bradshaw *et al.,* 1974), fibroblast growth factor (Jones and Addison, 1975), and erythropoietin (Jepson and McGarry, 1972).

4. Prolactin

The structure of human prolactin shows considerable homology with that of growth hormone, and from time to time the possibility that prolactin may be an anabolic hormone in man is raised. Phylogenetically, prolactin is the oldest of the pituitary hormones, and perhaps for this reason it displays an incredible diversity of actions among different vertebrates. Thus it plays a role in osmoregulation in teleosts, hepatic lipogenesis in birds, and gonadal function in rodents (Nicoll and Bern, 1972). The exact spectrum of activities in any species presumably depends upon the evolutionary distribution of specific prolactin receptors in various target organs (Turkington and Frantz, 1972).

In man prolactin clearly plays a role in lactation and may also have some modulating effects upon renal function and steroidogenesis (Turkington, 1972; McNeilly, 1974). Although some animal prolactins may mimic some of the metabolic effects of growth hormone in hypopituitary children, no definite effect of human prolactin upon linear growth has yet been demonstrated (Blizzard *et al.*, 1966; McGarry and Beck, 1972). But to confuse this area even more, a recent report has suggested that prolactin may stimulate production of somatomedin by rat liver (Francis and Hill, 1975), and thus the issue of some influence upon growth is not closed.

Serum concentrations of prolactin are high in neonates (approximately 200 μg/ liter), but fall during the first few weeks of life to basal levels of about 5 μg/liter (Guyda and Friesen, 1973; Aubert *et al.*, 1974). At all ages prolactin secretion is increased during sleep (Finkelstein *et al.*, 1974b; Parker *et al.*, 1974). In boys there is no change in mean prolactin levels with adolescence, but in girls there is a gradual rise through puberty to adult female concentrations of about 10 μg/liter (Lee *et al.*, 1974; Ehara *et al.*, 1975). Serum prolactin concentrations are not influenced by the presence of growth-hormone deficiency.

5. Thyroid Hormones

Two calorigenic hormones, L-thyroxine and L-triiodothyronine, are secreted by the thyroid gland. In addition a net increase in the metabolic effect of this secretion is achieved by the peripheral conversion of a significant fraction of the circulating thyroxine to the more potent triiodothyronine (Braverman *et al.*, 1970).

A deficiency of thyroid hormones results in a general retardation of growth and organ maturation with a striking reduction in rates of cell division in all proliferating tissues (Shellabarger, 1964; Brasel and Winick, 1970; Greenberg *et al.*, 1974). Replacement therapy with thyroxine is followed by enhanced protein synthesis and a recovery of statural growth (Graystone and Cheek, 1968). An example of the striking "catch-up" growth which can occur after treatment of hypothyroidism is shown in Figure 3. In some circumstances an excess of thyroid hormones may be associated with an acceleration of growth and skeletal maturation (Hung *et al.*, 1962). Hermosa and Sobel (1972) have suggested that pharmacologic amounts of thyroxine might stimulate growth in short euthyroid children, but the efficacy and complications of this approach have not been established.

Although the clinical correlates of thyroid deficiency or excess are well recognized, less is known about the cellular basis for the anabolic effects of thyroid hormones. Both thyroxine and triiodothyronine can stimulate cell growth in tissue culture, but excessive amounts inhibit growth (Siegel and Tobias, 1966; Bartfeld and Siegel, 1968). Most evidence supports a nuclear locus for these growth-related effects with an action at a transcriptional level, but there is also evidence for an additional effect of thyroid hormones directly upon mitochondria, concerned with energy metabolism and thermogenesis (Oppenheimer, 1973).

Control of thyroid function is mediated via hypothalamic secretion of thyrotropin-releasing hormone (TRH) which in turn stimulates release of pituitary thyroid-stimulating hormone (TSH). There is considerable opportunity for interaction between this system and other endocrine control systems, since TRH also stimulates secretion of prolactin (Foley *et al.*, 1972), while secretion of TSH may be inhibited not only by the feedback effects of thyroxine, but also by somatostatin

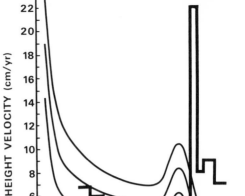

Fig. 3. Height velocity chart of a girl with chronic lymphocytic thyroiditis who became hypothyroid around age 5 years. Note the striking "catch-up" growth after initiation of thyroxine replacement.

(Siler *et al.*, 1974), growth hormone (Root *et al.*, 1973), and glucocorticoids (Faglia *et al.*, 1973).

There is some evidence that thyroid hormones enhance the growth-promoting effect of growth hormone (Bergenstal and Lipsett, 1960; Thorngren and Hansson, 1973), while in the absence of growth hormone thyroxine may only stimulate skeletal maturation without proportionate linear growth (Van den Brande *et al.*, 1973). Secretion of growth hormone is often impaired in hypothyroid children (Fink, 1967; MacGillivray *et al.*, 1968; Katz *et al.*, 1969), but it is not clear to what extent this may contribute to the growth failure that occurs. Surprisingly, growth-hormone secretion may also be impaired in hyperthyroidism, a condition in which linear growth may even be enhanced (Finkelstein *et al.*, 1974a).

During normal childhood one observes changes in growth velocity, basal metabolic rates, and various other metabolic parameters (Lewis *et al.*, 1943), but the relationship of these to alterations in thyroid function is not obvious. A transient period of thyroid hyperactivity in response to increased secretion of TRH and TSH is characteristic of the neonate (Erenberg *et al.*, 1974; Czernichow *et al.*, 1975). Thereafter, as can be seen in Table I, mean serum concentrations of thyroid hormones remain during childhood at or slightly above the levels seen in adults. Some evidence suggests a gradual decline in both serum free thyroxine concentrations and in fractional thyroxine turnover, with both reaching their lowest levels at the beginning of puberty (Malvaux *et al.*, 1966; Lamberg *et al.*, 1973b; Oddie *et al.*, 1966). Some authors have described a rise in mean serum TSH concentrations during early puberty (Golstein-Golaire and Delange, 1971; Lamberg *et al.*, 1973a), but the values reported in these studies suggest problems of assay specificity and require confirmation using the more sensitive and specific TSH assays that have recently become available. In spite of the known effects of sex steroids upon thyroxine-binding globulin, puberty does not appear to cause significant changes in the total serum binding and thus the metabolism of thyroid hormones (Braverman *et al.*, 1966).

JEREMY S. D. WINTER

Table I. Serum Concentrations of Thyroid Hormones in Healthy Subjects[a]

Age	TSH (mIU/liter)	Thyroxine[d] (μg/dl)	Triiodothyronine[d] (ng/dl)	T3 resin uptake (% of standard)	Free T4 (ng/dl)	Free T3 (pg/dl)
cord	8.5 (1.5–31.5)	11.0 (7.3–15.3)[b]	48 ± 3[c]	0.9 (0.75–1.05)	2.2 ± 0.1	130 ± 10
1–5 days	7.3 (<1.5–12.1)	15.5 (10.1–20.9)	140 ± 16	1.15 (0.9–1.4)	4.9 ± 0.3	410 ± 20
1–8 wk	<6	12.4 (7.1–16.6)	163 ± 6	0.95 (0.85–1.1)	2.1 ± 0.1	400 ± 20
2–12 mo	<6	9.6 (5.5–15.0)	124 ± 8	0.9 (0.8–1.1)		
1–6 yr	<6	9.1 (5.6–12.6)	138 ± 3	0.95 (0.8–1.1)		
6–10 yr	<6	8.3 (4.9–11.7)	141 ± 7	0.95 (0.8–1.1)		
10–16 yr	<6	7.2 (3.8–10.6)	131 ± 6	0.95 (0.8–1.1)		
16–20 yr	<6	7.5 (4.1–10.9)	120 ± 5	0.95 (0.8–1.1)		
Adult	<6	7.9 (4.7–11.1)	138 (80–170)	1.0 (0.85–1.15)	1.9 (1.5–4.0)	290 (230–450)

[a]Data summarized from: Abuid et al., 1974; Rubinstein et al., 1973; Avruskin et al., 1973; O'Halloran and Webster, 1972; Hays and Mullard, 1974; Fisher, 1973; Mayberry et al., 1971; Havek et al., 1973.

[b]Mean (range in brackets).

[c]Mean ± standard error.

[d]Levels of thyroxine and triiodothyronine are given in the units that are presently in general clinical use; to convert these to SI units multiply thyroxine values by 0.0129 (μmol/liter) and triiodothyronine values by 0.015 (nmol/liter). Recent data of Corcoran et al. (1977) suggest that serum concentration of thyroid hormones in children may be higher than has been previously reported, with mean serum thyroxine levels of 10.0 ± 2.5 μg/dl (SD) and serum triiodothyronine levels of 194 ± 35 ng/dl below the age of 10 years.

During severe illness or chronic malnutrition, TSH secretion may be reduced and peripheral conversion of thyroxine to triiodothyronine may be impaired (Carter et al., 1976; Wartofsky, 1977; Chopra and Smith, 1975; Miyai et al., 1975). The significance of these changes for the growth retardation associated with these conditions has not been established, however.

6. Sex Steroids

It is necessary to devote a proportionately larger segment of this review to a discussion of the sex steroids, both because of their very real significance for growth and development and because of the relative complexity of the changes which are observed in their secretion patterns through childhood and adolescence. The gonads, and to a lesser extent the adrenals, synthesize several sex steroids (Figure 4) which may be classified by structure and function as the progestins (21-carbon steroids), the androgens (19-carbon steroids), and the estrogens (18-carbon steroids). The ultimate concentrations of these steroids in serum at any moment depend upon their respective rates of gonadal and adrenal secretion plus the effects of peripheral interconversion, binding to serum proteins, and varying rates of metabolism and excretion.

In adult men the principal circulating androgen is testosterone, 95% of which is secreted by the Leydig cells of the testis (Baird et al., 1969). Most of the estradiol found in male serum is derived by peripheral conversion from testosterone, but some is also secreted by the testis. Androgen production in the male is primarily under the influence of pituitary luteinizing hormone (LH); follicle-stimulating hormone (FSH) may have a synergistic role in steroidogenesis, but its major actions are upon testicular growth and spermatogenesis. Control of both these pituitary gonadotropins appears to be mediated via hypothalamic neurosecretion of a decapeptide termed LH-releasing hormone (LH-RH).

In adult women the major estrogen is estradiol, which is secreted during the first half of the menstrual cycle by the theca interna of the ripening follicle and after

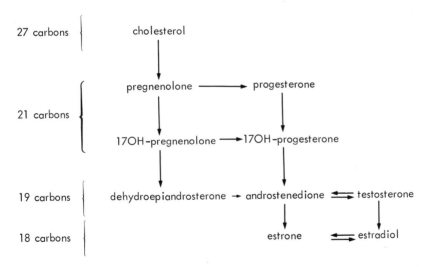

Fig. 4. A simplified scheme of sex steroid biosynthesis to show the sequential structural relationships among the progestins (C_{21} steroids), androgens (C_{19} steroids), and estrogens (C_{18} steroids).

ovulation by the corpus luteum. Most of the circulating testosterone and much of the estrone in women is derived from peripheral conversion of androstenedione, which is secreted by both the ovaries and the adrenals (Baird *et al.*, 1969). Progesterone, the major progestin, is produced in large amounts by the corpus luteum during the latter half of the cycle. FSH and LH appear to act synergistically to promote follicular estrogen production, but LH is the primary stimulus to ovulation and corpus luteum formation. The pattern of these hormone changes during one menstrual cycle is illustrated in Figure 5.

In the circulation the bulk of the sex steroids are bound to either albumin or to specific high-affinity binding globulins. Changes in the plasma concentrations of these globulins cause reciprocal effects on the metabolic clearance of steroids, an indication that it is primarily the unbound and albumin-bound fractions that are available for interaction with target tissues to induce growth, differentiation, or the synthesis of specific proteins (Baulieu *et al.*, 1971). However, except in the first few days of life, it appears that levels of binding of sex steroids remain relatively unchanged during childhood and puberty (Radfar *et al.*, 1976).

The ultimate physiologic effect of a sex steroid requires selective uptake and binding to specific cytoplasmic receptor proteins which occur in the appropriate hormone-responsive tissues. This steroid–protein complex is transported to the nucleus where it interacts with regions of the genome which in turn serve as a

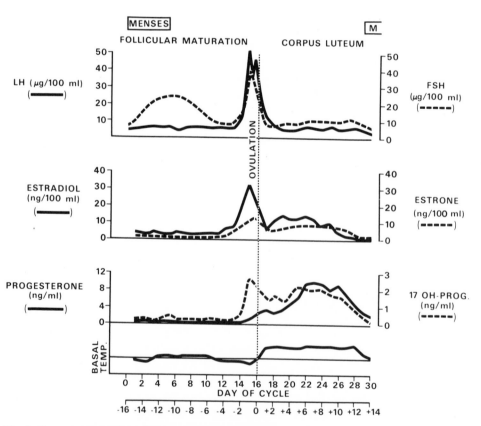

Fig. 5. Serum concentrations of FSH and LH (in μg LER907/dl) and of ovarian sex steroids through a menstrual cycle, in relation to the basal body temperature graph and menses.

template for the synthesis of new messenger RNA. In most androgen-dependent tissues testosterone itself is not transported to the nucleus but is first reduced to 5α-dihydrotestosterone (Wilson, 1972). Several excellent reviews have recently appeared describing in detail the mechanisms of action of the androgens (Minguell and Sierralta, 1975), the estrogens (Jensen and DeSombre, 1972), and the progestins (O'Malley and Strott, 1973).

Maturation of the pituitary–gonadal axis and of the neural mechanisms which modulate gonadotropin release begins before birth, when sex steroids play a fundamental role in the phenotypic expression of sex. By the time of birth the infant is capable of producing significant amounts of gonadotropins and sex steroids, but already basal gonadotropin secretion is under the control of a classic negative-feedback mechanism by which any increase in circulating sex steroid concentrations induces an inverse change in gonadotropin secretion, both through direct inhibition of release of LH-RH and through modulation of the effect of LH-RH upon the pituitary gland (Wang and Yen, 1975; Jaffe and Keye, 1975). Shortly after birth, presumably in response to the withdrawal of placental estrogens, a brisk surge of gonadotropin secretion occurs in both sexes (Winter *et al.,* 1975). This stimulates a transient rise in androgen production in male infants (Forest *et al.,* 1974; Winter *et al.,* 1976), but the negative-feedback axis is intact and remarkably sensitive, and serum levels of both gonadotropins and sex steroids return to barely detectable levels by a few months of age. In female infants an ovarian estrogen response to this gonadotropin rise may also occur but to a more variable degree (Bidlingmaier *et al.,* 1973; Winter *et al.,* 1976); for this reason serum FSH levels may not return to the basal range until as late as four years of age.

It is likely that some low-level gonadal steroid production continues through childhood, since in the absence of a normal gonad high levels of serum FSH and LH are seen at all ages (Winter and Faiman, 1972*a*). Although prepubertal testes do not show obvious Leydig cells, histochemical evidence of continuing steroidogenesis is apparent in some interstitial cells of boys (Wolfe and Cohen, 1964). In prepubertal girls there is also evidence that some gonadotropin-mediated estrogen secretion may continue through childhood, possibly with an influence upon uterine growth (Ross, 1974). In our laboratory we have preliminary data showing that castration of prepubertal monkeys results in a decline in serum concentrations of sex steroids and a corresponding rise in serum gonadotropin levels (Ellsworth *et al.,* 1975). However, it has as yet not been demonstrated that sex steroids play any important part in general body growth before the time of puberty.

The physical changes of adolescence described in Chapter 6 result from increased production of gonadotropins and sex steroids, which in turn reflects a series of complex maturational processes involving the hypothalamus, pituitary, gonads, and adrenal glands. Mean serum concentrations of pituitary gonadotropins and sex steroids during childhood and adolescence are shown in Table II. The first detectable rise in mean sex steroid levels occurs at about age 11 in girls and age 12 in boys. Similar evidence of increasing gonadotropin and sex steroid production through puberty has been obtained in studies of urinary hormone excretion (Raiti *et al.,* 1969; Baghdassarian *et al.,* 1970; Dalzell and El Attar, 1973; Fellier *et al.,* 1974; Waaler *et al.,* 1974; Gleispach, 1974; Widholm *et al.,* 1974). Unfortunately these cross-sectional studies obscure rapid incremental changes in hormone levels because of individual differences in the timing of the events of puberty.

In Figure 6 is shown the mean serum FSH, LH, and testosterone concentrations as a function of age as seen in a mixed longitudinal study of 56 healthy boys

Table II. Mean Serum Concentrations of Gonadotropins and Sex Steroids in Children and Adolescents [a,b]

Age	FSH (μg LER907/dl)	LH (μg LER907/dl)	Testosterone (ng/dl)	Dihydrotestosterone (ng/dl)	Androstenedione (ng/dl)	Dehydroepiandrosterone (ng/dl)	Estrone (ng/dl)	Estradiol (ng/dl)	Progesterone (ng/dl)	17OH-Progesterone (ng/dl)
Males										
5–7 days	3.1 (3–21.5)	4.2 (3.1–5.2)	20 (14–25)		36 (17–44)	230 (54–685)	1.2 (0–2.0)	1.9 (0–2.3)		98 (63–144)
8–60 days	8.6 (3–55.0)	5.2 (2.2–15.6)	165 (53–360)		27 (3–98)	70 (10–330)	1.3 (0–2.4)	<1 (0–3.1)		138 (22–210)
2–12 months	5.0 (3–35.0)	2.1 (0.8–4.0)	60 (3–308)		14 (2–68)		1.0 (0–1.6)	<1 (0–1.9)		51 (7–137)
1–4 years	6.0 (3–16.0)	1.5 (0.8–3.3)	8 (3–12)	2 (0–7)	5 (3–27)	31 (12–79)	3.0 (1.8–5.3)	<1 (0–2.7)		16 (4–76)
4–8	9.0 (3–13.5)	1.6 (0.8–2.9)	5 (3–13)	3 (0–7)	12 (7–19)	47 (12–233)	2.9 (1.7–4.8)	<0.5 (0–0.9)		30 (12–70)
8–10	10.3 (6.0–21.3)	1.9 (1.1–3.0)	9 (3–26)	3 (0–7)	18 (13–31)	85 (21–283)	3.3 (2.0–5.4)	<0.5 (0–1.1)		36 (18–75)
10–12	12.6 (6.1–22.6)	2.0 (1.0–3.0)	13 (3–48)	4 (3–7)	45 (31–65)	135 (46–505)	3.2 (2.1–4.9)	<0.5 (0–1.5)		39 (12–75)
12–14	13.6 (8.9–23.5)	2.5 (1.4–4.9)	78 (10–437)	13 (9–20)	62 (45–99)	229 (84–583)	2.7 (1.7–4.4)	1.0 (0–2.5)	39 (20–60)	42 (15–219)
14–16	15.8 (11.5–25.6)	3.1 (2.1–4.9)	340 (150–772)	32 (20–52)	80 (48–140)	333 (175–634)	3.9 (2.1–7.0)	1.5 (0–2.4)	34 (20–50)	80 (12–200)
16–18	15.7 (11.5–27.0)	4.1 (2.4–5.8)	532 (264–900)	50 (20–75)	108 (70–140)	380 (180–600)	4.6 (2.5–7.0)	1.8 (0.9–3.0)	56 (30–75)	97 (42–200)
Females										
5–7 days	5.3 (3–70.0)	2.3 (1.3–3.5)	14 (11–28)		26 (16–57)	344 (99–696)	1.2 (0–2.0)	1.0 (0–3.2)		105 (58–128)
8–60 days	18.1 (3–88.3)	3.3 (1.3–8.5)	11 (3–24)		25 (4–68)	95 (20–220)	1.2 (0–2.0)	1.5 (0–5.0)		86 (19–214)
2–12 months	13.3 (3–160.0)	2.8 (0.8–7.3)	5 (3–17)		13 (2–23)		1.0 (0–1.6)	1.1 (0–7.5)		43 (4–147)
1–4 years	9.3 (3–24.6)	1.7 (0.8–3.5)	8 (3–15)	1 (0–3)	8 (2–16)	29 (19–42)	3.0 (1.9–4.6)	1.1 (0–2.0)	*	26 (20–39)
4–8	6.9 (4.0–17.4)	2.0 (0.8–3.3)	6 (3–14)	3 (0–5)	18 (12–27)	54 (12–187)	2.7 (1.7–4.4)	<0.5 (0–1.7)	5 (3–23)	28 (9–63)
8–10	7.0 (4.1–13.1)	2.3 (1.2–5.0)	8 (4–20)	6 (4–10)	32 (22–47)	82 (24–289)	4.6 (3.1–7.0)	<0.5 (0–3.1)	6 (3–32)	35 (15–90)
10–12	11.2 (5.6–21.8)	2.5 (1.0–6.3)	18 (5–47)	8 (5–12)	65 (42–100)	261 (50–916)	4.4 (2.8–6.8)	1.6 (0–6.3)	8 (3–55)	42 (18–111)
12–14	13.8 (7.2–22.4)	4.5 (1.1–12.3)	26 (11–60)	12 (7–19)	123 (80–190)	473 (93–2000)	5.9 (3.7–14.0)	4.2 (0.5–12.2)	37 (10–450)	84 (24–234)
14–16	18.9 (6.5–51.3)	8.7 (2.4–45.0)	34 (16–72)	14 (10–30)	133 (77–224)	555 (250–2000)	8.9 (2.1–14.0)	8.6 (1.0–25.2)	189 (5–1400)	94 (27–234)
16–18	22.2 (8.7–53.0)	9.4 (5.3–45.0)	41 (20–65)	15 (10–32)	151 (80–300)	550 (250–2000)	6.7 (1.4–14.0)	8.4 (1.0–28.4)	161 (23–1500)	108 (34–240)

[a] Data summarized as mean (range in brackets). The data in this table and in Table III are derived mainly from cross-sectional studies of healthy children carried out in Winnipeg (Winter and Faiman, 1972b; Winter and Faiman, 1973a; Faiman and Winter, 1974; Winter et al., 1975, 1976; Hughes et al., 1976). These have been supplemented by data from the literature (Blizzard et al., 1972; Forest et al., 1974; Jenner et al., 1972; Bidlingmaier et al., 1973; Nankin et al., 1975; Sizonenko and Paunier, 1975; Hopper and Yen, 1975; Lee and Migeon, 1975; Gupta et al., 1975; Angsusingha et al., 1974; Penny et al., 1970; Korth-Schutz et al., 1976; de Peretti and Forest, 1976).
[b] Values of FSH and LH can be converted to IU/liter by multiplying by 0.5 and 4.5, respectively. Steroid levels can be converted to nmol/liter by multiplying by the following factors: testosterone (0.0347); dihydrotestosterone (0.0344); androstenedione (0.0349); dehydroepiandrosterone (0.0347); estrone (0.0370); estradiol (0.0367); progesterone (0.0318); and 17OH-progesterone (0.0303).

(Faiman and Winter, 1974). It can be seen that FSH and LH levels rose in parallel between age 10 and 15, while a more rapid increment in serum testosterone occurred between ages 12 and 14. Similar patterns were seen in a longitudinal study of serum FSH, LH, and estradiol concentrations in 58 healthy girls (Faiman and Winter, 1974) and are shown in Figure 7.

When serum hormone concentrations are analyzed in terms of pubertal staging (Table III), we can begin to see the relationship between the endocrine and the clinical phenomena of puberty. In boys (Figure 8), serum FSH levels begin to rise during the year before more rapid testicular enlargement signals the clinical onset of puberty. Shortly thereafter one sees the first increases in LH and testosterone levels; initially these occur only during nocturnal sleep (Boyar *et al.*, 1974, Judd *et al.*, 1974, Lee *et al.*, 1976a). In midpuberty serum concentrations of testosterone and other gonadal steroids rise steeply—this results in the appearance of sexual hair, phallic enlargement, the adolescent growth spurt with its associated increase in muscle mass, and eventual epiphyseal closure. Some gynecomastia occurs commonly in midpuberty, possibly as a result of increasing estrogen production (Lee, 1975).

In girls one can discern a similar general pattern, with an initial increase in serum FSH and LH levels, followed by an increase in ovarian estrogen secretion causing breast development, female fat distribution, vaginal and uterine develop-

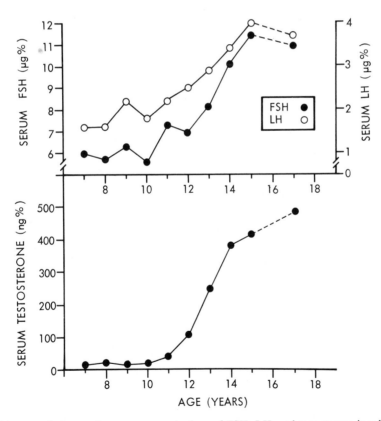

Fig. 6. Mean trends by age for serum concentrations of FSH, LH, and testosterone in a longitudinal study of 56 male subjects. The values for the 17-year-olds represent the means for 11 subjects aged between 16 and 18 years. (Reproduced with permission from Faiman and Winter, 1974.)

ment, and contributing to the adolescent enhancement of growth and skeletal maturation. But these events are complicated by the rhythmic character of hormonal secretion in females. Thus in early puberty, as the breasts begin to enlarge, one sees rhythmic fluctuations in serum FSH and estradiol levels (Winter and Faiman, 1973b), as ovarian follicles are first stimulated to secrete estrogen and then become atretic. Sometime in midpuberty (usually in breast stage 3–4, but with some variation—see Chapter 6), there is a sufficient decrement in estradiol levels during one of these fluctuations for the first withdrawal menstrual bleeding to occur. For several months after menarche most of the cycles are anovulatory, but soon one can observe the adult menstrual pattern in which FSH-mediated follicular estradiol production triggers a midcycle ovulatory surge of LH (and FSH), which is later followed by a sustained luteal rise in serum progesterone levels (Winter and Faiman, 1973b; Hayes and Johanson, 1972; Widholm and Kantero, 1974).

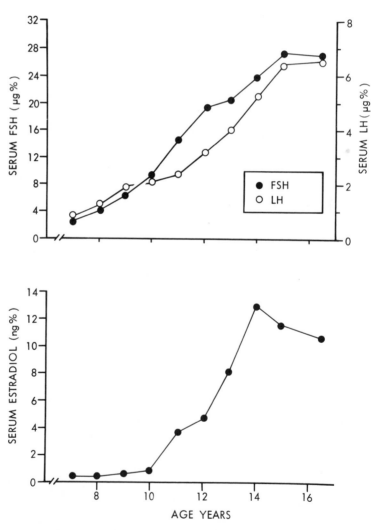

Fig. 7. Mean trends by age for serum concentrations of FSH, LH, and estradiol in a longitudinal study of 58 female subjects. (Reproduced with permission from Faiman and Winter, 1974.)

Table III. Mean Serum Concentrations of Gonadotropins and Sex Steroids by Pubertal Stage[a,b]

Stage of puberty	FSH		LH		Testosterone (ng/dl)	Dihydro-testosterone (ng/dl)	Andro-stenedione (ng/dl)	DHEA (ng/dl)	Estrone (ng/dl)	Estradiol (ng/dl)	Progesterone (ng/dl)	17OH-Progesterone (ng/dl)
	μg LER907/dl	IU/liter	μg LER907/dl	IU/liter								
Males												
G1	9.7 ± 0.4	4.5 (2.5–7.0)	1.7 ± 0.1	3.9 (2.5–5.8)	10 ± 1	8 (2–20)	20 ± 2	63 (10–280)	1.1 ± 0.3	0.8 ± 0.3	30	33 ± 3
G2	12.8 ± 0.3	5.9 (3.0–9.0)	2.2 ± 0.1	6.8 (4.0–12.0)	85 ± 5	16 (5–28)	40 ± 3	175 (50–560)	1.6 ± 0.3	1.1 ± 0.3	36	37 ± 3
G3	13.5 ± 0.7	8.1 (2.5–14.0)	2.5 ± 0.2	8.5 (6.0–11.0)	121 ± 17	22 (6–36)	66 ± 5	266 (120–590)	2.1 ± 0.2	1.6 ± 0.5	40	57 ± 5
G4	17.2 ± 1.0	8.5 (3.5–15.0)	3.5 ± 0.2	9.5 (4.0–15.5)	493 ± 42	36 (15–51)	111 ± 4	339 (180–640)	3.3 ± 0.6	2.2 ± 0.8	40	85 ± 8
G5	17.0 (8.6–25.5)	8.0 (2.9–14.3)	4.2 (2.7–6.5)	11.5 (4.4–19.0)	605 (260–1000)	52 (21–76)	107 (50–200)	370 (180–700)	4.2 (2.5–6.9)	2.5 (1.4–3.8)	35 (16–60)	97 (38–200)
Females												
B1/PH1	8.2 ± 0.6	4.2 (3.1–5.7)	2.1 ± 0.1	2.9 (2.0–7.5)	7 ± 1	5 (2–16)	26 ± 2	75 (19–300)	1.4 ± 0.3	0.8 ± 0.3	10	35 ± 3
B2/PH2	9.4 ± 0.9	5.5 (4.6–7.1)	2.1 ± 0.1	3.9 (2.5–11.5)	17 ± 3	8 (3–24)	60 ± 3	293 (45–1600)	2.0 ± 0.3	1.3 ± 0.5	16	45 ± 7
B3/PH3	14.2 ± 1.1	8.0 (5.0–12.0)	3.7 ± 0.3	8.4 (2.5–14.0)	30 ± 2	14 (9–25)	70 ± 5	465 (125–1700)	3.0 ± 0.5	2.5 ± 0.5	23	62 ± 7
B4/PH4	19.5 ± 1.2	8.0 (3.5–13.0)	7.6 ± 1.0	11.3 (3.0–29.0)	38 ± 2	16 (9–32)	130 ± 3	471 (153–1620)	4.9 ± 0.6	7.6 ± 0.8	161	114 ± 13
Adult												
Follicular	21.9 ± 2.2	12.3 (10.4–14.2)	6.9 ± 0.7	18.9 (15.2–22.6)	35 (15–65)	15 (10–32)	121 (80–300)	430 (200–1500)	2.7 (1.4–3.8)	8.7 ± 1.5	40 (29–75)	42 (11–80)
Midcycle	Peak 25–50	18.8 (14.3–22.8)	Peak 20–60	80.3 (62.4–98.2)	42	c	160	c	6.2 (3.8–18.0)	peak 15–35	—	112 (70–160)
Luteal	8.2 ± 1.0	7.4 (6.6–8.3)	5.3 ± 0.7	14.2 (11.9–16.5)	30	c	130	c	3.9 (1.8–7.6)	6.2 ± 1.4	940 (250–2000)	160 (30–285)

a Data shown as means ±SEM or means (range in brackets). Pubertal staging of males is based upon changes in testicular and penile size, independent of pubic hair growth. Pubertal staging of females is based upon breast development (B1 to B5) for FSH, LH, estrogen, and progestin results, but upon pubic hair growth (PH1 to PH5) for androgen results.
b Factors for conversion of steroid levels to the less commonly used nmol/liter are shown in Table II. Values of FSH and LH may be converted from μg LER907/dl to μg pure standard/liter by multiplying by 0.069 and 0.227, respectively.
c No significant change with menstrual cycle.

While the major physical manifestations of puberty in both sexes represent responses to increasing gonadal secretion of sex steroids, there is also a concomitant increase in adrenal secretion of the weak androgens dehydroepiandrosterone and androstenedione, and the weak estrogen estrone (Saez and Bertrand, 1968; Rosenfield *et al.*, 1971; Hopper and Yen, 1975; Ducharme *et al.*, 1976; Korth-Schutz *et al.*, 1976). This change is not accompanied by a parallel rise in adrenal glucocorticoid secretion. In some cross-sectional studies this rise in adrenal sex steroid production appeared to antedate the onset of clinical puberty, and this has led to a suggestion that the adrenal cortex might function in some way to trigger the pubertal rise in pituitary gonadotropin production. However, the only longitudinal study to examine this question suggests rather that the increase in adrenal sex steroid production coincides with the onset of puberty (Lee and Migeon, 1975). It seems more likely that this "adrenarche" represents an alteration in adrenal steroidogenesis that is induced at the time of early puberty, possibly by a small increment in gonadal sex steroid secretion. Thus it is known that estradiol, for example, will inhibit adrenal 3β-hydroxysteroid dehydrogenase and that its administration causes an increase in adrenal androgen production (Sobrinho *et al.*, 1971). The adrenal sex steroids themselves have only limited biologic activity, but they can be converted elsewhere to more potent steroids, particularly testosterone

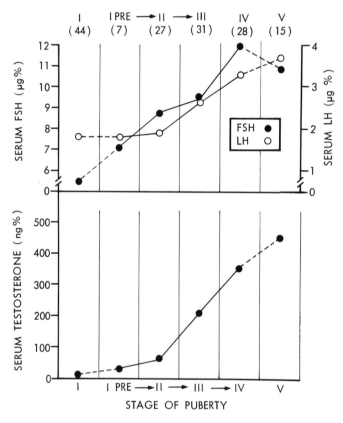

Fig. 8. Mean trends in serum FSH, LH, and testosterone during puberty in a longitudinal study of 56 male subjects. Pubertal stages are based upon testicular and phallic growth. Stage "1-PRE" refers to subjects who advanced to stage 2 (early testicular enlargement) within the subsequent year. (Reproduced with permission from Faiman and Winter, 1974.)

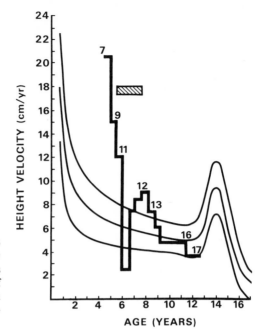

Fig. 9. Growth velocity chart of a boy who developed idiopathic precocious puberty before age 4. The figures superimposed on his line of growth velocity indicate his level of skeletal maturation (expressed in years) at each age. The shaded area indicates a trial of therapy with medroxyprogesterone acetate.

(Drucker *et al.,* 1972; Nieschlag and Kley, 1975). The adrenal androgens appear to contribute significantly to the growth of sexual hair and the adolescent growth spurt in girls, but their influence is overshadowed by the effect of testicular testosterone in boys.

The mechanism for the onset of the pubertal increase in pituitary gonadotropin secretion appears to begin with maturational changes in adrenergic hypothalamic neurons involved in LH-RH secretion (Ruf, 1973) which in turn leads to a reduction in the sensitivity of these neurons to feedback inhibition by sex steroids (Kulin and Reiter, 1973) and an increase in LH-RH production. This change is amplified at each level as the pituitary becomes increasingly responsive to LH-RH stimulation (Roth *et al.,* 1973; Garnier *et al.,* 1974) and the gonad becomes similarly more responsive to gonadotropin stimulation (Winter *et al.,* 1972). In severe illness or chronic malnutrition, puberty is delayed; in the postpubertal individual malnutrition decreases sex steroid secretion, both by impairment of LH-RH release and through a reduction in gonadal responsiveness to gonadotropin stimulation (Smith *et al.,* 1975*b*).

The striking effect that sex steroids have upon growth is seen best in children with precocious puberty. As can be seen in Figure 9, precocious puberty at first causes a marked enhancement of growth velocity, but as more rapid epiphyseal maturation occurs, there is a premature cessation of growth. Usually the eventual adult height is significantly reduced.

7. Adrenocorticosteroids

Normal glucocorticoid secretion is essential for survival, but it does not appear to be a major factor in human growth. However, when glucocorticoids are present in excess, as in Cushing's disease or because of steroid therapy, then retardation of both growth and skeletal maturation results. It is not clear at this time to what extent this inhibition of growth represents solely a direct action of glucocorticoids upon growing cartilage or to what degree suppression of growth-hormone secretion and antagonism of its effects may contribute. An example of the growth retardation which accompanies hypercortisolism is shown in Figure 10.

Cortisol is the major glucocorticoid in man. Of the total cortisol in serum, roughly 10% is available for metabolic effect, while the remainder is bound to corticosteroid-binding globulin and other serum proteins. In target tissues cortisol binds to specific cytoplasmic receptors; as with the sex steroids, this hormone–receptor complex is then transported to the nucleus to initiate synthesis of new mRNA, which in turn codes for the proteins that give rise to the hormone's metabolic effects. One of these effects is a net reduction in protein synthesis which results in muscle wasting and reduced bone formation (Peck *et al.*, 1967; Jowsey and Riggs, 1970). A relative deficiency of growth hormone may also contribute to the observed growth failure, since hypercortisolism can cause disappearance of the normal sleep-associated episodes of growth-hormone secretion (Evans *et al.*, 1973;

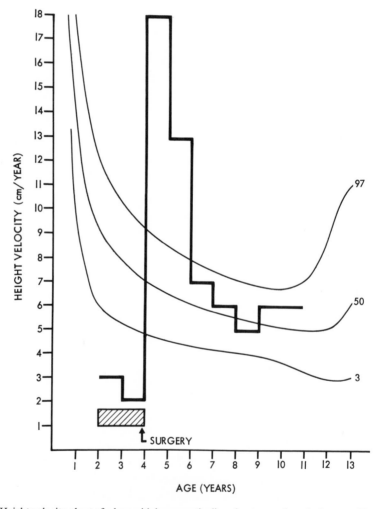

Fig. 10. Height velocity chart of a boy with hypercortisolism due to an adrenal adenoma. Note the catch-up growth which followed removal of the neoplasm. (Reprinted with permission from Winter, 1975.)

Krieger and Glick, 1974), variable suppression of response to stimuli such as hypoglycemia (Copinschi *et al.*, 1975), and a reduction in mean growth-hormone production rates (Thompson *et al.*, 1972*a*). In addition glucocorticoids in excess antagonize the anabolic actions of growth hormone, an effect which may involve both a reduction in somatomedin generation and inhibition of the cartilage response to somatomedin (Soyka and Crawford, 1965).

Secretion of ACTH and therefore of cortisol is episodic, with more frequent secretory episodes occurring in the early morning (Gallagher *et al.*, 1973; de Lacerda *et al.*, 1973). Throughout childhood and puberty cortisol production and corticoid excretion increase in proportion to body size (Kenny *et al.*, 1970; Franks, 1973; Savage *et al.*, 1975). Serum cortisol concentrations remain relatively constant, although mean levels may be slightly higher in males than in females after puberty (Zumoff *et al.*, 1974). In obese children cortisol production rises in proportion to the body weight (O'Connell *et al.*, 1973). Conversely in severe malnutrition cortisol production is reduced, but serum cortisol levels may be elevated because of an even greater reduction in cortisol metabolism (Smith *et al.*, 1975*a*).

8. An Overview of the Endocrinology of Growth

It is difficult, if not impossible, to extract from this gallimaufry a simplistic scheme for the hormonal control of growth in children. As with the instruments of a symphony orchestra, no hormone is more important than another. In the final analysis all hormones are primarily involved in the preservation of a stable internal environment, and a deficiency of any will in time lead to impaired growth. Clearly growth hormone, the somatomedins, and probably insulin act to maintain normal growth velocity from infancy onward, but we need to know more about the interactions of these hormones if only so we can understand how a growth-hormone-deficient child can grow normally before birth and occasionally for the first few months of postnatal life. It now appears that the sex steroids are produced in small amounts during childhood, but they do not influence growth to any degree until puberty, when they are the basis for the adolescent growth spurt, the related changes in body composition, the appearance of adult secondary sexual characteristics, and eventual epiphyseal fusion. The thyroid hormones and cortisol must be available in appropriate amounts, and any significant disruption in their normal secretion can result in failure of growth and maturation. Finally we must consider also the important permissive roles played by hormones such as aldosterone, vasopressin, and parathormone in the maintenance of the appropriate ionic environment for the orderly processes of cellular growth and differentiation.

ACKNOWLEDGMENTS

I would like to thank my colleagues, Doctors Charles Faiman and Francisco Reyes; my fellows, Doctors Ieuan Hughes and Garry Warne; and my secretary, Miss Emma Gulbis, for their assistance in composing this review. I am particularly grateful to Doctors Maria New, Peter Lee, and Jacques Ducharme for sharing with me their unpublished data. Support for our own studies was provided by the Medical Research Council of Canada and the Children's Hospital of Winnipeg Research Foundation.

9. References

Abrams, R. L., Grumbach, M. M., and Kaplan, S. L., 1971, The effect of administration of human growth hormone on the plasma growth hormone, cortisol, glucose and free fatty acid response to insulin: Evidence for growth hormone autoregulation in man, *J. Clin. Invest.* **50**:940.

Abuid, J., Klein, A. H., Foley, T. P., Jr., and Larsen, P. R., 1974, Total and free triiodothyronine and thyroxine in early infancy, *J. Clin. Endocrinol. Metab.* **39**:263.

Aceto, T., Jr., Frasier, S. D., Hayles, A. B., Meyer-Bahlburg, H. F. L., Parker, M. L., Munschauer, R., and Di Chiro, G., 1972, Collaborative study of the effects of human growth hormone in growth hormone deficiency. I. First year of therapy, *J. Clin. Endocrinol. Metab.* **35**:483.

Adamson, U., and Cerasi, E., 1975, Acute suppressive effect of human growth hormone on basal insulin secretion in man, *Acta Endocrinol. (Kbh)* **79**:474.

Alford, F. P., Baker, H. W. G., and Burger, H. G., 1973, The secretion rate of human growth hormone. I. Daily secretion rates, effect of posture and sleep, *J. Clin. Endocrinol. Metab.* **37**:515.

Almqvist, S., and Rune, I., 1961, Studies on sulfation factor (SF) activity of human serum: The variation of serum SF with age, *Acta Endocrinol. (Kbh)* **36**:566.

Angsusingha, K., Kenny, F. M., Nankin, H. R., and Taylor, F. H., 1974, Unconjugated estrone, estradiol and FSH and LH in prepubertal and pubertal males and females, *J. Clin. Endocrinol. Metab.* **39**:63.

Aubert, M. L., Grumbach, M. M., and Kaplan, S. L., 1974, Heterologous radioimmunoassay for plasma human prolactin (hPRL); values in normal subjects, puberty, pregnancy and in pituitary disorders, *Acta Endocrinol. (Kbh)* **77**:460.

Avruskin, T. W., Tang, S. C., Shenkman, L., Mitsuma, T., and Hollander, C. S., 1973, Serum triiodothyronine concentrations in infancy, childhood, adolescence and pediatric thyroid disorders, *J. Clin. Endocrinol. Metab.* **37**:235.

Baghdassarian, A., Guyda, H., Johanson, A., Migeon, C. J., and Blizzard, R. M., 1970, Urinary excretion of radioimmunoassayable luteinizing hormone (LH) in normal male children and adults according to age and stage of sexual development, *J. Clin. Endocrinol. Metab.* **31**:428.

Baird, D. T., Horton, R., Longcope, C., and Tait, J. F., 1969, Steroid dynamics under steady-state conditions, *Rec. Prog. Horm. Res.* **25**:611.

Bartfeld, H., and Siegel, S. M., 1968, Antioxidant activity of thyroxine and related substances. Effect on *in vitro* cell growth, *Exp. Cell Res.* **49**:25.

Baulieu, E. E., Alberga, A., Jung, I., Lebeau, M. C., Mercier-Bodard, C., Milgrom, E., Raynaud, J. P., Raynaud-Jammet, C., Rochefort, H., Truong, H., and Robel, P., 1971, Metabolism and protein binding of sex steroids in target organs: An approach to the mechanism of hormone action, *Rec. Prog. Horm. Res.* **27**:351.

Beaton, G. R., and Singh, V., 1975, Age-dependent variation in cartilage response to somatomedin, *Pediatr. Res.* **9**:683 (abstract).

Berganstal, D. M., and Lipsett, M. B., 1960, Metabolic effects of human growth hormone and growth hormone of other species in man, *J. Clin. Endocrinol. Metab.* **20**:1427.

Bidlingmaier, F., Wagner-Barnack, M., Butenandt, O., and Knorr, D., 1973, Plasma estrogens in childhood and puberty under physiologic and pathologic conditions, *Pediatr. Res.* **7**:901.

Bidlingmaier, F., Versmold, H., and Knorr, D., 1974, Plasma estrogens in newborns and infants, in: *Endocrinologie Sexuelle de la Periode Perinatale* (M. G. Forest and J. Bertrand, eds.), p. 299, Editions INSERM, Paris.

Birkbeck, J. A., 1972, Growth in juvenile diabetes mellitus, *Diabetologia* **8**:221.

Blizzard, R. M., Drash, A. L., Jenkins, M. E., Spaulding, J. S., Glick, A., Weldon, V. V., Powell, G. F., and Raiti, S., 1966, Comparative effect of animal prolactins and human growth hormone (HGH) in hypopituitary children, *J. Clin. Endocrinol. Metab.* **26**:852.

Blizzard, R. M., Penny, R., Foley, T. P., Jr., Baghdassarian, A., Johanson, A., and Yen, S. S. C., 1972, The pituitary–gonadal inter-relationships in relation to puberty, in: *Gonadotropins* (B. B. Saxena, C. G. Beling, and H. M. Gandy, eds.), pp. 502–523, Wiley, New York.

Boyar, R. M., Rosenfeld, R. S., Kapen, S., Finkelstein, J. W., Roffwarg, H. P., Weitzman, E. D., and Hellman, L., 1974, Human puberty. Simultaneous augmented secretion of luteinizing hormone and testosterone during sleep, *J. Clin. Invest.* **54**:609.

Bradshaw, R. A., Hogue-Angeletti, R. A., and Frazier, W. A., 1974, Nerve growth factor and insulin: Evidence of similarities in structure, function and mechanism of action, *Rec. Prog. Horm. Res.* **30**:575.

Brasel, J. A., and Winick, M., 1970, Differential cellular growth in the organs of hypothyroid rats, *Growth* **34**:874.

Braverman, L. E., Dawber, N. A., and Ingbar, S. H., 1966, Observations concerning the binding of thyroid hormones in sera of normal subjects of varying ages, *J. Clin. Invest.* **45**:1273.

Braverman, L. E., Ingbar, S. H., and Sterling, K., 1970, Conversion of thyroxine (T4) to triiodothyronine (T3) in athyreotic human subjects, *J. Clin. Invest.* **49**:855.

Cacciari, E., Tassoni, P., Cicognani, A., Pirazzoli, P., and Collina, A., 1972, Enteroinsular axis and relationship between insulin and growth hormone in the normal and obese child. Effects of oral lipidic and proteic load, *Helv. Paediatr. Acta* **27**:405.

Carnelutti, M., del Guercio, M. J., and Chiumello, G., 1970, Influence of growth hormone on the pathogenesis of obesity in children, *J. Pediatr.* **77**:285.

Carter, J. N., Corcoran, J. M., Eastman, C. J., and Lazarus, L., 1976, Serum T3 and T4 levels in sick children, *Pediatrics* **58**:776.

Chaussain, J. L., Binet, E., Schlumberger, A., and Job, J. C., 1976, Serum somatomedin activity in obese children, *Pediatr. Res.* **9**:667 (abstract).

Chochinov, R., and Daughaday, W. H., 1976, Current concepts of somatomedin and other biologically related growth factors, *Diabetes* **25**:994.

Chopra, I. J., and Smith, S. R., 1975, Circulating thyroid hormones and thyrotropin in adult patients with protein–calorie malnutrition, *J. Clin. Endocrinol. Metab.* **40**:221.

Clemmons, D. R., Hintz, R. L., Underwood, L. E., and Van Wyk, J. J., 1974, Common mechanism of action of somatomedin and insulin on fat cells, *Isr. J. Med. Sci.* **10**:1254.

Cohen, K. L., and Nissley, S. P., 1976, The serum half-life of somatomedin activity: Evidence for growth hormone dependence, *Acta Endocrinol. (Kbh.)* **83**:243.

Cohen, S., and Savage, C. R., Jr., 1974, Recent studies on the chemistry and biology of epidermal growth factor, *Rec. Prog. Horm. Res.* **30**:551.

Cole, H. S., 1973, Carbohydrate metabolism in children: (1) Blood sugar, serum insulin and growth hormone during oral glucose tolerance tests; (2) blood sugar during the cortisone glucose tolerance test, *Metabolism* **22**:289.

Copinschi, G., L'Hermite, M., Lelercq, R., Golstein, J., Vanhaelst, L., Virasoro, E., and Robyn, C., 1975, Effects of glucocorticoids on pituitary hormonal response to hypoglycemia. Inhibition of prolactin release, *J. Clin. Endocrinol. Metab.* **40**:442.

Corcoran, J. M., Eastman, C. J., Carter, J. N., and Lazarus, L., 1977, Circulating thyroid hormone levels in children, *Arch. Dis. Childh.* **52**:716.

Costin, G., Kogut, M. D., Phillips, L. S., and Daughaday, W. H., 1976, Craniopharyngioma: The role of insulin in promoting postoperative growth, *J. Clin. Endocrinol. Metab.* **42**:370.

Curry, D. L., and Bennett, L. I., 1973, Dynamics of insulin release by perfused rat pancreases: Effect of hypophysectomy, growth hormone, adrenocorticotrophic hormone and hydrocortisone, *Endocrinology* **93**:602.

Czernichow, P., Oliver, C., and Friedman, R., 1975, TRH and prolactin (HPr) in plasma of newborns during the first hours of life, *Pediatr. Res.* **9**:668 (abstract).

Dalzell, D. P., and El Attar, T. M. A., 1973, Gas chromatographic determination of urinary excretion of testosterone, epitestosterone and androstenedione in pre-adolescent and adolescent children, *J. Clin. Endocrinol. Metab.* **36**:1237.

Daughaday, W. H., and Parker, M. L., 1965, Human pituitary growth hormone, *Annu. Rev. Med.* **16**:47.

Daughaday, W. H., Hall, K., Raben, M., Salmon, W. D., Jr., Van den Brande, J. L., and Van Wyk, J. J., 1972, Somatomedin: A proposed designation for the "sulfation factor," *Nature* **235**:107.

Daughaday, W. H., Herington, A. C., and Phillips, L. S., 1975, The regulation of growth by endocrines, *Annu. Rev. Physiol.* **37**:211.

Daughaday, W. H., Phillips, L. S., and Mueller, M. C., 1976, The effects of insulin and growth hormone on the release of somatomedin by the isolated rat liver, *Endocrinology* **98**:1214.

de Lacerda, L., Kowarski, A., and Migeon, C. J., 1973, Integrated concentration and diurnal variation of plasma cortisol, *J. Clin. Endocrinol. Metab.* **36**:227.

de Peretti, E., and Forest, M. G., 1976, Unconjugated dehydroepiandrosterone plasma levels in normal subjects from birth to adolescence in human: The use of a sensitive radioimmunoassay, *J. Clin. Endocrinol. Metab.* **43**:982.

D'Ercole, A. J., Foushee, D. B., and Underwood, L. E., 1976, Somatomedin-C receptor ontogeny and levels in porcine fetal and human cord serum, *J. Clin. Endocrinol. Metab.* **43**:1069.

Drucker, W. D., Blumberg, J. M., Gandy, H. M., David, R. R., and Verde, A. L., 1972, Biologic activity of dehydroepiandrosterone sulfate in man, *J. Clin. Endocrinol. Metab.* **35**:48.

Ducharme, J.-R., Forest, M. G., De Peretti, E., Sempe, M., Collu, R., and Bertrand, J., 1976, Plasma adrenal and gonadal sex steroids in human pubertal development, *J. Clin. Endocrinol. Metab.* **42:**468.

Eddy, R. L., Gilliland, P. F., Ibarra, J. D., Jr., McMurry, J. F., Jr., and Thompson, J. Q., 1974, Human growth hormone release: Comparison of provocative test procedures, *Am. J. Med.* **56:**179.

Ehara, Y., Yen, S. S. C., and Siler, T. M., 1975, Serum prolactin levels during puberty, *Am. J. Obstet. Gynecol.* **121:**995.

Ellsworth, L. R., Hobson, W. C., Reyes, F. I., Faiman, C., and Winter, J. S. D., 1975, Evidence for gonadal steroid secretion and pituitary–gonadal interaction before puberty in a primate, *Clin. Res.* **23:**615A (abstract).

Erenberg, A., Phelps, D. L., Lam, R., and Fisher, D. A., 1974, Total and free thyroid hormone concentrations in the neonatal period, *Pediatrics* **53:**211.

Evans, H. M., and Long, J. A., 1921, The effect of the anterior lobe of the hypophysis administered intraperitoneally upon growth and the maturity and oestrus cycles of the rat, *Anat. Rec.* **21:**61.

Evans, H. M., Simpson, M. E., Marx, W., and Kibrick, E., 1943, Bioassay of pituitary growth hormone. Width of proximal epiphyseal cartilage of tibia in hypophysectomized rats, *Endocrinology* **32:**13.

Evans, J. I., Glass, D., Daly, J. R., and McLean, A. W., 1973, The effect of Zn-tetracosactrin on growth hormone release during sleep, *J. Clin. Endocrinol. Metab.* **36:**36.

Faglia, G., Ferrari, C., Beck-Peccoz, P., Spada, A., Travaglini, P., and Ambrosi, B., 1973, Reduced plasma thyrotropin response to thyrotropin releasing hormone after dexamethasone administration in normal subjects, *Horm. Metab. Res.* **5:**289.

Fahrer, M., Gruneiro, L., Rivarola, M., and Bergada, C., 1974, Levels of plasma growth hormone in children with congenital heart disease, *Acta Endocrinol. (Kbh.)* **77:**451.

Faiman, C., and Winter, J. S. D., 1974, Gonadotropins and sex hormone patterns in puberty: Clinical data, in: *Control of the Onset of Puberty* (M. M. Grumbach, G. D. Grave, and F. E. Mayer, eds.), pp. 32–61, Wiley, New York.

Fellier, H., Frisch, H., and Gleispach, H., 1974, Untersuchung der LH-, Östrogen- und Testosteronausscheidung im Vergleich zur Testesgrösse, *Z. Kinderheilk.* **116:**319.

Fink, C. W., 1967, Thyrotropin deficiency in a child resulting in secondary growth hormone deficiency, *Pediatrics* **40:**881.

Finkelstein, J. W., Anders, T. F., Sachar, E. J., Roffwarg, H. P., and Hellman, L., 1971, Behaviorial state, sleep stage and growth hormone levels in human infants, *J. Clin. Endocrinol. Metab.* **32:**368.

Finkelstein, J. W., Kream, J., Ludan, A., and Hellman, L., 1972a, Sulfation factor (somatomedin): An explanation for continued growth in the absence of immunoassayable growth hormone in patients with hypothalamic tumors, *J. Clin. Endocrinol. Metab.* **35:**13.

Finkelstein, J. W., Roffwarg, H. P., Boyar, R. M., Kream, J., and Hellman, L., 1972b, Age-related change in the twenty-four-hour spontaneous secretion of growth hormone, *J. Clin. Endocrinol. Metab.* **35:**665.

Finkelstein, J. W., Boyar, R. M., and Hellman, L., 1974a, Growth hormone secretion in hyperthyroidism, *J. Clin. Endocrinol. Metab.* **38:**634.

Finkelstein, J. W., Wu, R. H. K., and Boyar, R. M., 1974b, 24 hour secretory pattern of prolactin (hPRL) and its response to stimulation in hypopituitary patients, Program 56th Annual Meeting, Endocrine Society, Atlanta, Abstract 363.

Fisher, D. A., 1973, Advances in the laboratory diagnosis of thyroid disease, part 1, *J. Pediatr.* **82:**1.

Foley, T. P., Jr., Jacobs, L. S., Hoffman, W., Daughaday, W. H., and Blizzard, R. M., 1972, Human prolactin and thyrotropin concentrations in the serums of normal and hypopituitary children before and after the administration of synthetic thyrotropin-releasing hormone, *J. Clin. Invest.* **51:**2143.

Forest, M. G., Sizonenko, P. C., Cathiard, A. M., and Bertrand, J., 1974, Hypophyso-gonadal function in humans during the first year of life. 1. Evidence for testicular activity in early infancy, *J. Clin. Invest.* **53:**819.

Francis, M. J. O., and Hill, D. J., 1975, Prolactin-stimulated production of somatomedin by rat liver, *Nature* **255:**167.

Franks, R. C., 1973, Urinary 17-hydroxycorticosteroid and cortisol excretion in childhood, *J. Clin. Endocrinol. Metab.* **36:**702.

Frasier, S. D., 1974, A review of growth hormone stimulation tests in children, *Pediatrics* **53:**929.

Frasier, S. D., Hilburn, J. M., and Smith, F. G., Jr., 1970, Effect of adolescence on the serum growth hormone response to hypoglycemia, *J. Pediatr.* **77:**465.

Froesch, E. R., Bürgi, H., Ramseier, E. B., Bally, P., and Labhart, A., 1963, Antibody-suppressible and nonsuppressible insulin-like activities in human serum and their physiologic significance. An insulin assay with adipose tissue of increased precision and specificity, *J. Clin. Invest.* **42:**1816.

Froesch, E. R., Bürgi, H., Müller, W. A., Humbel, R. E., Jakob, A., and Labhart, A., 1967, Nonsuppressible insulin-like activity of human serum: Purification, physicochemical and biological properties and its relation to total serum ILA, *Rec. Prog. Horm. Res.* **23**:565.

Gallagher, T. F., Yoshida, K., Roffwarg, H. D., Fukushima, D. K., Weitzman, E. D., and Hellman, L., 1973, ACTH and cortisol secretory patterns in man, *J. Clin. Endocrinol. Metab.* **36**:1058.

Garn, S. M., and Haskell, J. A., 1960, Fat thickness and developmental status in childhood and adolescence, *Am. J. Dis. Child.* **99**:746.

Garnier, P. E., Chaussain, J. L., Binet, E., Schlumberger, A., and Job, J-C., 1974, Effect of synthetic luteinizing hormone-releasing hormone (LH-RH) on the release of gonadotrophins in children and adolescents. VI. Relations to age, sex and puberty, *Acta Endocrinol. (Kbh)* **77**:422.

Girard, F., Schimpff, R. M., Lassare, C., and Donnadieu, M., 1975, Somatomedin activity in the human hepatic vein, *Pediatr. Res.* **9**:683 (abstract).

Gleispach, H., 1974, The urinary steroid excretion in girls during puberty, *Horm. Metab. Res.* **6**:325.

Glick, S. M., Roth, J., Yalow, R. S., and Berson, S. A., 1965, The regulation of growth hormone secretion, *Rec. Prog. Horm. Res.* **21**:241.

Golstein-Golaire, J., and Delange, F., 1971, Serum thyrotropin level during growth in man, *Eur. J. Clin. Invest.* **1**:405.

Goodman, A. D., Tanenbaum, R., Wright, D. R., Trimble, K. D., and Rabinowitz, D., 1974, Existence of "big" and "little" forms of immunoreactive growth hormone in human plasma, *Isr. J. Med. Sci.* **10**:1230.

Gorden, P., Lesniak, M. A., Hendricks, C. M., and Roth, J., 1973, "Big" growth hormone components from human plasma: Decreased reactivity demonstrated by radioreceptor assay, *Science* **182**:829.

Grant, D. B., 1967, Fasting serum insulin levels in childhood, *Arch. Dis. Child.* **42**:375.

Grant, D. B., Hambley, J., Becker, D., and Pimatone, B. L., 1973, Reduced sulphation factor in undernourished children, *Arch. Dis. Child.* **48**:596.

Graystone, J. E., and Cheek, D. B., 1968, Connective tissue growth, in: *Human Growth* (D. B. Cheek, ed.), pp. 221–241, Lea and Febiger, Philadelphia.

Greenberg, A. H., Najjar, S., and Blizzard, R. M., 1974, Effects of thyroid hormone on growth, differentiation and development, in: *Handbook of Physiology,* Section 7 (Endocrinology), Vol. 3 (Thyroid) (M. A. Greer and D. H. Solomon, eds.), p. 377, Williams and Wilkins, Baltimore.

Guillemin, R., 1973, Hypothalamic peptides regulating the secretion of growth hormone: growth hormone releasing factor, in: *Advances in Growth Hormone Research* (S. Raiti, ed.), DHEW Publication No. (NIH) 74-612, p. 139.

Gupta, D., Attanasio, A., and Roaf, S., 1975, Plasma estrogen and androgen concentrations in children during adolescence, *J. Clin. Endocrinol. Metab.* **40**:636.

Guyda, H. J., and Friesen, H. G., 1973, Serum prolactin levels in humans from birth to adult life, *Pediatr. Res.* **7**:534.

Guyda, H., Friesen, H., Bailey, J. D., Leboeuf, G., and Beck, J. C., 1975, Medical Research Council of Canada therapeutic trial of human growth hormone: First 5 years of therapy, *Can. Med. Assoc. J.* **112**:1301.

Hall, K., and Filipson, R., 1975, Correlation between somatomedin A in serum and body height development in healthy children and children with certain growth disturbances, *Acta Endocrinol. (Kbh)* **78**:239.

Hall, K., and Luft, R., 1974, Growth hormone and somatomedin, in: *Advances in Metabolic Disorders* (R. Levine, ed.), Vol. 7, pp. 1–36, Academic Press, New York.

Hansen, A. P., and Johansen, K., 1970, Diurnal patterns of blood glucose, serum free fatty acids, insulin, glucagon and growth hormone in normals and juvenile diabetics, *Diabetologia* **6**:27.

Hayek, A., Maloof, F., and Crawford, J. D., 1973, Thyrotropin behavior in thyroid disorders of childhood, *Pediatr. Res.* **7**:28.

Hayes, A., and Johanson, A., 1972, Excretion of follicle-stimulating hormone (FSH) and luteinizing hormone (LH) in urine by pubertal girls, *Pediatr. Res.* **6**:18.

Haymond, M. W., Karl, I., Weldon, V. V., and Pagliara, A. S., 1976, The role of growth hormone and cortisone on glucose and gluconeogenic substrate regulation in fasted hypopituitary children, *J. Clin. Endocrinol. Metab.* **42**:846.

Hays, G. C., and Mullard, J. E., 1974, Normal free thyroxine index values in infancy and childhood, *J. Clin. Endocrinol. Metab.* **39**:958.

Hepp, K. D., 1972, Adenylate cyclase and insulin action, *Eur. J. Biochem.* **31**:266.

Hermosa, B. D., and Sobel, E. H., 1972, Thyroid in the treatment of short stature, *J. Pediatr.* **80**:988.

Hershko, A., Mamont, P., Shields, R., and Tomkins, G. M., 1971, "Pleiotypic response": Hypothesis relating growth regulation in mammalian cells to stringent controls in bacteria, *Nature (London) New Biol.* **232**:206.

Hintz, R. L., Clemmons, D. R., Underwood, L. E., and Van Wyk, J. J., 1972, Competitive binding of somatomedin to the insulin receptors of adipocytes, chondrocytes, and liver membranes, *Proc. Natl. Acad. Sci. U.S.A.* **69**:2351.

Hopper, B. R., and Yen, S. S. C., 1975, Circulating concentrations of dehydroepiandrosterone and dehydroepiandrosterone sulfate during puberty, *J. Clin. Endocrinol. Metab.* **40**:458.

Hughes, I. A., and Winter, J. S. D., 1976, The application of a serum 17OH-progesterone radioimmunoassay to the diagnosis and management of congenital adrenal hyperplasia, *J. Pediatr.* **88**:766.

Hung, W. L., Wilkins, L., and Blizzard, R. M., 1962, Medical therapy of thyrotoxicosis in children, *Pediatrics* **30**:17.

Imura, H., Kato, Y., Ikeda, M., Morimoto, M., and Yawata, M., 1971, Effect of adrenergic-blocking or -stimulating agents on plasma growth hormone, immunoreactive insulin, and blood free fatty acid levels in man, *J. Clin. Invest.* **50**:1069.

Jaffe, R. B., and Keye, W. R., Jr., Modulation of pituitary response to hypothalamic releasing factors, *J. Steroid Biochem.* **6**:1055.

Jenner, M. R., Kelch, R. P., Kaplan, S. L., and Grumbach, M. M., 1972, Hormonal changes in puberty: IV. Plasma estradiol, LH, and FSH in prepubertal children, pubertal females, and in precocious puberty, premature thelarche, hypogonadism, and in a child with a feminizing ovarian tumor, *J. Clin. Endocrinol. Metab.* **34**:521.

Jensen, E. V., and DeSombre, E. R., 1972, Mechanism of action of the female sex hormones, *Annu. Rev. Biochem.* **41**:203.

Jepson, J. H., and McGarry, E., 1972, Hemopoiesis in pituitary dwarfs treated with human growth hormone and testosterone, *Blood* **39**:238.

Jones, K. L., and Addison, J., 1975, Pituitary fibroblast growth factor as a stimulator of growth in cultured rabbit articular chondrocytes, *Endocrinology* **97**:359.

Jowsey, J., and Riggs, B. L., 1970, Bone formation in hypercortisolism, *Acta Endocrinol. (Kbh)* **63**:21.

Judd, H. L., Parker, D. C., Siler, T. M., and Yen, S. S. C., 1974, The nocturnal rise of plasma testosterone in pubertal boys, *J. Clin. Endocrinol. Metab.* **38**:710.

Kantero, R.-L., Wide, L., and Widholm, O., 1975, Serum growth hormone and gonadotrophins and urinary steroids in adolescent girls, *Acta Endocrinol.* **78**:11.

Katz, H. P., Youlton, R., Kaplan, S. L., and Grumbach, M. M., 1969, Growth and growth hormone. III. Growth hormone release in children with primary hypothyroidism and thyrotoxicosis, *J. Clin. Endocrinol. Metab.* **29**:346.

Kenny, F. M., Richards, C., and Taylor, F. H., 1970, Reference standards for cortisol production and 17-hydroxycorticosteroid excretion during growth: Variation in the pattern of excretion of radio-labelled cortisol metabolites, *Metabolism* **19**:280.

Kenny, F. M., Guyda, H. J., Wright, J. C., and Friesen, H. G., 1973, Prolactin and somatomedin in hypopituitary patients with "catch-up" growth following operations for cranio-pharyngioma, *J. Clin. Endocrinol. Metab.* **36**:378.

Knobil, E., and Greep, R. O., 1959, The physiology of growth hormone with particular reference to its action in the rhesus monkey and the "species specificity" problem, *Rec. Prog. Horm. Res.* **15**:1.

Kogut, M. D., Kaplan, S. A., and Shimizu, C. S. N., 1963, Growth retardation: use of sulfation factor as a bioassay for growth hormone, *Pediatrics* **31**:538.

Korth-Schutz, S., Levine, L. S., and New, M. I., 1976, Serum androgens in normal prepubertal and pubertal children and in children with precocious adrenarche, *J. Clin. Endocrinol. Metab.* **42**:117.

Kostyo, J. L., Uthne, K., Reagan, C. R., and Gimpel, L. P., 1973, Comparison of the effects of growth hormone and somatomedin on skeletal muscle metabolism, in: *Advances in Human Growth Hormone Research* (S. Raiti, ed.), pp. 127–134, DHEW Publication No. (NIH) 74-612.

Kram, R., and Tomkins, G. M., 1973, Pleiotypic control by cyclic AMP: Interaction with cyclic GMP and possible role of microtubules, *Proc. Natl. Acad. Sci. U.S.A.* **70**:1659.

Kram, R., Mamont, P., and Tomkins, G. M., 1973, Pleiotypic control by adenosine 3',5'-cyclic monophosphate: A model for growth control in animal cells, *Proc. Natl. Acad. Sci. U.S.A.* **70**:1432.

Krieger, D. T., and Glick, S. M., 1974, Sleep EEG stages and plasma growth hormone concentration in states of endogenous and exogenous hypercortisolemia or ACTH elevation, *J. Clin. Endocrinol. Metab.* **39**:986.

Kulin, H. E., and Reiter, E. O., 1973, Gonadotropins during childhood and adolescence—a review, *Pediatrics* **51**:260.

Lamberg, B.-A., Kantero, R.-L., Saarinen, P., and Widholm, O., 1973*a*, Endocrine changes before and after the menarche. IV. Serum thyrotrophin in female adolescents, *Acta Endocrinol. (Kbh)* **74**:695.

Lamberg, B.-A., Kantero, R.-L., Saarinen, P., and Widholm, O., 1973*b*, Endocrine changes before and after the menarche. III. Total thyroxine and "free thyroxine index" and the binding capacity of thyroxine binding proteins in female adolescents, *Acta Endocrinol. (Kbh)* **74**:685.

Laron, Z., 1974, Syndrome of familial dwarfism and high plasma immunoreactive growth hormone, *Isr. J. Med. Sci.* **10**:1247.

Laron, Z., Hochman, I. H., and Keret, R., 1972*a,* The effect of methandrostenolone on pituitary growth hormone secretion in prepubertal children, *Clin. Endocrinol.* **1**:91.

Laron, Z., Karp, M., Pertzelan, A., and Kauli, R., 1972*b,* Insulin, growth and growth hormone, *Isr. J. Med. Sci.* **8**:440.

Lee, P. A., 1975, The relationship of concentrations of serum hormones to pubertal gynecomastia, *J. Pediatr.* **86**:212.

Lee, P. A., and Migeon, C. J., 1975, Puberty in boys: Correlation of plasma levels of gonadotropins (LH, FSH), androgens (testosterone, androstenedione, dehydroepiandrosterone and its sulfate), estrogens (estrone and estradiol) and progestins (progesterone and 17-hydroxyprogesterone), *J. Clin. Endocrinol. Metab.* **41**:556.

Lee, P. A., Jaffe, R. B., and Midgley, A. R., Jr., 1974, Serum gonadotropin, testosterone and prolactin concentrations throughout puberty in boys: A longitudinal study, *J. Clin. Endocrinol. Metab.* **39**:664.

Lee, P. A., Plotnick, L. P., Steele, R. E., Thompson, R. G., and Blizzard, R. M., 1976*a,* Integrated concentrations of luteinizing hormone and puberty, *J. Clin. Endocrinol. Metab.* **43**:168.

Lee, P. A., Xenakis, T., Winer, J., and Matsenbaugh, S., 1976*b,* Puberty in girls: Correlation of serum levels of gonadotropins, prolactin, androgens, estrogens and progestins with physical changes, *J. Clin. Endocrinol. Metab.* **43**:775.

Lenko, H. L., and Perheentupa, 1975, Effect of HGH treatment on growth of non-GH-deficient short children, *Pediatr. Res.* **9**:671 (abstract).

Lewis, R. C., Duval, A. M., and Iliff, A., 1943, Standards for the basal metabolism of children from 2 to 15 years of age, inclusive, *J. Pediatr.* **23**:1.

Li, C. H., and Papkoff, H., 1956, Preparation and properties of growth hormone from human and monkey pituitary glands, *Science* **124**:1293.

Lippe, B., Wong, S.-L. R., and Kaplan, S. A., 1971, Simultaneous assessment of growth hormone and ACTH reserve in children pretreated with diethylstilbestrol, *J. Clin. Endocrinol. Metab.* **33**:949.

Loeb, H., 1966, Variations in glucose tolerance during infancy and childhood, *J. Pediatr.* **68**:237.

Lunn, P. G., Whitehead, R. G., Hay, R. W., and Baker, B. A., 1973, Progressive changes in serum cortisol, insulin and growth hormone concentrations and their relationship to the distorted amino acid pattern during the development of kwashiorkor, *Br. J. Nutr.* **29**:399.

MacGillivray, M., Aceto, T., Jr., and Frohman, L., 1968, Plasma growth hormone response and growth retardation of hypothyroidism, *Am. J. Dis. Child.* **115**:273.

MacGillivray, M. H., Frohman, L. A., and Doe, J., 1970, Metabolic clearance and production rates of human growth hormone in subjects with normal and abnormal growth, *J. Clin. Endocrinol. Metab.* **30**:632.

Malvaux, P., De Nayer, Ph., Beckers, C., Van den Schrieck, H. G., and De Visscher, M., 1966, Serum free thyroxine and thyroxine binding proteins in male adolescents, *J. Clin. Endocrinol. Metab.* **26**:459.

Manchester, K. L., 1972, Effect of insulin on protein synthesis, *Diabetes* **21**:447.

Martin, J. B., 1973, Neural regulation of growth hormone secretion, *N. Engl. J. Med.* **288**:1384.

Mayberry, W. E., Gharib, H., Bilstad, J. M., and Sizemore, G. W., 1971, Radioimmunoassay for human thyrotropin. Clinical value in patients with normal and abnormal thyroid function, *Ann. Intern. Med.* **74**:471.

McConaghey, P., 1972, The production of "sulphation factor" by rat liver, *J. Endocrinol.* **52**:587.

McGarry, E. E., and Beck, J. C., 1972, Biological effects of non-primate prolactin and human placental lactogen, in: *Lactogenic Hormones* (G. E. W. Wolstenholme and J. Knight, eds.), p. 361, Churchill Livingstone, Edinburgh.

McNeilly, A. S., 1974, Prolactin and human reproduction, *Br. J. Hosp. Med.* **12**:57.

Megyesi, K., Kahn, C. R., Roth, J., and Gorden, P., 1975, Circulating NSILA-s in man: Preliminary studies of stimuli *in vivo* and of binding to plasma components, *J. Clin. Endocrinol. Metab.* **41**:475.

Merimee, T. J., and Fineberg, S. E., 1974, Growth hormone secretion in starvation: A reassessment, *J. Clin. Endocrinol. Metab.* **39**:385.

Merimee, T. J., and Rabin, D., 1973, A survey of growth hormone secretion and action, *Metabolism* **22**:1235.

Milman, A. E., de Moor, P., and Lukens, F. D. W., 1951, Relation of purified pituitary growth hormone and insulin in regulation of nitrogen balance, *Am. J. Physiol.* **166**:354.

Minguell, J. J., and Sierralta, W. D., 1975, Molecular mechanism of action of the male sex hormones, *J. Endocrinol.* **65**:287.

Miyai, K., Yamamoto, T., Azukizawa, M., Ishibashi, K., and Kumahara, Y., 1975, Serum thyroid hormones and thyrotropin in anorexia nervosa, *J. Clin. Endocrinol. Metab.* **40**:334.

Morell, B., and Froesch, E. R., 1973, Fibroblasts as an experimental tool in metabolic and hormone studies, *Eur. J. Clin. Invest.* **3**:119.

Morris, H. G., 1975, Growth and skeletal maturation in asthmatic children: Effect of corticosteroid treatment, *Pediatr. Res.* **9**:579.

Muller, E. E., 1973, The nervous control of human growth hormone release, in: *Advances in Growth Hormone Research* (S. Raiti, ed.), DHEW Publication No. (NIH) 74-612, p. 192.

Nankin, H. R., Sperling, M., Kenny, F. M., Drash, A. L., and Troen, P., 1975, Correlation between sexual maturation and serum gonadotropins: Comparison of black and white youngsters, *Am. J. Med. Sci.* **268**:139.

Niall, H. D., Hogan, M. L., Tregear, G. W., Segre, G. V., Hwang, P., and Friesen, H., 1973, The chemistry of growth hormone and the lactogenic hormones, *Rec. Prog. Horm. Res.* **29**:387.

Nicoll, C. S., and Bern, H. A., 1972, On the actions of prolactin among the vertebrates: Is there a common denominator? in: *Lactogenic Hormones* (G. E. W. Wolstenholme and J. Knight, eds.), p. 299, Churchill Livingstone, Edinburgh.

Nieschlag, E., and Kley, H. K., 1975, Possibility of adrenal–testicular interaction as indicated by plasma androgens in response to HCG in men with normal suppressed and impaired adrenal function, *Horm. Metab. Res.* **7**:326.

O'Connell, M., Danforth, E., Jr., Horton, E. S., Salans, L., and Sims, E. A. H., 1973, Experimental obesity in man. III. Adrenocortical function, *J. Clin. Endocrinol. Metab.* **36**:323.

Oddie, T. H., Meade, J. H., Jr., and Fisher, D. A., 1966, An analysis of published data on thyroxine turnover in human subjects, *J. Clin. Endocrinol. Metab.* **26**:425.

O'Halloran, M. T., and Webster, H. L., 1972, Thyroid function assays in infants, *J. Pediatr.* **81**:916.

O'Malley, B. W., and Strott, C. A., 1973, The mechanism of action of progesterone, in: *Handbook of Physiology,* Section 7, Vol. 2 (R. O. Greep and E. D. Astwood, eds.), pp. 591–602, American Physiological Society, Washington, D.C.

Oppenheimer, J. H., 1973, Possible clues in the continuing search for the subcellular basis of thyroid hormone action, *Mt. Sinai J. Med.* **40**:491.

Otten, J., Johnson, G. S., and Patan, I., 1972, Regulation of cell growth by cyclic adenosine 3',5'-monophosphate, *J. Biol. Chem.* **247**:7082.

Parker, D. C., Rossman, L. G., and Vanderlaan, E. F., 1974, Relation of sleep-entrained human prolactin release to REM–nonREM cycles, *J. Clin. Endocrinol. Metab.* **38**:646.

Parlow, A. F., 1973, Human pituitary growth hormone, adrenocorticotropic hormone, follicle stimulating hormone, and luteinizing hormone concentrations in relation to age and sex, as revealed by biological assays, in: *Advances in Growth Hormone Research* (S. Raiti, ed.), p. 658, DHEW Publication No. (NIH) 74-612.

Peck, W. A., Brandt, J., and Miller, I., 1967, Hydrocortisone-induced inhibition of protein synthesis and uridine incorporation in isolated bone cells in vitro, *Proc. Natl. Acad. Sci. U.S.A.* **57**:1599.

Penny, R., Guyda, H., Baghdassarian, A., Johanson, A. J., and Blizzard, R. M., 1970, Correlation of serum follicular stimulating hormone (FSH) and luteinizing hormone (LH) as measured by radioimmunoassay in disorders of sexual development, *J. Clin. Invest.* **49**:1847.

Phillips, L. C., and Young, H. S., 1976, Nutrition and somatomedin. II. Serum somatomedin activity and cartilage growth activity in streptozotocin-diabetic rats, *Diabetes* **25**:516.

Pickens, J. M., Burkeholder, J. N., and Womack, W. N., 1967, Oral glucose tolerance test in normal children, *Diabetes* **16**:11.

Pierson, R. W., Jr., and Temin, H. M., 1972, The partial purification from calf serum of a fraction with multiplication-stimulating activity for chicken fibroblasts in tissue culture and with nonsuppressible insulin-like activity, *J. Cell Physiol.* **79**:319.

Pimstone, B. L., Barbezat, G., Hansen, J. D. L., and Murray, P., 1968, Studies on growth hormone secretion in protein–calorie malnutrition, *Am. J. Clin. Nutr.* **21**:482.

Plotnick, L. P., Thompson, R. G., Beitins, I., and Blizzard, R. M., 1974, Integrated concentrations of growth hormone correlated with stage of puberty and estrogen levels in girls, *J. Clin. Endocrinol. Metab.* **38**:436.

Plotnick, L. P., Thompson, R. G., Kowarski, A., Lacerda, L., Migeon, C. L., and Blizzard, R. M., 1975, Circadian variation of integrated concentration of growth hormone in children and adults, *J. Clin. Endocrinol. Metab.* **40**:240.

Raben, M. S., Murakawa, S., and Matute, M., 1972, Some observations concerning serum thymidine factor, in: *Growth and Growth Hormone* (A. Pecile and E. E. Müller, eds.), p. 124, Excerpta Medica, Amsterdam.

Radfar, N., Ansusingha, K., and Kenny, F. M., 1976, Circulating bound and free estradiol and estrone during normal growth and development and in premature thelarche and isosexual precocity, *J. Pediatr.* **89**:719.

Raiti, S. M. B., Light, C., and Blizzard, R. M., 1969, Urinary follicle-stimulating hormone excretion in boys and adult males as measured by radioimmunoassay, *J. Clin. Endocrinol. Metab.* **29**:884.

Rao, K. S. J., and Raghuramulu, 1972, Insulin secretion in kwashiorkor, *J. Clin. Endocrinol. Metab.* **35**:63.

Root, A. W., 1972, *Human Pituitary Growth Hormone,* p. 247, Charles C Thomas, Springfield, Illinois.

Root, A. W., Snyder, P. J., Rezvani, I., Di George, A. M., and Utiger, R. D., 1973, Inhibition of thyrotropin-releasing hormone-mediated secretion of thyrotropin by human growth hormone, *J. Clin. Endocrinol. Metab.* **36**:103.

Rosenfield, R. L., Grossman, B. J., and Ozoa, N., 1971, Plasma 17-ketosteroids and testosterone in prepubertal children before and after ACTH administration, *J. Clin. Endocrinol. Metab.* **33**:249.

Ross, G. T., 1974, Gonadotropins and preantral follicular maturation in women, *Fertil. Steril.* **25**:522.

Roth, J. C., Grumbach, M. M., and Kaplan, S. L., 1973, Effect of synthetic luteinizing hormone-releasing factor on serum testosterone and gonadotropins in prepubertal, pubertal and adult males, *J. Clin. Endocrinol. Metab.* **37**:680.

Roth, J. C., Kahn, R., Lesniak, M. A., Gorden, P., De Meyts, P., Megyesi, K., Neville, D. M., Jr., Gavin, J. R., III, Soll, A. H., Feychet, P., Goldfine, I. D., Bar, R. S., and Archer, J. A., 1975, Receptors for insulin, NSILA-S, and growth hormone: Applications to disease states in man, *Rec. Prog. Horm. Res.* **31**:95.

Rubinstein, H. A., Butler, V. P., Jr., and Werner, S. C., 1973, Progressive decrease in serum triiodothyronine concentrations with human aging: Radioimmunoassay following extraction of serum, *J. Clin. Endocrinol. Metab.* **37**:247.

Rudman, D., Patterson, J. H., Gibbas, D. L., Richardson, T. J., Awrich, A. I., Anthony, A. E., Bixler, T. J., II, and Giansanti, J. S., 1974, Responsiveness of human subjects to human growth hormone: Relation to bone age, *J. Clin. Endocrinol. Metab.* **38**:848.

Ruf, K. B., 1973, How does the brain control the process of puberty? *Z. Neurol.* **204**:96.

Saenger, P., Wiedemann, E., Schwartz, E., Korth-Schutz, S., Lewy, J. E., Riggio, R. R., Rubin, A. L., Stenzel, K. H., and New, M. I., 1974, Somatomedin and growth after renal transplantation, *Pediatr. Res.* **8**:163.

Saez, J. M., and Bertrand, J., 1968, Studies on testicular function in children: Plasma concentrations of testosterone, dehydroepiandrosterone and its sulfate before and after stimulation with human chorionic gonadotrophin, *Steroids* **12**:749.

Salmon, W. D., Jr., and Daughaday, W. H., 1957, A hormonally controlled serum factor which stimulates sulfate incorporation by cartilage *in vitro, J. Lab. Clin. Med.* **49**:825.

Salmon, W. D., Jr., and DuVall, M. R., 1970, A serum fraction with "sulfation factor activity" stimulates *in vitro* incorporation of leucine and sulfate into protein–polysaccharide complexes, uridine into RHA and thymidine into DHA of costal cartilage from hypophysectomized rats, *Endocrinology* **86**:721.

Samuel, A. M., and Deshpande, U. R., 1972, Growth hormone levels in protein calorie malnutrition, *J. Clin. Endocrinol. Metab.* **35**:863.

Savage, D. C. L., Forsyth, C. C., McCafferty, E., and Cameron, J., 1975, The excretion of individual adrenocortical steroids during normal childhood and adolescence, *Acta Endocrinol. (Kbh)* **79**:551.

Schwartz, A. G., and Amos, H., 1968, Insulin dependence of cells in primary culture: Influence of ribosome integrity, *Nature* **219**:1366.

Shellabarger, C. J., 1964, The effect of thyroid hormones on growth and differentiation, in: *The Thyroid Gland,* Vol. 1 (R. Pitt-Rivers and W. R. Trotter, eds.), pp. 187–198, Butterworths, London.

Shaywitz, B. A., Finkelstein, J. W., Hellman, L., and Weitzman, E. D., 1971, Growth hormone in newborn infants during sleep–wake periods, *Pediatrics* **48**:103.

Siegel, E., and Tobias, C. A., 1966, Actions of thyroid hormones on cultured human cells, *Nature* **212**:1318.

Siler, T. M., Yen, S. S. C., Vale, W., and Guillemin, R., 1974, Inhibition by somatostatin on the release of TSH induced in man by thyrotropin-releasing factor, *J. Clin. Endocrinol. Metab.* **38**:742.

Sims, E. A. H., and Danforth, E., Jr., 1974, Role of insulin in obesity, *Isr. J. Med. Sci.* **10**:1222.

Sizonenko, P. C., and Paunier, L., 1975, Hormonal changes in puberty. III. Correlation of plasma dehydroepiandrosterone, testosterone, FSH and LH with stages of puberty and bone age in normal boys and girls and in patients with Addison's disease or hypogonadism or premature or late adrenarche, *J. Clin. Endocrinol. Metab.* **41**:894.

Sizonenko, P. C., Rabinovitch, A., Schneider, P., Paunier, L., Wollheim, C. B., and Zahnd, G., 1975,

Plasma growth hormone, insulin, and glucagon responses to arginine infusion in children and adolescents with idiopathic short stature, isolated growth hormone deficiency, panhypopituitarism, and anorexia nervosa, *Pediatr. Res.* **9**:733.

Slone, D., Soeldner, J. S., Steinke, J., and Crigler, J. F., Jr., 1966, Serum insulin measurements in children with idiopathic spontaneous hypoglycemia and in normal infants, children and adults, *N. Engl. J. Med.* **274**:820.

Smith, G. L., and Temin, H. M., 1974, Purified multiplication-stimulating activity from rat liver cell conditioned medium: Comparison of biological activities with calf serum, insulin and somatomedin, *J. Cell. Physiol.* **84**:181.

Smith, P. E., 1930, Hypophysectomy and replacement therapy in the rat, *Am. J. Anat.* **45**:205.

Smith, S. R., Bledsoe, T., and Chhetri, M. K., 1975a, Cortisol metabolism and the pituitary–adrenal axis in adults with protein–calorie malnutrition, *J. Clin. Endocrinol. Metab.* **40**:43.

Smith, S. R., Chhetri, M. K., Johanson, A. J., Radfar, N., and Migeon, C. J., 1975b, The pituitary–gonadal axis in men with protein–calorie malnutrition, *J. Clin. Endocrinol. Metab.* **41**:60.

Sneid, D. S., Jacobs, L. S., Weldon, V. V., Trivedi, B. L., and Daughaday, W. H., 1975, Radioreceptor-inactive growth hormone associated with stimulated secretion in normal subjects, *J. Clin. Endocrinol. Metab.* **41**:471.

Sobrinho, I. G., Kase, N. G., and Grunt, J. A., 1971, Changes in adrenocortical function of patients with gonadal dysgenesis after treatment with estrogen, *J. Clin. Endocrinol. Metab.* **33**:110.

Soyka, L. F., and Crawford, J. D., 1965, Antagonism by cortisone of the linear growth induced in hypopituitary patients and hypophysectomized rats by human growth hormone, *J. Clin. Endocrinol. Metab.* **25**:469.

Sperling, M. A., Kenny, F. M., and Drash, A. L., 1970, Arginine-induced growth hormone responses in children: Effect of age and puberty, *J. Pediatr.* **77**:462.

Talwar, G. P., Pandian, M. R., Kumar, N., Hanjan, S. N. S., Saxena, R. K., Krishnaraj, R., and Gupta, S. L., 1975, Mechanism of action of pituitary growth hormone, *Rec. Prog. Horm. Res.* **31**:141.

Tato, L., Marc, V. L., Du Caju, C. P., and Rappaport, R., 1975, Early variations of plasma somatomedin activity in the newborn, *J. Clin. Endocrinol. Metab.* **40**:534.

Tell, G. P. E., Cuatrecasas, P., Van Wyk, J. J., and Hintz, R. L., 1973, Somatomedin inhibition of adenylate cyclase activity in subcellular membranes of various tissues, *Science* **180**:312.

Thompson, R. G., Rodriguez, A., Kowarski, A., and Blizzard, R. M., 1972a, Growth hormone: Metabolic clearance rates, integrated concentrations, and production rates in normal adults and the effect of prednisone, *J. Clin. Invest.* **51**:3193.

Thompson, R. G., Rodriguez, A., Kowarski, A., Migeon, C. J., and Blizzard, R. M., 1972b, Integrated concentrations of growth hormone correlated with plasma testosterone and bone age in preadolescent and adolescent males, *J. Clin. Endocrinol. Metab.* **35**:334.

Thorngren, K.-G., and Hansson, L. I., 1973, Effect of thyroxine and growth hormone on longitudinal bone growth in the hypophysectomized rat, *Acta Endocrinol. (Kbh.)* **74**:24.

Trygstad, O., 1969, Human growth hormone and hypopituitary growth retardation, *Acta Paediatr. Scand.* **58**:407.

Turkington, R. W., 1972, Human prolactin, *Am. J. Med.* **53**:389.

Turkington, R. W., and Frantz, W. L., 1972, The biochemical action of prolactin, in: *Prolactin and Carcinogenesis* (A. R. Boyns and K. Griffiths, eds.), p. 39, Alpha Omega Alpha, Cardiff.

Utiger, R. D., Parker, M. L., and Daughaday, W. H., 1962, Studies on human growth hormone. I. A radioimmunoassay for human growth hormone, *J. Clin. Invest.* **41**:254.

Vale, W., Brazeau, P., Rivier, C., Rivier, J., and Guillemin, R., 1973, Biological studies with somatostatin, in: *Advances in Growth Hormone Research* (S. Raiti, ed.), p. 159, DHEW Publication No. (NIH) 74-612.

Van den Brande, J. L., and Du Caju, M. V. L., 1973, Plasma somatomedin activity in children with growth disturbances, in: *Advances in Growth Hormone Research* (S. Raiti, ed.), p. 98, DHEW Publication No. (NIH) 74-612.

Van den Brande, J. L., Van Wyk, J. J., French, F. S., Strickland, A. L., and Radcliffe, W. B., 1973, Advancement of skeletal age of hypopituitary children treated with thyroid hormone plus cortisone, *J. Pediatr.* **82**:22.

Van den Brande, J. L., Du Caju, M. V. L., Visser, H. K. A., Schopman, W., Hackeng, W. H. L., and Degenhart, H. J., 1974, Primary somatomedin deficiency, *Arch. Dis. Child.* **49**:297.

Van Wyk, J. J., Underwood, L. E., Lister, R. C., and Marshall, R. N., 1973, The somatomedins: A new class of growth-regulating hormones? *Am. J. Dis. Child.* **126**:705.

Van Wyk, J. J., Underwood, L. E., Baseman, J. B., Hintz, R. L., Clemmons, D. R., and Marshall, R. N., 1975, Exploration of the insulin-like and growth-promoting properties of somatomedin by membrane receptor assays, *Adv. Metab. Disord.* **8**:128.

Vigneri, R., and D'Agata, 1971, Growth hormone release during the first year of life in relation to sleep–wake periods, *J. Clin. Endocrinol. Metab.* **33:**561.

Waaler, P. E., Thorsen, T., Stoa, K. F., and Aarskog, D., 1974, Studies in normal male puberty, *Acta Paediatr. Scand. Suppl.* **249:**1–36.

Wang, C. F., and Yen, S. S. C., 1975, Direct evidence of estrogen modulation of pituitary sensitivity to luteinizing hormone-releasing factor during the menstrual cycle, *J. Clin. Invest.* **55:**201.

Wartofsky, L., Burman, K. D., Dimond, R. C., Noel, G. L., Frantz, A. G., and Earll, J. M., 1977, Studies on the nature of thyroidal suppression during acute Falciparum malaria: Integrity of pituitary response to TRH and alterations in serum T3 and reverse T3, *J. Clin. Endocrinol. Metab.* **44:**85.

Westermak, B., and Wasteson, A., 1975, The response of cultured human normal glial cells to growth factors, *Adv. Metab. Disord.* **8:**85.

Widholm, O., and Kantero, R.-L., 1974, Daily urinary excretion of follicle stimulating (FSH) and luteinizing hormone (LH) immediately after menarche, *Int. J. Fertil.* **19:**1.

Widholm, O., Kantero, R.-L., Axelson, E., Johansson, E. D. B., and Wide, L., 1974, Endocrine changes before and after the menarche. I. Urinary excretion of estrogen, FSH and LH, and serum levels of progesterone FSH and LH, *Acta Obstet. Gynecol. Scand.* **53:**197.

Wiedemann, E., and Schwartz, E., 1972, Suppression of growth hormone-dependent human serum sulfation factor by estrogen, *J. Clin. Endocrinol. Metab.* **34:**51.

Wilson, J. D., 1972, Recent studies on the mechanism of action of testosterone, *N. Engl. J. Med.* **287:**1284.

Winter, J. S. D., 1975, Cushing's syndrome in childhood, in: *Endocrine and Genetic Diseases of Childhood and Adolescence,* 2nd ed. (L. I. Gardner, ed.), p. 501, W. B. Saunders, Philadelphia.

Winter, J. S. D., and Faiman, C., 1972a, Serum gonadotropin levels in agonadal children and adults, *J. Clin. Endocrinol. Metab.* **35:**561.

Winter, J. S. D., and Faiman, C., 1972b, Pituitary–gonadal relations in male children and adolescents, *Pediatr. Res.* **6:**126.

Winter, J. S. D., and Faiman, C., 1973a, Pituitary–gonadal relations in female children and adolescents, *Pediatr. Res.* **7:**948.

Winter, J. S. D., and Faiman, C., 1973b, The development of cyclic pituitary–gonadal function in adolescent females, *J. Clin. Endocrinol. Metab.* **37:**714.

Winter, J. S. D., Taraska, S., and Faiman, C., 1972, The hormonal response to HCG stimulation in male children and adolescents, *J. Clin. Endocrinol. Metab.* **34:**348.

Winter, J. S. D., Faiman, C., Hobson, W. C., Prasad, A. V., and Reyes, F. I., 1975, Pituitary–gonadal relations in infancy. I. Patterns of serum gonadotropin concentrations from birth to four years of age in man and chimpanzee, *J. Clin. Endocrinol. Metab.* **40:**545.

Winter, J. S. D., Hughes, I. A., Reyes, F. I., and Faiman, C., 1976, Pituitary–gonadal relations in infancy: 2. Patterns of serum gonadal steroid concentrations in man from birth to two years of age, *J. Clin. Endocrinol. Metab.* **42:**679.

Wolfe, H. J., and Cohen, R. B., 1964, Glucose-6-phosphate dehydrogenase activity in the human fetal and prepubertal testis: A histochemical study, *J. Clin. Endocrinol. Metab.* **24:**616.

Wu, A., Grant, D. B., Hambley, J., and Levi, A. J., 1974, Reduced serum somatomedin activity in patients with chronic liver disease, *Clin. Sci. Mol. Med.* **47:**359.

Zapf, J., Waldvogel, M., and Froesch, E. R., 1975, Binding of nonsuppressible insulin-like activity to human serum, *Arch. Biochem. Biophys.* **168:**638.

Zingg, A. E., and Froesch, E. R., 1973, Effects of partially purified preparations with nonsuppressible insulin-like activity (NSILA-S) on sulfate incorporation into rat and chicken cartilage, *Diabetologia* **9:**472.

Zumoff, B., Fukushima, D. K., Weitzman, E. D., Kream, J., and Hellman, L., 1974, The sex difference in plasma cortisol concentration in man, *J. Clin. Endocrinol. Metab.* **39:**805.

8

The Central Nervous System and the Onset of Puberty

MELVIN M. GRUMBACH

1. Introduction

The onset of puberty is a consequence of a complex sequence of maturational changes that are incompletely understood (Grumbach *et al.*, 1974; Grumbach, 1975; Odell and Swerdloff, 1976). The development of secondary sex characteristics, the adolescent growth spurt, the attainment of fertility, and the psychosocial changes entrain from the maturation of the gonads and the increase in sex steroid secretion. Two independent but associated processes, controlled by different mechanisms but closely linked temporally, are involved in the increased secretion of sex steroids in the peripubertal and pubertal period. One has been designated "adrenarche," the increase in adrenal androgen secretion (reviewed in Grumbach *et al.*, 1978) which precedes by two years or so the second event, "gonadarche" or the pubertal activation of the hypothalamic–pituitary gonadotropin–gonadal apparatus (Grumbach *et al.*, 1974). These two events and their role in puberty shall be considered separately.

The development of gonadal function may be viewed as a continuum extending from sexual differentiation and the ontogeny of the hypothalamic–pituitary gonadotropin–gonadal system in the fetus, through puberty, to the attainment of full sexual maturation and fertility, and then to senescence (Grumbach, 1975; Grumbach *et al.*, 1974). In this light, the pubertal process is not an isolated *de novo* event but rather a critical stage—a developmental milestone that involves the reactivation of the hypothalamic–pituitary gonadotropin system and gonadotropin secretion. This complex system, which differentiates and operates to a degree during fetal life and infancy (Kaplan *et al.*, 1976), is suppressed to a low level of activity in childhood, as exemplified by the small amount of gonadotropin secretion.

MELVIN M. GRUMBACH • Department of Pediatrics, University of California San Francisco, San Francisco, California.

Let us consider the mechanisms involved in the restraint of gonadotropin secretion and, hence, gonadal function during childhood and the factors that may be important in the onset of puberty. It has long been appreciated that certain central nervous system lesions involving the hypothalamus and nearby structures advance or delay the onset of human puberty (see Donovan and van der Werff ten Bosch, 1965; Critchlow and Bar-Sela, 1967; Wilkins, 1965; Grumbach *et al.*, 1974; Bierich, 1975). For example, true precocious puberty (including cyclic ovulation in girls and spermatogenesis in boys) can occur secondary to a central nervous system tumor or an inflammatory lesion that involves the hypothalamus. Thus, the hypothalamic–pituitary gonadotropin–gonadal complex can be activated prematurely.

A large body of experimental and clinical studies support the hypothesis that the central nervous system, not the pituitary gland or gonads, restrains activation of the hypothalamic–pituitary gonadotropin–gonadal system (reviewed in Grumbach *et al.*, 1974; Donovan and van der Werff ten Bosch, 1965; Critchlow and Bar-Sela, 1967). This inhibitory effect appears to be mediated through the hypothalamus and its neurosecretory neurons that synthesize and secrete luteinizing hormone-releasing factor (LRF)—the hypothalamic hypophysiotropic factor that is essential for the release of both FSH and LH. (This factor also is known as the gonadotropin-releasing hormone.) The cell bodies of the LRF neurosecretory neurons are mainly, but not exclusively, located in the medial basal hypothalamus in the region of the arcuate nucleus (Bugnon *et al.*, 1976; Okon and Koch, 1976); the axons terminate in the central portion of the basal hypothalamus, the median eminence. At this site, the chemical transmitter is released into the primary plexus of the hypothalamic–hypophyseal portal circulation and transmitted by the portal vessels to the anterior pituitary gland.

Recent evidence, derived from experiments in the rhesus monkey (Knobil and Plant, 1978), strongly suggests that the neural component of the system which controls gonadotropin secretion is located in the medial basal hypothalamus. However, in certain experimental animals, especially the rat, extrahypothalamic central nervous system structures (Critchlow and Bar-Sela, 1967; Gorski, 1974), including the limbic system (hippocampus and amygdala), influence gonadotropin secretion. Further, the secretion of LRF is modified by the neurotransmitter catecholaminergic and serotonergic neurons and their effect on hypothalamic norepinephrine, dopamine, and serotonin. For example, α-adrenergic blocking agents inhibit the release of gonadotropins in the monkey (Bhattacharya *et al.*, 1972). Whether the influence of extrahypothalamic factors on LRF release is mediated via these biogenic amines is unclear. Thus, the hypothalamic–pituitary gonadotropin unit is not influenced solely by sex steroids [and inhibin (Baker *et al.*, 1976)] but by the complex neural influences that integrate a variety of intrinsic and extrinsic stimuli.

1.1. The Pattern of Gonadotropin Secretion

There are three patterns of gonadotropins: tonic, cyclic, and pulsatile or episodic. Tonic or basal secretion is regulated by the negative or inhibitory feedback mechanism, by which changes in the concentration of circulating sex steroids, and possibly "inhibin," result in reciprocal changes in the secretion of pituitary gonadotropins. This is the pattern of secretion in the male and one of the control mechanisms in the female. Cyclic secretion involves a positive or stimulatory feedback mechanism in which an increment in circulating estrogens, to a critical level and of sufficient duration, initiates the synchronous release of LH and FSH (the preovulatory LH surge) which is characteristic of the pattern in the normal adult female before menopause. Quite likely the secretion of FSH and LH is always pulsatile or episodic, at a frequency of about 1–2 hr under normal conditions, irrespective of whether the secretion is tonic or cyclic. However, this is difficult to detect when the plasma concentration of gonadotropins is low (as in prepubertal individuals) because of methodologic limitations. The inherent oscillatory characteristic of gonadotropin secretion is a consequence of the pulsatile release of LRF (Boyar *et al.*, 1972). Hence, intrinsic central nervous system influences, which are not dependent upon fluctuations in gonadal steroid levels, mediate episodic gonadotropin secretion. Striking changes in the pattern of the pulsatile gonadotropin spikes and in their circadian rhythm occur in the peripubertal period and during puberty (Boyar *et al.*, 1972, 1974).

In a previous section (p. 193), the changing pattern of gonadotropin and sex steroid secretion was considered in relation to age. In the human fetus the fetal gonad is affected by a placental gonadotropin and by fetal pituitary FSH and LH. Early in gestation the placental gonadotropin, chorionic gonadotropin, has an important influence on the secretion of testosterone by the Leydig cells of the fetal testes during the critical period of sex differentiation of the genital ducts and the external genitalia; later in gestation, fetal FSH and LH influence the growth and maturation of the fetal testis and ovary (Grumbach and Kaplan, 1973; Kaplan *et al.*, 1976).

FSH and LH are detectable in the human fetal pituitary gland by 10 weeks of gestation and the content increases sharply until about 25–29 weeks of gestational age. The fetal pituitary gland has not only the capacity to synthesize and store FSH and LH, but it is capable of secreting these hormones by 11–12 weeks (Kaplan *et al.*, 1976). The concentrations of fetal serum LH and FSH rise to peak levels by mid-gestation, then decrease; the values in umbilical venous blood at term are low (Figure 1). The mean FSH content of fetal pituitary glands and the concentration of fetal serum FSH are greater in female than male fetuses at mid-gestation. The sex difference in FSH has been ascribed to the higher concentration of plasma testosterone between 11 and 24 weeks in the male fetus and the decrease in both serum FSH and LH toward term, due to the maturation of the negative feedback mechanism and the development of steroid receptors in the hypothalamus and pituitary gland for the sex steroids (Grumbach and Kaplan, 1973; Kaplan *et al.*, 1976).

LRF has been detected in the human fetal hypothalamus early in gestation (by about 6 weeks); further, the fetal pituitary gonadotropes are responsive to LRF (reviewed in Kaplan *et al.*, 1976). The pattern of changes in FSH and LH concentration in both the fetal pituitary glands and serum is consistent with a sequence of increasing synthesis and secretion, in which peak serum concentrations reach castrate levels, followed by a decline after mid-gestation that persists to term (Kaplan *et al.*, 1976). The high serum concentrations have been attributed to autonomous secretion of FSH and LH or to the relatively unrestrained secretion and stimulation of the fetal pituitary gland by LRF (Grumbach and Kaplan, 1974; Kaplan *et al.*, 1976) (Figure 2). As fetal development advances, the negative feedback mechanism matures and the hypothalamus secretes less LRF which, in turn, leads to decreased secretion of FSH and LH. This inhibition of hypothalamic LRF release and pituitary gonadotropin secretion appears to be a consequence of the progressive acquisition of increased sensitivity of the hypothalamus and its "gonadostat" (and the pituitary gland) to the inhibitory effects of the high concentration of sex steroids in the fetal circulation (Grumbach and Kaplan, 1974; Kaplan

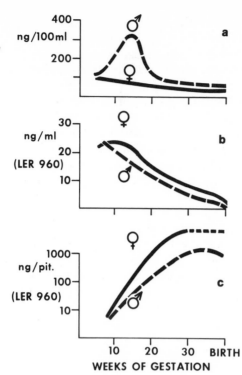

Fig. 1. The concentration of (a) serum testosterone (ng/dl), (b) serum human luteinizing hormone–chorionic gonadotropin in LH equivalents (ng/ml LER 960), and (c) the content of pituitary LH (plotted on logarithmic scale) against gestational age in days for female and male human fetuses. Note the correlation of serum testosterone with serum LH-HCG in the male fetus. (From Grumbach and Kaplan, 1974.)

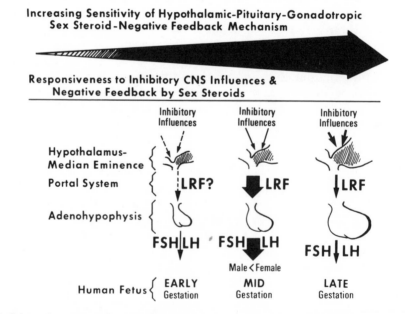

Fig. 2. Schematic representation of the development of regulatory mechanisms for the control of follicle-stimulating hormone (FSH) and luteinizing hormone (LH) secretion in the human fetus. LRF, LH-releasing factor; CNS, central nervous system. (From Grumbach and Kaplan, 1974.)

et al., 1976). The increasing hypothalamic control of gonadotropin secretion may involve the maturation of sex steroid receptors in the LRF neurosecretory neurons and of inhibitory central nervous system pathways.

The hypothalamic regulatory mechanism, like that of other pituitary hormones, is not fully developed at birth (Grumbach and Kaplan, 1974; Grumbach *et al.*, 1974). After the fall in circulating sex steroids, especially estrogens, during the first days after birth, the concentration of serum FSH and LH increases and exhibits wide perturbations during the first few months of age; high gonadotropin concentra-

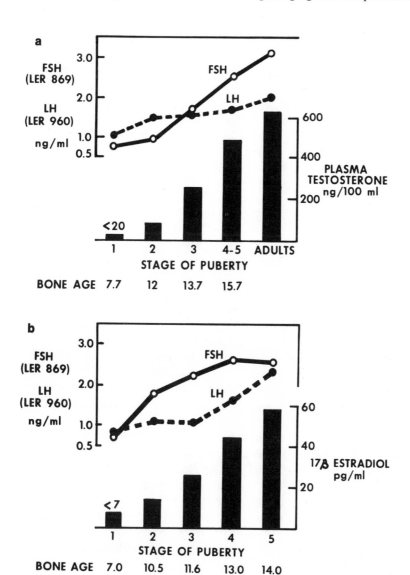

Fig. 3. (a) Mean plasma testosterone, LH, and FSH in prepubertal and pubertal boys by stages of maturation (1 = prepubertal) and the mean bone age for each stage. (From Grumbach, 1975.) (b) Mean plasma estradiol, FSH, and LH concentrations in prepubertal and pubertal females by stage of maturation (1 = prepubertal, 5 = menstruating adolescents) and the mean bone age for each stage. (From Grumbach, 1975.)

tions are associated with increased serum testosterone values in male infants and estradiol levels in females (Winter *et al.*, 1975; Forest *et al.*, 1976). FSH tends to be higher in females than males during the first few years of age. By about age 2 years, the concentration of plasma gonadotropins decreases to the low levels present during childhood until the onset of puberty (Figure 3a,b). Thus, the hypothalamic–pituitary gonadotropin–sex steroid feedback loop does not appear to attain its greatest sensitivity to sex steroid feedback until late in infancy or early childhood (Grumbach *et al.*, 1974; Grumbach, 1975).

1.2. The Neural Control of Puberty

The neural control of puberty includes two major aspects: (1) the timing of puberty and (2) the mechanisms involved in the control of the transition from the prepubertal or sexually infantile state through complete sexual maturation. The time of onset of puberty and its course are strongly influenced by genetic factors and inheritance and are modified by a variety of environmental factors (Grumbach *et al.*, 1974; Grumbach, 1975) operating through the CNS. These encompass socioeconomic factors which influence nutrition and general health, including their effect on the fetus and newborn infant; chronic disease, geography, altitude, and light perception.

1.3. Timing of Puberty

The specific mechanisms involved in the timing of puberty are complex and poorly understood. Frisch and Revelle (1970; Frisch, 1976) have suggested that, in healthy girls, despite differing ages, there is an "invariant mean weight" for the initiation of the pubertal spurt in weight, the maximum rate of weight gain, and menarche (48 kg). Using cross-sectional data and derived measurements, they postulated that a minimum percent of body fat correlates with the onset of menarche and is a "critical" factor in the signal that triggers menarche (Frisch, 1976). In support of the role of nutritional factors and body composition in the onset of menarche is the earlier age of menarche in moderately obese girls (Zacharias *et al.*, 1976), delayed menarche in states of malnutrition and chronic disease, in twins, and the relationship of weight to changes in gonadotropin secretion and amenorrhea in women with anorexia nervosa (Frisch, 1976).

The proposed causal relationship of "critical body weight" and "critical metabolic rate" and body fat to the time of onset of puberty is speculative and controversial (Johnston *et al.*, 1975; Billewicz *et al.*, 1976; Cameron, 1976) and has not been substantiated by direct, rather than derived, measurements. Moreover, menarche is a late event in the pubertal process and may be far removed from the factors that affect the time of onset of the hormonal events and the first physical features of sexual maturation. Nevertheless, the possibility that some component of body composition can provide a metabolic signal to the CNS for the timing of puberty warrants further study.

The timing of puberty has been linked to the vague but generally accepted concept of "maturation" of the CNS; the "maturation" is the outcome or consequence of the totality of environmental and genetic factors that retard or accelerate the onset of puberty. It is a provocative but unproved hypothesis that a metabolic signal related to body composition is an important factor in the maturation or activation of the CNS trigger mechanism for puberty that involves, as the "final common pathway,'' the hypothalamic–pituitary gonadotropin unit and is not a result of the early hormonal changes in puberty. In either event, a large body of clinical and experimental data support the contention that the factors influencing the timing of puberty are expressed finally through the CNS regulation of the onset of puberty (Donovan and van der Werff ten Bosch, 1965; Critchlow and Bar-Sela, 1967; Ramirez, 1973; Davidson, 1974; Grumbach *et al.*, 1974; Bierich, 1975; Odell and Swerdloff, 1976).

1.4. The Onset of Puberty

There is good evidence, although still incomplete, that the hypothalamic–pituitary gonadotropin–gonadal complex develops in two stages. Recent data support the concept that the negative, or inhibitory, feedback mechanism is operative by mid- to late fetal life (Grumbach *et al.*, 1974; Grumbach, 1975; Kaplan *et al.*, 1976). During childhood this tonic control mechanism is exquisitely sensitive to the suppressive effects of small amounts of circulating sex steroids. Coincident with the onset of puberty, the hypothalamic gonadostat becomes progressively less sensitive to the suppressive effects of sex steroids on LRF release.

The second aspect of the mechanism of puberty is maturation of the positive, or stimulatory, feedback response to estrogen. In normal menstruating women, about every 28 days, the negative feedback mechanism is interrupted by the midcycle LH surge, which induces ovulation. The LH surge entrains from the prior increase in plasma estradiol and its positive action on LRF release and in enhancing the sensitivity of the pituitary gonadotrope to the action of LRF. The positive feedback mechanism does not appear to mature until mid-puberty in girls (Kelch *et al.*, 1973; Grumbach *et al.*, 1974; Reiter *et al.*, 1974).

Species differences in the ontogenesis and regulation of gonadotropin secretion need emphasis. In lower species, for example the rat, exteroceptive factors (Donovan and van der Werff ten Bosch, 1965; Critchlow and Bar-Sela, 1967; Ramirez, 1973; Gorski, 1974), including light, olfaction, and pheromones, are important; the neural control mechanism is more sensitive to the environment, and the pattern of gonadotropin secretion is different. In contrast to rodents, there is some evidence that male as well as female primates have the capacity to exhibit an estrogen-provoked LH surge and appear not to exhibit sex-specific differentiation of the hypothalamus (Karsch *et al.*, 1973; Kulin and Reiter, 1976), but data are limited.

1.5. The Negative Feedback Mechanism and the Hypothalamic Control of Puberty

The principal evidence for an operative and highly sensitive negative feedback mechanism in prepubertal children is summarized as follows (Grumbach *et al.*, 1974):

1. The pituitary gland of the prepubertal child secretes small amounts of FSH and LH, which suggests that the hypothalamic–pituitary gonadotropin–gonadal complex is functional in childhood, but at a low level of activity.

2. The absence of functional gonads in the prepubertal child, e.g., in patients with the syndrome of gonadal dysgenesis or other congenital or postnatal agonadal disorders, is associated with increased secretion of FSH and, to a lesser degree, LH. The elevated gonadotropin concentrations in infancy and early childhood in patients with gonadal dysgenesis are evidence that hormones secreted by the

prepubertal gonad, despite their low level, inhibit gonadotropin secretion, supporting the hypothesis that a sensitive, functional tonic negative feedback mechanism is operative in infants and prepubertal children. Moreover, the development of this mechanism is not dependent upon the presence of the testis or ovary (Figure 4a,b). The diphasic pattern of FSH and LH secretion from infancy to adulthood, in patients with gonadal dysgenesis (Grumbach *et al.*, 1974; Conte *et al.*, 1975), is qualitatively similar to that in normal individuals but the levels are strikingly higher except during the mid-childhood nadir.

3. The low level of gonadotropin secretion in childhood is shut off by the administration of small amounts of sex steroids, which indicates a high degree of sensitivity of the hypothalamus and probably the pituitary gland to the feedback effect of sex steroids (Kelch *et al.*, 1973; Grumbach *et al.*, 1974).

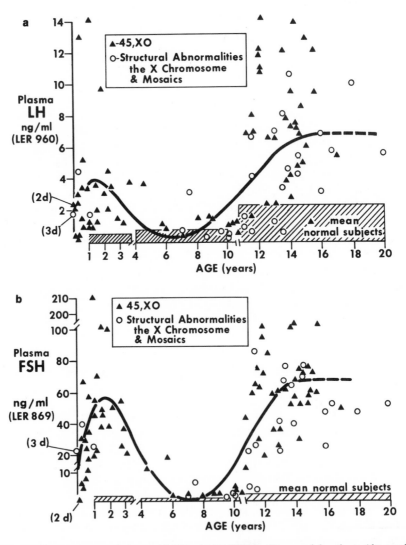

Fig. 4. The diphasic pattern of plasma (a) LH and (b) FSH in 58 agonadal patients (the syndrome of gonadal dysgenesis). The curve is a polynomial regression plot of the data. The hatched lines indicate the mean plasma FSH and LH values in normal females. Longitudinal studies in these patients confirm the patterns obtained from cross-sectional data. (From Conte *et al.*, 1975.)

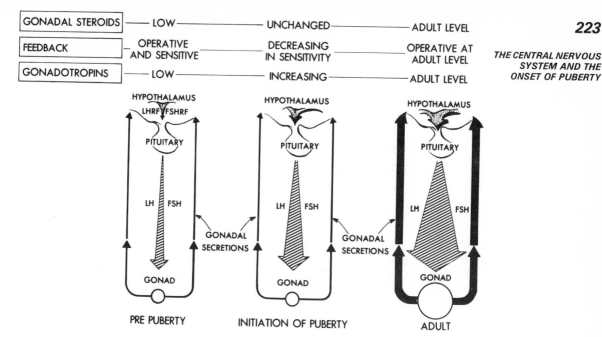

Fig. 5. A schematic diagram of the changes in sensitivity of the hypothalamic gonadostat. In the prepubertal state the concentration of sex steroids and gonadotropins is low; the hypothalamic "gonadostat" is functional but highly sensitive to low levels of sex steroids. With the onset of puberty there is decreased sensitivity of the hypothalamus to negative feedback by sex steroids, increased release of LRF, and enhanced secretion of gonadotropins. In adults the negative feedback mechanism in the hypothalamus is less sensitive to feedback by sex steroids (adult set point), and adult levels of gonadotropins and sex steroids are present. (From Grumbach *et al.,* 1974.)

1.5.1. Change in Sensitivity of Negative Feedback Mechanism

Many years ago Hohlweg and Dohrn (1932), from experiments in the rat, advanced the concept of a change at puberty in sensitivity to circulating sex steroids of a CNS *Sexualzentrum* that regulates gonadotropin secretion. There is now a substantial body of data in experimental animals and the human to support the hypothesis that the hypothalamic gonadotropin-regulating mechanism of the prepubertal individual is more sensitive to the negative feedback effects of circulating androgens and estrogens than that of the adult (Donovan and van der Werff ten Bosch, 1965; Critchlow and Bar-Sela, 1967; Ramirez, 1973; Davidson, 1974; Grumbach *et al.,* 1974). The low levels of sex steroids in the prepubertal individual suppress the release of LRF (and possibly directly inhibit the pituitary gonadotropes as well), and thus the secretion of FSH and LH.

With the approach of puberty and maturation of the CNS, there is a progressive decrease in sensitivity (higher set point) of the hypothalamic negative feedback receptors to sex steroids. This results in an increased secretion of pituitary gonadotropins, stimulation of sex steroid output, and development of secondary sex characteristics (Donovan and van der Werff ten Bosch, 1965; Critchlow and Bar-Sela, 1967; Ramirez, 1973; Davidson, 1974; Grumbach *et al.,* 1974).

The concept of decreasing sensitivity of the hypothalamus to gonadal steroid feedback as a critical occurrence in the initiation of human puberty is illustrated in Figure 5. According to this hypothesis, LRF, gonadotropins, and sex steroids interact at low levels on a highly sensitive negative feedback. The initiation of

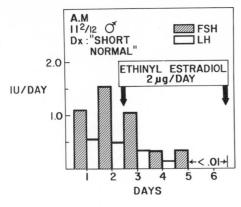

Fig. 6. The effect of administration of ethinyl estradiol (2 μg/day) to an 11²/₁₂-year-old prepubertal normal male on the urinary excretion of LH and FSH. Note the rapid and significant decrease in LH and FSH by the third day following treatment with estradiol. (From Kelch *et al.*, 1973.)

puberty is associated with an increase in the set point of the hypothalamic negative feedback mechanism, so that the low concentrations of sex steroids are no longer effective in suppressing secretion of FSH and LH. This results in an increased release of LRF, FSH, and LH, as well as stimulation of the gonads, and finally the attainment of an adult set point of the gonadostat. The change in sensitivity is probably gradual rather than acute and may exhibit narrow fluctuations before the pubertal increase in set point is well established. It seems to correlate with the attainment of a critical level of CNS and somatic maturation.

Figure 6 shows the striking sensitivity of the hypothalamic–pituitary gonadotropin complex to the administration of small amounts of ethinyl estradiol, as manifested by suppressed gonadotropin output. An estimate of the differential sensitivity of the prepubertal and adult set point to estrogen feedback suggests that the prepubertal hypothalamic gonadostat is about 6–15 times more sensitive than the adult negative feedback mechanism (Grumbach *et al.*, 1974). These studies indicate that the hypothalamic–pituitary gonadotropic–gonadal negative feedback mechanism in the human is operative and sensitive to sex steroids in prepubertal children, exhibits decreased sensitivity to sex steroid feedback with the onset and progression of puberty, and provides support for the concept that decreased sensitivity is a major determinant of the increased gonadotropin secretion at puberty. These are consistent with studies in experimental animals (Donovan and van der Werff ten Bosch, 1965; Critchlow and Bar-Sela, 1967; Ross *et al.*, 1970; Ramirez, 1973).

1.5.2. Location of the Negative Feedback Mechanism

Recent studies in the monkey by Knobil and Plant (1978), using surgical isolation of the hypothalamus from the rest of the brain, indicate that the neural component of the negative feedback system (the gonadostat) resides in the medial basal hypothalamus and seems to require the integrity of the arcuate nucleus. The feedback action of estradiol and testosterone appears to act at both the level of the medial basal hypothalamus and the pituitary gland.

1.5.3. Ontogeny of the Negative Feedback Mechanism

The development of the hypothalamic–pituitary gonadotropin unit has been discussed. The data (Grumbach *et al.*, 1974; Winter *et al.*, 1975; Kaplan *et al.*, 1976) support the concept that the negative feedback mechanism becomes operative in the fetus during mid- to late gestation (Figure 7). The lower pituitary content of FSH and LH, the lower serum concentration of FSH in the male fetus, and the decrease in serum FSH and LH in late gestation in female and male fetuses can be explained by the maturation of the negative feedback mechanism. The sex differences in FSH and LH values during mid-gestation appear to be a consequence of the high testosterone concentration in the circulation in the male fetus during the gestational period. A sex dichotomy has not been found for fetal estrogen levels, which increase with advancing gestation in both sexes. However, in the newborn infant there is a sharp decrease in circulating estrogen and other sex steroids during the first week of life. Coincidentally, the plasma concentration of FSH and LH increases from the low levels at birth in response to the diminished feedback suppression of the hypothalamic–pituitary gonadotropin unit (Winter *et al.*, 1975; Forest *et al.*, 1976). The rise in circulating gonadotropins in the second week is associated with an increase in the plasma concentration of testosterone in male infants and of estradiol in the female. The sex steroid values fall to prepubertal levels by about 6 months of age, and by about 1–2 years of age the gonadotropin concentrations finally diminish to those characteristic of the prepubertal child (later in girls than boys) (Winter *et al.*, 1975; Forest *et al.*, 1976). The hypothalamic–pituitary gonadotropin unit does not appear to attain its maximum sensitivity to sex steroid feedback until about 2–3 years of age (Grumbach *et al.*, 1974; Grumbach, 1975).

Apparently, the onset of puberty is characterized by a gradual decrease in the sensitivity of the hypothalamus and pituitary to circulating sex steroids, which results in the increased secretion of pituitary gonadotropins. This concept of the maturation of the negative feedback system and the change in set point of the gonadostat is illustrated in Figures 5 and 7.

The decrease in circulating FSH and LH in agonadal children between 4 and 10 years of age suggests that the change in sensitivity can occur independent of the gonads. The diminished gonadotropin output (Conte *et al.*, 1975) appears to be a

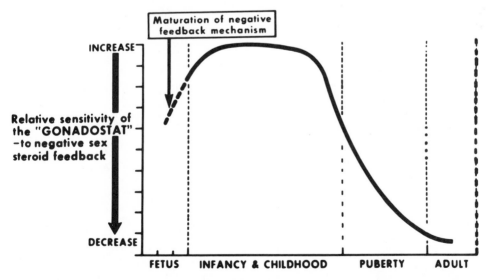

Fig. 7. The ontogenesis of the maturation of the negative feedback system and the changes in set point of the hypothalamic gonadostat extending from the fetus through puberty to the adult are illustrated schematically. (Modified from Grumbach *et al.*, 1974.)

consequence of two factors: (1) CNS factors which lead to increased sensitivity of the hypothalamic–pituitary gonadotropin unit, and (2) in addition to the augmented CNS restraint of LRF release, the onset of adrenarche and the rise in the secretion of adrenal androgens (Grumbach *et al.*, 1978).

2. LRF and the Pituitary and Gonadal Sensitivity to Tropic Stimuli

Puberty appears to encompass orderly maturational changes that involve, sequentially, the suprahypothalamic CNS and hypothalamus, pituitary gland, gonads, and the sex steroid target organs (Grumbach *et al.*, 1974). At each level these structures may exhibit differences in responsiveness to neural or tropic stimuli, depending on their sensitivity or a particular hormonal milieu. If the increased secretion of gonadotropins with the approach of puberty is a consequence of a change in the neural and hormonal restraints on synthesis and secretion of LRF, the decreased sensitivity of the gonadostat should lead to increased release of LRF initially, followed by increased gonadotropic secretion by the pituitary and, finally, to augmented output of sex steroids by the gonad (Figure 8).

With the availability of synthetic LRF, the hypothalamic hypophysiotropic hormone which stimulates LH and FSH release, it has been possible to study the pituitary sensitivity to LRF and the dynamic reserve or readily releasable pool of pituitary gonadotropins. The effect of LRF on the release of FSH and LH has been studied at different stages of sexual maturation (Grumbach *et al.*, 1974; Roth *et al.*, 1972; Job *et al.*, 1972) and in disorders involving the hypothalamic–pituitary gonadal system.

The release of LH following the administration of LRF (Figure 9a) is minimal in prepubertal children beyond infancy, increases strikingly during the peripubertal period and puberty (Grumbach *et al.*, 1974), and is still greater in adult males and females [depending upon the phase of the menstrual cycle (Yen *et al.*, 1975)]. This

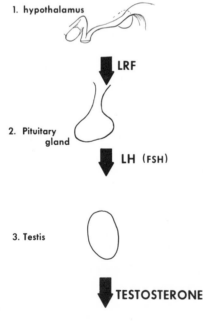

1. hypothalamus

LRF

2. Pituitary gland

LH (FSH)

3. Testis

TESTOSTERONE

Fig. 8. Schematic representation of the sequential events in the maturation of the hypothalamic–pituitary–gonadal system in the male. At puberty increased secretion of LRF leads to increased stimulation of the pituitary gland and increased responsiveness to LRF; both factors lead to enhanced release of pituitary LH. The increased secretion of LH (in the presence of FSH) stimulates the Leydig cells of the testes, which results in increased synthesis and release of testosterone. (From Grumbach *et al.*, 1974.)

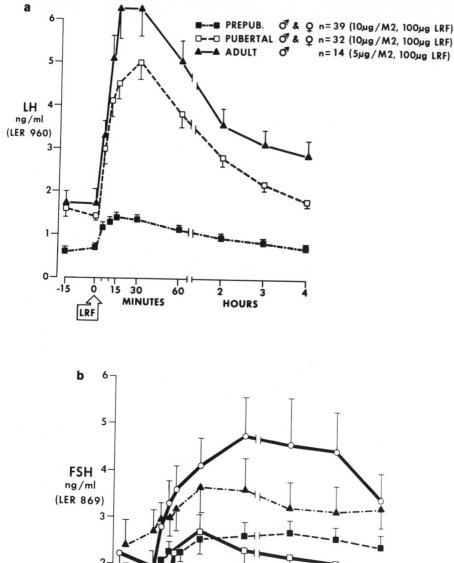

Fig. 9. The change in plasma LH (a) and FSH (b) elicited by LRF in prepubertal, pubertal, and adult subjects. Note the limited LH response in prepubertal children when compared with that of pubertal and adult subjects. The FSH response to LRF is similar in prepubertal, pubertal, and adult males. In females the FSH response is significantly greater than that of prepubertal, pubertal, or adult males. (From Grumbach *et al.*, 1974.)

change at puberty leads to a striking reversal of the FSH/LH ratio of gonadotropin released after LRF between prepuberty and puberty (Grumbach *et al.*, 1974). However, FSH release after the exhibition of LRF is comparable in prepubertal, pubertal, and adult males, indicating similar pituitary sensitivity to LRF (Figure 9b). Moreover, there is a sex difference in the FSH response; prepubertal and pubertal females release more FSH than males at all stages of sexual maturation (Job *et al.*, 1972; Grumbach *et al.*, 1974). These observations suggest a striking change in pituitary sensitivity to LRF in prepubertal and pubertal individuals, as well as a sex difference in the "dynamic reserve" of pituitary FSH (Grumbach *et al.*, 1974).

The sex difference in LH and FSH response to LRF suggests that the pituitary gonadotropes of prepubertal females are more sensitive to LRF than those of prepubertal males, even though there is no apparent difference in the concentration of circulating sex steroids at this stage of maturation. Prepubertal girls have a larger readily releasable pool of pituitary FSH than prepubertal or pubertal males (Figure 9b). This sex difference in sensitivity to LRF and releasable FSH may be a factor in the higher frequency of idiopathic precocious puberty in girls. The data are consistent with the hypothesis that less LRF is required for FSH than LH synthesis and storage in a readily releasable form. These findings also point out the difference between pituitary sensitivity and the actual secretory rate of FSH and LH.

The response to LRF in peripubertal children who do not yet exhibit physical signs of sexual maturation provides further evidence of increased pituitary responsiveness to LRF as a major factor in the increased gonadotropin secretion at puberty, a change in responsiveness that is thought to be mediated by increased endogenous secretion of LRF (Roth *et al.*, 1972, 1973; Grumbach *et al.*, 1974).

The degree of previous exposure of gonadotropes to endogenous LRF appears to affect both the magnitude and quality of FSH and LH responses to a single intravenous dose of LRF. Studies of the effects of acute and chronic administration of synthetic LRF in hyper- and hypogonadism, hypogonadotropic hypogonadism, constitutional delayed adolescence, and idiopathic precocious puberty provide indirect but strong support for this concept (Grumbach *et al.*, 1974; Reiter *et al.*, 1975*a*). Before puberty, the low set point of the gonadostat and the inhibitory feedback effect of low concentrations of plasma sex steroids suppress secretion of LRF. As a consequence, the prepubertal pituitary gland has a significantly smaller pool of releasable LH and exhibits decreased responsiveness to the acute administration of synthetic LRF. With the approach of puberty, increased release of LRF augments pituitary sensitivity to LRF and enlarges the reserve of LH. The explanation for the discordance in FSH and LH release prepubertally is not clear, but it may reside, at least in part, at the level of the pituitary gland.

Gonads also increase in their responsiveness to gonadotropins during puberty. For example, the augmented testosterone secretion in response to the administration of human chorionic gonadotropin at puberty (Winter *et al.*, 1972) is probably a consequence of the priming effect of the increase in endogenous secretion of LH [in the presence of FSH (Sizonenko *et al.*, 1973; Odell and Swerdloff, 1976)] on the Leydig cell.

3. Sleep-Associated LH Release and the Onset of Puberty

Discrete episodic or pulsatile bursts of LH (about 12 episodes over a 24-hr period) are superimposed on the tonic negative feedback control of gonadotropin secretion; the episodic character apparently is a consequence of the pulsatile release of endogenous LRF. In adult men and women, no difference in the amplitude and frequency of these pulses is apparent during a 24-hr period (Boyar *et al.*, 1972). In prepubertal children, there is evidence of episodic secretion (Penny *et al.*, 1977), but the plasma concentrations of FSH and LH are low and it is difficult to demonstrate for methodologic and statistical reasons. Boyar *et al.* (1972) found a striking change in the pattern of gonadotropin secretion in boys and girls beginning in the peripubertal period and extending through puberty. In late prepubertal and pubertal children, enhanced secretion of LH occurs during sleep (Boyar *et al.*, 1972; Judd *et al.*, 1977). The magnitude of the nighttime sleep pulses, and possibly their frequency, are increased (Figure 10). In late puberty the daytime LH secretory episodes increase in amplitude but are still less than during sleep, until finally the adult pattern is achieved. In boys, the augmented LH release during sleep leads to increased testosterone secretion and a rise in the plasma concentration of testosterone at night (Boyar *et al.*, 1974). This pattern of enhanced sleep-associated LH secretion occurs in agonadal patients during the pubertal age period (Boyar *et al.*, 1973*a*), which suggests that this pattern is not dependent upon gonadal function. Further sleep-related gonadotropin release is demonstrable in children with idiopathic precocious puberty (Boyar *et al.*, 1973*b*).

Sleep-enhanced LH secretion can be viewed as a maturational phenomenon related to changes in the CNS and the hypothalamic restraint of LRF release.

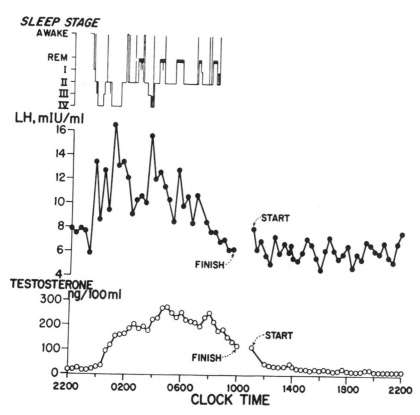

Fig. 10. Plasma LH and testosterone every 20 min in a 14-year-old boy in pubertal stage 2. The histogram displaying sleep stage sequence is depicted above the period of nocturnal sleep. Sleep stages are REM (graph) with stages I–IV shown by depth of line graph. Plasma LH (●—●—●) is expressed as mIU/ml 2nd IRP-HMG. Plasma testosterone (○—○—○) is expressed as ng/100 ml. (From Boyar *et al.*, 1974.)

Episodic release of gonadotropins is suppressed by anti-LRF serum, and by the administration of sex steroids or certain catecholaminergic agonists and antagonists. Sleep-associated LH release correlates with the increased sensitivity of the pituitary gonadotropes to the administration of LRF in the peripubertal period and puberty. We have suggested previously that an increase in endogenous LRF secretion at puberty has a priming effect on the gonadotropes (Roth *et al.*, 1972; Grumbach *et al.*, 1974) and leads to increased sensitivity of the pituitary gland to LRF (either endogenous or exogenous). The augmented LH release at night in both sexes is evidence that the increased secretion of LRF with the onset of puberty occurs initially and principally during sleep.

4. Maturation of the Positive Feedback Mechanism

In normal women the mid-cycle surge in LH and FSH secretion (Figure 11) is attributed to the positive feedback effect of an increased, critical concentration of estradiol during the latter part of the follicular phase (Ross *et al.*, 1970). Estradiol has both a negative and positive feedback effect on the hypothalamic–pituitary gonadotropin system. While its suppressive effect is probably present from late fetal life on, the positive action of estradiol on gonadotropin release has not been

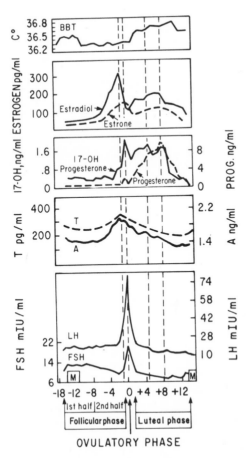

Fig. 11. Hormonal changes in the normal menstrual cycle. A representation of the changes in basal body temperature (BBT) and in the concentrations of plasma estrogens, 17-hydroxyprogesterone (T), androstendione (A), FSH, and LH. The menstrual cycle is dated from the day of the ovulatory surge in plasma LH and FSH (M = menses). The rise in plasma estradiol late in the follicular phase has a *positive feedback* effect on gonadotropin release and leads to the sharp increase in FSH and LH (the ovulatory surge). (From S. S. C. Yen, and R. B. Jaffe, 1978, in: *Reproductive Endocrinology: Physiology, Pathophysiology and Clinical Management*, W. B. Saunders Co.)

demonstrated in prepubertal and early pubertal children (Kelch *et al.*, 1973; Grumbach *et al.*, 1974; Reiter *et al.*, 1974). Hence, acquisition of positive feedback is a late maturational event in puberty, and from the present evidence probably does not occur before mid-puberty in girls (Kulin *et al.*, 1972; Grumbach *et al.*, 1974; Reiter *et al.*, 1974).

Among the prerequisites for a positive feedback action of estradiol on gonadotropin release at puberty are (1) ovarian follicles primed by FSH to secrete sufficient estradiol to reach and maintain a critical level in the circulation, (2) a pituitary gland which is already sensitive to LRF and contains a large enough pool of releasable LH to support an LH surge, and (3) sufficient LRF stores for the LRF neurosecretory neurons to respond with an acute increase in LRF release. Whether the main site of estradiol is at the level of the medial basal hypothalamus or of the anterior pituitary is uncertain. Knobil and Plant (1978) have shown in the rhesus monkey that positive as well as negative feedback can occur in adult ovariectomized females in whom the medial basal hypothalamus has been surgically disconnected from the remainder of the CNS. Further, prior to and during the LH surge, there is enhanced secretion of LRF. However, estradiol also has a positive feedback effect directly on the pituitary gland, and prolonged administration of estradiol to women is accompanied by an augmented LH response to LRF administration (Keye and Jaffe, 1975). The failure to elicit a positive feedback action of estradiol (in high dosage) could be related to either the immaturity of the CNS or the pituitary, or both components.

Even though estradiol-induced positive feedback can be demonstrated by mid-puberty and prior to menarche, this does not imply that the positive feedback loop is complete. Indeed, the modulating action of the pubertal ovary and its output of estradiol on the hypothalamic–pituitary gonadotropin unit may be insufficient to induce an ovulatory LH surge even when there is an adequate pituitary store of readily releasable LH and FSH. The ovary, either from lack of sufficient gonadotropin stimulation or decreased responsivity or other local factors does not secrete estradiol at a high level or long enough to induce an ovulatory surge. We visualize the process leading to an ovulation as a gradual one in which the ovary [the "zeitgeber" for ovulation (Knobil and Plant, 1978)] and the hypothalamic–pituitary gonadotropin complex progressively become more integrated and synchronous until finally an ovary primed for ovulation secretes sufficient estradiol to induce an ovulatory LH surge.

During the first 2 years postmenarche, studies of basal body temperature changes (Doring, 1963) and of plasma progesterone concentrations (Winter and Faiman, 1973; Apter and Vihko, 1977) suggest that as many as 55–90% of cycles are anovulatory, the latter decreasing to less than 20% of cycles by 5 years after menarche (Apter and Vihko, 1977). There is evidence of a cyclic surge of LH during at least some of the anovulatory cycles in adolescence, but the mechanism of ovulation seems unstable and immature and has not yet attained the fine tuning and synchronization that is a requisite for maintenance of regular ovulatory cycles.

4.1. Summary of Present Concept

Our present concept of the ontogeny of the hypothalamic–pituitary gonadotropin–gonadal system in relation to the control of the onset of puberty is illustrated in Figure 12 and summarized in Table I. Clearly, the understanding of these complex maturational processes is deficient, but much illuminating information has emerged over the past decade.

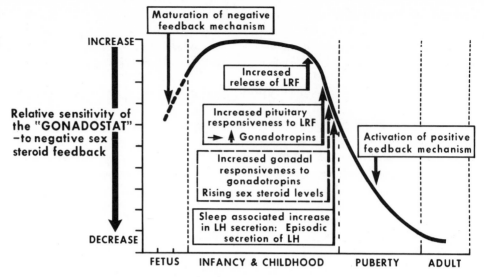

Fig. 12. Schematic illustration of the change in set point of the hypothalamic gonadostat (denoted by the dashed and solid lines) and the maturation of the negative and positive feedback mechanism from fetal life to adulthood in relation to the normal changes of puberty. (From Grumbach *et al.*, 1974.)

Table I. Postulated Ontogeny of Hypothalamic–Pituitary Gonadotropin–Gonadal Circuit [a,b]

Fetus
    1. Secretion of pituitary FSH and LH by 80 days gestation
    2. "Unrestrained" secretion of LRF (100–150 days)
    3. Maturation of negative sex steroid feedback mechanisms after 150 days gestation—sex difference
    4. Low level of LRF secretion at term

Infancy and childhood
    1. Negative feedback control of FSH and LH secretion becomes highly sensitive to sex steroids (low set point), maximum sensitivity attained by age 2–4 years
    2. Higher mean serum FSH and LH levels in female infants and transient increase in testosterone in males and estradiol in females during first 6 months of life owing to "immaturity" of gonadostat

Late prepubertal period
    1. Decreasing sensitivity of hypothalamic "gonadostat" to sex steroids (increased set point)
    2. Increased secretion of LRF
    3. Increased responsiveness of gonadotropes to LRF
    4. Increased secretion of FSH and LH
    5. Increased responsiveness of gonad to FSH and LH
    6. Increased secretion of gonadal hormones

Puberty
    1. Further decrease in sensitivity of negative feedback mechanism to sex steroids
    2. Sleep-associated increase in episodic secretion of LH
    3. Progressive development of secondary sex characteristics
    4. Mid- to late puberty—maturation of *positive* feedback mechanism and capacity to exhibit an estrogen-induced LH surge
    5. Spermatogenesis in male; ovulation in female

[a]FSH, follicle-stimulating hormone; LH, luteinizing hormone; LRF, LH-releasing factor.
[b]From Grumbach *et al.* (1974).

For many years, considerable speculation has focused on the adrenal component of puberty and the potential role of adrenal androgens in the onset of puberty. Recent studies have shed new light on these questions.

5.1. The Nature and Regulation of Adrenal Androgens

The major adrenal androgens secreted by the adrenal cortex are dehydroepiandrosterone (DHA), its sulfate (DHAS), and androstenedione. By extraglandular metabolism, the adrenal androgens contribute to circulating and physiologically active testosterone and estrone. In normal adult females only androstenedione is an important precursor and DHA and DHAS contribute little to plasma testosterone and estrone. However, scant information is available on the metabolism and kinetics of DHA and DHAS in prepubertal children. Androstenedione is the major androgen secreted by the ovary during and after puberty. It has considerably more androgenic activity than DHA and DHAS.

Cross-sectional and mixed longitudinal studies have demonstrated a progressive increase in the plasma concentration of DHA and DHAS in boys and girls by 7–8 years chronological age (6–8 years skeletal age) and continuing to 13–15 years (Hopper and Yen, 1974; Sizonenko and Paunier, 1975; Reiter *et al.*, 1977; reviewed in Grumbach *et al.*, 1978). It is accompanied by an increased excretion of urinary 17-ketosteroids, especially 11-deoxy C_{19} steroids. This increase, which serves as a mark of the onset of adrenarche, occurs about 2 years before the increase in gonadotropin and gonadal sex steroid secretion. It is not associated with increased sensitivity to exogenous LRF (Reiter *et al.*, 1975a) or sleep-associated LH secretion. Hence, it occurs at an age when the hypothalamic–pituitary gonadotropin–gonadal complex is functioning at a low level of activity (Grumbach *et al.*, 1978).

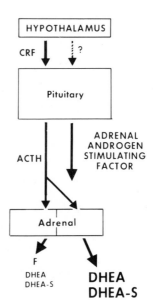

Fig. 13. Schematic representation of an hypothesis of the control of pituitary adrenal androgen secretion by a putative separate adrenal androgen-stimulating hormone acting on an ACTH-primed adrenal cortex. The zona reticularis appears to be the major site of DHAS secretion. (Modified from Grumbach *et al.*, 1978.)

MELVIN M. GRUMBACH

a **CHANGE OF SERUM DHAS WITH AGE RELATED TO GROWTH OF ZONA RETICULARIS**

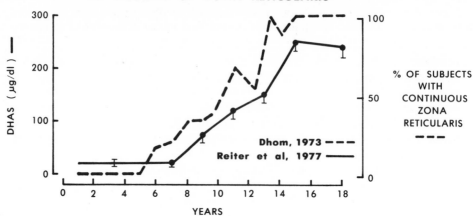

b **DEVELOPMENT OF THE ZONA RETICULARIS**

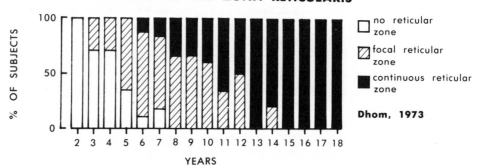

c **GROWTH OF THE ADRENAL**

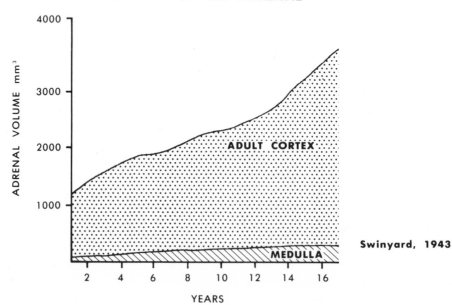

Recently hypotheses advanced for the control of adrenal androgen secretion have been reviewed (Grumbach *et al.*, 1978). The evidence, while incomplete and indirect, suggests that "the regulation of adrenal androgen secretion is based on a dual control mechanism: (a) ACTH is obligatory for (b) the action of an unidentified pituitary adrenal androgen stimulating hormone. ACTH is required to prime the adrenal for the second pituitary adrenal stimulating factor to be effective" (Grumbach *et al.*, 1978). This concept is illustrated in Figure 13. Neither the known pituitary hormones (such as ACTH, prolactin, FSH or LH, or growth hormone), an extrapituitary factor, nor a *de novo* or estrogen-induced maturational change in adrenal biogenesis with reactivation of C_{17-20} lyase and the C_{19} steroid pathway were viewed as more supportable alternatives to a separate pituitary adrenal androgen-stimulating hormone.

The hypothesis of a distinct pituitary adrenal androgen-stimulating hormone affords an explanation of the following observations: (1) the spurt in adrenal growth and the increase in the size of the zona reticularis at adrenarche that occurs independently of an increase in ACTH or cortisol secretion but correlates with the increase in plasma DHAS (Figure 14); (2) cortisol and adrenal androgen secretion vary independently with age, in normal, as well as premature adrenarche, starvation, malnutrition, anorexia nervosa and chronic disease; and (3) unlike cortisol, the secretion of DHA and DHAS in response to ACTH administration varies with age.

5.2. Adrenal Androgens and Puberty

The earlier onset of adrenarche than "gonadarche," the contribution of adrenal androgens to the growth of pubic and axillary hair, and experiments in the rat (recently interpreted in a different light) have led to the hypothesis that in normal children adrenal androgens are an important factor in the age of onset of puberty and the maturation of the hypothalamic–pituitary gonadotropin–gonadal complex. While true precocious puberty may occur in circumstances in which the prepubertal child previously had been exposed to excessive androgens from either an endogenous or exogenous source (e.g., after the initiation of glucocorticoid therapy in congenital virilizing adrenal hyperplasia) (Reiter *et al.*, 1975*b*; Grumbach *et al.*, 1978), there is little evidence to suggest that adrenal androgens play an important qualitative or rate-limiting role in the onset of puberty in normal children (Grumbach *et al.*, 1978).

Most patients with premature adrenarche, who precociously secrete excessive amounts of adrenal androgens for their age, enter puberty and experience menarche within the normal age range (Grumbach *et al.*, 1978). Moreover, prepubertal children who have congenital or acquired chronic adrenal insufficiency (Addison's disease) and, consequently deficient or absent adrenal androgen secretion, when given appropriate glucocorticoid and mineralocorticoid replacement therapy usually have a normal onset of and progression through puberty. Thus, early activation of

---

Fig. 14. Relation of plasma dehydroepiandrosterone sulfate (DHAS) to growth of the zona reticularis and the increase in adrenal volume with age. Part (a) shows the close correlation of the development of the zona reticularis and the increase in plasma DHAS. Part (b) indicates the age at which either focal islands of reticular tissue or a continuous reticular zone were found in children and adolescents with sudden death who had not had an antecedent illness. Part (c) demonstrates the increase in adrenal volume at the time of puberty. (From Grumbach *et al.*, 1978.)

adrenal androgen secretion does not commonly lead to precocious puberty nor is deficient or absent adrenal androgen output usually associated with delayed puberty. Further, growth studies in children with chronic adrenal insufficiency, isolated gonadotropin deficiency, and hypergonadotropic hypogonadism suggest that in girls and boys adrenal androgens are not essential for the adolescent growth spurt, whereas sex steroids secreted by the testis and ovary (Grumbach *et al.*, 1978) are in concert with growth hormone.

6. References

Apter, D., and Vihko, R., 1977, Serum pregnenolone, progesterone, 17-hydroxyprogesterone, testosterone and 5α-dihydrotestosterone during female puberty, *J. Clin. Endocrinol. Metab.* **45**:1039.

Baker, H. W. G., Bremner, W. J., Burger, H. G., de Kretser, D. M., Dulmanis, A., Eddie, L. W., Hudson, B., Keogh, E. J., Lee, V. W. K., and Rennie, G. C., 1976, Testicular control of follicle-stimulating hormone secretion, *Rec. Prog. Horm. Res.* **32**:429.

Bhattacharya, A. N., Dierschke, D. J., Yamaji, T., and Knobil, E., 1972, The pharmacologic blockade of the circhoral mode of LH secretion in the ovariectomized rhesus monkey, *Endocrinology* **90**:778.

Bierich, J. R. (ed.), 1975, Disorders of puberty, *Clin. Endocrinol. Metab.* **4**:1–222.

Billewicz, W. Z., Fellowes, H. M., and Hytten, C. A., 1976, Comments on the critical metabolic mass and the age of menarche, *Ann. Hum. Biol.* **3**:51.

Boyar, R., Finkelstein, J., Roffwarg, H., Kapen, S., Weitzman, E., and Hellman, L., 1972, Synchronization of augmented luteinizing hormone secretion with sleep during puberty, *N. Engl. J. Med.* **287**:582.

Boyar, R. M., Finkelstein, J. W., Roffwarg, H., Kapen, S., Weitzman, E. D., and Hellman, L., 1973*a*, Twenty-four-hour luteinizing hormone and follicle-stimulating hormone secretory patterns in gonadal dysgenesis, *J. Clin. Endocrinol. Metab.* **37**:521.

Boyar, R. M., Finkelstein, J. W., David, R., Roffwarg, H., Kapen, S., Weitzman, E. D., and Hellman, L., 1973*b*, Twenty-four hour patterns of plasma luteinizing hormone and follicle-stimulating hormone in sexual precocity, *N. Engl. J. Med.* **289**:282.

Boyar, R. M., Rosenfeld, R. S., Kapen, S., Finkelstein, J. W., Roffwarg, H. P., Weitzman, E. D., and Hellman, L., 1974, Simultaneous augmented secretion of luteinizing hormone and testosterone during sleep, *J. Clin. Invest.* **54**:609.

Bugnon, C., Bloch, B., and Fellman, D., 1976, Mise en évidence cyto-immunologique de neurones à LH-RH chez le foetus humain, *C. R. Acad. Sci.* **282**(D):1625.

Cameron, N., 1976, Weight and skinfold variation at menarche and the critical body weight hypothesis, *Ann. Hum. Biol.* **3**:279.

Conte, F. A., Grumbach, M. M., and Kaplan, S. L., 1975, A diphasic pattern of gonadotropin secretion in patients with the syndrome of gonadal dysgenesis, *J. Clin. Endocrinol. Metab.* **40**:670.

Critchlow, V., and Bar-Sela, M. E., 1967, Control of the onset of puberty, in: *Neuroendocrinology,* Vol. 2 (L. Martini and W. F. Ganong, eds.), pp. 101–162, Academic Press, New York.

Davidson, J. M., 1974, Hypothalamic–pituitary regulation of puberty: Evidence from animal experimentation, in: *Control of the Onset of Puberty* (M. M. Grumbach, G. D. Grave, and F. E. Mayer, eds.), p. 79, John Wiley & Sons, New York.

Donovan, B. T., and van der Werff ten Bosch, J. J., 1965, *Physiology of Puberty,* Williams and Wilkins, Baltimore.

Doring, G. K., 1963, Uber die relative sterilitat in den Jahren nach der menarche, *Geburtschilfe Frauenheilkd.* **23**:30.

Forest, M. G., de Peretti, E., and Bertrand, J., 1976, Hypothalamic–pituitary–gonadal relationships in man from birth to puberty. *Clin. Endocrinol.* **5**:551.

Frisch, R. E., 1976, Fatness of girls from menarche to age 18 with a nomogram, *Hum. Biol.* **48**:353.

Frisch, R. E., and Revelle, R., 1970, Height and weight at menarche and a hypothesis of critical body weights and adolescent events, *Science* **169**:97.

Gorski, R. A. 1974, Extrahypothalamic influences on gonadotropin secretion, in: *Control of the Onset of Puberty* (M. M. Grumbach, G. D. Grave, and F. E. Mayer, eds.), p. 182, John Wiley & Sons, New York.

Grumbach, M. M., 1975, Onset of Puberty, in: *Puberty, Biologic and Psychosocial Components* (S. R. Berenberg, ed.), pp.1–21, H. E. Stenfert Kroese B.V., Leiden.

Grumbach, M. M., and Kaplan, S. L., 1973, Ontogenesis of growth hormone, insulin, prolactin, and gonadotropin secretion in the human foetus, in: *Foetal and Neonatal Physiology,* Proc. Sir Joseph Barcroft Centenary Symposium, pp. 462–497, Cambridge University Press, Cambridge.

Grumbach, M. M., and Kaplan, S. L., 1974, Fetal pituitary hormones and the maturation of central nervous system regulation of anterior pituitary function, in: *Modern Perinatal Medicine* (L. Gluck, ed.), pp. 247–271, Year Book Medical Publishers, Chicago.

Grumbach, M. M., Roth, J. C., Kaplan, S. L., and Kelch, R. P., 1974, Hypothalamic–pituitary regulation of puberty in man: Evidence and concepts derived from clinical research, in: *Control of the Onset of Puberty* (M. M. Grumbach, G. D. Grave, and F. E. Mayer, eds.), pp. 115–166, John Wiley & Sons, New York.

Grumbach, M. M., Richards, G. E., Conte, F. A., and Kaplan, S. L., 1978, Clinical disorders of adrenal function and puberty: An assessment of the role of the adrenal cortex in normal and abnormal puberty in man and evidence for an ACTH-like pituitary adrenal androgen stimulating hormone, in: *The Endocrine Function of the Human Adrenal Cortex* (M. Serio, ed.), Serono Symposium, 1977, Academic Press, New York (in press).

Hohlweg, W., and Dohrn, M., 1932, Uber die Beziehungen zwischen Hypophysenvorderlappen und Keimdrusen, *Klin. Wochenschr.* **11**:233.

Hopper, B. R., and Yen, S. S. C., 1974, Circulating concentrations of dehydroepiandrosterone and dehydroepiandrosterone sulfate during puberty, *J. Clin. Endocrinol. Metab.* **40**:458.

Job, J. C., Garnier, P. E., Chaussain, J. L., and Milhaud, G., 1972, Elevation of serum gonadotropins (LH and FSH) after releasing hormone (LH-RH) injection in normal children and in patients with disorders of puberty, *J. Clin. Endocrinol. Metab.* **35**:473.

Johnston, F. E., Roche, A. F., Schell, L. M., and Wettenhall, N. B., 1975, Critical weight at menarche, *Am. J. Dis. Child.* **129**:19.

Judd, H. L., Parker, D. C., and Yen, S. S. C., 1977, Sleep-wake patterns of LH and testosterone in prepubertal boys, *J. Clin. Endocrinol. Metab.* **44**:865.

Kaplan, S. L., Grumbach, M. M., and Aubert, M. L., 1976, The ontogenesis of pituitary hormones and hypothalamic factors in the human fetus: Maturation of central nervous system regulation of anterior pituitary function, *Rec. Prog. Horm. Res.* **32**:161.

Karsch, F. J., Dierschke, D. J., and Knobil, E., 1973, Sexual differentiation of pituitary function: Apparent difference between primates and rodents, *Science* **179**:484.

Kelch, R. P., Kaplan, S. L., and Grumbach, M. M., 1973, Suppression of urinary and plasma follicle-stimulating hormone by exogenous estrogens in prepubertal and pubertal children, *J. Clin. Invest.* **52**:1122.

Keye, W. R., Jr., and Jaffe, R. B., 1975, Strength-duration characteristics of estrogen effects on gonadotropin response to gonadotropin-releasing hormone in women. I. Effects of varying duration of estradiol administration, *J. Clin. Endocrinol. Metab.* **41**:1003.

Knobil, E., and Plant, T. M., 1978, The neuroendocrine control of gonadotropin secretion in the female rhesus monkey, in: *Frontiers in Neuroendocrinology,* Vol. 5 (W. F. Ganong and L. Martini, eds.), Raven Press, New York (in press).

Kulin, H. E. and Reiter, E. O., 1976, Gonadotropin and testosterone measurements after estrogen administration to adult men, prepubertal and pubertal boys, and men with hypogonadotropism: Evidence for maturation of positive feedback in the male, *Pediatr. Res.* **10**:46.

Kulin, H. E., Grumbach, M. M., and Kaplan, S. L., 1972, Gonadal–hypothalamic interaction in prepubertal and pubertal man: Effect of clomiphene citrate on urinary follicle-stimulating hormone and luteinizing hormone and plasma testosterone, *Pediatr. Res.* **6**:162.

Odell, W. D., and Swerdloff, R. S., 1976, Etiologies of sexual maturation: A model system based on the sexually maturing rat, *Rec. Prog. Horm. Res.* **32**:245.

Okon, E., and Koch, Y., 1976, Localisation of gonadotropin-releasing and thyrotropin-releasing hormones in human brain by radioimmunoassay, *Nature* **263**:345.

Penny, R., Olambiwonnu, N. O., and Frasier, S. D., 1977, Episodic fluctuations of serum gonadotropins in pre- and post-pubertal girls and boys, *J. Clin. Endocrinol. Metab.* **45**:307.

Presl, J., Horejsi, J., Strouflova, A., and Herzmann, J., 1976, Sexual maturation in girls and the development of estrogen-induced gonadotropic hormone release, *Ann. Biol. Anim. Bioch. Biophys.* **16**:377.

Ramirez, V. D., 1973, Endocrinology of puberty, in: *Handbook of Physiology,* Vol. 2, Part 1, Female Reproductive System (R. O. Greep, ed.), pp. 1–28, American Physiological Society, Washington, D.C.

Reiter, E. O., Kulin, H. E., and Hamwood, S. M., 1974, The absence of positive feedback between estrogen and luteinizing hormone in sexually immature girls, *Pediatr. Res.* **8:**740.

Reiter, E. O., Kaplan, S. L., Conte, F. A., and Grumbach, M. M., 1975*a,* Responsivity of pituitary gonadotropes to luteinizing hormone-releasing factor in idiopathic precocious puberty, precocious thelarche, precocious adrenarche, and in patients treated with medroxyprogesterone acetate, *Pediatr. Res.* **9:**111.

Reiter, E. O., Grumbach, M. M., Kaplan, S. L., and Conte, F. A., 1975*b,* The response of pituitary gonadotropes to synthetic LRF in children with glucocorticoid-treated congenital adrenal hyperplasia: Lack of effect of intrauterine and neonatal androgen excess, *J. Clin. Endocrinol. Metab.* **40:**318.

Reiter, E. O., Fuldauer, V. G., and Root, A. W., 1977, Secretion of the adrenal androgen, dehydroepiandrosterone sulfate, during normal infancy, childhood, and adolescence, in sick infants, and in children with endocrinologic abnormalities *J. Pediatr.* **90:**766.

Ross, G. T., Cargille, C. M., Lipsett, M. B., Rayford, P. L., Marshall, J. R., Strott, C. A., and Rodbard, D., 1970, Pituitary and gonadal hormones in women during spontaneous and induced ovulatory cycles, *Rec. Prog. Horm. Res.* **26:**1.

Roth, J. C., Kelch, R. P., Kaplan, S. L., and Grumbach, M. M., 1972, FSH and LH response to luteinizing hormone-releasing factor in prepubertal and pubertal children, adult males and patients with hypogonadotropic and hypergonadotropic hypogonadism, *J. Clin. Endocrinol. Metab.* **35:**926.

Roth, J. C., Grumbach, M. M., and Kaplan, S. L., 1973, Effect of synthetic luteinizing hormone-releasing factor on serum testosterone and gonadotropins in prepubertal, pubertal and adult males, *J. Clin. Endocrinol. Metab.* **37:**680.

Sizonenko, P. C., and Paunier, L., 1975, Hormonal changes in puberty. III. Correlation of plasma dehydroepiandrosterone, testosterone, FSH, and LH with stages of puberty and bone age in normal boys and girls and in patients with Addison's disease or hypogonadism or with premature or late adrenarche, *J. Clin. Endocrinol. Metab.* **41:**894.

Sizonenko, P. C., Cuendet, A., and Paunier, L., 1973, FSH. I. Evidence for its mediating role on testosterone secretion in cryptorchidism, *J. Clin. Endocrinol. Metab.* **37:**68.

Wilkins, L., 1965, *The Diagnosis and Treatment of Endocrine Disorders in Childhood and Adolescence,* 2nd ed., Charles C Thomas, Springfield, Illinois.

Winter, J. S. D., and Faiman, C., 1973, Pituitary–gonadal relations in female children and adolescents, *Pediatr. Res.* **7:**948.

Winter, J. S. D., Taraska, S., and Faiman, C., 1972, The hormonal response to HCG stimulation in male children and adolescents, *J. Clin. Endocrinol. Metab.* **34:**348.

Winter, J. S. D., Faiman, C., Hobson, W. C., Prasad, A. V., and Reyes, F. I., 1975, Pituitary–gonadal regulations in infancy. I. Patterns of serum gonadotropin concentrations from birth to four years of age in man and chimpanzee, *J. Clin. Endocrinol. Metab.* **40:**545.

Yen, S. S. C., and Jaffe, R. B., 1978, *Reproductive Endocrinology:* W. B. Saunders, Philadelphia.

Yen, S. S. C., Lasley, B. L., Wang, C. F., Leblanc, H., and Siler, T. M., 1975, The operating characteristics of the hypothalamic–pituitary system during the menstrual cycle and observations of biological action of somatostatin, *Rec. Prog. Horm. Res.* **31:**321.

Zacharias, L., Rand, W. M., and Wurtman, R. J., 1976, A prospective study of sexual development in American girls: The statistics of menarche, *Obstet. Gynecol. Survey,* **31:**325.

9

Body Composition in Adolescence

GILBERT B. FORBES

1. Introduction

Anatomists showed many years ago that some organs grow at rates which differ from that of the body as a whole. In recent decades we have witnessed the development of techniques which permit the assessment of certain features of body composition in the living subject. Humans rank among the fattest of mammals, and it is this fact, among others, which has stimulated the development of methods for estimating the relative proportions of lean and fat in our species. These methods make it possible to define the contribution of each component to the adolescent growth process. Body composition data can also help in estimating certain nutritional requirements for growth. Although long a favorite of nutritionists, the metabolic balance method is not equal to this task (Wallace, 1959; Forbes, 1973), and one can look forward to more realistic estimates based on these newer techniques.

1.1. The Concept of Chemical Maturity

For more than a century biologists and chemists have known that young tissues differ from old in chemical composition (Bezold, 1857). The young body has a higher proportion of water and a lower proportion of ash; young bones contain less calcium, young muscle less potassium, and the ratio of extracellular fluid (ECF) volume to intracellular fluid (ICF) volume declines during growth (Friis-Hansen, 1957; Nichols *et al.*, 1968); indeed the average adult (of today) is fatter than the child.

A half century ago Moulton (1923) put forth the concept of chemical maturity, namely that at some point during growth body composition approaches that of the adult, and hence is considered "mature." (Incidentally, aging reverses this process: for example, relative ECF volume increases, bone loses calcium.) Moulton concluded that mammals, including man, reach chemical maturity at about 4% of their

---

GILBERT B. FORBES • The University of Rochester, School of Medicine and Dentistry, Rochester, New York.

life span. The age for man is not known precisely, and it is likely it varies for different tissues. Roentgenograms show that the skeleton is not completely ossified until after adolescence; on the other hand skeletal muscle composition achieves values in the adult range for potassium, chloride, and water by about age six years (Nichols *et al.*, 1968). Cureton *et al.* (1975) suggest that "chemical maturity for potassium and the density of the fat-free body" has been achieved by the age of eight years. Data on composition of the newborn and adult, as determined by chemical analysis, are summarized in a paper by Forbes (1962).

1.2. The Fat-Free Body

Early workers noted that tissue samples were not always of uniform composition. These discrepancies were finally explained in large part by the work of such individuals as Pfeiffer (1887) and Magnus-Levy (1906), and later Hastings and Eichelberger (1937), who pointed out the necessity of accounting for the fat content of the tissue sample. Neutral fat does not bind water, nitrogen, or electrolyte, and it is now common practice to express the results of tissue analysis on a fat-free basis. It is this relative constancy of the composition of fat-free tissues, and of the fat-free body as a whole (Pace and Rathbun, 1945; Forbes and Lewis, 1956) that forms the basis for some of the methods now in use for estimating body composition in living man. Figure 1 shows values for chemically determined content of water, potassium, and calcium in man when these are calculated on a fat-free weight basis. It is obvious that major changes take place during growth: water content declines, as does Cl; K and Ca contents increase, as do N, P, and Mg (Forbes, 1962). The fact that body Ca content increases more rapidly than K means that the ossified skeleton grows relatively more rapidly than soft tissues. It is also evident that there is reasonable compositional consistency for each age group when values are calculated on a fat-free basis.

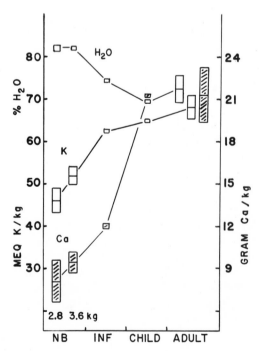

Fig. 1. Composition of fat-free human body by chemical analysis. Means ±SD for H_2O and K (clear boxes, scale at left) and Ca (filled boxes, scale at right). Data from Widdowson *et al.* (1951), Garrow *et al.* (1965), Forbes (1962).

The terms "fat-free weight" (FFW) and "lean body mass" (LBM) are both in use. The author interprets these terms as synonymous, and prefers the latter as the more delicate of the two. As used in this chapter LBM means body mass minus ether-extractable fat and hence includes the stroma of adipose tissue.

Behnke and Wilmore (1974) have characterized the LBM as possessing 3–5% of its weight as "essential lipid," i.e., structural as opposed to storage lipid. Garrow (1974) defines LBM as total weight minus adipose tissue, i.e., fat together with its framework of adipocytes and connective tissue. Brozek *et al.* (1963) prefer to use an entity called "the reference man," designated to contain a certain amount of fat, to which can be added variable amounts of "obesity tissue." Womersley *et al.* (1976) have produced a sophisticated analysis of the contribution made by various techniques to the field of body composition.

Moore *et al.* (1963) have introduced the term "body cell mass" (BCM) to designate an entity whose mass (in grams) is calculated by multiplying total exchangeable K (mEq) by 8.33. Burmeister and Bingert (1967) also make use of this concept (although they assume BCM contains 92.5 mEq K/kg) which in essence defines that portion of the body which contains K at a given concentration.

2. Techniques for Assessment of Body Composition

2.1. Dilution Procedures

These make use of a relationship long used for the analysis of static systems, namely $C_1 V_1 = C_2 V_2$, where C is concentration and V is volume. Since quantity (Q) of solute $= CV$, the volume of a solution (V_2) can be determined by adding a known quantity of solute and then, after mixing has occurred, measuring the final concentration (C_2):

$$V_2 = \frac{Q}{C_2} \tag{1}$$

This general principle has been used to estimate the volume of various body fluid compartments. However, the body is not a static system, and several assumptions have to be made: (1) the administered material enters the compartment in question quickly and becomes thoroughly mixed within a reasonable period of time; (2) the apparent volume of distribution coincides with that of the compartment in question; (3) if a steady-state concentration is not achieved, by virtue of metabolism and/or excretion of the administered material, the change in plasma concentration should be some simple function of time so that the concentration at zero time can be calculated; (4) the compartment in question is not undergoing changes in volume during the mixing period; and (5) the material administered does not significantly alter normal physiologic processes or homeostatic mechanisms. Obviously, allowance must be made for excretion of the administered material; if this is significant, equation (1) must be modified to read: $V_2 = (Q_{\text{administered}} - Q_{\text{excreted}})/C_2$. It is evident, therefore, that the prerequisites for such fluid volume assays are rigorous.

In the last analysis, it is difficult to conceive of a material which would completely satisfy assumption (2) above. Plasma and ECF represent the only avenues of communication of body cells with the outside world, so materials of all sorts are continuously being transported in and out of these compartments and at a very rapid rate. Such transfer rates are generally assumed to represent first-order reactions, and the mathematical technique known as kinetic analysis has been developed to a high art.

The materials which have been used for estimation of body fluid volumes are listed in Figure 2 and Table I. The list is fairly long. Some materials can be given orally; some must be given intravenously. Not all yield the same result: for example, sucrose and inulin both yield lower values for ECF volume than does bromide.

Intracellular fluid (ICF) volume is estimated by difference (ICF = total H_2O − ECF), and it is clear that the calculated ICF volume will depend on the materials used to estimate ECF volume.

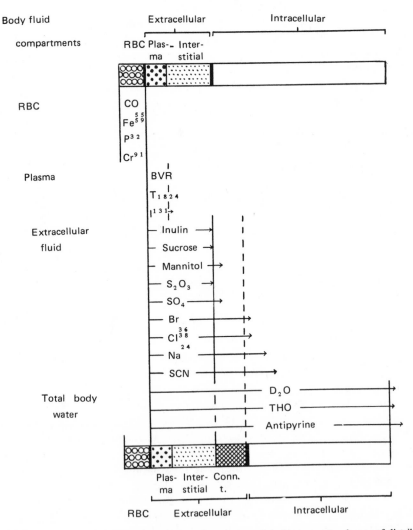

Fig. 2. Materials used to estimate body fluid compartments and their apparent volumes of distribution. T-1824 = Evans blue dye; [131]I = labeled serum albumin. (From Elkinton and Danowski, 1955, with permission.)

Table I. Additional Methods for Estimation of Body Fluid Volumes

Compartment	Material	Remarks
Extracellular fluid volume	^{82}Br ($t_{1/2}$ 35 hr)	Stable Br analysis by fluorescent excitation (Kaufman and Wilson, 1973)
Total body water	^{35}SO$_4^{2-}$ ($t_{1/2}$ 87 days)	Can be given p.o. (Bauer, 1976)
	N-Acetyl-4-amino antipyrine	Not bound to plasma proteins (Brodie *et al.*, 1951)
	Urea	McCance and Widdowson, 1951; Bradbury, 1961
	Alcohol	Give p.o., breath analysis (Grüner and Salmen, 1961)
Total exchangeable K	^{42}K ($t_{1/2}$ 12.4 hr)	
	^{43}K ($t_{1/2}$ 22 hr)	Davies and Robertson (1973)
Total exchangeable Na	^{24}Na ($t_{1/2}$ 15 hr)	^{22}Na ($t_{1/2}$ 2.6 yr) has long residence time in bone
Total exchangeable Cl	Br, ^{82}Br	

2.1.1. Isotopic Dilution

The advantage of radioisotopes lies in the fact that extremely small quantities can be measured with reasonable accuracy; hence the mass of administered material is so small as to not appreciably add to the *net* amount already present in the body, or to alter physiologic mechanisms. As an example, 50 μCi of ^{24}Na, an amount sufficient to measure total exchangeable Na in an adult, weighs only 6×10^{-9} mg. There is, of course, a penalty for this tremendous advantage in the form of the accompanying radioactive emissions.

The dilution principle is the same, except that one attempts to measure a quantity rather than a volume. A known amount of the radioisotope is administered, and a period allowed for mixing to occur with its sister isotope in the body. At the end of the equilibration period, during which urine has been collected, a sample of urine or plasma is obtained for analysis. Then

$$\text{Total body H}_2\text{O (ml)} = \frac{^3\text{H}_{\text{adm}} - {}^3\text{H}_{\text{exc}}}{(^3\text{H/ml serum H}_2\text{O})}$$

(2)

$$\text{Total exchangeable Na (mEq)} = \frac{^{24}\text{Na}_{\text{admin}} - {}^{24}\text{Na}_{\text{exc}}}{(^{24}\text{Na/mEq Na in serum})}$$

In like manner total exchangeable K and Cl can be determined (Br is used for the latter since there is no suitable radioisotope of Cl).

The assumptions here are (1) that the radioisotope is handled by the body in the same way as its sister stable isotope; and (2) that mixing is complete, i.e., specific activity (X^*/X) is the same in all tissues. In the case of tritium and deuterium, mixing is complete in about 3 hr; in point of fact these isotopes also exchange with the hydrogens of amino and carboxyl groups and so lead to an overestimation of body water content by a few percent (Foy and Schnieden, 1960; Bradbury, 1961; Tisavipat *et al.*, 1974). Radiosodium dilution underestimates total body Na because of incomplete exchange with Na of bone (Forbes and Perley, 1951). Recently neutron activation analysis has been compared with isotopic dilution (Rudd *et al.*, 1972; Ellis *et al.*, 1976), to yield an underestimate of about 25%.

Administered ^{42}K mixes rather slowly and not completely with body K. Comparisons of total body K by ^{40}K counting and by isotopic dilution indicate that the latter method underestimates body K content by 3–10% (Rundo and Sagild, 1955; Surveyor and Hughes, 1968; Talso *et al.,* 1960; Jasani and Edmonds, 1971; Davies and Robertson, 1973; Boddy *et al.,* 1974; Lye *et al.,* 1976). Administered radiochloride mixes with body Cl in most tissues (brain is a clear exception) within a few minutes (Manery and Haege, 1941), but neither ^{36}Cl or ^{38}Cl is particularly suitable for human use. Bromide, either stable or radioactive, is therefore used; its volume of distribution is slightly larger than that of isotopic Cl and continues to increase for at least 21 hr (Gamble *et al.,* 1953). In the rat, however, total exchangeable Cl estimated by Br dilution has been found to equal chemically determined carcass Cl (Cheek and West, 1955).

It is customary to speak of the results of isotopic dilution assays as total exchangeable Na, K, and Cl. The radiation dose delivered to the body from the tracer amounts used in such studies is about 2 mrad/μCi.

2.2. Potassium-40

Nature has been kind to those studying body composition in devising an isotope of potassium which is long-lived ($t_{1/2} = 1.3 \times 10^9$ yr), emits a strong gamma ray (1.46 MeV), and is sufficiently abundant (0.012% of K) to permit its detection and quantitation in human subjects. The human adult body contains about 0.1 μCi

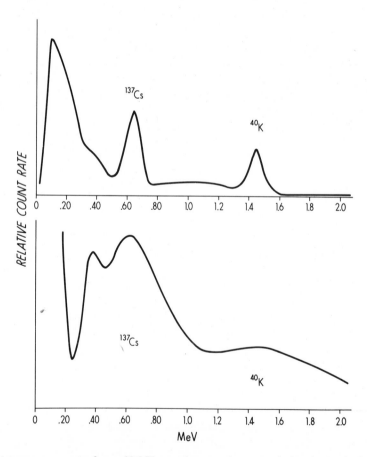

Fig. 3. Net gamma ray spectra from a NaI-Tl crystal counter (upper section) and a 4π plastic scintillator (lower section)—for subject containing 10 nCi ^{137}Cs (fallout level *circa* 1965) and 130 gK.

(or 17 mg) of ^{40}K, which produces about 200,000 beta rays and 25,000 gamma rays each minute. The resultant radiation dose is about 0.016 rem/yr, which is 16% of the total radiation background to man from all natural sources.

The body content of ^{40}K can be assayed in specially constructed counters, which essentially consist of a large shielded room (to reduce background radiation from cosmic and terrestrial sources) containing a gamma-ray detector connected to a suitable recording apparatus. The detectors are of two types: large thallium-activated NaI crystals, one or more of which are positioned near the subject; and large hollow cylinders, or half cylinders, the walls of which contain liquid or plastic scintillation material and into which the subject is placed so as to be completely or partially surrounded by the detector. These are referred to as having 4π or 2π geometry. The crystal type has a very good energy resolution, a low background rate, and low efficiency, while the plastic or liquid systems have very poor resolution, high background rate, and high efficiency.

Figure 3 illustrates the net gamma-ray spectra for ^{137}Cs and ^{40}K obtained from a 4π and a crystal counter. The International Atomic Energy Agency (1970) has published a directory describing the features of the whole-body monitors now in use throughout the world. Rather elaborate instrumentation is needed to properly generate and interpret the gamma-ray spectra obtained from human subjects. Most importantly, each instrument must be properly calibrated so that the recorded ^{40}K activity can be accurately translated into terms of body K content. The calibration must take into account subjects of varying size, the position of the subject relative to the detector, and the extent to which ^{40}K gamma rays are absorbed by the subject's body. The whole-body counter provides a noninvasive means for estimating body composition; the only hazard is claustrophobia.

2.3. Body Density

Archimedes is credited with discovering that one can estimate the relative proportions of a two-component mixture, each component being of known density, by measuring the density of the whole system. Now it is known that fat has a density of 0.900 g/cm^3, and it has been estimated that lean has one of about 1.100 g/cm^3; hence if the density of the body is determined, the relative proportions of lean and fat can be calculated. The formulation proceeds as follows: let Wf represent weight of fat, Wl weight of lean, and Wb weight of body, and Vf, Vl, and Vb their respective volumes. Since density $(D) = W/V$, the relationship $Vb = Vf + Vl$ can be written as

$$\frac{Wb}{Db} = \frac{Wf}{Df} + \frac{Wl}{Dl} \qquad (3)$$

and since $Wl = Wb - Wf$, it follows that

$$\frac{Wb}{Db} = \frac{Wf}{Df} + \frac{Wb - Wf}{Dl}$$

Rearranging, one has

$$\frac{Wb}{Db} - \frac{Wb}{Dl} = \frac{Wf}{Df} - \frac{Wf}{Dl}$$

whence

$$Wb\left(\frac{1}{Db} - \frac{1}{Dl}\right) = Wf\left(\frac{1}{Df} - \frac{1}{Dl}\right)$$

which can be written as

$$\frac{Wf}{Wb} = \frac{\left(\dfrac{1}{Db} - \dfrac{1}{Dl}\right)}{\left(\dfrac{1}{Df} - \dfrac{1}{Dl}\right)}$$

A final rearrangement yields the following

$$\frac{Wf}{Wb} = \left[\frac{1}{Db} \times \overbrace{\left(\frac{1}{\dfrac{1}{Df} - \dfrac{1}{Dl}}\right)}^{a}\right] - \left[\frac{1}{Dl} \times \overbrace{\left(\frac{1}{\dfrac{1}{Df} - \dfrac{1}{Dl}}\right)}^{a'}\right] \tag{4}$$

Now if Df and Dl are known, the constants a and a' can be calculated, and equation (4) becomes

$$\text{Fraction fat}\left(\frac{Wf}{Wb}\right) = \frac{a}{Db} - a' \tag{5}$$

and fraction lean is

$$1 - \frac{Wf}{Wb}$$

Body density can be measured by underwater weighing:

$$Db = \frac{W(\text{air})}{W(\text{air}) - W(\text{water})} \tag{6}$$

Appropriate corrections have to be made for the density of water and air and for the volume of air in the lungs (gastrointestinal-tract air is neglected). Lung volume can be estimated by nitrogen washout. This method requires a fair degree of subject cooperation and is obviously unsuited to young children.

Body volume can be estimated by water displacement, by helium dilution, by air displacement, and even by special photographic techniques, whence density is simply W/V. Serious technical problems have limited the use of the latter two methods, and the presence of variable amounts of air in the gastrointestinal tract constitutes a real problem in using the helium dilution method for infants. The underwater weighing technique is the only one in common use. Details of these various techniques can be found in Whipple and Brozek (1963) and in Brozek and Henschel (1961).

2.4. Neutron Activation

Many elements respond to neutron bombardment by capturing the neutron, whereupon the atomic nucleus becomes unstable and emits particulate or electromagnetic radiation of characteristic energy. It is this property together with the decay rate which permits identification of the particular isotope produced by the neutron bombardment. This technique was first used some years ago for the analysis of tissue samples. The sample is exposed to a neutron beam of known flux, and the gamma rays emanating from the induced isotopes are detected by means of

Table II. Nuclear Reactions

Substance	Activation product half-life
^{48}Ca (0.18%)[a] (n, γ)[b] ^{49}Ca	8.8 min
^{23}Na (100%) (n, γ) ^{24}Na	15 hr
^{37}Cl (25%) (n, γ) ^{38}Cl	37 min
^{31}P (100%) (n, α) ^{28}Al	2.3 min
^{14}N (99.6%) $(n, 2n)$ ^{13}N	10 min
^{26}Mg (11%) (n, γ) ^{27}Mg	9.5 min

[a]Isotopic abundance.
[b]Nuclear reaction.

a scintillation counter connected to a multichannel pulse-height analyzer; hence the isotopes can be identified by their characteristic gamma-ray spectra and quantitated. This method is extremely sensitive; for example, as little as 10^{-9} mEq Na can be assayed.

During the past decade this technique has been applied to whole animals, including man. The subject is irradiated with neutrons, and then placed in a whole-body scintillation counter for detection of the induced radioactivity (Anderson *et al.*, 1964; Harvey *et al.*, 1973; Nelp *et al.*, 1970; Cohn and Dombrowski, 1971; Dombrowski *et al.*, 1973).

The reactions which have been employed are listed in Table II.

A survey of the technique, including *in vivo* estimation of thyroid iodine content, liver cadmium, and body hydrogen content can be found in Cohn and Fairchild (1976).

Thus the total body content of Ca, P, Na, Cl, and N can be determined in the living subject and with reasonable accuracy.* But there are formidable problems associated with this truly elegant technique: The equipment is very expensive and intricate (indeed there are only a few laboratories in the world capable of assaying human subjects); the radiation dose from neutrons and the neutron energy spectra must be precisely known; and the subject must be exposed to radiation. Although the radiation dose is very small ($\sim$ 30 mrad), it is this feature which has thus far deterred the use of this technique for children.

2.5. Uptake of Fat-Soluble Gases

Both cyclopropane and radioactive krypton are much more soluble in fat than in lean tissues. A subject placed in an environment containing a known amount of such a gas will progressively take up the gas into his body, and the rate of uptake will depend almost entirely on the amount of fat contained therein. This method has been used successfully in estimating body fat content in adults and in laboratory animals (Lesser *et al.*, 1960, 1971; Hytten *et al.*, 1966), but one attempt to use it for infants failed (Halliday, 1971). It has the advantage of being the only technique available for *direct* estimation of body fat, lean weight being determined by difference. The disadvantages are the elaborate apparatus needed, the very long equilibration time (several hours of uninterrupted exposure), the multitude of

---

*Boddy *et al.* (1976) have used this technique to monitor changes in body composition of animals given a low Ca diet, and Cohn *et al.* (1974) have documented the reduced body Ca content in patients with osteoporosis.

corrections (temperature, pressure, etc.) necessary, and the sophisticated mathematical treatment which must be employed.

There is one report of the use of nitrogen washout for estimating body fat content (Behnke, 1945).*

2.6. Urinary Creatinine Excretion

This compound is formed in muscle from phosphocreatine, and provided the diet does not contain excessive amounts of preformed creatinine, the quantity which appears in the urine is generally considered to be an index of muscle mass (Cheek, 1968). A recent study of subjects of both sexes and of widely varying body size revealed an excellent correlation ($r = 0.988$) between LBM and creatinine excretion when two ^{40}K counts and three consecutive 24-hr collections of urine were made (Forbes and Bruining, 1976). The equation describing this relationship is LBM (kg) = 0.0291 Cr (mg/day) + 7.38.

Certain technical considerations must be kept in mind. Despite Shaffer's (1908) early dictum, creatinine excretion does vary somewhat from day to day, even in face of a constant diet or during fasting. The urine collection period must be accurately timed, since an error as small as 15 min represents 1% of a 24-hr period. It is advisable to make three consecutive 24-hr collections.

Picou *et al.* (1976) estimated muscle mass in infants and children by a complicated technique involving the administration of [^{15}N]creatine (^{15}N is a stable isotope). The time course of the urinary excretion of labeled creatine is determined, together with urinary creatinine and the creatine concentration in a sample of muscle. Kreisberg *et al.* (1970) used this same procedure in adults by administering [^{14}C]creatine. While this technique is of theoretical interest, it is very complicated and requires a muscle biopsy.

The regression equations derived from the data of Forbes and Bruining (1976), as well as many others, possess positive intercepts on the y axis. Hence the ratio of muscle mass, or LBM,† to urinary creatinine excretion declines hyperbolically as creatinine excretion increases, so that one cannot speak of a constant ratio.

3-Methylhistidine, a nonmetabolizable amino acid, is known to be formed in muscle and to be linked to actin and myosin. Attempts are now being made to use the urinary excretion of this compound as an index of muscle protein turnover and of muscle mass (Long *et al.*, 1975).‡

2.7. External Radiation Techniques

2.7.1. Bone

Roentgenograms of the hand are taken, in a manner to minimize parallax, and measurements are made of the cortex of the second metacarpal, either with a special caliper§ or with a magnifying glass fitted with a reticule. Garn (1970) and Gryfe *et al.* (1971) prefer to measure cortex thickness at a point equidistant from the

---

*Mettau *et al.* (1977) have successfully estimated body fat in young infants with nonradioactive xenon.

† Analyses of human adults and of a large number of mammalian species reveal that fat-free muscle constitutes about half the total fat-free body weight (see Forbes and Bruining, 1976).

‡ A new method for estimating lean weight is now undergoing trials in animals (Domermuth *et al.*, 1976). It is based on the change in electrical conductivity when the subject is placed in an electromagnetic field; the change is proportional to the amount of electrolyte-containing material present.

§ Helios Caliper fitted with needle points and a dial which can be read to the nearest 0.05 mm.

two ends of the bone, while Bonnard (1968) takes the reading where the cortex is thickest and includes the 3rd and 4th metacarpals as well as the 2nd. This measurement yields a value for cortex thickness [periosteal diameter (OD) minus endosteal diameter (ID)], from which one can calculate cross-sectional cortical bone area [$\pi/4$ $(OD^2 - ID^2)$], assuming that the cortex is a regular cylinder at the point of measurement. Garn (1970) has postulated that metacarpal cortex thickness serves as an index of total skeletal mass.

Maresh (1966) has published data on bone, muscle, and subcutaneous fat widths measured radiographically at various points on the extremities of children and adolescents.

The technique of photon densitometry has been used by several investigators. The bone, usually distal radius and ulna, is scanned transversely by a low-energy photon beam (^{125}l, 27 KeV; or ^{241}Am, 60 keV), and the transmission is monitored by a scintillation detector. The change in transmission as the beam is moved across the extremity is a function of the bone density in that region, and there is evidence that this is related to total skeletal mass, at least in adults (Manzke *et al.*, 1975; Christiansen and Rödbro, 1975*a*; Cohn and Ellis, 1975).

On the other hand, Mazess and Cameron (1972) claim that the correlation ($r = 0.84$) which they found for radius cortex cross-sectional area and photon densitometry is not high enough to produce confidence in either measurement as a predictor of skeletal mass.

Isherwood *et al.* (1976) have presented preliminary observations on the use of computer-assisted tomography for estimating bone mineral content. A claimed advantage for this technique is that trabecular as well as cortical bone mass can be determined and that it is suitable for studying vertebrae as well as long bones.

2.7.2. Muscle

Garn (1961) and Tanner (1962) discuss the technique involved in making satisfactory measurements of muscle widths from roentgenograms of the extremities (see Chapter 10). Since the entire width is measured, this procedure overlooks the possibility that there may be variable amounts of fat between the muscle layers.

2.7.3. Subcutaneous Fat Layer

With appropriate X-ray exposures, the subcutaneous fat layer can be identified and measured. The favored sites are the lateral chest wall and the region over the greater trochanter of the femur (Stuart *et al.*, 1940; Comstock and Livesay, 1963; Garn, 1961) (see Chapter 3).

Mention should be made here of the use of ultrasound for estimation of subcutaneous fat thickness (Stouffer, 1963; Booth *et al.*, 1966).

2.8. Anthropometry

In contrast to the techniques described above, anthropometry requires little in the way of apparatus. Calipers of various sorts and a tape measure are all one needs.

2.8.1. Thickness of Skin plus Subcutaneous Tissue

This is variously called skinfold thickness and fatfold thickness. Since a double layer of skin at the usual sites is only 1.8 mm thick, the subcutaneous fat layer contributes the bulk of the measured value (Durnin and Womersley, 1974). The two most commonly used sites are over the triceps midway between elbow and shoulder and at the lower tip of the scapula. Others are the midbiceps region, at the lower rib margin in the anterior axillary line, the periumbilical region, and over the iliac crest.

The measurement is made by grasping the subcutaneous tissue between thumb and forefinger, shaking it gently to (hopefully) exclude underlying muscle, and stretching it just far enough to permit the jaws of the spring-actuated caliper* to impinge on the tissue. Since the jaws of the caliper compress the tissue, the caliper reading diminishes for a few seconds and then the dial is read (see Chapter 3). In subjects with moderately firm subcutaneous tissue the measurement is easy to make, but those with flabby, easily compressible tissue and those with very firm tissue not easily deformable, present somewhat of a problem. In these latter two situations it may be difficult to achieve consistent readings.

The assumptions underlying the use of this method for estimating body fat content are two in number: first, that the thickness of the subcutaneous fat mantle reflects the total amount of fat in the body, and second, that the sites chosen for the measurement, either singly or in combination, represent the average thickness of the entire mantle. Neither assumption has been proven true. Despite the oft-repeated statement that subcutaneous fat makes up about half of the total body fat, there are no data to support such a contention. There are only two analyses of humans which bear on this point: a neonate in whom subcutaneous fat represented 42% of total body fat (Forbes, 1962) and an adult woman for whom the value was 32% (Moore et al., 1968). Pitts and Bullard (1968) analyzed the bodies of a number of mammalian species and found a wide variation, from 4 to 43%.

Booth et al. (1966) have criticized the caliper measurement of skinfold thickness as inaccurate and conclude that ultrasonic measurements are inherently superior.

A number of investigators have correlated skinfold thickness at one or more sites with body density or with body fat content estimated by total body water, ^{40}K counting, or densitometry (Parizková, 1961a; Heald et al., 1963; Nagamine and Suzuki, 1964; Sloan, 1967; Young et al., 1968; Burmeister and Fromberg, 1970; Forbes and Amirhakimi, 1970; Hermansen and Döbeln, 1971; Brook, 1971; Durnin and Womersley, 1974; Ward et al., 1975; Lohman et al., 1975). For the most part the correlations are not very high, although some authors claim that body fat estimates are reasonably accurate; others, such as Shephard et al. (1973), were very disappointed. While of theoretical interest, equations relating body density to skinfold thickness are of little use unless a ready means is at hand for converting this value to actual fat content.

2.8.2. Arm Circumference

By combining skinfold thickness (SF) with arm circumference at the same level, one can calculate a "corrected" arm circumference which represents the muscle and bone (M + B) component of the extremity. Novak et al. (1973) and others have used only the triceps skinfold for this measurement, yet for most subjects the subcutaneous tissue layer of the arm is not of uniform thickness, so a

---

*Three are recommended: the Harpenden Caliper, H.E. Morse Co., Holland, Michigan; the Holtain-Harpenden, Holtain Ltd., Brynberian, Crymmych, Pembrokeshire, Wales; and the Lange caliper, Cambridge Scientific Industries, Inc., Cambridge, Maryland.

more appropriate formula would be

$$\text{M} + \text{B circumference} = \text{arm circ.} - \frac{\pi}{2}(\text{triceps} + \text{biceps SF}) \tag{7}$$

$$\text{M} + \text{B area} = \frac{1}{4\pi}[\text{circ.} - \frac{\pi}{2}(T + \text{B SF})]^2 \tag{8}$$

Data on cross-sectional area of arm M + B and arm fat for 6- to 17-year-old American children have been published by the National Center for Health Statistics (1974). It is unfortunate that these values were based on triceps skinfold only.

2.8.3. Other Anthropometric Measurements

Behnke and Wilmore (1974) present a complicated formula for estimating lean weight. Body diameters are measured at six locations (biacromial, biiliac, upper chest, bitrochanteric, wrist, and ankle), and these together with height are entered in the following formula:

$$\text{LBM} = \frac{(\text{sum of diameters})^2}{a} \times \text{height}^b \tag{9}$$

where a and b are constants, which differ for the sexes. Huenemann *et al.* (1974) used this method on a group of 16-year-old school children and compared the results with those obtained by the densitometric and ^{40}K techniques. The correspondence for boys was very good: average of 55.2 kg lean weight by anthropometry, 55.7 by ^{40}K counting, and 55.9 kg by densitometry. The values for girls were, respectively, 47.3, 40.9, and 41.1 kg; hence for them the anthropometric estimates were clearly out of line.

Steinkamp *et al.* (1965) were able to predict body fat content in adults (by the H_2O-density method) rather well from multiple regression formulas involving various combinations of body circumferences, diameters, and skinfold thicknesses. Edwards and Whyte (1962) estimated body fat from skinfolds and height. Pollock *et al.* (1976) used equations involving skinfold thicknesses, body circumferences, and diameters to predict body density in young men, and Young *et al.* (1968) have used this approach in adolescent girls. Others include Michael and Katch (1968), Katch and Michael (1968), Kjellberg and Reizenstein (1970), Döbeln (1959), Delwaide and Crenier (1973), and Lohman *et al.* (1975). Crook *et al.* (1966) combined skinfold thickness with percentage overweight.

2.8.4. Functions of Weight and Height

Physiologists commonly use weight$^{0.75}$ as an index of "active metabolic" weight. Moore *et al.* (1963) claim that body fat is a linear function of weight in adults; certainly the two should be well correlated in series where the weight range is large enough to include obese and underweight subjects. Since fat is a component of weight, they must perforce be related. While it is obvious that a man of average height weighing 150 kg is obese, one who weighs 100 kg at a height of 195 cm may well not be. It was this simple observation that led Behnke *et al.* (1942) to develop the densitometric method some three decades ago and to emphasize the importance of considering body weight as a two-compartment system—lean and fat.

Cheek (1968) presented a formula for predicting LBM from height and weight ($aW + bH + c$) for children, and Hume (1966) and Hume and Weyers (1971) have done the same for the adult. Of more than passing interest here are the facts that: (1) the equations for children differ from those for the adult; (2) those for males differ from those for females; and (3) Hume's 1966 equations differ appreciably, and inexplicably, from the ones published by him and his associate in 1971. Burmeister and Bingert (1967) attempted to predict body cell mass from the formula: $A^a W^b H^c$ (A is age; a, b, and c are constants); four such equations were needed to encompass the entire age range for each sex. Wilmore and Behnke's (1970) admonition that anthropometric equations are valid only for the subject sample from which they were derived seems to have been borne out.

Flynn *et al.* (1972) and Forbes (1972*a*, 1974) have related body K and LBM, respectively, to stature. However, the regression slopes and intercepts vary with age and sex, and although the correlation coefficients are statistically significant, they are not high enough to provide satisfactory predictability. Nevertheless, the effect of stature is great enough to make control of this variable mandatory in any study involving comparisons of LBM among groups of subjects.

2.9. Cell Number and Cell Size

2.9.1. Muscle

A biopsy is done and the tissue analyzed for protein, intracellular H_2O, and DNA; an estimate of total muscle mass is derived from urinary creatinine excretion. Under the assumptions that each muscle cell has one nucleus, and that each nucleus contains 6.2 pg DNA, one can calculate total muscle number and relative muscle cell size.

Cheek (1968) has investigated the growth changes in muscle cell number and size in a small series of children and adolescents. Cell number increases from 0.2×10^{12} in the newborn to about 1.1×10^{12} at age 9 years, then rises to about 3.0×10^{12} at age 16 in the boy and 2.0×10^{12} in the girl. Cell size increases from about 150 mg protein/mg DNA in the baby to about 300 mg/mg in the adolescent (see Chapter 10).

2.9.2. Adipose Tissue

A biopsy is done, the tissue is analyzed for lipid, and the cells are counted. Total body fat is estimated. From these, adipocyte size (μg lipid per cell) and total adipocyte number can be calculated. Knittle's data (1976) show that cell size increases from about 0.35 μg lipid/cell in young children to about 0.50 in late adolescence, and cell number from 8×10^9 to about 32×10^9.

3. Calculation of Lean Body Mass and Fat

The three most widely used methods—total body H_2O, ^{40}K counting, and densitometry—all involve the assumption that the LBM is constant in composition. Water and K contents have actually been determined, but density cannot be measured and hence must be calculated from the densities and relative weights of the various tissues.

Data obtained by chemical analysis were shown in Figure 1 (also see Forbes, 1962). The adult body contains less H_2O but more K and Ca per unit fat-free weight, and also has a higher Ca/K ratio than the infant. Questions have been raised about the absolute constancy of fat-free body composition quite apart from the known effect of age (Wedgewood, 1963). In studies of pigs in whom body fat varied from 16 to 38% of body weight, Fuller et al. (1971) found that the K content of the fat-free body declined from 69.1 mEq/kg to 64.8 as H_2O content increased from 73.5 to 75.5%. The data of Ellis and Cohn (1975, and personal communication) show that the K/Ca ratio (g/g) of adult humans varies with stature, from about 0.10 in a 153-cm woman to 0.13 in a 188-cm man. Since bone has a lower H_2O and K content and a greater density than do soft tissues, this means that neither H_2O nor K form a constant fraction of the LBM, nor is its density absolutely constant.

A few sex differences have been noted. Data cited above on the influence of stature on body K/Ca ratio imply that this ratio will on average be lower in women than in men. Compilation of data from several sources (see Forbes et al., 1968) shows that the ratio of total exchangeable K to total body water is 5.8% less in adult females than in males. No such studies have been done on children and adolescents, so the scanty available data on ICF/total H_2O ratios were reviewed (see Forbes and Amirhakimi, 1970). These show that the female–male ratio for ICF/total H_2O is about 1.0 at age 9–12, about 0.98 at age 14, and about 0.95 at 17–18 years. Thus, a graduated correction factor was used for adolescent girls (see footnotes to equations 10 and 11). Incidentally, it has been shown that the ratio of ECF to ICF volume tends to increase in old age (Norris et al., 1963).*

Despite these diversions the weight of evidence is that LBM composition does not vary greatly among subjects of specified age (Cheek and West, 1955; Talso et al., 1960; compilation by Forbes and Hursh, 1963).

The formulas used for the older child, adolescent, and adult are based on the results of chemical analysis:

$$LBM(kg) = \text{total body } H_2O \text{ (liter)}/0.73 \qquad (10)$$
$$LBM(kg) = \text{total body K (mEq)}/68.1 \text{ (males, and girls} < 13 \text{ yr)}\dagger \qquad (11)$$
$$LBM(kg) = \text{total body K (mEq)}/64.2 \text{ (females 18+ yr)}\dagger \qquad (12)$$

whence

$$\text{Body fat} = \text{weight} - LBM \qquad (13)$$

The equations used to calculate LBM and fat from body density are derived from measurements of the density of human depot fat and the major tissue components of the LBM: at 37°C fat has a density of 0.900 g/cm³, water 0.993, protein 1.340, and mineral 3.000 (Brozek et al., 1963). As noted in Section 2.3, the formula for calculating fraction fat is $Wf/Wb = (a/D) - a'$. Using values for fat and lean of 0.900 and 1.10, respectively, the formula is

$$\frac{Wf}{Wb} = \frac{4.950}{D} - 4.500 \qquad (14)$$

whence $Wl/Wb = 1 - Wf/Wb$.

*Claims that the K/LBM ratio drops markedly with advancing age (Pierson et al., 1974) rest on the premise that LBM can be satisfactorily estimated by anthropometric methods, a premise which lacks validation.

†Based on data referred to above. For adolescent girls the author has used a graduated value for the K content of the LBM: 68.1 mEq/kg at age 12, 67.4 at 13, 66.7 at 14, 66.1 at 15, 65.4 at 16, 64.7 at 17, and 64.2 at 18 and above. If total exchangeable K is determined rather than ⁴⁰K content, the numerator of equations (11) and (12) should be multiplied by 1.05 or 1.10, depending on the extent to which one believes the ⁴²K dilution method underestimates body K content.

The constants a and a' have been assigned various values by various investigators, and there is a bewildering array of formulas from which to choose (Pearson *et al.*, 1968). The calculated fat content will obviously depend on the particular formula selected by the investigator.

Siri (1961) has devised a formula utilizing data from both total body H_2O and density, which offers the advantage of being unaffected by the state of hydration, although it does require the assumption of a fixed mineral/protein ratio of the body:

$$\frac{Wf}{Wb} = \frac{2.1366}{D} - \left(0.780 \times \frac{H_2O}{W} \right) - 1.374 \tag{15}$$

3.1. Precision of Various Techniques

The ultimate precision is limited by the fact that the human body is not a static system: body weight varies by a percent or so from day to day; a long drink of water can easily increase body H_2O content by 1%; an overnight thirst depletes body H_2O by about 0.5%.

Table III gives the variations found by investigators who have assayed their subjects on repeated occasions. It is evident that LBM cannot be estimated with a precision of much better than 5%.

Tyson *et al.* (1970) fed one subject a high-K diet and found a good correspondence between change in body K as estimated by K balance and ^{40}K counting.

4. Growth of Lean Body Mass

There are a number of reports on body K content of children and adolescents (Allen *et al.*, 1960; Meneely *et al.*, 1963; Oberhausen *et al.*, 1965; Burmeister and Bingert, 1967; Forbes, 1972*b*; Novak *et al.*, 1973; Flynn *et al.*, 1972), on total body H_2O (Young *et al.*, 1968; Heald *et al.*, 1963), and on density (Parizková, 1961*b*). Estimates of LBM can be made from these data.

Table III. Reproducibility Data

Technique	Coefficient of variation	Reference
^{40}K counting	1.9–2.9%	Shukla *et al.* (1973)
	2–3%	Forbes *et al.* (1968)
	1.4–4.1%	Johny *et al.* (1970)
	1.2–4.8%	Pierson *et al.* (1974)
Total exchangeable K	6.1%	Price *et al.* (1969)
	2.9%	Haxhe (1963)
	2.5%	Davies and Robertson (1973)
THO, D_2O, antipyrine dilution	2.5%	Price *et al.* (1969)
	1.8%	Haxhe (1963)
	9%	Greenway *et al.* (1965)
Density	0.0023 g/ml (SE)	Durnin and Taylor (1960)
	0.0004–0.0043	Buskirk (1961)
	0.0029	Norris *et al.* (1963)
Cyclopropane, ^{85}Kr uptake	7–8% "uncertainty"	Lesser *et al.* (1971)
Skinfold thickness	16%	Greenway *et al.* (1965)
	6–24%	Nat'l Center for Health Statistics (1974)
Creatinine excretion	2–19%	Forbes and Bruining (1976)

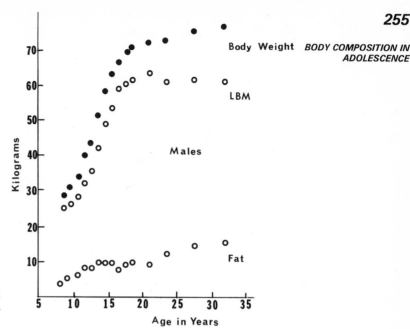

Fig. 4. Mean weight, LBM, and fat for 604 white males aged 8–35 years (author's data).

All are in agreement that the adolescent spurt in LBM is more intense in the adolescent boy than in the girl and that he achieves a mature value which is considerably larger. A collation of data obtained by various techniques showed a remarkable degree of agreement when correction was made for stature differences among the various groups (Forbes, 1972b).

Figures 4 and 5 present the author's data on normal white subjects by ^{40}K counting,* and compare the growth of the LBM with that of body weight. The average preadolescent boy has a slightly larger LBM than the girl, but it is only with the onset of puberty that the difference achieves real significance. The up-swing in the boy is rapid and sustained, and maximum values are achieved by about age 20. [The word "maximum" is appropriate here, since there is a slow decline during the adult years (Forbes, 1976a).] The LBM spurt in girls is not as intense, and values close to the maximum are attained by age 18.

Figure 6 represents an attempt to conceptualize the adolescent LBM spurt as being superimposed on the preadolescent growth pattern. Viewed in this manner the male spurt is seen to be much more intense and more prolonged than that of the female.

Figure 7 depicts changes in the LBM/height ratio during adolescence. Boys finally achieve a much higher value for this ratio than do girls. This means that the spurt in LBM is relatively greater than the spurt in height, a result which follows naturally from the known effects of androgens.

The variability in LBM is shown in Figure 8, where the hatched areas enclose the 95% confidence limits for our subjects. A number of studies have shown that

*Meneely *et al.* (1963) found the black females in their group to have a larger LBM than whites, while the reverse was true for adolescent males. As might be expected, Japanese and Filipino young adults have a lower LBM than Caucasians (Nagamine and Suzuki, 1964; Novak, 1970). Eskimo women, on the other hand, have a slightly larger LBM than Caucasians, while the reverse is true for men (Shephard *et al.*, 1973). As will be seen later, some of these racial differences could be due to stature.

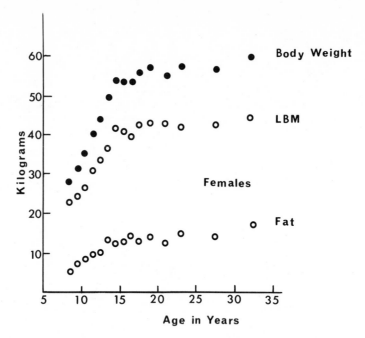

Fig. 5. Mean weight, LBM, and fat for 467 white females aged 8–35 years (author's data).

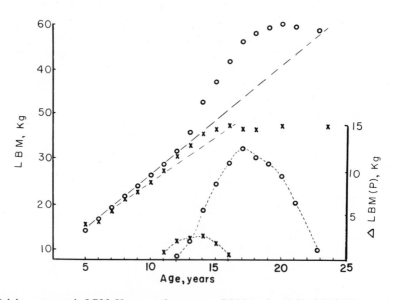

Fig. 6. Adolescent spurt in LBM. Upper portion: average LBM (scale at left), data from several sources. Dashed lines represent calculated regression of LBM on age for preadolescent years and extrapolation of same; males—o, females—x. Lower portion: difference between actual LBM and extrapolated line (scale at right). (From Forbes, 1976b, with permission.)

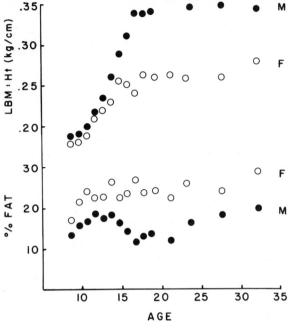

Fig. 7. Mean LBM/height ratio (kg/cm) (upper portion), and percent body fat (lower portion) (author's data).

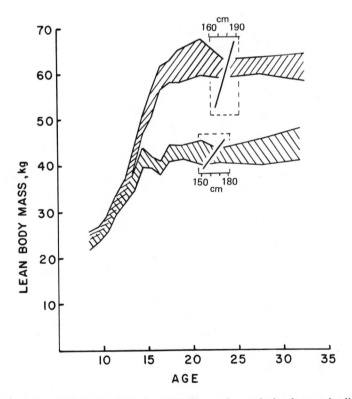

Fig. 8. Ninety-five percent confidence limits for LBM. Inserts show calculated regression lines for LBM against height in subjects 22–25 years old: slope for males ($N = 41$) is much steeper (0.69 kg/cm) than for females ($N = 29$) (0.29 kg/cm) (author's data).

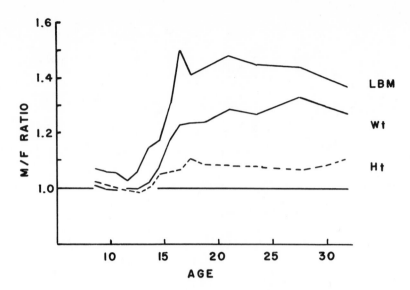

Fig. 9. Male/female ratios for LBM, weight, and height (author's data).

LBM is a function of stature (Forbes, 1974), and the inserts in Figure 8 show the calculated regression lines for 22- to 25-year-old males and females. The former has a slope of 0.69 kg/cm, the latter one of 0.29 kg/cm. On average, then, tall individuals have a larger LBM than short individuals, an attribute which undoubtedly confers an advantage in athletics (Khosla, 1968). Data are also available to show that the regression slopes of LBM on height are rather gentle in childhood, and gradually steepen as adolescence progresses (Forbes, 1972a).

Tanner (1974) has published curves depicting increments in radiographically determined muscle widths during adolescence. The peak velocity for males coincides with peak height velocity, as do those for biacromial and biiliac diameters; there is a lag of about a year for females. Parizková (1976) found that peak velocity in LBM growth coincides with peak height velocity in adolescent boys.

Malina (1969) has summarized the literature on growth of muscle (and bone and subcutaneous fat) as estimated by radiography for children and adolescents.

Figure 9 illustrates the sex ratios for body weight, height, and LBM as functions of age in our subjects. The latter clearly exhibits the greatest degree of sexual dimorphism; on average the 15-year-old girl has an LBM which is 81% as large as the boy's, and by age 20 this has dropped to 68%. Since the LBM comprises the active metabolic tissues of the body, it is likely that certain nutritional needs and doses of certain medications possess greater sex differentials than would be anticipated from differences in total body weight.

5. Growth of Body Fat

Girls tend to accumulate more fat than boys. Slight sex differences in estimated body fat content have been detected in children (Novak, 1966a; Flynn et al., 1970) and even in infants (Owen et al., 1962), but it is not until adolescence that the difference becomes striking. Figures 4 and 5 show the age curve for body fat content, and Figure 7 depicts fat as a percent of body weight. Boys and girls both gain body fat early in adolescence; later the gain stops, even reverses temporarily,

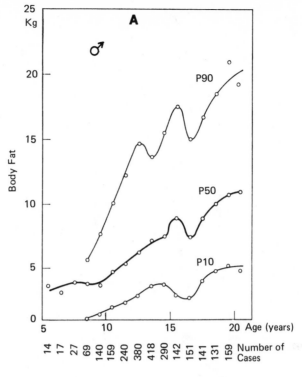

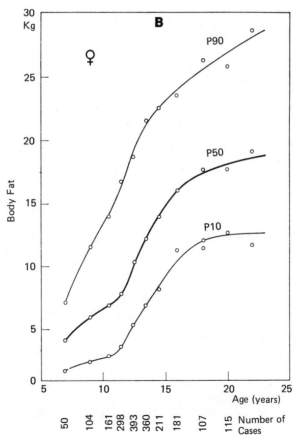

Fig. 10. A and B: Variation in body fat estimated by ⁴⁰K counting. (From Burmeister and Bingert, 1967, with permission.)

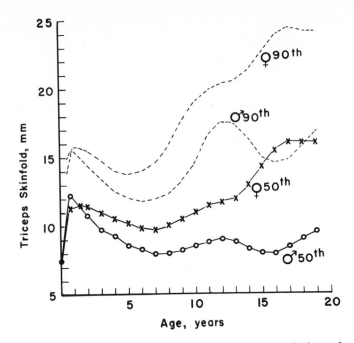

Fig. 11. Skinfold thickness at the midtriceps region for British children. (Redrawn from Tanner and Whitehouse, 1975.)

in boys, while girls continue to put on fat as adolescence proceeds. The adult years see a further accumulation in both sexes.

Body fat content tends to be more variable than lean. Figure 10 is taken from the work of Burmeister and Bingert (1967), who assayed ^{40}K content in several thousand German children. The range between the 10th and 90th percentiles of body fat is wide indeed, and the skewness in the data is evident from the fact that the distance from the 50th and the 90th percentile exceeds that from the 10th to the 50th percentile. These authors rightfully conclude that "The nongaussian distribution of weight (which is seen in most population studies) is thought to be due to the skewness in distribution of fat."

These findings are mirrored by data on skinfold thickness, for here one sees similar age and sex trends and a distinct tendency for skewness (Figure 11). Tanner and Whitehouse (1975) attempted to eliminate the skewness by plotting the logarithm of skinfold thickness, but achieved only partial success. Data from American adolescents do not differ greatly (National Center for Health Statistics, 1974).

Of considerable interest are the data on Shephard *et al.* (1973) on young Eskimo women, who contrary to popular belief proved to be no fatter than Caucasians.

A publication from the National Center for Health Statistics (1974) contains some interesting graphs relating the yearly increments in triceps skinfold (SF) thickness to height velocity for children and adolescents. For boys velocity for triceps SF is positive in the earlier years, with a peak at about age 10; then velocity declines and actually becomes negative, with a nadir which nicely coincides with peak height velocity at about age 13½ years. The velocity curve for girls is quite similar in shape, the entire curve being shifted upwards, so that nadir of the velocity dip just touches zero; but here, too, the nadir coincides in time with peak height velocity at about age 12 years. The later years of adolescence witness a positive triceps SF velocity once again. Tanner's (1974) published curves for radiographically determined subcutaneous fat thicknesses at various sites are similarly related to peak height velocity.

This interesting coincidence, in which it appears that triceps SF and height velocity curves resemble mirror images of each other *in both sexes,* raises the possibility that a single hormonal influence is responsible for both phenomena.

6. Changes in Total Body Calcium

An earlier illustration (Figure 1) depicted the Ca enrichment of the LBM which takes place during growth, as determined by chemical analysis. Trotter and Hixon's (1974) data on skeletal weights show a progressive increase in skeletal weights during childhood and adolescence, to a maximum of about 4400 g (dry fat-free weight) in males and 3100 g in females (once again the word "maximum" is appropriate, for skeletal weight declines during the adult years).

Figure 12 shows changes in estimated total body Ca during childhood and adolescence. One set of curves is from the data of Garn (1970), who estimated skeletal weight from measurements of the thickness of the mid-cortex of the second metacarpal bone, whence skeletal weight × 0.25 = body Ca (over 99% of body Ca is in bone). The second is based on data derived from photon densitometry of the

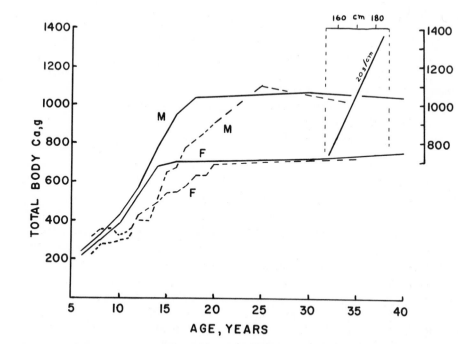

Fig. 12. Total body caclcium as a function of age. Solid lines—from Garn's (1970) estimates of skeletal weight derived from metacarpal cortex thickness; dotted lines—data of Christiansen and Rödbro (1975*b*) and Christiansen *et al.* (1975) based on photon scans of the forearm; insert—data of Ellis and Cohn (1975, and personal communication) by neutron activation: regression of body Ca on stature for 6 women, 9 men aged 30–39 years.

distal radius and ulna (Christiansen and Rödbro, 1975a,b; Christiansen et al., 1975). The two sets of estimates agree rather well for preadolescent children and for adults, but there is a marked difference for the adolescent years. Garn's subjects were American, Christiansen's were Danish. The basis for Garn's (1970) extrapolation of metacarpal cortex thickness to total skeletal weight is not clear—indeed he has been criticized (Mazess and Cameron, 1972)—while Christiansen and Rödbro (1975a) checked their photon measurements against the weight and Ca content of various human bones. On the other hand, the latter authors' subjects showed some irregularities in height growth.

This discrepancy between the two sets of estimates is unfortunate, for now any estimate of the rapidity of Ca accretion during adolescence is open to doubt, and this is just the time in life when such an estimate would be most helpful from a nutritional point of view. Schuster (1970) did photon measurements on large numbers of children, but made no attempt to extrapolate these values to skeletal weight.

It is evident that the sex difference in total body Ca is appreciable and of the same order of magnitude as the sex difference in LBM (Figure 8). At age 20 years the male–female ratio is 1.47/1 for Garn's data and 1.31/1 for Christiansen's; the ratio for Trotter and Hixon's (1974) adult skeletons (dry weight) is 1.4/1, and for a series of adult Russian skeletons (wet weight) it is about 1.2/1 (Borisov and Marei, 1974); by neutron activation the ratio for body Ca is 1.4/1. These are consistent with the sex ratio for LBM (Figure 9).

As is the situation for LBM, stature exerts an influence on body Ca content, and indeed this is one plausible reason for the larger body Ca content of men. The insert in Figure 12 shows that, on average, body Ca in young adults increases by 20 g for each cm increment in height. Hence a 186-cm male would be expected to contain 640 g more Ca than a 154-cm woman (1370 − 730 g Ca), a remarkable difference and one that certainly is of nutritional importance. The series of skeleton weights reported by Borisov and Marei (1974) also demonstrate the effect of stature: the tallest men had skeletons weighing about 12 kg, the shortest women ones weighing only about 8 kg (net weight).

Cohn et al. (1974) also determined total body phosphorus in adults by neutron activation. The ratio of body P/Ca (g/g) averaged 0.48 in males and 0.50 in females. No assays are available for children and adolescents.

7. Effect of Exercise and Physical Training

It is well known that the athletically inclined are apt to have a larger LBM than sedentary individuals (Parizková, 1963; Novak, 1966b, Behnke and Wilmore, 1974). Yet the former may be naturally better endowed.

Table IV lists the results of a number of experiments designed to actually test the effects of various exercise regimes on body composition. Many were disappointing in that LBM, whether estimated by total body H_2O, densitometry, or ^{40}K counting, did not change appreciably. The first four groups in the table are of interest, since it appears that the superimposition of diet control resulted in a greater loss of weight but prevented a gain of LBM. Intense training is needed to appreciably affect the LBM, and the best results are obtained by concomitant use of androgens.

Table IV. Effect of Physical Training

Number	Age	Subjects	Period of training	Change in Wt	Change in LBM	Remarks	Reference
33	12½ yr	Obese girls, boys	7 wk	−7 kg	0	+1700-cal diet	Parízková et al. (1962)
7	10–12	Overweight boys	7 wk	−6	−1 kg	+1700-cal diet	Šprynarová and Parízková (1965)
8	17–19	Obese students	9 wk	−3.2	+2.7		Boileau et al. (1971)
11	17–19	Lean students	9 wk	−1.0	+1.4		Boileau et al. (1971)
31	6–13	Boys	10 wk		0		Lohman et al. (1975)
55	17–59	Men	10 wk	−1.0	0		Wilmore et al. (1970)
12	High school	Normal girls	15 wk	−0.6	0		Moody et al. (1972)
28	High school	Obese girls	15 wk	−1.1	+1.1		Moody et al. (1972)
23	14–18	Overweight girls	9 mo	−0.9	+1.6	?LBM increase due to growth	Wilmore et al. (1970)
22	16–28	Soccer players	½ season	+1	+1.5		Boddy et al. (1974)
8		Female gymnasts	16 wk	0	+1.2[a]	Training for Olympics	Parízková (1963)
7		Male gymnasts	16 wk	0	+1.8[a]	Training for Olympics	Parízková (1963)
1	24	Male student	10 wk	+3.9	+3.4	4200-mile bicycle trip	Forbes, unpublished
11	19–25	Athletes, male	6 wk	0	−1.1		Hervey et al. (1976)
11	19–25	Athletes, male	6 wk	+3.3	+6.3	+ androgens	Hervey et al. (1976)
1	26	Body builder	7 mo	+12.2	+16.5	+ androgens	Forbes, unpublished
1	31	Weight lifter	7 mo	+6.0	+16.0	+ androgens	Forbes, unpublished
8		College students	5 wk	+1.0	+1.1		Ward (1973)
8		College students	5 wk	+1.3	+3.1	+ androgens	Ward (1973)
14		College students	9 wk	+1.4	+1.4		Fahey and Brown (1973)
		College students	9 wk	+2.1	+2.5	+ androgens	Fahey and Brown (1973)

[a]Calculated from density data.

GILBERT B. FORBES

8.1. Obesity

It is plainly obvious that obese individuals carry a lot of excess fat, and assays reveal that in some the fat burden exceeds 50%. What is less well known is that they also tend to have a modest increase in LBM when compared to nonobese individuals of similar height and age. In assessing the results of overfeeding experiments, Brozek *et al.* (1963) estimated that "obesity tissue" was only ⅔ fat, the remainder being ECF and "cell residue." Of the weight gained by the subjects of Goldman *et al.* (1975), 69% was fat and 31% was lean; furthermore, calculations from their data show that the increment in LBM, as well as total weight, was directly proportional to the *total excess calories* consumed. Table V lists some data on obese subjects, either in terms of percent of the excess weight accounted for by LBM or as the ratio of observed to expected LBM, i.e., expected on the basis of the average nonobese subject of the same age, height, and sex. It would appear, therefore, that the human LBM can respond positively to caloric overnutrition. Knittle (1976) finds that obese children have an excessive fat cell size and an even greater increase in total fat cell number.

Both Forbes (1964) and Cheek *et al.* (1970) report that children who become obese early in life tend to have a larger increment in LBM together with an advance in bone age and stature when compared to those whose obesity is of more recent onset.

The usual form of human obesity thus differs from certain types of experimental obesity in animals, for hypothalamic lesions in the rat are accompanied by a reduction in carcass nitrogen and in body length as well as an increase in body fat.

8.2. Muscular Dystrophy

Children with this disorder have a markedly reduced body K content (Blahd *et al.*, 1963; Borgstedt *et al.*, 1970). The former authors report a slight reduction in body K in unaffected relatives, while the latter authors found no difference when K content was compared to normal subjects of the same age and height.

Table V. Body Composition in Obesity

Subjects	Number	Excess weight due to LBM (%)	Ratio Obs./ Exp. LBM	Reference
Boys and girls	27		1.09	Forbes (1964)
14-yr-old boys	12	25		Parizková *et al.* (1971)
14-yr-old girls	9	32		Parizková *et al.* (1971)
Boys and girls	22		1.29	Cheek *et al.* (1970)
Adult women	17	30		Ljunggren *et al.* (1957)
Adult women	13		1.41	Linquette *et al.* (1969)
Adult men	8		1.18	Linquette *et al.* (1969)
Adult men	10	36		Brozek *et al.* (1963)— deliberate overfeeding
Adult men	13	31		Goldman *et al.* (1975)— deliberate overfeeding

East *et al.* (1976) report that males with a 47,XXY chromosome complement tend to have a moderately reduced LBM as determined by ^{40}K counting, to the point where the values were about midway between the usual male and female values. The situation for the 47,XYY male is less clear; the values were slightly reduced but the reduction was of questionable statistical significance.

8.4. Growth-Hormone Deficiency

Tanner and Whitehouse (1967) and Collipp *et al.* (1973) have documented a decrease in subcutaneous fat thickness in hypopituitary dwarfs given growth hormone, an effect which would be anticipated from the known metabolic actions of this hormone. The latter authors also noted an increase in body K content. Cheek's (1968) data show an increase in muscle mass with treatment.

9. Pregnancy

Hytton and Leitch (1971) have reviewed the changes in body composition which occur during normal pregnancy: blood volume, ECF volume, and total body water all show an increase. Seitchik (1967) studied 99 women, some of whom were teenagers: their average weight gain was 16 kg, of which 8 kg was water; fat-free body mass calculated by the combined H_2O–density method increased by 9 kg. The 40 women studied by Godfrey and Wadsworth (1970) gained an average of 19 kg weight and 6.7 g K. MacGillivray and Buchanan (1958) found a mean increase of 770 mEq in total exchangeable Na and one of 560 mEq in total exchangeable K in eight pregnant women. King *et al.* (1973) studied ten teenage girls during the last trimester of their pregnancy: average weight gain was 77 g/day and average K accretion was 3.4 mEq/day. It is likely, therefore, that body fat is also gained, but the exact interpretation of these data with respect to the magnitude of the lean and fat components is difficult, for in such studies one is measuring two subjects simultaneously, one of which is known to differ markedly in body composition from the other.

10. Heredity

Brook *et al.* (1975) in a study of mono- and dizygotic twins found a high coefficient of heredibility for skinfold thickness, and metacarpal cortex thickness also appears to be strongly influenced by heredity (Smith *et al.*, 1973). It remains to be determined whether other aspects of body composition have a genetic basis.

11. Nutritional Implications

The LBM comprises the bulk of the active metabolic tissue, so it is no accident that BMR is strongly correlated with the size of the lean body (Döbeln, 1956), as is blood volume (Muldowney, 1961). The marked sex difference in the intensity and duration of the adolescent spurt in LBM dictates that caloric and certain nutrient needs are higher for adolescent boys than for girls. For example, Hawkins (1964) has made some calculations to show that the boy requires more iron at this time of life than does the menstruating girl. Increments in body Ca, P, N, Mg, and Zn should exhibit a similar degree of sexual dimorphism. Note should be taken on the documented LBM–stature and body Ca–stature relationships, which suggest that nutrient needs for growth are considerably higher in tall adolescents.

Many adolescent boys carry rather small amounts of fat, so one must decry the efforts of some high school wrestling coaches at encouraging them to lose weight in preparation for contests.

ACKNOWLEDGMENT

The author's work has been supported by grants from the National Institute of Child Health and Human Development and under a contract with the U.S. Energy Research and Development Administration at the University of Rochester Biomedical and Environmental Research Project, and has been assigned Report No. UR-3490-1121.

12. References*

Allen, T. H., Anderson, E. C., and Langham, W. H., 1960, Total body potassium and gross body composition in relation to age, *J. Gerontol.* **15**:348.

Anderson, J., Osborn, S. B., Tomlinson, R. W. S., Newton, D., Rundo, J., Salmon, L., and Smith, J. W., 1964, Neutron-activation analysis in man *in vivo:* A new technique in medical investigation, *Lancet* **2**:1201.

Bauer, J. H., 1976, Oral administration of radioactive sulfate to measure extracellular fluid space in man, *J. Appl. Physiol.* **40**:648.

Behnke, A. R., 1945, The absorption and elimination of gases of the body in relation to its fat and water content, *Medicine* **24**:359.

Behnke, A. R., and Wilmore, J. H., 1974, *Evaluation and Regulation of Body Build and Composition,* Prentice-Hall, Englewood Cliffs, New Jersey.

Behnke, A. R., Feen, B. G., and Welham, W. C., 1942. The specific gravity of healthy men, *J. Am. Med. Assoc.* **118**:495.

Bezold, A. von, 1857, Untersuchungen über die Vertheilung von Wasser, organischer Materie und anorganische Verbindungen im Thierreiche, *Z. Wiss. Zool.* **8**:486.

Blahd, W. H., Cassen, B., and Lederer, M., 1963, Body potassium content in patients with muscular dystrophy, *Ann. N. Y. Acad. Sci.* **110**:282.

Boddy, K., Hume, R., King, P. C., Weyers, E., and Rowan, T., 1974, Total body plasma and erythrocyte potassium and leucocyte ascorbic acid in "ultra-fit" subjects, *Clin. Sci. Mol. Med.* **46**:449.

Boddy, K., Lindsay, R., Holloway, I., Smith, D. A. S., Elliott, A., Robertson, I., and Glaros, D., 1976, A study of changes in whole-body calcium, phosphorus, sodium and nitrogen by neutron activation analysis *in vivo* in rats on a calcium-deficient diet, *Clin. Sci. Mol. Med.* **51**:399.

Boileau, R. A., Buskirk, E. R., Horstman, D. H., Mendez, J., and Nicholas, W. C., 1971, Body composition changes in obese and lean men during physical conditioning, *Med. Sci. Sports* **3**:183.

Bonnard, G. D., 1968, Cortical thickness and diaphyseal diameter of the metacarpal bones from the age of three months to eleven years, *Helv. Paediatr. Acta* **30**:445.

Booth, R. A. D., Goddard, B. A., and Paton, A., 1966, Measurement of fat thickness in man: A comparison of ultrasound, Harpenden calipers, and electrical conductivity, *Br. J. Nutr.* **20**:719.

*References preceded by an asterisk constitute a general bibliography.

Borgstedt, A., Forbes, G. B., and Reina, J. C., 1970, Total body potassium and lean body mass in patients with Duchenne dystrophy and their female relations, *Neuropädiatrie* **1**:447.

Borisov, B. K., and Marei, A. N., 1974, Weight parameters of adult human skeleton, *Health Phys.* **27**:224.

Bradbury, M. W. B., 1961, Urea and deuterium-oxide spaces in man, *Br. J. Nutr.* **15**:177.

Brodie, B. B., Berger, E. Y., Axelrod, J., Dunning, M. F., Porosowska, Y., and Steele, J. M., 1951, Use of N-acetyl-4-aminoantipyrine (NAAP) in measurement of total body water, *Proc. Soc. Exp. Biol. Med.* **77**:794.

Brook, C. G. D., 1971, Determination of body composition of children from skinfold measurements, *Arch. Dis. Child.* **46**:182.

Brook, C. G. D., Huntley, R. M. C., and Slack, J., 1975, Influence of heredity and environment in determination of skinfold thickness in children, *Br. Med. J.* **2**:719.

*Brozek, J., 1961, Body composition, *Science* **134**:920.

*Brozek, J. (ed.), 1965, *Human Body Composition: Approaches and Applications,* Pergamon Press, Oxford.

*Brozek, J., and Henschel, A. (eds.), 1961, *Techniques for Measuring Body Composition,* Nat'l Acad. Sci.–Nat'l Res. Council, Washington, D.C.

Brozek, J., Grande, F., Anderson, J. T., and Keys, A., 1963, Densitometric analysis of body composition: Revision of some quantitative assumptions, *Ann. N. Y. Acad. Sci.* **110**:113.

Burmeister, V. W., and Fromberg, G., 1970, Depotfet, bestimmt nach der Kalium-40 Methode, und seine Beziehung zur Hautfaltendicke, *Arch. Kinderh.* **180**:228.

Burmeister, W., and Bingert, A., 1967, Die quantitativen Veränderungen der menschlichen Zellmasse zwischen dem 8 und 90 Lebensjahr, *Klin. Wochenschr.* **45**:409.

Buskirk, E. R., 1961, Underwater weighing and body density: A review of procedures, in: *Techniques for Measuring Body Composition* (J. Brozek and A. Henschel, eds.), pp. 90–107, National Academy of Science, Washington, D.C.

*Cheek, D. B., 1968, *Human Growth,* Lea and Febiger, Philadelphia.

Cheek, D. B., and West, C. D., 1955, An appraisal of methods of tissue chloride analysis: The total carcass chloride, exchangeable chloride, potassium and water of the rat, *J. Clin. Invest.* **34**:1744.

Cheek, D. B., Schultz, R. B., Parra, A., and Reba, R. C., 1970, Overgrowth of lean and adipose tissues in adolescent obesity, *Pediatr. Res.* **4**:268.

Christiansen, C., and Rödbro, P., 1975a, Estimation of total body calcium from the bone mineral content of the forearm, *Scand. J. Clin. Lab. Invest.* **35**:425.

Christiansen, C., and Rödbro, P., 1975b, Bone mineral content and estimated total body calcium in normal adults, *Scand. J. Clin. Lab. Invest.* **35**:433.

Christiansen, C., Rödbro, P., and Nielsen, C. T., 1975, Bone mineral content and estimated total body calcium in normal children and adolescents, *Scand. J. Clin. Lab. Invest.* **35**:507.

Cohn, S. H., and Dombrowski, C. S., 1971, Measurement of total-body calcium, sodium, chlorine, nitrogen and phosphorus in man, *J. Nucl. Med.* **12**:499.

Cohn, S. H., and Ellis, K. J., 1975, Predicting radial bone mineral content in normal subjects, *Int. J. Nucl. Med. Biol.* **2**:53.

Cohn, S. H., and Fairchild, R. G., 1976, Application of neutron activation analysis in nuclear medicine, in: *CRC Handbook of Nuclear Medicine* (R. P. Spencer, ed.), CRC Press, Cleveland.

Cohn, S. H., Ellis, K. J., Wallach, S., Zanzi, I., Atkins, H. L., and Aloia, J. R., 1974, Absolute and relative deficit in total-skeletal calcium and radial bone mineral in osteoporosis, *J. Nucl. Med.* **15**:428.

Collipp, P. J., Curti, V., Thomas, J., Sharma, R. K., Maddaiah, V. T., and Cohn, S. H., 1973, Body composition changes in children receiving human growth hormone, *Metabolism* **22**:589.

Comstock, G. W., and Livesay, V. T., 1963, Subcutaneous fat determinations from a community-wide chest X-Ray survey in Muscogee County, Georgia, *Ann. N. Y. Acad. Sci.* **110**:475.

Crook, G. H., Bennett, C. A., Norwood, W. D., and Mahaffey, J. A., 1966, Evaluation of skinfold measurements and weight chart to measure body fat, *J. Am. Med. Assoc.* **198**:39.

Cureton, K. J., Boileau, R. A., and Lohman, T. G., 1975, A comparison of densitometric, potassium-40 and skinfold estimates of body composition in prepubescent boys, *Hum. Biol.* **47**:321.

Davies, D. L., and Robertson, J. W. K., 1973, Simultaneous measurement of total exchangeable potassium and sodium using ^{43}K and ^{24}Na, *Metabolism* **22**:133.

Delwaide, P. A., and Crenier, E. J., 1973, Body potassium as related to lean body mass measured by total water determination and by anthropometric method, *Hum. Biol.* **45**:509.

Döbeln, W. von, 1956, Human standard and maximal metabolic rate in relation to fat-free body mass, *Acta Physiol. Scand.* **37**(Suppl):126.

Döbeln, W. von, 1959, Anthropometric determination of fat-free body weight, *Acta Med. Scand.* **165**:37.

Dombrowski, C. S., Wallach, S., Shukla, K. K., and Cohn, S. H., 1973, Determination of whole body magnesium by *in vivo* neutron activation, *Int. J. Nucl. Med. Biol.* **1**:15.

Domermuth, W., Veum, T. L., Alexander, M. A., Hedrick, H. B., Clark, J., and Eklund, D., 1976, Prediction of lean body composition of live market weight swine by indirect methods. *J. Anim. Sci.* **43**:966.

Durnin, J. V. G. A., and Taylor, A., 1960, Replicability of measurements of density of the human body as determined by underwater weighing, *J. Appl. Physiol.* **15**:142.

Durnin, J. V. G. A., and Womersley, J., 1974, Body fat assessed from total body density and its estimation from skinfold thickness: Measurements on 481 men and women aged from 16 to 72 years, *Br. J. Nutr.* **32**:77.

East, B. W., Boddy, K., and Price, W. H., 1976, Total body potassium content in males with X and Y chromosome abnormalities, *Clin. Endocrinol.* **5**:43.

Edwards, K. D. G., and Whyte, H. M., 1962, The simple measurement of obesity, *Clin. Sci.* **22**:347.

Elkinton, J. R., and Danowski, T. S., 1955, *The Body Fluids,* Williams and Wilkins, Baltimore.

Ellis, K. J., and Cohn, S. H., 1975, Correlation between skeletal calcium mass and muscle mass in man, *J. Appl. Physiol.* **38**:455.

Ellis, K. J., Shukla, K. K., Cohn, S. H., and Pierson, R. N., Jr., 1974, A predictor for total body potassium in man based on height, weight, sex and age: Application in metabolic disorders, *J. Lab. Clin. Med.* **83**:716.

Ellis, K. J., Vaswani, A., Zanzi, I., and Cohn, S. H., 1976, Total body sodium and chlorine in normal adults, *Metabolism* **25**:645.

Fahey, T. D., and Brown, C. H., 1973, The effects of an anabolic steroid on the strength, body composition, and endurance of college males when accompanied by a weight training program, *Med. Sci. Sports* **5**:272.

Flynn, M. A., Murthy, Y., Clark, J., Comfort, G., Chase, G., and Bentley, A. E. T., 1970, Body composition of Negro and white children, *Arch. Environ. Health* **20**:604.

Flynn, M. A., Woodruff, C., and Chase, G., 1972, Total body potassium in normal children, *Pediatr. Res.* **6**:239.

Forbes, G. B., 1962, Methods for determining composition of the human body, *Pediatrics* **29**:477.

Forbes, G. B., 1964, Lean body mass and fat in obese children, *Pediatrics* **34**:308.

Forbes, G. B., 1972*a*, Relation of lean body mass to height in children and adolescents, *Pediatr. Res.* **6**:32.

Forbes, G. B., 1972*b*, Growth of the lean body mass in man, *Growth* **36**:325.

Forbes, G. B., 1973, Another source of error in the metabolic balance method, *Nutr. Rev.* **31**:297.

Forbes, G. B., 1974, Stature and lean body mass, *Am. J. Clin. Nutr.* **27**:595.

Forbes, G. B., 1976*a*, The adult decline in lean body mass, *Hum. Biol.* **48**:161.

Forbes, G. B., 1976*b*, Biological implications of the adolescent growth process: Body composition, in: *Nutrient Requirements in Adolescence* (J. I. McKigney and H. N. Munro, eds.), pp. 57–66, M.I.T. Press, Cambridge, Massachusetts.

Forbes, G. B., and Amirhakimi, G. H., 1970, Skinfold thickness and body fat in children, *Hum. Biol.* **42**:401.

Forbes, G. B., and Bruining, G. J., 1976, Urinary creatinine excretion and lean body mass, *Am. J. Clin. Nutr.* **29**:1359.

Forbes, G. B., and Hursh, J. B., 1963, Age and sex trends in lean body mass calculated from K[40] measurements: With a note on the theoretical basis for the procedure, *Ann. N. Y. Acad. Sci.* **110**:255.

Forbes, G. B., and Lewis, A., 1956, Total sodium, potassium, and chloride in adult man, *J. Clin. Invest.* **35**:596.

Forbes, G. B., and Perley, A. M., 1951, Estimation of total body sodium by isotopic dilution, *J. Clin. Invest.* **30**:558, 566.

Forbes, G. B., Schultz, F., Cafarelli, C., and Amirhakimi, G. H., 1968, Effects of body size on potassium-40 measurement in the whole body counter (tilt-chair technique), *Health Phys.* **15**:435.

Foy, J. M., and Schnieden, H., 1960, Estimation of total body water (virtual tritium space) in the rat, cat, rabbit, guinea pig and man, and of the biological half-life of tritium in man, *J. Physiol.* **154**:169.

Friis-Hansen, B., 1957, Changes in body water compartments during growth, *Acta Paediatr.* **110**(Suppl):46.

Fuller, M. F., Houseman, R. A., and Cadenhead, A., 1971, The measurement of exchangeable potassium in living pigs and its relation to body composition, *Br. J. Nutr.* **26**:203.

Gamble, J. L., Jr., Robertson, J. S., Hannigan, C. A., Foster, C. G., and Farr, L. F., 1953, Chloride, bromide, sodium, and sucrose spaces in man, *J. Clin. Invest.* **32**:483.

Garn, S. M., 1961, Radiographic analysis of body composition, in: *Techniques for Measuring Body Composition* (J. Brozek and A. Henschel, eds.), pp. 36–58, National Academy of Science, Washington, D.C.

Garn, S. M., 1970, *The Earlier Gain and the Later Loss of Cortical Bone in Nutritional Perspective*, Charles C Thomas, Springfield, Illinois.

Garrow, J. S., 1974, *Energy Balance and Obesity in Man*, American Elsevier, New York.

Garrow, J. S., Fletcher, K., and Halliday, D., 1965, Body composition in severe infantile malnutrition, *J. Clin. Invest.* **44**:417.

Godfrey, B. E., and Wadsworth, G. R., 1970, Total body potassium in pregnant women, *J. Obstet. Gynaecol. Br. Commonw.* **77**:244.

Goldman, R. F., Haisman, M. F., Bynum, G., Horton, E. S., and Sims, E. A. H., 1975, Experimental obesity in man, in: *Obesity in Perspective* (G. A. Bray, ed.), pp. 165–186, DHEW Pub. No. (NIH) 75-708, Washington, D.C.

Greenway, R. M., Littell, A. S., Houser, H. B., Lindan, O., and Weir, D. R., 1965, An evaluation of the variability in measurement of some body composition parameters, *Proc. Soc. Exp. Biol. Med.* **120**:487.

Grüner, O., and Salmen, A., 1961, Vergleichende Körperwasserbestimmungen mit Hilfe von N-Acetyl-4-amino Antipyrin und Alkohol, *Klin. Wochenschr.* **39**:92.

Gryfe, C. I., Exton-Smith, A. N., Payne, P. R., and Wheeler, E. F., 1971, Pattern of development of bone in childhood and adolescence, *Lancet* **1**:523.

Halliday, D., 1971, An attempt to estimate total body fat and protein in malnourished children, *Br. J. Nutr.* **26**:147.

Harvey, T. C., Dykes, P. W., Chen, N. S., Ettinger, K. V., Jain, S., James, H., Chettle, D. R., Fremlin, J. H., and Thomas, B. J., 1973, Measurement of whole-body nitrogen by neutron-activation analysis, *Lancet* **2**:395.

Hastings, A. B., and Eichelberger, L., 1937, The exchange of salt and water between muscle and blood, *J. Biol. Chem.* **117**:73.

Hawkins, W. W., 1964, Iron, copper, and cobalt, in: *Nutrition: A Comprehensive Treatise* (G. H. Beaton and E. W. McHenry, eds.), Vol. 1, pp. 309–372, Academic Press, New York.

Haxhe, J. J., 1963, *La Composition Corporelle Normale*, Librairie Maloine S.A., Paris.

Heald, F. P., Hunt, E. E., Jr., Schwartz, R., Cook, C. D., Elliot, O., and Vajda, B., 1963, Measures of body fat and hydration in adolescent boys, *Pediatrics* **31**:226.

Hermansen, L., and Döbeln, W. von, 1971, Body fat and skinfold measurements, *Scand. J. Clin. Lab. Invest.* **27**:315.

Hervey, G. R., Hutchinson, I., Knibbs, A. V., Burkinshaw, L., Jones, P. R. M., Norgan, N. G., and Levell, M. J., 1976, "Anabolic" effects of methandienone in men undergoing athletic training, *Lancet* **2**:699.

Huenemann, R. L., Hampton, M. C., Behnke, A. R., Shapiro, L. R., and Mitchell, B. W., 1974, *Teenage Nutrition and Physique*, Charles C Thomas, Springfield, Illinois.

Hume, R., 1966, Prediction of lean body mass from height and weight, *J. Clin. Pathol.* **19**:389.

Hume, R., and Weyers, E., 1971, Relationship between total body water and surface area in normal and obese subjects, *J. Clin. Pathol.* **24**:234.

Hytten, F. E., and Leitch, I., 1971, *The Physiology of Human Pregnancy*, 2nd ed., Blackwell, Oxford.

Hytten, F. E., Taylor, K., and Taggart, N., 1966, Measurement of total body fat in man by absorption of [85]Kr, *Clin. Sci.* **31**:111.

International Atomic Energy Agency, 1970, *Directory of Whole Body Radioactivity Monitors*, Vienna.

Isherwood, I., Rutherford, R. A., Pullan, B. R., and Adams, P. H., 1976, Bone-mineral estimation by computer-assisted transverse axial tomography, *Lancet* **2**:712.

Jasani, B. M., and Edmonds, C. J., 1971, Kinetics of potassium distribution in man using isotope dilution and whole-body counting, *Metabolism* **20**:1099.

Johny, K. V., Worthey, B. W., Lawrence, J. R., and O'Halloran, M. W., 1970, A whole body counter for serial studies of total body potassium, *Clin. Sci.* **39**:319.

Katch, F. I., and Michael, E. D., Jr., 1968, Prediction of body density from skinfold and girth measurements of college females, *J. Appl. Physiol.* **25**:92.

Kaufman, L., and Wilson, C. J., 1973, Determination of extracellular fluid volume by fluorescent excitation analysis of bromine, *J. Nucl. Med.* **14**:812.

Khosla, T., 1968, Unfairness of certain events in the Olympic Games, *Br. Med. J.* **4**:111.

King, J. C., Calloway, D. H., and Morgen, S., 1973, Nitrogen retention, total body ^{40}K and weight gain in teenage pregnant girls, *J. Nutr.* **103**:772.

Kjellberg, J., and Reizenstein, P., 1970, Body composition in obesity, *Acta Med. Scand.* **188**:161.

Knittle, J. L., 1976, Discussion, in: *Nutrient Requirements in Adolescence* (J. I. McKigney and H. N. Munro, eds.), pp. 76–83, M.I.T. Press, Cambridge, Massachusetts.

Kreisberg, R. A., Bowdoin, B., and Meador, C. K., 1970, Measurement of muscle mass in humans by isotopic dilution of creatine-^{14}C, *J. Appl. Physiol.* **28**:264.

Lesser, G. T., Perl, W., and Steele, J. M., 1960, Determination of total body fat by absorption of an inert gas: measurements and results in normal human subjects, *J. Clin. Invest.* **39**:1791.

Lesser, G. T., Deutsch, S., and Markofsky, J., 1971, Use of independent measurement of body fat to evaluate overweight and underweight, *Metabolism* **20**:792.

Linquette, M., Fossati, P., Lefebvre, J., and Checkau, C., 1969, The measurement of total water, exchangeable sodium and potassium in obese persons, in: *Pathophysiology of Adipose Tissue* (J. Vague, ed.), pp. 302–316, Excerpta Medica Foundation, Amsterdam.

Ljunggren, H., Ikkos, D., and Luft, R., 1957, Studies on body composition, *Acta Endocrinol.* **25**:187.

Lohman, T. G., Boileau, R. A., and Massey, B. H., 1975, Prediction of lean body mass in young boys from skinfold thickness and body weight, *Hum. Biol.* **47**:245.

Long, C. L., Haverberg, L. N., Young, V. R., Kinney, J. M., Munro, H. N., and Geiger, J. W., 1975, Metabolism of 3-methyl histidine in man, *Metabolism* **24**:929.

Lye, M., May, T., Hammick, J., and Ackery, D., 1976, Whole body and exchangeable potassium measurements in normal elderly subjects, *Eur. J. Nucl. Med.* **1**:167.

MacGillivray, I., and Buchanan, T. J., 1958, Total exchangeable sodium and potassium in non-pregnant women and in normal and pre-eclamptic pregnancy, *Lancet* **2**:1090.

Magnus-Levy, A., 1906, Physiologie des Stoffwechsels, in: *Handbuch der Pathologie des Stoffwechsels* (C. von Noorden, ed.), p. 446, Hirschwald, Berlin.

*Malina, R. M., 1969, Quantification of fat, muscle, and bone in man, *Clin. Orthop.* **65**:9.

Manery, J. F., and Haege, L. F., 1941, The extent to which radioactive chloride penetrates tissues, and its significance, *Am. J. Physiol.* **134**:83.

Manzke, E., Chestnut, C. H., III, Wergedal, J. E., Baylink, D. J., and Nelp, W. B., 1975, Relationship between local and total bone mass, *Metabolism* **24**:605.

Maresh, M. M., 1966, Changes in tissue widths during growth, *Am. J. Dis. Child.* **111**:142.

Mazess, R. B., and Cameron, J. R., 1972, Growth of bone in school children: Comparison of radiographic morphometry and photon absorptiometry, *Growth* **36**:77.

McCance, R. A., and Widdowson, E. M., 1951, A method of breaking down the body weights of living persons into terms of extracellular fluid, cell mass and fat, and some applications of it to physiology and medicine, *Proc. R. Soc. Ser. B* **138**:115.

Meneely, G. R., Heyssel, R. M., Ball, C. O. T., Weiland, R. L., Lorimer, A. R., Constantinides, C., and Meneely, E. U., 1963, Analysis of factors affecting body composition determined from potassium content in 915 normal subjects, *Ann. N. Y. Acad. Sci.* **110**:271.

Mettau, J. W., Degenhart, H. J., Visser, H. K. A., and Holland, W. P. S., 1977, Measurement of total body fat in newborns and infants by absorption and desorption of nonradioactive xenon, *Pediatr. Res.* **11**:1097.

Michael, E. D., Jr., and Katch, F. I., 1968, Prediction of body density from skinfold and girth measurements of 17-year-old boys, *J. Appl. Physiol.* **25**:747.

Moody, D. L., Wilmore, J. H., Girandola, R. N., and Royce, J. P., 1972, The effects of a jogging program on the body composition of normal and obese high school girls, *Med. Sci. Sports* **4**:210.

*Moore, F. D., Olesen, K. H., McMurray, J. D., Parker, H. V., Ball, M. R., and Boyden, C. M., 1963, *The Body Cell Mass and Its Supporting Environment*, W. B. Saunders Co., Philadelphia.

Moore, F. D., Lister, J., Boyden, C. M., Ball, M. R., Sullivan, N., and Dagher, F. J., 1968, The skeleton as a feature of body composition: values predicted by isotope dilution and observed by cadaver dissection in an adult female, *Hum. Biol.* **40**:135.

Moulton, C. R., 1923, Age and chemical development in mammals, *J. Biol. Chem.* **57**:79.

Muldowney, F. P., 1961, Lean body mass as a metabolic reference standard, in: *Techniques for Measuring Body Composition* (J. Brozek and A. Henschel, eds.), pp. 212–222, National Academy of Science, Washington, D.C.

Nagamine, S., and Suzuki, S., 1964, Anthropometry and body composition of Japanese young men and women, *Hum. Biol.* **36**:8.

*National Academy of Sciences, 1968, Body Composition in Animals and Man, Pub. 1598, Washington, D.C.

National Center for Health Statistics, 1974, Skinfold thickness of youths 12–17 years, United States, DHEW Pub. No. (HRA) 74-614, Series 11, #132, Rockville, Maryland.

Nelp, W. B., Palmer, H. E., Murano, R., Pailthorp, K., Hinn, G. M., Rich, C., Williams, J. L., Rudd, T. G., and Denney, J. D., 1970, Measurement of total body calcium (bone mass) *in vivo* with the use of total body neutron activation, *J. Lab. Clin. Med.* **76**:151.

Nichols, B. L., Hazlewood, C. F., and Barnes, D. J., 1968, Percutaneous needle biopsy of quadriceps muscle: Potassium analysis in normal children, *J. Pediatr.* **72**:840.

Norris, A. H., Lundy, T., and Shock, N. W., 1963, Trends in selected indices of body composition in men between the ages 30 and 80 years, *Ann. N. Y. Acad. Sci.* **110**:623.

Novak, L. P., 1966a, Total body water and solids in six- to seven-year-old children: Differences between the sexes, *Pediatrics* **38**:483.

Novak, L. P., 1966b, Physical activity and body composition of adolescent boys, *J. Am. Med. Assoc.* **197**:891.

Novak, L. P., 1970, Comparative study of body composition of American and Filipino women, *Hum. Biol.* **42**:206.

Novak, L. P., Tauxe, W. N., and Orvis, A. L., 1973, Estimation of total body potassium in normal adolescents by whole body counting: age and sex differences, *Med. Sci. Sports* **5**:147.

Oberhausen, E., Burmeister, W., and Huycke, E. J., 1965, Das Wachstrum des Kaliumbestandes im Menschen gemessen mit dem Ganzkörperzähler, *Ann. Paediatr.* **205**:381.

Owen, G. M., Jensen, R. L., and Fomon, S. J., 1962, Sex-related difference in total body water and exchangeable chloride during infancy, *J. Pediatr.* **60**:858.

Pace, N., and Rathbun, E. N., 1945, Studies on body composition, *J. Biol. Chem.* **158**:685.

Parizkova, J., 1961a, Total body fat and skinfold thickness in children, *Metabolism* **10**:794.

Parizkova, J., 1961b, Age trends in fat in normal and obese children, *J. Appl. Physiol.* **16**:173.

Parizkova, J., 1963, Impact of age, diet, and exercise on man's body composition, *Ann. N. Y. Acad. Sci.* **110**:661.

Parizkova, J., 1976, Growth and growth velocity of lean body mass and fat in adolescent boys, *Pediatr. Res.* **10**:647.

Parizkova, J., Vaněčková, M., and Vamberová, M., 1962, A study of changes in some functional indicators following reduction of excessive fat in obese children, *Physiol. Bohemoslov.* **11**:351.

Parizkova, J., Vaněčková, M., Šprynarová, S., and Vamberová, M., 1971, Body composition and fitness in obese children before and after special treatment, *Acta Paediatr. Scand. Suppl.* **217**:6.

Pearson, A. M., Purchas, R. W., and Reineke, E. P., 1968, Theory and potential usefulness of body density as a predictor of body composition, in: *Body Composition in Animals and Man* (J. T. Reid, ed.), pp. 153–185, National Academy of Science, Pub. #1598, Washington, D.C.

Pfeiffer, L., 1887, Über den Fettgehalt des Körpers, *Z. Biol.* **23**:340.

Picou, D., Reeds, P. J., Jackson, A., and Poulter, N., 1976, The measurement of muscle mass in children using [^{15}N]creatine, *Pediatr. Res.* **10**:184.

Pierson, R. N., Jr., Lin, D. H. Y., and Phillips, R. A., 1974, Total-body potassium in health: Effects of age, sex, height, and fat, *Am. J. Physiol.* **226**:206.

Pitts, G. C., and Bullard, T. R., 1968, Some interspecific aspects of body composition in mammals, in: *Body Composition in Animals and Man,* pp. 45–70, National Academy of Science, Pub. #1598, Washington, D.C.

Pollock, M. L., Hickman, T., Kendrick, A., Jackson, A., Linnerud, A., and Dawson, G., 1976, Prediction of body density in young and middle-aged men, *J. Appl. Physiol.* **40**:300.

Price, W. F., Hazelrig, J. B., Kreisberg, R. A., and Meador, C. K., 1969, Reproducibility of body composition measurements in a single individual, *J. Lab. Clin. Med.* **74**:557.

Rudd, T. G., Pailthorp, K. G., and Nelp, W. B., 1972, Measurement of non-exchangeable sodium in normal man, *J. Lab. Clin. Med.* **80**:442.

Rundo, J., and Sagild, U., 1955, Total and "exchangeable" potassium in humans, *Nature* **175**:774.

Schuster, W., 1970, Über Methoden und Ergebnisse quantitativer Mineralsalz-bestimmungen am kindlichen Skelett, *Arch. Kinderh.* **180**:256.

Seitchik, J., 1967, Total body water and total body density of pregnant women, *Obstet. Gynecol.* **29**:155.

Shaffer, P., 1908, The excretion of kreatinin and kreatin in health and disease, *Am. J. Physiol.* **23**:1.

Shephard, R. J., Hatcher, J., and Rode, A., 1973, On the body composition of the Eskimo, *Eur. J. Appl. Physiol.* **32**:3.

Shukla, K. K., Ellis, K. J., Dombrowski, C. S., and Cohn, S. H., 1973, Physiological variation of total-body potassium in man, *Am. J. Physiol.* **224**:271.

Siri, W. E., 1961, Body composition from fluid spaces and density: Analysis of methods, in: *Techniques for Measuring Body Composition* (J. Brozek and A. Henschel, eds.), pp. 223–244, National Academy of Science, Washington, D. C.

Sloan, A. W., 1967, Estimation of body fat in young men, *J. Appl. Physiol.* **23**:311.

Smith, D. M., Nance, W. E., Kang, K. W., Christian, J. C., and Johnston, C. C., Jr., 1973, Genetic factors in determining bone mass, *J. Clin. Invest.* **52**:2800.

Šprynarová, S., and Parizková, J., 1965, Changes in the aerobic capacity and body composition in obese boys after reduction, *J. Appl. Physiol.* **20**:934.

Steinkamp, R., Cohen, N. L., Gaffey, W. R., McKey, T., Bron, G., Siri, W. E., Sargent, T. W., and Isaacs, E., 1965, Measures of body fat and related factors in normal adults—II: A simple clinical method to estimate body fat and lean body mass, *J. Chron. Dis.* **18**:1291.

Stouffer, J. R., 1963, Relationship of ultrasonic measurements and X-rays to body composition, *Ann. N. Y. Acad. Sci.* **110**:31.

Stuart, H. C., Hill, P., and Shaw, C., 1940, The growth of bone, muscle and overlying tissue as revealed by studies of roentgenograms of the leg area, *Monogr. Soc. Res. Child Dev.* **V**:(No. 3, serial No. 26) 1–190.

Surveyor, I., and Hughes, D., 1968, Discrepancies between whole-body potassium content and exchangeable potassium, *J. Lab. Clin. Med.* **71**:464.

Talso, P. J., Miller, C. E., Carballo, A. J., and Vasquez, I., 1960, Exchangeable potassium as a parameter of body composition, *Metabolism* **9**:456.

*Tanner, J. M., 1962, *Growth at Adolescence,* 2nd ed., Blackwell, Oxford.

Tanner, J. M., 1974, Sequence and tempo in the somatic changes in puberty, in: *Control of the Onset of Puberty* (M. M. Grumbach, G. D. Grave, and F. E. Mayer, eds.), pp. 448–470, John Wiley & Sons, New York.

Tanner, J. M., and Whitehouse, R. H., 1967, The effect of human growth hormone on subcutaneous fat thickness in hyposomatotrophic and panhypopituitary dwarfs, *J. Endocrinol.* **39**:263.

Tanner, J. M., and Whitehouse, R. H., 1975, Revised standards for triceps and subscapular skinfolds in British children, *Arch. Dis. Child.* **50**:142.

Tisavipat, A., Vibulsreth, S., Sheng, H.-P., and Huggins, R. A., 1974, Total body water measured by desiccation and by tritiated water in adult rats, *J. Appl. Physiol.* **37**:699.

Trotter, M., and Hixon, B. B., 1974, Sequential changes in weight, density, and percentage ash weight of human skeletons from an early fetal period through old age, *Anat. Rec.* **179**:1.

Tyson, I., Genna, S., Jones, R. L., Bikerman, V., and Burrows, B. A., 1970, Body potassium measurements with a total-body counter, *J. Nucl. Med.* **11**:255.

Wallace, W. M., 1959, Nitrogen content of the body and its relation to retention and loss of nitrogen, *Fed. Proc.* **18**:1125.

Ward, G. M., Krzywicki, H. J., Rahman, D. P., Quaas, R. L., Nelson, R. A., and Consolazio, C. F., 1975, Relationship of anthropometric measurements to body fat as determined by densitometry, potassium-40, and body water, *Am. J. Clin. Nutr.* **28**:162.

Ward, P., 1973, The effect of an anabolic steroid on strength and lean body mass, *Med. Sci. Sports* **5**:277.

Wedgewood, R. J., 1963, Inconstancy of the lean body mass, *Ann. N. Y. Acad. Sci.* **110**:141.

*Whipple, H. E., and Brozek, J. (eds.), 1963, Body composition, *Ann. N. Y. Acad. Sci.* **110**:1.

Widdowson, E. M., McCance, R. A., and Spray, C. M., 1951, The chemical composition of the human body, *Clin. Sci.* **10**:113.

Wilmore, J. H., and Behnke, A. R., 1970, An anthropometric estimation of body density and lean body weight in young women, *Am. J. Clin. Nutr.* **23**:267.

Wilmore, J. H., Girandola, R. N., and Moody, D. L., 1970, Validity of skinfold and girth assessment for predicting alterations in body composition, *J. Appl. Physiol.* **29**:313.

Womersley, J., Durnin, J. V. G. A., Boddy, K., and Mahaffy, M., 1976, Influence of muscular development, obesity, and age on the fat-free mass of adults, *J. Appl. Physiol.* **41**:223.

Young, C., Sipin, S. S., and Roe, D. A., 1968, Body composition of pre-adolescent and adolescent girls, *J. Am. Diet. Assoc.* **53**:25, 357, 469, 579.

10

Growth of Muscle Tissue and Muscle Mass

ROBERT M. MALINA

1. Growth of Muscle Tissue

1.1. Histological Considerations

The basic unit of muscle tissue is the muscle cell or muscle fiber, a cylindrical, multinucleated unit. A skeletal muscle contains many long fibers which have the ability to contract. This contractile property resides in specialized and interacting proteins localized in myofibrils which comprise the muscle fiber. Myofibrils have a transverse banding pattern which gives skeletal muscle its striated appearance. The units comprising each band are sarcomeres.

Muscle fibers are multinucleate, the number of cylindrical nuclei varying with the length of a fiber and with age. Nuclei are located at the periphery of a muscle fiber, just beneath the plasma or cell membrane, which immediately encloses the fiber's contents. Some evidence (Landing *et al.*, 1974) suggests that nuclei of a muscle fiber are arranged in a hexagonal array, with each nucleus covering a fiber area that is uniform in size and shape.

Closely related to muscle fibers are satellite cells, located at the periphery of muscle fibers, between the plasma and basement membranes. The latter is an outer coating of the fiber. Satellite cells appear to be important in muscle growth (see below).

Muscle fibers have their origin in the middle germinal layer of the embryo, the mesenchyme. They arise from mononucleated myoblasts, which fuse to form myotubes. Myotubes are multinucleated syncitia and can be viewed as immature muscle fibers. Their myofibrils are distributed circumferentially with the nuclei at the central area of a syncitium. The nuclei then migrate to the periphery beneath the plasma membrane, and the myofibrils become more evenly distributed throughout

ROBERT M. MALINA • Department of Anthropology, University of Texas, Austin, Texas.

the cytoplasm so that the cell now has the appearance of a proper muscle fiber (Fischman, 1972).

Goldspink (1972) suggests that the postembryonic growth of muscle tissue can be divided into two stages: (1) an early postembryonic stage during which myotubes develop into proper muscle fibers and increase in girth and length; and (2) a subsequent period of further growth in girth and length. The latter stage continues postnatally, and early postnatally mature biochemical and physiological characteristics of the muscle develop.

In humans the number of muscle fibers increases prenatally and for a short period postnatally. Montgomery (1962a) noted that the number of muscle fibers approximately doubles between the 32nd week of gestation and 4 months of age. Goldspink (1972), however, cautions that the degree of postnatal increase in muscle fibers seems dependent on the organism's maturity status at birth and that the postnatal fiber increase shortly after birth should be considered an extension of the embryonic differentiation of muscle tissue.

Muscle fibers grow in length by an increase in the number of sarcomeres in series along myofibrils, and possibly by an increase in length of individual sarcomeres (Goldspink, 1972). The former, i.e., increase in sarcomere number, is the main contributing factor to muscle length. Growth in length occurs primarily at the musculotendinous junction, and the number of sarcomeres along a muscle fiber is apparently adjusted to the functional length of a muscle (Goldspink, 1972). Needless to say, the differential growth of limb segments postnatally requires similar length increases in individual muscles.

Growth of a muscle in length also entails an increase in muscle fiber nuclei, an increase which continues after growth in fiber length has ceased (Goldspink, 1972; Burleigh, 1974; Cheek et al., 1971). Since larger-diameter fibers generally have more nuclei, it can be concluded that growth of a muscle fiber in both girth and length is associated with an increase in the number of nuclei. Thus, the number of nuclei observed in fiber cross-sections or along a fiber's length increases with age. The additional nuclei are apparently derived from satellite cells (Montgomery, 1962a; Goldspink, 1972; Burleigh, 1974; Cheek et al., 1971). The incorporation of these nuclei into the muscle fiber seemingly assures sufficient nuclear control for the new fiber components added through growth in length and girth. Note, however, that alternative explanations for the increase in nuclei with growth have been proposed (Burleigh, 1974).

The marked postnatal increase in muscle girth is due almost entirely to hypertrophy or continued growth of existing muscle fibers, and not to hyperplasia (Goldspink, 1972). Muscle fibers increase in diameter linearly with age (Figure 1) and body size postnatally (Bowden and Goyer, 1960; Aherne et al., 1971). Increase in fiber diameter varies with muscle studied, and is apparently related to function or intensity of workload. During infancy and childhood, muscle fibers of boys and girls do not consistently differ in diameter, and Brooke and Engel (1969b) note that adult diameters are attained between 12 and 15 years of age. However, the data shown in Figure 1, as well as other childhood data cited, indicate a paucity of "normal" data in middle childhood and adolescence.

The preceding consideration of change in muscle diameters with growth is limited in that there is a need for information on the distribution of fiber types within muscle during growth. Mature muscle has two primary histochemical fiber types: type I, red, slow-twitch fibers, characterized by high mitochondrial oxidative enzyme activity, low phosphorylase and ATPase reactions, and low glycogen concentration; and type II, white, fast-twitch fibers, characterized by high phosphorylase and ATPase reactions, high glycogen content, and low activity of most dehydrogenases (Jennekens *et al.*, 1970; Brooke and Kaiser, 1970). Type I fibers have larger and more numerous mitochondria than type II fibers (Payne *et al.*, 1975). Type II fibers are larger in adult males than in females (Jennekens *et al.*, 1970). Type II fibers are also larger than type I fibers in adult males (Brooke and Engel, 1969*a;* Jennekens *et al.*, 1970; Polgar *et al.*, 1973). The reverse is generally true in adult females, although there is variation among specific muscles (Brooke and Engel, 1969*a;* Jennekens *et al.*, 1970). For example, type II fibers are larger in the deltoid and biceps muscles in both men and women, but are smaller in the rectus femoris and gastrocnemius (Jennekens *et al.*, 1970). Note that sample sizes in such studies as those cited are characteristically very small, thus contributing to the observed variation. In addition, some of the variation between muscles might reflect sampling variation within the muscle tissue studied.

Some data suggest that predominantly phasic muscles have a high percentage of type II fibers, while predominantly tonic muscles have a high percentage of type I fibers. However, many muscles perform both functions and show no predominance of either fiber type (Johnson *et al.*, 1973).

Although there is variation between individual muscles sampled, type I fibers comprise approximately 37%, and type II approximately 63% of adult human muscle (Brooke and Engel, 1969*a;* Brooke and Kaiser, 1970). Limited data for children's muscle indicates a slightly higher proportion of type I fibers, about 45% (Brooke and Engel, 1969*b*). Data for age-associated variation in type I and type II fibers are not available. Comparison of childhood and adult values would seem to indicate a relative increase in type II fibers during growth.

1.2. Chemical Composition of Muscle

Changes in the chemical composition of muscle during growth and development reflect histological changes. In the fetus, fibers are small, few in number and widely separated by extracellular material. At term, fibers are still small, but are greater in number and more closely packed in the muscle. In adults, muscle fibers are larger in diameter with little space between them (Widdowson, 1969). These histological changes are illustrated in the decreasing concentrations of extracellular ions (sodium and chloride) and increasing concentrations of intracellular constituents (potassium and phosphorus) with growth and development (Table I). In addition, there is a fall in the percentage of water. The size of the extracellular component of muscle with growth and development is first reduced by the increase in fiber number prenatally, and then by the increase in fiber size postnatally.

The decrease in the relative water content of muscle tissue is accompanied by an increase in total nitrogen with development (Table II). Both nonprotein nitrogen and cellular protein nitrogen (sarcoplasmic and fibrillar) increase with growth and development from early in prenatal life to adulthood. The sarcoplasmic and fibrillar protein components change independently during growth and development. The relative contribution of sarcoplasmic protein decreases during fetal development and then increases postnatally. The relative contribution of fibrillar protein, on the other hand, does not change much prenatally, but increases postnatally. Dickerson and Widdowson (1960) suggest that the rate at which fibrillar protein develops postnatally may be influenced by the functional activity of the muscle.

The absolute extracellular protein nitrogen increases through infancy and then decreases to adulthood (Table II). As a percentage of total nitrogen, the extracellular fraction, however, increases to a maximum at about the time of birth, and then decreases postnatally to a lower level in the adult. In a general way, the increase in relative contribution of extracellular protein nitrogen appears to parallel the increase in muscle fiber numbers prenatally, and its decreasing relative contribution apparently occurs when muscle fibers are increasing in size (Dickerson and Widdowson, 1960; Widdowson, 1969).

The lack of information for children and adolescents is especially apparent in the chemical composition data. Dickerson and Widdowson (1960), however, note

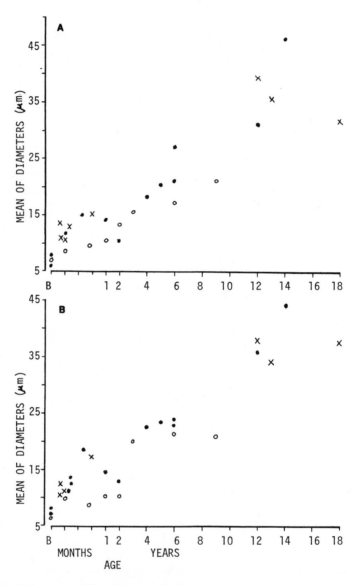

Fig. 1. Postnatal increase in fiber diameter of selected muscles. (A, biceps brachii; B, deltoid; C, gastrocnemius; D, vastus lateralis). x, girls, and ●, boys from Aherne *et al.* (1971); ○, combined, from Bowden and Goyer (1960).

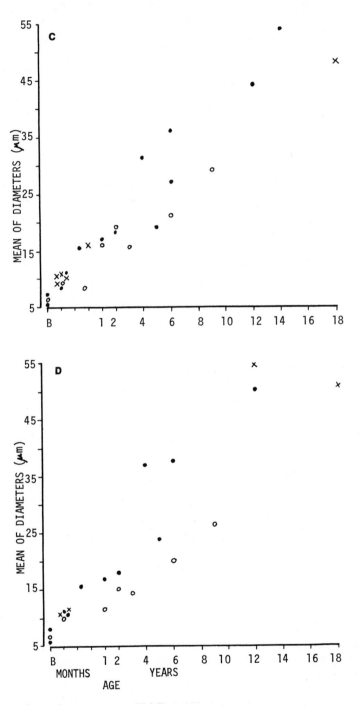

Fig. 1. (*Continued*)

Table I. Water and Electrolyte Composition of Human Skeletal Muscle Pre- and Postnatally [a]

	Fetus		Infant		
	13–14 weeks	20–22 weeks	Full-term newborn	4–7 months	Adult
Water, g/100 g [b]	91	89	80	79	79
Na, mEq/kg	101	91	60	50	36
Cl, mEq/kg	76	66	43	35	22
K, mEq/kg	56	58	58	89	92
P, m mol/kg	37	40	47	65	59
Na space, g/100 g	80	71	43	35	26
Cl space, g/100 g	67	58	35	29	18

[a] Adapted from Dickerson and Widdowson (1960).
[b] Results are expressed per unit of fresh muscle.

Table II. Concentration of Nitrogen in Various Fractions of Human Skeletal Muscle (g/100 g Fresh Muscle) Pre- and Postnatally [a]

	Fetus		Infant		
	14 weeks	20–22 weeks	Full-term newborn	4–7 months	Adult
Total N	1.09	1.52	2.09	2.90	3.05
Nonprotein N	0.12	0.17	0.24	0.32	0.30
Sarcoplasmic protein N	0.36	0.37	0.39	0.50	0.67
Fibrillar protein N	0.57	0.87	1.09	1.70	1.99
Extracellular protein N	0.06	0.18	0.38	0.46	0.14
Sarcoplasmic protein, %	33.0	24.3	18.7	17.2	22.0
Fibrillar protein, %	52.3	57.2	52.2	58.6	65.2
Extracellular protein, %	5.5	11.8	18.2	15.9	4.6

[a] Adapted from Dickerson and Widdowson (1960).

that figures for the composition of muscle in an 11- and a 16-year-old boy were similar to those for adults, and thus combined them with adult values. This would seem to suggest that muscle tissue attains chemical maturity during childhood, at least in terms of the variables studied.

1.3. Muscle Cell Size and Nuclear Number

Growth of muscle tissue after the first few months of postnatal life is characterized by constancy in number of muscle fibers, an increase in fiber size, and a considerable increase in number of muscle nuclei (Montgomery, 1962a). Age changes in the number of nuclei during growth have received much attention in both experimental animals and humans. The number of nuclei in developing muscles has been estimated from measurements of DNA. Since DNA per nucleus is relatively constant at 6.2 pg/diploid nucleus, total number of nuclei in a muscle tissue sample can be determined by dividing the total DNA in the tissue by 6.2. Cheek and colleagues (Cheek, 1968, 1974, 1975; Cheek and Hill, 1970; Cheek et al., 1971) have applied this method to assess muscular growth in a small sample of normal children. Using DNA content of a sample of gluteal muscle tissue as an index of nuclear or "cell" number, and the ratio of protein to DNA as an index of "muscle cell size," Cheek and colleagues have shown an increase in nuclear number or in DNA and an increase in amounts of protein or cell mass per nucleus or per unit DNA during growth and development. Assuming that the gluteal biopsy sample is representative of body musculature in general and using creatinine excretion per day (see below) as an estimate of total muscle mass in the body, Cheek and colleagues have generalized nuclear number and cell size estimates to total muscle mass in growing children.

Based on the preceding, boys show a 14-fold increase in muscle nuclear number from infancy through adolescence, while girls show a tenfold increase. In terms of limits for muscle nuclear number, Cheek's (1975) estimates suggest a limit of 3.8×10^{12} nuclei for boys and 2.0×10^{12} nuclei for girls, approximately half the number in males. Relative to age, males increase in muscle nuclear number especially prior to 2 years of age and again after 9 years of age. The latter comprises the adolescent spurt in muscle mass. Prior to the adolescent spurt, sex differences in estimated nuclear numbers are not obvious. On the other hand, increase in estimated "cell size" or protein or cell mass per nucleus or per unit of DNA is only twofold for boys and girls. Cheek's data indicate larger protein/DNA ratios in girls at an earlier age, implying accelerated growth of "muscle cell size" in females. Males eventually catch up in this muscle tissue estimate and may eventually surpass females. Cheek's data, however, do not extend past 17 years of age.

The preceding estimates of nuclear number and "cell size," although interesting, should be viewed with care. The sample from which they are derived is small, 33 boys and 19 girls. Of the former, 17 boys are clustered between 0.18 and 1.37 years, while 16 range from 5.27 to 16.05 years. Only five of the boys are over 12 years of age. One can inquire, therefore, as to the accuracy of projections through male adolescence based on such a limited data base. The 19 girls range from 3.50 to 17.08 years of age, only eight are over 11 years of age, and only two are sexually mature. In addition, generalizations from a sample of gluteal muscle tissue to the body as a whole needs clarification.

Methodological concerns relative to estimates of "cell" or nuclear number are also apparent. Muscle cells or fibers are multinucleate, and because of this, the amount of DNA in a muscle shows nothing of the number of fibers in it. In addition, the ratio of protein or cell mass per nucleus or per unit DNA reveals nothing about the size of the muscle cell (Widdowson, 1970). What is shown, however, is the amount of cytoplasm associated with each nucleus, and Widdowson (1970) emphasizes that this ratio increases in human muscle tissue (as in other organs) during postnatal growth until the proper functional relationship is attained. The amount of protein in the nucleus per unit DNA of skeletal muscle in the human newborn is 175 mg/mg and rises to 300 mg/mg in adult muscle (Widdowson, 1970).

Cheek and colleagues have referred to the increase in nuclear number as an increase in "muscle cell" number, choosing a working definition which assumes that each nucleus has jurisdiction over a finite volume of cytoplasm (see Cheek, 1968; Cheek et al., 1971). Although this concept can be criticized and has added a certain amount of confusion to the literature, it should be noted that the term hyperplasia does not necessarily have to be applied to whole muscle fibers (Goss, 1966).

The source of additional nuclei during growth is not clear. Satellite cells are commonly viewed as responsible for the increase in nuclei or DNA of muscle during growth (Cheek et al., 1971; Goldspink, 1972; Burleigh, 1974). However, alternative possibilities for the increase in nuclei with growth must be considered. Burleigh (1974), for example, indicates that increments in total DNA can also be the result of replication of nuclei in the connective tissue between muscle fibers, in the capillaries of the blood supply, and perhaps in fat cells. This would thus imply other explanations for the increase in total muscle DNA or nuclear number. Interpretations of gross DNA estimates and generalizations from such estimates to the total muscle mass can also be obscured by other factors. Burleigh (1974), for example, indicates two such considerations: first, muscle fibers differ in the extent to which their fibers elongate with age, and second, fibers may vary in the number of nuclei within them and in the number of dividing satellite cells that occur along their length. Thus, although there are problems evident in the use and interpretation of the DNA unit and associated ratios, data derived from such estimates have provided insights into muscle tissue growth.

2. Growth of Muscle Mass

2.1. Dissection Studies

There is no verified method by which total muscle mass can be measured *in vivo* (Garn, 1963; Durnin, 1969; Kreisberg *et al.,* 1970). Estimates of muscle mass derived from dissection studies (Table III) indicate that muscle comprises about 23–25% of body weight at birth and about 40% of body weight in adults, although the range of reported values for adults varies. Further, the adult dissection specimens are generally older adults so that one can perhaps assume a slightly higher percentage contribution of muscle mass to body weight in young adulthood, probably about 45% or more. Dissection data for the growing years are lacking. Relative to fat-free body weight, i.e., body weight minus total fat, muscle mass comprises approximately 45–50% of the fat-free body mass (Durnin, 1969).

Table III. Muscle Mass as a Percentage of Body Weight[a]

Age	Sex	Body weight	Muscle mass (%)
Fetus, 6 months		401 g	22.75
Fetus, 6 months		491 g	22.71
Fetus, 6 months			24.60
Newborn		2.36 kg	23.30
Newborn		2.91 kg	24.03
Newborn		3.10 kg	25.05
Newborn			24.80
Adult		66.20 kg	43.40
Adult			43.07
Adult	M	69.67 kg	41.77
Adult	M	55.75 kg	41.36
Adult	M	76.51 kg	42.07
Adult, 35 years	M	70.55 kg	31.56
Adult, 46 years	M	53.80 kg	39.76
Adult, 60 years	M	73.50 kg	40.22
Adult, 48 years	M	62.00 kg	42.53
Adult	F	55.40 kg	35.83
Adult, 67 years	F	43.40 kg	23.13

[a]Based on dissection data collated from various sources (after Malina, 1969).

Urinary creatinine excretion is frequently used as an index of total muscle mass in the body. Although creatinine is not solely a byproduct of muscle metabolism, the volume of creatinine excreted is a function of the muscle mass. Creatinine excretion can be expressed simply in grams per day or relative to body weight (the creatinine coefficient). The former serves as an index of muscle mass, while the latter estimates the relative amount of muscle in the body (Novak, 1963; Cheek, 1968).

Muscle tissue development, however, is affected by a variety of factors, including age, sex, degree of physical training, hormonal states, and metabolic states. Further, dietary intake directly influences creatinine excretion values (Garn, 1963; Grande, 1963; Cheek, 1968), making creatinine difficult to control in man and perhaps limiting its utility in muscle mass estimates. Nevertheless, within the limitations imposed by the complexities of creatinine metabolism and excretion, urinary levels of creatinine, when collected under properly controlled conditions, offer the advantage of ready availability and thus provide an estimate of muscle mass.

The amount of creatinine excreted in the urine over a 24-hr period is used as an estimate of muscle mass. Mean urinary creatinine excretion increases with age, especially during adolescence, and is greater in boys than in girls (Reynolds and Clark, 1947; Clark *et al.,* 1951; Novak, 1963; Zorab, 1969). From 10 through 18 years of age, the amount of creatinine excreted more than doubles in boys and almost doubles in girls.

A constant relationship between creatinine excreted per unit of muscle mass has been suggested. Talbot (1938) estimated that 1 g of creatinine excreted per 24 hr was derived from 17.9 kg of muscle tissue. Cheek and colleagues (see Cheek, 1968), however, recommend a conversion factor of 20 as being more appropriate, i.e., each gram of creatinine excreted in the 24-hr urine sample is derived from 20 kg of muscle tissue. Using this conversion factor, absolute muscle mass has been estimated from published creatinine excretion and body weight data (Figure 2). Absolute muscle mass increases with age, sex differences being minor prior to the adolescent spurt but marked during adolescence. The estimates from Clark *et al.* (1951, mixed longitudinal), Novak (1963, cross-sectional), and Young *et al.* (1968, cross-sectional) agree rather closely. The muscle mass estimates from Cheek's (1968, 1975) small sample are consistently lower. This perhaps represents sampling and methodological variation, as Cheek's data are derived over a 3-day period in the presence of a low-creatinine diet. However, the small sample above 3 years of age for whom creatinine excretion data are available (N = 19 boys and 19 girls) is somewhat unusual. Among the boys, 16 of the 17 for whom skeletal age (SA) data were available had skeletal ages less than their respective chronological ages (CA). The average SA–CA difference was −1.45 years. In contrast, 14 of the 19 girls had skeletal ages in advance of their chronological ages, an average SA–CA difference of +0.46 years. Note that muscular development and skeletal and/or sexual maturation are positively related (see below), so that the lesser estimated muscle mass values might reflect maturity differences, especially in the boys.

Estimated muscle mass expressed as a percentage of body weight increases from 5 through 17 years of age in boys, from about 42% to 54% (Table IV). In girls, it increases from 5 through 13 years of age, from about 40% to 45%, decreasing somewhat after 13 years of age (Malina, 1969). The decrease in relative contribution of muscle mass to body weight is probably associated with the accumulation of fat during female adolescence. At all ages during childhood and adolescence, muscle mass contributes a greater percentage to body weight in boys than in girls. In young adults (20–29 years), the relative contribution of estimated muscle mass to body weight decreases somewhat relative to the late adolescent values, about 51% in males and 40% in females (Malina, 1969). This reflects a relative increase in fatness in both sexes during the third decade of life.

2.3. Potassium Concentration

Muscle tissue is rich in potassium, and the muscle mass of the body accounts for 50–70% of body potassium (Pierson *et al.*, 1974). As such, measures of potassium offer a marker for muscle tissue. Quite commonly, though, measures of potassium are used to derive estimates of lean body mass (LBM) and/or body cell mass (BCM) (Moore *et al.*, 1963; Malina, 1969; Forbes, 1972; Pierson *et al.*, 1974).

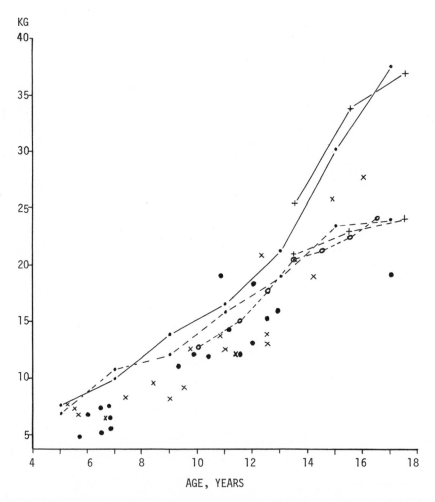

Fig. 2. Age changes and sex differences in muscle mass estimated from creatinine excretion and body weight. ●—●, boys, and ●----●, girls, from Clark *et al.* (1951); +——+, boys, and +----+, girls, from Novak (1963); x, boys, and ●, girls from Cheek (1968); ○----○, girls, from Young *et al.* (1968).

Table IV.   Estimated Muscle Mass as a
Percentage of Body Weight[a]

Age (years)	Muscle (%)	
	Males	Females
5	42.0	40.2
7	42.5	46.6
9	45.9	42.2
11	45.9	44.2
13	46.2	43.1
13.5	50.2	45.5
15	50.3	43.2
15.5	50.6	44.2
17	52.6	42.0
17.5	53.6	42.5
20–29	51.5	39.9

[a]Data are derived from average values for body weight and creatinine excretion, g/24 hr, and are collated from several sources (Malina, 1969).

The radioisotope ^{40}K concentration provides the basis from which LBM and BCM estimates of body composition are derived.

Potassium concentration of the body increases from infancy throughout childhood, the slope of increase being the same for both sexes until the adolescent spurt. Boys have consistently higher concentrations of potassium in the body than girls in infancy and childhood, with the sex difference becoming especially marked during adolescence (Burmeister, 1965; Flynn et al., 1970; Novak, 1973). Boys, however, do not have a greater concentration of potassium in muscle tissue (Cheek, 1968).

Although potassium concentration or LBM estimates derived from them are frequently expressed relative to body weight, several analyses suggest stature as the standard for expressing potassium concentration or LBM, as body weight is too variable (Forbes, 1972, 1974; Flynn et al., 1975). Potassium concentration per unit of stature relative to age is shown in Figure 3. There appears to be a rather marked increase in potassium per unit of height during infancy, followed by a slower increase during childhood. During adolescence, boys show a sharp increase in potassium per unit stature, while girls show only a slight gain. This increase in potassium reflects the male spurt in muscle mass during adolescence. Estimates of LBM derived from other in vivo methods, for example, total body water and densitometry, show general correspondence among themselves although differing in quantitative estimates of lean tissue, and show a general correspondence to age- and sex-associated variation in muscle tissue (Malina, 1969; Forbes, 1974).

2.4. Radiographic Studies

Muscle can be measured radiographically in a manner similar to that used to assess fat and bone. Arm, forearm, thigh, and calf are the most commonly measured areas, and the data are reported as summated measurements for "total limb muscle-width" (Tanner, 1965, 1968; Maresh, 1966, 1970), or as limb-specific curves (Reynolds, 1944, 1948; Lombard, 1950; Baker et al., 1958; Johnston and Malina, 1966; Malina and Johnston, 1967a). Qualitatively similar age and sex profiles during childhood and adolescence are provided by both the composite muscle-width

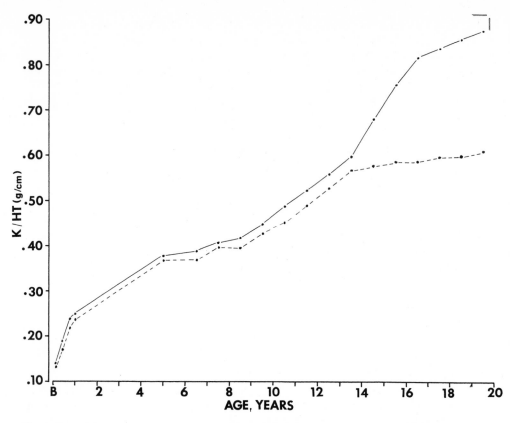

Fig. 3. Potassium concentration per unit stature in boys (—) and girls (---). Data for infancy are from Novak (1973), for 5 years from Flynn *et al.* (1970), and for 6–20 years from Burmeister (1965).

curves and the limb-specific curves. Nevertheless, correlations between muscle widths measured on different limb segments indicate at best only moderate relationships which are slightly higher in males (Table V).

Muscles of the extremities increase in size with age from infancy through adolescence, and the growth curve generally resembles that for other estimates of muscle mass. Sex differences, although apparent, are small during childhood, boys having slightly wider muscles. By 10 or 11 years of age, girls begin their adolescent

Table V.   Reported Correlations between Muscle Widths Measured in the Arm, Thigh, and Calf[a]

Areas compared	Age	Correlations	
		Males	Females
Arm/calf	6.0–15.5 years	0.53	0.42
Arm/calf	Young adult	0.42	0.36
Arm/thigh	Young adult	0.54	0.49
Calf/thigh	Young adult	0.51	0.43
Arm/calf	Adult, athletes	0.42	
Arm/thigh	Adult, athletes	0.30	
Calf/thigh	Adult, athletes	0.45	

[a]Adapted from Malina (1969).

spurt and have slightly wider muscles than boys for a year or two. This temporary size advantage seems more apparent in muscle width measures of the lower extremity (Maresh, 1970). Boys then have their adolescent spurt in muscle mass and have considerably wider extremity muscles than girls. Sex differences in muscle widths, established during adolescence, persist into adulthood (see Malina, 1969).

Velocities of growth in muscle widths are similar in boys and girls prior to adolescence. Males have a later and greater adolescent spurt. When longitudinal data for radiographic muscle widths are aligned on peak height velocity (Tanner, 1965), the sex difference in the magnitude and perhaps timing of the adolescent muscle tissue spurt is apparent. Boys have a spurt in muscle width that is approximately $1\frac{1}{2}$ times the magnitude of that in females. The male peak is coincident with peak height velocity or slightly after if the data are smoothed, and the female peak occurs slightly after peak height velocity (Tanner, 1965).

2.5. Estimated Arm Musculature

Muscle is the body's main protein reserve, and as such estimates of muscle mass are important to field studies of nutritional status (Jelliffe and Jelliffe, 1969). Arm circumference and triceps skinfold, both measured at the level midway between the acromial and olecranon processes, are most commonly used for nutritional surveys. Estimated mid-arm muscle circumference can then be derived from the equation

$$C_m = C_a - \pi S_t$$

where C_m is the estimated mid-arm muscle circumference; C_a is arm circumference (relaxed); and S_t is the triceps skinfold.

Gurney and Jelliffe (1973) recommend the use of cross-sectional fat and muscle areas in nutritional anthropometry. The equation for estimated muscle area then is

$$\text{Muscle area} = \frac{(C_a - S_t)^2}{4\pi}$$

Nomograms for the estimation of arm muscle circumference and area from arm circumference and triceps skinfold in children and in adolescents and adults are provided in Gurney and Jelliffe (1973). Although the circumference or area is indicated as being an estimated muscle measurement, it should be noted that bone tissue, the humerus, is also included in the estimate. The pattern of age- and sex-associated variation illustrated in curves for estimated arm muscle circumference or area are qualitatively similar to those for height and weight and for radiographic measures of muscle widths.

2.6. Interrelationships and Limitations

Dissection studies, creatinine excretion, and potassium concentration provide only estimates of total muscle mass in the body. They do not indicate regional variation in distribution of muscle tissue, or possible age-, sex-, or race-associated variation in muscularity. The head and trunk, for example, constitute about 40% of the total weight of the musculature at birth, but only 25–30% at maturity. The muscles of the lower extremity show an increase from approximately 40% at birth to 55% of the total weight of the muscle mass at maturity, while the muscles of the upper extremity remain rather constant, comprising about 18–20% of the musculature during life (Scammon, 1923). Thus, body regions contribute differentially to the total muscle mass of the body.

Radiographic measures of muscle widths provide only estimates of relative muscularity, and the problem of muscle patterning, analogous to fat patterning, needs to be explored. This is significant in assessing racial differences in muscularity. There are, for example, differences in calf muscle mass among races (Garn, 1963), and blacks have substantially smaller calf muscles in relation to muscle development in the arm and thigh (Tanner, 1964). Estimated mid-arm muscle circumferences also tend to be larger in American black than in white children (Malina *et al.*, 1974). Such observations would seem to suggest differences in muscle patterning.

Creatinine excretion and radiographic measures of muscle widths are moderately and significantly correlated (Garn, 1961; Reynolds and Clark, 1947). Both estimates of muscularity are significantly related to total body potassium, while creatinine excretion is highly correlated with other estimates of lean body mass (Malina, 1969). Lean body mass, however, is heterogeneous, and thus cannot be regarded as an absolute indicator of muscle mass.

3. Development of Muscular Strength

The composition and size of muscle is functionally manifest in muscular strength, the capacity to exert force against a resistance. Strength of a muscle is related to its cross-sectional area.

Muscular strength increases linearly with chronological age from early childhood to approximately 13–14 years in males, when there is a marked acceleration in strength development through 20 years of age. In girls, strength improves linearly with age through about 15 years of age, followed by a tendency to level off (Figure 4). Boys demonstrate, on the average, greater strength than girls at all ages. Sex differences throughout childhood are consistent, although small. The marked acceleration of strength development during male adolescence magnifies the preadolescent sex difference (Metheny, 1941; Jones, 1949; Asmussen, 1962; Bouchard, 1966; Malina, 1974). With increasing age during adolescence, the percentage of girls whose performance on strength tests equals or exceeds that of boys drops considerably, so that after 16 years of age, few girls perform as high as the boys' average, and, conversely, practically no boys perform as low as the girls' average (Jones, 1949).

The relationship between strength development and general growth and maturation during male adolescence is such that the strength spurt is frequently considered as a maturity indicator. Maximum strength development occurs after peak velocity of growth in height and weight, the relationship being better with body weight (Stolz and Stolz, 1951; Tanner, 1962; Carron and Bailey, 1974). The adolescent peak in strength development occurs about nine months to one year after peak weight gain.

These data should be related to those for growth of muscle tissue during adolescence. Peak growth in muscle width occurs slightly after peak height velocity, while peak weight gain occurs about six months after peak height velocity. Thus, peak growth in muscle tissue occurs prior to peak weight gain, emphasizing the role of muscle mass in the adolescent weight gain. It is also clear that muscle tissue increases first in mass and then in strength. This would seem to suggest a qualitative change in muscle tissue as adolescence progresses or perhaps a neuro-muscular maturation affecting the volitional demonstration of strength.

Strength is related to body size and estimates of lean body mass (Malina, 1975; Lamphiear and Montoye, 1976), so that sex differences in strength might relate to a size advantage in boys. Using height as an index of size, Asmussen and Heebøll-Nielsen (1955) noted that strength in boys and girls increases more than predicted from height alone. In boys, strength improves out of proportion to gains in height or age after about 14 years. Carron and Bailey (1974) also noted an increase in strength greater than that predicted from statural growth in boys followed longitudinally from 10 to 16 years. The predicted average yearly increase in strength was approximately 12%, while the actual average yearly strength increase was about 23%, or about twice the predicted value. These observations thus emphasize the magnitude

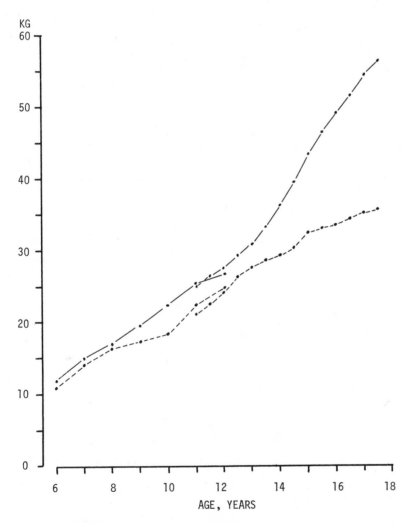

Fig. 4. Changes in muscular strength (right grip) during growth (—, boys, and ---, girls). Data from 6 to 12 years from Malina (unpublished) and from 11 to 18 years from Jones (1949).

of the male adolescent strength spurt. It might be plausible to hypothesize a relationship between the observed strength increments and the doubling of muscle nuclear number between 9 and 14 years of age in boys (see above).

The disproportionate strength increase in male adolescence is more apparent in strength of the upper extremities than in the trunk or lower extremities (Asmussen, 1962; Carron and Bailey, 1974). It should be noted, in this regard, that the brachial muscle mass practically doubles during male adolescence (Baker *et al.,* 1958; Hunt and Heald, 1963; Malina and Johnston, 1967*b*).

Sex differences in strength after adjusting for height differences are not evident in lower extremity strength from 7 to 17 years of age. However, from 7 years on, boys are significantly stronger in upper extremity and trunk strength even after adjusting for sex differences in height (Asmussen, 1962).

4. Muscularity and Maturity Status

Maturity-associated variations in size and physique (Tanner, 1962; Malina and Rarick, 1973) are also evident for measures of muscle mass and strength. Correlations between skeletal age and radiographic muscle width measurements are low during early childhood and only slightly higher at older ages (Table VI). However, when correlations are limited to the circumpubertal years, 9–14 in girls and 11–16 years in boys, correlations between skeletal maturity and muscle widths of the arm and calf are moderately high, especially in boys.

The inference to be made from the correlations in Table VI is that the more mature child in any age group has larger muscles, especially during male adolescence. This is evident also in comparisons of children grouped as early, average, and late maturers (Reynolds, 1946; Johnston and Malina, 1966). The early maturers have larger muscle measurements of the arm and calf, and this, of course, reflects their larger body size.

Data on maturity differences in other measures of muscle mass are limited. The effects of sexual maturation of girls on creatinine excretion and muscularity estimates are shown in Table VII. There are increasing creatinine excretion values and

Table VI. Correlations between Skeletal Maturity and Muscle Widths in the Arm and Calf[a]

Age	Area	Males	Females
6–60 months	Calf[b]	0.19	0.07
7–16 years	Calf[c]	0.35	0.36
7–16 years	Arm[d]	0.43	0.37
11–16 years	Calf[c]	0.51	
11–16 years	Arm[d]	0.61	
9–14 years	Calf[c]		0.39
9–14 years	Arm[d]		0.45

[a]Average correlations derived from z-transformed age-specific correlations weighted for sample size.
[b]Adapted from Hewitt (1958).
[c]Malina and Johnston (unpublished data).
[d]Adapted from Malina and Johnston (1967*b*).

Table VII. *Creatinine Excretion and Estimated Muscle Mass in Girls Grouped According to Maturity Status*[a]

Maturity group[b]	N	Age, mean (range) (yr)	Creatinine excretion (g/24 hr)	Estimated muscle mass (kg)	Relative muscle mass (%)
I	18	10.2 (9.3–11.9)	0.635	12.70	39.3
II	25	11.3 (9.5–13.1)	0.793	15.86	40.8
III	12	12.9 (11.7–15.6)	0.928	18.56	39.5
IV	14	13.6 (10.7–15.3)	1.087	21.74	44.1
V	32	14.6 (11.9–17.0)	1.150	23.00	42.1

[a]Adapted from Young *et al.* (1968). Absolute and relative muscle mass are estimated from reported mean values using the conversion factor, 1 g creatinine = 20 kg muscle tissue.
[b]The maturity groups are defined by Young *et al.* (1968) as follows: I—premenarche, no secondary sex development; II—premenarche, some secondary sex development; III—premenarche at time of study, but reaching menarche within 6 months; IV—postmenarche, some secondary sex development; V—postmenarche, complete secondary sex development.

thus absolute muscle mass with age and advancing degrees of sexual maturity from premenarcheal to postmenarcheal stages. Sample size within specific age groups in the data of Young *et al.* (1968) are too small to illustrate maturity-associated variation within a single age category. When estimated muscle mass is viewed relative to body weight, there is no gradient with increasing maturity, although girls in the more advanced maturity categories (IV and V, postmenarcheal) have a slightly greater relative muscle mass. Similar analysis of sexual maturity relative to muscle mass development of boys is lacking. Nevertheless, increased muscularity with advanced maturity status can be inferred from the elevated circulating levels of testosterone, which is specific for muscle tissue anabolism, with advancing stages of sexual maturity (Frasier *et al.*, 1969; August *et al.*, 1972).

Correlations between strength and skeletal maturity are moderate (+0.35 to 0.63) in primary-grade boys (Rarick and Oyster, 1964). Thus, the more mature boys are generally stronger. However, when the effects of body size are statistically controlled, correlations between skeletal maturity and strength are considerably reduced. When boys were divided into the skeletally mature and immature, the former were stronger, taller, and heavier (Rarick and Oyster, 1964). Thus, strength differences among young boys reflect primarily size differences.

During adolescence maturity relationships with strength are more apparent for boys than they are for girls. Early maturing boys 10–17 years of age are stronger age for age than their average and late-maturing peers (Jones, 1949; Clarke, 1971; Carron and Bailey, 1974). Strength differences between early and late maturers are especially apparent between 13 and 16 years of age, and the strength advantage for the early-maturing boys reflects their larger body size and muscle mass. When the effects of body weight are removed in comparing early- and late-maturing boys, strength differences between the contrasting maturity groups are eliminated (Carron and Bailey, 1974).

Early-maturing girls are stronger than their late-maturing peers during early adolescence. They do not, however, maintain this advantage as adolescence progresses (Jones, 1949). Early- and late-maturing girls attain comparable strength levels in later adolescence by apparently different routes. The early maturer shows rapid strength development through 13 years of age and then improves only slightly thereafter. The late maturer, on the other hand, improves in strength gradually between 11 and 16 years, so that at 16 years of age, the contrasting maturity categories do not differ in strength (Jones, 1949).

5. Selected Factors Influencing Muscle Growth and Function

Growth of muscle tissue is affected by a number of factors, including physical activity and inactivity, nerve supply, hormones, and nutrition. Detailed review of these is beyond the scope of this chapter (see Goldspink, 1972; Edgerton, 1973), and only a brief overview of physical activity and undernutrition effects on muscle will be made here.

5.1. Training

Physical training results in muscular hypertrophy, and the intensity of muscular work is the critical factor underlying the hypertrophy observed in normal muscle subjected to regular training. Hypertrophy is accompanied by an increase in contractile proteins, myofibrils, and functional capacity (e.g., strength). Some evidence indicates longitudinal splitting of fibers in hypertrophying muscle (Goldspink, 1972; Edgerton, 1973; Goldberg et al., 1975). Whether fiber splitting is an important component of work-induced hypertrophy or a pathological effect of increased muscle load remains to be established (Goldberg et al., 1975).

Studies of the DNA content of muscle in growing animals undergoing regular training indicate a significant rise in DNA content during growth and suggest that training during growth may be a significant factor determining adult levels of DNA in muscle tissue (see Chapter 17).

Muscle changes with training also include improved functioning as reflected, for example, in an increase in the percentage of fibers having high oxidative enzyme activity (Edgerton, 1973). In a small sample of 11-year-old boys ($N = 5$) trained for six weeks, Eriksson (1972) observed similar changes. There was a marked increase in the oxidative potential of both slow- and fast-twitch fibers (as indicated by enhanced staining by DPNH-diaphorase). A 30% increase was also noted in succinate dehydrogenase (SDH) activity. In contrast, max $\dot{V}_{O_2}$ increased by only 8% after six weeks of training, suggesting that the oxidative capacity of muscle tissue may not limit oxygen consumption. In addition, phosphofructokinase (PFK) activity, although low compared to adult values, increased markedly with the short training program (83%). Since PFK is regarded as the rate-limiting enzyme for glycolysis, the increased PFK activity with training indicates enhanced glycolytic potential in this small sample of young boys.

5.2. Undernutrition

Studies of severe protein–calorie malnutrition indicate a reduction of muscle mass, muscle potassium, and muscle cell size (Montgomery, 1962b; McFie and Welbourn, 1962; Alleyne et al., 1969; Nichols et al., 1969, 1972; Widdowson, 1969; Cheek et al., 1970), while studies of severely malnourished children who have undergone subsequent nutritional rehabilitation do not show complete recovery or catch-up in muscle tissue measurements, for example, muscle cell size (Cheek et *al.*, 1970). Muscle tissue of children with protein–calorie malnutrition is further characterized by reduced energy metabolism (Nichols *et al.*, 1972).

291

GROWTH OF MUSCLE TISSUE AND MUSCLE MASS

One can inquire about possible changes in muscle metabolism and mass in children who develop under chronic mild to moderate protein–calorie malnutrition. These children are generally shorter and lighter than better nourished children, and also show less muscle mass (Jelliffe and Jelliffe, 1969; Viteri, 1971). Since strength of a muscle is proportional to its cross-sectional area, reduced muscle mass in undernourished individuals should result in a corresponding reduction in muscular strength. Data relating strength to muscle mass in malnourished children are not available. Previously well-nourished adults who voluntarily underwent an acute period of semistarvation, however, showed reductions in muscle mass and in strength (Keys *et al.*, 1950). This Minnesota experiment demonstrated a marked deterioration of grip strength as the semistarvation continued. The decrease in strength after 24 weeks represented about 28% of the control value. After a 12-week nutritional rehabilitation period, grip strength showed minimal recovery (approximately one third of the strength lost). After 20 weeks of rehabilitation, recovery in strength was not yet complete, approximating only 64% of the starvation loss. Although this experiment was conducted on individuals who had not been previously exposed to undernutrition, implications for children living under conditions of chronic protein–calorie malnutrition are obvious.

6. References

Aherne, W., Ayyar, D. R., Clarke, P. A., and Walton, J. N., 1971, Muscle fibre size in normal infants, children and adolescents: An autopsy study, *J. Neurol. Sci.* **14**:171.

Alleyne, G. A. O., Halliday, D., and Waterlow, J. C., 1969, Chemical composition of organs of children who died from malnutrition, *Br. J. Nutr.* **23**:783.

Asmussen, E., 1962, Muscular performance, in: *Muscle as a Tissue* (K. Rodahl and S. M. Horvath, eds.), pp. 161–175, McGraw-Hill, New York.

Asmussen, E., and Heebøll-Nielsen, K., 1955, A dimensional analysis of physical performance and growth in boys, *J. Appl. Physiol.* **7**:593.

August, G. P., Grumbach, M. M., and Kaplan, S. L., 1972, Hormonal changes in puberty. III. Correlations of plasma testosterone, LH, FSH, testicular size and bone age with male pubertal development, *J. Clin. Endocrinol. Metab.* **34**:319.

Baker, P. T., Hunt, E. E., Jr., and Sen, T., 1958, The growth and interrelations of skinfolds and brachial tissues in man, *Am. J. Phys. Anthropol.* **16**:39.

Bouchard, C., 1966, Les différences individuelles en force musculaire statique, *Mouvement* **1**:49.

Bowden, D. H., and Goyer, R. A., 1960, The size of muscle fibers in infants and children, *Arch. Pathol.* **69**:188.

Brooke, M. H., and Engel, W. K., 1969*a*, The histographic analysis of human muscle biopsies with regard to fiber types. I. Adult male and female, *Neurology* **19**:221.

Brooke, M. H., and Engel, W. K., 1969*b*, The histographic analysis of human muscle biopsies with regard to fiber types. 4. Children's biopsies, *Neurology* **19**:591.

Brooke, M. H., and Kaiser, K. K., 1970, Muscle fiber types: How many and what kind?, *Arch. Neurol.* **23**:369.

Burleigh, I. G., 1974, On the cellular regulation of growth and development in skeletal muscle, *Biol. Rev.* **49**:267.

Burmeister, W., 1965, Body cell mass as the basis of allometric growth functions, *Ann. Paediatr.* **204**:65.

Carron, A. V., and Bailey, D. A., 1974, Strength development in boys from 10 through 16 years, *Monogr. Soc. Res. Child Dev.* **39**(4).

Cheek, D. B., 1968, *Human Growth: Body Composition, Cell Growth, Energy, and Intelligence*, Lea and Febiger, Philadelphia.

Cheek, D. B., 1974, Body composition, hormones, nutrition, and adolescent growth, in: *Control of the Onset of Puberty* (M. M. Grumbach, G. D. Grave, and F. E. Mayer, eds.), pp. 424–442, John Wiley & Sons, New York.

Cheek, D. B., 1975, Growth and body composition, in: *Fetal and Postnatal Cellular Growth: Hormones and Nutrition* (D. B. Cheek, ed.), pp. 389–408, John Wiley & Sons, New York.

Cheek, D. B., and Hill, D. E., 1970, Muscle and liver cell growth: Role of hormones and nutritional factors, *Fed. Proc.* **29:**1503.

Cheek, D. B., Hill, D. E., Cordano, A., and Graham, G. G., 1970, Malnutrition in infancy: Changes in muscle and adipose tissue before and after rehabilitation, *Pediatr. Res.* **4:**135.

Cheek, D. B., Holt, A. B., Hill, D. E., and Talbert, J. L., 1971, Skeletal muscle cell mass and growth: the concept of the deoxyribonucleic acid unit, *Pediatr. Res.* **5:**312.

Clark, L. C., Thompson, H. L., Beck, E. I., and Jacobson, W., 1951, Excretion of creatine and creatinine by children, *Am. J. Dis. Child.* **81:**774.

Clarke, H. H., 1971, *Physical and Motor Tests in the Medford Boys' Growth Study,* Prentice-Hall, Englewood Cliffs, New Jersey.

Dickerson, J. W. T., and Widdowson, E. M., 1960, Chemical changes in skeletal muscle during development, *Biochem. J.* **74:**247.

Durnin, J. V. G. A., 1969, Muscular and adipose tissue and the significance of increase in body weight with age, in: *Physiopathology of Adipose Tissue* (J. Vague, ed.), pp. 387–389, Excerpta Medica, Amsterdam.

Edgerton, V. R., 1973, Exercise and the growth and development of muscle tissue, in: *Physical Activity: Human Growth and Development* (G. L. Rarick, ed.), pp. 1–31, Academic Press, New York.

Eriksson, B. O., 1972, Physical training, oxygen supply and muscle metabolism in 11–13 year old boys, *Acta Physiol. Scand., Suppl.* 384.

Fischman, D. A., 1972, Development of striated muscle, in: *The Structure and Function of Muscle,* 2nd ed., Vol. I, Structure, Part 1 (G. H. Bourne, ed.), pp. 75–148, Academic Press, New York.

Flynn, M. A., Murthy, Y., Clark, J., Comfort, G., Chase, G., and Bentley, A. E. T., 1970, Body composition of Negro and white children, *Arch. Environ. Health* **20:**604.

Flynn, M. A., Clark, J., Reid, J. C., and Chase, G., 1975, A longitudinal study of total body potassium in normal children, *Pediatr. Res.* **9:**834.

Forbes, G. B., 1972, Relation of lean body mass to height in children and adolescents, *Pediatr. Res.* **6:**32.

Forbes, G. B., 1974, Stature and lean body mass, *Am. J. Clin. Nutr.* **27:**595.

Frasier, S. D., Gafford, F., and Horton, R., 1969, Plasma androgens in childhood and adolescence, *J. Clin. Endocrinol. Metab.* **29:**1404.

Garn, S. M., 1961, Radiographic analysis of body composition, in: *Techniques for Measuring Body Composition* (J. Brozek and A. Henschel, eds.), pp. 36–58, National Academy of Sciences–National Research Council, Washington, D.C.

Garn, S. M., 1963, Human biology and research in body composition, *Ann. N.Y. Acad. Sci.* **110:**429.

Goldberg, A. L., Etlinger, J. D., Goldspink, D. F., and Jablecki, C., 1975, Mechanism of work-induced hypertrophy of skeletal muscle, *Med. Sci. Sports* **7:**248.

Goldspink, G., 1972, Postembryonic growth and differentiation of striated muscle, in: *The Structure and Function of Muscle,* 2nd ed., Vol. I, Structure, Part 1 (G. H. Bourne, ed.), pp. 179–236, Academic Press, New York.

Goss, R. J., 1966, Hypertrophy versus hyperplasia, *Science* **153:**1615.

Grande, F., 1963, Comment, *Ann. N.Y. Acad. Sci.* **110:**576.

Gurney, J. M., and Jelliffe, D. B., 1973, Arm anthropometry in nutritional assessments: Nomogram for rapid calculation of muscle circumference and cross-sectional muscle and fat areas, *Am. J. Clin. Nutr.* **26:**912.

Hewitt, D., 1958, Sib resemblance in bone, muscle and fat measurements of the human calf, *Ann. Hum. Genet.* **22:**213.

Hunt, E. E., Jr., and Heald, F. P., 1963, Physique, body composition, and sexual maturation in adolescent boys, *Ann. N.Y. Acad. Sci.* **110:**532.

Jelliffe, E. F., and Jelliffe, D. B., 1969, The arm circumference as a public health index of protein–calorie malnutrition of early childhood, *J. Trop. Pediatr.* **15:**177.

Jennekens, F. G. I., Tomlinson, B. E., and Walton, J. N., 1970, The sizes of the two main histochemical fibre types in five limb muscles in man: An autopsy study, *J. Neurol. Sci.* **13:**281.

Johnson, M. A., Polgar, J., Weightman, D., and Appleton, D., 1973, Data on the distribution of fibre types in thirty-six human muscles: An autopsy study, *J. Neurol. Sci.* **18:**111.

Johnston, F. E., and Malina, R. M., 1966, Age changes in the composition of the upper arm in Philadelphia children, *Hum. Biol.* **38:**1.

Jones, H. E., 1949, *Motor Performance and Growth,* University of California Press, Berkeley.

Keys, A., Brozek, J., Henschel, A., Mickelsen, O., and Taylor, H. L., 1950, *The Biology of Human Starvation,* University of Minnesota Press, Minneapolis.

Kreisberg, R. A., Bowdoin, B., and Meador, C. K., 1970, Measurement of muscle mass in humans by isotopic dilution of creatine-^{14}C, *J. Appl. Physiol.* **28:**264.

Lamphiear, D. E., and Montoye, H. J., 1976, Muscular strength and body size, *Hum. Biol.* **48:**147.

Landing, B. H., Dixon, L. G., and Wells, T. R., 1974, Studies on isolated human skeletal muscle fibers, *Hum. Pathol.* **5:**441.

Lombard, O. M., 1950, Breadth of bone and muscle by age and sex in childhood, *Child Dev.* **21:**229.

Malina, R. M., 1969, Quantification of fat, muscle and bone in man, *Clin. Orthop. Rel. Res.* **65:**9.

Malina, R. M., 1974, Adolescent changes in size, build, composition and performance, *Hum. Biol.* **46:**117.

Malina, R. M., 1975, Anthropometric correlates of strength and motor performance, *Exercise Sport Sci. Rev.* **3:**249.

Malina, R. M., and Johnston, F. E., 1967*a,* Relations between bone, muscle and fat widths in the upper arms and calves of boys and girls studied cross-sectionally at ages 6 to 16 years, *Hum. Biol.* **39:**211.

Malina, R. M., and Johnston, F. E., 1967*b,* Significance of age, sex, and maturity differences in upper arm composition, *Res. Q.* **38:**219.

Malina, R. M., and Rarick, G. L., 1973, Growth, physique, and motor performance, in: *Physical Activity: Human Growth and Development* (G. L. Rarick, ed.), pp. 125–153, Academic Press, New York.

Malina, R. M., Hamill, P. V. V., and Lemeshow, S., 1974, Body dimensions and proportions, White and Negro children 6–11 years, United States, *Vital Health Stat.* Ser. 11 (143).

Maresh, M. M., 1966, Changes in tissue widths during growth, *Am. J. Dis. Child.* **111:**142.

Maresh, M. M., 1970, Measurements from roentgenograms, heart size, long bone lengths, bone, muscle and fat widths, skeletal maturation, in: *Human Growth and Development* (R. W. McCammon, ed.), pp. 155–200, Charles C Thomas, Springfield, Illinois.

McFie, J., and Welbourn, H. F., 1962, Effect of malnutrition in infancy on the development of bone, muscle and fat, *J. Nutr.* **76:**97.

Metheny, E., 1941, The present status of strength testing for children of elementary school and preschool age, *Res. Q.* **12:**115.

Montgomery, R. D., 1962*a,* Growth of human striated muscle, *Nature* **195:**194.

Montgomery, R. D., 1962*b,* Muscle morphology in infantile protein malnutrition, *J. Clin. Pathol.* **15:**511.

Moore, F. D., Oleson, K. H., McMurrey, J. D., Parker, H. V., Ball, M. R., and Boyden, C. M., 1963, *The Body Cell Mass and Its Supporting Environment,* W. B. Saunders, Philadelphia.

Nichols, B. L., Alleyne, G. A. O., Barnes, D. J., and Hazelwood, C. F., 1969, Relationship between muscle potassium and total body potassium in infants with malnutrition, *J. Pediatr.* **74:**49.

Nichols, B. L., Alvarado, J., Hazelwood, C. F., and Viteri, F., 1972, Clinical significance of muscle potassium depletion in protein–calorie malnutrition, *J. Pediatr.* **80:**319.

Novak, L. P., 1963, Age and sex differences in body density and creatinine excretion of high school children, *Ann. N.Y. Acad. Sci.* **110:**545.

Novak, L. P., 1973, Total body potassium during the first year of life determined by whole-body counting of ^{40}K, *J. Nucl. Med.* **14:**550.

Payne, C. M., Stern, L. Z., Curless, R. G., and Hannapel, L. K., 1975, Ultrastructural fiber typing in normal and diseased human muscle, *J. Neurol. Sci.* **25:**99.

Pierson, R. N., Jr., Lin, D. H. Y., and Phillips, R. A., 1974, Total-body potassium in health: Effects of age, sex, height, and fat, *Am. J. Physiol.* **226:**206.

Polgar, J., Johnson, M. A., Weightman, D., and Appleton, D., 1973, Data on fibre size in thirty-six human muscles: an autopsy study, *J. Neurol. Sci.* **19:**307.

Rarick, G. L., and Oyster, N., 1964, Physical maturity, muscular strength, and motor performance of young school-age boys, *Res. Q.* **35:**523.

Reynolds, E. L., 1944, Differential tissue growth in the leg during childhood, *Child Dev.* **15:**181.

Reynolds, E. L., 1946, Sexual maturation and the growth of fat, muscle and bone in girls, *Child Dev.* **17:**121.

Reynolds, E. L., 1948, Distribution of tissue components in the female leg from birth to maturity, *Anat. Rec.* **100:**621.

Reynolds, E. L., and Clark, L. C., 1947, Creatinine excretion, growth progress and body structure in normal children, *Child Dev.* **18:**155.

Scammon, R. E., 1923, A summary of the anatomy of the infant and child, in: *Pediatrics* (I. A. Abt, ed.), pp. 257–444, W. B. Saunders, Philadelphia.

Stolz, H. R., and Stolz, L. M., 1951, *Somatic Development of Adolescent Boys,* Macmillan, New York.

Talbot, N. B., 1938, Measurement of obesity by the creatinine coefficient, *Am. J. Dis. Child.* **55**:42.

Tanner, J. M., 1962, *Growth at Adolescence,* 2nd ed., Blackwell Scientific Publications, Oxford, England.

Tanner, J. M., 1964, The Physique of the Olympic Athlete, Allen and Unwin, London. Issued by Department of Growth and Development, Institute of Child Health, University of London.

Tanner, J. M., 1965, Radiographic studies of body composition in children and adults, *Symp. Soc. Study Hum. Biol.* **7**:211.

Tanner, J. M., 1968, Growth of bone, muscle and fat during childhood and adolescence, in: *Growth and Development of Mammals* (G. A. Lodge and G. E. Lamming, eds.), pp. 3–18, Plenum Press, New York.

Viteri, F. E., 1971, Considerations on the effect of nutrition on the body composition and physical working capacity of young Guatemalan adults, in: *Amino Acid Fortification of Protein Foods* (N. S. Scrimshaw and A. M. Altschul, eds.), pp. 350–375, MIT Press, Cambridge, Massachusetts.

Widdowson, E. M., 1969, Changes in the extracellular compartment of muscle and skin during normal and retarded development, *Biblioth. Nutr. Dieta* **13**:60.

Widdowson, E. M., 1970, Harmony of growth, *Lancet* **1**:901.

Young, C. M., Bogan, A. D., Roe, D. A., and Lutwak, L., 1968, Body composition of pre-adolescent and adolescent girls. IV. Total body water and creatinine excretion, *J. Am. Diet. Assoc.* **53**:579.

Zorab, P. A., 1969, Normal creatinine and hydroxyproline excretion in young persons, *Lancet* **2**:1164.

11

Adipose Tissue Development in Man

JEROME L. KNITTLE

1. Introduction

Adipose tissue in man can be divided into two major categories. One is the deep-seated part spread around the various visceral spaces of the trunk, neck, and limbs; the other is subcutaneous, which appears to develop later. It is this latter portion that plays the most prominent role in the development of obesity.

There are many modes of distributions of subcutaneous fat accumulation in the obese state, and each individual has his own characteristic features. The most common sites are: the nape of the neck; the cheeks; under the chin; the region overlying the seventh cervical vertebra where the spiny apophysis is prominent (the so-called ''buffalo hump''); the supraclavicular area which bulges in extreme obesity (pseudolipoma of Potain and Verneuil); the back of the upper limbs along the triceps; abdominal fat (above and/or below the umbilicus), and in the female is especially prominent in the supra pubic areas (mons veneris); the buttocks, which appears to have three distinct areas (inferiointernal, posterior, and high external); and finally, a characteristic distribution of fat along the hips and lower extremity. Some authors have attempted to further classify such distributions according to gynoid (female) vs. android (male) types.

As early as 1749, Buffon noted that the adipose distribution in the buttock was related to sex and the erect position and, although found among lower anthropoids, only reaches its fullest development in man. It is usually dominated by muscular features in the male and adipose tissue in females.

A number of other anatomic traits have been postulated to contrast the two types. Gynoid adipose tissue is supposedly larger; thus, the fat-to-muscle ratio is greater in this form. It tends to develop over the lower half of the body while the android is predominantly in the upper half. The localizations of the android type are typically the nape of the neck, the deltoid area, and epigastrium. For the gynoid it is the breasts, hypogastrium, pelvis, and anterior aspect of the thigh. This is obviously

JEROME L. KNITTLE • Division of Nutrition and Metabolism, Mt. Sinai School of Medicine, New York, New York.

an oversimplification, and the details of factors that increase adipocity and determine its specific subcutaneous distribution are poorly understood.

Given these variations in types of adipose tissue and distribution, the question arises as to what extent a particular area of adipose tissue is representative of the general behavior of the body fat stores. Obviously, during weight gain and loss, various portions of the body fat respond at differing times. A number of metabolic differences have been demonstrated in man when mesenteric is compared to subcutaneous fat derived from the same individual. Variations are also apparent in the degree of neural innervation, volume of vascular supply, and the concentration of enzymes relative to substrate. All of these could determine rate-limiting steps of metabolism in different areas. Broadly speaking, it may be stated that when such variables as the uptake of glucose and its stimulation by insulin, the extent of glycogen deposition after refeeding, and the rates of free fatty acid esterification and release are considered, mesenteric and omental depots are more active than subcutaneous fat. The mainly supportive fat depots such as orbital and duct fats are the most sluggish. Obviously, more information is needed in this area. The physiology, number of adipocytes, and their size and distribution, are probably a result of numerous factors which include familial, genetic, and metabolic nutritional components, many of which will be explored in the following text, but to date remain poorly understood.

Two other nonhepatic fat-synthesizing tissues should also be mentioned for the sake of completeness: the mammary gland and brown adipose tissue.

During lactation, mammalian females synthesize an important part of the fatty acids of the milk for their young in the mammary gland itself. These include short-chain acids and those of medium length which the other lipogenic tissues do not form in any appreciable amount. The mammary gland also has the ability to remove ingested triglycerides from the circulation. It has been demonstrated that, in the lactating rat, lipoprotein lipase activity increases in the mammary gland while decreasing in the adipose depot, thus directing exogenous fat to the gland. However, the mammary gland does not represent a major focus of fat synthesis in nonlactating females or in males.

Brown fat is found in many hibernating animals. In man it has been demonstrated in some newborns, but is especially prominent in the premature infant. The physiologic role of brown adipose tissue in hibernating animals appears to be that of heat production during arousal from hibernation. During each of the periodic occurrences of acute wakening from deep hibernation, the animal rapidly warms itself from temperatures of 5°C to that of a normal warm-blooded animal.

Brown fat differs from white adipose tissue in its higher content of mitochondria, cytochromes, and ubiquinones. This parallels the high rate of oxygen consumption found *in vitro* when compared to white adipose tissue. Thus, brown adipose tissue appears to contribute to heat production by means of oxidative processes occurring within the cell itself. However, as in other fat, one also sees release of stored lipid into the circulation as free fatty acid and glycerol. This latter phenomenon occurs during the early stages of arousal and is roughly equal in magnitude to that of white adipose tissue while the oxidative processes are ten times greater in brown fat. The role of brown adipose tissue in man, and its possible relation to cold adaptation and energy balance in the premature and newborn, is still a moot point and an area worthy of further investigation, since it could shed important light on the energy needs and ability for survival in the neonatal period.

Since subcutaneous fat is the most easily and widely studied in man, and most conspicuously enlarged in the obese state, the remainder of our discussion will be devoted to this depot. However, despite the universally accepted fact of enlarged fat depots in obesity, immediate disagreement arises when one attempts to define what degree of expansion of this organ should be considered pathologic. Normal values for adipose tissue growth and development are not readily available. Within any population, wide variations can be found depending upon age, sex, genetic endowment, and nutritional and other environmental factors. To complicate the matter further, techniques for direct measurement of body fat in large populations are not currently available, so that one must rely on less accurate and indirect indices.

Ideally, one would like an absolute measure of the growth and development of the fat depot in relation to the growth of lean body mass so that normal values could be developed. The utilization of total body weight is not satisfactory since this variable is made up of a number of components, each of which can contribute differentially to weight for differences in age, sex, or height. Thus, an individual can be overweight relative to some arbitrary standard by virtue of increase in bony structure, musculature, or adipose tissue.

The term obesity should be limited to an excessive deposition and storage of fat and is to be distinguished from "overweight," which does not have any direct implication of fatness. It is essential, therefore, that a study of the obese state include some estimate of total body fat. Although total body fat can be determined directly by the use of inert gases or indirectly by the measurement of body density, total body water, or total body potassium, these techniques are not readily available to the clinician in practice (Lesser and Zak, 1963; Brozek *et al.,* 1963; Owen *et al.,* 1962; Forbes, 1964). He must rely on less exact methods such as skinfold thickness, body weight for height and sex, or weight charts in which obese subjects are defined as those who exceed mean values by more than two standard deviations (Heald *et al.,* 1963; Wolff, 1955; Quaade, 1964; Jackson and Kelly, 1945; Stuart and Meredith, 1946). Thus, while most investigators agree that obesity is a major nutritional problem, accurate figures as to its prevalence, especially in children less than five years of age, are not available.

Until recently simple and accurate measurements of fat cell size and number have not been available. Such studies are of importance since merely knowing the total mass of the fat organ does not give one insight into its developmental patterns. Like other organs, the fat depot must grow and develop by some combination of cellular enlargement and division.

Enlargement of fat cells in animals has been demonstrated by many investigators utilizing cross-sectional anatomic preparations. Between birth and 3 months the inguinal fat cell of the guinea pig has been shown to enlarge some seven- to tenfold in size (Wasserman and McDonald, 1960). However, cell enlargement alone cannot conceivably account for the enormous masses of fat that may accumulate in certain regions of obese bodies. The limited amount of cytoplasm that covers the intracellular fat droplet, no matter how large, must place a limit on the extensibility of the fat cell. Thus, some degree of cellular division must occur. The queston is, how does this proceed in man?

The knowledge of morphology and function of adipose tissue in subjects of varying ages could, therefore, provide bases for an understanding of the complex pathogenesis of obesity, including the mechanisms involved in the growth and development of the enlarged fat depots characteristic of most obese patients. There is need for classification of the obese state, which would be useful in assessing the prognosis of obesity. Indeed, much of the confusion in the current literature regarding the pathogenesis of obesity is due to lack of agreement concerning the definition of the obese state and to variations in the methods used in determining body composition. By combining morphologic and metabolic data, one could obtain a rational basis for a better understanding of the factors which contribute to the development of fat depots in normal and pathologic states.

2. Morphology of Fat Depot Development

Although small droplets of neutral fat may occur in any cell, there are cells which have a special fat-storing and fat-releasing function and only these should be termed true adipocytes. These cells are found scattered singly or in groups in the loose connective tissue and along blood vessels. The first step in the formation of an adipocyte is the appearance of a few small droplets of fat in the cytoplasm of presumably undifferentiated mesenchymal fibroblast-like cells. The droplets then increase in size and gradually fuse, causing the cell body to swell.

Three major theories of adipose tissue development have been postulated. According to Toldt (1870) the adipose cell develops from a specific primitive cell which is not mesenchymatous. Fleming (1871) suggested that adipose cells are fibroblasts that progressively become laden with lipid, and Hammar (1895) united both theories, stating that adipocytes develop from nondifferentiated cells of a primitive mesenchymatous type. This last theory is currently the most widely accepted, and Wasserman (1926) showed that adipose cells first appear in the perivascular reticulum referred to as the primitive fat organ.

The cells of the fat organ are in all respects like other mesenchymal cells except for the presence of small fat droplets which are represented by the vacuoles in the cytoplasm. Prenatally fat cells contain not only lipids but a considerable amount of glycogen. In the embryo fat or glycogen is only found in the primitive fat organ and in fibroblasts of the connective tissue. This latter fact lends support to the theory of Hammar and Wasserman as to the non-connective-tissue origin of adipocytes.

The first adipose cells appear in the stromal areas called mesenchymatous lobules. These contain mostly ground substance with scattered star-shaped cells and argentaffin fibrils. This primitive structure is not specific for adipose tissue but can be seen in hair follicles and sebaceous gland development. The mesenchymatous lobule is not vascularized, and the beginning of adipogenesis occurs with the penetration of a capillary bud into the lobule. The primitive fat cells appear to differentiate from reticular cells, not from preexisting fibroblasts, since the latter do not become laden with lipid.

Wasserman (1926) thought that the primitive fat organ is formed by the growth of both endothelial and adventitial cells. Thus, the first step in adipogenesis is a type of vascular growth at various places in connective tissue. Each lobule is penetrated by a single capillary which proliferates along the axis of the lobule, then ramifies, and sends branches to the periphery. One sees all stages of adipocyte development in the lobule; the youngest cells are at the periphery, the most differentiated at the hilus.

In the human fetus, adipose tissue appears between the 26th and 30th weeks. The primitive cell is of the reticular type, the nucleus is elongated, and the protoplasm is basophilic without extensions and is difficult to distinguish from endothelial cells with the light microscope. The nucleus soon becomes rounded, and the cytoplasm fills with sudanophilic droplets which are insoluble in acetone and chloroform. Once droplets are formed, it appears that the cell loses its capacity to divide. Other changes that occur are an increase in the number and size of mitochondria as well as an increase in free ribosomes and smooth vesicles of the endoplasmic reticulum.

As the cell develops, the lipid inclusions increase in size (1–2 μm) and become acetone- and chloroform-soluble. At this stage the lipid droplets, surrounded by free ribosomes, begin to fuse. As more and more lipid accumulates, the nucleus is surrounded by fused droplets; the larger droplets appear to be surrounded by a fine membrane, and the central droplets are more osmophilic. The large droplets then continue to fuse, forcing the nucleus to the periphery where it becomes oval shaped.

Thus at birth a well-defined fat organ with adipocytes in varying degrees of maturity can be found. Formation of primitive fat lobules appears to cease in late fetal and early postnatal life. The infant has little fat in the omental or perirenal areas, but subcutaneous fat is well established in the normal full-term baby. The fat content of the body increases from birth to about 6 months of age, but the rate of increase rapidly falls off during the latter part of the first year. At this time a plateau is reached and thereafter a slight decrement occurs until approximately 7 years of age, when fat once again increases. The changes during the adolescent growth spurt are described by Malina (See Chapter 10). In females the early fat loss is less, as is the development of lean body mass; hence the average female finishes as an adult with a greater amount of body fat (Falkner, 1966). The growth of the fat depot is not linear, and the relative contribution of cell number and size to the growth of this organ remains unclear.

Studies in the rat indicate that there is a finite time for cellular division. Early in life the epididymal fat depot of the rat increases by cell division alone. This stage is followed by a combined increase in cell number and size and finally, at 7–8 weeks of age, cell division appears to cease and further growth of the fat depot is accomplished by the enlargement of fat cells while cell number remains constant (Hirsch and Gallian, 1969). However, a second possibility in this last stage of development may also account for the growth of fat depots. Cells that look like histiocytes and fibroblasts can be seen among mature fat cells. It is conceivable that they represent a stock from which new fat cells could be supplied as needed, or that new "embryonic growths" of recently divided cells occur to form new fat cells. Studies utilizing tritrated thymidine incorporation into DNA indicate that at 5 months of age there is no appreciable cell division in mature adipocytes of the rat (Hollenberg and Vost, 1968; Greenwood and Hirsch, 1974).

Studies of adipose depot development in man have been much stimulated by the methodology of Hirsch and Gallian (1968). The method is limited first by the fact that only cells with some lipid content can be identified. Those cells that are either depleted of their fat stores or destined to store lipid at some future date cannot be counted since the sole identifying characteristic of the adipocyte is the presence of fat. Secondly, the method is not readily adapted to large epidemiologic studies. However, given these shortcomings, the technique has been useful in defining the cellular characteristics of obese and nonobese subjects and in clarifying data relating metabolic function to cellularity.

In our laboratory we have been interested in cellular development of the fat depot and have instituted longitudinal studies of the fat depot in children ranging from 4 months to 16 years of age (Knittle *et al.,* 1977). The technique of needle aspiration was used (Hirsch and Gallian, 1968). The procedure is simple and safe, even in children. One anesthetizes the subcutaneous area with 1–2 ml of 1% xylocaine, then inserts a 15-gauge needle attached to a siliconized 50-ml glass syringe. When one pulls back on the plunger, the fat tissue issues forth by suction. As much as 500 mg of fat can be removed with each sampling. Only one sample was obtained for the following studies. Fat cell number was determined by dividing the average fat cell size into a value for total body fat which was derived from the measurement of total body K^+. This technique is limited by the fact that we were unable to measure total body K^+ in children under 2 years of age because of the lack of sensitivity of the counter. Thus, results of children younger than 2 are based on height–weight relationships. Two groups of subjects were studied: obese subjects over 130% of ideal weight and nonobese subjects between 90% and 120% of ideal weight. From 2 to 16 years of age no significant differences in cell size were observed between children of different ages in obese subjects. In nonobese subjects, cell size also showed no difference with age from 2 to 10; however, after this time cell size was significantly higher. From 2 to 10, obese children had significantly larger fat cells than nonobese children of the same age; however, the differences in size were not statistically significant after 12 years of age. Furthermore, at 2 years, obese children displayed cell sizes within the adult range of 0.4–0.7 µg lipid/cell, while nonobese subjects did not achieve adult levels until age 12 (Figure 1).

Differences in adipose cell number were also found. Significant differences in cell number were detected as early as 2 years of age and persisted throughout the

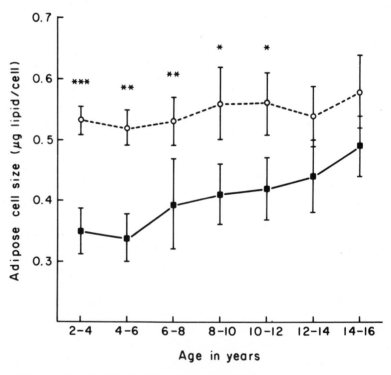

Fig. 1. Adipose cell size of obese and nonobese subjects as a function of age. Obese subjects, ○; nonobese subjects, ■. ***, $P < 0.001$; **, $P < 0.01$; *, $P < 0.05$.

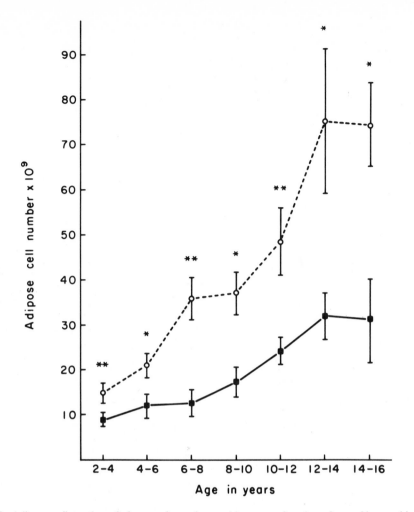

Fig. 2. Adipose cell number of obese and nonobese subjects as a function of age. Obese subjects, ○; nonobese subjects, ■. $**P < 0.01$; $*P < 0.05$.

entire age range studied. Obese subjects displayed increases in cell number at all ages. However, no change in cell number was observed in nonobese subjects between 2 and 10 years of age. Once again increments in this factor were found only after age 10. Both groups appeared to plateau somewhere between ages 14 and 16 (Figure 2). Differences in cellular development were also observed in the course of longitudinal studies where the development of cell number in the same child was followed over a 1-year period. While obese subjects displayed significant increments at all ages, nonobese subjects did not show any significant change until age 10. Studies were also repeated at 2-year intervals, and no change in cellularity was observed in the nonobese group prior to age 10, while obese children continued to display significant increases at all ages.

Thus, after 2 years of age one can detect significant quantitative as well as qualitative differences in adipose tissue growth and development between obese and nonobese children. Obese children attained adult levels for fat cell size by age 2 and thereafter increased their fat stores almost exclusively by increase in cell number. Nonobese subjects displayed a relative quiescence in adipose depot enlargement from 2 until the prepubescent and adolescent periods. At that time both cell size and number appeared to increase.

We have also followed children from 4 months to 4 years of age. Fat aspirations were repeated at 6- to 8-month intervals prior to age 12 months and at yearly intervals thereafter. Eight of the subjects studied clinically developed overt obesity. The data obtained from these individuals during the course of our observations were matched with subjects of comparable age and sex who did not become obese. No significant differences in weight were observed prior to and up to 12 months of age. Indeed, although not statistically significant, the mean weight of the subsequently obese children was less than that observed in the nonobese at birth and 4 months of age. However, by 24 months of age, the obese group was significantly heavier with a mean weight of 18.8 kg compared to 13.0 kg in the nonobese, and they remained heavier until 4 years of age. In both groups cell number continued to increase throughout the study. Although obese subjects did display a greater number of cells at all ages, the differences were not statistically significant until 24 months of age. Thus, one could not clearly distinguish between obese and nonobese children on the basis of either total cell number or rate of cellular development at any time under 24 months of age. From 4 to 12 months of age both groups showed significant increments in lipid content and one could not distinguish the groups on the basis of size at either of these ages. However, from 12 to 24 months of age, while obese children continued to display significant increases in lipid content, nonobese subjects displayed a decrease and by 24 months had a mean cell size that was significantly less than the obese child and smaller than that observed at 12 months of age.

Studies in nonobese adults indicate that, as in the rat, the cell number remains fairly constant. Thus, overfeeding results in an increase in cell size without alteration in number. In the obese adult marked degrees of weight loss result in a depletion of fat cell lipid content without altering cell number. Studies in children also reveal the same general phenomenon. Once a particular cell number is attained it appears that it cannot be decreased. At what age cell division ends in man is still a moot point. This is particularly true in the case of obese subjects who become obese in later life and display hypercellularity of their adipose depot. Were the cells always there in a preadipocyte form or did new cells arise *de novo* from undifferentiated embryonal forms as seen in fetal development? Current techniques do not allow one to clearly resolve this dilemma. Techniques for distinguishing potential adipose cells prior to lipid deposition are not available at present.

Thus, although concrete evidence for ''critical periods'' or ''set points'' in adipose tissue development has not been demonstrated, the cellular data compiled to date suggest that two intervals in the development of adipose tissue may have important consequences for the development of obesity in man. The first period extends from birth to 2 years of age, and the second occurs somewhere in the prepubescent–pubescent period, stretching from about age 8–10 to 16. In both periods alterations in both cell number and size occur in nonobese subjects, some of whom may later become obese. Since alterations in cell size are intimately related to nutritional and hormonal factors, one could, perhaps, develop techniques to identify enzymatic markers or histochemical techniques that identify important precursor cells.

There is abundant literature describing the metabolic interactions of adipose tissue and relating a number of parameters to adipose cell size and number.

Unfortunately, to date, none of these has been identified prior to the obese state or in precursor cells, and in almost every case the metabolic aberration found in the obese subject returns to "normal" after weight reduction is achieved. Thus, a marker of the preobese or obesity-prone subject still eludes the investigator. It might be useful, however, to review the metabolic development of fat depots in order to delineate future paths of investigation which would be useful in defining normal as well as abnormal development of fat depots.

3. Metabolic Studies Related to Fat Depot Development

What little animal and human evidence there is indicates that marked variations in metabolic function can be observed when one compares the fetal, suckling, and postweaning periods. Delivery entails a sudden change from an essentially high-carbohydrate fetal diet to a short period (12–24 hr) of relative starvation followed by a relatively high-fat diet during suckling and a gradual increase in carbohydrate again after weaning. The hormonal and enzymatic changes evoked by delivery lead to alterations in glucose and fat metabolism. The major changes in fat metabolism occurring in each of these periods are in fatty acid synthesis, oxidation, and lipolysis. In addition, marked alterations in glucose metabolism are reflected in altered gluconeogenesis and glycolysis (Walker, 1971; Hahn, 1972).

In rats the rate of fatty acid synthesis in the liver is high in the fetus and decreases rapidly after birth. Synthesis remains low for the entire period of suckling but increases again when the postweaning period commences. The changes in lipogenesis are reflected in enzymatic activity, in particular malic enzyme and ATP citrate lyase. One can also demonstrate parallel changes in *de novo* synthesis from glucose, pyruvate, citrate, and acetate (Hahn, 1972; Hahn and Novak, 1975). Little is known about liver lipogenesis in man; however, in one study of human neonatal adipose tissue a similar pattern was found (Novak and Monkus, 1972).

During fetal life lipogenesis and fat accumulation are favored while fatty acid oxidation is low. If this fetal condition were to persist after birth, it could lead to fat accumulation. The fetus is not equipped for the oxidation of fatty acids or for the formation of ATP by fatty acid oxidation. Studies of rat heart indicate decreased carnitine and acylcarnitine transferase at birth (Wittels and Bressler, 1965). Ten days after birth the carnitine system becomes active, and in the neonate a larger portion of fatty acid is oxidized via this pathway than in the adult. In rabbits the capacity for beta-oxidation has also been shown to be higher in newborns than in adults. In addition, beta-hydroxyacyl CoA dehydrogenase has been shown to be low in human isolated mitochondria of newborn subcutaneous adipose tissue but to increase during the first week of life (Wolf *et al.*, 1974; Novak *et al.*, 1972). Thus it appears that in most species fatty acid oxidation, while absent or low *in utero*, increases postnatally, then decreases again with age after weaning.

Immediately after birth free fatty acid levels increase markedly, presumably released from fat stores by the effect of catecholamines. These hormones are secreted in response to the stress of birth and altered environmental temperatures. In human newborns glycerol release is high during the first hours of life and decreases with subsequent feeding (Novak *et al.*, 1965). Furthermore, the rate of glycerol release *in vitro* can be enhanced by phentolamine (an alpha blocker) and decreased by propanolol (a beta blocker), without prior addition of catecholamines which are necessary in adult tissues (Monkus *et al.,* 1974; Novak *et al.,* 1968; Burns and Langley, 1970; Burns *et al.,* 1972). This fact, in conjunction with a documented decreased sensitivity of neonatal adipose tissue to catecholamines, suggests that hormone receptor sites are activated at birth and less sensitive to additional catecholamine stimulation. If this insensitivity persisted, one could visualize an increase in fat cell size due to a lack of lipolytic response to normal catecholamine stimulation.

We have detected a decreased response to the lipolytic effect of epinephrine in obese children as early as 2 years of age (Knittle *et al.,* 1977). It is tempting, therefore, to postulate that alterations in adipose cell catecholamine receptor sites and responses to alpha and beta blockers, which normally occur in nonobese children, are altered in the obese child.

The importance of glucose as a fetal food has been cited above. Immediately after birth there is an initial absence of all exogenous nutrition during the first hours of life. After this time the carbohydrate content of food gradually increases. Thus studies of the developmental aspects of glucose metabolism of fat cells could provide important data related to subsequent growth. We have studied hexokinase I and II in adipocytes of varying size and have demonstrated differences in total quantity between large and small adipocytes (Katz *et al.,* 1976). In addition, adipocytes from subjects with abnormal glucose tolerance display alterations in the relative ratios of hexokinase I and II (Belfiore *et al.,* 1975). Varying patterns with age have also been demonstrated in this isozyme system in the rat, as well as in the rate of gluconeogenesis and glycolysis (Katzen and Schimke, 1965; Bernstein *et al.,* 1975).

Studies of glucose metabolism in fat cells of adults also demonstrate important differences in metabolic function in cells of differing size from obese and nonobese subjects. When adipose tissue is obtained from subjects during periods of weight maintenance and ingestion of diets of similar composition, increasing adipose cell size is associated with unchanging rates of glucose oxidation in the absence of insulin but decreasing stimulation by insulin. On the other hand the rate of glucose incorporation into triglyceride glycerol and lypolysis of triglyceride to FFA and glycerol increases with increasing cell size. These differences can be altered within a given subject by gaining and/or losing weight with parallel changes in cell size (Salans *et al.,* 1968; Smith, 1971).

In addition, one can alter responses to insulin by altering diet. Thus basal glucose metabolism and the response to insulin in large adipose cells obtained from obese individuals ingesting weight-maintenance diets high in carbohydrate may equal or exceed that of smaller cells from nonobese subjects ingesting isocaloric diets low in carbohydrate. Furthermore, basal glucose metabolism and the response to insulin in large adipocytes from obese rats actively gaining weight may be greater than that observed in smaller cells obtained from nonobese rats at a constant body weight. These data indicate the importance of distinguishing between enlarging and enlarged fat cells (Salans *et al.,* 1972).

Since the alterations in glucose metabolism in the obese are not only reversible through weight loss, but also inducible in nonobese subjects through weight gain, they may represent adaptive rather than primary changes in adipose cell function. However, the exact role of glucose metabolism and insulin effects on maintaining and developing increased cell size remain to be more fully examined.

Recently interest has been generated in insulin receptor sites in adipocytes and monocytes in obese and nonobese subjects. Although the exact mechanism remains unknown, evidence has come from several sources indicating that insulin may act through a simple binding of the hormone to a specific site on the plasma membrane. Thus the ability to directly measure and study the interaction of insulin with its membrane receptor could provide insight to a variety of conditions in which insulin resistance exists, such as the obese state. In obese hyperglycemic mice (*ob/ob*) a decrease in binding of the larger fat cells is found when compared to their lean littermates. Similar decreases were found in liver and thymic lymphocytes (Freychet *et al.*, 1972; Goldfine *et al.*, 1973; Kahn *et al.*, 1972). Other animal models of obesity studied (*db/db* mice and New Zealand rats) have yielded similar results (Kahn *et al.*, 1973; Baxter *et al.*, 1973).

However, others have not found any difference in insulin binding of large vs. small adipocytes from rats of different ages (Livingston *et al.*, 1972). Furthermore, the finding that high-carbohydrate or high-calorie intake acutely enhances the insulin response of enlarged adipose cells suggests that a change in insulin binding may not be fundamental to altered metabolic processes or development. Further studies of binding in preobese states are needed to determine the role of abnormal binding in the development and enlargement of fat stores. Studies to date demonstrate that the insulin receptor is not a static structure but may be altered by metabolic, hormonal, dietary, and perhaps genetic factors. Like most indications of altered glucose metabolism in obese states, it appears to be a secondary rather than a primary attribute.

In vivo studies also reveal that glucose tolerance, plasma insulin, and free fatty acid levels are changed by weight reduction (Knittle and Ginsberg-Fellner, 1972; Kalkhoff *et al.*, 1971; Newburgh, 1952). Thus abnormalities of glucose metabolism and the hyperinsulinemia found in the obese adult may be secondary to, rather than the cause of, the obese state.

Little is known about the causal relationship in children and adolescents since the finding of similar abnormalities at these ages have only been demonstrated in overtly obese subjects (Parra *et al.*, 1971; Paulsen *et al.*, 1968). Prospective studies will be needed before one can state definitely that a hyperinsulinemic state and/or abnormal glucose tolerance precedes the obese state. However, recent studies in our laboratory of newly diagnosed diabetic children, before and after therapy, indicate that treatment with insulin affects adipose cell lipid content without significantly altering cell number (Ginsberg-Fellner and Knittle, 1970). Thus one might predict that if the hyperinsulinemic state precedes obesity, it will lead to an increased lipid content of cells rather than to hypercellularity. Also related to metabolic function are growth hormone, adrenal function, and the thyroid. The effect of each will be briefly summarized.

Growth Hormone. Studies of the effect of growth hormone on the cellular development of a variety of organs other than fat have shown that this hormone increases cellular proliferation (Cheek and Hill, 1974; Cheek *et al.*, 1966; Beach and Kostyo, 1968). *In vitro* studies of adipose tissue in animals suggest that the main metabolic action of growth hormone is to increase the rate of lipolysis (Vaughn and Steinberg, 1963). However, little is known about the effect of human growth hormone on human adipose tissue *in vitro*. *In vivo* studies indicate that the administration of human growth hormone increases free fatty acid release and decreases skinfold thickness, presumably due to its effect on adipose tissue lipolysis (Raben and Hollenberg, 1959). Preliminary studies in our laboratory indicate that the treatment of growth-hormone-deficient children produces significant increases in total body potassium and adipose cell number 6 to 8 months after onset of therapy

(Knittle *et al.*, 1972). These findings plus the fact that obese subjects have abnormalities in serum growth hormone levels suggest that growth hormone may also play a role in the development of the hypercellular state of obesity by virtue of its effect on cell division (Parra *et al.*, 1971).

Furthermore, we have found that infants of diabetic mothers studied at 6 months of age or less are relatively hypercellular in regard to adipocytes (Ginsberg-Fellner and Knittle, 1971). Indeed, three of the five subjects studied developed clinically evident obesity one year after the initial studies. It is tempting to postulate that frequent fluctuations of blood sugar with periods of hypoglycemia in diabetic mothers may serve as the stimulus for increased secretion of growth hormone in the fetus. This in turn could lead to an increase in cellularity of adipose tissue.

At present the role of growth hormone in the development of the obese state remains conjectural, but its importance in cellular division indicates that further studies of its action on the development of adipose tissue are warranted.

Adrenal Function. Increased adrenal cortical function has been found in 30–60% of obese subjects, but it is not clear how this may relate to the development or maintenance of the obese state (Migeon *et al.*, 1963; Schteingart and Conn, 1965). Weight reduction is accompanied by a decrease in 17-hydroxycorticosteroids, but little is known about the effect of corticosteroids on human adipose tissue metabolism other than its effect on lipolysis.

Studies of *in vivo* responses to exogenous epinephrine are conflicting; some authors suggest that obese subjects fail to increase free fatty acids after an epinephrine load, while others maintain that the response is normal (Gordon *et al.*, 1962; Orth and Williams, 1960). These differences may be due to the heterogeneity of obese persons, and no data related to adipose cell number or size are available.

In vitro studies of epinephrine-stimulated glycerol release indicate that the increase over basal values is diminished in larger adipocytes (Goldrick and McLoughlin, 1970; Zinder and Shapiro, 1971). As with insulin, it is not clear whether this phenomenon precedes or follows the development of obesity. However, it has been shown that weight reduction does not alter the *in vitro* epinephrine response in obese adults or adolescents. No data relative to weight reduction are presently available in obese children who have diminished epinephrine response (Knittle and Ginsberg-Fellner, 1969).

One could speculate that the lack of epinephrine response serves as a stimulus for the development of new adipose cells. Thus a decrease in fatty acid release secondary to epinephrine stimulation could be overcome by providing a greater number of less responsive cells to meet energy needs. Since epinephrine acts via adenylcyclase, one could also postulate a defect in this system. Studies of the levels of $3',5'$-cyclic adenosine monophosphate and adenylcyclase in human and rat adipose tissue are currently in progress in our laboratory to explore this hypothesis.

Thyroid. At present there are few data to support any role for thyroid malfunction in the pathogenesis of obesity. Serum levels of protein-bound iodine or thyroxine all appear to be normal in obese subjects. Bray (1969) has recently identified a small group of obese women with exceptionally large cells that respond to T3 therapy. However, one should not use this drug routinely in the treatment of simple exogenous obesity. Furthermore, there is no evidence that similar subjects exist in the pediatric or adolescent age group.

From the above it is clear our understanding of the metabolic and endocrine factors involved in fat cell development is far from complete. The extent to which any of these factors play a role in the pathogenesis of the obese state is not clear.

Indeed, the fact that weight reduction results in a more normal metabolic and endocrine profile suggests that these phenomena are secondary to the obese state rather than causal. Nonetheless, it does appear that hormonal actions can markedly affect the adipose depot by virtue of their effect on metabolism and cellular proliferation. Further studies of the complex interaction of hormonal control and early nutritional factors could provide important clues about the development and maintenance of the ultimate size of the fat depot in man.

The development of the fat depot in man is the result of many morphological, metabolic, and presumably genetic factors acting in varying degrees at different times in the life of an individual. At present it is impossible to determine the relative role of genetics and environmental factors. No specific genetic effect has been identified in man.

4. Future Areas of Research

Obviously, many questions remain. Too few facts and all too many "theories and myths" can be found regarding the relative importance of a variety of early nutritional interventions on the development of adipose depots. Thus we do not know to what extent such commonly implicated factors as maternal diet and weight gain during pregnancy and early infant intake contribute to the pathogenesis and development of fat depots in later life.

4.1. Maternal Weight Gain

Maternal nutrition prior to pregnancy and weight gain during pregnancy have been frequently implicated as one of several environmental factors capable of affecting birth weight. In general, studies related to this phenomenon have focused their attention on the effect of undernutrition and extreme protein and calorie deprivation in animals, or on that found in lower socioeconomic classes in Third-World countries (Lechtig *et al.*, 1975*a;* Ebbs, 1941; Burke *et al.*, 1943). Natural experiments involving humans under acute starvation conditions, such as famine, seige during wars, and climatic disasters, have also been cited as evidence for a consistently lower birth weight in infants born during times of maternal deprivation (Antonov, 1947; Smith, 1947; Gruenwald and Funakawa, 1967). A correlation between birth weight and maternal weight prior to and during pregnancy has been found by some in the general population and has been interpreted as evidence of a nutritional effect on birth weight of the child. However, studies utilizing dietary surveys and/or food supplementation programs in more affluent societies have produced varying and sometimes contradictory results (Thomson, 1959; McGanity *et al.*, 1954; Dieckman *et al.*, 1944). In addition, inferring a cause-and-effect relationship based on current evidence is a bit risky since numerous factors other than nutritional deprivation can be cited. These include poor sanitation, ill health, smoking, and the use of drugs, all of which, in and of themselves, produce alterations in birth weight.

Even if one accepts all the positive data related to the effect of severe maternal undernutrition on birth weight, one is still left with the question of the role of maternal overnutrition on not only birth weight, but the development of obesity in later life. Indeed, there is conflicting evidence regarding the correlation between maternal weight gain, birth weight, and subsequent obesity. In general, most investigators agree that there is no correlation between an infant's weight and subsequent adult weight until approximately five years of age, implying that in obesity early neonatal factors are more important than genetic or maternal nutritional influences (Abrahams and Nordsieck, 1960; Asher, 1966; Eid, 1970).

However, a recent study by Fisch *et al.* (1975) indicates that the weight of the mother is important. They found a high positive correlation between maternal weight and weight gain during pregnancy and the development of obesity in children at age 7. It is also known that the probability of obesity occurring in a child increases remarkably if one parent is obese and even more so if both are. The degree to which this is under genetic control or modifiable by environmental factors is not presently known.

Studies of food supplementation in deprived areas suggest that there is a limit to the effectiveness that an increase in food intake has on birth weight. It has been shown in these studies that an increment of greater than 5 kg in maternal weight during the latter part of pregnancy did not produce a further increase in birth weight. These authors felt a threshhold effect exists and that the relationship between calories and birth weight is curvilinear, not rectilinear, and depends upon the previous nutritional status of the mother (Lechtig *et al.,* 1975*b*). The effect of food intake and weight increase on subsequent infant body weight will obviously vary depending upon the time it occurred, its duration, what the limiting or excessive nutrient is, the functional capacity of the placenta, and the degree to which maternal nutrients are available to and utilized by the fetus.

Metabolic studies related to fetal development indicate that despite a high degree of safety, physiologic factors can be altered to produce a change in birth weights. In general, the hormonal system of the pregnant woman is modified in the direction of increased nitrogen retention for growth and deposition of fat for use as a subsequent source of energy needs. Adequate provision of amino acids and glucose is also made available to the fetus and is usually maintained independent of changes in the maternal diet. Even with low levels of amino acids in maternal blood, the placenta can selectively maintain adequate concentrations in fetal blood presumably for protein synthesis and growth. However, severe degrees of maternal undernutrition have been shown to retard fetal development (Oldham and Sheft, 1951; Pearse and Sornson, 1969; Page, 1957; Callard and Leathen, 1970).

On the other hand, metabolic factors can also contribute to excessive adiposity. Infants born to diabetic mothers are heavier at birth, are relatively fatter, and have a higher rate of subsequent obesity than infants of nondiabetic mothers of equal gestational age (Osler, 1960; Baker, 1969). It has been suggested that this phenomenon may be the result of long exposure of the fetus to hyperglycemic states resulting in the deposition of excessive adipose tissue. Adipose tissue cellular studies in our laboratory have also demonstrated an increased cell number in these children (Ginsberg-Fellner and Knittle, 1971).

It is also well known that the deposition of fat in the fetus occurs in the last trimester of pregnancy and is accelerated particularly in the final 37th to 40th weeks (Widdowson, 1950). In infants of birth weights between 3.0 and 3.2 kg, the difference in weight is mainly a function of the deposition of fat (Lechtig *et al.,* 1975). Thus a rise in blood sugar and hormones related to glucose and fat metabolism due to ingested calories by the mother could serve as an important substrate for the synthesis of fat during this critical time period. This is particularly true in obese women, many of whom have normal glucose tolerance prior to but become class A diabetics during pregnancy.

Although it is tempting to postulate a relationship between maternal nutrition and weight gain during pregnancy and obesity in the child, the current data are not sufficient to make any causal correlations. The above-mentioned studies suffer from the fact that in assessing maternal nutrition, investigators relied on either weight gain alone or height–weight indices, with little if any reference to the type of weight gain (i.e., fat, lean body mass, or water) or diet. To date little is known about the effect of maternal diet on the deposition of fat in either the mother or neonate. Hytten (1964) has estimated that of the average maternal weight gain in pregnancy (i.e., 27.5 lb), 9 lb is deposited mostly as subcutaneous fat around the trunk and hips. However, the range of weight gain is huge in "normal" pregnancies, varying from a loss to a gain of 60 lb or more (Hytten, 1964). Thus one could postulate that marked variations in fat deposition can occur in different women and that this could have important consequences for the fetus and neonate. Indeed, the probability of demonstrating causal relationships between maternal fat deposits and obesity in the neonate would increase if data related to the degree and type of weight gain are observed in an "at-risk" population. Much confusion exists in the study of the effect of maternal nutrition on the development of obesity because we lack precise definitions and classification of the population under investigation. This information is urgently needed since the effects of weight gain prior to or during pregnancy may vary drastically in terms of their effect on metabolic function and development of the fetus in women with similar weights but vastly different body compositions and metabolic function. Another area of concern is the proportion and type of dietary fat in the maternal diet. In general pregnant women are told to increase their caloric intake about 200 calories over their prepregnant intake and to ingest approximately 30% of calories as fat. Little is known about the metabolic consequences related to such diets.

4.2. Early Infant Feeding

The question of the effect of overnutrition during early neonatal life on subsequent obesity in adult life has recently received a great deal of attention. Vociferous advocates can be found both pro and con. As with the study of maternal nutrition, the arguments are based on retrospective studies of the general population and are filled with numerous correlations but little in the way of direct causal prospective studies.

Although the cellular data generated in our laboratory indicate that the cellular development of fat stores of obese children does increase earlier and at a more rapid rate than nonobese subjects, the exact mechanism involved (i.e., nutrition vs. genetic endowment) has not been determined. It is also true that our longitudinal data further indicate that once adipose tissue hypercellularity is achieved it cannot be altered by current dietary regimens. Many investigators utilizing retrospective analyses of the onset of obesity have popularized the concept of an early critical period when an immutable endowment of an increased number of adipocytes is produced. Unfortunately, this has caused a great deal of confusion and distress regarding the issue of early infant feeding and the statements made have been premature. More data based on prospective studies of carefully delineated populations are necessary before concrete recommendations can be made to mothers of newborns.

Studies of caloric intake during infancy in America, and New York City in particular, indicate that infants are ingesting more calories and protein than the recommended dietary allowances and that the introduction of solid food is occurring at earlier and earlier ages (Grossi, personal communication; Filer and Martinez, 1964). It has been postulated, therefore, that these practices and trends in infant feeding may be contributing to the growing prevalence of obesity in industrialized Western nations. It is even further postulated that the prevention of obesity may require alterations of eating habits, especially in the earliest neonatal period. However, this postulate still remains to be proved. Recent studies in England indicate that there is a remarkable degree of self-regulatory control within the infant and that metabolic balance of small-for-dates (SFD) infants differs from large-for-dates (LFD) infants in early postnatal months (Ounsted and Sleigh, 1975). Thus, SFD infants took significantly more milk per kilogram of body weight than LFD infants. Furthermore, within each group there was a negative correlation between actual weight at two months and milk intake per kilogram body weight. In addition, breast-fed babies tended to have a lower weight gain per day in the LFD group, whereas in the SFD they had a greater gain (Foman *et al.,* 1964). These data indicate that there are self-regulatory controls within the infant that vary markedly from one child to the other. The authors further suggest that this control mechanism is allowed its greatest effect in breast-fed children as compared to those who are bottle fed. Indeed, human breast milk varies during a single feed, and Hall (1975) has stated that the breast-fed baby controls its feeding in response to certain biochemical changes not present in bottle feeding. The exact biochemical mechanisms are not known at present. However, a number of differences in fat content between human and cow's milk have been found (Freundenberg, 1953; Breckenridge and Kuksis, 1967; Carey and Dils, 1972). Human milk has little or no short-chain (C_4–C_6) and less saturated and methyl-branched saturated acids than cow's milk; on the other hand, there is a greater amount of linoleic acid isomers, total organic phosphorus, cholesterol, 7-dehydrocholesterol, vitamin A, tocopherol, and phospholipid in human milk. There is also a preference for the C_2 position in palmitic acid distribution in human milk. This last fact may be important for palmitic acid absorption since lipase activity attacks the 1 and 3 position forming the C_2 palmitic monoglyceride which is more readily absorbed than the free palmitic acid. Finally, human milk contains two lipases compared to one in cow's milk.

It is of interest that diet can alter the composition of milk. When fat is eliminated from the diet, the C_{16} and C_{18} composition of milk simulated that found in adipose tissue and the relative amounts of C_{12} and C_{14} acids increased compared to diets containing fat. When caloric intake was increased from 1800 to 3800/day, the C_{16} and C_{18} acids were further decreased by more than 30%, and C_{12} and C_{14} acids increased by 100% (Insull *et al.,* 1959).

Variations in human milk have also been related to the age of milk. Thus colostrum (1–5 days) has a lower total lipid short-chain and C_{12} fatty acid concentration and an increased concentration of vitamin A, carotenoids, vitamin E, total phospholipid, and arachidonic acid compared to mature (over 30 days) milk.

The metabolic and hormonal mechanism involved in milk production during lactation has been examined by Scow *et al.* (1973) in the rat. Their results indicate that an inverse relationship exists between lipoprotein lipase activity when the mammary gland is compared to adipose tissue. During lactation lipoprotein lipase activity is increased in mammary tissue and decreased in adipose. Lipoprotein lipase activity decreases in mammary tissue during the first 20 days of pregnancy and increases several days prior to parturition. It then decreases abruptly during parturition and immediately increases and remains high during lactation. The inverse relationship between these two systems suggests that hormonal control in milk secretion may direct dietary fatty acids to mammary gland for milk production and away from adipose tissue deposition. Indeed many women who breast feed report a decrease in weight, and the studies in the rat also show no appreciable weight gain during suckling when the lipoprotein lipase activity in adipose tissue is down. Furthermore, when the rat is not permitted to suckle for 9 hr, there is an increase in plasma triglyceride and a decrease in lipoprotein lipase activity in the mammary gland. After 18 hr complete inhibition of lipoprotein lipase in mammary glands occurs with an increase of 55% activity in adipose tissue.

Thus, in the nonsuckling postpartum animal (and perhaps man as well) dietary fat is directed into the fat depot rather than into milk production. One could postulate that the postpartum weight gain described by some obese women is due to the fact that they do not breast feed, thus causing an increase in lipoprotein lipase of their adipose tissue, resulting in excessive fat accumulation.

Variations in the intake of maternal total calories, carbohydrate, and fat content can influence not only total milk production but produce qualitative changes in fat composition that may have important consequences for the food intake and development of adipose depot in both the mother and neonate. Similarly, differences in fat content of breast milk vs. formula feeding may also be involved in the ability of the infant to modify and regulate food intake.

Investigations of normal bottle-fed babies in the general population have revealed a wide range of intake when infants are allowed *ad libitum* feedings (Filer and Martinez, 1964). Food intake varies and depends upon a number of factors such as total energy requirements, appetite, food composition and taste, disease states, physical activity, growth, and the extent of fat deposition present as a reserve for energy needs. Thus, a normal placid infant can consume as little as 70 kcal/kg of body weight, whereas a more active child can ingest as much as 130 kcal/kg or more (Holt and Snyderman, 1964).

In addition to total caloric intake and type of calories supplied, the caloric density of infant formulas appears to be critical. Recent evidence has accrued that suggests that in the normal full-term neonate, ingesting a normal mean caloric intake of 525 kcal, 185 kcal are used for growth and 340 kcal for nongrowth; 45.6% of calories used for growth in this "normal" diet represented weight gain as fat. Infants fed diets of "normal" caloric density during the first 8–41 days of life (i.e., 100 kcal/ml of formula) gain a greater percentage of body weight as fat as compared to those on lower concentrations (54 kcal/ml). In neonates fed 133 kcal/ml from day 8 to day 111, the percentage of weight gain as fat was 59.3% as compared to the 45.6% previously calculated on normal diets (Foman *et al.,* 1971, 1975).

Unfortunately, the calculations utilized height–weight relationships and were based on the fact that for similar increases in body length greater increases in total body weight were observed. No attempt was made to measure changes in total fat. Furthermore, no measurements were made of maternal prepregnant weight or weight gain during pregnancy.

In summary, the current available data indicate that one can alter the degree of weight gain and adiposity in neonates by a number of manipulations of early food intake. However, the ingestion of food and its subsequent disposal as a source of energy and/or growth as protein or fat may vary markedly from one infant to another. Merely measuring the route of delivery, type, quantity, and/or caloric density of the formula does not generate the data necessary to determine the final route the ingested calories take. To date, our knowledge of the hormonal and enzymatic developmental patterns and changes induced in them by diet are rather meager. More information is necessary regarding the dietary, metabolic, and genetic mechanisms involved in fat deposition during the antenatal, neonatal, and adult period in man. It is still not certain to what degree differences in diet during these time intervals influence subsequent adipose cellularity and metabolism. Finally, techniques must be developed to ascertain whether or not preadipocytes do exist in man which form the basis of future obesity in later life.

5. References

Abrahams, S., and Nordsieck, M., 1960, Relationship of excess weight in children and adults, *Public Health Rep.* **75:**263.

Antonov, A. N., 1947, Children born during the siege of Leningrad in 1942, *J. Pediatr.* **30:**250.

Asher, P., 1966, Fat babies and fat children: The prognosis of obesity in the very young, *Arch. Dis. Child.* **41:**672.

Baker, G. L., 1969, Human adipose tissue composition and age, *Am. J. Clin. Nutr.* **22:**829.

Baxter, D., Gates, R. J., and Lazarus, N. R., 1973, Insulin receptor of the New Zealand obese mouse (NZO): Changes following the implantation of islets of Langerhans. Congress of the VIII International Diabetes Federation, Brussels, July 15–20, 1973, *Excerpta Med. Int. Congr. Ser.* **280:**74.

Beach, R. F., and Kostyo, J. L., 1968, Effect of growth hormone on the DNA content of muscles of young hypophysectomized rats, *Endocrinology* **32:**882.

Belfiore, F., Rabuazzo, A., Napoli, E., Borzi, V., and LoVecchio, L., 1975, Enzymes of glucose metabolism and of the citrate cleavage pathway in adipose tissue of normal and diabetic subjects, *Diabetes* **24:**865.

Bernstein, R. S., Grant, N., and Kipnis, D. M., 1975, Hyperinsulinemia and enlarged adipocytes in patients with endogenous hyperlipoproteinemia without obesity or diabetes mellitus, *Diabetes* **24:**207.

Bray, G. A., 1969, Effect of diet and triiodothyronine on the activity of SN-glycerol-3-phosphate dehydrogenase and on the metabolism of glucose and pyruvate by adipose tissue of obese patients, *J. Clin. Invest.* **48:**1413.

Breckenridge, W. C., and Kuksis, A., 1967, Molecular weight distributions of milk fat triglycerides from seven species, *J. Lipid Res.* **8:**473.

Brozek, J., Grande, F., Anderson, J. T., and Keys, A., 1963, Densitometric analysis of body composition, *Ann. N.Y Acad. Sci.* **110:**113.

Burke, B. S., Beal, V. A., Kinkwood, B., and Stuart, H. C., 1943, The influence of nutrition during pregnancy upon the condition of the infant at birth, *J. Nutr.* **26:**569.

Burns, T. W., and Langley, P. E., 1970, Lipolysis by human adipose tissue: The role of 3'5' cyclic AMP and andrenergic receptor sites, *J. Lab. Clin. Med.* **75:**983.

Burns, T. W., Langley, P. E., and Robinson, G. A., 1972, Studies on the role of cAMP in human lipolysis, *Adv. Cyclic Nucleotide Res.* **1:**63.

Callard, I. P., and Leathen, 1970, Pregnancy maintenance in protein deficient rats, *Acta Endocrinol.* **63:**539.

Cary, E. M., and Dils, R., 1972, The pattern of fatty acid synthesis in lactating rabbit mammary gland studied *in vivo, Biochem. J.* **126:**1005.

Cheek, D. B., and Hill, D. E., 1974, The effect of growth hormone on cell and somatic growth, in: *Handbook of Physiology,* Section 7 (R. O. Greep and E. B. Atwood, eds.), Endocrinology IV, Part 2, Chapter 28, p. 159, Williams and Wilkins, Baltimore.

Cheek, D. B., Brasel, J. A., Elliott, D., and Scott, R., 1966, Muscle cell size and number in normal children and in dwarfs (pituitary cretins and primordial), preliminary observations, *Bull. Johns Hopkins Hosp.* **119:**46.

Dieckmann, W. J., Adair, F. L., Michel, H., Kramer, S., Dunkle, F., Arthur, B., Costin, M., Campbell, A., Wensley, A. C., and Lorang, E., 1944, Calcium, phosphorus, iron and nitrogen balance in pregnant women, *Am. J. Obstet. Gynecol.* **47:**357.

Ebbs, J., 1941, The influence of prenatal diet on the mother and child, *J. Nutr.* **22:**515.

Eid, E. E., 1970, Follow-up study of physical growth of children who had excessive weight gain in first six months of life, *Br. Med. J.* **2:**74.

Falkner, F., 1966, General considerations in human development, in: *Human Development* (F. Falkner, ed.), p. 10, W. B. Saunders, Philadelphia.

Filer, L. J., and Martinez, B. S., 1964, Intake of selected nutrients by infants in the U. S., *Clin. Pediatr.* **3**:633.

Fisch, R. O., Bilek, M. S., and Ustrom, 1975, Obesity and leanness at birth and their relationship to body habitus in later childhood, *Pediatrics* **56**:521.

Flemming, W., 1871, On the formation and regression of fat cells in connective tissue with comment on the structure of the latter, *Arch. Mikrosk. Anat.* **7**:32.

Foman, S. J., Owen, G. M., and Thomas, L. N., 1964, Milk or formula volume ingested by infants fed ad libitum, *Am. J. Dis. Child.* **108**:601.

Foman, S. J., Thomas, L. N., Filer, L. J., Ziegler, G. G., and Leonard, M. T., 1971, Food consumption and growth of normal infants fed milk-based formulas, *Acta Paediatr. Scand. Suppl.* 223.

Foman, S. J., Filer, L. J., Thomas, L. N., Anderson, T. A., and Nelson, S., 1975, Influence of formula concentration on caloric intake and growth of normal infants, *Acta Paediatr. Scand.* **64**:172.

Forbes, G. B., 1964, Lean body mass and fat in obese children, *Pediatrics* **34**:308.

Freundenberg, E., 1953, *Jahrb. Kinderheilkd.* **54**:1.

Freychet, P., Luadat, M. H., Luadat, P., *et al.,* 1972, Impairment of insulin binding to the fat cell membrane in the obese hyperglycemic mouse, *FEBS Lett.* **25**:339.

Ginsberg-Fellner, F., and Knittle, J. L., 1970, Adipose tissue cellularity and metabolism in newly diagnosed juvenile diabetics, *Soc. Pediatr. Res.* **40**:134.

Ginsberg-Fellner, F., and Knittle, J. L., 1971, Maternal diabetes as a factor in the development of childhood obesity, *Soc. Pediatr. Res.* **41**:197.

Goldfine, I. D., Soll, A., Kahn, C. R., Neville, D. M., Jr., Gardner, J. D., and Roth, J., 1973, The isolated thymocyte: A new cell for the study of insulin receptor concentrations, *Clin. Res.* **21**:492.

Goldrick, R. B., and McLoughlin, G. M., 1970, Lipolysis and lipogenesis from glucose in human fat cells of different sizes: Effects of insulin, epinephrine and theophylline, *J. Clin. Invest.* **49**:1213.

Gordon, E. S., Goldberg, E. M., Brandabur, J. J., Gee, J. B., and Rankin, J., 1962, Abnormal energy metabolism in obesity, *Trans. Assoc. Am. Physicians* **75**:118.

Greenwood, M. R. C., and Hirsch, J., 1974, Postnatal development of adipocyte cellularity in the normal rat, *J. Lipid Res.* **15**:474.

Grossi, M., Bureau of Child Health, New York City Department of Health, personal communication.

Gruenwald, P., and Funakawa, I. L., 1967, Influence of environmental factors on fetal growth in man, *Lancet* **1**:1026.

Hahn, P., 1972, Lipid metabolism and nutrition in the prenatal and postnatal period, in: *Nutrition and Development* (M. Winick, ed.), Chapter IV, John Wiley & Sons, New York.

Hahn, P., and Novak, M., 1975, Development of brown and white adipose tissue, *J. Lipid Res.* **16**:79.

Hall, B., 1975, *Lancet* 1779.

Hammar, J. A., 1895, Contribution to our knowledge of adipose tissue, *Arch. Mikrosk. Anat.* **45**:512.

Heald, F. P., Hunt, E. E., Schwartz, R., Cook, C. D., Elliott, O., and Vajda, B., 1963, Measures of body fat and hydration in adolescent boys, *Pediatrics* **31**:226.

Hirsch, J., and Gallian, E., 1968, Methods for the determinatio.: of adipose cell size in man and animals, *J. Lipid Res.* **9**:110.

Hirsch, J., and Gallian, E., 1969, Cellularity of rat adipose tissue: Effects of growth, starvation, and obesity, *J. Lipid Res.* **10**:77.

Hollenberg, C. H., and Vost, A., 1968, Regulation of DNA synthesis in fat cells and stromal elements from rat adipose tissue, *J. Clin. Invest.* **47**:2485.

Holt, L. E., Jr., and Snyderman, S., 1964, Nutrition in infancy and adolescence, in: *Modern Nutrition in Health and Disease,* Chapter 37, p. 1124, Lea and Febiger, Philadelphia.

Hytten, F. E., 1964, Nutritional aspects of fetal growth, Proc. VI International Congress of Nutrition, p. 59.

Insull, W., Jr., Hirsch, J., James, T., and Ahrens, E. H., Jr., 1959, The fatty acids of human milk. II Alterations produced by manipulation of caloric balance and exchange of dietary fats, *J. Clin. Invest.* **38**:443.

Jackson, R. L., and Kelly, H. G., 1945, Growth charts for use in pediatric practice, *J. Pediatr.* **27**:215.

Kahn, C. R., Neville, D. M., Jr., Gorden, P., Freychet, P., and Roth, J., 1972, Insulin receptor defect in insulin resistance: Studies in the obese hyperglycemic mouse, *Biochem. Biophys. Res. Commun.* **48**:135.

Kahn, C. R., Soll, A., Neville, D. M., Jr., and Roth, J., 1973, Severe deficiency in insulin receptor: A common denominator in the insulin resistance of obesity, *Clin. Res.* **21**:628.

Kalkhoff, R. K., Kim, H. D., Cerletty, J., and Ferrou, G. A., 1971, Metabolic effects of weight loss in obese subjects; changes in plasma substrate levels, *Diabetes* **20**:83.

Katz, D. P., Dixon-Shanies, D., and Knittle, J., 1976, The effects of cell size on the hexokinase multiple enzyme system in human adipocytes, *Clin. Res.* **24:**362A(abstract).

Katzen, H. M., and Schimke, R. T., 1965, Multiple forms of hexokinase in the rat: Tissue distribution, age dependency, and properties, *Proc. Natl. Acad. Sci. U.S.A.* **54:**1218.

Knittle, J. L., and Ginsberg-Fellner, F., 1969, Adipose tissue cellularity and epinephrine stimulated lipolysis in obese and non-obese children, *Clin. Res.* **17:**387.

Knittle, J. L., and Ginsberg-Fellner, F., 1972, The effect of weight reduction on *in vitro* adipose tissue lipolysis and cellularity in obese adolescents and adults, *Diabetes* **21:**754.

Knittle, J. L., Sussman, L., Collipp, P. J., and Gertner, M., 1972, The effect of treatment with growth hormone on glucose tolerance and adipose tissue cellularity in ateliotic dwarfism, American Diabetes Association Meeting, June 1972.

Knittle, J. L., Ginsberg-Fellner, F., and Brown, R., 1977, Adipose tissue development in man, *Am. J. Clin. Nutr.* **30:**762.

Lechtig, A., Delgado, H., and Lasky, R., 1975a, Maternal nutrition and fetal growth in developing countries, *Am. J. Dis. Child.* **129:**553.

Lechtig, A., Yarbrough, C., Delgado, H., Habicht, J., Matorell, R., and Klein, R., 1975b, Influence of maternal nutrition on birth weight, *Am. J. Clin. Nutr.* **28:**1223.

Lesser, G. T., and Zak, G., 1963, Measurement of total body fat in man by simultaneous adsorption of two inert gases, *Ann. N.Y. Acad. Sci.* **110:**40.

Livingston, J. N., Cuatrecasas, P., and Lockwood, D., 1972, Insulin insensitivity of large fat cells, *Science* **177:**626.

McGanity, W. J., Cannon, R. O., Bridgforth, E. B., Martin, M. P., Densen, P. M., Newbill, J. A., McClellan, G. S., Christie, A., Peterson, J. C., and Darby, W. J., 1954, The Vanderbilt cooperative study of maternal and infant nutrition, *Am. J. Obstet. Gynecol.* **67:**501.

Migeon, C. J., Green, O. C., and Eckert, J. P., 1963, Study of adrenocortical function in obesity, *Metabolism* **12:**718.

Monkus, E., Penn-Walker, D., and Novak, M., 1974, Effects of altered gestation on adipose tissue metabolism in the neonate, 4th European Congress Perinatal Medicine, Prague, Czechoslovakia.

Newburgh, L. H., 1952, Control of the hyperglycemia of obese "diabetics" by weight reduction, *Ann. Intern. Med.* **17:**935.

Novak, M., and Monkus, E., 1972, Metabolism of subcutaneous adipose tissue in the immediate postnatal period of human newborns, *Pediatr. Res.* **6:**73.

Novak, M., Melichar, V., and Hahn, P., 1965, Changes in the content and release of glycerol in subcutaneous adipose tissue of newborn infants, *Biol. Neonate* **8:**253.

Novak, M., Melichar, V., and Althan, P., 1968, Changes in the reactivities of human adipose tissue *in vitro* to epinephrine and norepinephrine during postnatal development, *Biol. Neonate* **13:**175.

Novak, M., Monkus, E., Wolf, H., and Stave, U., 1972, The metabolism of subcutaneous tissue in the immediate postnatal period of human newborns, *Pediatr. Res.* **6:**211.

Oldham, H., and Sheft, 1951, Effect of caloric intake on nitrogen utilization during pregnancy, *J. Am. Med. Assoc.* **27:**847.

Orth, R. D., and Williams, R. H., 1960, Response of plasma nefa levels to epinephrine infusions in normal and obese women, *Proc. Soc. Exp. Biol. Med.* **104:**119.

Osler, M., 1960, Body fat of newborn infants of diabetic mothers, *Acta Endocrinol.* **34:**277.

Ounsted, M., and Sleigh, G., 1975, The infant's self-regulation of intake and weight gain, *Lancet* 1393.

Owen, G. M., Jensen, R. L., and Fomon, S. J., 1962, Sex-related difference in total body water and exchangeable chloride during infancy, *J. Pediatr.* **60:**858.

Page, E. W., 1957, Transfer of materials across the human placenta, *Am. J. Obstet. Gynecol.* **74:**705.

Parra, A., Schultz, R. B., Graystone, J. E., and Cheek, D. B., 1971, Correlative studies in obese children and adolescents concerning body composition and plasma insulin and growth hormone levels, *Pediatr. Res.* **5:**605.

Parra, A., Schultz, R. B., Graystone, J. E., and Cheek, D. B., 1971, Correlative studies in obese children and adolescents concerning body composition and plasma insulin and growth hormone levels, *Pediat. Res.* **5:**605.

Paulsen, B. L., Richenderfer, F., and Ginsberg-Fellner, F., 1968, Plasma glucose, free fatty acids and immunoreactive insulin in sixty obese children, *Diabetes* **17:**261.

Pearse, W. H., and Sornson, 1969, Free amino acids of normal and abnormal human placenta, *Am. J. Obstet. Gynecol.* **105:**696.

Quaade, F., 1964, Prevention of overnutrition, in: Symposia of the Swedish Nutrition Foundation (G. Blix, ed.), p. 25, Almqvist and Wiksells, Uppsala.

Raben, M. S., and Hollenberg, C. H., 1959, Effect of growth hormone on plasma free fatty acids, *J. Clin. Invest.* **38:**484.

Salans, L. B., Knittle, J. L., and Hirsch, J., 1968, The role of adipose cell size and adipose tissue insulin sensitivity in the carbohydrate intolerance of obesity, *J. Clin. Invest.* **47**:153.

Salans, L. B., Zrnowski, M. J., and Segal, R., 1972, The effect of insulin upon the cellular character of rat adipose tissue, *J. Lipid Res.* **13**:616.

Schteingart, D., and Conn, J., 1965, Characteristics of the increased adrenocortical function observed in many obese patients, *Ann. N.Y. Acad. Sci.* **131**:388.

Scow, R. U., Mendelson, C. R., Zinder, O., Hamosh, M., and Blanchette-Mackie, E. J. 1973, Role of lipoprotein lipase in delivery of dietary fatty acids to lactating mammary tissue, in: *Dietary Lipids and Postnatal Development,* p. 91, Raven Press, New York.

Smith, C. A., 1947, Effects of maternal undernutrition upon the newborn infant in Holland (1944–45), *J. Pediatr.* **30**:229.

Smith, U., 1971, Effect of cell size on lipid synthesis by human adipose tissue *in vitro, J. Lipid Res.* **12**:65.

Stuart, H. D., and Meredith, H. V., 1946, Use of body measurements in the school health program, *Am. J. Public Health* **36**:1365.

Thomson, A. M., 1959, Diet in pregnancy: II. Assessment of the nutritive value of diets, especially in relation to differences between social classes, *Br. J. Nutr.* **13**:509.

Toldt, C., 1870, Contribution to the histology and physiology of adipose tissue, *Sitzungsber. Akad. Wiss. Wien. Math. Naturwiss. Kl.* **62**:445.

Vaughn, M., and Steinberg, D., 1963, Effect of hormones in lipolysis and esterification of free fatty acids during incubation of adipose tissue *in vitro, J. Lipid Res.* **4**:193.

Walker, D. G., 1971, Development of enzymes for carbohydrate metabolism, in: *The Biochemistry of Development,* Chapter IV (P. Benson, ed.), J. B. Lippincott, Philadelphia.

Wasserman, F., 1926, The fat organs of man: Development, structure and systematic place of the so-called adipose tissue, *Z. Zellforsch. Mikrosk. Anat. Abt. Histochem.* **3**:325.

Wasserman, F., 1927, On the formation of appendices epiploicae in man, with consideration of the development of the fat organs therein, *Ergeb. Anat. Anz.* **63**:155.

Wasserman, F., and McDonald, T. F., 1960, Electron microscopic investigation of the surface membrane structures of the fat cell and of their changes during depletion of the cell, *Z. Zellforsch. Mikrosk. Anat. Abt. Histochem.* **52**:778.

Widdowson, W. M., 1950, Chemical composition of newly born mammals, *Nature* **166**:626.

Wittels, B., and Bressler, R., 1965, Lipid metabolism in the newborn heart, *J. Clin. Invest.* **44**:1639.

Wolf, H., Stave, U., Novak, M., and Monkus, E. F., 1974, Recent investigation on neonatal fat metabolism, *J. Perinat. Med.* **2**:75.

Wolff, O. H., 1955, Obesity in childhood, *Q. J. Med.* **24**:109.

Zinder, O., and Shapiro, B., 1971, Effect of cell size on epinephrine and ACTH-induced fatty acid release from isolated fat cells, *J. Lipid Res.* **12**:91.

12

Bone Growth and Maturation

ALEX F. ROCHE

1. Introduction

"Growth" is often defined as an increase in size, but more specificity is needed. An increase in size of the whole body, or of an organ such as a bone, may be due to one or a combination of three processes: (1) Hyperplasia or an increase in cell number. This involves duplication of DNA and cell division. (2) Hypertrophy or an increase in cell size. True hypertrophy implies an increase in the size of the active functional elements of a cell as occurs in skeletal muscle with exercise. (3) Storage of organic or nonorganic materials within or among cells. Each of these three processes occurs during bone growth, but the extent to which any one of them dominates depends upon age and the part of the bone considered. Furthermore, these processes are often reversed in localized areas. While the over-all change is an increase in size, some parts may become smaller or be removed completely. In such areas, there is a reduction in cell number or cell size or a reduction in the amounts of organic or nonorganic material stored, or, more commonly, a combination of all these processes.

Maturation is more difficult. A widely used American dictionary defines maturation as: "the process of becoming mature"; unfortunately it defines mature as "having undergone maturation." In regard to bone, maturation refers to the sequential alterations in this tissue as it changes from an embryonic anlage of a bone until each of its parts attains adult form and functional level.

The growth and maturation of a typical long bone and an irregularly shaped bone will be considered from the early prenatal period to early adulthood. The changes during the growth and maturation of these areas occur in most other parts of the skeleton also, with modifications involving the extent of the changes rather than their type or sequence. Separate descriptions are necessary for two complex regions: the craniofacial area and the pelvis. The former area is considered else-

---

ALEX F. ROCHE • The Fels Research Institute, and Department of Pediatrics, Wright State University School of Medicine, Yellow Springs, Ohio.

where in this treatise (Chapter 13); the latter has been described in detail by several authors (Reynolds, 1945, 1947; Harrison, 1958*a,b*, 1961; Coleman, 1969). Attention will not be given to factors that regulate bone growth—these have received adequate attention elsewhere (Bourne, 1971).

2. Prenatal Growth and Maturation of a Long Bone

The following description is appropriate for most long bones, although the timing of changes differs among these bones. Timing before birth is expressed as postconceptional age. Alternative schemes are based either on crown–rump length or the maturity horizons of Streeter (1951); all these classification criteria are closely correlated.

Where a long bone will develop, the first visible change is a condensation of mesenchyme. This packing of cells that were at first loosely arranged occurs at about 7 weeks of fetal life; soon after this the cells secrete compounds characteristic of cartilage. This occurs first in the central part of the condensed area; it is associated with a separation of the central cells and compression of the marginal cells into a membrane called the perichondrium. The inner cells of this membrane remain undifferentiated and retain the potential to form cartilage. The cells in the outer, more obvious layer change to fibroblasts and sharply define the shape and size of the model of the future bone (Olivier, 1962). Even when first formed, this model resembles the adult bone in shape (Lewis, 1902; Hesser, 1926), although its ends are relatively wider (Olivier and Pineau, 1959; Gardner and Gray, 1970) and there are differences in structural details. For example, the linea aspera of the femur is not present until the third prenatal month (Gardner and Gray, 1970). With chondrification, there is an indication of future joints in areas where the mesenchyme persists as homogeneous interzones.

The model of a long bone enlarges by apposition from the deep layer of the perichondrium that surrounds it completely. This apposition involves the formation of chondrocytes (cartilage cells). The bone grows also by the division of chondrocytes within the model and by an increase in the intercellular material. Newly formed chondrocytes are concentrated at each end of the model where the cells are densely packed. Rapid growth in these areas results in elongation of the model and increase in its width. The increase in width is more rapid at the ends than in the central part of the model, partly because the central cells do not divide. These central cells grow old, enlarge, develop vesicles and later vacuoles in their cytoplasm, and finally disintegrate. While these changes occur, the older chondrocytes secrete alkaline phosphatase and the intercellular substance around them becomes calcified (Niven and Robison, 1934; Fell and Robison, 1934) (Figure 1A). Usually this calcified area is too small to be visible radiographically, but exceptions occur in the vertebral bodies (Hadley, 1956).

Ossification begins around the central part of the model after this area of cartilage is calcified (Figure 1B). At first, osteoid tissue (uncalcified bone matrix) forms deep to the perichondrium as a collar. This calcifies quickly to form periosteal bone in contact with the cartilage. The membrane external to this bone is called the periosteum; the deep cells in the periosteum are osteoblasts (bone-forming cells) that can become osteocytes (bone cells).

The calcified cartilage in the central part of the model is now vascularized by masses of cells that extend quickly from the periosteum through the periosteal bone. These masses contain osteoclasts (bone-removing cells) and undifferentiated cells. The cartilage cells are destroyed and blood vessels form over an area that elongates in the major axis of the bone. Some of the cells differentiate to become osteoblasts; others become blood-forming cells. The osteoblasts form bone around the remains of the calcified cartilage (Rambaud and Renault, 1864; Gray *et al.,* 1957; Brunk and Sköld, 1962; Gardner, 1971) (Figure 1B,C). Some, e.g., Ham (1974), illogically restrict the term ''center of ossification'' to this early endochondral ossification; the term should be applied to each discrete site of ossification. For example, in the femur, there are subperiosteal and endochondral centers near the midshaft and other endochondral centers form much later near the ends of the bone. It is important that these centers of ossification should not be called ''growth centers'' because the latter term connotes control over subsequent growth (Koski, 1968).

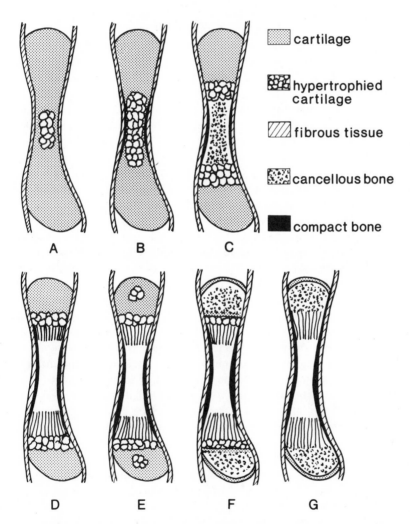

cartilage

hypertrophied cartilage

fibrous tissue

cancellous bone

compact bone

Fig. 1. A diagram of the maturation of a long bone in which the length of the bone has been kept constant. The approximate age scale is: (A) sixth prenatal week, (B) seventh prenatal week, (C) twelfth prenatal week, (D) sixteenth prenatal week to two years, (E) 2–6 years, (F) 6–16 years, and (G) adulthood. The clear area in D–G represents the marrow cavity (based on Roche, 1967).

The ossified area extends toward the ends of the model both within the model (endochondral) and on its surface (subperiosteal; Figure 1C,D). The endochondral bone consists of trabeculae (spicules) with cores of calcified cartilage. These trabeculae are separated from each other by vascular tissue. The subperiosteal bone is more dense and is called "compact bone," although it is penetrated by numerous channels that contain blood vessels (Haversian systems). Near each end of the model, just beyond the ossified area, there is a narrow zone of calcified cartilage and next to it, still nearer the end, a zone in which the chondrocytes are hypertrophied.

Periosteal ossification extends more rapidly than endochondral ossification until it reaches the level of the zone of hypertrophied chondrocytes. While the ossified area is extending to occupy relatively more of the cartilaginous model, the model itself is elongating and becoming wider. Elongation is mainly due to the division of chondrocytes near the ends of the model and an increase in the intercellular substance. In the humerus, for example, the length of the ossified part of the model increases from 19% of the total at 10 weeks to 46% at 12 weeks, 71% at 17 weeks, and 79% at term (Gray and Gardner, 1969).

Soon the central trabeculae are absorbed with the formation of a marrow cavity that elongates toward the ends of the bone and enlarges laterally as the bone becomes wider (Figure 1D). The trabeculae around this marrow cavity widen as bone is deposited on their surfaces. By about the third month of fetal life, the bone has reached a level of maturation that is relatively fixed until birth or later (Figure 1D). This is in agreement with the generalization that organogenesis occurs mainly during the first three prenatal months.

The ossified shaft (diaphysis) now has a dense cortex that surrounds trabeculae and a marrow cavity. The ends of the bone are cartilaginous and, at the junctions between cartilage and bone (the future epiphyseal discs), there are transverse zones of hypertrophied chondrocytes that cover both the dense outer cortex and the loose cancellous bone (trabeculae) over the ends of the marrow cavity. These hypertrophic cells are in columns that have their long axes approximately parallel to that of the bone. The columns are separated by calcified cartilage strips around which bone is deposited. The importance of this area in bone elongation will be considered later. The radiographic appearance of the hand–wrist area in the early prenatal period is illustrated in Figure 2.

Bone growth is dependent not only on apposition but resorption. Without the latter, a normal adult shape would not be attained. One obvious aspect of remodeling concerns the width of the shaft. This flares near the cartilaginous end of the bone, and its width at this level is reduced as the bone elongates and the wider area is incorporated into a more central, narrower part of the shaft. This reduction in width involves the resorption of bone from the external surface of the cortex and concomitant apposition of bone on its deep surface. Due to these processes, trabeculae formed in cartilage are incorporated into the cortex which is of periosteal origin. In some areas, this change is so marked that the whole periosteal cortex is replaced by endochondral bone.

Remodeling, which occurs by a combination of apposition and resorption, is necessary also to retain the relative positions of muscular prominences and other features on the external surfaces of bones (Amprino and Cattaneo, 1937; Lacroix, 1951; Enlow, 1963; Gardner and Gray, 1970). Without this remodeling, features nearer the midpoint of the diaphysis than the epiphyseal zone would become relatively closer to this midpoint. There is also considerable internal remodeling

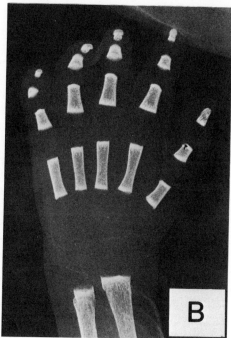

Fig. 2. Radiographs of silver-impregnated hand–wrists in fetal life. (A) 12 weeks: All diaphyses have ossified except middle phalanx V, but proportionately large areas are not yet radiopaque. (B) 16 weeks: Note the considerable increase in the proportion of the hand–wrist area that is radiopaque and the progression toward adult proportions in the diaphyses.

throughout life that involves a rearrangement of the Haversian systems in the cortex and of the trabeculae. Further discussion of this important aspect will be omitted, but the reader is referred to the early work of Amprino and Bairati (1936) and a recent review by Lacroix (1971).

3. Postnatal Growth and Maturation of a Long Bone

The major early change from the prenatal state is the onset of ossification in the cartilaginous ends of the bone. These endochondral centers develop separately from the ossified shaft and are called "epiphyses" or "epiphyseal centers of ossification" (Figure 1F). In some bones, several centers develop at one end. For example, in the cartilage at the proximal end of the humerus there are separate centers for the head, greater tuberosity, and lesser tuberosity; these enlarge and fuse, forming a compound epiphysis.

At the site where ossification will occur, the chondrocytes enlarge and become vesicular. The intercellular substance in this area calcifies and is invaded by vascular buds from adjacent cartilage canals (Figure 1E). Most of the calcified cartilage is removed; bone forms around the remnants (Figure 1F). Histologically, these changes are the same as those that occur when an endochondral center of ossification forms in the shaft. The ossified epiphyseal area enlarges rapidly within the cartilage, after the cartilage around the ossified area has passed through the same sequential changes that occurred in the first part of the cartilage to be ossified. Ossification does not extend throughout the cartilage at the end of the bone. The extreme end remains cartilaginous throughout life as articular cartilage and, for a long time, a disc of cartilage remains between the shaft and the ossified epiphysis. This disc is essential for further elongation; it persists until the bone reaches its adult length (Figure 1G).

At first, the ossified epiphysis is spherical and therefore appears circular in radiographs. Later its shape changes to approximate that of the cartilaginous end of the bone (Todd, 1930a; Greulich and Pyle, 1959; Pyle et al., 1961; Gardner, 1971) when it is curved on all its surfaces except for the approximately flat aspect facing the shaft. This flat surface is covered by a thin layer of bone called the terminal plate, which was noted in radiographs as early as 1910 (Hasselwander, 1910). Only for a brief period after the onset of ossification does the ossified area enlarge, to any marked extent, on the surface adjacent to the epiphyseal zone (Payton, 1933; Siegling, 1941; Haines, 1975).

Soon after ossification begins, the epiphysis widens much more rapidly than the shaft. Consequently, the ratio between epiphyseal width and shaft width (the latter being measured at the end of the shaft, i.e., the metaphysis) is very useful in assessing skeletal maturity (Mossberg, 1949; Scheller, 1960; Murray et al., 1971; Tanner et al., 1975; Roche et al., 1975b), although it may be misleading in some pathological conditions, especially rickets (Acheson, 1966). While the ossified epiphysis widens, the epiphyseal disc also increases in width by the proliferation of chondrocytes deep in the cartilaginous epiphysis and in the margin of the epiphyseal disc and by the hypertrophy of chondrocytes in the disc (Heřt, 1972).

4. The Epiphyseal Disc

The layer of the disc near the ossified epiphysis consists of resting cartilage. A little nearer the diaphysis, the chondrocytes are arranged in columns that have their long axes parallel to that of the bone. The chondrocytes in the columns divide and later mature before they hypertrophy, thus expanding the disc lengthwise. Soon after their maturation is complete, the nearby cartilage is calcified and the chondrocytes die. Most of the calcified cartilage is resorbed and ossification occurs on the surface of the remnant. These latter changes occur on the diaphyseal aspect of the epiphyseal disc. After the ossified epiphysis of a long bone is well developed, elongation of the shaft occurs only by the formation of bone on the shaft side of the cartilage of the epiphyseal disc. The disc remains approximately constant in thickness despite the growth changes in this area. The rate at which chondrocytes divide and enlarge is approximately balanced by the rate at which they die and are replaced by bone.

A short bone, e.g., metacarpal II, has an epiphysis at one end only. The nonepiphyseal end of the shaft is covered by articular cartilage and chondrocytes divide at the junction between this cartilage and the bone. Consequently, some elongation of the shaft occurs at the nonepiphyseal end. In a short bone, between 2 and 11 years, about 20% of total elongation occurs at the nonepiphyseal end (Roche, 1965; Lee, 1968).

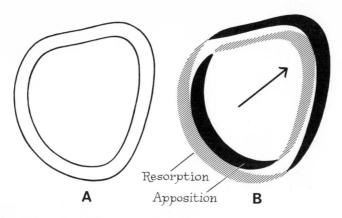

Fig. 3. A diagrammatic representation of remodeling of the shaft of a long bone resulting in cortical "drift" in the direction of the arrow.

5. Increase in Shaft Width

Increase in shaft width occurs by the apposition of bone on the external surface due to the activity of osteoblasts in the deep layer of the periosteum. At the same time, bone is resorbed from the internal surface of the periosteum. As a result, the external diameter of the shaft and the diameter of the marrow cavity enlarge, but the cortex does not necessarily increase in thickness. In general, however, cortical thickness does increase with growth (Garn, 1970). In most bones, the processes of external apposition and internal resorption are unevenly distributed on the various aspects of the shaft (medial, lateral, anterior, posterior) with resultant changes in the position and shape of the cortical bone. Such a change is shown diagrammatically in Figure 3A, which represents the earlier cross section of a bone in which apposition and resorption occur on different aspects of the external surface of the shaft with the opposite changes occurring similarly on the internal surface of the cortex. As a result, the shaft drifts in the direction of the arrow while retaining the same shape. In general, the areas of apposition and resorption are recognized from histological sections of dried bones (Enlow, 1963), but limited inferences can be made from serial radiographs if bone scars are present (Garn *et al.*, 1968).

6. Epiphyseal Fusion

At early ages, the end of the shaft consists of the surrounding cortex and the ends of trabeculae (Figure 4). Later, the end of the shaft is covered by a thin transverse layer of bone that separates it from the cartilage of the epiphyseal disc (Hasselwander, 1910; Hellman, 1928; Todd, 1930*b;* Moss and Noback, 1958; Silberberg and Silberberg, 1961). This layer of bone, together with the calcified cartilage on its epiphyseal aspect, is visible radiographically, and its thickness is related to the rate of growth in stature (Park, 1954; Edlin *et al.*, 1976).

The division of chondrocytes in the disc becomes slower and finally ceases. During this phase, the epiphyseal disc becomes thinner and some of the elongation

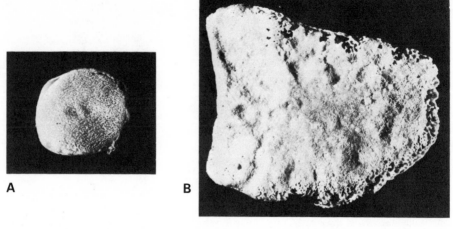

Fig. 4. The distal end of the diaphysis of the radius, viewed from the "epiphyseal" aspect, at 14 weeks of intrauterine life (A) and at 10 years (B). Before birth, the trabeculae of endochondral bone are not covered by a transverse layer of bone.

of the shaft occurs at the expense of the epiphyseal cartilage (Noback *et al.,* 1960). Finally all the cartilage of the epiphyseal disc is replaced by bone. This occurs first peripherally or eccentrically—not necessarily in the central part of the disc (Stevenson, 1924; Todd, 1930*b;* McKern and Stewart, 1957; Haines and Mohiuddin, 1959; Haines, 1975). The last part of the epiphyseal disc to be replaced by bone is the circumferential margin. Consequently, in dried bones at the stage of nearly complete fusion, the epiphysis is fused to the shaft over a large area, but a circumferential groove remains at the level of fusion where the cartilage has not been replaced by bone.

In the area where fusion occurs, a transverse layer of bone forms that is a combination of the terminal plate of the epiphysis and the plate of bone that covered the end of the shaft. This compound plate is visible radiographically. Usually it is resorbed soon after the completion of fusion but it may persist for years later (Paterson, 1929; Sahay, 1941). After fusion is complete, the bone is adult in maturity although there may be a radiopaque line at the level of fusion. The articular cartilage remains as the only remnant of the original model.

7. Growth and Maturation of an Irregularly Shaped Bone

In an irregularly shaped bone such as a carpal, ossification usually begins after birth. The early stages of chondrification and ossification are histologically the same as those that occur within the models of long bones. Each carpal bone develops first as a condensation of embryonic connective tissue. Subsequently, cavitation occurs in this connective tissue at the sites of future joints between the carpal and neighboring bones. After this occurs, the cartilaginous model, now resembling the future bone in shape, articulates with its neighbors on adjacent surfaces and is covered by a well-defined perichondrium on other surfaces (Figure 5B). It is invaded by blood vessels within cartilage canals, but further changes do not occur until ossification begins. Ossification starts in this cartilaginous model (Figure 5C)

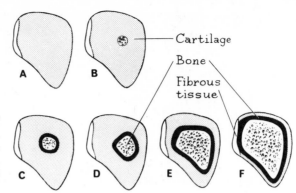

Fig. 5. A diagram of the maturation of a carpal bone in which size has been kept constant. The approximate age scale is: (A) prenatal, (B) 1–3 years, (C) 4–5 years, (D) 6 years, (E) 10 years, and (F) adulthood.

with the same histological processes as those described for endochondral ossification of long bones (Siffert, 1966). Occasionally, some carpals and tarsals have two primary centers, and the calcaneum always develops an ossified epiphysis. With the exception of the latter bone, the carpals and tarsals lack epiphyseal zones. At first, the ossified area expands rapidly in all directions (Gardner, 1971); later growth is more rapid in some directions than others. The changes as the ossified area gradually matures to match the shape of an adult carpal (Figure 5D–F) are used to assess the skeletal maturity levels of children from radiographs. In adulthood, some surfaces of these bones are covered by periosteum and the others by articular cartilage which is the only remnant of the cartilaginous model.

8. Radiographic Changes

Calcified cartilage and bone are radiopaque; thus very small radiopaque areas may be calcified cartilage or bone surrounded by calcified cartilage. The only reliable radiographic evidence that bone is present is the recognition of trabeculae. As the ossified area enlarges, the changes in size and shape are visible radiographically in a two-dimensional view, but only those relating to shape are useful for the assessment of maturity. Measurements of size can be made with accuracy from radiographs, although the circumstances under which the radiographs were taken must be known so that appropriate corrections for enlargement can be made (Tanner, 1962). The changes that occur during remodeling are not directly visible radiographically, although their occurrence can be inferred. Except insofar as they are reflected in changes of shape, they are not useful in the assessment of skeletal maturity.

9. Reference Data for Size and Shape

Bone lengths can be measured in various ways. Measurements of dried bones are inappropriate as a source of reference data because of obvious problems due to "sample" selection. Radiographic measurements are more reliable than those made on the living ("classical anthropometry") because of the difficulty of identifying landmarks and because measurements on the living have to be made through

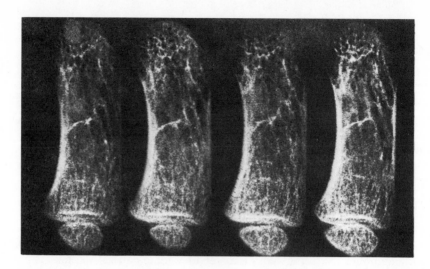

Fig. 6. Serial radiographs at 6-monthly intervals of a first metacarpal with the photographic enlargement adjusted to keep the total length constant. Note the fixed trabecular patterns that become relatively more distal with growth.

overlying soft tissues. This difference in reliability has been demonstrated convincingly by Maresh and Deming (1939) and by Day and Silverman (1952). One exception may concern the ulna; it has been reported that the length of this bone can be measured accurately without radiographs using the apparatus described by Valk (1971, 1972).

The generalization that measurements on radiographs are preferable to those in the living is true only when the radiographs are taken under standardized conditions in which orientation is fixed and enlargement is known. A description of appropriate radiographic techniques and discussions of the principles involved have been given by Goff (1960) and Tanner (1962).

The recommended reference values for the lengths of the major long bones are those of Maresh (1970). Dial calipers with sharp points should be used to measure radiographs and the distances should be recorded to the nearest 0.1 mm.* This level of accuracy is necessary for many research purposes. In other circumstances, useful data can be obtained using a transparent rule graduated to 1.0 mm; such a rule is adequate for the measurements required in the TW2 and RWT methods of assessing skeletal maturity. Serial standardized radiographs can be used also to determine the amounts of elongation at particular sites. For this purpose, natural bone markers, e.g., growth arrest lines, notches, or fixed trabecular patterns, are used as landmarks (Roche, 1965; Lee, 1968) (Figure 6).

Reference values for the lengths of the bones of the hand from birth to 15 months have been reported by Gefferth (1972) and from 2 years to adulthood by Garn *et al.* (1972*a*). Gefferth included the epiphyses whenever they were ossified; consequently, his reported lengths are dependent upon levels of maturity. Garn and his associates included the epiphyses in the distances measured. These were the

---

*One suitable instrument is Helios caliper, No. 4; $99.95; Mager Scientific Inc., 209 East Washington, Ann Arbor, Michigan 48108.

maximum lengths in the long axis of each diaphysis. The data of Garn *et al.* (1972*a*) have been used to construct profiles of lengths within individual hands, and the deviations from these reference profiles in several diseases, e.g., hand–foot–uterus syndrome, Holt-Oram syndrome, and Turner's syndrome, have been reported (Poznanski *et al.*, 1972*a,b*).

Some proportions involving the lengths of the short bones of the hand are important also in the clinical diagnosis of Turner's syndrome because the fourth metacarpal is short in many of these XO individuals. This can be detected without measurement either by placing a ruler across the medial two knuckles of the clenched fist as described by Archibald *et al.* (1959) or by drawing a line on a radiograph corresponding to the plane of the ruler. As illustrated in Figure 7, in a normal individual, a tangent to the most distal points on the heads of the fifth and fourth metacarpals passes distal to the head of the third metacarpal. In those with Turner's syndrome and in some normal individuals, the tangent passes across the head of the third metacarpal. The shortness of the fourth metacarpal in Turner's syndrome can be identified also by comparing the length of the fourth metacarpal with the combined lengths of the fourth proximal and distal phalanges in the same hand (Kosowicz, 1965).

Comparisons between lengths are used to determine whether brachymesophalangia (shortness) of the fifth finger is present (Figure 8). Usually this can be recognized subjectively, but it is better to use measurements. The best choice is the ratio between the lengths of the fifth and fourth middle phalanges (Hewitt, 1963; Wetherington and Haurberg, 1974). This is preferable to criteria based on comparisons between the lengths of the fifth middle phalanx and other bones of the fifth ray because the correlation coefficients between the lengths of the bones of the hand are higher among bones in the same row, e.g., metacarpals, than among bones in the same ray, e.g., metacarpal II, proximal phalanx II, middle phalanx II, and distal phalanx II (Roche and Hermann, 1970). Reference values are given in Table I.

Reference values for bone widths have been reported for metacarpal II (external diameter, internal diameter, cortical thickness, and calculated cortical area) in white American children by Garn *et al.* (1971). The measurements were made at the midpoint of the bones, including the epiphyses; cortical areas were calculated on

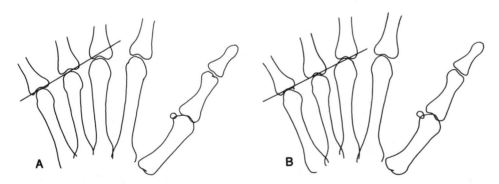

Fig. 7. Diagrams of the hand–wrist: (A) a tangent to the heads of the fifth and forth metacarpals usually passes distal to the head of the third metacarpal (negative metacarpal sign) and (B) a similarly constructed tangent passing across the head of the third metacarpal because the fourth metacarpal is short (positive metacarpal sign).

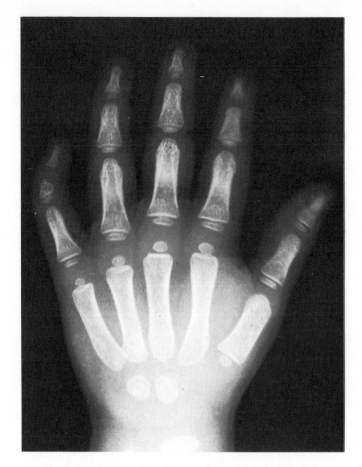

Fig. 8. An example of brachymesophalangia of the fifth digit.

Table 1. Values for the Ratio (Middle Phalanx V/Middle Phalanx IV)
Lengths[a] in the Hands of United States White Children[b]

Age (years)	Males		Females	
	Mean	S.D.	Mean	S.D.
1	0.66	0.06	0.65	0.06
4	0.67	0.05	0.66	0.04
9	0.69	0.04	0.68	0.04
Adult	0.73	0.03	0.71	0.03

[a] Lengths include epiphyses except at 1 year.
[b] Data from Garn et al., quoted by Poznanski, 1974.

the assumption that the bones and the medullary cavities were circular. Corresponding data for Swiss children from 3 months to 11 years for metacarpal II and for metacarpals II–IV combined have been reported by Bonnard (1968). The influence of age and nutrition on cortical thickness and other measurements of the second metacarpal have been described fully by Garn (1970).

In general, the reference data for shapes of bones are unsatisfactory. In part this is due to the use of inappropriate methods. These complex shapes cannot be described accurately using a single ratio, e.g., length/width. Recently, bone shapes (judged from radiographic outlines) have been summarized by curve-fitting methods. For example, Fourier analyses have been made that provide efficient summations of shape for both ectocranial and endocranial lateral silhouettes and for the outline of the distal end of the femur using parameters that can be interpreted readily (Lestrel, 1974, 1975; Lestrel *et al.*, 1977; Lestrel and Roche, 1977). Such approaches are suitable for research purposes only—it would be unreasonable to suggest the reported parameters be used clinically as reference values.

Despite their limitations in regard to the description of skeletal shape, simple ratios between length and width can be useful clinically. If length/width ratios are calculated using data from metacarpals II–V combined, width being measured at the midpoint of the bone, the values increase from infancy to adulthood, being slightly lower in males (Sinclair *et al.*, 1960). As expected, the ratio is high in Marfan's syndrome. Kosowicz (1965) reported the ratio between the width of the subungual expansions and that of the shafts of the distal phalanges of the hand. This

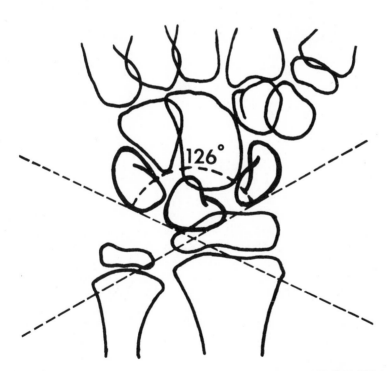

Fig. 9. A diagram of the carpal area to indicate the carpal angle (Kosowicz, 1962, 1965). This is measured between a tangent to the proximal margins of the triquetral and lunate and a tangent to the proximal margins of the scaphoid and lunate.

Table II.   Selected Skeletal Dimensions Measured on Radiographs with Their Clinical Relevance

Dimension	Literature sources	Use
Atlas–odontoid distance	Hinck and Hopkins, 1960; Locke et al., 1966	Subluxation
Spinal canal–A.P. diameter	Naik, 1970; Hinck et al., 1962, 1965	Trauma, tumors, pathology of disc
Vertebral column		
Interpedicular width	Hinck et al., 1966	Achondroplasia
Disc and body height	Brandner, 1970	Pathology of disc
Shoulder, width of joint space	Arndt and Sears, 1965	Dislocation
Acetabular angle	Caffey et al., 1956	Congenital dislocation of hip
Iliac angle and index	Caffey and Ross, 1958	Down's syndrome
Axial relations within the foot	Templeton et al., 1965	Congenital abnormalities

ratio is low in apical dystrophy, craniocleidodysostosis, and cretinism and high in Turner's syndrome (Macarthur and McCulloch, 1932; Brailsford, 1943, 1953; Kosowicz, 1965).

The shape of the carpal area is abnormal in Turner's syndrome. Usually this is determined by measuring the angle between tangents to the medial and lateral parts of the proximal margin of the carpus. The normal range for this "carpal angle" is often given as 124.3° to 138.7° (Kosowicz, 1962, 1965) (Figure 9). It is important to note that it is about 10° higher in American blacks than whites, that it is 5° higher in the right hand than the left, and that it increases 5–10° between 5 and 14 years (Harper et al., 1974). This angle is small in Turner's syndrome.

Many other skeletal dimensions are measured by radiologists. A partial list with the conditions in which they are useful is provided in Table II. Further details are available in the text of Lusted and Keats (1972).

10. Skeletal Weight

Data on skeletal weight have been reported for fetuses, infants, and children (Garrow and Fletcher, 1964; Trotter and Peterson, 1968, 1969a,b, 1970). Both before and after birth, the relationship between skeletal weight and body length approximates a sigmoid function (Figures 10 and 11), with larger increases in weight per centimeter as body length increases. This increase becomes very rapid after 170 cm of body length in males. This may be due to vagaries of sampling, but it could reflect later transverse growth of the trunk. The increase in skeletal weight at larger body lengths is much less marked in females than males. Before birth, the weight of the femur in relation to body length also has an approximately sigmoid function, following a curve that is almost parallel to that for total skeletal weight. This indicates that during fetal life the weight of the femur is a fixed proportion of the total skeletal weight. The relationship between the length of the femur and body length is essentially rectilinear for body lengths ranging from 23 to 176 cm. The means given graphically in Figures 10 and 11 should be accepted as showing general trends only—the sample sizes are very small for all variables in the fetal period and for skeletal weight after birth.

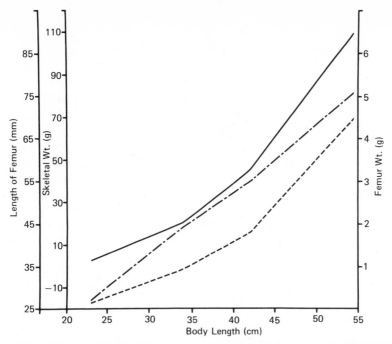

Fig. 10. The relationships of length of femur (—·—·), skeletal weight (——), and weight of femur (----) with body length during the prenatal period in white United States fetuses (based on Trotter and Peterson, 1969*a,b*).

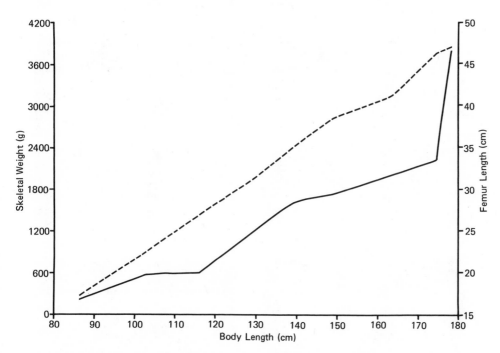

Fig. 11. The relationships of skeletal weight (——) and femur length (---) with body length after birth in white United States males (based on Maresh, 1970; Trotter and Peterson, 1970).

ALEX F. ROCHE

The assessment of skeletal maturation from radiographs is based on the recognition of "maturity indicators." A maturity indicator is a radiographically visible feature that assists the determination of the maturity level of a bone. In the atlases, they are described in the text accompanying each standard; in the method of Tanner *et al.* (1975) they are the "criteria" for stages; in the method of Roche *et al.* (1975*b*) they are called "indicators" and divided into grades. After they have been recognized, a skeletal age is assigned to the radiograph, using a subjective or objective method of "scoring." To be useful an "indicator" must occur during the maturation of every child (Pyle and Hoerr, 1969); that is, be universal. Because of limitations in the schedule of serial radiographic examinations, however, some indicator grades appear to be skipped in some children. Some of the appearances used as maturity indicators are shown in Figures 12 and 13. These illustrations of the distal end of the radius in boys have been enlarged so that the width of the end of the shaft is constant, thus removing over-all size differences. Even to the inexperienced observer, there are clear differences in maturity between epiphyses with widely separated skeletal ages, e.g., 2.5 vs. 3.5 years or 11.0 vs. 14.0 years. It is equally evident that the recognition of the differences between, e.g., 10.0 and 11.0

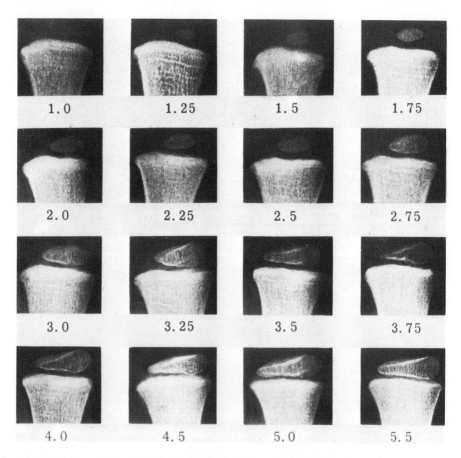

Fig. 12. The distal end of the left radius in boys 1.0–5.5 years with diaphyseal width kept constant. The figure under each insert is the Greulich–Pyle skeletal age.

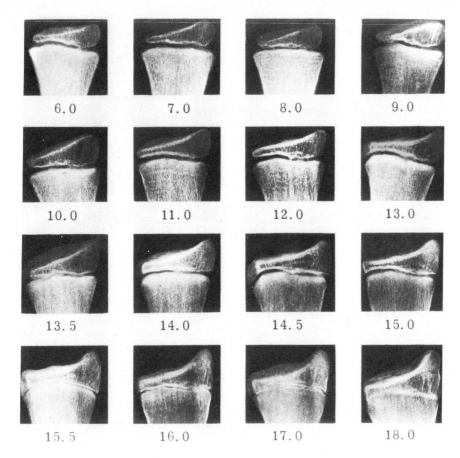

Fig. 13. The distal end of the left radius in boys 6.0–18.0 years with diaphyseal width kept constant. The figure under each insert is the Greulich–Pyle skeletal age.

requires training and care and that small differences, e.g., 7.0 vs. 8.0 have to be carefully distinguished from individual variations in shape that must, of course, be excluded from the system of rating.

Maturity indicators reflect the three-dimensional shapes of calcified or ossified areas. As bones become more mature, their external contours change due to differences in the rates of bone apposition and resorption at various surfaces. Parts of these surfaces that are approximately parallel to the central axis of the radiographic beam cause dense white zones on a radiograph that may be maturity indicators. If these zones are long and narrow, they are called "radiopaque lines." Radiographs used to assess skeletal age must be taken under standardized conditions because the radiographic outline of a bone, or the shadow cast by a radiopaque zone or line, depends on the inclination of the radiographic beam.

Maturity indicators provide information about the level of maturity in a particular bone at the time the radiograph was taken. Serial radiographs for an individual can show the duration of each indicator (or grade of an indicator), that is, the period the bone remained at the same maturity level. Information concerning the duration of indicators (or stages) is useful in developing a scoring system because indicators of long duration are less informative than those of brief duration (Healy and Goldstein, 1976; Roche *et al.*, 1975*b*). Because the schedule of radiography in longitudinal growth studies is related to chronological age, not the inception of maturity indicators, and because some scheduled visits are missed in all long-term series, this property of maturity indicators has been used only indirectly (Roche *et al.*, 1975*b*).

11.1. Number of Centers Method of Assessment

Some early methods for the "assessment" of skeletal maturity were based on the onset of ossification (Sontag *et al.*, 1939; Elgenmark, 1946). These methods have considerable appeal because radiographic positioning does not have to be controlled rigidly, the assessors need only minimal training, and observer errors are infrequent. While these methods are appropriate for preschool children and for areas that include many bones, e.g., hand–wrist, foot–ankle, if both the irregularly shaped bones and epiphyses are included (Yarbrough *et al.*, 1973), they are useless at older ages. One difficulty is that this method gives all centers the same weight despite their differences in representativeness and in the distributions of chronological ages over which the onset of ossification occurs. The problem is exemplified in Figure 14 which shows two hand–wrists with equal numbers of centers ossified that differ by 0.8 years in Greulich–Pyle skeletal age and by 35 points (0.7 years) on the TW2 scale.

11.2. Atlas Method of Assessment

The recognition and subjective grading of maturity indicators is the basis of the atlas method. This method was developed by the Cleveland "school," first by Todd (1937) and then by others who have provided atlases of standards for the hand–

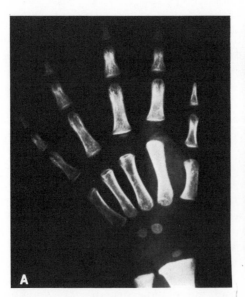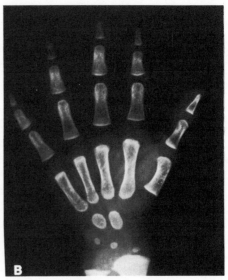

Fig. 14. Hand–wrist radiographs of two boys. Each has five centers ossified in the left hand–wrist area. In each boy, the capitate, hamate, and distal radius are ossified but the other two centers are those for proximal phalanges II and III in boy A, whereas they are the triquetral and lunate in boy B (from Yarbrough *et al.*, 1973).

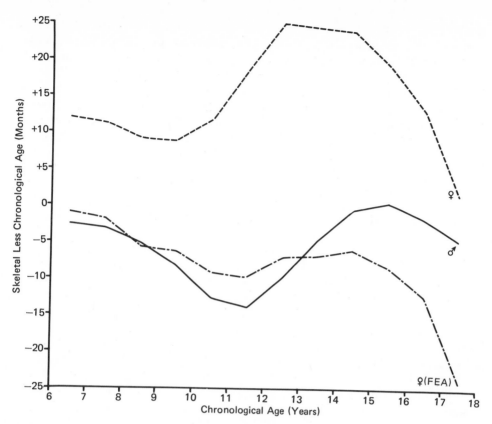

Fig. 15. Mean difference between skeletal age (Greulich–Pyle) and chronological age in a national probability sample of United States children and youth [girls, ----; boys, ———; and girls (FEA), —·—·] (Roche *et al.*, 1975*c*, 1976).

wrist, foot–ankle, and knee (Greulich and Pyle, 1959; Hoerr *et al.*, 1962; Pyle and Hoerr, 1969). The atlas standards were derived from a group of white children of very high socioeconomic status living in Cleveland, Ohio, who were enrolled in the Brush Foundation Study. These classic atlases differ both in the area of the skeleton considered and the manner in which the standards are presented. The Greulich–Pyle atlas for the hand–wrist has separate series of standards for boys and girls, each with their own set of corresponding skeletal ages expressed in years and months. The atlases for the knee and foot–ankle have only one set of standards for both sexes but sex-specific skeletal age equivalents are provided for each standard. The standards show the approximately median levels of maturity for Brush Foundation children with chronological ages matching the skeletal ages assigned to the standards. The children from whom the Cleveland atlases were derived were born between 1917 and 1942, and secular changes might be expected. In fact, the skeletal maturity standards for the hand–wrist in the Greulich–Pyle atlas are markedly in advance of the mean levels of maturity in a national probability sample of United States children between 9 and 13 years (Roche *et al.*, 1974, 1976). This advancement is most marked at 11.5 years when it is 13.8 months in boys and 10.7 months in girls (Figure 15). These data require explanation.

All radiographs taken in this survey were assessed, without the chronological age or the sex being known, against a set of male standards (Pyle *et al.*, 1971) that are essentially the same as the male standards in the Greulich and Pyle atlas. Consequently, the means of the skeletal ages assigned to the girls (Figure 15, ♀) are in advance of the chronological ages but those assigned to the boys (Figure 15, ♂) are less than the chronological ages. The skeletal ages assigned to the girls were transformed to what it was considered they would have been had they been assessed against female standards. These female equivalent ages were obtained using the bone-specific sex differences published by Pyle *et al.* (1971). The mean female equivalent ages (Figure 15, ♀ FEA) are less than the chronological ages, but the differences are smaller than those for the boys. These findings for American children are of great importance to all who use the Greulich–Pyle and other Cleveland atlases. Children in Finland and Denmark also are about 0.7 years retarded in comparison with the Greulich–Pyle standards (Koski *et al.*, 1961; Andersen, 1968; Mathiasen, 1973). Particularly from 9 to 13 years, population groups should not be expected to be equivalent to the atlas standards in maturity unless membership in the population groups is restricted to privileged children.

The unusual protocol that was followed in this national survey has advantages and disadvantages. A totally unbiased estimate of skeletal maturity levels was obtained for boys, but the female equivalent ages are dependent on the accuracy of the sex differences in skeletal maturity reported by Pyle and her associates (1971). Because of the procedure applied, highly accurate estimates of sex differences were obtained (Figure 16). These are close to those reported by Pyle *et al.* (1971), except after the age of 12 years when the actual sex differences are much smaller. In fact, at 17.5 years, the sex difference in skeletal age is only 3.9 months (Roche *et al.*, 1976). This is markedly at variance with previous reports (e.g., Roche, 1968; Pyle *et al.*, 1971; Tanner *et al.*, 1975), but it is what was observed. To repeat, when radiographs

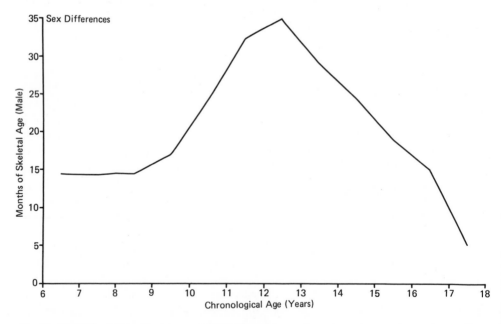

Fig. 16. Sex differences in skeletal maturity (Greulich–Pyle) from a national probability sample of United States children and youth (Roche *et al.*, 1975c, 1976).

were assessed without knowledge of chronological age or sex, against a single set of standards, the mean difference between the ages assigned to boys and girls aged 17.5 years was, in fact, only 3.9 months. The reasons for these variations among studies have been considered in full elsewhere (Roche *et al.*, 1976).

When the atlas method is applied, the radiograph to be assessed is compared with the standards until one is found that is at the same maturity level as the radiograph. The skeletal age recorded for the radiograph is that of the standard it matches. Often it is desirable to interpolate between standards, and the procedure should be applied to individual bones. In the latter case, bone-specific skeletal ages are recorded. The way in which these bone-specific ages are combined affects the over-all skeletal ages assigned to an area (Roche and Johnson, 1969). Few authors, other than some research workers, follow a definite scheme of combination— usually impressions are combined subjectively to a single-area skeletal age. Because the bones in the standards differ in maturity levels, there are differences among the bone-specific skeletal ages assigned to a single standard. This makes the atlas matching process less accurate if bone-specific skeletal ages are not recorded, but it is not known how much information is lost.

The most appropriate combination depends on two factors: the completeness with which the separate bones can be assessed and the purpose for which the area skeletal age is required. Skeletal ages cannot be assigned to bones until the appropriate parts are radiopaque nor can skeletal ages, expressed in years, be assigned to bones that have become adult. Consequently, a mean cannot be obtained for bone-specific hand–wrist skeletal ages except between about 5 and 13 years, but a median can be calculated with some reasonable assumptions if skeletal ages have been assigned to more than half the bones. In theory, one combination could be significantly better than another when the skeletal age assessment is needed to estimate an outcome variable. In regard to the prediction of adult stature, reported findings are not in agreement. One group of workers found the median of all Greulich–Pyle bone-specific skeletal ages more useful than various weighted means of these ages (Roche *et al.*, 1975*a*). Another group, however, reported that a combination of the maturity levels of the radius, ulna, and short bones was more useful in prediction than a combination of corresponding data from the carpals (Tanner *et al.*, 1975).

Some authors (Pryor and Carter, 1938; Francis, 1939; Francis and Werle, 1939; Sontag and Lipford, 1943; Greulich, 1954; Greulich and Pyle, 1959; Hansman and Maresh, 1961) have claimed that the range of bone-specific skeletal ages within a hand–wrist is related to the illness experience of a child; others have reported the contrary (Tiisala *et al.*, 1966; Acheson, 1966). Only recently have reference values for the range of bone-specific skeletal ages been made available (Roche *et al.*, 1977). These ranges, for a national probability sample of American children and youth, decrease in an approximately linear fashion with age except for a slight increase in girls from 8 to 11 years.

The accuracy of the atlas method, whether applied to an area or a bone, is unknown. Considerable data have been reported, however, concerning replicability and comparability among research workers (Mainland, 1953, 1954; Koski *et al.*, 1961; Moed *et al.*, 1962; Johnston, 1964; Acheson *et al.*, 1966; Andersen, 1968; Roche *et al.*, 1970, 1974, 1975*b,c*; Mazess and Cameron, 1971; Sproul and Peritz, 1971; Tanner *et al.*, 1975), among research workers and pediatric radiologists (Johnson *et al.*, 1973), and during the training of assessors (Roche *et al.*, 1970). In general, repeated ratings by experienced assessors of the hand–wrist using the atlas method differ by less than 0.5 years in 95% of cases; the differences between assessors are about 0.8 years in 95% of cases.

It has been claimed that maturity indicators appear in a fixed sequence for each bone (Pyle *et al.*, 1948; Greulich, 1954; Pyle and Hoerr, 1955; Greulich and Pyle, 1959; Tanner, 1962; Acheson, 1966) "no matter what the stock, parentage, economic standing, stature, weight or health of the child" (Todd, 1937). However, differences in sequence occur between parts of a bone, e.g., the distal end of the femur (Roche and French, 1970). If Todd's statement were true, Tanner and his co-workers (1962, 1972, 1975) would not have needed to describe multiple criteria for maturity stages.

The sequence of indicators is far from fixed in a given area (Pryor, 1907; Todd, 1937; Robinow, 1942; Kelly and Reynolds, 1947; Pyle *et al.*, 1948; Abbott *et al.*, 1950; Greulich and Pyle, 1959; Garn and Rohmann, 1960; Christ, 1961; Hewitt and Acheson, 1961; Acheson, 1966; Garn *et al.*, 1966; Poznanski *et al.*, 1971; Yarbrough *et al.*, 1973; Roche *et al.*, 1978). It is clear that not all these differences in sequence are associated with ill health (Sontag and Lipford, 1943; Garn *et al.*, 1961*b;* Acheson, 1966), although this has been suggested (Francis, 1940; Pyle *et al.*, 1948). There is convincing evidence that unusual sequences tend to be aggregated in families (Garn *et al.*, 1966); while this does not prove genetic factors are involved, such an explanation is plausible.

11.3. TW2 Method of Assessment

This alternative to the atlas method is based on assigning numerical scores to bones depending on their levels of maturity. This system was introduced by Acheson (1954), but the original method was unsatisfactory because the bone scores were ordinal and only limited inferences can be made from summing such values (Lord, 1953). The ratings can be recorded as 1, 2, 3, etc., but they cannot be added or subtracted as ordinary numbers. Partly to avoid the tendency to do this, the TW2 stages are given letters: A, B, C, etc. Tanner and his colleagues (1962, 1972, 1975) have improved and extended Acheson's method. Their TW2 method makes use of approximately nine maturity indicators (called "stages") for each of 20 hand–wrist bones. The bones of the second and fourth "rays"* are omitted because their skeletal maturity levels are correlated highly with those of corresponding bones in the other rays (Hewitt, 1963; Roche, 1970; Sproul and Peritz, 1971). However, if an underlying trait, such as skeletal maturity, is to be rated indirectly, some workers consider it preferable to use highly correlated indicators (Wilks, 1938; Wainer, 1976), as implied by Garn in his work on communality indices (Garn and Rohmann, 1959; Garn *et al.*, 1964). Not all agree. Tanner *et al.* (1975) adopted the principle of dropping bones with high correlations because they provide redundant information. They consider that the variations among bones in levels demonstrate differences in the rates of maturation, which is what they are trying to measure. Certainly, an assessment method restricted to highly correlated indicators would omit potentially important information from bones that tend to be divergent. The latter may be sensitive guides to environmental or genetic effects.

*A ray of the hand consists of a metacarpal and its associated phalanges, e.g., metacarpal II, proximal phalanx II, middle phalanx II, and distal phalanx II.

The scores assigned to the stages of each selected bone in the TW2 system were derived by assuming that the observed stages of each bone of the individual all reflect the same underlying quantity, the level of skeletal maturity for the hand–wrist, although the bones may be at different stages. Therefore the scores corresponding to stages were selected to minimize the disparity among the bone-specific scores for the individual, summed over all individuals in the standardizing group. This scoring system implies that the hand–wrist bones of an individual are normally at the same level of maturity, which is in agreement with Todd's (1937) concept of the evenly maturing skeleton. Although, as mentioned earlier, variations among bones are common, this concept receives strong support from a recent analysis of the structure of bone-specific skeletal ages in the hand–wrist, foot–ankle, and knee assessed by the atlas method (Roche *et al.,* 1975*a*).

The TW2 mathematical method required at least two points of scale to be fixed. This was achieved by requiring the average of all the initial stages to be zero and the average of all the final stages to be 100, thus defining a scale from zero to 100. Although the scores applied to each stage of each bone are objectively derived, as described above, an element of choice or subjectivity was introduced into some aspects of the system through the authors' feeling that the presence of many metacarpal and phalangeal bones in the hand–wrist would "swamp" the contribution of the carpals, radius, and ulna. Consequently, they promulgated two separate scores for maturity, one concerning the carpals only and the other the radius, ulna, and finger bones (RUS). In the RUS score, furthermore, they dropped the second and fourth rays to diminish the number of finger bones and then weighted the scores of the remaining finger bones so that they all together only equaled in importance the joint scores of radius and ulna. This "biological" weighting is wholly separate from the mathematical scoring technique, although for convenience of the user the final scores incorporate both. In practice, the RUS system seems more informative than the carpals, although insufficient experience has been accumulated for such a judgment to be certain.

The TW2 system assumes a fixed sequence of maturity indicators for each bone, but some stages have multiple criteria. A particular stage is allocated to a bone only after the first criterion of the preceding stage is present. For example, the maximum diameter of the radial epiphysis may be less than half the width of the radial diaphysis, and the epiphysis may not be thicker on its lateral side in an early radiograph of a child. Because the first criterion of stage D is not met, stage C would be assigned even if the criterion for stage E were present (the limit of the palmar surface visible as a radiopaque line).

The maturity indicators in the TW2 method are generally satisfactory, except those based on relationships between the diameters of bones or the distances between bones. With such indicators, a pair of bones is being graded. In addition, some raters have been confused by the instruction: "If at a particular stage a feature described in brackets is not present, this does not affect the rating: if it is, this confirms, but does not diagnose, the stage in question" (Tanner *et al.,* 1962, 1975). The descriptions in brackets could have been omitted without loss. In addition, there are problems at the upper end of the Tanner–Whitehouse scale where a difference of one grade for one bone can alter the skeletal age for the whole hand–wrist by as much as 0.8 years, even in the revised method (Tanner *et al.,* 1975).

There is convincing evidence that the TW2 system is at least slightly more reliable than the Greulich–Pyle method (Acheson *et al.,* 1963, 1966; Johnston, 1963; Johnston and Jahina, 1965; Roche *et al.,* 1970; Johnson *et al.,* 1973).

In the TW2 system, the bones are rated individually in a fixed order. Due to the objectivity of this system, this presents little problem, whereas difficulties occur with the more subjective atlas method because the skeletal ages assigned to the bones assessed first can influence the ages assigned to the remaining bones. This problem is not overcome by assessing in a random order as done by Sproul (Sproul and Peritz, 1971; Peritz and Sproul, 1971). The subject has been examined recently (Roche and Davila, 1976) using radiographs masked so that only one bone was visible at a time. In comparison with the findings when the whole of each radiograph was visible, both comparability and reliability of bone-specific skeletal ages were reduced; this reduction was particularly marked for the hamate and triquetral. Furthermore, when only one bone was visible at a time during assessment, the ranges of bone-specific skeletal ages within individual hand–wrist were increased.

11.4. RWT Method of Assessment

The Roche–Wainer–Thissen method is for the assessment of the knee (Roche *et al.*, 1975*b*); a corresponding system for the hand–wrist is being elaborated. The RWT method was developed using anteroposterior knee radiographs of "normal" southwestern Ohio children enrolled in the Fels Longitudinal Study. These radiographs had been taken between 1932 and 1972, but significant secular effects are absent. This was shown by correlating skeletal age (both RWT and Greulich–Pyle) with year of birth and by comparing the skeletal maturity levels of parents and children of the same sex at the same chronological age.

The RWT method, like all others, is based on maturity indicators. The authors listed all the maturity indicators reported, made the descriptions anatomically correct, and developed grading methods that were as objective as possible. Some indicators are graded by comparison with photographs and drawings, others by fitting standard curves to margins, and the remainder by measuring distances to obtain ratios.

The useful indicators that were retained were shown to be reliable, i.e., inter- and intraobserver differences were small. These indicators discriminated; i.e., they were present in some but not all children during a particular age range. This feature is age-limited: an indicator will discriminate at some ages but not others.

By definition, each grade of every indicator must occur during the maturation of an individual. Consequently, in an indicator with multiple grades, there must be ages when either the least mature or the most mature grade is universal although, because children mature at different rates, there may be no age when a particular intermediate grade is present in all the children. When this occurred, serial radiographs were reviewed to be sure that the intermediate grade did indeed occur in the sequence of changes in each child. The criterion of universality was applied to the ratios by determining whether they became approximately constant at older ages. Although serial radiographs were used for this and some other purposes, the raw data were recorded without reference to other radiographs of the same child. The aim was to develop a method that would be applicable when only one radiograph was available.

The final indicators were valid. In the present context, validity refers to the quality by which a radiographic feature indicates skeletal maturity. *A priori*, the bones of a child must become more mature with increasing chronological age. Therefore, as pointed out by Takahashi (1956), the prevalence of grades of valid maturity indicators must change systematically with age until the most mature grade is universal. Prevalence data can provide only a general guide to validity because some children miss visits. A better measure is obtained by reviewing serial data for individuals. "Reversals" were sought in these serially organized data, i.e., changes in indicators with increasing age that were in the reverse direction to those expected from the group trends. This procedure is similar to that adopted by Hughes and Tanner (1970) when developing an assessment method for the rat and the method used in constructing the TW2 stages.

Useful maturity indicators must have the quality of completeness. This refers to the extent to which the indicator can be graded in a series of radiographs. As with some other criteria, an indicator may have this quality at some ages but not others.

The parameters used to construct the RWT scale are the chronological age at which each indicator is present in 50% of children and the rate of change in each indicator's prevalence with age. These parameters were combined to a single continuous index using latent trait analysis (Birnbaum, 1968; Samejima, 1969, 1972). This index was scaled so that the mean and variance of skeletal age in the standardizing group was the mean and variance of chronological age.

This statistical method separates the within-age variance into two components. One is attributable to real variations in maturity level; the other is attributable to sampling error which is inherent in any scheme. Only the RWT method estimates this error component. The standard error tends to be small when many highly informative indicators are assessed that have grades with the age of 50% prevalence close to the chronological age considered. The standard error is large in the contrary circumstances and also when unusual combinations of indicator grades are observed.

The method is based on 34 maturity indicators for the femur, tibia, and fibula. At any particular age, only about half of these must be graded because there are limits to the age ranges during which each is useful. After the grades have been

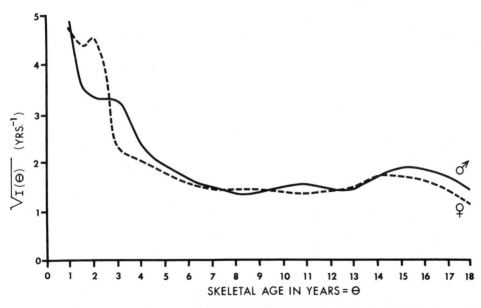

Fig. 17. The total information system curves for the RWT method (knee) in each sex. I = the inverse of the error variance at θ; θ = skeletal age (from Roche *et al.*, 1975*b*).

recorded, they are transferred to a computer. Published programs provide the skeletal age appropriate for the radiograph together with the standard error of the particular estimate. The errors tend to be larger at some ages than others as shown by the total information curve (Figure 17). The information available from a knee radiograph is relatively great at 2 years, decreases to 6 years, and then remains approximately constant.

The RWT system has advantages and disadvantages. It is very helpful to know the accuracy of each assessment, and the model allows the addition of corresponding parameters for other variables, e.g., hand–wrist indicators. Reliability and comparability are about the same as with the TW2 method and generally better than with the atlas method. Although an assessment does not require much time, there must be access to a computer.

11.5. Scale of Maturity

The scale applied in each method for the assessment of skeletal maturity is based on the assumption that skeletal maturity is zero at conception and that all individuals reach the same level of complete maturity in young adulthood. The endpoint of the scale is the completion of epiphyseodiaphyseal fusion in bones with epiphyses and the attainment of adult shape in bones that do not develop epiphyses. Thus, each individual achieves the same amount of maturation between conception and adulthood, although their rates of maturation differ. In practice, the scale begins with the onset of ossification in epiphyseal centers or in the primary centers of irregular bones.

All maturity scales are divided into years of skeletal age, except in the TW2 method. However, even the TW2 scores are usually transformed to skeletal ages using published tables or graphs. Skeletal-age "years" are not known to be equivalent to each other even when the same system of assessment is used for the same area of the skeleton. These ordinal scales do not provide a measure of skeletal maturity, but they allow a maturation level to be assigned to a radiograph relative to pictorial standards for a whole area, individual bones, or individual indicators. Typically, girls achieve adult levels of skeletal maturity at younger chronological ages than boys. Therefore, more maturation occurs during a typical female skeletal-age year than during a male skeletal-age year. Furthermore, bones differ in the chronological ages at which they reach adult maturity levels (Roche *et al.*, 1974); therefore the amount of maturity achieved per skeletal-age year differs among bones.

The sex differences in skeletal maturation rates differ among bones (Garn *et al.*, 1966; Roche, 1968; Thompson *et al.*, 1973; Tanner *et al.*, 1975), and their measurement is complicated by the fact that a given radiographic appearance does *not* necessarily indicate the same level of maturity in both sexes. Consequently, in the TW2 method, the sex differences between the scores for corresponding stages vary among bones. Similar variations occur with the RWT method in which each grade of each indicator has sex-specific parameters related to rate of change in prevalence with increasing age.

11.6. Variants and Assessment

Some problems arise if variations are present in the bones of the hand. When there are multiple centers of ossification in a single epiphyseal area, the width of the epiphysis should be regarded as the sum of the widths of its parts measured in the same straight line. Tanner and his associates (1975) give rules for assessing when some other variations are present. Marked variations, or even absence, of the hook of the hamate cause ratings of this bone to be based on other criteria. If the lunate develops from two centers of ossification, the general maturity of the bone is considered. In brachymesophalangia of the fifth finger, the epiphysis fuses early; such a bone is rated at the maturity level of the third middle phalanx in the same hand–wrist.

11.7. The Area to Assess

The accuracy of any estimate of the maturity of the whole skeleton obtained from the assessment of only one area is limited because skeletal areas vary in maturity levels within normal individuals (Crampton, 1908; Reynolds and Asakawa, 1951; Garn *et al.,* 1964; Roche and French, 1970). Ideally, either the whole skeleton or the hemiskeleton would be assessed. There are, however, several obvious difficulties, including the time needed for the assessment of so many bones, the expense of radiographic film, and, most importantly, the risk of excessive radiation. An area should be chosen that can be radiographed with minimal radiation, as few radiographic views as possible should be used, and all useful relevant information should be obtained from each radiograph. Bilateral assessments are unnecessary. Although lateral asymmetry occurs in all phases of skeletal maturation, the differences are small and not of practical importance (Roche, 1962*a*).

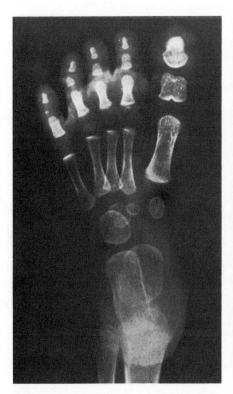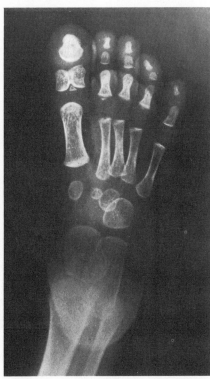

Fig. 18. Incompletely bifid first proximal phalanges in the feet of a child with Rubenstein–Taybi syndrome (from Rubenstein and Taybi, 1963).

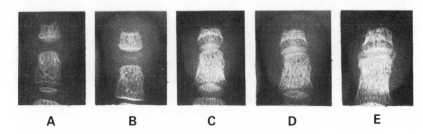

Fig. 19. Serial radiographs of an incomplete second distal phalanx in a girl at (A) 2.75 years, (B) 3 years, (C) 4.5 years, (D) 5.5 years, and (E) 8 years (from Roche, 1962*b*).

Only two areas are commonly assessed. The hand–wrist is the area most usually rated, partly because it was the first region for which atlases became available (Howard, 1928; Flory, 1936; Todd, 1937; Greulich and Pyle, 1950). Furthermore, the standards in the well-known Greulich and Pyle atlas are much clearer than those in the Cleveland atlases for the knee or the foot–ankle. The hand–wrist has other advantages: it can be positioned easily, the subject receives very little irradiation, and many bones are present in a relatively small area. However, there are disadvantages. Long age ranges occur during pubescence when the hand–wrist provides little information, and assessments of this area are not useful before one year (Sauvegrain *et al.*, 1962; Tanner *et al.*, 1975). During infancy, the knee can provide more information. Problems occur in assessments of the hand–wrist toward the end of maturation when some of the bones, e.g., distal phalanges, reach adult levels earlier than others (Todd, 1930*b;* Pyle *et al.*, 1961; Garn *et al.*, 1961*a;* Hansman, 1962). Only the designation "adult" can be applied to such bones.

The area to be assessed should be determined partly by the reason for the assessment. For example, if a measure of skeletal maturity is required to estimate

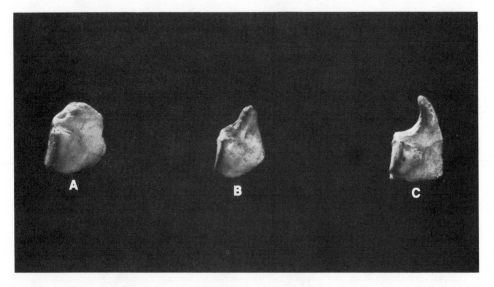

Fig. 20. Three left adult hamate bones viewed from the proximal aspect: (A) no hook, (B) small process in the position of the hook, and (C) typical adult hook.

the potential for elongation at each knee joint of an anisomelic child, the knee should be assessed. When an assessment of skeletal maturation is needed for a child whose stature deviates from the mean, the knee is preferable to the hand–wrist because it is an important site of growth in stature and can provide a slightly more accurate estimate of the potential for growth in stature than the hand–wrist (Roche *et al.*, 1975*a*), even when assessed using the atlas of Pyle and Hoerr (1969). The knee area includes only a few bones that are unusually informative (Roche *et al.*, 1975*b*). However, the knee is more difficult to position, and although the irradiation is very slight, it is double that required for the hand–wrist.

12. Minor Skeletal Variants

Major variants have been omitted from the present description; detailed accounts are available elsewhere (O'Rahilly, 1951; Schmorl and Junghans, 1971; Caffey, 1972). The minor variants considered have been described more fully by Köhler and Zimmer (1953) and Poznanski (1974).

Most duplications must be considered major variants, but minor expressions of this tendency occur as incomplete division of the subungual expansion of a terminal phalanx or complete or incomplete duplication of a bone (Figure 18). Some bones

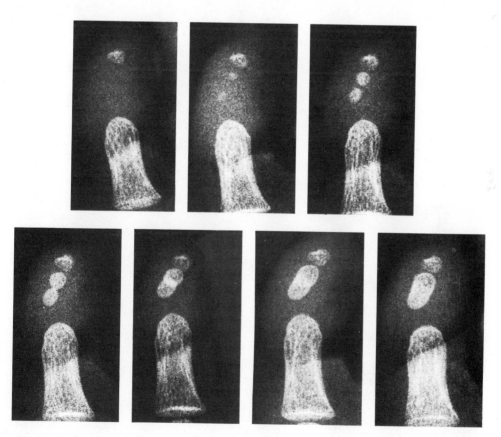

Fig. 21. Serial radiographs of a boy between the ages of 2.5 and 4 years showing ossification of the shaft of the fifth middle phalanx of the foot from two centers (from Roche and Sunderland, 1960).

may be unusually incomplete. For example, the subungual expansion of the second distal phalanx of the foot is incomplete in about 30% of children aged 2–8 years (Roche, 1962*b*) (Figure 19). Some bones in the hand–wrist or the foot may be "incomplete" even when fully mature. For example, an adult ulna may lack a styloid process or a hamate may lack its hook (Figure 20).

In about 25% of children the diaphysis of the fifth middle phalanx of the foot has two centers of ossification (Figure 21). This will be seen only in radiographs taken between 2 and 3 years. This bone may be short (brachymesophalangia): a condition often associated with clinodactyly and with an increase in the width of the fifth middle phalanx (Roche, 1961; Garn *et al.*, 1972*b*) (Figure 8). Brachymesophalangia is unusually common in Down's syndrome (Roche, 1961).

Variations occur at the proximal ends of metacarpals, particularly the second, where cortical notches, pseudoepiphyses, and supernumerary epiphyses are relatively common. A notch in the cortex commonly represents the last stage of fusion between the diaphysis and either an epiphysis or a pseudoepiphysis (Figure 22A). A

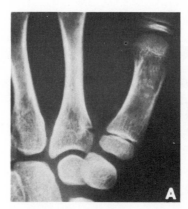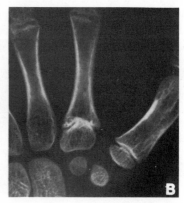

Fig. 22. Variations at the proximal end of the diaphysis of metacarpal II: (A) a notch in the cortex in a girl aged 7.5 years, and (B) a pseudoepiphysis in a girl aged 10 years.

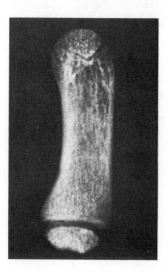

Fig. 23. An additional epiphysis at the distal end of the first metacarpal in a boy aged 7 years.

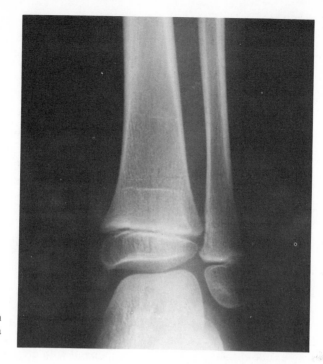

Fig. 24. Lines of arrested growth near the distal end of the tibia in a girl aged 3.5 years.

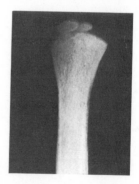

Fig. 25. Two centers of ossification in the epiphysis at the distal end of the ulna in a boy aged 7 years.

pseudoepiphysis differs from an epiphysis in that a separate center of ossification does not form in the cartilage at the end of the diaphysis. Instead, diaphyseal ossification extends in continuity into this cartilage over a small central area and then "mushrooms" within the cartilage (Figure 22B). The incidence of pseudoepiphyses on the second metacarpal has been reported as 14% in boys and 7% in girls, but supernumerary epiphyses are much less common (Iturriza and Tanner, 1969). Additional centers are common at the distal end of the first metacarpal also (Figure 23). It is not clear whether this platelike epiphysis should be classified as a pseudoepiphysis or a supernumerary epiphysis. The ossified area is flakelike, being concave on its diaphyseal aspect in all planes. Consequently, it is impossible to determine from radiographs whether this flake ossifies independently or by extension from the diaphysis.

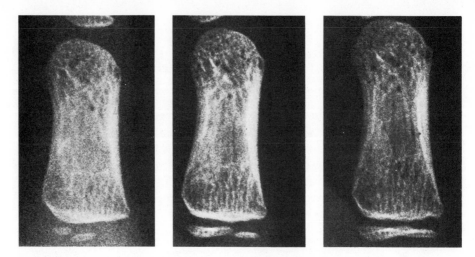

Fig. 26. Multiple centers of ossification at the proximal end of the first proximal phalanx in a girl at 3.5, 4.0, and 4.5 years.

Radiopaque lines are common (Figure 24) near the ends of the diaphyses of long bones. These are called "bone scars" or "lines of arrested growth." Commonly, but not always, these are associated with diseases during growth (Hewitt *et al.*, 1955; Marshall, 1966). They are useful as signs of past ill health and as natural bone markers from which amounts of diaphyseal elongation can be measured.

Some variations occur commonly in epiphyses also. There may be more than one center of ossification in a single epiphyseal cartilage (Figures 25 and 26). When this occurs, the separate centers enlarge until, about 2 years after the onset of ossification, they coalesce to form a single center within the cartilage that matches the shape of epiphyses formed from single centers.

The ossified epiphysis may be conical with its diaphyseal aspect extending into a depression on the end of the diaphysis (Figure 27). Conical epiphyses are

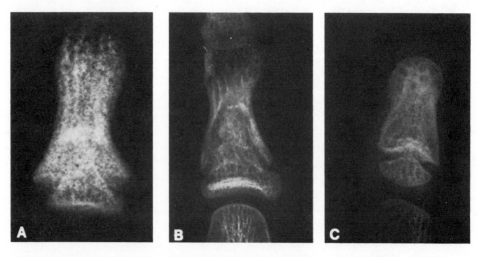

Fig. 27. Examples of conical epiphyses: (A) second proximal phalanx (foot), (B) third proximal phalanx (hand), and (C) first distal phalanx (hand).

characteristic of several clinical conditions, e.g., achondroplasia, cleidocranial dysostosis (Caffey, 1972; Gellis and Feingold, 1968), but they occur in normal children also.

13. Finale

The potential value of X rays in the study of child growth and development was realized very soon after their discovery. Much has been achieved in the intervening 80 years by the development of appropriate methods for the measurement of size and shape and for the estimation of levels of maturity of the skeleton. These advances required large collections of suitable radiographs—further advances will be similarly dependent on radiographic libraries.

As in almost every research field, the work is incomplete. Further information, for example, concerning bone mass or three-dimensional trabecular patterns, may soon be obtained accurately from radiographs. In planning for the future, collections of serial standardized radiographs must be maintained and extended. The study of trivia should be omitted, and future research should be limited to features that meaningfully discriminate among children or increase man's understanding of his own frame.

ACKNOWLEDGMENTS

This work was supported by the National Institutes of Health, Bethesda, Maryland, through grant HD-04629. Gratitude is expressed to Nancy Harvey for her help with the illustrations.

14. References

Abbott, O. D., Townsend, R. O., French, R. B., and Ahmann, C. F., 1950, Carpal and epiphyseal development. Another index of nutritional status of rural school children, *Am. J. Dis. Child.* **79**:69.

Acheson, R. M., 1954, A method of assessing skeletal maturity from radiographs. A report from the Oxford Child Health Survey, *J. Anat.* **88**:498.

Acheson, R. M., 1966, Maturation of the skeleton, in: *Human Development* (F. Falkner, ed.), pp. 465–502, W. B. Saunders Co., Philadelphia.

Acheson, R. M., Fowler, G., Fry, E. I., Janes, M., Koski, K., Urbano, P., and Van der Werff Ten Bosch, J. J., 1963, Studies in the reliability of assessing skeletal maturity from x-rays, Part I, Greulich–Pyle atlas, *Hum. Biol.* **35**:317.

Acheson, R. M., Vicinus, J. H., and Fowler, G. B., 1966, Studies in the reliability of assessing skeletal maturity from x-rays. Part III, Greulich–Pyle atlas and Tanner–Whitehouse method contrasted, *Hum. Biol.* **38**:204.

Amprino, R., and Bairati, A., 1936, Processi di recostruzione e di riassorbimento nella sostanza compatta delle ossa dell'uomo. Ricerche su cento soggetti dalla nascita sino a tarda età, *Z. Zellforsch. Mikrosk. Anat.* **24**:439.

Amprino, R., and Cattaneo, R., 1937, II. Substrato istologico delle varie modalitá di inserzioni tendinee alle ossa nell'uomo. Ricerche su individui di varia età, *Z. Anat. Entwicklungsgesch.* **107**:680.

Andersen, E., 1968, Skeletal maturation of Danish school children in relation to height, sexual development, and social conditions, *Acta Paediatr. Scand. Suppl.* **185**:133 pp.

Archibald, R. M., Finby, N., and de Vito, F., 1959, Endocrine significance of short metacarpals, *J. Clin. Endocrinol.* **19**:1312.

Arndt, J. H., and Sears, A. D., 1965, Posterior dislocation of the shoulder, *Am. J. Roentgenol.* **94**:639.

Birnbaum, A., 1968, Some latent trait models and their use in inferring an examinee's ability, in: *Statistical Theories of Mental Test Scores* (F. M. Lord and M. R. Novick, eds.), Chapters 17–20, Addison-Wesley, Reading, Massachusetts.

Bonnard, G. D., 1968, Cortical thickness and diaphysial diameter of the metacarpal bones from the age of three months to eleven years, *Helv. Paediatr. Acta* **23**:445.

Bourne, G. H., 1971, *The Biochemistry and Physiology of Bone,* 2nd ed., Vol. III, "Development and Growth," Academic Press, New York.

Brailsford, J. F., 1943, Variations in the ossification of the bones of the hand, *J. Anat.* **77**:170.

Brailsford, J. F., 1953, *The Radiology of Bones and Joints,* 5th ed., J. and A. Churchill, London.

Brandner, M. E., 1970, Normal values of the vertebral body and intervertebral disc index during growth, *Am. J. Roentgenol.* **110**:618.

Brunk, U., and Sköld, G., 1962, Length of foetus and ossification of the toe and finger phalanges, *Acta Histochem.* **14**:59.

Caffey, J., 1972, *Pediatric X-Ray Diagnosis,* Vol. 2, 6th ed., Year Book Medical Publishers, Chicago.

Caffey, J., and Ross, S., 1958, Pelvic bones in infantile mongoloidism. 3 Roentgenographic features, *Am. J. Roentgenol.* **80**:458.

Caffey, J., Ames, R., Silverman, W. A., Ryder, C. T., and Hough, G., 1956, Contradiction of the congenital dysplasia–predislocation hypothesis of congenital dislocation of the hip through a study of the normal variation in acetabular angles at successive periods in infancy, *Pediatrics* **17**:632.

Christ, H. H., 1961, A discussion of causes of error in the determination of chronological age in children by means of x-ray studies of carpal-bone development, *S. Afr. Med. J.* **35**:854.

Coleman, W. H., 1969, Sex differences in the growth of the human bony pelvis, *Am. J. Phys. Anthropol.* **31**:125.

Crampton, C. W., 1908, Physiological age—a fundamental principle, *Am. Phys. Educ. Rev.* **13**: reprinted in *Child Dev.* 1944, **15**:3.

Day, R., and Silverman, W. A., 1952, Growth of the fibula of premature infants as estimated in roentgen films: A method for assessing factors promoting or inhibiting growth, *Neo-Natal Studies* **1**:111.

Edlin, J. C., Whitehouse, R. H., and Tanner, J. M., 1976, Relationship of radial metaphyseal band width to stature velocity, *Am. J. Dis. Child.* **130**:160.

Elgenmark, O., 1946, The normal development of the ossific centres during infancy and childhood. A clinical, roentgenologic and statistical study, *Acta Paediatr.* **33**(Suppl. 1):79 pp.

Enlow, D. H., 1963, *Principles of Bone Remodeling. An Account of Post-Natal Growth and Remodeling Processes in Long Bones and the Mandible.* Charles C Thomas, Springfield, Illinois.

Fell, H. B., and Robison, R., 1934, The development of the calcifying mechanism in avian cartilage and osteoid tissue, *Biochem. J.* **28**:2243.

Flory, C. D., 1936, Osseous development in the hand as an index of skeletal development, *Monogr. Soc. Res. Child Dev.* **1**:1.

Francis, C. C., 1939, Factors influencing appearance of centers of ossification during early childhood, *Am. J. Dis. Child.* **57**:817.

Francis, C. C., 1940, The appearance of centers of ossification from 6–15 years, *Am. J. Phys. Anthropol.* **27**:127.

Francis, C. C., and Werle, P. P., 1939, The appearance of centers of ossification from birth to 5 years, *Am. J. Phys. Anthropol.* **24**:273.

Gardner, E., 1971, Osteogenesis in the human embryo and fetus, in: *The Biochemistry and Physiology of Bone,* Vol. III (G. H. Bourne, ed.), pp. 77–118, Academic Press, New York.

Gardner, E., and Gray, D. J., 1970, The prenatal development of the human femur, *Am. J. Anat.* **129**:121.

Garn, S. M., 1970, *The Earlier Gain and the Later Loss of Cortical Bone in Nutritional Perspective,* Charles C Thomas, Springfield, Illinois.

Garn, S. M., and Rohmann, C. G., 1959, Communalities of the ossification centers of the hand and wrist, *Am. J. Phys. Anthropol.* **17**:319.

Garn, S. M., and Rohmann, C. G., 1960, Variability in the order of ossification of the bony centers of the hand and wrist, *Am. J. Phys. Anthropol.* **18**:219.

Garn, S. M., Rohmann, C. G., and Apfelbaum, B., 1961a, Complete epiphyseal union of the hand, *Am. J. Phys. Anthropol.* **19**:365.

Garn, S. M., Rohmann, C. G., and Robinow, M., 1961b, Increments in hand–wrist ossification, *Am. J. Phys. Anthropol.* **19**:45.

Garn, S. M., Silverman, F. N., and Rohmann, C. G., 1964, A rational approach to the assessment of skeletal maturation, *Ann. Radiol.* **7**:297.

Garn, S. M., Rohmann, C. G., and Blumenthal, T., 1966, Ossification sequence polymorphism and sexual dimorphism in skeletal development, *Am. J. Phys. Anthropol.* **24**:101.

Garn, S. M., Silverman, F. N., Hertzog, K. P., and Rohmann, C. G., 1968, Lines and bands of increased density, *Med. Radiogr. Photogr.* **44**:58.

Garn, S. M., Poznanski, A. K., and Nagy, J. M., 1971, Bone measurement in the differential diagnosis of osteopenia and osteoporosis, *Radiology* **100**:509.

Garn, S. M., Hertzog, K. P., Poznanski, A. K., and Nagy, J. M., 1972a, Metacarpophalangeal length in the evaluation of skeletal malformation, *Radiology* **105**:375.

Garn, S. M., Poznanski, A. K., Nagy, J. M., and McCann, M. B., 1972b, Independence of brachymesophalangia-5 from brachymesophalangia-5 with cone mid-5, *Am. J. Phys. Anthropol.* **36**:295.

Garrow, J. S., and Fletcher, K., 1964, The total weight of mineral in the human infant, *Br. J. Nutr.* **18**:409.

Gefferth, K., 1972, Metrische Auswertung der kurzen Röhrenknochen der Hand von der Geburt bis zum Ende der Pubertät: Längenmasse, *Acta Paediatr. Acad. Sci. Hung.* **13**:117.

Gellis, S. S., and Feingold, M., 1968, Atlas of Mental Retardation Syndromes. Visual Diagnosis of Facies and Physical Findings, U.S. Dept. Health, Education, and Welfare Social and Rehabilitation Service, Rehabilitation Services Administration, Division of Mental Retardation, U. S. Government Printing Office, Washington, D.C.

Goff, C. W., 1960, *Surgical Treatment of Unequal Extremities,* Charles C Thomas, Springfield, Illinois.

Gray, D. J., and Gardner, E., 1969, The prenatal development of the human humerus, *Am. J. Anat.* **124**:431.

Gray, D. J., Gardner, E., and O'Rahilly, R., 1957, The prenatal development of the skeleton and joints of the human hand, *Am. J. Anat.* **101**:169.

Greulich, W. W., 1954, The relationship of skeletal status to the physical growth and development of children, in: *Dynamics of Growth Processes* (E. J. Boell, ed.), pp. 212–223, Princeton University Press, Princeton, New Jersey.

Greulich, W. W., and Pyle, S. I., 1950, *Radiographic Atlas of Skeletal Development of the Hand and Wrist,* 1st ed., Stanford University Press, Stanford, California.

Greulich, W. W., and Pyle, S. I., 1959, *Radiographic Atlas of Skeletal Development of the Hand and Wrist,* 2nd ed., Stanford University Press, Stanford, California.

Hadley, L. A., 1956, *The Spine; Anatomico-Radiographic Studies: Development and the Cervical Region,* Charles C Thomas, Springfield, Illinois.

Haines, R. W., 1975, The histology of epiphyseal union in mammals, *J. Anat.* **120**:1.

Haines, R. W., and Mohiuddin, A., 1959, A preliminary note on the process of epiphysial union, *J. Fac. Med. Baghdad* **1**:141.

Ham, A. W., 1974, *Histology,* 7th ed., J. B. Lippincott, Philadelphia.

Hansman, C. F., 1962, Appearance and fusion of ossification centers in the human skeleton, *Am. J. Roentgenol.* **88**:476.

Hansman, C. F., and Maresh, M. M., 1961, A longitudinal study of skeletal maturation, *Am. J. Dis. Child.* **101**:305.

Harper, H. A. S., Poznanski, A. K., and Garn, S. M., 1974, The carpal angle in American populations, *Invest. Radiol.* **9**:217.

Harrison, T. J., 1958a, The growth of the pelvis in the rat—a mensural and morphological study, *J. Anat.* **92**:236.

Harrison, T. J., 1958b, An experimental study of pelvic growth in the rat, *J. Anat.* **92**:483.

Harrison, T. J., 1961, The influence of the femoral head on pelvic growth and acetabular form in the rat, *J. Anat.* **95**:12.

Hasselwander, A., 1910, Ossifikation des menschlichen Fusskeletts, *Z. Morphol. Anthropol.* **12**:1.

Healy, M. J. R., and Goldstein, H., 1976, An approach to the scaling of categorised attributes, *Biometrika* **63**:219.

Hellman, M., 1928, Ossification of epiphyseal cartilages in the hand, *Am. J. Phys. Anthropol.* **11**:223.

Heřt, J., 1972, Growth of the epiphyseal plate in circumference, *Acta Anat.* **82**:420.

Hesser, C., 1926, Beitrag zur Kenntnis der Gelenkentwicklung beim Menschen, *Morphol. Jahrb.* **55**:489.

Hewitt, D., 1963, Pattern of correlations in the skeleton of the growing hand, *Ann. Hum. Genet.* **27**:157.

Hewitt, D., and Acheson, R. M., 1961, Some aspects of skeletal development through adolescence. I. Variations in the rate and pattern of skeletal maturation at puberty, *Am. J. Phys. Anthropol.* **19**:321.

Hewitt, D., Westropp, C. K., and Acheson, R. M., 1955, Oxford Child Health Survey: Effect of childish ailments on skeletal development, *Br. J. Prev. Soc. Med.* **9**:179.

Hinck, V. C., and Hopkins, C. E., 1960, Measurement of the atlanto-dental interval in the adult, *Am. J. Roentgenol.* **84**:945.

Hinck, V. C., Hopkins, C. E., and Savara, B. S., 1962, Sagittal diameter of the cervical spinal canal in children, *Radiology* **79**:97.

Hinck, V. C., Hopkins, C. E., and Clark, W. M., 1965, Sagittal diameter of the lumbar spinal canal in children and adults, *Radiology* **85**:929.

Hinck, V. C., Clark, W. M., Jr., and Hopkins, C. E., 1966, Normal interpediculate distances (minimum and maximum) in children and adults, *Am. J. Roentgenol.* **97**:141.

Hoerr, N. L., Pyle, S. I., and Francis, C. C., 1962, *Radiographic Atlas of Skeletal Development of the Foot and Ankle, A Standard of Reference,* Charles C Thomas, Springfield, Illinois.

Howard, C. C., 1928, The physiologic process of the bone centers of the hands of normal children between the ages of five and sixteen inclusive; also a comparative study of both retarded and accelerated hand growth in children whose general skeletal growth is similarly affected, *J. Orthod. Oral Surg. Radiogr.* **14**:948, 1041.

Hughes, P. C. R., and Tanner, J. M., 1970, The assessment of skeletal maturity in the growing rat, *J. Anat.* **106**:371.

Iturriza, J. R., de, and Tanner, J. M., 1969, Cone-shaped epiphyses and other minor anomalies in the hands of normal British children, *J. Pediatr.* **75**:265.

Johnson, G. F., Dorst, J. P., Kuhn, J. P., Roche, A. F., and Davila, G. H., 1973, Reliability of skeletal age assessments, *Am. J. Roentgenol.* **18**:320.

Johnston, F. E., 1963, Skeletal age and its prediction in Philadelphia children, *Am. J. Phys. Anthropol.* **21**:406.

Johnston, F. E., 1964, The relationship of certain growth variables to chronological and skeletal age, *Hum. Biol.* **36**:16.

Johnston, F. E., and Jahina, S. B., 1965, The contribution of the carpal bones to the assessment of skeletal age, *Am. J. Phys. Anthropol.* **23**:349.

Kelly, H. J., and Reynolds, L., 1947, Appearance and growth of ossification centers and increases in the body dimensions of white and Negro infants, *Am. J. Roentgenol.* **57**:477.

Köhler, A., and Zimmer, E. A., 1953, *Grenzen des Normalen und Anfänge des Pathologischen im Röntgenbilde des Skelettes,* Georg Thieme Verlag, Stuttgart.

Koski, K., 1968, Cranial growth centers: facts or fallacies? *Am. J. Orthod.* **54**:566.

Koski, K., Haataja, J., and Lappalainen, M., 1961, Skeletal development of hand and wrist in Finnish children, *Am. J. Phys. Anthropol.* **19**:379.

Kosowicz, J., 1962, The carpal sign in gonadal dysgenesis, *J. Clin. Endocrinol. Metab.* **22**:949.

Kosowicz, J., 1965, The roentgen appearance of the hand and wrist in gonadal dysgenesis, *Am. J. Roentgenol.* **93**:354.

Lacroix, P., 1951, *The Organization of Bones,* McGraw-Hill (Blakiston), New York.

Lacroix, P., 1971, The internal remodeling of bones, in: *The Biochemistry and Physiology of Bone,* Vol. III (G. H. Bourne, ed.), pp. 119–144, Academic Press, New York.

Lee, M. M. C., 1968, Natural markers in bone growth, *Am. J. Phys. Anthropol.* **29**:295.

Lestrel, P. E., 1974, Some problems in the assessment of morphological size and shape differences, *Yearb. Phys. Anthropol.* **18**:140.

Lestrel, P. E., 1975, Fourier analysis of size and shape of the human cranium: A longitudinal study from four to eighteen years of age, Unpublished Ph.D. dissertation, University of California, Los Angeles.

Lestrel, P. E., and Roche, A. F., 1976, Fourier analysis of the cranium in trisomy 21, *Growth* **40**:385.

Lestrel, P. E., Kimbel, W. H., Prior, W. F., and Fleischmann, M. L., 1977, Size and shape of the hominoid distal femur: Fourier analysis, *Am. J. Phys. Anthropol.* **46**(2):281.

Lewis, W. H., 1902, The development of the arm in man, *Am. J. Anat.* **1**:145.

Locke, G. R., Gardner, J. I., and Van Epps, E. F., 1966, Atlas-Dens Interval (ADI) in children. A survey based on 200 normal cervical spines, *Am. J. Roentgenol.* **97**:135.

Lord, F. M., 1953, On the statistical treatment of football numbers, *Am. Psychol.* **8**:750.

Lusted, L. B., and Keats, T. E., 1972, *Atlas of Roentgenographic Measurement,* 3rd ed., Year Book Medical Publishers, Chicago.

Macarthur, J. W., and McCulloch, E., 1932, Apical dystrophy. An inherited defect of hands and feet, *Hum. Biol.* **4**:179.

McKern, T. W., and Stewart, T. D., 1957, Skeletal age changes in young American males analysed from the standpoint of age identification, Technical Report EP-45, Headquarters Quartermaster Research & Development Command, Environmental Protection Research Division, U.S. Army, Natick, Massachusetts.

Mainland, D., 1953, Evaluation of the skeletal age method of estimating children's development. I. Systematic errors in the assessment of roentgenograms, *Pediatrics* **12**:114.

Mainland, D., 1954, Evaluation of the skeletal age method of estimating children's development. II. Variable errors in the assessment of roentgenograms, *Pediatrics* **13**:165.

Maresh, M. M., 1970, Measurements from roentgenograms, heart size, long bone lengths, bone, muscle and fat widths, skeletal maturation, in: *Human Growth and Development* (R. W. McCammon, ed.), pp. 157–200, Charles C Thomas, Springfield, Illinois.

Maresh, M. M., and Deming, J., 1939, The growth of the long bones in 80 infants: Roentgenograms versus anthropometry, *Child Dev.* **10**:91.

Marshall, W. A., 1966, Problems in relating radiopaque transverse lines in the radius to the occurrence of disease, *Symp. Soc. Hum. Biol.* **8**:245.

Mathiasen, M. S., 1973, Determination of bone age and recording of wrist or skeletal hand anomalies in normal children, *Dan. Med. Bull.* **20**:80.

Mazess, R. B., and Cameron, J. R., 1971, Skeletal growth in school children: Maturation and bone mass, *Am. J. Phys. Anthropol.* **35**:399.

Moed, G., Wight, B. W., and Vandergrift, H. N., 1962, Studies of physical disability: Reliability of measurement of skeletal age from hand films, *Child Dev.* **33**:37.

Moss, M. L., and Noback, C. R., 1958, A longitudinal study of digital epiphyseal fusion in adolescence, *Anat. Rec.* **131**:19.

Mossberg, H.-O., 1949, The x-ray appearance of the knee joint in obese overgrown children, *Acta Paediatr. Scand.* **38**:509.

Murray, J. R., Bock, R. D., and Roche, A. F., 1971, The measurement of skeletal maturity, *Am. J. Phys. Anthropol.* **35**:327.

Naik, D. R., 1970, Cervical spinal canal in normal infants, *Clin. Radiol.* **21**:323.

Niven, J. S. F., and Robison, R., 1934, The development of the calcifying mechanism in the long bones of the rabbit, *Biochem. J.* **28**:2237.

Noback, C. R., Moss, M. L., and Leszczynska, E., 1960, Digital epiphyseal fusion of the hand in adolescence. A longitudinal study, *Am. J. Phys. Anthropol.* **18**:13.

O'Rahilly, R., 1951, Morphological patterns in limb deficiencies and duplications, *Am. J. Anat.* **89**:135.

Olivier, G., 1962, *Formation du Squelettes des Membres chez l'Homme,* Vigot Frères, Paris.

Olivier, G., and Pineau, H. A., 1959, Embryologie de l'humerus, *Arch. Anat. Pathol.* **7**:121.

Park, E. A., 1954, Bone growth in health and disease, *Arch. Dis. Child.* **29**:269.

Paterson, R. S., 1929, A radiological investigation of the epiphyses of the long bones, *J. Anat.* **64**:28.

Payton, C. G., 1933, The growth of the epiphyses of the long bones in the madder-fed pig, *J. Anat.* **67**:371.

Peritz, E., and Sproul, A., 1971, Some aspects of the analysis of hand–wrist bone-age readings, *Am. J. Phys. Anthropol.* **35**:441.

Poznanski, A. K., 1974, *The Hand in Radiologic Diagnosis,* W. B. Saunders, Philadelphia.

Poznanski, A. K., Garn, S. M., Kuhns, L. R., and Sandusky, S. T., 1971, Dysharmonic maturation of the hand in the congenital malformation syndromes, *Am. J. Phys. Anthropol.* **35**:417.

Poznanski, A. K., Garn, S. M., Nagy, J. M., and Gall, J. C., Jr., 1972a, Metacarpophalangeal pattern profiles in the evaluation of skeletal malformation, *Radiology* **104**:1.

Poznanski, A. K., Garn, S. M., Gall, J. C., Jr., and Stern, A. M., 1972b, Objective evaluation of the hand in the Holt–Oram syndrome, *Birth Defects, Orig. Art. Ser.* **8**:125.

Pryor, J. W., 1907, The hereditary nature of variation in the ossification of bones, *Anat. Rec.* **1**:84.

Pryor, H. B., and Carter, H. D., 1938, Phases of adolescent development in girls, *Calif. West. Med.* **48**:89.

Pyle, S. I., and Hoerr, N. L., 1955, *Radiographic Atlas of Skeletal Development of the Knee. A Standard of Reference,* Charles C Thomas, Springfield, Illinois.

Pyle, S. I., and Hoerr, N. L., 1969, *A Radiographic Standard of Reference for the Growing Knee,* Charles C Thomas, Springfield, Illinois.

Pyle, S. I., Mann, A. W., Dreizen, S., Kelly, H. J., Macy, I. G., and Spies, T. D., 1948, A substitute for skeletal age (Todd) for clinical use: The red graph method, *J. Pediatr.* **32**:125.

Pyle, S. I., Stuart, H. C., Cornoni, J., and Reed, R. B., 1961, Onsets, completions, and spans of the osseous stage of development in representative bone growth centers of the extremities, *Monogr. Soc. Res. Child Dev.* **26**(1), 126 pp.

Pyle, S. I., Waterhouse, A. M., and Greulich, W. W., 1971, *A Radiographic Standard of Reference for the Growing Hand and Wrist. Prepared for the United States National Health Examination Survey,* Case Western Reserve University, Cleveland, Ohio.

Rambaud, A., and Renault, C., 1864, *Origine et Développement des Os,* F. Chamerot, Paris.

Reynolds, E. L., 1945, Bony pelvic girdle in early infancy, *Am. J. Phys. Anthropol.* **3:**321.

Reynolds, E. L., 1947, The bony pelvis in prepuberal childhood, *Am. J. Phys. Anthropol.* **5:**165.

Reynolds, E. L., and Asakawa, T., 1951, Skeletal development in infancy; standards for clinical use, *Am. J. Roentgenol.* **65:**403.

Robinow, M., 1942, Appearance of ossification centres: Grouping obtained from factor analysis, *Am. J. Dis. Child.* **64:**229.

Roche, A. F., 1961, Clinodactyly and brachymesophalangia of the fifth finger, *Acta Paediatr.* **50:**387.

Roche, A. F., 1962*a,* Lateral comparisons of the skeletal maturity of the human hand and wrist, *Am. J. Roentgenol.* **89:**1272.

Roche, A. F., 1962*b,* Incomplete distal phalanges in the foot during childhood, *Acta Anat.* **51:**369.

Roche, A. F., 1965, The sites of elongation of human metacarpals and metatarsals, *Acta Anat.* **61:**193.

Roche, A. F., 1967, The elongation of the mandible, *Am. J. Orthod.* **53:**79.

Roche, A. F., 1968, Sex-associated differences in skeletal maturity, *Acta Anat.* **71:**321.

Roche, A. F., 1970, Associations between the rates of maturation of the bones of the hand–wrist, *Am. J. Phys. Anthropol.* **33:**341.

Roche, A. F., and Davila, G. H., 1976, The reliability of assessments of the maturity of individual hand–wrist bones, *Hum. Biol.* **48:**585.

Roche, A. F., and French, N. Y., 1970, Differences in skeletal maturity levels between the knee and hand, *Am. J. Roentgenol.* **109:**307.

Roche, A. F., and Hermann, R., 1970, Associations between the rates of elongation of the short bones of the hand, *Am. J. Phys. Anthropol.* **32:**83.

Roche, A. F., and Johnson, J. M., 1969, A comparison between methods of calculating skeletal age (Greulich–Pyle), *Am. J. Phys. Anthropol.* **30:**221.

Roche, A. F., and Sunderland, S., 1960, Observations on the developing fifth toe in normal children, *Acta Anat.* **41:**261.

Roche, A. F., Rohmann, C. G., French, N. Y., and Davila, G. H., 1970, Effect of training on replicability of assessments of skeletal maturity, *Am. J. Roentgenol.* **108:**511.

Roche, A. F., Roberts, J., and Hamill, P. V. V., 1974, Skeletal Maturity of Children 6–11 Years, United States, Vital and Health Statistics Series 11, No. 140, U.S. Dept. Health, Education and Welfare, Washington, D.C.

Roche, A. F., Wainer, H., and Thissen, D., 1975*a,* Predicting adult stature for individuals, *Monogr. Pediatr.* **3,** 115 pp.

Roche, A. F., Wainer, H., and Thissen, D., 1975*b, Skeletal Maturity. The Knee Joint as a Biological Indicator,* Plenum Publishing, New York.

Roche, A. F., Roberts, J., and Hamill, P. V. V., 1975*c,* Skeletal Maturity of Children 6–11 Years: Racial, Geographic Area of Residence, Socioeconomic Differentials, United States, Vital and Health Statistics Series 11, No. 149, U.S. Dept. Health, Education and Welfare, Washington, D.C.

Roche, A. F., Roberts, J., and Hamill, P. V. V., 1976, Skeletal Maturity of Youths 12–17 Years, United States, Vital and Health Statistics Series 11, No. 160, U.S. Dept. Health, Education and Welfare, Washington, D.C., 81 pp.

Roche, A. F., Roberts, J., and Hamill, P. V. V., 1978, Skeletal Maturity of Youths 12–17 Years: Racial, Geographic Area, and Socioeconomic Differentials, Vital and Health Statistics Series 11, No. 167, U.S. Dept. Health, Education and Welfare, Washington, D.C.

Rubinstein, J. H., and Taybi, H., 1963, Broad thumbs and toes and facial abnormalities. A possible mental retardation syndrome, *Am. J. Dis. Child.* **105:**588.

Sahay, G. B., 1941, Determination of age by x-ray examination (with reference to epiphyseal union), *Indian Med. J.* **35:**37.

Samejima, F., 1969, Estimation of latent ability using a response pattern of graded scores, *Psychometrika* **34:** Monogr. Suppl. 1.

Samejima, F., 1972, General model for free response data, *Psychometrika* **37:** Monogr. Suppl. 18.

Sauvegrain, J., Nahum, H., and Bronstein, H., 1962, Etude de la maturation osseuse du coude, *Ann. Radiol.* **5:**542.

Scheller, S., 1960, Roentgenographic studies on epiphysial growth and ossification in the knee, *Acta Radiol.* Suppl. 195.

Schmorl, G., and Junghanns, H., 1971, *The Human Spine in Health and Disease,* 2nd ed., Grune and Stratton, New York.

Siegling, J. A., 1941, Growth of the epiphyses, *J. Bone Joint Surg.* **23:**23.

Siffert, R. S., 1966, The growth plate and its affections, *J. Bone Joint Surg.* **48-A:**546.

Silberberg, M., and Silberberg, R., 1961, Ageing changes in cartilage and bone, in: *Structural Aspects of Ageing* (G. H. Bourne, ed.), pp. 85–108, Pitman Medical, London.

Sinclair, R. J. G., Kitchin, A. H., and Turner, R. W. D., 1960, The Marfan syndrome, *Q. J. Med.* **53**:19.

Sontag, L. W., and Lipford, J., 1943, The effect of illness and other factors on the appearance pattern of skeletal epiphyses, *J. Pediatr.* **23**:391.

Sontag, L. W., Snell, D., and Anderson, M., 1939, Rate of appearance of ossification centers from birth to the age of five years, *Am. J. Dis. Child.* **58**:949.

Sproul, A., and Peritz, E., 1971, Assessment of skeletal age in short and tall children, *Am. J. Phys. Anthropol.* **35**:433.

Stevenson, P. H., 1924, Age order of epiphyseal union in man, *Am. J. Phys. Anthropol.* **7**:53.

Streeter, G. L., 1951, *Developmental Horizons in Human Embryos. Age Groups 11 to 23,* Embryology Reprint, Vol. 2, Carnegie Institute, Washington, D.C.

Takahashi, Y., 1956, Studies on skeletal development in Japanese children. Roentgenographic indicators of maturity in bones of the foot, ankle, knee and elbow areas, U.S. Atomic Energy Commission, Oak Ridge, Tennessee.

Tanner, J. M., 1962, *Growth at Adolescence; with a General Consideration of the Effects of Hereditary and Environmental Factors upon Growth and Maturation from Birth to Maturity,* 2nd edition, Blackwell, Oxford.

Tanner, J. M., Whitehouse, R. H., and Healy, M. J. R., 1962, A New System for Estimating Skeletal Maturity from the Hand and Wrist, with Standards Derived from a Study of 2,600 Healthy British Children, International Children's Centre, Paris.

Tanner, J. M., Whitehouse, R. H., and Goldstein, H., 1972, A Revised System for Estimating Skeletal Maturity from Hand and Wrist Radiographs, with Separate Standards for Carpals and Other Bones (TW II System), International Children's Centre, Paris.

Tanner, J. M., Whitehouse, R. H., Marshall, W. A., Healy, M. J. R., and Goldstein, H., 1975, *Assessment of Skeletal Maturity and Prediction of Adult Height (TW2 Method),* Academic Press, London.

Templeton, A. W., McAlister, W. H., and Zim, I. D., 1965, Standardization of terminology and evaluation of osseous relationships in congenitally abnormal feet, *Am. J. Roentgenol.* **93**:374.

Thompson, G. W., Popovich, F., and Luks, E., 1973, Sexual dimorphism in hand and wrist ossification, *Growth* **37**:1.

Tiisala, R., Kantero, R.-L., and Bäckström, L., 1966, The appearance pattern of skeletal epiphyses among normal Finnish children during the first five years of life, *Ann. Paediatr. Fenn.* **12**:64.

Todd, T. W., 1930a, The roentgenographic appraisement of skeletal differentiation, *Child Dev.* **1**:298.

Todd, T. W., 1930b, The anatomical features of epiphyseal union, *Child Dev.* **1**:186.

Todd, T. W., 1937, *Atlas of Skeletal Maturation,* C. V. Mosby, St. Louis, Missouri.

Trotter, M., and Peterson, R. R., 1968, Weight of bone in the fetus: A preliminary report, *Growth* **32**:83.

Trotter, M., and Peterson, R. R., 1969a, Weight of bone during the fetal period, *Growth* **33**:167.

Trotter, M., and Peterson, R. R., 1969b, Weight of bone in the fetus during the last half of pregnancy, *Clin. Orthop.* **65**:46.

Trotter, M., and Peterson, R. R., 1970, Weight of the skeleton during postnatal development, *Am. J. Phys. Anthropol.* **33**:313.

Valk, I. M., 1971, Accurate measurement of the length of the ulna and its application in growth measurements, *Growth* **35**:297.

Valk, I. M., 1972, Ulnar length and growth in twins with a simplified technique for ulnar measurement using a condylograph, *Growth* **36**:291.

Wainer, H., 1976, Estimating coefficients in linear models: It don't make no nevermind, *Psychol. Bull.* **82**:213.

Wainer, H., and Thissen, D., 1975, Multivariate semi-metric smoothing in multiple prediction, *J. Am. Stat. Assoc.* **70**:568.

Wetherington, R. K., and Haurberg, S., 1974, Study of brachymesophalangia in a Hong Kong sample, *Am. J. Phys. Anthropol.* **41**:509.

Wilks, S. S., 1938, Weighting systems for linear functions of correlated variables when there is no dependent variable, *Psychometrica* **3**:23.

Yarbrough, C., Habicht, J.-P., Klein, R. E., and Roche, A. F., 1973, Determining the biological age of the preschool child from a hand–wrist radiograph, *Invest. Radiol.* **8**:233.

13

The Fundamentals of Cranial and Facial Growth

HARRY ISRAEL, III

1. Introduction

Time plays its hand nowhere more expressively than in the human face. It performs this task so explicitly that we seldom misjudge any person's age by more than a few years. The cherubic and large-headed infant gives way to the full-faced adolescent and gradually the years add the special embellishment of maturity. In that face is expressed health and well-being, aspirations and inspirations, and happiness or sorrow. To that face is drawn the glance of the stranger, the acquaintance, the relative, and the health professional; no one is excluded. In a sense, this makes diagnosticians of us all. To some this implies over-all health of the infant; to others, the status of the teeth; and to still others, the state of the mind. There is no limit to the list of interested parties because any person-to-person association initiates with a glance to the face.

This discussion of growth in the face and skull is focused on the skeletal portion which forms the scaffolding upon which the respective parts rest. As a formative and protective mechanism, it contributes not a little to countenance, but it is after all only the lattice-work. The mental and physical complexities of man are veiled by the face; and often this is a transparent veil.

Many disciplines have an interest in this region of the body. Each seeks out different facets that are most applicable to its own needs. Take, for example, the orthodontist and radiologist. The former depends heavily upon radiocephalometrics for the assessment of growth patterns, to seek predictive aids, and formulate dentoskeletal diagnoses (Ricketts, 1960; Broadbent *et al.,* 1975; L. E. Johnston, 1975). The radiologist usually views the image for diagnostic and therapeutic purposes unrelated to that of the orthodontist (Henderson and Sherman, 1946; Dorst, 1964; Caffey, 1973). Both overlap, but one has only to read descriptions of

HARRY ISRAEL, III • Dental Research Section, Children's Medical Center, Dayton, Ohio.

embryology and growth in the respective writings of both specialists to be aware of their varying interests and intentions.

The object of this chapter is to explain some of the mechanisms that have been proposed in craniofacial development and then to translate these into morphologic consequences. Since growth is a *continuum,* the attempt here is to carry in sequential fashion the observable change from conception to older age. Amounts of growth may diminish, shape and form may appear stationary, but the capacity for cells and tissues to produce change ceases when the organism dies and not at some deduced period along the way to that end.

2. Taking Shape

Generally, the germ layers begin to form tissues and organs in the embryonic period around four weeks (Patton, 1968; Langman, 1969). The nervous system progresses rapidly following the appearance of the neural plate. High cellular activity converts the plate into an ectodermal groove; ultimately this groove disappears as constricting ectoderm fuses above, and this pinching together of the groove results in the formation of the neural tube. This gives a separate ectodermal surface and ectodermal neural tube and from the latter both spinal cord and brain vesicles are derived. The junction area between groove and surface ectoderm prior to fusion and tube formation is known as the neural crest and from here come the important neural crest cells (Figure 1). It is believed that some of the cranium and all of the skeletal and connective tissues of the face are of crest-cell origin (M. C. Johnston, 1975). This excludes portions of the eye, endothelium, epithelium, and skeletal muscle. Below the neural tube is the notochord, and placed laterally are the mesodermal-derived specializations known as somites. As the cell types differentiate they become the sclerotome composed of so-called mesenchyme; they have the ability to develop into blast cells which form connective tissues, cartilage, and bone. Also, an associated myotome arises that is responsible for muscular construc-

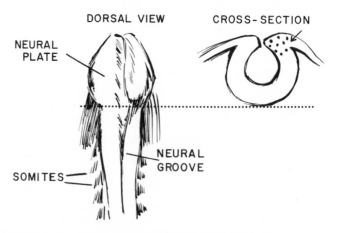

Fig. 1. Neural crest cells (arrow) are located at the junction of the neural groove and surface ectoderm of the early embryo. These cells migrate to form much of the skeleton, musculature, and connective tissue of the facial processes and visceral arches.

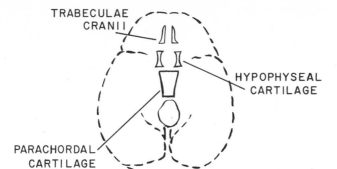

Fig. 2. The cranial base is formed from mesenchymal cartilage, and three plates are depicted here in this simplified illustration of chondrocranial development.

tion. With the principal participants identified, the developing facial structures become recognizable around the stomodeum.

It is usually an advantage to consider the skull in two portions although they are not necessarily totally segregated embryologically. These portions are the neurocranium, with its association to the central nervous system and organs of special sense, and the viscerocranium forming the facial segment connected with alimentation and respiration. At the earliest formative stage, the phylogenetically oldest portion of the viscerocranium consists of cartilage (the chondrocranium). This intermediary between neuro- and viscerocranium is known as the cranial base. It is a midline cartilaginous structure extending nasally from the foramen magnum and contributes to all three cranial cavities. Initially, cartilage forms in the noto-chordal mesenchyme (Figure 2). Paranotochordal cartilage extends from the sella turcica to the occipital sclerotomes. Anterior to the parachordal cartilages are the hypophyseal cartilages and trabeculae cranii. The former forms the body of the sphenoid and the latter the ethmoid bone. The skull base then develops out of the confluence of the occipital, sphenoid, and ethmoid bones. The occipital bone contributes to foramen magnum construction and supports much of the rhomben-cephalon consisting of medulla, pons, and cerebellum. It is formed from the parachordal cartilage and arises in a number of centers. These are the basiocciput anteriorly, the paired exoccipital laterally, and posteriorly the four-part supraoccipi-tal (Figure 3). These endochondral ossification centers are joined with a pair of intramembraneous centers above the superior nuchal line (interparietal). Hard tissue formation begins to appear about the ninth week. The Mendosal suture separates the supraoccipital from interparietal and thus delineates the two ossification types. Full confluence of the occipital bone is protracted in that the Mendosal suture disappears in radiographs in the first few weeks after birth, the separation of exoccipital from supraoccipital disappears during the second or third year, and the sphenooccipital synchrondosis between basisphenoid and basiocciput is obliterated in the second decade of life.

The sphenoid bone has a number of paried endochondral ossification centers (Figure 4). There is one pair for the lesser wings (orbitosphenoids), one for the greater wings (alisphenoids), and two for the body of the sphenoid (presphenoid anteriorly and basisphenoid posteriorly). Some other small centers also occur. Intramembranous centers contribute laterally to the greater wings and posteriorly to the medial pterygoid process. Some centers appear before ten weeks and at birth the sphenoid is still in three parts. The ethmoid connected with limbic system is formed in cartilage both medially and laterally. Ossification brings forth the perpendicular plate and crista galli. As the lateral cartilages ossify into wide-spaced trabeculae, they form ethmoid air cells and the superior and middle conchae. Further coalescence of the medial and lateral cartilage awaits the formation of olfactory nerves before full endochondral ossification completes the cribriform plate. Ossification begins in the fourth month after fertilization.

The fourth bone having principally endochondral components is the temporal. It is formed from the outer capsule of the inner ear (periotic cartilage) which is part of the primitive chondrocranium. The squamous portion is membrane bone, and the temporal has a contribution from the second branchial arch (Reichert's cartilage) for formation of the stylohyoid process. Endochondral ossification in the ear capsule encasing the internal ear occurs within what is called petrous temporal. The portions of the temporal bone that are phylogenetically most recent calcify earliest, in the fifth or sixth fetal month. This is followed in succession by the intramembranous ossification of the squamous portion and the tympanic ring. The developing temporal bone is readily followed on radiographs and these developing milestones are well known to the radiologist. For similar reasons, the occipital bone often holds diagnostic significance. The membraneous bones that comprise the lateral and rostral portions of the cranium appear to begin calcifying before ten weeks. Customarily, frontal and interparietal centers appear followed by rapidly progressing enlargement of the parietal region. By birth, the coronal suture is demarcated with a wide gap known as the anterior fontanelle bordered by the frontal and parietal bones. The posterior fontanelle lies between the parietal and occipital bones. The saggital suture connects both of these openings. Laterally, sphenoid and mastoid fontanelles exist at either end of the squamosal suture. The sphenoid fontanelle represents incompletion in expansive growth of the temporal, great wings of the sphenoid, and frontal and parietal bones, while the mastoid fontanelle indicates the

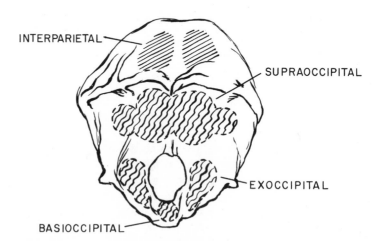

Fig. 3. Ossification centers in occipital bone. The occipital bone is fashioned from separate ossification centers and includes both endochondral and intramembraneous elements. The interparietals, located above the superior nuchal line, are membraneous while the remainder ossify from cartilaginous centers (modified after Patton, 1968).

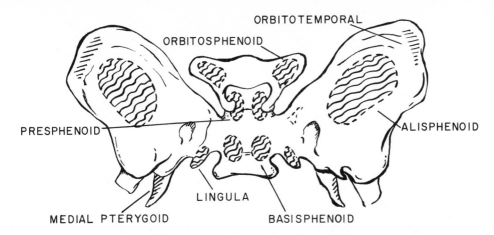

Fig. 4. Ossification centers in sphenoid bone. The bulk of the sphenoid bone arises from endochondral elements with two small exceptions. At the tip of the greater wings and in the medial pterygoid plate, the bone has membraneous origins. This is depicted by the straight lines while all endochondral centers are represented by the irregular lines (modified after Patton, 1968).

same situation involving parietal, squamous temporal, and the separate parts of occipital.

In its over-all aspects, the skull forms as a membrane capsule during the first few weeks of postfertilization. Chondrification develops at the base beginning at around nine weeks. This is followed by discrete areas of ossification in both the membranous vault and cartilaginous base. At birth, the separate parts are not joined but are appropriately positioned. Some rather striking radiologic features can be seen during and in the first few hours after birth, as a demonstrable record of the molding that has taken place in the skull to accommodate it in the journey through the birth canal (Figure 5).

Having reviewed the early development of the neurocranium consequent on rapid brain formation, we go on to consider the visceral section whose foundation stems partly from the branchial arches. A full discussion is not possible here, but this does not intimate that it is insignificant. On the contrary, total comprehension is impossible without it.

The branchial arches are critical to understanding both the development and anatomy of facial structures. Each arch forms its own cartilaginous and muscular framework. The associated vascular and neural components can also be easily identified. Most of the cartilage parts are lost, but some persist as bony or cartilaginous structures. Muscles sometimes migrate from the framework in their own arch, yet their origins are never obscured because they carry with them the nerve supply of each respective arch. The posterior belly of the digastric muscle, for example, extends from the mastoid notch of the temporal, passes through the stylohyoid muscle, and attaches by an aponeurotic sling to the hyoid. Innervation is by the facial nerve (hyoid or second arch). The muscle then swings up and forward to attach at the digastric fossa of the mandibular symphysis. This anterior belly is innervated through the trigeminal nerve (mandibular or first arch). Obviously, this muscle arises from two separate arches.

The first or mandibular branchial arch consists of a maxillary and mandibular process. The latter contains Meckel's cartilage. It largely disappears except for those portions which persist as the incus and malleus. Both the maxilla and

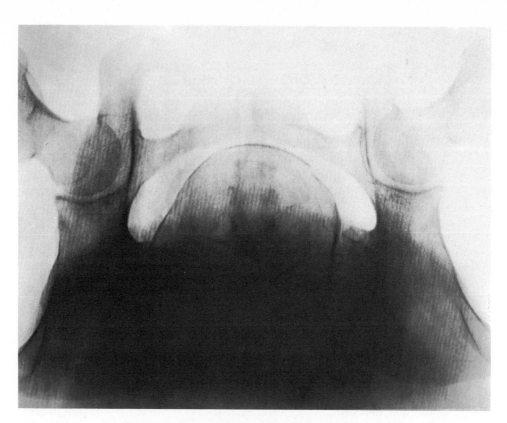

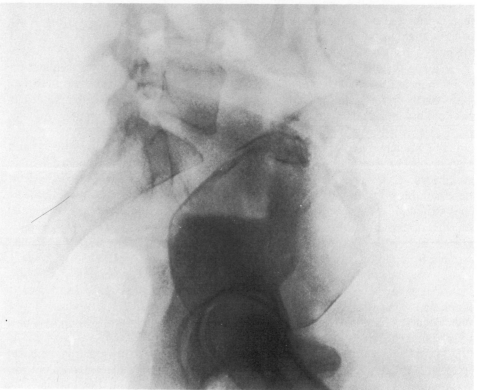

mandible ossify from membranous elements. The Meckel's cartilage serves only as temporary scaffolding. Arch II, or the hyoid, contains Reichert's cartilage forming the stapes, styloid process of temporal bone, the ligament connecting styloid to hyoid, and portions of the hyoid itself. The nerve of the second arch is the facial, and it innervates, naturally, the stapedius and stylohyoid muscles, along with those of facial expression. The digastric has already been noted. Most interesting, from the standpoint of arch development, is the tongue. It arises from the lateral and medial swellings of the mandibular arch, and is joined by elements from arches II, III, and IV. This, then, helps to explain the curious neural pattern of the tongue. Cranial nerves V and IX, anterior to posterior, with the superior laryngeal branch of X near the epiglottis, comprise the sensory component. Taste is derived from VII (chorda timpani traveling with the lingual nerve of V) and IX and X posteriorly. The intrinsic and extrinsic muscles of the tongue (except palatoglossus) are innervated by nerve XII, and this fact supports the supposition that they originate from the occipital somites.

The third through sixth branchial arches and respective aortic arches belong to another discussion, but, as stated before, knowledge about them will serve the student admirably when attempting to comprehend the characteristics of the tongue, floor of the mouth, pharynx, and associated glossopharyngeal and vagus nerves. Clearly, the structures arising from adjacent pharyngeal pouches are equally worth the effort to understand them.

The rostral portion of the developing embryo has one salient characteristic and that is early and swift neurocranial growth. The branchial arches develop slightly later, as already described. This leaves the integral facial structures, which grow out of the tissues surrounding the stomodeum. A series of elevations in the thickening mesenchyme known colloquially as ''swellings or processes'' appear in the fourth or fifth week. The upper face and palate develop from the frontal prominence and a series of paired swellings about the nasal placodes. Caudal to the oral cavity is the mandibular component of branchial arch I, and this seems to present less complexity than the maxillary arch. A more complete discussion should be sought elsewhere (Patton, 1968). By the sixth and seventh weeks, the midface is rapidly taking form. The swellings grow toward the midline and ultimately merge. Often these surface changes get disproportionate emphasis in embryologic discussions, and the underlying transformation of tissue into muscle, connective tissue, cartilage, blood vessels, and nerves is ignored. Nevertheless, the processes can become difficult to comprehend if one overlooks the fact that these swellings are less than distinct entities and that high activity is occurring deep within them. Whether the processes come together and merge either by fusion or by some other alternative seems unresolved (Poswillo, 1975). Unification of the processes (probably median and lateral–nasal) above the primitive oral cavity results in the philtrum of the lips, the jaw part containing four incisor teeth, and a palatal portion known as the primary palate. The secondary palate is formed by a separate process. The deeper parts of the maxillary

Fig. 5. A lateral and posterior–anterior view of an infant with the head engaged in the birth canal prior to birth. The modeling necessary to accommodate the skull is apparent. Note the displacement of the parietal bone and overriding of the occipital bone at lambda in the PA projection (posterior fontanelle). The anterior fontanelle changes are not so easily viewed. Over-all lateral compression and separation at the junction of the frontal, parietal, sphenoid, and temporal bones is demonstrable in the lateral radiograph (anterior lateral fontanelle). (Supplied through the courtesy of Robert C. Winslow, M.D., Dayton, Ohio.)

swellings in conjunction with extension of the head by the fetus results in tongue displacement which in turn allows the vertical palatal shelves to move horizontally and merge with each other and the nasal system. This alignment of the secondary and primary palate results in separation of the nasal and oral cavities through formation of the hard and soft palates.

Certainly by the eighth week, morphogenesis is complete and calcification of the maxillary bone begins. The nasal chambers and nasolacrimal ducts are in continuing development at this same time. Overall, craniofacial contour has seen a dramatic change. During the first month after fertilization, the frontal area is prominent. In subsequent weeks, the middle third of the face undergoes rapid morphological development. The early and explosive growth in the maxillary region gives the fetus decided humanoid features. Up to this time, the mandible has not kept pace and appears relatively small and weak by comparison. The nasofrontal lines become more symmetric and free flowing, fat pads in the cheeks are full, and, finally, the mandible accelerates (free of a prominent chin) bringing the fetus to its cherubic facial contours by birth.

3. Control Mechanisms for Craniofacial Growth

The continuing search for an explanation of the fundamental mechanisms which guide craniofacial growth has for the moment been satisfied by what has come to be known as the functional matrix hypothesis. It holds a paramount position in terms of trying to understand the force or forces which direct the over-all pattern of development in both neural and visceral components. As stated by Moss (1975) " . . . the origin, growth and maintenance in being of all skeletal tissues (cartilage as well as bone) are all secondary, compensatory, and mechanically obligatory responses to the prior demands of related organs, tissues and functioning spaces, these latter being subsumed under the term of functional matrices." Moss has applied a tenable concept which should help in comprehending those factors that guide craniofacial formation. This hypothesis will be continuously tested in the birth-defect field since aberrations in or absence of these so-called matrices should correspondingly alter the respective development fields that each is reputed to control.

Moss considers that every purposeful function in the craniofacial region controls development. Respiration, alimentation, vision, hearing, smell, the entire central nervous system organ, etc., are the respective functions which exert a field of influence over the hard structures which form the scaffolding of that particular function. These are termed skeletal units and include building materials other than bone, such as cartilage and tendon. Taking the calvarium as an example, the bony material develops within a capsular matrix consisting of dura within and scalp outside. The latter includes skin, subcutaneous and aponeurotic layers, and periosteum. It is this complex which is hypothetically sensitive to expansion of the central nervous system. The bone within the capsule is then stimulated by the enlarging brain and this is translated into expansive growth. In this same context, other stimulii must exert additional and very different forces. For example, the dorsal cervical muscle group comprising part of the external capsule must certainly produce an operational force unlike anything in the frontal region. The added posterior pressures should be translated into modification of the occipital bone.

This all seems logical and convincing, but it is nonetheless essential that control mechanisms for a functional matrix hypothesis be explained on a chemical basis. Their pathways of mediation must be elucidated. To satisfy this, Moss (1971) offers the nervous system as the vehicle and a group of so-called neurotrophic substances as the power source. He believes that the epithelial components of any matrix are excited chemically by cytoplasmic substances conveyed within afferent or sensory neurons. Skeletal muscle is similarly stimulated through efferent or motor neurons. Through these peripheral nerve pathways the soft tissue matrix is considered to control the skeletal unit. This concept is made more plausible when one acknowledges the fact that muscular tissue attaches to the periosteum. Albeit teleological, the intimate envelopment of bone by muscle-inserted periosteum helps make neurotrophic mediation an attractive hypothesis.

4. Birth to Maturity

We have considered embryogenesis and the mechanisms of craniofacial development. However, postnatal growth considered as the sum of size increase and shape alteration is the principle interest for most people. Following growth patterns by comparison with the norm gives clear insight into the course of development. It assists the observer in his task of uncovering aberrations early in the course of growth. The ability to predict precise final adult form during the formative years is still an unreached goal, and can be considered only in broad terms. Enlow (1968) gives a notably lucid description of the principles governing maturation of the human face. Since the cranium follows general neural patterns of growth, the major mechanics are complete soon after birth. On the other hand, the face has a prolonged course until adult form is acquired, and major interest centers on it and the cranial base.

Cartilage and bone are the hard structures upon which the craniofacial complex is built. Cartilage has both an appositional and interstitial growth mechanism. It serves the purpose of providing a scaffolding upon which bone forms (endochondral ossification), for example, the chondrocranium. In the developing craniofacial system, cartilaginous accumulations in the mandibular condyles, nasal system, and basal cranium have enjoyed a past reputation as centers of primary expansive growth. This concept is in disfavor at present as little evidence exists for concluding that growth focuses at these sites. The Meckel's and Reichert's cartilages formed out of the first and second branchial arches comprise a special situation again, but they disappear very early in embryogenesis.

Bone has no interstitial component and is appositional in its growth. Osteoprogenitor cells occupy a surface position and, for the most part, periosteal and endosteal surfaces are usually considered as the two bone-producing sites. Added to these are the Haversian surfaces as explained by Frost (1966). Although not continuous in the same recognizable sense as periosteum or endosteum, it comprises a large and important surface area within the two other envelopes.

It is extremely important to recognize that facial and cranial growth is not a single incremental process where new elements are laid upon an existing facade. The "growing crystal" concept is not applicable here. Differential surface remodeling is the guiding process through which the skull and face enlarge and alter in form. Here, growth is the sum of size increase and shape modification. The excellent and exhaustive investigations of Enlow (1968) provide profound insight into general, as well as craniofacial, bone growth. As bone is added to one section and deleted from another, quantitative differences will result in a product that is not only larger, but drastically altered in shape. For example, bone deposited on the lateral surfaces of the mandibular ramus may ultimately comprise the internal surface simply by continued addition on the facial surface and progressive resorption of its oral aspect. That segment of bone has traveled from outside to within and becomes relocated due to contiguous bone deposition and absorption. In sculpture, one may create a figure by continually adding molding material. Unlike the artist's method, human skeletal dynamics has the dimension of resorption complementing the additive process.

Craniofacial remodeling occurs not only on the surface but at other locations known as major growth sites. These include the sutures between bones or parts of bones, synchrondoses between cartilage, junctions of bone with cartilage, and the condylar cartilages. In understanding craniofacial development, it is imperative to appreciate the function of these centers. All too often, persons tend to extrapolate their understanding of postcranial growth to growth of the skull and face. The assumption is made that epiphyseal regions of long bones are synonymous with sutural areas. This is incorrect and more complete knowledge of both these regions and thus their differences can be obtained from the work of Moss (1954), Enlow (1968), and Ham (1974). The endochondral growth site in the long bone provides a mechanism for expansion or lengthening through a growth plate interposed between epiphysis and the contiguous segment of the diaphysis known as the metaphysis. Differential growth and/or death and replacement of cartilage by bone contribute to increased total length. Apparently, the sutural area in the skull comprised of bone edges and interposed soft tissue does not produce expansion through the exertional pressure of growing parts. Rather than serving as major growth sites, the sutures allow for movement between bones as the skull grows by surface opposition and resorption. Functional demands seem to dictate the type of simple or complicated interdigitation in the morphology of the suture line. Suture areas are rather serpentine at the outer table and more linear endocranially. Moss (1959) believes that premature cranial synostosis and skull deformation is not primarily a sutural defect. He contends that the cranial base is spatially altered in position, and it is the pressure transmitted through the aberrant attachment of the dura which inhibits proper suture function: ''The spatial malpositioning of the basal points of dural attachment causes the development of abnormal tensile stresses (hoop tension) within these tissues. This increased tension is transmitted directly by the dura to the tissues of the sutural areas'' (Moss, 1959). Thus, in the face of early life neural growth pressures, the osseous container of the brain cannot function in the usual manner and deformation results. The dolichocephalys, brachycephalys, oxycephalys, and so forth, become prominent not because of sutural area defects but because the abnormal cranial base acts to transmit abnormal pressure on the sutures through the dura.

In addition to the sutures separating individual bones or segments of bone, there are synchondroses which are the growth centers in cartilage. These major sites of developmental activity exist in the largely endochondral cranial base, and actually share a small association with the long bone epiphysis in their manner of operation, with a major exception. A long bone has a growth pattern which proceeds in one direction away from the bone's center, and the result is size increase or lengthening. The synchondrosis, however, is designed for bidirectional growth; by pointing toward a common center, the vectors are offsetting. The sum of this activity is compression and obliteration; therefore, these regions are contained and incapable of enormous expansive growth. There is obviously great contrast between epiphyses and synchondroses relative to over-all effect even though they share cellular similarity.

Since the sutural regions at the sites of both intramembranous and intracartilaginous ossification in the craniofacial area are probably not highly important as areas of expansive growth, there is no reason to expect that closure necessarily occurs at or near attainment of adult size. This is actually the case and patency often persists into the fourth or fifth decade in prescribed regions both anatomically and radiographically. This includes the saggital, coronal, and lambdoidal sutures. Conventional radiography shows the sphenooccipital synchondrosis closure in the later part of the second decade. Fusion of the frontal bones along the so-called metopic suture occurs much earlier, during the first decade of life. Suture closure begins endocranially and proceeds toward the ectocranium.

In summary, suture closure in the craniofacial skeleton area is not similar to that in postcranial areas. Closure of the centers in a bone such as the femur signals the end of extreme statural growth. Conversely, the reductions of the posterior and anterior fontanelles at two months and one year, respectively, indicate developmental milestones, but since the fontanelles function as slip mechanisms rather than primary growth sites, they must not be viewed as indications of growth cessation. The same might be said for the mandibular condyle cartilages where it was long assumed that thrust applied against the condylar fossae by cell growth resulted in a forward and downward enlargement of the lower jaw. This conclusion seems no longer tenable.

5. Skeletal Change and Countenance

The greatest difference between the appearance of the neonate and the adult is in proportionality. The face is small relative to the calvarium. It is generally stated that early and rapid central nervous system development produces a configuration in which the volume of the skull is eight times that of the face as opposed to an adult where the volume is only twice as great. By one year the brain has attained nearly 50% of its total size (in terms of weight). Two years later, this figure is 75%, then 90% by about age 7, and it is essentially complete by 10–12 years of age. Cranial growth follows the dictates of the brain, and the later programmed facial growth patterns proceed along the lines of general somatic development. Therefore, skull and facial proportions vary accordingly. The skull as depicted in the lateral radiograph has an area which is six times the size of the face at one month of age, $4\frac{1}{2}$ by one year, 4 by 4 years, $3\frac{1}{2}$ by 6 years, 3 by adolescence, and reduces thereafter to $2\frac{1}{2}$ by adulthood (Israel, 1978a).

Stated thus, these figures present a rather sterile approach to a dynamic and intricate biological process. Any change stated only in a metrical sense cannot be fully descriptive, but at least it gives a tangible base from which craniofacial growth can be elaborated. Visual experience is the essential ingredient, and it cannot be substituted by numbers. Added to these proportions are those more subtle maturing factors. We anticipate the frontal and parietal prominence in the newborn. The orbits appear large, and they are, in order to accommodate the globes. The cranial base is well developed but flatter than it will be later in life. The mastoids are small, the incomplete temporal bone at the acoustic meatus appears as a ringlike structure, and the sinus cavities of the skull are not well formed. The mandible appears diminutive, the chin is not prominent, and the lack of developed alveolar processes further contributes to the relative diminutiveness of the lower third of the face.

The frontal bone is divided by a metopic suture at birth. The occipital bone is not united and surrounds the foramen magnum as four parts. The sphenoid exists as three segments, and the mandible is separated at the midline. As the infant ages, cranial growth progresses slowly. The bones thicken, skull tables and the diploic space appear, and vascular markings become noticeable.

Shifting attention to the face, the differential growth patterns and displacement processes produce continuing change in shape and form. Patterns of expansion in an area like the chin are easily observed while the intricate adding or subtracting of bone above and below the external oblique line, or above and below the internal oblique line, is less apparent separately. The net effect of these acrobatics is the accomplishment of adult size and contour of the mandible.

To appreciate craniofacial growth fully it is worth recognizing some of its components. These parts have to be elucidated in more than one way or they cannot be fully comprehended. Mensurational methods of assessment as depicted here (Figure 6) demonstrate net effect, and invariably the curve takes an S-shaped configuration when size is plotted against age. Scammon (1930) demonstrated this type of somatic development many years ago; there is rapid pre- and postnatal activity, followed by slowing until adolescent acceleration which continues to a tapering off near adulthood. Contrast this with Scammon's demonstration of the so-called neural type of growth with explosive development early in life, and these are the ingredients which separately are neurocranial and viscerocranial growth. One should be cognizant of the fact that this type of assessment elucidates craniofacial development in a collective sense, and one has simply viewed expansive growth as

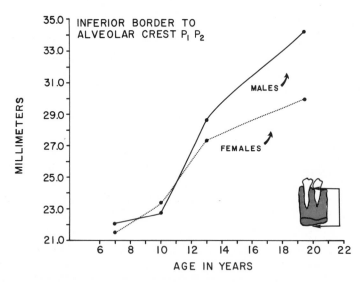

Fig. 6. Surface-to-surface measurements depict enlargement and when plotted relative to age usually have a sigmoid configuration. It is especially important to recognize that the differential patterns of apposition and/or resorption are not elucidated by these methods. There is little more to infer from this type of evaluation than over-all expansion. (Reprinted with permission from Israel, H., 1969, *Arch. Oral Biol.*, **14**:583, Pergamon Press Ltd., Oxford.)

the net effect of surface activity alone. Very obviously, these methods fail to elucidate the differential apposition and resorption factors producing the displacement effect. Microanatomical methods, such as those forming the basis of Enlow's work (1968), must be utilized to depict surface action. At any given point in time, apposition, resorption, or resting inactivity might be observed at the precise location tested.

The important principle to recognize here is that change has its component portions producing a net effect. Understanding skeletal growth requires recognition that enlargement is usually the over-all effect, but within that process is selective apposition and resorption. The surface that adds today may be subtracting tomorrow. Cumulatively, net gain is the result, but changes in form and shape signify that there has been an enormous interplay of addition and subtraction. Without this, growth would be simply a continuing enlargement of the existing image.

6. The Influence of Aging

Even though growth is appropriately measured through its performance early in life, there is no reason to assume that it terminates at any ongoing period. Progressive change in facial contour occurs throughout adulthood to later life. Discounting skin and fatty tissue alteration, which is a matter for another discussion, skeletal transformation is a continual process. The mechanisms and rates of change may differ between preadulthood and postmaturity, but, nevertheless, a reduction and renewal process should not necessarily result in an assumption that the net effect between the two is equal. In the skull, union or near union of sutures inhibits free movement of bones, and therefore displacement type of remodeling is inconceivable. Persistent surface activity of viable bone suggests that skeletal aging produces some change exclusive of drastic shape alteration. A standard anatomy text (for example, Cunningham) portrays a sorry plight for the craniofacial skeleton (Inkster, 1964). First, no teeth are shown. The bony alveolar ridges are absorbed, and the angles of the mandible are obtuse. With the drastic reduction in the lower third of the face, the mandible appears protruding and the maxilla is diminutive— gloomy thoughts for one's own future. None of these processes are inexorable; the loss of the dentition is a product of dental disease and not age. The mandibular angle does not change drastically, and the mandibular condyles hardly atrophy to any degree (Israel, 1973b,c). In short, aging is inevitable, but these depicted skeletal changes are the result of indiscriminantly considering dental disease as a concomitant of age (Figure 7). Obviously, the longer one lives, the more exposure there is to a disease process, and, therefore, the more likely one is to be affected. But this does not assure that everyone will ultimately yield to every disease during their lifetime.

In later life the craniofacial skeletal changes that seem to be occurring relate to bone surface activity. Continuing bone deposition beneath the periosteum produces modest enlargement which is reflected in a measurably larger skull and face over the passing years (Figure 8) (Israel, 1968, 1973a, 1978b). Precise figures are yet unavailable but mean values in the 1% per decade range are probably not exaggerated. Since sutures are rigid and individual bone displacement is inconceivable, expansion proceeds in such a way as generally to reflect uniform enlargement. Change occurs in the image of that which is already present. The skull tables and air sinuses seem to be segments which belie this fact (Israel, 1977). The skull thickens at about twice the pace as the entire skull expands over the same extended time

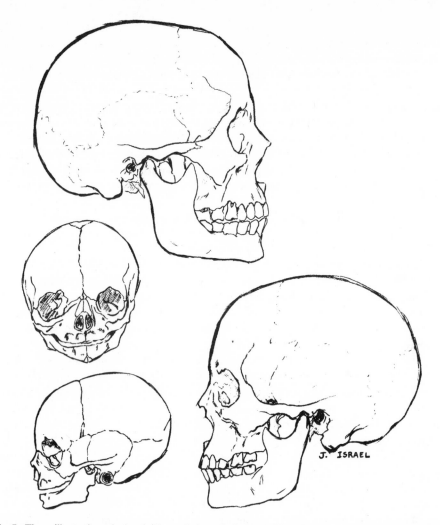

Fig. 7. These illustrations depict skull morphology during early life, adulthood, and older age. The lower right demonstrates later-life configuration. The dentition and supporting structures are represented because there is no evidence to incriminate age as responsible for loss of teeth or supporting bone. Dental caries and periodontal disease eliminate the dentition, not age. Continuing surface apposition of bone produces over-all growth throughout life.

period (Israel, 1973*a*). The bony outline of the sella turcica also fails to subscribe to the generalized over-all cranial pattern of expansion, as it is also somewhat more accelerated (Israel, 1970). How craniofacial aging bone apposition relates to systematic enlargement in the postcranial skeleton remains obscure at present, but it is likely that common associations exist (Garn *et al.*, 1967; Israel, 1967; Smith, 1967).

7. Nutritional Consequences

There is no abundance of information dealing with the effects of mal- or undernutrition in the total craniofacial complex. Short of prominent growth failure in protein–calorie malnutrition, there is little substantive material in biologic writing that reports anything but rudimentary investigation into the skull region. Discounting individual nutritional problems such as rickets and its skull-deforming effects, scant attention has been devoted to craniofacial development among disadvantaged groups in the industrialized nations such as urban or rural poor, or in undernourished individuals in the developing nations. In all likelihood, there are serious potential problems relative to the nutrition factor (Berg, 1973). Inferring from the rapidly accumulating information about the effects of nutrition and brain development, one would certainly expect the calvarium to reflect any underlying reduction in neural tissue. The fact that brain development is 75% complete by two years of age suggests that if brain development were restricted, skull formation would also lag. Therefore, this fact should be apparent early in life, and a persistence of it might be expected to exist for the life of an individual. This general area is discussed fully in Chapter 19.

Information about possible size reduction in the skull with undernutrition exists, and this further attests to the unexplored facial relationships. Two examples will be considered. Using the case of iodine nutriture as a starting point, a deficiency of this mineral has created a serious medical problem in many parts of the world. This is painfully apparent in the Andean region of South America (Stanbury, 1972). Rural Indian villages are burdened with endemic goiter, and a sizable number of offspring are cretinous. Endemic cretinism in Ecuador implies mental deficiency, deaf–mutism, goiter, dwarfism, and neuromuscular aberrations reflected in visible gait disturbances and muscular incoordination (Querido, 1969). The problem of characterizing cretinism is discussed in Chapter 14. The craniofacial skeleton of a young dwarfed cretin matched against a nonaffected contemporary is presented in

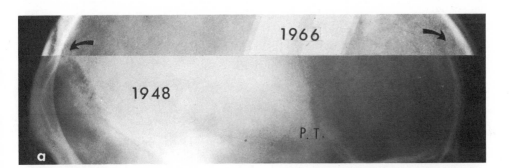

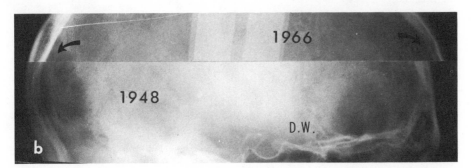

Fig. 8. The extent and configuration of aging enlargement in the skull is demonstrated here in two subjects. Over an elapsed 18 years, longitudinal changes are depicted by cropping these skull radiographs through the calvarium and then opposing the older segment with one from younger age. Skull diameter and thickness are demonstrably greater (Reprinted with permission from Israel, H., 1968, *Arch. Oral Biol.*, **13:**133, Pergamon Press Ltd., Oxford.)

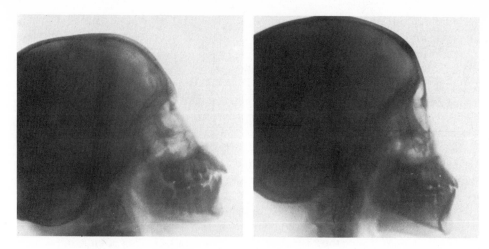

Fig. 9. The 11-year-old cretinous female on the left demonstrates the forward position of the face, a lack of frontal prominence, and conspicuous absence of the frontal sinus. Her noncretinous contemporary on the right emphasizes the defective development in cretinism. These radiographic plates were exposed under field conditions with a portable X-ray unit.

Figure 9. The cretin on the left clearly demonstrates the anterior positioning of the face relative to the neurocranium. The frontal sinus is not conspicuous, and the backward sweeping frontal region produces an altered craniofacial contour. In this obviously dwarfed individual the ratio of facial to skull size is not drastically altered. The demeanor of two so-called nervous cretins is clearly demonstrated in Figure 10. Their appearance must certainly be the cumulative effect of intellectual level, motor control, and the aberrant pattern of skeletal development. Further skeletal and dental characterizations are available elsewhere (Israel *et al.*, 1969*a,b*).

Fig. 10. These photographs of two adult cretins emphasize the unusual associations in the neural and visceral cranial structures (Supplied through the courtesy of John B. Stanbury, M.D., Department of Nutrition and Food Science, Massachusetts Institute of Technology, Cambridge, Massachusetts).

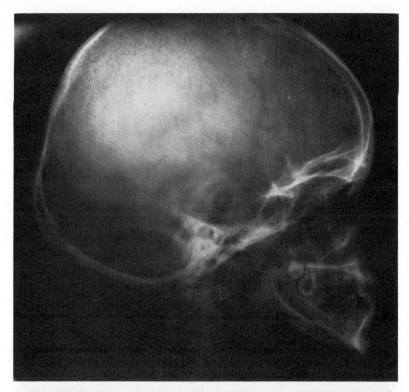

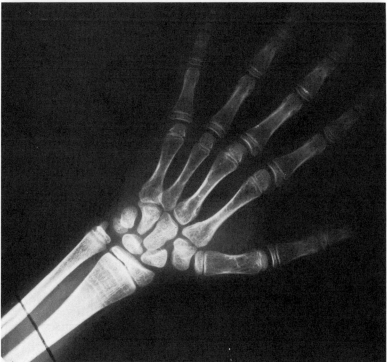

Fig. 11. This zinc-deficient male is from Egypt, and the facial structures belie his age of 20 years. The associated hand radiograph appears to be from a child of about 11 years old, and thus restricted somatic development is obvious. (Supplied through the courtesy of Harold H. Sandstead, M.D., Human Nutrition Research Laboratory, USDA, Grand Forks, North Dakota.)

Another problem of mineral nutrition is dwarfism and zinc deficiency (Ronaghy *et al.*, 1969; Halsted *et al.*, 1972; Patwardhan and Darby, 1972). In the Middle East, a syndrome of dwarfism, hypogonadism, iron-deficiency, and anemia are purported to be related to zinc and geophagia. The craniofacial system in these dwarfs is remarkable not only for its diminutive size but for the immature appearance. The face of the 20-year-old shown in Figure 11 remains small relative to the mature-appearing calvarium. The associated hand radiograph adjudged to have a skeletal age of 11 years (Greulich and Pyle, 1959) emphasizes what the skull film demonstrates. Somatic development has been severely restricted, and this is nowhere more apparent than in the face and skull. Accelerated body growth and sexual maturation can be induced at even the late age of 20 years with nutritional therapy including zinc (Halsted *et al.*, 1972).

These two examples of abnormal development help to illustrate some of the relationships of growth between the neural and visceral cranium. The defect in cretinism obviously has an intrauterine component which is reflected in the face and calvarium. The non-mentally-retarded zinc-deficient dwarf has little skull affliction, it is just that the somatic pattern of facial development has been restricted. The former cannot be benefited by replacement therapy, while the latter still retains some potential for improvement into the third decade of life.

8. Aberrant Facial Formation

8.1. Birth Defect

A segment about craniofacial malformation can hardly appear as more than an afterthought when tacked onto a discussion about normative development because of the enormous expansion of knowledge in the birth-defect field. Formerly, it was customary to find a short discussion with a few drawings or photographs about malformation at the termination of any discourse on growth. Usually malunion of facial planes would occupy some space just prior to the bibliography. This is no longer the case, but it is appropriate to point out some generalizations about the normal which might assist in the comprehension of the abnormal. Good physical diagnosis demands four basics: inspection, percussion, palpation, and auscultation, and inspection begins with the face. Family resemblances are strong; even when members tend not to look alike there are still some characteristics, although indescribable, which lay bare the common genetic source. Congenital malformation can wipe this away, and simply by looking at the face a "syndrome" can be identified, meaning that the defective individual bears a greater resemblance to those who share his or her developmental anomaly than to his own family. Down's syndrome, Crouzon's disease, and others explain this readily. There are many syndromes which have traits in common; yet they are not as profoundly obvious as the defects mentioned.

To explore normative development relative to malformation, knowledge about the formation of the face is helpful (DeMyer, 1975). This includes the processes from the mesenchyme about the stomodeum of the first branchial arch. The frontal, median–nasal, and lateral–nasal swellings are the precursors of the upper third and median portion of the middle third of the face. These include frontal region, nose, philtrum of the lip, the jaw component of the four incisor teeth, and the primary palate. The mandibular and maxillary processes of the first branchial arch are responsible for the secondary palate, middle third portion lateral to the alae of the nose, and the entire lower third. In a most simplified form, the forehead portion is exclusively the product of the frontal prominence. The maxillary or midportion is shared in origin by the maxillary processes of branchial arch I and the swellings of the nasal placode. The lower is of solitary beginnings by way of the mandibular processes of the first branchial arch. Failure of unification results in clefts. In broad context, knowing the source of each respective portion of the face helps one understand the resulting defective configuration of the affected segment.

The birth-defect problem often involves multiple organ systems, and skull and facial anomalies occur with a high rate of frequency under innumerable causative processes and varied physical appearances. This has been the impetus behind the exhaustive nomenclature used for identification purposes (Gorlin *et al.*, 1971; Pruzansky, 1975). To this end, the environmental effects, chromosomal aberrations, metabolic errors, and the rest deserve extensive coverage in their own right and none can have justice served them as part of a treatise on normative development. Suffice it to say that understanding the usual events of development can measurably aid the elucidation of that gone awry—whether this be thalidomide etiology, chromosome trisomy, or problems yet to be appreciated.

8.2. Improving Facial Form

If, as is almost certainly the case, the functional demands of growing soft parts profoundly influence osseous form, then it is essential to understand that each organ system develops at a rate that is not necessarily similar to every other. Rapid progressive growth or relative cessation of growth in one segment may be succeeded or overlapped by the same process in one or more other parts.

The maxilla is used as an example here. Early in life it is small relative to the calvarium, the orbits are large, and the alveolar ridges are not well developed. Looking at the constituent influencing factors, the brain expands rapidly at a young age and promotes skull growth, the orbits are large to accommodate the globes which nearly attain adult size by 7 years of age, and the alveolar ridges are diminutive since the dental apparatus has yet to require the space that will become necessary later on. This essentially demonstrates the dynamics of facial configuration based upon demand. One region may have maximally expanded, another might be in the midst of its greatest input, while still another has yet to produce its contribution. To illustrate this further, muscle mass following a pattern of general somatic growth will play a continuing major role before "final form" is realized. An area such as the zygomatic complex will respond to masseteric needs, and the infratemporal fossa has to accommodate the fleshy mass of the expanding temporal muscle.

Selective influences on the ultimate product point out some of the challenges facing investigators and clinicians alike. Understanding the facets of growth, recognizing the temporal sequences, and predicting final form are aspects that would allow the therapist to know whether or not to intervene. In a more subtle sense, there is the question of when one should attempt any intervention and, finally, what effect it will have upon the product. Not only does one want to know the effects, but whether or not they are lasting.

Dentocraniofacial form may be influenced by both mechanical and surgical means. The orthodontist attempts to arrange the dentition in an orderly fashion and thereby influence the associated skeletal form. Using the mechanical means at his disposal, teeth are moved within bones for purposes of establishing a healthy dentition and improving over-all facial contours.

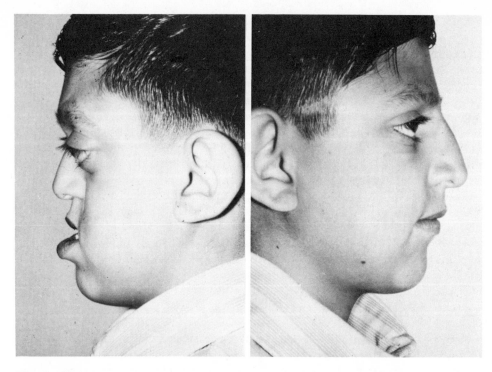

Fig. 12. This is an 11-year-old boy with Crouzon's disease. The posttreatment photograph on the right demonstrates the pleasing symmetrical arrangement that can be accomplished with combined surgical and orthodontic therapy. (This and the succeeding two figures were supplied through the courtesy of Bruce N. Epker, D.D.S., Ph.D., Fort Worth, Texas).

The past three or four decades have also seen advances in craniofacial surgery. Improvements in the use of prolonged anesthesia and postoperative care have enabled surgeons to alter dramatically the whole of the midface. This orbitocraniofacial surgery now offers rather broad and sweeping assistance to individuals with birth defects who carry the indelible mark of what has come to be known as "facial Gestalt" in the syndrome field. Many of these individuals are of high intelligence, and their position in society might be greatly influenced by the stigma of their appearance. Assistance for them has established a new surgical discipline for the facially deformed. The kind efforts of Dr. Bruce N. Epker* demonstrate the surgical methods that are now available to assist these people (Figures 12–14). Both orthodontic and surgical means were utilized. The midface was actually separated from its contiguous structures and relocated anteriorly, and bone grafts relieved the orbital rim deficiences.

Preventing the problem of craniofacial malformation is an obvious ultimate goal. In the meantime, treating deformity is desirable both as a corrective and preventive measure. The example above demonstrates what can be accomplished by surgically repositioning the component portions and amplifying those segments that are deficient. The ideal measure of accomplishment is to meaningfully realign the bony structures so that the growth process will not be impeded, and in that way continuing improvement of the craniofacial contours may be anticipated.

*DDS, PhD, Director, Department of Oral-Maxillofacial Surgery, Tarrant County Hospital District, Fort Worth, Texas.

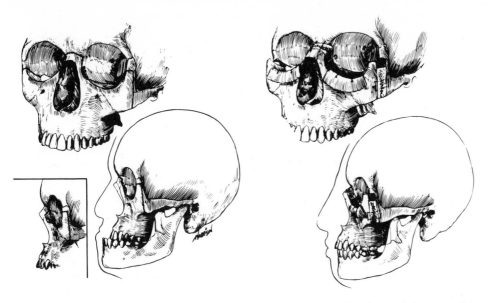

Fig. 13. This illustrates the surgical fracture of the midface, its anterior relocation, and the appropriate bone grafting methods to maintain placement and construct definitive orbits.

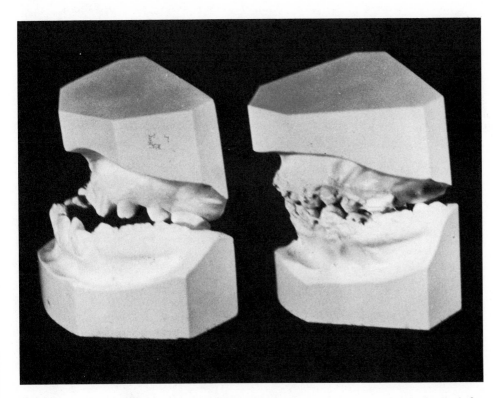

Fig. 14. The extent of the alteration in interarch relationships can be seen with this before-and-after treatment arrangement of the boy's dental casts.

It is imperative that growth be recognized as a lifelong function because a more profound understanding of personal health and individual well-being cannot be accomplished without considering life as a continuous spectrum. Many problems of aging certainly must be imprinted upon the organism early in life. Trying to unravel these nuances might be fruitless unless our biological vision is panoramic. Toward the intention to achieve a broader view of health-related factors, Robbins (1974) stated, " . . . the greatest impact upon the quality of life is to be achieved through preventive measures applied at an early age. This applies no matter whether one is concerned with the role of infectious agents, nutrition, or the general environment. Those of us who deal with children must better inform ourselves on the life-long spectrum of human illness. We must have a greater sensitivity for the long-term implications of the problems with which we deal and the measures we prescribe with the intent of improving health."

The requisite for an integrated comprehension of disease is an understanding of all temporal periods of life. We designate growth as an early affair and call this childhood. There is no quarrel with the term childhood, rather the hope that this discourse has helped to extend our view of the growing period from beyond spring and summer and into autumn and winter.

ACKNOWLEDGMENTS

Jane B. Israel deserves special thanks for contributing the illustrations. Appreciation also goes to Donna M. Sanders and Trudy J. Lawrence for technical assistance and manuscript preparation. Partial financial support of this work was through research grants DE 01294 and HD 09154 from the National Institutes of Health, Bethesda, Maryland. The Fels Research Institute also contributed continuously and generously.

10. References

Berg, A. D., 1973, *The Nutrition Factor: Its Role in National Development,* The Brookings Institution, New York.

Broadbent, B. H., Sr., Broadbent, B. H., Jr., and Golden, W. H., 1975, *Bolton Standards of Dentofacial Developmental Growth,* C. V. Mosby, St. Louis, Missouri.

Caffey, J., 1973, *Pediatric X-Ray Diagnosis,* 6th ed., Year Book Medical Publishers, Chicago.

DeMyer, W., 1975, Median facial malformations and their implications for brain malformations, in: *Morphogenesis and Malformation of Face and Brain* (D. Bergsma, ed.), pp. 155–181, The National Foundation–March of Dimes, Birth Defects: Original Article Series, Vol. II, No. 7, Alan R. Liss, New York.

Dorst, J. P. 1964, Functional craniology: An aid in interpreting roentgenograms of the skull, *Radiol. Clin. North Am.* **2:**347–366.

Enlow, D. H., 1968, *The Human Face: An Account of the Postnatal Growth and Development of the Craniofacial Skeleton,* Hoeber Medical Division, Harper and Row, New York.

Frost, H. M., 1966, Bone dynamics in metabolic bone disease, *J. Bone Joint Surg.* **48A:**1192–1203.

Garn, S. M., Rohmann, C. G., and Wagner, B., 1967, Bone loss as a general phenomenon in man, *Fed. Proc.* **26:**1729–1736.

Gorlin, R. J., Cervenka, J., and Pruzansky, S., 1971, Facial clefting and its syndromes, in: *The Third Conference on the Clinical Delineation of Birth Defects, Part II, Orofacial Structures,* held at the Johns Hopkins Hospital, Baltimore, Maryland, June 15–19, 1970, sponsored by the Johns Hopkins

Medical Institutions and The National Foundation–March of Dimes (D. Bergsma, ed.), Birth Defects: Original Article Series, Vol. 7, No. 7, Williams and Wilkins, Baltimore.

Greulich, W. W., and Pyle, S. I., 1959, *Radiographic Atlas of Skeletal Development of the Hand and Wrist,* 2 ed., Stanford University Press, Stanford, California.

Halsted, J. A., Ronaghy, H. A., Abadi, P., Haghshenass, M., Amirhakemi, G. H., Barakat, R. M., and Reinhold, J. G., 1972, Zinc deficiency in man: The Shiraz experiment, *Am. J. Med.* **53:**277–284.

Ham, A. W., 1974, *Histology,* 7th ed., p. 431, J. B. Lippincott, Philadelphia.

Henderson, S. G., and Sherman, L. S., 1946, The roentgen anatomy of the skull in the newborn infant, *Radiology* **46:**107–118.

Inkster, R. G., 1964, Osteology, in: *Cunningham's Textbook of Anatomy,* 10th ed. (G. J. Romanes, ed.), p. 126, Oxford University Press, London.

Israel, H., 1967, Loss of bone and remodeling–redistribution in the craniofacial skeleton with age, *Fed. Proc.* **26:**1723–1728.

Israel, H., 1968, Continuing growth in the human cranial skeleton, *Arch. Oral Biol.* **13:**133–137.

Israel, H., 1969, Pubertal influence upon the growth and sexual differentiation of the human mandible, *Arch. Oral Biol.* **14:**583–590.

Israel, H., 1970, Continuing growth in sella turcica with age, *Am. J. Roentgenol. Radium. Ther. Nucl. Med.* **108:**516–527.

Israel, H., 1973*a,* Age factor and the pattern of change in craniofacial structures, *Am. J. Phys. Anthropol.* **39:**111–128.

Israel, H., 1973*b,* Recent knowledge concerning craniofacial aging, *Angle Orthod.* **43:**176–184.

Israel, H., 1973*c,* The failure of aging or loss of teeth to drastically alter mandibular angle morphology, *J. Dent. Res.* **52:**83–90.

Israel, H., 1977, The dichotomous pattern of craniofacial expansion during aging, *Am. J. Phys. Anthropol.* **47:**47–52.

Israel, H., 1978*a,* Skull and face relationships among malformed children, *J. Dent. Res.* **57:**187 (Abstr. No. 452).

Israel, H., 1978*b,* Evidence for continued apposition of adult mandibular bone from skeletalized materials, *J. Prosthet. Dent.* (in press).

Israel, H., Fierro-Benítez, R., and Garcés, J., 1969*a* Skeletal and dental development in the endemic goiter and cretinism areas of Ecuador, *J. Trop. Med. Hyg.* **72:**105–113.

Israel, H., Fierro-Benítez, R., and Garcés, J., 1969*b,* Iodine therapy for endemic goiter and its effect upon skeletal development of the child, in: *Endemic Goiter,* Report of the meeting of the PAHO scientific group on research in endemic goiter held in Puebla, Mexico, June 27–29, 1968 (J. B. Stanbury, ed.), pp. 360–372, Scientific publication number 193, Pan American Health Organization, Washington, D.C.

Johnston, L. E., 1975, A simplified approach to prediction, *Am. J. Orthod.* **67:**253–257.

Johnston, M. C., 1975, The neural crest in abnormalities of the face and brain, in: *Morphogenesis and Malformation of Face and Brain* (D. Bergsma, ed.), pp. 1–18, The National Foundation–March of Dimes, Birth Defects: Original Article Series, Vol. II, No. 7, Alan R. Liss, New York.

Langman, J., 1969, *Medical Embryology: Human Development—Normal and Abnormal,* 2nd ed., Williams and Wilkins, Baltimore.

Moss, M. L., 1954, Growth of the calvaria in the rat, *Am. J. Anat.* **94:**333–362.

Moss, M. L., 1959, The pathogenesis of premature cranial synostosis in man, *Acta Anat.* **37:**351–370.

Moss, M. L., 1971, Neurotrophic processes in orofacial growth, *J. Dent. Res.* **50:**1492–1494.

Moss, M. L., 1975, New studies of cranial growth, in: *Morphogenesis and Malformation of Face and Brain* (D. Bergsma, ed.), pp. 283–295, The National Foundation–March of Dimes, Birth Defects: Original Article Series, Vol. II, No. 7, Alan R. Liss, New York.

Patton, B. M., 1968, *Human Embryology,* 3rd ed., pp. 224–373, McGraw-Hill, New York.

Patwardhan, V. N., and Darby, W. J., 1972, The zinc-deficiency syndrome, in: *The State of Nutrition in the Arab Middle East,* pp. 122–140, Vanderbilt University Press, Nashville, Tennessee.

Poswillo, D., 1975, Causal mechanisms of craniofacial deformity, *Br. Med. Bull.* **31:**101–106.

Pruzansky, S., 1975, Anomalies of face and brain, in: *Morphogenesis and Malformation of Face and Brain* (D. Bergsma, ed.), pp. 183–204, The National Foundation–March of Dimes, Birth Defects, Original Article Series, Vol. II, No. 7, Alan R. Liss, New York.

Querido, A., 1969, Endemic cretinism: A search for a tenable definition, in: *Endemic Goiter,* Report of the meeting of the PAHO scientific group on research in endemic goiter held in Puebla, Mexico, June 27–29, 1968 (J. B. Stanbury, ed.,), pp. 85–90, Scientific publication number 193, Pan American Health Organization, Washington, D.C.

Ricketts, R. M., 1960, Cephalometric synthesis: An exercise in stating objectives and planning treatment with tracings of the head roentgenogram, *Am. J. Orthod.* **46:**647–673.

Robbins, F. C., 1974, The long term effects of infection in early life (long view II), *Pediatr. Res.* **8:**927–976.

Ronaghy, H., Spivey Fox, M. R., Garn, S. M., Israel, H., Harp, A., Moe, P. G., and Halsted, J. A., 1969, Controlled zinc supplementation for malnourished school boys: A pilot experiment, *Am. J. Clin. Nutr.* **22:**1279–1289.

Scammon, R. E., 1930, The measurement of the body in childhood, in: *The Measurement of Man* (J. A. Harris, C. M. Jackson, D. G. Paterson, and R. E. Scammon, eds.), pp. 173–215, University of Minnesota Press, Minneapolis.

Smith, R. W., Jr., 1967, Dietary and hormonal factors in bone loss, *Fed. Proc.* **26:**1737–1746.

Stanbury, J. B. 1972, The clinical pattern of cretinism as seen in highland Ecuador, in: *Human Development and the Thyroid Gland: Relation to Endemic Cretinism* (J. B. Stanbury and R. L. Kroc, eds.), pp. 3–17, Plenum Press, New York.

14

Skull, Jaw, and Teeth Growth Patterns

PATRICK G. SULLIVAN

1. Introduction

The growth of the human skull has been recorded in many ways over the centuries, ranging from the drawings found in the tombs of ancient Egypt to the computer-drawn plots of modern times. Although both the advent of radiography and the data-processing power of computer technology have permitted a much improved understanding of the over-all patterns of skull growth, the stimuli and control mechanisms which govern the growth of an individual are still subject to debate and controversy.

Developmentally the structure of the skull can be divided into two components: one, the neurocranium, concerned with the support and protection of the brain, and the other, the viscerocranium (nasofacial complex), concerned with the mechanisms for respiration, mastication, and speech. The relationship and evolution of these components has to be appreciated for a full understanding of skull growth. The configuration of the human skull, together with that of the other members of the suborder Anthropoidea, differs markedly from the remainder of the animal kingdom. Perhaps the most obvious and certainly the most significant difference is in the relative proportions of the neuro- and viscerocranium. In man evolutionary development has been accompanied by a marked increase in brain capacity, which has resulted in enlargement of the neurocranium. Furthermore as the neural component has increased in volume, it has also become progressively repositioned relative to the remainder of the cranial architecture. During the same evolutionary time span other changes have taken place. Thus with the adoption of an upright posture, the opening of the foramen magnum has been relocated from the posterior to the inferior surface of the skull. The summation of these changes is seen

PATRICK G. SULLIVAN • London Hospital Medical College, Dental School, London, England.

in the progressive movement of the viscerocranium from in front of the neural component to a position underneath the forebrain (Figure 1).

The growth pattern of the human skull in its passage from neonatal to adult form is a measure of the difference in growth characteristics between the neural and visceral components. The neurocranium ceases development at a comparatively earlier age, and the pattern of skull growth reflects the dynamic relationship

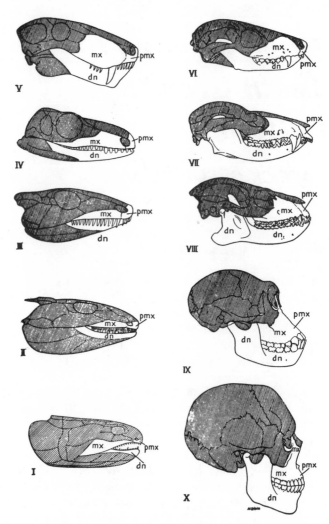

Fig. 1. Evolution of the maxillary bones from lobe-finned fish to man (pmx, premaxilla; mx, maxilla; and dn, dentary). In the ascending series, consisting of primitive amphibian, stem reptile, and three stages of mammal-like reptiles, the dentary and maxilla increase greatly in size. In the earlier adult mammals the dentary becomes the sole bone of each half of the lower jaw, and the premaxilla and maxilla together form the subocular surface of the face. The transition from this stage to that of man involves chiefly the deepening and anteroposterior shortening of the maxillary bones (including the premaxilla, maxilla, and dentary). (I) Lobe-finned fish, Devonian age (essentially Rhizodopsis). After Traquair, Watson, Bryant. (II) Primitive amphibian (Palaeogyrinus), lower Carboniferous. After Watson. (III) Primitive cotylosaurian reptile (Seymouria), Permocarboniferous. After Broili, Williston, Watson. (IV) Primitive theromorph reptile (Mycterosaurus), Permocarboniferous. After Williston. (V) Gorgonopsian reptile (Scymnognathus). After Broom. (VI) Primitive cynodont reptile (Ictidopsis), Triassic. After Broom, Haughton. (VII) Recent opossum. (VIII) Primitive primate (Notharctus), Eocene. After Gregory. (X) Anthropoid (female chimpanzee), Recent. (X) Man, Recent. (From Gregory, 1927.)

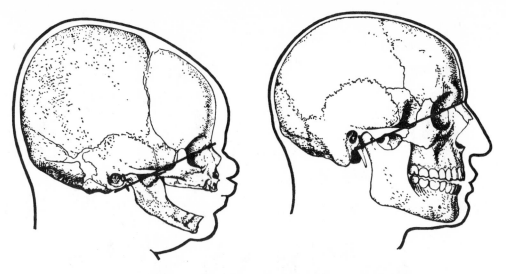

Fig. 2. The jaws at birth compared to the adult (After Tulley and Campbell, 1975.)

between the later-developing facial complex and the base of the skull. The change in shape between the infant and the adult skull shown in Figure 2 is related to the relative increase in size of the nasofacial complex (viscerocranium) when compared to the calvaria (neurocranium). Between birth and adult life the volume of the calvaria increases four times, and the volume of the facial region about 12 times. Furthermore as 80% of the postnatal growth of the calvaria occurs during the first two years, the major part of calvarial growth is completed long before facial growth is terminated.

Although the skull may be viewed developmentally as being capable of division into two component regions, this is not demonstrated by the bone architecture. The demands of evolution have resulted in the development of a complex structure made up from 22 individual bones. Thus change in size and shape is achieved by the integrated activity of many component bones, each bone having its own individual growth pattern, which may differ from the others in the complex in both direction and rate of growth.

2. Methods Employed to Record Skull Growth

Three mechanisms are available to bring about changes in shape of a bone. These are: ossification of cartilage, new bone addition at a suture (Pritchard *et al.*, 1956), and remodeling by deposition and resorption at the periosteal and endosteal surfaces. All three mechanisms are involved in skull growth, each making a varying contribution to different stages in the achievement of adult skull form.

The growth pattern followed by a point on a given bone surface can only be appreciated by comparison with a fixed point, which may be taken as an unchanging fiducial marker, constant in all three spatial planes. One of the complicating factors in the appreciation of skull growth is that there are no absolute fixed points from which growth can be said to have taken place. Any selected anatomic or radiographic landmark is itself liable to movement by the growth activity of both the bone on which the landmark is situated and that of the adjacent bones.

PATRICK G. SULLIVAN

A

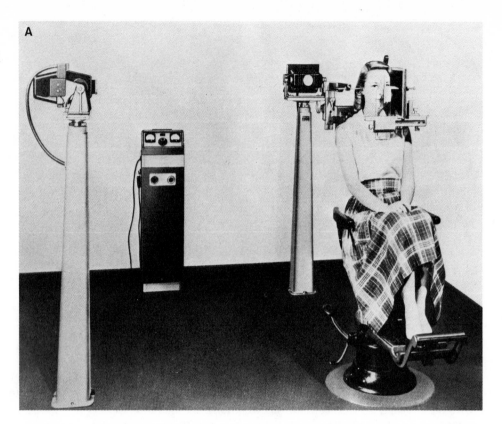

B

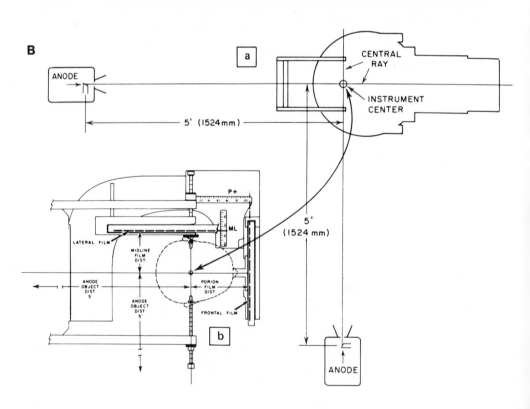

A landmark on the surface of a bone of the skull may move in space relative to a fixed point by the following mechanisms: (1) Growth by surface bone deposition. (2) Growth at a suture between the bone bearing the landmark and an adjacent bone. In this case because of the new bone formation at the sutural interface between the bones, the landmark is moved away from the fixed point by the bodily movement of the bone on which the landmark is located. (3) The landmark-bearing bone may be carried away from the fixed point by the movement of an intervening bone nearer to the fixed point.

An ideal recording method should be able to register not only change in position of selected landmarks, but also the three-dimensional shape of a constituent skull bone. The method should also permit the measurement of small changes in the same individual to allow longitudinal growth analysis (Tanner, 1951). The nearest approach to these requirements is found in the use of intravital dyes in experimental animals (Hunter, 1798; Harris, 1960; Suzuki and Matthews, 1966), where the detailed relationship between the successive dye-marked contours and the surface of the bone may be established and measured by the use of a three-dimensional reconstruction method (Sullivan, 1972).

In the study of human skull growth, where the examination must obviously be limited to the use of noninjurious methods, these techniques are not applicable, and longitudinal growth recording is limited to the use of direct craniometry and radiography.

Several workers have employed methods of direct craniometry on the living subject; for instance, Goldstein (1936) examined 100 Jewish males at biannual intervals between 2 and 21 years of age. All measurement techniques employed in face and jaw growth studies are liable to error, the source of the error residing in the characteristics of the measurement method. In the case of direct craniometry, the error lies in the accurate location of surface landmarks through the overlying tissues.

Other investigators have made extensive use of radiography, and this approach confers the ability to measure the relative positions of anatomical structures within the skull as well as on the surface. There are two basic drawbacks to the use of a radiographic method: first, the necessity of achieving the same head orientation at successive recordings, and secondly the distortion encountered when a "shadow" of a complex three-dimensional object is thrown onto a two-dimensional film.

A variety of methods have been suggested to obtain the same head position on successive radiographs. Lynsholm (1931) employed an optical method whereby the head was positioned by means of images in a mirror system. However, Broadbent (1931) adapted an instrument previously employed for anthropological measurements for use as a radiographic craniostat (Figure 3), and instruments based on this design have become widely used for skull-growth recording.

The instrument employs two adjustable ear rods, inserted into the external auditory meati, and a pointer is adjusted to contact the left infraorbital foramen. The ear rods are positioned such that they are coincident in a beam of X rays, and a lateral view of the skull is recorded.

Fig. 3 A. Broadbent–Bolton radiographic cephalometer used in the Bolton Study, Case Western Reserve University, Cleveland, Ohio (Broadbent *et al.*, 1975). B. Bolton cephalometric floor plan. (a) Relation of the anodes, the central rays, and the instrument center. (b) Relation of the instrument center to the lateral and frontal films.

A magnification error is inherent in recording different parts of the same skull in the lateral skull view. Those structures farthest from the film will be magnified to a greater degree than those nearest to the film. This error is minimized by setting the X-ray source at a distance from the subject to obtain as near a parallel beam of X rays as possible. The remaining magnification error may be overcome by measurement of a frontal radiograph of the same skull. From this view the distance of a structure from the X-ray film may be calculated and an appropriate correction factor applied to the measurement taken from the lateral film. A further means of calibrating these errors was described by Adams (1940), who used a radiopaque scale projected onto the radiograph.

The relative movement of all parts of the craniofacial complex during growth has already been stressed. Thus, if cephalometric radiographs are used, either in the case of a cross-sectional survey on different individuals or in a longitudinal survey on the same individual, a uniform superimposition method is required to provide a true representation of the growth changes. Broadbent evolved a method based on the identification of certain anatomical landmarks associated with the skull base which could be identified with most certainty on a cephalometric radiograph. These landmarks were employed as the points of origin for the reference planes used for the superimposition of successive radiographs. He described a reference plane, the Bolton plane, drawn between the suture between the frontal and nasal bones and the highest point in the notch posterior to the occipital condyle. This plane was used to demonstrate the growth pattern of the facial complex (Figure 4). Since 1931 many other planes and landmarks have been used for fiducial purposes (Krogman and Sassouni, 1957). The selection of these landmarks was made on the assumption that as the base of the skull has completed most of its growth at a relatively early stage, it may be considered as a fixed base against which the growth of the facial skeleton takes place. The statistical reliabilities of some of these planes and points were examined by Bjork and Solow (1962).

Bjork (1955) contributed a major advance in the use of radiography by developing a method for the insertion of small metallic pins into the bones of the jaws. The pins are inserted beneath the periostal surface and are unaffected by subsequent changes in bone shape. When radiographs are taken, the pins appear as small radiopaque points. If tracings taken from successive radiographs are superimposed on the radiopaque points, the true change in bone shape is demonstrated (Figure 5). An alternative superimposition method, based on the position of the auditory canals, was published by Delattre and Fenart (1958). They suggested that in view of the function of the semicircular canals in the appreciation of spatial position, they might be used for the superimposition of skull radiographs.

Enlow (1966) adopted a method whereby tracings taken from radiographs were superimposed according to growth patterns derived from histological studies of areas of bone deposition and resorption found in postmortem skulls. Although not a quantitative method, this approach has proved valuable in identifying the growth mechanisms contributing to regional skull growth. Several investigators have used radiographs taken from views of the skull at right angles to each other to examine the three-dimensional positions of points in the radiographic image (Broadbent 1931; Ricketts et al., 1972).

In 1972, Savara (1972) described a method whereby the spatial positions of selected landmarks were abstracted from radiographs taken simultaneously at right angles. By a process of triangulation the landmarks are positioned in space by means of three measurements. However, the difficulty in identifying the same point on the two radiographs is such as to render this approach of limited accuracy.

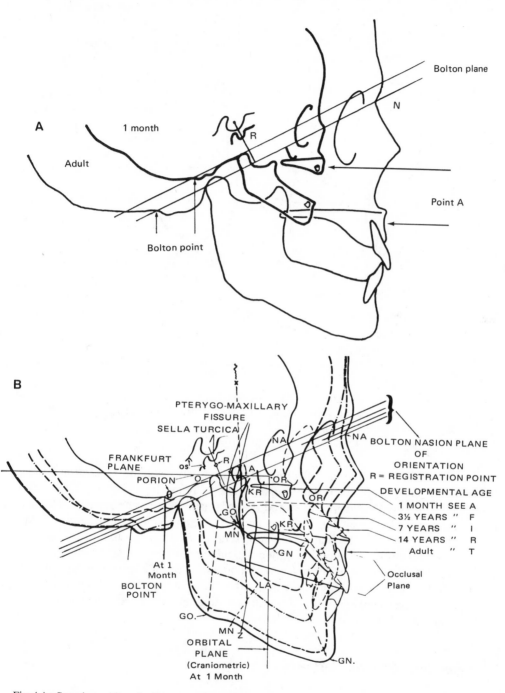

Fig. 4 A. Superimposition of radiograph tracings, using the Bolton plane for orientation. The two tracings are superimposed on the R point, which is found halfway along a perpendicular to the Bolton plane drawn from the midpoint of the sella turcica. B. Composite tracing showing mean growth trends (Broadbent, 1931).

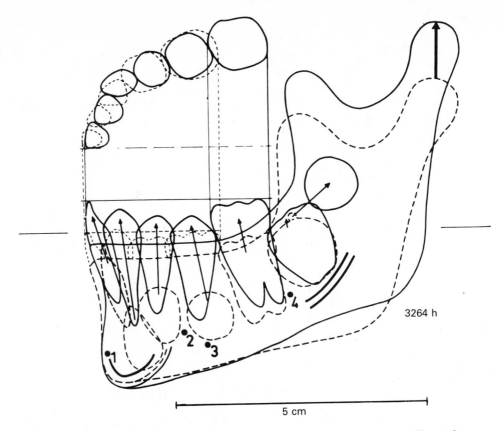

3264 h

5 cm

Fig. 5. Growth of the mandible analyzed by superimposition on implant markers. The tracings are superimposed on the images of four pins inserted into the bone. Broken line, age 5 years, 8 months; solid line, age 10 years, 8 months (Bjork, 1963).

The ability of a computer to store and process large volumes of information has led to the establishment of data banks based on recordings obtained from radiographic surveys. Walker (1972) stored tracings using a computer-based technique, and as part of his survey developed an atlas of the mean stages for progress of skull growth at given ages throughout the adolescent period (Figure 6).

The use of radiography, the only practical approach to the study of human jaw growth, has played an important part in understanding the trends of human skull growth. However, the study of the detailed movements of forming bone surfaces is limited both by the resolution of the clinical radiograph and by its inability to record the shape of a contour in three dimensions.

3. Regional Growth Patterns

The skull is a complex structure formed from many component bones which articulate along an intricate pattern of junctions. Thus any individual change in size of a component bone must be accompanied by a balanced readjustment in the adjacent bones. Similarly more distant bones must also adjust in size such that an integrated, progressive change in shape is seen throughout the growth period. It is

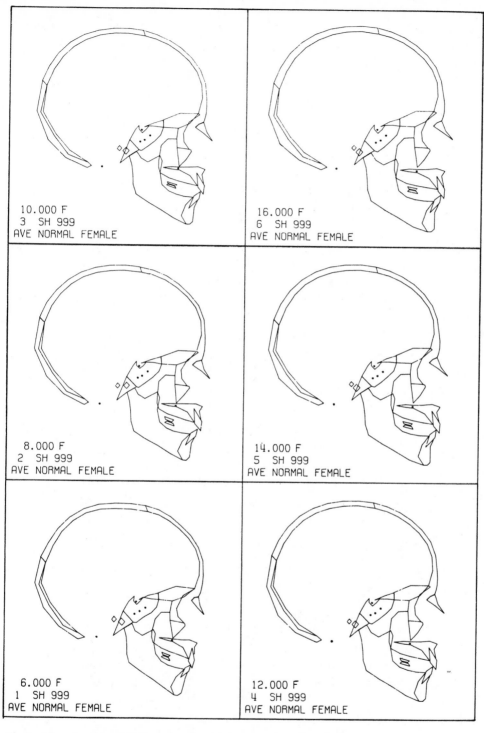

Fig. 6. Atlas of average skull dimensions for girls from 6 to 16 years of age. Sample size for each group is approximately 100 (Walker, 1972).

thus misleading to analyze skull growth on a regional basis, as if each region grows in isolation without regard to the over-all skull-growth pattern. However, the different regions of the skull do vary one from another, not only in the obvious morphologic and functional characteristics, but also in the balance between the growth mechanisms employed to bring about the integrated change in size and shape.

3.1. Calvaria

The cranial vault is constructed from the frontal and parietal bones and parts of the temporal, occipital, and sphenoid bones. These bones are platelike in structure and are formed by ossification in membrane.

At birth the bones are separated anteriorly, posteriorly, and at other locations in the calvarium by six fontanelles. Each fontanelle is bridged by a fibrous membrane, which becomes progressively reduced in size by bone formation around the periphery of the adjacent bones. The anterior fontanelle is the last to close, at about 18 months, leaving the bones of the cranial vault in contact along a complex pattern of suture lines.

The increase in size of the cranial vault necessary to accommodate the progressively enlarging brain is accomplished by bone deposition along the lines of the sutural system. The diagram in Figure 7 demonstrates the new bone formation

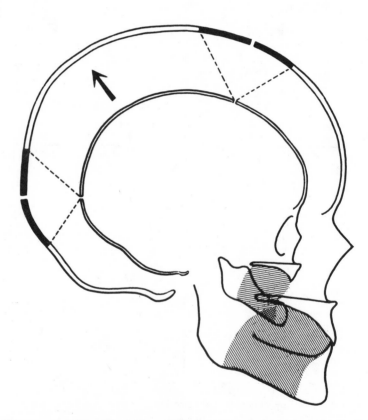

Fig. 7. Radial movement of the calvarial bones following expansion of the brain. Compensatory sutural growth shown in black. (After Salzmann, 1966.)

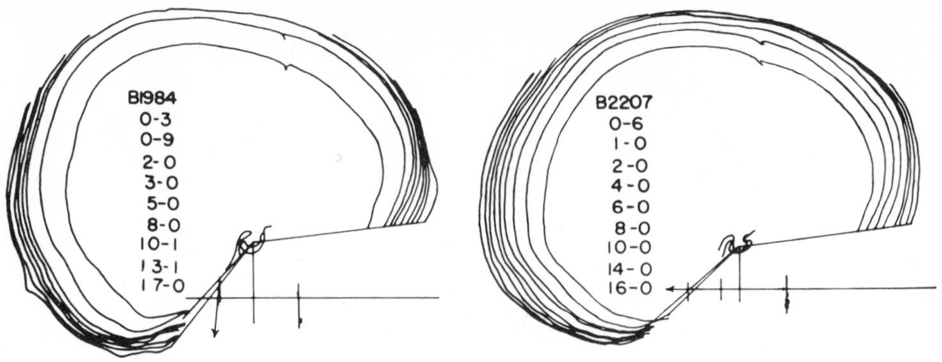

Fig. 8. Progressive growth of the calvaria. Tracings from radiographs at ages (years and months) shown inside the composite drawings (Brodie, 1953).

which has taken place at the coronal and lambdoidal sutures in order to accommodate the increase in size of the cranial contents.

The curvature of the surface of a large sphere is less than that on a small sphere. The adult calvarium is larger than the infant's and shows a corresponding reduction in curvature. Thus, in addition to increase in circumference by sutural deposition, the bones of the cranial vault undergo a progressive flattening. The reduction in curvature is carried out by selective bone deposition and resorption at areas over both the inner and outer surfaces of the bones.

Originally the sutures were regarded as primary growth centers (Weinmann and Sicher, 1955), that is the new bone laid down at the suture interface was thought to be responsible for the movement apart of the adjacent bones. This has been questioned by several authors (Moss, 1954; Scott, 1956), and it is now more generally accepted that sutural growth is a secondary phenomenon, involved with roofing over the separation between the bones created by the effect of increase in size of the enclosed soft tissues. Thus increase in brain volume will cause a potential separation at, for instance, the lambdoidal and coronal sutures, which is filled in by bone formation at the suture.

The progressive growth pattern of the calvaria has been demonstrated by Brodie (1953), using the superimposition of tracings taken from lateral skull radiographs (Figure 8). The spacing of the tracings over the period from 6 months to 4 years indicates a rapid phase of growth. After this age the tracing lines become more coincident, indicative that only a relatively small increase in sagittal cranial growth occurs after 4 years, shown mainly by an increase in the anteroposterior dimension.

3.2. Cranial Base

The cranial base is largely composed of bones formed by the ossification of cartilage precursors. The base is formed from the basal part of the occipital, the sphenoid, the petrous part of the temporal, and the ethmoid bones. The main postnatal growth mechanism found in this region is provided by endochondral ossification at the sphenooccipital synchondrosis (Figure 9) which is active until about 12–15 years and fuses at about 17–20 years of age. Cartilage replacement at this site is related to forward growth of the whole anterior cranial segment.

Anterior movement in this manner causes an increase in the anteroposterior dimension of the nasopharynx and carries the upper facial skeleton forward. At the

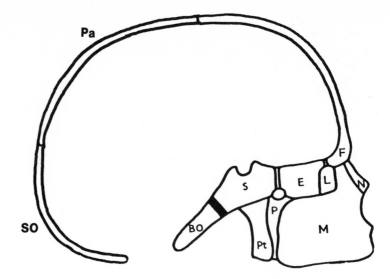

Fig. 9. The bones of the calvaria, cranial base, and upper face. Position of the sphenooccipital synchondrosis shown in black. M, maxilla; N, nasal bone; F, frontal bone; L, lacrimal bone; E, ethmoid; P, vertical plate of palatine bone; S, body of the sphenoid; Pt, Pterygoid plate; BO, basioccipital; Pa, Parietal.

same time the glenoid fossae (the sites of articulation of the mandibular condyles) are carried posteriorly and inferiorly (Figure 13). Thus the anteroposterior growth of the mandible must be greater, when compared to the maxilla, if the bony bases carrying the dentition are to be related in a harmonious fashion. The direction of growth also has a vertical component which tends to open the space between the maxilla and mandible associated with the vertical development of the alveolar processes.

The process of endochondral ossification in a synchondrosis appears to be similar to that found in the epiphyseal cartilage of a long bone, except that, unlike the epiphyseal cartilage, ossification occurs on two surfaces. In the case of the sphenooccipital synchondrosis ossification takes place on both the sphenoidal and the occipital faces of the cartilage.

The presence of an area of cartilage with interstitial growth potential, which has been shown to be active during the period of skull growth, has led to its implication as a primary growth center for the base of the skull. Evidence for this view is obtained by examination of the effects of developmental disorders which affect the formation and growth of cartilage. For instance in achondroplasia, the base of the skull is shortened, with a flattened palate, sunken bridge to the nose, and a general reduction in the development and size of the facial region. Conversely the growth of the vault of the skull, which does not rely on the ossification of cartilage for growth, is unhindered (Brash, 1956).

The growth characteristics of cartilage have been studied experimentally by Koski and Ronning (1970), using an intracerebral transplantation technique in the rat. They found that the growth capacity of the transplanted cartilage depended on the origin of the cartilage, that from the epiphysis showing a continuing ability to proliferate while the growth of the synchondral cartilage was much reduced. Nevertheless synchondral cartilage grows normally *in situ*. Thus, unlike an epiphyseal plate, it appears that the sphenooccipital synchondrosis requires an additional extrinsic factor for the stimulation of normal growth.

Although the deepening of the middle cranial fossae on either side of the midline is related to increase in the temporal lobes of the brain, Scott (1967) suggested that growth of the midline cranial base is largely independent of the growth of the brain. He drew his evidence from the finding that patients suffering from microcephaly show a normal range of cranial-base dimensions despite the marked reduction of brain volume.

Whether the synchondrosis acts as a primary growth center or not, it must not be regarded as the only growth mechanism participating in cranial base growth. If independent endochondral ossification were to occur in the sphenooccipital synchondrosis, the skull would be split into two parts, anterior and posterior to the synchondrosis. No splitting takes place because a synchronized process of sutural growth and surface remodeling takes place in the adjoining bones, with the effect of producing an integrated change in size and shape of the whole bone complex. In this instance all three mechanisms of endochondral ossification, sutural growth, and surface remodeling participate in the achievement of an harmonious change in skull form.

3.3. Upper Face

The bones of the upper face ossify in membrane and are developmentally associated with the capsules of the eyes, nose, and mouth cavity.

The upper facial region consists of the nasal, lacrimal, maxilla, zygomatic, vomer, palatine, and pterygoid bones. It will be seen that this regional complex is closely related to the anterior segment of the cranium formed by frontal, ethmoid, and sphenoid bones (Figure 9). Thus any relative forward growth of the anterior cranial base will carry the upper facial region with it into a more anterior position.

The growth of the nasomaxillary complex takes place in a downward and forward direction and is related to the growing eyeballs and nasal septum (Figure 10). The increase in height of this region contributes to the relative increase in height of the face compared with the neural part of the cranium and is related to the enlargement of the maxillary sinus and the development and eruption of the dentition.

The nasofacial region is attached to the anterior skull base by a complex series of sutures, including the frontomaxillary, zygomaticotemporal, zygomaticosphenoidal, zygomaticomaxillary, ethmomaxillary, ethmofrontal, nasofrontal, frontolacrimal, palatine, and vomerine sutures. The orientation of this circumfacial system is such that sutural growth will result in the downward and forward movement of the facial complex when compared with the base of the skull (Scott, 1967).

It is suggested by Enlow (1975) that more vertical displacement takes place by sutural growth in the posterior part of the facial complex than in the anterior region. The balance is said to be maintained anteriorly by surface deposition and resorbtion. This involves the concept of sutural slippage, the relative movement of adjacent bones past one another by adjustment at the sutural interface, described by Latham (1968).

The marked increase in vertical height of the maxilla, which is associated both with increase in the size of the maxillary sinus and with development of the alveolar bone, is brought about by surface deposition and internal resorption (Figure 11).

A measure of the increase in size of the maxilla is given by the change in size of the maxillary sinus. At birth the sinus occupies a small depression in the lateral wall of the nasal cavity, but by adult life the small depression has expanded to a large cavity.

The growth pattern shown by the maxilla, the major bone in the upper face, provides an example of the operation of the three mechanisms of bone movement described on p. 383. If the movement of a landmark on the anterior surface of the maxilla from soon after birth to adult life is examined (point A in Figure 4A), all three growth mechanisms are found to participate in the movement.

First, the landmark (point A) will move in a mainly downward direction by bone deposition on the inferior-facing surfaces of the maxilla (Figure 11). An equivalent, integrated degree of bone resorption is found on the superiorly facing surfaces. This remodeling is associated with the increase in maxillary height which occurs with the enlargement of the maxillary sinus and the development and eruption of the dentition.

Secondly, the landmark will move following sutural growth at the circumnasal suture system. The effect on point A of growth at these sutures is in a downward and forward direction, but in this instance the movement is achieved by the bodily movement of the whole maxilla.

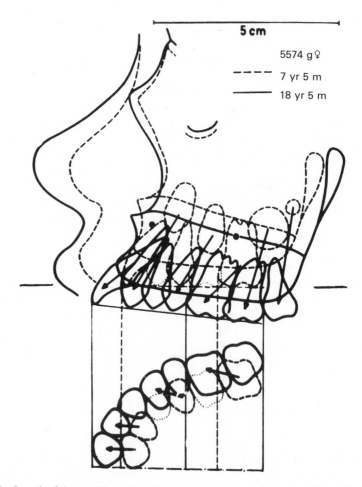

5 cm

5574 g ♀

— — — 7 yr 5 m

——— 18 yr 5 m

Fig. 10. Growth of the maxilla analyzed by superimposition on implant markers (Bjork, 1964).

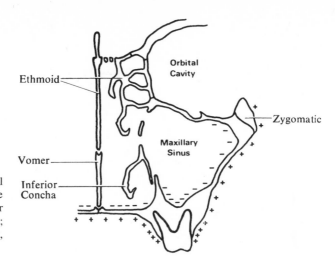

Fig. 11. Growth of the upper facial skeleton. Coronal section at the level of the first permanent molar tooth. Surface deposition, +; resorption, −. (After Enlow, 1975.)

Finally, the landmark is repositioned in a forward direction by endochondral bone formation at the sphenooccipital synchondrosis. In this instance the whole anterior cranial base has been repositioned forward by an enlargement of the middle cranial fossa.

Furthermore, it will be appreciated that the movement of one point on the surface of the maxilla in a downward and forward direction does not mean that all the points in the maxilla are growing in the same direction at the same rate. For instance, at stages during the development of the molar teeth, the posterior aspect of the maxilla is growing rapidly in a distal direction.

This increase in anteroposterior length is obtained by deposition at the posterior borders of the alveolar arch to form the maxillary tuberosities. In the tuberosity is found developing the next molar tooth to be added to the dental arch. Thus at 3 years, the tuberosity contains the developing first permanent molar, at 6 years the developing second permanent molar, and at 12 years the developing third permanent molar. As the current molar tooth erupts the posterior development of the tuberosity continues to accommodate the next tooth in the series.

The part played in maxillary growth by sutural deposition and surface remodeling is well recognized; the contribution made by ossification of cartilage is more controversial. In the midline of the nasomaxillary complex is found the septal cartilage which articulates in infancy with the ethmoid sphenoid, vomer, and premaxillary bones (Figure 12). This structure is derived from the cartilage of the embryonic nasal capsule and is thus formed from primary cartilage. At the septoethmoidal junction the cartilage is replaced by bone in a zone of endochondral ossification (Baume, 1961). Furthermore the cartilage is attached anteriorly to the premaxilla by a septopremaxillary ligament (Latham, 1969). The growth potential of cartilage and the anatomy of this region have prompted several authors (e.g., Scott, 1967) to suggest that the nasal septum acts as a primary growth center for the normal development of the upper facial region. This view is supported by experiments involving the resection of all or part of the nasal septum in experimental animals, with a resulting distortion in the normal growth pattern of the facial region (Wexler and Sarnat, 1961; Sarnat and Wexler, 1966, 1967). A similar result was found by Gange and Johnston (1974), who severed the septopremaxillary attachment in rats.

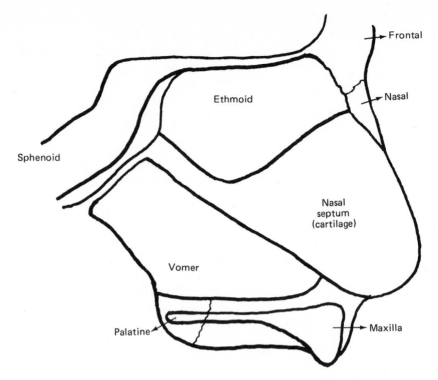

Fig. 12. The ansal septum and adjacent bones in a young child.

These results are contradicted by the work of Moss *et al.* (1968) in the rat, and by Stenstrom and Thilander (1972) in the guinea pig. These papers suggest that the nasal septum acts solely as a structural component and has no primary determining effect on facial growth. Moss *et al.* (1968) also claim that in humans, congenital absence of the cartilaginous nasal septum leads to little reduction in facial growth.

4. Mandible

The human mandible is a membrane bone which forms in close association with Meckel's cartilage, the first branchial arch cartilage. At birth the bone is in two parts, joined in the midline by the symphysis menti, which closes by the end of the first year of life.

No other bone of the skull has been more extensively studied, the basic growth pattern being first described by John Hunter in the 18th century. The mandible is unusual in having a secondary growth cartilage located posteriorly under the surface of the articular condyle. The secondary growth cartilage differs from a synchondrosis or epiphyseal cartilage in that it is derived from a membrane condensation and is not a remnant of a primary cartilage precursor. Not only does the cartilage differ in origin but also in structure. It is covered with a layer of fibrous tissue which forms the articular surface, unlike other examples of cartilage which are bounded by bone on both sides of the area of interstitial cartilaginous growth.

The over-all change in shape is shown in Figure 2. The most striking difference is found in the relative increase in height of the ramus when compared with the body of the mandible. This change in proportion is reflected by the decrease in the angle between the posterior and the inferior borders of the bone.

The growth to adult dimensions is brought about by a combination of endochondral ossification of the condylar cartilage and by remodeling over the bone surface. The larger part of the increase in the height of the mandibular body is brought about by addition to the alveolar bone on the superior surface, and similarly most of the increase in length occurs through a process of progressive distal relocation of the ramus (Figure 13). The movement of the ramus is effected by deposition on the posterior border and resorption from the anterior border of the bone. This process was first described by Hunter (1798) and subsequently confirmed in a classical experiment by Humphrey (1863). He inserted wire rings into both the anterior and posterior borders of the mandibular ramus in the pig. After growth the anterior ring became extruded from the bone and was found in the soft tissues, while the posterior ring became more deeply embedded.

The remodeling changes seen in the development of an individual mandible, over a 5-year period are shown in Figure 5. In this case the two mandibular tracings have been superimposed on the images of metal pins inserted into the bone, and the change in outline represents the true contour change experienced by the individual.

As with other sites of cartilage proliferation, the mandibular secondary growth cartilage has been suggested as a primary growth center for mandibular development. This view is supported by the observation that damage to the condyle at an early age may be followed by a reduction in mandibular growth (Rushton, 1944). Subsequent studies in experimental animals have provided results which suggest that following condylectomy only minor changes (Sarnat, 1963) or no change (Giannelly and Moorrees, 1965) in mandibular growth pattern occurs. It may well be that modification of the growth pattern following early damage may be due to

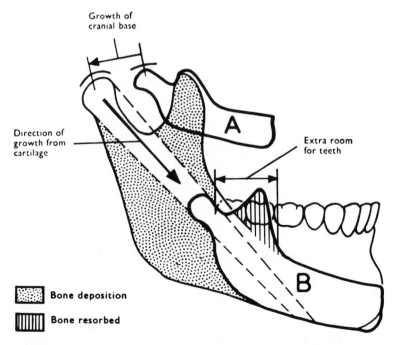

Fig. 13. Growth changes in the mandible. (After Weinmann and Sicher, 1955.)

contraction or fixation by scar tissue rather than loss of the growth potential of the condylar cartilage. Moss and Rankow (1968) claim that careful surgical resection of the condyle in childhood produces little reduction in mandibular growth.

The matter is further complicated by the effects of pathological conditions on the secondary growth cartilage. In eosinophilic adenoma of the pituitary gland (which causes gigantism before and acromegaly after the fusion of the epiphyses) the mandibular growth cartilage reacts like epiphyseal cartilage, and a characteristic overgrowth of the mandible occurs. However, in the hereditary disease achondro-plasia, unlike the long bones (epiphyseal cartilage), the mandible continues to grow with a normal growth pattern. Furthermore, transplantation studies have shown that, unlike epiphyseal cartilage, transplanted condylar cartilage has very little independent growth potential (Koski and Ronning, 1965). Thus it would seem that on balance the evidence is against the role of the condylar cartilage as a primary growth center.

5. Dentition

In man two sets of teeth are formed. The deciduous dentition (20 teeth) starts to erupt at 6 months and is completed by about 2½ years. Eruption of the permanent dentition (32 teeth) commences at 6 years and the last of the deciduous teeth are shed by 12–14 years. The permanent dentition is usually completed by the end of the second decade.

The timing of both the development and eruption of the teeth has been studied by several authors, for instance Moorrees *et al.* (1963*a,b*). The increase in arch length necessary to accommodate the increase in number of teeth is achieved both by expansion of the dental arch and by the addition of bone at the ends of the arch. The shape of the dental arch is established with the eruption of the deciduous dentition, and the arch expands in size until about 8 years (Figure 14) (Sillman, 1964). After this age only slight increase in transverse arch width is found, the major enhancement in length being provided by increase at the posterior ends of the arch. In the upper jaw arch length is increased by bone deposition in the tuberosity region of the maxilla. In the lower jaw, space for eruption of the posterior teeth is provided by the distal relocation of the mandibulas ramus.

The positions of the teeth in the vertical dimension are influenced by several factors: (1) Movements associated with the initial eruption of the teeth. (2) Move-ment associated with surface deposition and remodeling of the alveolar bone after tooth eruption. Confirmation of this form of vertical movement is found in the misnamed clinical condition of tooth "submergence" (Figure 15). In this case a single tooth becomes ankylosed to the surrounding bone and does not experience the vertical development shown by the adjacent teeth. Although originally in occlusion with the teeth of the opposing arch, the tooth is overtaken by the vertical development of the neighboring teeth, and is subsequently found below the surface of the bone. (3) Bodily movement of the bone on which the teeth and alveolar bone are based.

The growth and remodeling changes of both the mandibular ramus and the middle cranial fossa produce a progressive lowering of the mandibular arch. The space created by this aspect of growth is filled by the vertical development of the craniofacial complex. During this process the teeth are maintained in occlusion by the operation of the factors mentioned above, but to a differing degree in the two

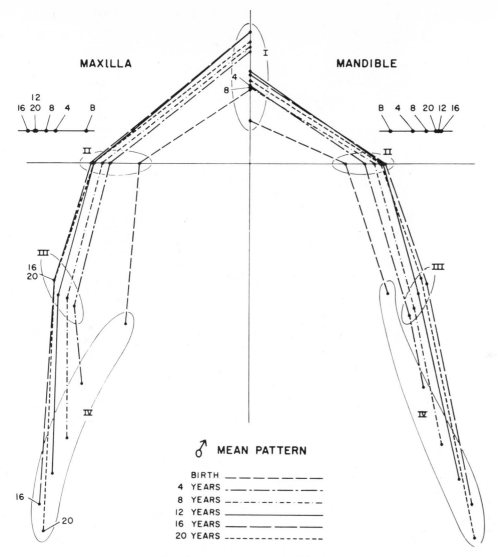

MAXILLA   MANDIBLE

♂ MEAN PATTERN

BIRTH	
4 YEARS	
8 YEARS	
12 YEARS	
16 YEARS	
20 YEARS	

Fig. 14. Composite mean growth pattern of male maxilla and mandible from birth to 20 years of age (Sillman, 1964).

arches. For instance, the vertical movement of the maxillary teeth is effected by surface remodeling to a greater degree than that of the mandibular teeth (Enlow, 1975).

6. Analysis of the Patterns of Skull Growth

The over-all growth of the facial complex in a downward and forward direction was demonstrated by Broadbent as part of the Bolton Study involving the measurement of radiographs from 5000 children (Figure 4). The progressive change in facial form is a cross-sectional study and represents the mean pattern of facial growth.

A similar approach to skull growth, by the derivation of the mean growth

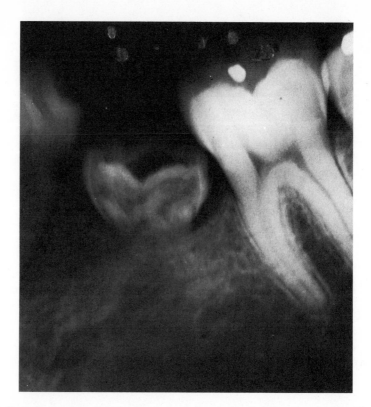

Fig. 15. Radiograph showing relative submergence of a previously fully erupted tooth. The tooth has become ankylosed to the supporting alveolar bone and has not experienced the occlusal movement shown by the adjacent teeth.

pattern, has been illustrated (Figure 6), where Walker has constructed the mean profile derived from the recording of 100 subjects at each stage in development.

Many other workers (Wylie, 1947; Bjork, 1947; Downs, 1948) have developed analytical strategies for the assessment of facial shape and form which have been extensively used in planning orthodontic and surgical treatment. For instance, the Steiner analysis (1953, 1959, 1960) relates the positions of the teeth and jaws to the cranial base and permits a comparison of these relationships with the population average.

Other authorities have developed methods which have sought to illustrate the discrepancy between the individual and the mean by the use of a graphical representation.

Voorhies and Adams (1951) drew out a polygon, based on Downs (1948), representing the range of variation about the mean of 10 selected measurements taken from cephalometric radiograph tracings (Figure 16). The measurements record the configuration of both the skeletal and the dental elements in the skull tracing. The value for each measurement taken from any individual radiograph is entered on the chart, and a straight line is drawn between the points. The "shape" described by this line is characteristic of the overall skull configuration of the individual concerned.

Moorrees and Yem (1955) described a method to illustrate the skull growth changes based on the coordinate analysis method described by D'Arcy Thompson

(1917). The changes in shape and position are demonstrated by the derivation of a rectangular mesh which is superimposed on the tracings of cephalometric radiographs taken before and after treatment (Figure 17). The relationships of certain radiographic landmarks to the horizontal and vertical mesh lines in the pretreatment tracing are taken as a basis of reference. The relationships of the same landmarks to the mesh lines in the posttreatment tracing are then compared to those in the pretreatment tracing. The horizontal or vertical mesh lines in the posttreatment tracing are changed, if necessary, in order to achieve a similar proportionate distance of landmarks to mesh lines as observed in the tracing made before treatment. The resulting distortion of the mesh lines calls attention to differences in the position of landmarks within various rectangles of the diagram.

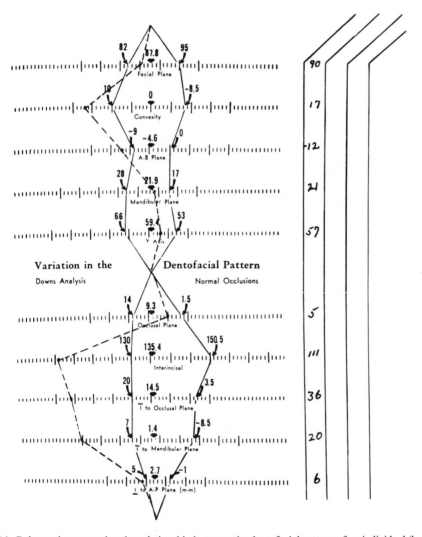

Fig. 16. Polygon demonstrating the relationship between the dentofacial pattern of an individual (broken line) and that of the population (solid line). The upper half of the polygon records the range of skeletal measurements, and the lower half refers to the range of dental measurements (Voorhies and Adams, 1951).

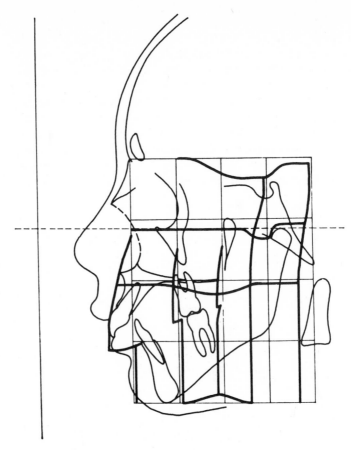

Fig. 17. Mesh analysis of craniofacial growth in an 11-year-old boy. See text for an outline of mesh construction procedure (Moorrees and Kean, 1958).

This approach was extended by Moorrees and Kean (1958) to examine the relationship between head orientation at rest and certain of the intracranial reference lines. All their radiographs were taken using an optical method to ensure a "natural head position." A true vertical marking registered on the radiograph was employed as a reference line in the construction of the rectangular mesh. The relationship between the cranial reference lines and the mesh was used to orient the cranial base to the vertical and to study the relationship between anatomical structure and the posture and profile of the head. In addition they derived the average facial pattern of 50 female students. The ranges of individual variation about the mean are shown in Figure 18.

With the completion of longitudinal growth studies came the realization of the individual nature of jaw growth. It was realized that individual variation within the group is just as normal a characteristic of the phenomena as a whole as is the fact of the mean value. The possible range of the individual growth pattern was particularly well demonstrated by the use of the fiducial marker-pin technique developed by Bjork (Figure 19). In a study of mandibular growth in 45 males, Bjork (1963) found mandibular growth could vary widely between a predominantly downward and backward to a predominantly downward and forward pattern (Figure 20).

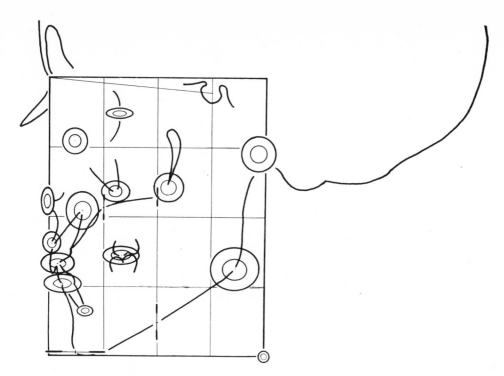

Fig. 18. The average facial pattern, based on 50 North American females. The concentric ovals show the ranges of individual variation around the mean at the one and two standard deviation limits (Moorrees and Kean, 1958).

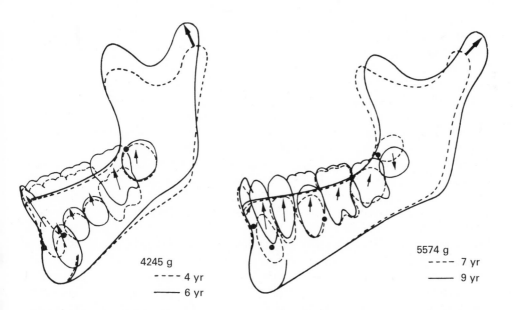

Fig. 19. Variation in mandibular growth pattern. Examples illustrated show extremes of difference in pattern. (Bjork, 1955).

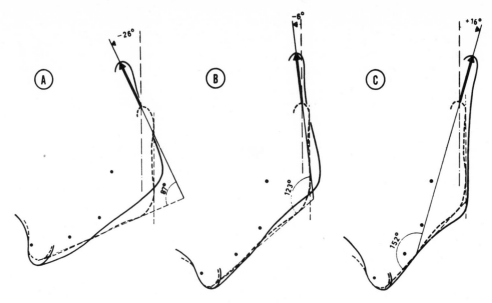

Fig. 20. Diagram illustrating range of mandibular growth pattern in 45 males. B illustrates the mean; and A and C show the extremes of condylar growth direction (Bjork, 1963).

An approach published by Solow (1973) suggested the use of a theoretical model system to represent the pattern of skull growth (Figure 21). He divided the skull into five regions, cranial base, maxilla, maxillary alveolar bone, etc. Each region is represented diagrammatically on a computer-controlled visual display. The simulated growth of each region may be varied in magnitude, direction, and timing by the operation of a computer program. With the use of this model system, the effect that variation in regional growth has on the over-all skull growth pattern may be tested and compared with that found in radiographic studies.

The use of a computer-based technique was extended by Ricketts (1973; Ricketts *et al.*, 1972) to make predictions of individual growth. He collected and stored information on skull growth both from his own and other published studies. The growth patterns found in this data are analyzed according to ethnic type, age, sex, and size. The appropriate pattern is applied to the state of skull formation shown in a radiograph tracing to obtain an estimation of the skull form to be expected after a selected time interval. The predicted form is demonstrated by a computer-drawn plot of the skull. The prediction may be further modified by the anticipated effect of orthodontic treatment on the positions of the teeth. On the basis of these data, Ricketts suggested that the basic growth pattern of the mandible follows a circular path, the particular curve for an individual being derived from a geometric construction based on tracings taken from the lateral skull radiograph (Figure 22).

The accuracy of this method of growth prediction was examined by Mitchell *et al.* (1975). They studied eight cases using the position of metal implants in the jaws to compare the predicted tracing and the tracing taken at the end of the growth period. They claimed an acceptable accuracy of prediction in six of the eight uses studied.

The growth pattern of the mandible was examined by Moss and Selentijn,

(1970) and Moss *et al.* (1974), using a metal pin superimposition method. These authors also suggested a curving path for mandibular growth, in this case based on a logarithmic spiral (Figure 23). They claimed that the radiographic positions of three neurovascular foramina, the mandibular foramen, mental foramen, and foramen ovale, are found to lie, during growth, along the path of a logarithmic spiral.

There is some support for the concept that the nerve foramina and canals might in some way act as orienting points or pathways during skull growth. Koski (1973) has drawn attention to the relatively unchanging orientation of the inferior dental and other nerve canals during growth. He found that if tracings of lateral skull radiographs were superimposed and scaled to overcome differences in size, the orientation of the inferior dental canal in the mandible was relatively unchanging during growth.

The ''stability'' of the inferior dental canal has also been mentioned by Bjork (1963). He suggested, based on his metal pin measurements in the mandible, that three radiographic landmarks are relatively unchanging during growth and may be used as fiducial markers. In addition to the outline of the inferior dental canal, the other reference contours comprised the lower margin of a molar tooth germ prior to root formation and the inner cortical structure of the inferior border of the symphysis.

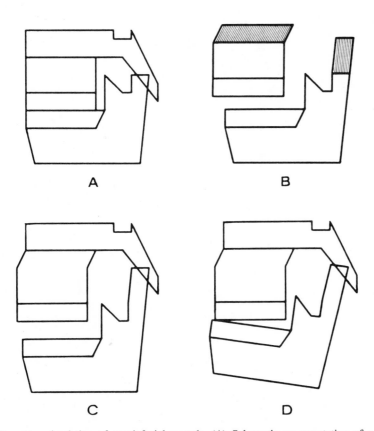

Fig. 21. Computer simulation of craniofacial growth. (A) Schematic representation of craniofacial complex. (B) Maxilla and mandible with growth addition shaded. (C) First phase of simulation, displacement of maxilla and mandible. (D) Second phase of simulation, occlusal surfaces of maxilla and mandible brought into contact (Solow, 1973).

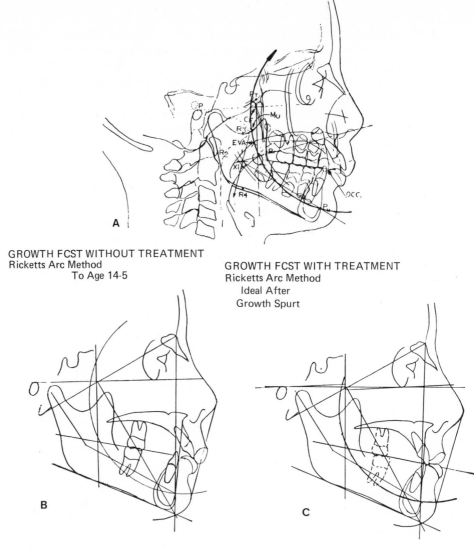

GROWTH FCST WITHOUT TREATMENT
Ricketts Arc Method
To Age 14-5

GROWTH FCST WITH TREATMENT
Ricketts Arc Method
Ideal After
Growth Spurt

Fig. 22. Ricketts' arcial method for the prediction of facial growth. (A) Tracing of a girl at 8 years of age with an arc drawn through the protuberance menti or suprapogonion and up through a point called EVA at the center of the upper anterior quadrant of the ramus. A third point formed with the distance EVA to PM as its radius will produce the arc as demonstrated. It is then hypothesized that this is the direction in which the mandible grows. (B) The prediction without treatment at the age of 14½ years after principal growth has ceased. (C) The idealized positioning of the teeth after treatment (Ricketts *et al.,* 1972).

The contributions made by the different mechanisms for the increase in bone size, whether sutural growth, cartilage replacement, or surface remodeling, to the growth of the regions of the skull have been mentioned previously in this chapter. However, the control system by which the effects of these mechanisms are integrated into the over-all growth of the skull are still little understood. Whereas transplantation studies suggest that epiphyseal cartilage (the growth mechanism for long bones) appears to possess an independent growth potential, which would perhaps explain the elongation of a long bone, both the sphenooccipital synchondrosis and the condylar cartilage appear to require an additional stimulus to achieve their full potential.

Following studies in experimental animals (Moss, 1954; Watanabe *et al.,* 1957; Sarnat, 1963), it is now generally accepted that sutural growth at a bone interface is secondary to the increase in size of the tissues contained within the bone. For example the increase in size of the neural tissue associated with growth of the brain tends to separate the bones of the cranial vault, which provides the stimulus for bone formation along the suture systems of the calvaria. This concept of the adaptation of the bone elements to the requirements of the contained tissues has been greatly extended by Moss (1968) in his functional matrix theory. Moss suggested that the head could be considered as an aggregation of functional units, each unit responsible for a separate activity. The functional unit is divided into a skeletal component and a functional matrix which includes all the other tissues, organs, or spaces necessary to carry out the function.

The functional matrices may be of two types, depending on whether they contain muscles attached to the bone via a perosteum (periosteal matrix) or consist of functional cavities (capsular matrix). The capsular matrices may include tissue masses, such as are found in the orbital and intracranial cavities, or contain functional spaces as in the nose, mouth, and pharynx. Moss has suggested that the capsular matrix acts by producing displacement of the surrounding skeletal units, examples being the growth of calvaria or circumorbital bones. Evidence for this

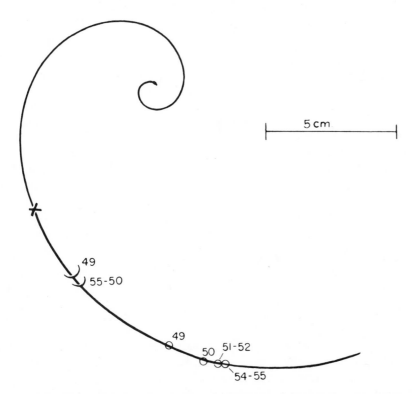

Fig. 23. Analysis of mandibular growth by the use of a unitary logarithmic spiral. With constant registration at foramen ovale (×) and with the spiral, held in an arbitrary position, the absolute "movement" of the sites of the mandibular (∪) and mental (○) foramina are demonstrated at annual intervals over a 6-year period. The numerals indicate calendar years, e.g., 49 = 1949 (Moss *et al.,* 1974).

view has been supplied by experiments in the rat carried out by Young (1959), who found that reduction in intracranial contents led to a decrease in the size of the calvaria, whereas an enlargement followed the induction of experimental hydrocephalus.

The periosteal matrix is taken to operate by the stimulation of bone deposition and resorption. This view is supported by the experiments of Washburn (1947) and Avis (1961), who found that the extirpation of certain of the muscles of mastication can lead to an almost complete absence of the associated area of bone into which the muscle would normally have been inserted, i.e., absence of the related skeletal component.

Thus the theory would appear to suggest that the stimulus to bone growth is transmitted by the mechanical forces applied by the functional matrix. The effect of pressure causing bone resorption, and of tension causing deposition, is well established (Reitan, 1951); were this not so the orthodontic movement of teeth would not be possible. Furthermore Frost (1964) put forward theories based on the tendency to deformation of the surface of a bone when placed under conditions of stress. He suggested that bone areas which tend to a more concave contour under load were associated with deposition, whereas areas tending to a more convex contour lead to resorption.

The investigation of Bassett (1972) may form the basis of an explanation of how the effects on bone structure are mediated. He found that bone exhibits piezoelectric properties, i.e., deformed bone can produce weak electric potentials. It has also been shown experimentally that small electric potentials are capable of stimulating bone formation (Bassett *et al.*, 1964). The sign of the potential is associated with the nature of the change: bone deposition accompanies a negative potential, and resorption is associated with a positive potential. Perhaps a feedback system relying on minute deformation-induced potentials is responsible for the change of bone contour.

Facial growth is controlled by the inherent genetic constitution and the environmental influences upon an individual. Similarly, the development of a malocclusion is subject to the effects of both hereditary and environmental factors. The correct relationship of the two dental arches depends upon three basic characteristics. The first is whether there is a disproportion in the relative positions of the bone arches in which the teeth develop; if there is a disproportion, the teeth will not erupt into a harmonious interdigitated relationship.

Second, the molding effect of the oral musculature affects the position and shape of the dental arches. The morphology and activity of the tongue inside and the circumoral musculature of the lips and cheeks outside dictate a balance position for the teeth on eruption. If the skeletal bony base elements are malrelated but the soft tissue environment is favorable, the teeth, on eruption, are molded into angulations which tend to compensate for the skeletal disproportion. However, this compensatory effect is limited and capable of compensating satisfactorily for only minor degrees of skeletal malrelation.

Third, the degree of dental crowding affects the positions of the teeth. During the process of evolution there has been a reduction in the size of the bone arches supporting the teeth. Despite the human dentition consisting of less teeth than the usual mammalian dentition (32 compared with 44), the bone arches are frequently too short to accommodate the complete dentition in a smooth arch. Thus on development and eruption of the teeth there occurs a competition for space resulting in crowding and displacement of individual teeth. In some populations a degree of dental crowding is a majority finding, thus constituting the "normal" state of affairs.

In addition to these three basic factors, other influences of a more local nature may operate in individual cases. These influences are many in number and include the failure of teeth to develop, the loss of teeth due to decay, the development of supernumerary teeth, thumb- or finger-sucking habits causing tooth displacement, etc.

The interplay between hereditary and environmental factors is complex and presents formidable problems for investigation. In this respect several authors have drawn attention to the effects that some forms of orthopedic or orthodontic treatment can have on the facial growth pattern. Logan (1962), Alexander (1966), Rock and Baker (1972), and others have described the effect of the Milwaukee brace on the facial growth pattern. The Milwaukee brace is used in the treatment of scoliosis (abnormal lateral curvature of the spine) and consists of a metal braced caliper or corset. The appliance is designed to provide traction to the vertebral column, yet still permit the patient to stand and walk. The bottom end of the brace rests against the iliac crests of the pelvis, and the upper end against the underside of the mandible and occipital bones of the skull. The pressure exerted on the mandible is transmitted to the whole facial region and over a period of time distorts the normal growth pattern.

Similarly, changes in facial growth have also been reported due to orthodontic treatment techniques. These include the use of orthodontic appliances to cause rapid expansion of the mid-palatal suture (Haas, 1961) and the use of headgear harnesses to provide extraoral traction to orthodontic appliances (Klein, 1957; Graber *et al.*, 1967). Thus it would appear that the normal growth pattern is capable of modification by the operation of external environmental factors.

A classical approach to the differentiation between hereditary and environmental effects lies in the study of pairs of twins, and this form of investigation has been applied to the patterns of skull growth (Lundstrom, 1954; Hunter, 1965). Differences in growth pattern found between the members of monozygous pairs must be due to environmental factors. Dudas and Sassouni (1973) studied the growth of 12 pairs of monozygotic twins and 10 pairs of dizygotic twins. They compared skull growth between members of each pair of twins and between the two groups. The authors concluded that of the 15 skeletal measurements recorded in the investigation, only four appeared to be under strong hereditary (or weak environmental) influence. The four measurements record the relative anteroposterior position of two points on the mandible compared with the base of the skull and also the total and lower part of the anterior facial height.

7. Conclusions

Thus the present state of knowledge concerning the subject of skull growth presents a confusing picture. The mechanisms that have evolved to bring about change in bone shape have been explored in some detail, but the stimuli which initiate and control their operation are still subject to controversy. Whether the requirements of the essential activities of the skull provide the control, or whether there exists a more direct genetic control mechanism, has yet to be conclusively established. The studies on twins appear to implicate a mixture of hereditary and environmental factors, but more investigation is required into this and all other aspects of skull growth.

What is certain is the wide range of variation of facial form. It would appear from the world-wide variety in craniofacial contours that to be different is the norm. Perhaps we should rejoice in the difficulty, or more likely impossibility, of reducing facial growth to a simple mathematical model. For if it were possible to describe facial growth by one simple all-encompassing formula, would the resulting face provide the variation and charm of that bequeathed to us by evolution?

8. References

Adams, J. W., 1940, Correction of errors in cephalometric roentgenograms, *Angle Orthod.* **10**:3–13.

Alexander, R. B., 1966, The effects on both position and maxillo facial vertical growth during treatment of scoliosis with the Milwaukee brace, *Am. J. Orthod.* **52**:161–189.

Avis, V., 1961, The significance of the angle of the mandible: An experimental and comparative study, *Am J. Phys. Anthropol.* **19**:55–61.

Bassett, C. A. L., 1972, Biophysical properties affecting bone structure, in: *The Biochemistry and Physiology of Bone,* Vol. III (G. H. Bourne, ed.), pp. 1–76, Academic Press, New York.

Bassett, C. A. L., Pawluk, R. J., and Becker, R. O., 1964, Effects of electric currents on bone *in vivo, Nature (London)* **202**:652–654.

Baume, L. J., 1961, The postnatal growth activity of the nasal cartilage septum, *Helv. Odontol. Acta* **47**:881–901.

Bjork, A., 1947, The face in profile, *Sven. Tandlaek. Tidskr.* **40**:56.

Bjork, A., 1955, Facial growth in man, studies with the aid of metallic implants, *Acta Odontol. Scand.* **13**:9–34.

Bjork, A., 1963, Variations in the growth pattern of the human mandible: Longitudinal radiographic study by the implant method, *J. Dent. Res.* **42**:400–411.

Bjork, A., 1964, Sutural growth of the upper face studied by the implant method, *Trans. Eur. Orthod. Soc.* **49**:49–65.

Bjork, A., and Solow, B., 1962, Measurements on radiographs, *J. Dent. Res.* **41**:672–683.

Brash, J. C., 1956, The aetiology of irregularity and malocclusion of the teeth, Part 1, Public Dental Board of U.K.

Broadbent, B. H., Sr., Broadbent, B. H., Jr., and Golden, W. H., 1975, *Bolton Standards of Dentofacial Developmental Growth,* C. V. Mosby, St. Louis, Missouri.

Broadbent, H., 1931, A new X-ray technique and its application to orthodontics, *Angle Orthod.* **1**:45–66.

Brodie, A. G., 1953, Late growth changes in the human face, *Angle Orthod.* **23**:146–157.

D'Arcy Thompson, W., 1917, *On Growth and Form,* Cambridge University Press, Cambridge.

Delattre, A., and Fenart, R., 1958, La méthode vestibulaire appliqué à l'étude du crâne. Son champ d'application, *Z. Morphol. Anat.* **1**:90–114.

Downs, W. B., 1948, Variations in facial relationships, their significance in treatment and prognosis, *Am. J. Orthod.* **34**:812–840.

Dudas, M., and Sassouni, V., 1963, The hereditary components of mandibular growth, a longitudinal twin study, *Angle Orthod.* **43**:314–323.

Enlow, D. H., 1966, A morphogenetic analysis of facial growth, *Am. J. Orthod.* **52**:283–299.

Enlow, D. H., 1975, *Handbook of Facial Growth,* W. B. Saunders, Philadelphia.

Frost, H. M., 1964, *The Laws of Bone Structure,* Charles C Thomas, Springfield, Illinois.

Gange, R. J., and Johnston, L. E., 1974, The septo-premaxillary attachment and mid-facial growth, *Am. J. Orthod.* **66**:71–81.

Giannelly, A. A., and Moorrees, C. F. A., 1965, Condylectomy in the rat, *Arch. Oral Biol.* **10**:101–106.

Goldstein, M. S., 1936, Changes in dimension and form of the face and head with age, *Am. J. Phys. Anthropol.* **22**:37–89.

Graber, T., Chung, D., and Aoba, J., 1967, Dentofacial orthopedics versus orthodontics, *J. Am. Dent. Assoc.* **75**:1145–1166.

Gregory, W. K., 1927, The palaeomorphology of the human head: Ten structural stages from fish to man: Part 1—The skull in norma lateralis, *Q. J. Biol.* **2**:267.

Haas, A. J., 1961, Rapid expansion of the maxillary dental arch and nasal cavity by opening the mid-palatal suture, *Angle Orthod.* **31**:73–90.

Harris, W. H., 1960, A microscopic method of determining rates of bone growth, *Nature (London)* **188**:1038–1039.

Humphrey, G. M., 1863, Results of experiments on the growth of the jaws, *Br. J. Dent. Sci.* **6**:548–550.

Hunter, J., 1798, Experiments and observations on the growth of bones, in: *Hunter's Works,* 1837 (D. F. Palmers, ed.), Longman, London.

Hunter, W. S., 1965, Study of inheritance of craniofacial characteristics as seen in lateral cephalograms of 72 like-sexed cases, *Trans. Eur. Orthod. Soc.* **41**:59–69.

Klein, P. L., 1957, An evaluation of cervical traction on the maxilla and the upper first permanent molar, *Angle Orthod.* **27**:61.

Koski, K., 1973, Variability of the facial skeleton: An exercise in roentgen-cephalometry, *Am. J. Orthod.* **64**:188–196.

Koski, K., and Ronning, O., 1965, Growth potential of transplanted components of the mandibular ramus of the rat III, *Suom. Hammaslaak Seur Toim* **61**:292–297.

Koski, K., and Ronning, O., 1970, Growth potential in intracerebrally transplanted cranial base synchondrosis in the rat, *Arch. Oral Biol.* **15**:1107–1108.

Krogman, W. M., and Sassouni, V., 1957, *A Syllabus in Roentgenographic Cephalometry,* Philadelphia Center for Research in Child Growth, Philadelphia.

Latham, R. A., 1968, The sliding of cranial bones at sutural surfaces during growth, *J. Anat. (London)* **102**:593.

Latham, R. A., 1969, The septopremaxillary ligament and maxillary development, *J. Anat.* **104**:584–586.

Logan, W. R., 1962, The effect of the Milwaukee brace on the developing dentition, *Dent. Pract.* **12**:447–452.

Lundstrum, A., 1954, The importance of genetic and non-genetic factors in the facial skeleton studied in 100 pairs of twins, *Trans. Eur. Orthod. Soc.* **30**:92–106.

Lynsholm, E., 1931, Apparatus and technique for roentgen examination of the skull, *Acta Radiol. Suppl.* **12.**

Mitchell, D. L., Jordan, J. F., and Ricketts, R. M., 1975, Arcial growth with metallic implants in mandibular growth prediction, *Am. J. Orthod.* **68**:655–659.

Moorrees, C. F. A., and Kean, M. R., 1958, Natural head position a basic consideration in the interpretation of cephalometric radiographs, *Am. J. Phys. Anthropol.* **16**:213–234.

Moorrees, C. F. A., and Yem, P. K.-J., 1955, An analysis of changes in the dentofacial skeleton following orthodontic treatment, *Am. J. Orthod.* **41**:526–538.

Moorrees, C. F. A., Fanning, E. A., and Hunt, E. E., 1963*a*, Age variation of formation stages for ten permanent teeth, *J. Dent. Res.* **42**:1490–1502.

Moorrees, C. F. A., Fanning, E. A., and Hunt, E. E., 1963*b*, Formation and resorbtion of three deciduous teeth in children, *Am. J. Phys. Anthropol.* **21**:205.

Moss, M. L., 1954, The growth of the calvaria in the rat, *Am. J. Anat.* **94**:333.

Moss, M. L., 1968, The primacy of functional matrices in orofacial growth, *Dent. Pract. Dent. Rec.* **19**:65–73.

Moss, M. L., and Rankow, R. M., 1968, The role of the functional matrix in mandibular growth, *Angle Orthod.* **39**:95–103.

Moss, M. L., and Salentijn, L., 1970, The logarithmic growth of the human mandible, *Acta Anat.* **77**:341–360.

Moss, M. L., Bromberg, B. E., Song, I. C., and Eisenman, G., 1968, The passive role of nasal septal cartilage in mid-facial growth, *Plast. Reconstr. Surg.* **41**:536–542.

Moss, M. L., Salentijn, L., and Ostreicher, H. P., 1974, The logarithmic properties of active and passive mandibular growth, *Am. J. Orthod.* **66**:645–664.

Pritchard, J. J., Scott, J. H., and Girgis, F. G., 1956, The structure of cranial and facial sutures, *J. Anat.* **90**:73–86.

Reitan, K., 1951, The initial tissue reaction incident to orthodontic tooth movement as related to the influence of function, *Acta Odontol. Scand. Suppl. 6.*

Ricketts, R. M., 1973, New findings and concepts emerging from the clinical use of the computer, *Trans. Eur. Orthod. Soc.* **1973**:507–515.

Ricketts, R. M., Bench, R. W., Hilgers, J. I., and Schulhof, A. B., 1972, An overview of computerized cephalometrics, *Am. J. Orthod.* **61**:1–28.

Rock, W. P., and Baker, R., 1972, The effect of the Milwaukee brace upon dentofacial growth, *Angle Orthod.* **42**:96–102.

Rushton, M. A., 1944, Growth at the mandibular condyle in relation to some deformities, *Br. Dent. J.* **76**:57–68.

Salzmann, J. A., 1966, *Practice of Orthodontics,* Vol. 1, Lippincott, Philadelphia.

Sarnat, B. G., 1963, Postnatal growth of the upper face; some experimental considerations, *Angle Orthod.* **33**:139–161.

Sarnat, B. G., and Wexler, M. R., 1966, Growth of the face and jaws after resection of the septal cartilage in the rabbit, *Am. J. Anat.* **118**:755–760.

Sarnat, B. G., and Wexler, M. R., 1967, Rabbit snout growth after resection of central linear segments of nasal septal cartilage, *Acta Oto-Laryngol.* **63**:467–478.

Savara, B. S., 1972, The role of computers in dentofacial research and the development of diagnostic aids, *Am. J. Orthod.* **61**:231–245.

Scott, J. H., 1954, The growth of the human face, *Proc. R. Soc. Med. (London)* **47**:91.

Scott, J. H., 1956, Growth at facial sutures, *Am. J. Orthod.* **42**:381.

Scott, J. H., 1967, *Dento-facial Development and Growth,* Pergamon, London.

Sillman, J. H., 1964, Dimensional changes of the dental arches: Longitudinal study from birth to 25 years, *Am. J. Orthod.* **50**:824–842.

Solow, B., 1973, Graphical simulation of craniofacial growth, *Trans. Eur. Orthod. Soc.* **1973**:516–521.

Steiner, C. C., 1953, Cephalometrics for you and me, *Am. J. Orthod.* **39**:729–755.

Steiner, C. C., 1959, Cephalometrics in clinical practice, *Angle Orthod.* **29**:8–29.

Steiner, C. C., 1960, The use of cephalometrics as an aid to planning and assessing orthodontic treatment, *Am. J. Orthod.* **46**:721–735.

Stenstrom, S. J., and Thilander, B. L., 1972, Effects of nasal septal cartilage resections on young guinea pigs, *Plast. Reconstr. Surg.* **45**:160–170.

Sullivan, P. G., 1972, A method for the study of jaw growth using a computer-based three-dimensional recording technique, *J. Anat.* **112**:457–470.

Suzuki, H. K., and Matthews, A., 1966, Two colour fluorescent labelling of mineralising tissues with tetracycline and 2,4-bis(N,N'-di-(carboxymethyl)aminomethyl) fluorescein, *Stain Technol.* **41**:57–60.

Tanner, J. M., 1951, Some notes on the reporting of growth data, *Hum. Biol.* **23**:93–159.

Tulley, W. J., and Campbell, A. C., 1975, *A Manual of Practical Orthodontics,* 3rd ed., Wright and Sons, Bristol.

Voorhies, J. M., and Adams, J. W., 1951, Polygonic interpretation of cephalometric findings, *Angle Orthod.* **21**:194–197.

Walker, G. F., 1972, A new approach to the analysis of craniofacial morphology, *Am. J. Orthod.* **61**:221–230.

Washburn, S. L., 1947, The relation of the temporal muscle to the form of the skull, *Anat. Rec.* **99**:239–248.

Watanabe, M., Laskin, D. M., and Brodie, A., 1957, The effect of antotransplantation on growth of the zygomaticomaxillary suture, *Am. J. Anat.* **10**:319.

Weinmann, J. P., and Sicher, H., 1955, *Bone and Bones: Fundamentals of Bone Biology,* 2nd ed., Kimpton, London.

Wexler, M. R., and Sarnat, B. G., 1961, Rabbit snout growth: Effect of injury to the septovomeral region, *Arch. Oto-Laryngol.* **74**:305–313.

Wylie, W. L., 1947, Assessment of antero-posterior dysplasia, *Angle Orthod.* **17**:97.

Young, R. W., 1959, The influence of cranial contents on postural growth of the skull of the rat, *Am. J. Anat.* **103**:383–415.

15

Dentition

ARTO DEMIRJIAN

1. Introduction

The dental system is an integral part of the human body and its growth and development can, and should be, studied in parallel with other physiological maturity indicators, such as bone age, menarche, and height. Knowledge about the development of dentition can be applied not only in dentistry and orthodontics, but also in such diverse fields as physical anthropology, endocrinology, nutrition, and forensic odontology.

There is considerable latitude in the study of dental maturation, since various criteria are available for evaluation. Dental maturity can be based on not one, but two, complete sets of dentition. Both sets can be observed from the beginning, or at least from very early stages of calcification (formation). Furthermore, the clinical emergence of both sets of teeth and the shedding of the deciduous teeth can yet serve as other maturity indicators. However, the study of dental development, unlike that of skeletal development, has unfortunately not progressed to the stage where norms for dental maturity have been wholly clarified, standardized, and universally accepted. It is expected that this obstacle will be overcome in the near future, as improved techniques for the measurement of dental maturation are developed.

The most commonly used parameter for assessing dental maturity has always been dental "eruption," because it is both rapid and convenient. The term "eruption" is used almost universally, albeit wrongly, to indicate the clinical appearance of the cusps of a tooth through the gum. In fact eruption is not an instantaneous phenomenon, corresponding to the moment when a tooth pierces the gum; it is rather the continuous upward movement of the dental bud, from the depth to the edge of the alveolar bone, and further to the occlusal level.

Historically, however, "eruption" data refer to the clinical appearance of a tooth in the oral cavity. The first use of eruption as an indicator of physical maturity occurred in England (Saunders, 1837), when the Factory Act stipulated that a child without a second permanent molar would not be allowed to work in the factories. In

ARTO DEMIRJIAN • Université de Montréal, Montréal, Quebec, Canada.

the first quarter of the 20th century, Beik (1913), Bean (1914), and Cattell (1928) used dental criteria as a maturity indicator for school entrance purposes.

Up to the 1940s, data concerned with dental development were mostly descriptive and were based on histological sections from very small samples. Errors from this unreliable approach persist to the present day, for most eruption and other data are still compared to standards derived from histological studies. For example, Legros and Magitot (1893) first reported on the chronology of the formation of tooth buds; their study was based on histological sections. Regardless of the fact that "they based their conclusions about the onset of calcification on sections from a single foetus" (Kraus, 1959), their results were used for many decades, even in textbooks of histology and embryology.

In their well-known article, Logan and Kronfeld (1933) criticized the Legros and Magitot material and its unchallenged use for 40 years. Yet their new material was hardly better, having been derived from a heterogenous sample composed of 3 newborns, 13 infants up to 12 months, and 9 children from 1 to 15 years of age. Moreover, they stated that some of their subjects were pathological. Obviously, the results of this study, also based on histological sections, are too biased and unreliable to be of any use as a guide to the chronology of dental development.

It is not very edifying to trace the history of repetitive error in this field; however, it is interesting to note that as late as 1940, Schour and Massler (1940) published the "Chronology of Growth of Human Teeth," a table modified from that of Logan and Kronfeld. The authors described the continuous development not only of the permanent teeth, but of the deciduous teeth as well. Yet examination of the Logan and Kronfeld article reveals no reference at all to the deciduous dentition, and it was difficult to find reference to any other sample in Schour and Massler's paper. We suspect that histological and roentgenographic material was used without distinction. Schour and Massler did, however, caution their readers that "since the X-ray film is a record of density, the stage of tooth development, as indicated in histological sections, will be slightly earlier than that indicated by X-rays." Although roentgenographic representations and anatomical findings are not identical, the distinction has sometimes been disregarded.

Today, studies of dental maturity try to use and correlate histological, radiological, and clinical data. The only approach to the study of the prenatal phase of the deciduous dental development is still the histological approach; however, in postnatal growth studies, radiological (formation) and clinical (emergence) data can be used, either independently or in parallel. It must be remembered that formation and eruption involve two essentially different processes, which can be variously influenced by genetic, environmental, and hormonal factors. The study of dental development can be even further complicated by such factors as the prenatal environment and its influence on the formation of the deciduous teeth, the possible physiological or pathological influences on one or both sets of teeth, the shedding of the deciduous dentition, and the effect of the deciduous teeth on the timing of the emergence of their permanent successors.

Over the years, we have become more aware of such experimental pitfalls and have developed more precise and scientific methods of studying dental development. The results, although not always accepted universally, provide a far more reliable account of the chronology of the human dentition. In the following discussion of dental emergence and development, the most prominent research in the field of dental maturity assessments will be described and illustrated.

As indicated above, the word "eruption" is an erroneous term used to denote the appearance of a tooth in the oral cavity. The correct term for the piercing of the gum is "clinical emergence," or simply "emergence." The piercing of the alveolar bone, seen in a radiographic image and referred to as the bony or alveolar eruption, should similarly be called "alveolar emergence." From here on, the use of the word eruption will be confined to the dynamic movement of the tooth, both before and after the actual emergence. This process resembles the pre- and postnatal growth of an infant, where the actual birth of the child corresponds to emergence and is an event of very short duration during the uninterrupted development of the child.

The literature is replete with reports of research into the chronology and timing of dental emergence. These studies have been conducted in order to establish population standards for clinical, anthropological, medicolegal, and other purposes. Several such standards have also been used as criteria for the determination of biological (dental) age.

Emergence has been analyzed in many different ways. Individual teeth, specific groups of teeth, and the entire dentition (deciduous and permanent) have been studied in cross-sectional population samples, in single-age groups, and also longitudinally, with groups of subjects closely monitored throughout the crucial years of their growth and development. Groups of children from many parts of the world have been observed in an effort to trace any development differences which may be related to race, sex, nutritional standards, socioeconomic levels, hormone conditions, or other intrinsic or extrinsic factors. All such studies have contributed valuable information about dentition and its applicability as a measure of maturity.

A thorough review of the subject was presented by Tanner (1962). We have here an excellent opportunity to bring Tanner's review up to date; as much as possible, the deciduous dentition will be treated separately from the permanent dentition. In discussions of both, however, information gained about the various intrinsic and extrinsic influences mentioned above will be emphasized.

3. Emergence of Deciduous Dentition

The emergence of deciduous dentition takes place between the sixth and the 30th months of postnatal life. Although this is a relatively short period of time, the study of dental maturity in these age groups is useful in growth studies concerned with environmental and hereditary factors. For the evaluation of maturity, radiographic assessment of the deciduous dentition is a more meaningful approach, since the development of every stage of each tooth can be followed regularly between birth and the third year of life, rather than during the more limited period of actual emergence, from the sixth to the 30th months. However, the difficulty of developing adequate radiological techniques for infants, as well as the hazards of the radiation itself, represent two significant disadvantages to this method. Nevertheless, Moorrees *et al.* (1963*b*) studied the development of the mandibular deciduous canines and molars and presented normative data, using the Fels material.

Many authors, having searched to find *sex differences* in the emergence of the deciduous dentition, were unanimous in reporting no difference between boys and girls in the timing of emergence (Falkner, 1957; Yun, 1957; Lysell *et al.,* 1962;

Bailey, 1964; Roche *et al.*, 1964; McGregor *et al.*, 1968; Friedlaender and Bailit, 1969; Brook and Barker, 1972, 1973; Billewicz *et al.*, 1973; Robinow, 1973; Trupkin, 1974; El Lozy *et al.*, 1975). Other investigators, however, found an earlier emergence pattern in boys (Doering and Allen, 1942; Robinow *et al.*, 1942; Meredith, 1946; MacKay and Martin, 1952; Ferguson *et al.*, 1957). Only one group of investigators (Bambach *et al.*, 1973) reported a delayed emergence among boys as compared to girls, but the difference was not statistically significant.

The emergence of deciduous teeth has also been investigated in different *racial and ethnic groups*. In 1957, Ferguson *et al.* (1957) reported that the emergence of deciduous teeth was delayed in white Americans, as compared to blacks, during the first year of life. Yun (1957) noted that the emergence of the deciduous teeth was slower in Korean children than in American children. McGregor *et al.* (1968) noted that American white, English, and French children were ahead of Gambian children with respect to the emergence of deciduous teeth. The Gambian children caught up, though, by the time they reached 2 years of age. This latter trend was more firmly established through the works of Mukherjee (1973) and Billewicz *et al.* (1973); the former compared Indian children, and the latter compared Hong Kong children to the American, British, and French samples and, like McGregor, they found a delay in both the Indian and Hong Kong children. Bambach and co-workers (1973) compared results from a sample of Tunisian children to that of McGregor *et al.* (1968) and also found a delay in the emergence of deciduous teeth of the African children as compared to the European and American groups. Still, in every case, it was reported that the delay was overcome by the age of 18 months to 2 years. Friedlaender and Bailit (1969) came to the conclusion that the emergence of deciduous dentition is complete by the end of the 29th month of life, regardless of ethnic origin or race, and independent of the age at which the first tooth appeared.

Most studies which attempted to correlate *socioeconomic status* to dental emergence have concentrated on the permanent dentition. Bambach *et al.* (1973) were the only workers who attempted to relate deciduous tooth emergence to socioeconomic conditions, but they found no significant correlations. The emergence of deciduous dentition seems to be more influenced by genetic and hereditary factors than by environmental or socioeconomic conditions. The latter seems to exert a more direct influence on body growth, as distinct from dental development.

Nutrition has also been investigated extensively (McGregor *et al.*, 1968; Cifuentes and Alvarado, 1973; Infante and Owen, 1973; Trupkin, 1974; Delgado *et al.*, 1975). In general terms, no correlation has been found between nutrition and the emergence of the deciduous teeth. Dental emergence might be delayed only in cases of severe protein–calorie malnutrition (PCM); even in those cases, the effect of malnutrition is greater on other parameters, namely height, weight, and bone age. El Lozy *et al.* (1975) reported that malnutrition of a lesser degree than PCM leads to a delayed emergence of deciduous teeth in Tunisian children; this fact might probably be true for the developing societies, even though dental development is less affected than somatic and skeletal growth (Boutourline-Young, 1969; Habicht *et al.*, 1974).

Thus, the next logical step would be to investigate severe maternal malnutrition in poor communities, in an effort to trace not only hypoplasia of the enamel, but also developmental and eruptive patterns of the deciduous teeth. With the information currently available, emergence and formation patterns of the deciduous dentition seem to be "programmed" in fetal life (Jelliffe and Jelliffe, 1973).

Low birth weight and *prematurity* are two other factors which have been linked to delayed emergence of the deciduous teeth. Some correlations between birth weight and the number of teeth in the mouth have been reported (Billewicz *et al.,* 1973; Infante and Owen, 1973; Trupkin, 1974; Delgado *et al.,* 1975). Children with higher birth weight tend to have more teeth in the mouth at any given age. Although birth weight seems to be closely related to the emergence of the deciduous teeth, there is no evidence as yet that this is a result of the length of gestation. Advanced emergence has also been correlated to height in both sexes and to head circumference in boys (Infante and Owen, 1973); this relation between the length of an infant's body and the emergence of deciduous dentition was corroborated by Falkner (1957).

No correlation has yet been found between deciduous dental emergence and *skeletal maturity* (Robinow, 1973; Billewicz *et al.,* 1973). McGregor *et al.* (1968) came to the same conclusion that there is no connection between emergence and/or the number of teeth in the mouth, and the general physiological growth of the child. Slight trends suggesting such relationships have been observed, but none have proven to be statistically significant. Since there are so few definitive correlations between deciduous tooth emergence and other physiological parameters, such as skeletal maturation, size, sex, etc. (Falkner, 1957), it appears that the deciduous dentition is remarkably independent of other morphological processes.

4. Emergence of Permanent Dentition

Cross-sectional surveys have been the major vehicle for studying the emergence of the permanent dentition. Means and standard deviations of the time of emergence of each tooth can be derived from the data gathered from these studies. Cumulative incidence curves can also be constructed from the cross-sectional data, which show the percentage of children, at a given age, in whom a certain tooth has emerged (Cattell, 1928; Boas, 1933; Klein *et al.,* 1937; Hellman, 1943; Hurme, 1948, 1949; C. Dahlberg and Maunsbach, 1948; Gödény, 1951; Leslie, 1951; Clements *et al.,* 1953*b;* Tanner, 1962). In later studies (Clements *et al.,* 1953*a;* Kihlberg and Koski, 1954; A. A. Dahlberg and Menegaz-Bock, 1958; Gates, 1964; Miller *et al.,* 1965; Perreault *et al.,* 1974), probit transformation has been adopted as the best method for analyzing these types of data.

The pattern and timing of the emergence of permanent dentition have been studied both cross-sectionally and longitudinally. In longitudinal studies, when subjects are only seen at yearly intervals, it is likely that the exact timing of emergence will be missed. For this reason, A. A. Dahlberg and Menegaz-Bock (1958) suggested that the age, at subsequent examinations, be reduced by half the time elapsed since the previous examination. If such an adjustment is not made, data resulting from longitudinal studies will always give higher values for the age of the emergence.

It is now a widely accepted fact that there is a *symmetry* in the emergence times between the teeth of the right and left sides (Clements *et al.,* 1957*b;* Gates, 1964; Lee *et al.,* 1965; Nanda and Chawla, 1966; Lysell *et al.,* 1969; Krumholt *et al.,* 1971; Perreault *et al.,* 1974; Helm and Seidler, 1974; Billewicz and McGregor, 1975). The emergence of the upper and lower teeth varies according to the individual tooth: the first to emerge is always the lower central incisor. The mandibular molars usually emerge earlier than their maxillary counterparts, while the upper premolars precede the lower premolars (A. A. Dahlberg and Menegaz-Bock, 1958; Nanda, 1960; Gates, 1964; Lee *et al.*, 1965; Nanda and Chawla, 1966; Houpt *et al.*, 1967; Lysell *et al.*, 1969; Garn *et al.*, 1973; Helm and Seidler, 1974).

Several authors have found significant differences between the two *sexes* in the emergence of the permanent dentition (Hurme, 1949; A. A. Dahlberg and Menegaz-Bock, 1958; Garn *et al.*, 1959; Fanning, 1961; Halikis, 1961; Lee *et al.*, 1965; Miller *et al.*, 1965; Houpt *et al.*, 1967; Krumholt *et al.*, 1971; Brook and Barker, 1972; Robinow, 1973, Helm and Seidler, 1974). Girls are usually 1–6 months ahead of boys; for certain teeth, such as the canines, this difference can be as great as 11 months in some populations. Other teeth vary individually.

Hereditary factors and *ethnic origin* exert considerable control over the emergence of the permanent teeth. Dental emergence is earlier in American blacks than in American whites (Garn *et al.*, 1973); likewise, it is earlier in African blacks than in Europeans generally, sometimes by as much as 1½ years (Houpt *et al.*, 1967; Krumholt *et al.*, 1971). New Guinea children are also ahead of American and British children (Voors and Metselaar, 1958; Brook and Barker, 1972). Lee *et al.* (1965) compared a group of Hong Kong children to samples of American whites, British, New Zealand, Hawaiian, and Chinese children; differences in the timing of the emergence were attributed to the ethnic origins of the various groups of children.

Not only does the timing of the emergence of permanent teeth come under ethnic influences, but also individual teeth emerge according to certain genetically controlled patterns. The timing of the emergence of the anterior and the posterior segments of dentition is different for Pima Indians, Hong Kong children, American, and British children (Leslie, 1951; A. A. Dahlberg and Menegaz-Bock, 1958; Lee *et al.*, 1965); A. A. Dahlberg and Menegaz-Bock (1958) offered a possible explanation for this phenomenon, that the "posterior teeth emerge earlier in a population subject to premature shedding of the deciduous molars." Leslie (1951) concluded that the later the deciduous molar is lost, the more accelerated is the emergence of the succeeding premolars. Racial differences have been noted to have a more significant effect on the incisors and molars, and a less important one on the molar and canine groups (Garn *et al.*, 1973).

Socioeconomic conditions have been variously defined as the profession of the father, family income, parents' education, or the family housing and environment. These criteria have been used either simultaneously or individually to establish the social level of a family. Socioeconomic conditions are known to have a definite effect on general body growth, but several investigators (Lee *et al.*, 1965; Houpt *et al.*, 1967; Garn *et al.*, 1973; Mukherjee, 1973) have concluded that either there is no such impact on permanent tooth emergence, or at most only an insignificant one as compared to the effect on general somatic growth. Clements *et al.* (1953a, 1957a,b) found that dental emergence was earlier among English children whose fathers were placed, by the registrar general's list, in the upper occupational levels. Yet, in the Ten-State Nutrition Survey (1972), in the United States, white children from high- and low-income states did not differ significantly in the age of emergence of the permanent teeth; among black children, the emergence was earlier in the higher-income states.

In some studies comparing rural and urban populations (Adler, 1958), early emergence of teeth was attributed to urbanization. Others (Clements *et al.*, 1975a,b) found a tendency of early emergence in children from rural schools compared to those from urban elementary schools. There is no agreement on this point, however, as Helm and Seidler (1974) found only a small and insignificant difference in the tooth emergence patterns of rural and urban Danish children. Consequently, we agree with Garn *et al.* (1973) that "population differences in tooth emergence exceed socio-economic differences."

The study of *nutrition,* in relation to dental development, has been of special interest to many investigators. Both the emergence and the formation of the deciduous and permanent teeth have been studied in the developing and Western countries (Voors, 1957; Niswander, 1963; Garn *et al.,* 1973; Billewicz and McGregor, 1975). As nature–nurture studies in dentistry are quite variable from one population to another, the results have always been compared and correlated to such anthropometric parameters as height, weight, and arm circumference, which very likely come under genetic influences as well. The literature is full of statements like: "children with fewer teeth are, in general, shorter" or "long-term undernutrition may have some effect on eruption," etc. It is true that severe undernutrition will affect the skeletal and dental systems, but the latter to a lesser degree; even statistically significant correlations always remain low.

There are very few studies of the role of *hormones* in the process of dental emergence or dental development. This is due to the failure, on the part of the clinicians, to extend the systematic diagnosis of their patients to the area of dentition. There has never been any consistent dental follow-up in cases of endocrinopathies. Such studies, when they are undertaken, should adopt radiographic techniques rather than emergence, in order to obtain a complete picture of the developmental history of the teeth.

In their longitudinal study of two children with pituitary insufficiency Cohen and Wagner (1948) came to the conclusion that, due to different development patterns and different origins (ectodermal and mesodermal), the bony structures (maxilla and mandible) and teeth showed independent growth patterns and were affected differently. Because of the lack of synchronization between bone and dental tissues, positional arrangements, such as changes in the vertical dimension, overbite, etc., become more complicated. Wagner *et al.* (1963) conducted a longitudinal study of 11 children with idiopathic sexual precocity, androgenic virilism, or adrenocortical hyperplasia. Their dental development was recorded radiologically; odontogenesis was advanced in the majority of children with either congenital adrenocortical hyperplasia or androgenic virilism. It is worthwhile to stress again here that sexual and skeletal maturity on the one hand, and dental development on the other, are dissociated. Garn *et al.* (1965a) studied several children with various endocrinopathies, such as thyroid abnormalities, athyreosis, hypopituitary dwarfism, etc. They separated the nonendocrine developmental delays from the endocrine-related cases and found that in congenital hypothyroidism and hypopituitarism, tooth emergence was slightly delayed. By contrast, emergence was delayed by 10% and skeletal development by 60%, in athyrotic children. In constitutional and endocrine precocities, dental advancement was noted along with skeletal advancement, but the degree of dental advancement remained low. Dental development also tends to be advanced in children with Turner's syndrome, but this phenomenon cannot be explained on endocrine grounds alone.

We can see that even serious endocrinopathies, which severely retard somatic growth and maturation, exert only a minor effect on dentition. Garn *et al.* (1962, 1965a) also attributed the movement of the PM_2 and M_2 to steroid hormones; they found little or no correlation between dental, sexual, and bone maturation before puberty. These authors also supposed that sex-linked genes might be concerned in tooth-eruption timings, but Mather and Jinks (1963) pointed out that this did not follow from their data.

5. Secular Trend in Dental Maturity

Several researchers have investigated secular trends in puberty, height, weight, and general growth, over a period of several decades. During recent decades, stature has increased more than other measurements, such as head or upper-arm circumference, weight, etc. (Tanner, 1962).

It has been assumed that the dental system, being an inherent part of the body, must be influenced by this trend, although perhaps to a lesser degree. Yet, as has already been mentioned, the development of deciduous dentition is quite independent of morphological processes (Falkner, 1957), and no secular trends have been detected in the development of this set of teeth (Lysell *et al.,* 1962). The timing of the shedding of deciduous teeth has a definite effect on the emergence of the permanent teeth (Clements *et al.,* 1957*a*). The early shedding of deciduous teeth advances the emergence of the permanent ones, or perhaps we could reverse this statement and say that an earlier development of the permanent teeth causes early shedding of the deciduous teeth. This latter statement remains to be proven. However, the extraction of a deciduous tooth, when very premature, can retard the emergence of its permanent successor (A. A. Dahlberg and Menegaz-Bock, 1958; Brook, 1973).

In disputing this apparent relationship, Butler (1962) argued that there is no evidence that the early shedding of the deciduous teeth is the major cause for the early emergence of permanent teeth; rather, he attributed the phenomenon to advanced human growth and development generally, as observed during this century. He thus supports a secular-trend theory in dentition. Brook (1973) favors the theory that the earlier emergence of permanent teeth parallels early puberty and an increase in height and weight noted in the last two decades. Miller *et al.* (1965) compared a sample of 2000 children to the previous results of Ainsworth (1925) and also concluded that emergence is earlier today.

The secular trend in general growth is attributed to improved environmental conditions, including better nutrition. We have already seen that, although teeth are influenced by environmental factors, they do not follow the same pattern as general growth, and their maturational process is relatively independent. To prove the existence of a secular trend in dentition, we must conduct studies in which the ethnic origin of the sample, methodology, and statistical treatment of the data are carefully controlled. Only then can we positively detect the existence of such a trend in dental development.

6. Concept of Dental Age

Several approaches have been used to give a child a "dental age," based on the number of teeth present in the mouth at each chronological age. Although the use of dental emergence seems to be a simple approach for prediction or age-determination purposes, there are certain inherent limitations to the method: (1) while this application may be useful for population groups, it must be recognized that there is a high probability of inaccuracies for any individual child; (2) little information is available during periods when no variation in the number of teeth occurs. If predictions are to be made for individuals, low birth weight, local pathological conditions, supernumerary teeth, the early or late loss of deciduous teeth, and some general pathological conditions should be taken into consideration.

Cattell (1928) was the first to conduct such a study and assign a dental maturity age according to the number of emerged teeth in the mouth. Later authors (Voors and Metselaar, 1958; Moorrees *et al.,* 1963*a;* McGregor *et al.,* 1968) have also worked in the same line. It is understandable that teeth, being less prone to environmental changes, can be regarded as the ideal instrument for chronological age determination. Bailey (1964) devised a linear regression formula, $Y + A = BX$ (where Y is the number of emerged teeth and X is the age in months), to predict age from dental emergence of New Guinean children. This approach was later criticized by several authors (Meredith, 1971; Brook and Barker, 1972; Billiewicz *et al.,* 1973).

Malcolm and Bue (1970), using regression equations, developed a system of predicting height, according to the number of teeth in the mouth. Following studies in six different New Guinean populations, they based their system on four different stages, where, respectively, 4, 4–10, 12, or 14–26 teeth were present in the mouth. A similar approach was used by Filipsson (1975), for dental-age determination, by using the curve of the total number of emerged permanent teeth. In a later paper, Filipsson and Hall (1975) attempted to predict final body height from the age at which the first 12 permanent teeth emerged, and the corresponding height for that age.

But the concept of "dental age," even though it is simpler to visualize and compare to either the chronological age or to other developmental ages, should be replaced by a scale of maturity, between "immaturity" and "complete maturity," which can be expressed by figures, say between 0 and 100, and could later on be transformed and presented in terms of age, if needed.

7. Evaluation of Dental Maturity by Developmental Stages

Up to now, we have reviewed dental maturity according to the criterion of emergence. It was Gleiser and Hunt (1955) who first stated, in their well-known study, that "the calcification of a tooth may be a more meaningful indication of somatic maturation than is its clinical emergence." Since then, this concept has steadily been gaining acceptance among workers in the dental field of human biology (Garn *et al.,* 1958; Nolla, 1960; Fanning, 1961; Moorrees *et al.,* 1963*a;* Demirjian *et al.,* 1973) for the following reasons:

1. While the emergence of a tooth is a fleeting event and its exact time very difficult to determine, its calcification is a continuous process which can be assessed by permanent records, such as X rays. Although several authors agree on the definition of clinical emergence as being the first tip of the cusp piercing the gum, we can still find in the literature different definitions of the word "eruption," ranging all the way from the moment the alveolar bone is broken to the attainment of the occlusal level.

2. The emergence of a tooth is disturbed by different exogenous factors, such as infection or the premature extraction of the deciduous predecessor. It is known that the early extraction of a deciduous tooth has a delaying effect on the emergence

of its permanent successor, and late extraction favors early emergence (A. A. Dahlberg and Menegaz-Bock, 1958; Fanning, 1961; Brook, 1973). We can also cite the influence of crowding, as the emergence of the permanent teeth is subject to the available space in the arch.

3. If we take the emergence of a tooth as a maturity indicator, we have to consider each tooth individually. We cannot assess the over-all picture of the dentition, unless we deal with the total count of the teeth. This means the evaluation of maturity becomes a very crude measurement.

4. The use of clinical emergence as a maturity indicator is limited to certain ages. The emergence of the deciduous teeth can only be used between the ages of 6 and 30 months, after which time there is no activity until the first permanent tooth (first molar) emerges, at the age of 6 years. The last permanent tooth emerges at approximately 12 years of age. The third molar is not taken into consideration because of the inconsistency and great variation in the timing of its emergence. Besides, the third molar is the only tooth to emerge between the ages of 14 and 20 years, and it would be too imprecise to base estimates of dental maturity for such an age-span on just one tooth. Thus we see that it is impossible, with emergence data alone, to assess the dental maturity of a child between 2½ and 6 years, and after 12 years of age, as there is no clinical emergence of teeth at those intervals.

For all the reasons just stated, dental formation or calcification, which is a continuous developmental process, should be considered as a better measure of physiological maturity than a short-lived and environmentally dependent phenomenon such as dental emergence.

The process of assessing dental maturity from X rays differs from that of other techniques used in physical anthropology, in which length measurements are the basis of the assessment. For example, both stature or the bigonial diameter are measures of the distance between two well-defined points. On the other hand, the assessment of dental maturity, like skeletal maturity, should be based on biological criteria rather than on length or width measurements alone. The latter are really estimates of growth, as distinct from maturity.

Like the bones of the hand–wrist area, teeth undergo different sequences of maturational stages. The first stage is the actual formation of the crypt, and the final stage is the fully mature tooth, defined as the "closure of the apex." During this maturational process, one can see continuous changes in the size and shape of the teeth. Each tooth follows the same sequence and in order to study the entire process, arbitrary stages must be selected and fully described, which trace the entire developmental process from beginning to end. These stages should: (1) describe the major developmental stages of the tooth; (2) be clearly defined (not merely by length increases); and (3) be objective enough to be reproducible. Unlike the skeletal system, the dental system has two overlapping developmental periods, for two sets of teeth. The developmental period for the deciduous teeth extends from the third month of intrauterine life to the third year of postnatal life. For the permanent teeth, it is from the age of 6 months (postnatal) to 14–15 years, excluding the third molars.

The radiographic study of the development of both dentitions, along with the resorption of the deciduous teeth, is very effective in the assessment of dental maturity. The study of a single tooth or a certain group of teeth can be useful in some clinical situations, but when evaluating maturity, it is, at least in theory, preferable to evaluate the entire dentition, either deciduous or permanent, or both, as every single tooth contributes to the dental maturity process (Prahl-Andersen and Van der Linden, 1972). Nevertheless, to study every tooth is prohibitively time-consuming and expensive, and there are certain technical difficulties involved as well. Therefore, acceptable modifications have been devised. For example, several authors (Nolla, 1960; Moorrees *et al.*, 1963a; Liliequist and Lundberg, 1971; Demirjian *et al.*, 1973) have demonstrated that a high degree of correlation exists between the developmental stages of the right and left sides of the maxilla and mandible. As a result, it is possible to study only one side of the arch. This type of selection is also made in bone-age assessments, where the hand–wrist area is usually selected for evaluation.

Most investigators have also preferred to limit their observations to the mandibular teeth (Gleiser and Hunt, 1955; Demisch and Wartmann, 1956; Demirjian *et al.*, 1973) and to the maxillary incisors and cuspids (Fanning, 1961; Moorrees *et al.*, 1963; Liliequist and Lundberg, 1971), because the radiographic view of most maxillary teeth is obstructed during the early developmental period of the permanent teeth (between 1 and 6 years): the tooth buds are still in the maxillary bone and the bony structure superimposed in the bicuspid and molar area.

Because the radiographic view of the mandibular teeth is so clear, it has been proposed that, for purposes of standardization, all mandibular teeth (excluding the third molars) on the left side be selected for study. Once a group of teeth is agreed upon for assessment, a maturity scale can be constructed. The concept of a maturity scale is different from that of a scale of height, for example, in that all subjects pass through the same series of points on the scale, starting with a stage designated as "0" for "complete immaturity," and concluding with "100," corresponding to "complete maturity." Scores between these two extremes are assigned to the degree of development of the specific teeth under consideration.

The first study found in the literature to assess dental formation using radiographic techniques was conducted by Hess *et al.* (1932), in order to assess the physiological maturity of children. Gleiser and Hunt (1955) studied 25 boys and 25 girls longitudinally, from birth to age 10, and traced the development of the first permanent mandibular molar. They described 13 different stages of maturity, from the calcification of the top of the cusp to closure of the apex. The description of the stages is based on length criteria, corresponding to, for example, $1/2$ of the crown or $3/4$ of the root completed. They found that although "calcification and eruption are not identically synchronized in all children, on the average clinical emergence occurs soon after one third of the root is completed."

In a longitudinal study of 255 boys and girls, Garn *et al.* (1958) defined five stages of the development of the second premolar (P_2) and the second molar (M_2): three for calcification and two for eruption. All assessments of the calcification stages, as well as of the alveolar emergence and occlusal attainment, were made from radiographs. In this study, clinical emergence was not assessed. The authors found that the sequences of calcification and emergence alternated in 55% of the subjects. Thus, they concluded that a stage of calcification, for example, cannot be a prediction criterion for a stage of eruption. Once again, the lack of synchronization between different developmental stages and the variability among individual teeth were demonstrated.

Nolla (1960) described another technique for the evaluation of the development of the permanent dentition, based on 10 length stages, such as $1/3$ of the crown or $2/3$ of the root. A score of 1–10 was assigned to each stage, starting from the presence of the crypt without calcification, to the completion of the apical end. These developmental stages were studied longitudinally for both the maxillary and mandibular teeth in 25 boys and 25 girls.

Fanning (1961) applied Gleiser and Hunt's 13 stages of dental development to all mandibular and maxillary incisors. For greater precision, she added a total of seven more stages: two initial stages, two for cleft formation in the molars, and three to describe the different degrees of apex closure. In fact, these added stages served only to detract from the precision of the method. It proved to be too difficult to differentiate between successive stages, and measurements became little more than subjective estimates.

Moorrees *et al.* (1963*b*) used the longitudinal records from the Fels sample to study the formation and resorption of the deciduous canines and molars. Weighted

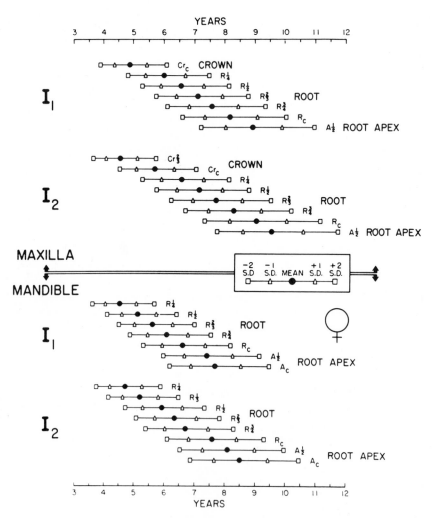

Fig. 1. Norms of the formation of permanent maxillary and mandibular incisor roots of females, including terminal stages of crown formation of the maxillary incisors. (From Moorrees, C. F. A., Fanning, E. A., and Hunt, E. E., Jr., 1963*a*, Age variation of formation stages for ten permanent teeth, *J. Dent. Res.* **42**:1490.)

Table I. Comparative Table of "Stages of Dental Formation" According to Different Authors

	Fanning (1961)	Moorrees (1963)	Hunt (1955)	Nanda (1966)	Nolla (1960)	Demirjian (1973)	Garn (1958)
Presence of crypt	1	—	—	—	1	—	—
Initial cusp formation	2	1	—	—	2	1 (A)	1
Coalescence of cusps	3	2	1		—	—	—
Occlusal surface completed	4	3	2	1	—	2 (B)	—
Crown $\frac{1}{3}$	—	—	—	—	3	—	—
Crown $\frac{1}{2}$	5	4	3	2	—	3 (C)	—
Crown $\frac{2}{3}$	6	—	4	—	4	—	—
Crown $\frac{3}{4}$	—	5	—	3	5	—	—
Crown formation completed	7	6	5	4	6	4 (D)	—
Initial radicular formation	8	7	6	5($\frac{1}{8}$)	—	—	2
Initial radicular bifurcation	8 A,B	8	—	—	—	—	—
Root $\frac{1}{4}$	9	9	7	6	—	5(E)	—
Root $\frac{1}{3}$	10	—	8	7($\frac{3}{8}$)	7	—	—
Root $\frac{1}{2}$	11	10	9	8	—	—	—
Root $\frac{2}{3}$	12	—	10	9($\frac{5}{8}$)	8	6(F)	—
Root $\frac{3}{4}$	13	11	11	10	—	—	—
Root completed	14	12	12	11($\frac{7}{8}$)	9	7(G)	—
Apex $\frac{1}{2}$ closed	—	13	—	—	—	—	—
Apex closed[a]	15	14	13	12	10	8(H)	3

[a]Apex closure: $\frac{1}{4}$; $\frac{1}{2}$, $\frac{3}{4}$.

estimates of the mean attainment age for each stage were obtained by using a standard deviation of 2.042 log conceptional-age units. A graphic representation of the chronology of tooth formation and root resorption was provided for each tooth. However, the authors cautioned that "standard score ratings for the various teeth should be presented preferably as a range of maximal and minimal deviate. They should not be averages, in recognition of the individuality of each tooth."

Moorrees et al. (1963a) also studied the developmental pattern of eight mandibular and two maxillary incisor permanent teeth. The incisors were assessed from intraoral radiographs of 134 Boston children, while uniradicular and multiradicular posterior teeth were assessed from lateral radiographs of 246 children from the Fels sample. The results of this study are presented in chart form. They are "designed specifically for determining the dental maturity of an individual child, for each tooth separately." The usefulness of this system can be very well appreciated in orthodontics clinics, where formative stages of individual teeth and their emergence are very important for treatment-planning purposes. The developmental chart of the maxillary and mandibular incisors, for the females, is given as an example in Figure 1.

Following a study of 287 children, Liliequist and Lundberg (1971) published a system of evaluation for dental development, based on the radiographic assessment of the mandibular teeth (excluding the third molars), upper incisors, and cuspids. The upper bicuspids and molars were not examined because of the difficulty in

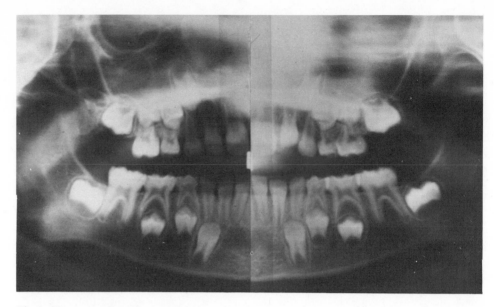

Fig. 2. The panoramic radiograph of the maxilla and the mandible gives a complete picture of the whole dentition. The posterior teeth of the upper jaw are superimposed with the palative and the zygomatic bones, while in the lower jaw there is a clearer view for a better assessment.

obtaining a clear image of these teeth on X rays. Eight developmental stages were defined between the noncalcified crown and the completion of root development. In this system, the intervals between each successive stage were considered to be equal. In other words, scores of 1 to 8 were assigned to each stage, respectively.

All the systems and stages of dental development described in the previous paragraphs are illustrated in Table I. It should be noted that all these systems have been based on absolute values, for the lengths of the teeth, crowns, or roots. If dental radiographs from longitudinal studies are not available, however, such a system is difficult to apply. Even an experienced investigator can have trouble distinguishing between such criteria as ¼ or ⅓ of a root length. Moreover, absolute lengths of teeth can be highly variable from individual to individual.

An acceptable alternative for these systems of assessing dental maturity should be both convenient and specific and should contain just enough stages to avoid confusion and subjectivity. A new approach has been adopted by Demirjian *et al.* (1973), which is based on the same principle as the bone-age assessment of Tanner *et al.* (1975). By this system, which utilizes a new technique of panoramic radiographs (Figure 2), weighted scores are assigned to each of the seven left mandibular teeth. Eight stages of development, from calcification of the tip of the cusp to the closure of the apex, were designated by a 0 (for no calcification), and letters A to H, corresponding to the eight stages (Figure 3). Letters rather than numbers were selected so as not to leave the false impression that each stage is equidistant from the other.

7.1. Assigning the Ratings

1. The mandibular permanent teeth are rated in the following order: 2nd molar, 1st molar, 2nd bicuspid, 1st bicuspid, canine, lateral incisor, central incisor.

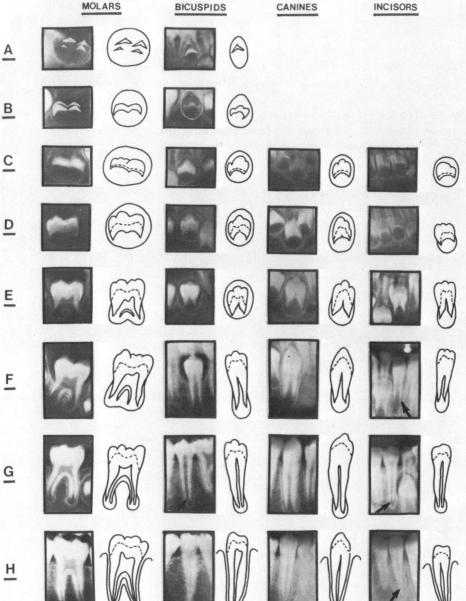

Fig. 3. The developmental status of each group of teeth (molars, bicuspids, canines, incisors) is defined for stages A to H. The definition of each stage is based on biological criteria (see text).

2. The rating is assigned by following carefully the written criteria for each stage and by comparing the tooth with the diagrams and X-ray pictures given in Figure 3. The illustrations should only be used as an aid, not as the sole source of comparison. For each stage, there are one, two, or three written criteria marked a, b, and c. If only one criterion is given, this must be met for the stage to be taken as reached; if two criteria are given, then it is sufficient if the first one of them is met for the stage to be recorded as reached; if three criteria are given, the first two of

them must be met for the stage to be considered reached. At each stage, in addition to the criteria for that stage, the criteria for the previous stage must be satisfied. In borderline cases, the earlier stage is always assigned.

3. There are no absolute measurements to be taken. A pair of dividers is sufficient to compare the relative length (crown/root). To determine apex closure stages, no magnifying glass is necessary. The ratings should be made with the naked eye.

4. The crown height is defined as being the maximum distance between the highest tip of the cusps and the cementoenamel junction. When the buccal and lingual cusps are not at the same level, the midpoint between them is considered as the highest point.

7.2. Dental Formation Stages

If there is no sign of calcification, the rating 0 is given. The crypt formation is not taken into consideration.

Each stage was then given a numerical score, according to the mathematical

Stage		Description
A		In both uniradicular and multiradicular teeth, a beginning of calcification is seen at the superior level of the crypt, in the form of an inverted cone or cones. There is no fusion of these calcified points.
B		Fusion of the calcified points forms one or several cusps, which unite to give a regularly outlined occlusal surface.
C	a	Enamel formation is complete at the occlusal surface. Its extension and convergence toward the cervical region is seen.
	b	The beginning of a dentinal deposit is seen.
	c	The outline of the pulp chamber has a curved shape at the occlusal border.
D	a	The crown formation is completed down to the cementoenamel junction.
	b	The superior border of the pulp chamber in uniradicular teeth has a definite curved form, being concave toward the cervical region. The projection of the pulp horns, if present, gives an outline like an umbrella top. In molars, the pulp chamber has a trapezoidal form.
	c	Beginning of root formation is seen in the form of a spicule.
E		*Uniradicular teeth*
	a	The walls of the pulp chamber now form straight lines, whose continuity is broken by the presence of the pulp horn, which is larger than in the previous stage.
	b	The root length is less than the crown height.
		Molars
	a	Initial formation of the radicular bifurcation is seen in the form of either a calcified point or a semilunar shape.
	b	The root length is still less than the crown height.
F		*Uniradicular teeth*
	a	The walls of the pulp chamber now form a more or less isosceles triangle. The apex ends in a funnel shape.
	b	The root length is equal to or greater than the crown height.

(Continued)

Stage		Description
		Molars
	a	The calcified region of the bifurcation has developed further down from its semilunar stage to give the roots a more definite and distinct outline, with funnel-shaped endings.
	b	The root length is equal to or greater than the crown height.
G	a	The walls of the root canals are now parallel (distal root in molars).
	b	The apical ends of the root canals are still partially open (distal root in molars).
H	a	The apical end of the root canal is completely closed (distal root in molars).
	b	The periodontal membrane has a uniform width around the root and the apex.

technique by Tanner *et al.* (1975) and Healy and Goldstein (1976), the sum of which provides an estimate of an individual's dental maturity on a scale from 0 to 100. The scores and percentile standards were calculated separately for boys and girls between the ages of 3 to 16 years, based on a standard sample of 1446 boys and 1482 girls, aged 2–20 years, of French-Canadian origin.

Although we have no absolute standard by which to judge the variability of a dental maturity system, we can require it to have two general properties: first, maturity scores must reflect the continuous nature of biological development and succeed each other in a smooth logical progression; second, the variability in the maturity score, at each age, should be sufficient to reflect the natural variability of the general population.

The system just described has two main shortcomings: (1) It is necessary to rate all 7 teeth. In many children, however, one or more teeth may be missing, and it may not be possible to substitute the corresponding tooth from the right side of the mandible. Also, for practical reasons, it is often simpler, using a standard X-ray

Table II. Self-Weighted Scores for Dental Stages, 7 Teeth (Mandibular Left Side)

Tooth	Stage								
	0	A	B	C	D	E	F	G	H
Boys									
M_2	0.0	1.7	3.1	5.4	8.6	11.4	12.4	12.8	13.6
M_1				0.0	5.3	7.5	10.3	13.9	16.8
PM_2	0.0	1.5	2.7	5.2	8.0	10.8	12.0	12.5	13.2
PM_1		0.0	4.0	6.3	9.4	13.2	14.9	15.5	16.1
C				0.0	4.0	7.8	10.1	11.4	12.0
I_2				0.0	2.8	5.4	7.7	10.5	13.2
I_1				0.0	4.3	6.3	8.2	11.2	15.1
Girls									
M_2	0.0	1.8	3.1	5.4	9.0	11.7	12.8	13.2	13.8
M_1				0.0	3.5	5.6	8.4	12.5	15.4
PM_2	0.0	1.7	2.9	5.4	8.6	11.1	12.3	12.8	13.3
PM_1		0.0	3.1	5.2	8.8	12.6	14.3	14.9	15.5
C				0.0	3.7	7.3	10.0	11.8	12.5
I_2				0.0	2.8	5.3	8.1	11.2	13.8
I_1				0.0	4.4	6.3	8.5	12.0	15.8

machine, to take a radiograph of fewer than 7 teeth; thus a system based on fewer teeth would be preferable. Naturally, some information and precision would be lost, with the result that slightly different components of dental maturity may be measured. (2) There was an insufficient number of children in the highest and lowest age groups in the standard sample. This meant that the earlier stages of some teeth could not be included, since they were not adequately represented in the sample. Consequently, centile standards could not be provided for the extreme age groups (3 and 14 years).

For these reasons, the standard sample was enlarged to include a total of 2047 boys and 2349 girls, aged 2–20 years (Demirjian and Goldstein, 1976). Two stages, which were excluded in the initial study, could not be included, namely stage A of the PM_1 and stage C for the I_1. The 3rd and 97th percentile estimates for the maturity standards could also now be included. The new scores and centile curves for the seven teeth are provided in Table II and Figures 4 and 5. The comparative curves of the maturity scores for boys and girls is given in Figure 6.

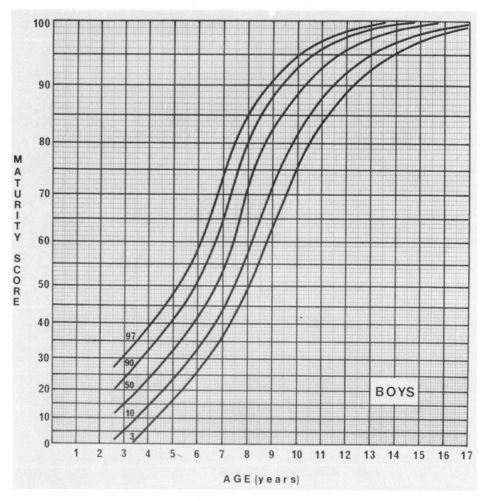

Fig. 4. Dental maturity percentiles given in scores, for boys from 3 to 17 years of age (based on seven left mandibular teeth).

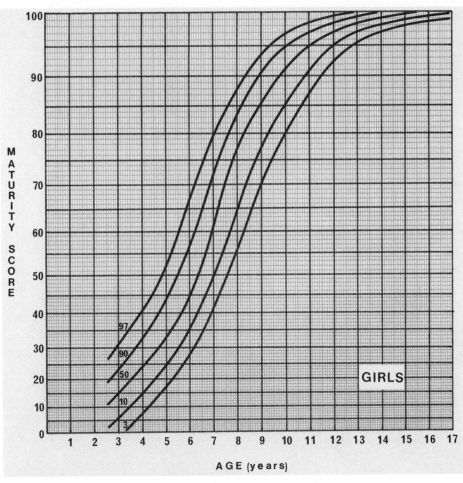

Fig. 5. Dental maturity percentiles given in scores, for girls from 3 to 17 years of age (based on seven left mandibular teeth).

For practical purposes, when panoramic radiographs cannot be used, it is convenient to take two periapical radiographs of the molar and the premolar areas. Therefore, the use of two molars and premolars (M_2, M_1, PM_2, and PM_1) has been considered as the basis of a second system of dental assessment, for which scores and standards have also been established (Table III and Figures 7 and 8).

Since the development of the lower incisor is chronologically almost the same as that of the first molar, the central incisor can be substituted for the molar in older age groups where the latter is frequently missing. This principle forms the basis of a third system of dental assessment, which assigns separate scores and standards for another group of four teeth (M_2, PM_2, PM_1, I_1). This third system requires an additional periapical film of the incisor area (see Table IV, Figures 9 and 10).

In all three systems just described, equal "biological" weights for each tooth have been used. The correlation among the three systems is quite high, the correlation coefficient being between 0.7 and 0.9 for each age group between 6 and 16 years. This could be anticipated since most of the same teeth are involved in all three scoring systems. What we have done is to devise two subsystems, each comprising convenient groups of teeth for rating purposes. The results of comparing

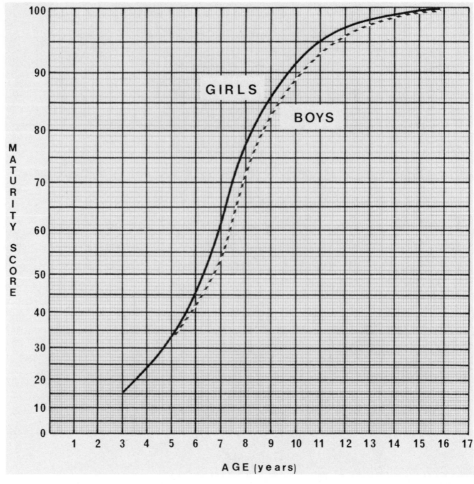

Fig. 6. Comparative curves (median) of dental maturity for boys and girls between ages 3 and 17 (based on seven left mandibular teeth).

Table III. Self-Weighted Scores for Dental Stages, 4 Teeth M₂, M₁, PM₂, PM₁ (Mandibular Left Side)

Tooth	| Stages								
	0	A	B	C	D	E	F	G	H
Boys									
M_2	0.0	3.2	6.2	9.9	14.4	18.4	20.7	21.9	23.3
M_1				0.0	8.0	12.6	16.9	21.8	27.4
PM_2	0.0	3.1	5.6	9.5	13.7	17.4	20.1	21.4	22.5
PM_1		0.0	5.9	10.7	15.7	20.7	23.8	25.4	26.8
Girls									
M_2	0.0	3.6	6.1	9.9	15.3	19.2	21.7	23.0	24.2
M_1				0.0	5.4	9.8	14.3	20.1	25.9
PM_2	0.0	3.7	5.8	9.8	14.7	18.1	20.8	22.3	23.3
PM_1		0.0	4.6	9.2	15.1	20.2	23.3	25.1	26.6

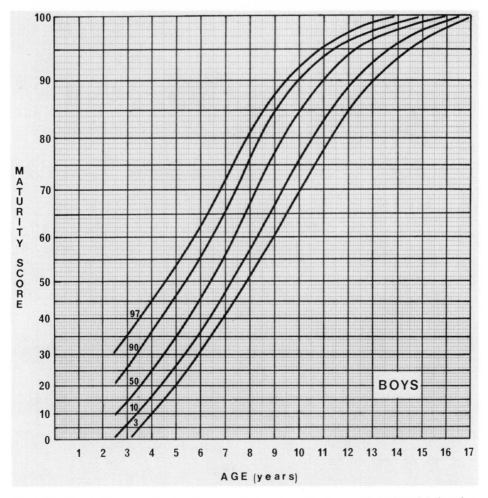

Fig. 7. Dental maturity percentiles given in scores, for boys from 3 to 17 years of age (4 teeth I) (based on M_2, M_1, PM_2, PM_1).

these systems with the seven-tooth system, and with each other, raises the possibility that somewhat different aspects of maturity are being measured. In order to study whether they are, in fact, measuring different maturity parameters, we need to compare children with longitudinal records, using the different systems. If the systems prove to be comparable, one or the other may be selected, according to whether a full panoramic radiograph is available, whether a particular tooth is missing, etc. The reasons for selecting one system or the other should be recorded when reporting a maturity score.

The centile curves enable us to assess the centile position of an individual's maturity score. If required, the maturity score can be converted to "dental age," by finding the age at which the 50th centile value equals the maturity score. However, since dental development is a maturational process and the scores reflect this maturation, we feel it more appropriate to use maturity scores, rather than dental age, when we compare the development of individuals or trace the progress of a particular child.

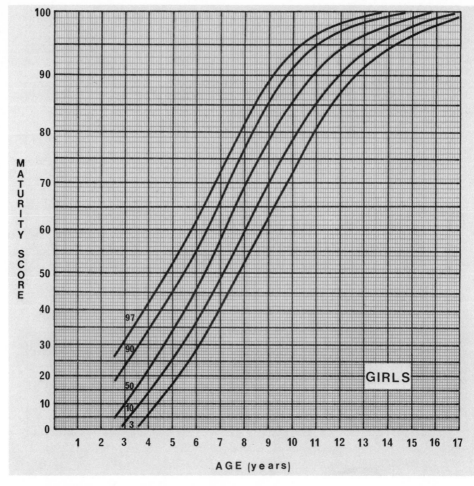

Fig. 8. Dental maturity percentiles given in scores, for girls from 3 to 17 years of age (4 teeth I) (based on M_2, M_1, PM_2, PM_1).

Table IV. Self-Weighted Scores for Dental Stages, 4 Teeth M_2, PM_2, PM_1, I_1 (Mandibular Left Side)

Tooth	Stages								
	0	A	B	C	D	E	F	G	H
Boys									
M_2	0.0	3.3	6.1	9.9	15.0	19.7	21.3	22.1	23.5
PM_2	0.0	3.2	5.6	9.6	14.2	18.8	20.9	21.7	22.8
PM_1		0.0	7.1	11.6	16.9	22.8	25.8	26.8	27.9
I_1				0.0	7.4	11.5	14.6	18.9	25.7
Girls									
M_2	0.0	3.4	6.3	10.2	15.7	20.0	21.5	22.3	23.5
PM_2	0.0	3.7	6.2	10.3	15.1	19.1	21.0	21.7	22.8
PM_1		0.0	5.9	10.2	16.2	21.9	24.6	25.6	26.8
I_1				0.0	8.1	12.2	15.6	20.7	27.0

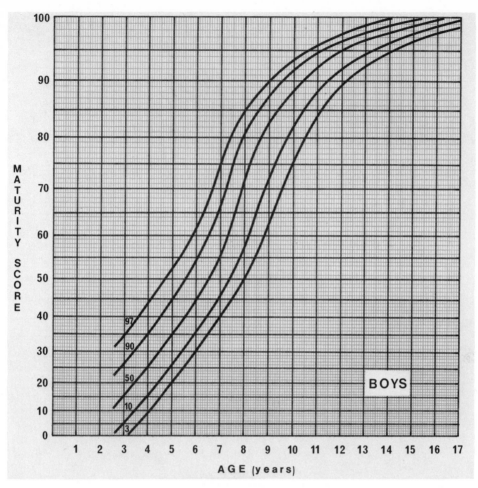

Fig. 9. Dental maturity percentiles given in scores, for boys from 3 to 17 years of age (4 teeth II) (based on M_2, PM_2, PM_1, I_1).

When using the scoring system and standards presented here, it should be remembered that the sample is entirely of French-Canadian origin. We do not know as yet how representative the results are of children generally. We would conjecture, however, that the scores for the eight stages will not vary much between populations, but that the maturity standards may change appreciably in different populations. It is possible to study differences in the average maturity of different populations, using the present scoring systems, with relatively small samples.

8. Correlations between Dental Development and Other Maturity Indicators

Many cross-sectional and longitudinal studies have been conducted to establish possible correlations between the development of the dental system and other physiological indicators, such as bone age, height, menarche, or circumpubertal growth. Most such studies have used dental emergence of both the deciduous and/

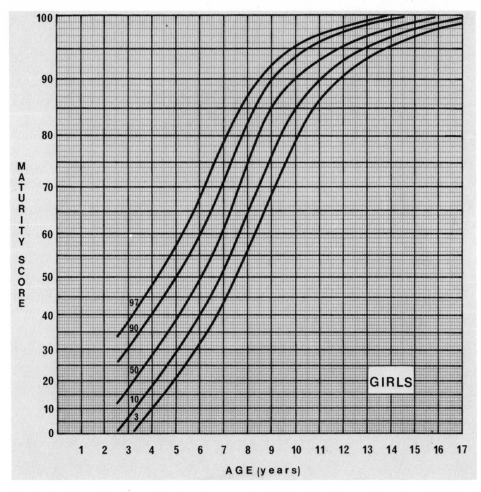

Fig. 10. Dental maturity percentiles given in scores, for girls from 3 to 17 years of age (4 teeth II) (based on M_2, PM_2, PM_1, I_1).

or permanent teeth as the maturity indicator; however, some have investigated the formation or calcification pattern of the teeth in relation to certain physiological measures.

Correlation studies have been undertaken for a variety of reasons. Usually, investigators are looking for a simple means of establishing correlations between bone age and dental maturity, or for a method which might replace bone age as a maturity indicator in child growth and development. Other investigators seek comparative growth data in an attempt to establish the etiology of malocclusions in orthodontic practice. Because of the different methodologies and approaches used in these studies, it is very instructive to look at them in some detail.

8.1. Correlations Using Dental Emergence Criteria

As early as 1942, Robinow *et al.* (1942) used factor analysis to establish correlations between the emergence of the deciduous teeth and the appearance of ossification centers, height, and the onset of walking. They found no correlation among any of these parameters and concluded that the development of the dentition is a relatively independent growth process. This should not be too surprising, since bones and teeth have different origins. Teeth are at least partially of epithelial origin, while bone is derived from the mesoderm.

The relationship of the emergence of deciduous dentition to birth weight has already been reported in an earlier section of this review. Falkner (1957) reported that small babies have a faster rate of dental emergence than larger babies, but he found no correlation between skeletal maturation and dental emergence. These results have been corroborated by Robinow (1973), Billewicz *et al.* (1973), and McGregor *et al.* (1968).

Doubtless because of the greater availability of data from older children (school and medical records), the emergence of permanent dentition and/or the number of permanent teeth in the mouth have been more thoroughly investigated in relation to somatic growth, skeletal development, and puberty. Earlier studies employed the Greulich and Pyle atlas (1950) as the reference base of skeletal development. The reference most frequently used for dental emergence was a series of dental charts compiled by Schour and Massler (1940).

Sutow *et al.* (1954) investigated the correlation between the number of permanent teeth in the mouth and skeletal development in Japanese children, between the ages of 6 and 14 years. They found that the children who had the greatest number of teeth in each group were also the most skeletally advanced. Nevertheless, differences within any one age group were not statistically significant. This illustrates the importance of distinguishing age-specific correlations between different maturity indicators from correlations using pooled ages, which are confounded by the maturation going on with time in both systems.

In a longitudinal study, Lamons and Gray (1958) came to the same conclusion, namely that there is a greater correlation between dental emergence and chronological age, than between dental emergence and bone age. In his longitudinal study of 43 girls and 29 boys, Meredith (1959) assessed dental development by the emergence of the canine and the first and second mandibular molars. A very low degree of correlation was found between the emergence of these teeth and age of maximum circumpubertal height growth. Nanda (1960), using the longitudinal records of the Denver study, came to the same conclusion and found a low and nonsignificant correlation (0.2 for boys and 0.3 for girls) between dental emergence and age of the maximum rate of the circumpubertal spurt in height. The correlation was slightly higher (0.6) between the completion of the permanent dentition and age at menarche. Another interesting corollary to this finding was that no close relationship has been found between the growth of the face, which follows the general somatic pattern, and the emergence of the teeth. This knowledge is of importance in the diagnosis and treatment of malocclusions.

Eveleth (1960) compared dental emergence and age at menarche in American children living in Rio de Janeiro, Brazil, to their counterparts living in the United States. Again, only a very low correlation was found between these two indices in both populations. Björk and Helm (1967) also failed to find any significant correlation between dental age and skeletal age, height, or age at menarche.

In concluding the review of the relationships between dental emergence and other maturational parameters, we can state that the evidence, so far, indicates that the skeletal system, as well as height and the onset of puberty, develop largely independently of the dental system. For further information on this, we now turn to studies using the newer, quantitative measures of dental formation.

ARTO DEMIRJIAN

As there are few studies in this field, and these have varied in the exact measures used and in the way the results have been presented, we will review them individually in some detail.

In one of the first correlation studies using dental formation as one of the criteria, Brauer and Bahador (1942) compared the calcification process to emergence, for both sets of teeth, in 415 children. They found that calcification age and emergence age do not necessarily correspond to chronological age. They also stressed that without radiographic data, a diagnosis based on dental maturity by tooth emergence could be inaccurate in at least 50% of the patients. The conclusions of Brauer and Bahador were corroborated by Garn *et al.* (1960), who found that tooth calcification and eruption processes are independent of each other.

Since it appeared well established that the two dental processes are autonomous, most authors have elected to try to correlate tooth calcification with other physiological maturity indicators. In one of the earliest of such studies, Demisch and Wartmann (1956) studied the calcification of the third molar, in relation to bone age, in 151 children. Their correlation coefficient was always between 0.73 and 0.92, between bone age, chronological age, and dental calcification, for all age groups pooled. However, within any one chronological age group, the correlation coefficient between bone age and dental calcification was about 0.45 for both boys and girls.

Hotz (1959) and Green (1961) concluded independently that there is a high correlation between dental and chronological age, but not between dental age and skeletal age. Both authors found large variations in skeletal age and as a result, Green suggested that dental age should be adopted as a more reliable criterion of somatic growth evaluation.

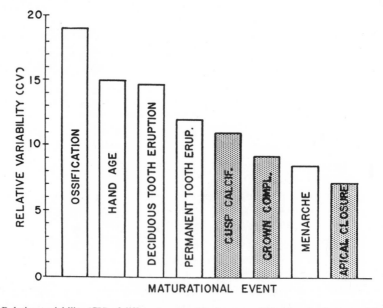

Fig. 11. Relative variability (CV) of different maturational events. (From Lewis, A. B., and Garn, S. M., 1960, The relationship between tooth formation and other maturational factors, *Angle Orthod.* **30**:70.)

Lewis and Garn (1960) also found less variability, as judged by coefficient variation, in dental development than skeletal development, in equivalent age groups, when they applied their five stages of tooth development to 255 children. This suggests that dental development can be a better criterion of biological maturation than ossification (Figure 11). They went a step further by suggesting that dental age could be used to determine an unknown chronological age. It is also interesting to note that in the same study, Lewis and Garn found that there was a weaker correlation between dental calcification stages and dental emergence, than among different stages of calcification.

Garn *et al.* (1962) studied the calcification and emergence patterns of the third molar in relation to bone age and chronological age. Due to its inconsistent development, they found a low correlation between the development of this tooth and menarche and skeletal development. A higher correlation was subsequently found between the development of the second molar and the other maturity indicators.

Grøn (1962) compared dental emergence, calcification of the root, and skeletal development in 874 children. She found a better correlation between dental emergence and root formation than between dental emergence and chronological or skeletal age. Thus, she suggested that emergence and root formation data be used for clinical and treatment purposes in orthodontics.

Steel (1965) investigated the relationships between dental and physiological maturation in a group of 12-year-old children. Each child was evaluated within a week of his twelfth birthday. No direct interdependence was detected between dental and skeletal maturation, and furthermore, the maturity stages of all three teeth investigated (mandibular second premolar and the second and third molars) varied independently. The author concluded that this problem might be overcome through the creation of a system using the combined results of the maturity of all teeth.

Gyulavari (1966) confirmed Steel's findings concerning the variability of individual tooth development and attributed the contradictory results of different authors to this fact. Following his study of low-birth-weight Hungarian children, he corroborated the view that dental and skeletal systems are independent of each other and stressed the necessity of establishing dental standards for each population.

Lacey (1973) investigated 40 children whose heights were below the third percentile. Skeletal age was assessed by Tanner's method (Tanner *et al.,* 1972) and dental development by the appearance of the root development of the second molar. Certain associations between dental and skeletal development were found, but Lacey concluded that, in normal children, the influences that control dental development are different from those that determine bone age. More recently, Anderson *et al.* (1975) found that dental development, in both sexes, was more strongly related to morphological development than to skeletal age. They also pointed out the sex specificity and individuality in the development of the teeth.

At the Growth Centre of the University of Montreal, some 200 boys and girls have been followed longitudinally since 1967. Their dental development and emergence have been studied in detail, along with their anthropometric measurements, skeletal maturation, and nutritional habits. We analyzed the correlations among these different parameters within each age group; very low correlation coefficients have been found, especially between dental age and skeletal age and/or stature

Table V. Correlation Coefficients for Dental Age (DA)[a] with Skeletal Age (SA)[b] and with Stature (STAT)[c], and for Skeletal Age with Stature[d]

Age	Girls				Boys			
	No.	DA/SA	DA/STAT	SA/STAT	No.	DA/SA	DA/STAT	SA/STAT
7	87	0.18	0.24	0.51	97	0.29	0.27	0.39
8	118	0.11	0.18	0.66	127	0.25	0.29	0.42
9	121	0.20	0.19	0.59	128	0.28	0.32	0.55
10	141	0.26	0.35	0.57	160	0.24	0.21	0.51
11	198	0.17	0.20	0.59	213	0.15	0.20	0.56
12	165	0.18	0.17	0.57	187	0.28	0.23	0.58
13	65	0.09	0.25	0.32	84	0.14	0.19	0.66

[a]Dental age calculated by unrevised 7-tooth weighting system and converted to ages, according to Montreal standards.
[b]Skeletal age according to System I of Tanner et al. (1972).
[c]As the data comes from an unfinished longitudinal study, it has not been possible to express stature in maturity form, that is, as a proportion of final adult height.
[d]Unpublished data from Montreal Growth Centre.

(0.09–0.35). The correlations between stature and skeletal age are always higher between the ages of 7 and 13 (Table V). A decrease is noted in girls at age 13, due to the beginning of puberty.

9. Conclusion

Since the beginning of the century, the clinical emergence of teeth, which was referred to as "eruption," has been used as the only criterion of dental development. Population standards have been established on a very few and unsatisfactory samples obtained from fetuses or living subjects; deciduous and permanent dentitions have been studied separately for this purpose, as one might expect. However, their study is limited to the time of appearance of each tooth in the mouth: for example, for the deciduous dentition, this period is between the sixth and the thirtieth months, while for the permanent dentition, it is between the sixth and twelfth years (if we omit the third molar). As for the study of dental calcification by means of X rays, its use is extended between birth and the fifteenth year, without interruption. While an early emergence pattern has been established for girls for permanent dentition, this point is still debated, as far as the deciduous dentition is concerned. In both the deciduous and permanent dentitions, hereditary and genetic factors seem to be more influential than the environmental factors, such as nutrition or socioeconomic status.

Biologists throughout the world started in the middle of this century to use dental calcification, rather than clinical emergence, as a maturity indicator of dental development. This latter approach has many advantages over the previous one; the most obvious is that the dental system can be assessed as a whole, like the skeletal system, instead of assessing the emergence of individual teeth, as is done for clinical purposes. Different methods are described for the evaluation of dental calcification as a maturity indicator.

As with the studies using dental emergence criteria, the evidence indicates that dental and skeletal maturation are substantially independent. As the two have different embryologic origins, with possible differences in genetic control, this is not too surprising. The same environmental factors, such as nutrition, affect both processes, but again to differing degrees. With the possible exception of gross nutritional deprivation, not generally found in Western countries, dental maturation has been found relatively impervious to environmental factors.

To the human biologist, the study of the development of dentition adds another dimension to the growing body of knowledge of human growth and development. Dental maturity should be considered, along with skeletal development and stature, as a major component of the developing physiological unit. To the clinical orthodontist, the developmental life of individual teeth may be of prime importance, but he also needs a method of evaluating the over-all developmental status of the mouth he is treating: this is provided by a dental maturity measure. If this tool is to have scientific use, meticulous longitudinal studies are required to elucidate the physiological relationships involved and to produce the necessary measurement standards.

10. References

Adler, P., 1958, Studies on the eruption of the permanent teeth. IV. The effect upon the eruption of the permanent teeth of caries in the deciduous dentition, and of urbanisation, *Acta Genet.* **8**:78.

Ainsworth, N. J., 1925, The incidence of dental disease in children, *Med. Res. Counc. Spec. Rep. Ser.* **97**.

Anderson, D. L., Thompson, G. W., and Popowich, F., 1975, Interrelationships of dental maturity, skeletal maturity, height and weight from age 4 to 14 years, *Growth* **39**:453.

Bailey, K. V., 1964, Dental development in New Guinean infants, *J. Pediatr.* **64**:97.

Bambach, M., Saracci, R., and Young, H. B., 1973, Emergence of deciduous teeth in Tunisian children in relation to sex and social class, *Hum. Biol.* **45**:435.

Bean, R. B., 1914, Eruption of teeth as physiological standard for testing development, *Pedagog, Sem.* **21**:596.

Beik, A. K., 1913, Physiological age and school entrance, *Pedagog. Sem.* **20**:283.

Billewicz, W. Z., 1973, A note on estimation of calendar age on the basis of development of primary teeth, *J. Trop. Pediatr. Environ.* **19**:243. (monograph no. 28).

Billewicz, W. Z., and McGregor, I. A., 1975, Eruption of permanent teeth in West African (Gambian) children in relation to age, sex and physique, *Ann. Hum. Biol.* **2**:117.

Billewicz, W. Z., Thomson, A. M., Baber, F. M., and Field, C. E., 1973, The development of primary teeth in Chinese (Hong Kong) children, *Hum. Biol.* **45**:229.

Björk, A., and Helm, S., 1967, Prediction of the age of maximum puberal growth in body height, *Angle Orthod.* **37**:134.

Boas, F., 1933, Studies in growth - II, *Hum. Biol.* **5**:429.

Boutourline-Young, H., 1969, The arm circumference as a public health index of protein–calorie malnutrition of early childhood. XII. Arm measurements as indicators of body composition in Tunisian children, *J. Trop. Pediatr.* **15**:222.

Brauer, J. C., and Bahador, M. A., 1942, Variation in calcification and eruption of the deciduous and permanent teeth, *J. Am. Dent. Assoc.* **29**:1373.

Brook, A. H., 1973, The secular trend in permanent tooth eruption times, *J. Trop. Pediatr. Environ. Child Health* **19**:206 (monograph no. 28).

Brook, A. H., and Barker, D. K., 1972, Eruption of teeth among the racial groups of Eastern New Guinea: A correlation of tooth eruption with calendar age, *Arch. Oral Biol.* **17**:751.

Brook, A. H., and Barker, D. K., 1973, The use of deciduous tooth eruption for the estimation of unknown chronological age, *J. Trop. Pediatr. Environ. Child Health* **19**:234 (monograph no. 28).

Butler, D. J., 1962, The eruption of teeth and its association with early loss of the deciduous teeth, *Br. Dent. J.* **112**:443.

Cattell, P., 1928, Dentition as a measure of maturity, *Harvard Monographs in Education,* No. 9, Harvard University Press, Cambridge. 91 pp.

Cifuentes, E., and Alvarado, J., 1973, Assessment of deciduous dentition in Guatemalan children, *J. Trop. Pediatr. Environ. Child Health* **19**:211 (monograph no. 28).

Clements, E. M. B., Davies-Thomas, E., and Pickett, K. G., 1953a, Time of eruption of permanent teeth in Bristol children in 1947–48, *Br. Med. J.* **1**:1421.

Clements, E. M. B., Davies-Thomas, E., and Pickett, K. G., 1953b, Order of eruption of the permanent human dentition, *Br. Med. J.* **1**:1425.

Clements, E. M. B., Davies-Thomas, E., and Pickett, K. G., 1957a, Age at which the deciduous teeth are shed, *Br. Med. J.* **1**:1508.

Clements, E. M. B., Davies-Thomas, E., and Pickett, K. G., 1957b, Time of eruption of permanent teeth in British children at independent, rural and urban schools, *Br. Med. J.* **1**:1511.

Cohen, M. M., and Wagner, R., 1948, Dental development in pituitary dwarfism, *J. Dent. Res.* **27**:445.

Dahlberg, A. A., and Menegaz-Bock, R. M., 1958, Emergence of the permanent teeth in Pima Indian children, *J. Dent. Res.* **37**:1123.

Dahlberg, G., and Maunsbach, A. B., 1948, The eruption of the permanent teeth in the normal population of Sweden, *Acta Genet (Basel)* **1**:77.

Delgado, H., Habicht, J.-P., Yarbrough, C., Lechtig, A., Martorell, R., Malina, R. M., and Klein, R. E., 1975, Nutritional status and the timing of deciduous tooth eruption, *Am. J. Clin. Nutr.* **28**:216.

Demirjian, A., and Goldstein, H., 1976, New systems for dental maturity based on seven and four teeth, *Ann. Hum. Biol.* **3**:411.

Demirjian, A., Goldstein, H., and Tanner, J. M., 1973, A new system of dental age assessment, *Hum. Biol.* **45**:211.

Demisch, A., and Wartmann, P., 1956, Calcification of the mandibular third molar and its relation to skeletal and chronological age in children, *Child Dev.* **27**:459.

Doering, C. R., and Allen, M. F., 1942, Data on eruption and caries of the deciduous teeth, *Child Dev.* **13**:113.

El Lozy, M., Reed, R. B., Kerr, G. R., Boutourline, E., Tesi, G., Ghamry, M. T., Stare, F. J., Kallal, Z., Turki, M., and Hemaidan, N., 1975, Nutritional correlates of child development in southern Tunisia. IV. The relation of deciduous dental eruption to somatic development, *Growth* **39**:209.

Eveleth, P. B., 1960, Eruption of permanent dentition and menarche of American children living in the tropics, *Hum. Biol.* **39**:60.

Falkner, F., 1957, Deciduous tooth eruption, *Arch. Dis. Child.* **32**:386.

Fanning, E. A., 1961, A longitudinal study of tooth formation and root resorption, *N.Z. Dent. J.* **57**:202.

Ferguson, A. D., Scott, R. B., and Bakwin, H., 1957, Growth and development of Negro infants, *J. Pediatr.* **50**:327.

Filipsson, R., 1975, A new method for assessment of dental maturity using the individual curve of number of erupted permanent teeth, *Ann. Hum. Biol.* **2**:13.

Filipsson, R., and Hall, K., 1975, Prediction of adult height of girls from height and dental maturity at ages 6–10 years, *Ann. Hum. Biol.* **2**:355.

Friedlaender, J. S., and Bailit, H. L., 1969, Eruption times of the deciduous and permanent teeth of natives of Bougainville Island, territory of New Guinea: A study of racial variation, *Hum. Biol.* **41**:51.

Garn, S. M., and Rohmann, C. G., 1962, X-linked inheritance of developmental timing in man, *Nature* **196**:695.

Garn, S. M., Lewis, A. B., Koski, P. K., and Polacheck, D. L., 1958, The sex difference in tooth calcification, *J. Dent. Res.* **37**:561.

Garn, S. M., Lewis, A. B., and Polacheck, D. L., 1959, Variability of tooth formation, *J. Dent. Res.* **38**:135.

Garn, S. M., Lewis, A. B., and Polacheck, D. L., 1960, Interrelations in dental development. 1. Interrelationships within the dentition, *J. Dent. Res.* **39**:1049.

Garn, S. M., Lewis, A. B., and Bonne, B., 1962, Third molar formation and its development course, *Angle Orthod.* **32**:270.

Garn, S. M., Lewis, A. B., and Blizzard, R. M., 1965a, Endocrine factors in dental development, *J. Dent. Res.* **44**:243.

Garn, S. M., Lewis, A. B., and Kerewsky, R. S., 1965b, Genetic, nutritional and maturational correlates of dental development, *J. Dent. Res.* **44**:228.

Garn, S. M., Sandusky, S. T., Nagy, J. M., and Trowbridge, F. L., 1973, Negro–Caucasoid differences in permanent tooth emergence at a constant income level, *Arch. Oral Biol.* **18**:609.

Gates, R. E., 1964, Eruption of permanent teeth of New South Wales school children, Part I. Ages of eruption, *Aust. Dent. J.* **9**:211.

Gleiser, I., and Hunt, E. E., Jr., 1955, The permanent mandibular first molar: Its calcification, eruption and decay, *Am. J. Phys. Anthropol.* **13**:253.

Gödény, E., 1951, Studies on the eruption of the permanent teeth. I. The age at the eruption of the different teeth in the normal school population in Hungary, *Acta Genet. (Basel)* **2**:331.

Green, L. J., 1961, The interrelationships among height, weight, and chronological, dental and skeletal ages, *Angle Orthod.* **31**:189.

Greulich, W. W., and Pyle, S. I., 1950, *Radiographic Atlas of Skeletal Development of the Hand and Wrist,* 190 pp., Stanford University Press, Stanford, California.

Grøn, A. M., 1962, Prediction of tooth emergence, *J. Dent. Res.* **41**:573.

Gyulavari, O., 1966, Dental and skeletal development of children with low birth weight, *Acta Paediatr. Acad. Sci. Hung.* **7**:301.

Habicht, J. P., Martorell, R., Yarbrough, C., Malina, R. M., and Klein, R. E., 1974, Height and weight standards for pre-school children. How relevant are ethnic differences in growth potential? *Lancet* **1**:611.

Halikis, S. E., 1961, Variability of eruption of permanent teeth and loss of deciduous teeth in western Australian children. I. Times of eruption of permanent teeth, *Aust. Dent. J.* **6**:137.

Healy, M. J. R., and Goldstein, H., 1976, An approach to the scaling of categorized attributes, *Biometrika* **63**:219.

Hellman, M., 1943, The phase of development concerned with erupting the permanent teeth, *Am. J. Orthod.* **29**:507.

Helm, S., and Seidler, B., 1974, Timing of permanent tooth emergence in Danish children, *Community Dent. Oral Epidemiol.* **2**:122.

Hess, A. F., Lewis, J. M., and Roman, B., 1932, Radiographic study of calcification of teeth from birth to adolescence, *Dent. Cosmos* **74**:1053.

Hotz, R., 1959, The relation of dental calcification to chronological and skeletal age, *Eu. Orthod. Soc.* **1959**:140.

Houpt, M. I., Adu-Aryee, S., and Grainger, R. M., 1967, Eruption times of permanent teeth in the Brong Ahafo region of Ghana, *Am. J. Orthod.* **53**:95.

Hurme, V. O., 1948, Standards of variation in the eruption of the first six permanent teeth, *Child Dev.* **19**:211.

Hurme, V. O., 1949, Ranges of normalcy in the eruption of the permanent teeth, *J. Dent. Child.* **16**:11.

Infante, P. F., and Owen, G. M., 1973, Relation of chronology of deciduous tooth emergence to height, weight, and head circumference in children, *Arch. Oral Biol.* **18**:1411.

Jelliffe, E. F. P., and Jelliffe, D. B., 1973, Deciduous dental eruption, nutrition and age assessment, *J. Trop. Pediatr. Environ. Child Health* **19**:194 (monograph no. 28).

Kihlberg, J., and Koski, K., 1954, On the properties of the tooth eruption curve, *Fin. Tandläk. Sällak. Forh.* **50**:6.

Klein, H., Palmer, C. E., and Kramer, M., 1937, Studies on dental caries. II. The use of the normal probability curve for expressing the age distribution of eruption of the permanent teeth, *Growth* **1**:385.

Kraus, B. S., 1959, Calcification of the human deciduous teeth, *J. Am. Dent. Assoc.* **59**:1128.

Krumholt, L., Roed-Petersen, B., and Pindborg, J. J., 1971, Eruption times of the permanent teeth in 622 Ugandan children, *Arch. Oral Biol.* **16**:1281.

Lacey, K. A., 1973, Relationship between bone age and dental development, *Lancet* **2**:736.

Lamons, F. F., and Gray, S. W., 1958, A study of the relationship between tooth eruption age, skeletal development age, and chronological age in sixty-one Atlanta children, *Am. J. Orthod.* **44**:687.

Lee, M. M. C., Chan, S. T., Low, W. D., and Chang, K. S. F., 1965, Eruption of the permanent dentition of southern Chinese children in Hong Kong, *Arch. Oral Biol.* **10**:849.

Legros, C. H., and Magitot, E., 1893, Chronologie des follicles dentaires chez l'homme, Congrès de Lyon.

Leslie, G. H., 1951, A Biometrical Study of the Eruption of the Permanent Dentition of New Zealand Children, Government Printing Office, Wellington, 72 pp.

Lewis, A. B., and Garn, S. M., 1960, The relationship between tooth formation and other maturational factors, *Angle Orthod.* **30**:70.

Liliequist, B., and Lundberg, M., 1971, Skeletal and tooth development, *Acta Radiol.* **11**:97.

Logan, W. H. G., and Kronfeld, R., 1933, Development of the human jaws and surrounding structures from birth to the age of fifteen years, *J. Am. Dent. Assoc.* **20**:379.

Lysell, L., Magnusson, B., and Thilander, B., 1962, Time and order of eruption of the primary teeth, *Odontol. Rev.* **13**:217.

Lysell, L., Magnusson, B., and Thilander, B., 1969, Relations between the times of eruption of primary and permanent teeth—A longitudinal study, *Acta Odontol. Scand.* **27**:271.

MacKay, D. H., and Martin, W. J., 1952, Dentition and physique of Bantu children, *J. Trop. Med. Hyg.* **55**:265.

Malcolm, L. A., and Bue, B., 1970, Eruption times of permanent teeth and the determination of age in New Guinean children, *Trop. Geogr. Med.* **22**:307.

Mather, K., and Jinks, J. L., 1963, Correlations between relatives arising from sex-linked genes, *Nature* **198:**314.

McGregor, I. A., Thomson, A. M., and Billewicz, W. Z., 1968, The development of primary teeth in children from a group of Gambian villages, and critical examination of its use for estimating age, *Br. J. Nutr.* **22:**307.

Meredith, H. V., 1946, Order and age of eruption for the deciduous dentition, *J. Dent. Res.* **25:**43.

Meredith, H. V., 1959, Relation between the eruption of selected mandibular permanent teeth and the circumpuberal acceleration in stature, *J. Dent. Child.* **26:**75.

Meredith, H. V., 1971, Growth in body size: A compendium of findings on contemporary children living in different parts of the world, *Adv. Child Dev. Behav.* **6:**153.

Miller, J., Hobson, P., and Gaskell, T. J., 1965, A serial study of the chronology of exfoliation of deciduous teeth and eruption of permanent teeth, *Arch. Oral Biol.* **10:**805.

Moorrees, C. F. A., Fanning, E. A., and Hunt, E. E., Jr., 1963a, Age variation of formation stages for ten permanent teeth, *J. Dent. Res.* **42:**1490.

Moorrees, C. F. A., Fanning, E. A., and Hunt, E. E., Jr., 1963b, Formation and resorption of three deciduous teeth in children, *Am. J. Phys. Anthropol.* **21:**205.

Mukherjee, D. K., 1973, Deciduous dental eruption in low income group Bengali Hindu children, *J. Trop. Pediatr. Environ. Child Health* **19:**207 (monograph no. 28).

Nanda, R. S., 1960, Eruption of human teeth, *Am. J. Orthod.* **46:**363.

Nanda, R. S., and Chawla, T. N., 1966, Growth and development of dentitions in Indian children. I. Development of permanent teeth, *Am. J. Orthod.* **52:**837.

Niswander, J. D., 1963, Effects of heredity and environment on development of dentition, *J. Dent. Res.* **42:**1288.

Nolla, C. M., 1960, The development of the permanent teeth, *J. Dent. Child.* **27:**254.

Perreault, J. G., Demirjian, A., and Jenicek, M., 1974, Emergence des dents permanentes chez les enfants canadiens-français, *J. Assoc. Dent. Can.* **40:**306.

Prahl-Andersen, B., and Van der Linden, F. P. G. M., 1972, The estimation of dental age, *Eur. Orthod. Soc.* **1972:**535.

Robinow, M., 1973, The eruption of the deciduous teeth (factors involved in timing), *J. Trop. Pediatr. Environ. Child Health* **19:**200 (monograph No. 28).

Robinow, M., Richards, T. W., and Anderson, M., 1942, The eruption of deciduous teeth, *Growth* **6:**127.

Roche, A. F., Barkla, D. H., and Maritz, J. S., 1964, Deciduous eruption in Melbourne children, *Aust. Dent. J.* **9:**106.

Saunders, E., 1837, *The Teeth a Test of Age, Considered with Reference to the Factory Children,* H. Renshaw, London.

Schour, I., and Massler, M., 1940, Studies in tooth development: The growth pattern of human teeth, Part I and Part II, *J. Am. Dent. Assoc.* **27:**1778, 1918.

Steel, G. H., 1965, The relation between dental maturation and physiological maturity, *Dent. Pract.* **16:**23.

Sutow, W. W., Terasaki, T., and Ohwada, K., 1954, Comparison of skeletal maturation with dental status in Japanese children, *Pediatrics* **14:**327.

Tanner, J. M., 1962, *Growth at Adolescence,* 2nd ed., pp. 143–155, Blackwell, Oxford.

Tanner, J. M., Whitehouse, R. H., Healy, M. J. R., and Goldstein, H., 1972, *Standards for Skeletal Age,* Centre International de l'Enfance, Paris.

Tanner, J. M., Whitehouse, R. M., Marshall, W. A., Healy, M. J. R., and Goldstein, H., 1975, *Assessment of Skeletal Maturity and Prediction of Adult Height: TW2 Method,* Academic Press, London.

Ten-State Nutrition Survey 1968–1970, 1972, Part III, Center for Disease Control, Atlanta, Georgia, U.S. DHEW Publ. No. (HSM)72-8131.

Trupkin, D. P., 1974, Eruption patterns of the first primary tooth in infants who were underweight at birth, *J. Dent. Child.* **41:**279.

Voors, A. W., 1957, The use of dental age in studies of nutrition in children, *Doc. Med. Geogr. Trop.* **9:**137.

Voors, A. W., 1973, Can dental development be used for assessing age in underdeveloped communities? *J. Trop. Pediatr. Environ. Child Health* **19:**242 (monograph no. 28).

Voors, A. W., and Metselaar, 1958, The reliability of dental age as a yardstick to assess the unknown calendar age, *Trop. Geogr. Med.* **10:**175.

Wagner, R., Cohen, M. M., and Hunt, E. E., Jr., 1963, Dental development in idiopathic sexual precocity, congenital adrenocortical hyperplasia and adrenogenic virilism, *J. Pediatr.* **63:**506.

Yun, D. J., 1957, Eruption of primary teeth in Korean rural children, *Am. J. Phys. Anthropol.* **15:**261.

16

Secular Growth Changes

J. C. VAN WIERINGEN

1. The Concept of Secular Changes in Growth and Maturation

The pattern of growth and somatic development of children in a particular population proves not to be static, but changing with time. Some investigators discussed the possibility of such changes, called secular growth changes, as early as the beginning of the 19th century, before the phenomenon had actually been observed (Virey, 1816; Villermé, 1829). Secular growth shift comprises a combination of the following phenomena: (1) decrease or increase in age at which a particular height or weight is attained, particular characteristics of maturation are developed, and linear growth is stopped; (2) increase or decrease in adult height. Increased growth velocity, earlier maturation (probably at an increased height), earlier cessation of linear growth, and a greater adult stature are changes that usually are associated. When these changes prevail, the adjective used is "positive," whereas the adjective for the reverse situation is "negative."

In this particular context the word "secular" means "during a prolonged period." Other authors have used the word in the sense of "lasting a century" or "for the last century." However, secular shifts may last less as well as more than a century. Changes do not occur uninterruptedly in one direction; they show positive as well as negative fluctuations, as well as large geographic differences, and the phenomenon is not exclusively connected to the last hundred years or the present century. It is therefore preferable not to use "the secular trend" and "the positive trend in industrialized countries" as synonyms.

The term *Akzeleration,* introduced by Koch (1953) in the German medical literature, is not very suitable either. The term is used to indicate an individually advanced growth and maturation, as well as an increased growth velocity and an earlier-occurring maturation of an entire population as compared to the past (Hagen, 1962). Because *sekuläre Akzeleration* is a term that chiefly emphasizes increase in growth velocity and earlier occurrence of maturation, indicates less

J. C. VAN WIERINGEN • University Children's Hospital, Het Wilhelmina Kinderziekenhuis, Utrecht, The Netherlands.

clearly the increment in adult height, and actually ignores the possibility of negative trends, it is a less suitable term than secular growth shift. The last-mentioned expression now has become an established term that is frequently encountered in medical literature. According to Moore (1966) interest in the phenomenon increased after World War II.

It has been suggested that measures of the skeleton other than height—head circumference (Nellhaus, 1970) and thorax width, for example—are likewise liable to secular trends, but data about this are lacking. Secular changes may also occur in hemoglobin level (Owen *et al.*, 1970), blood chemistry, skeletal age (de Wijn and Rusbach, 1961), and dental development (Garn and Russell, 1971). However, as secular changes are clearly perceptible in height, weight, and characteristics of sexual maturation, important parameters for the assessment of growth, these are the characteristics that usually represent the material for surveys on secular growth shifts.

2. Medical and Social Importance

The significance of secular changes of growth, and, therefore, the importance of studying this phenomenon, is threefold:

1. Many years ago a close interrelationship was established between shifts in growth and in patterns of morbidity and mortality. Within the framework of epidemiology (the study of characteristics and determinants of health, morbidity, and mortality as collective phenomena), secular growth trends can be used as an indicator of public health by study of both nationwide and regional growth patterns. For international comparisons this health indicator should be used as a variable akin to the conventional indices of vital statistics (Chase, 1972; Habicht *et al.*, 1974).

2. For the assessment of an individual child's growth, one has to be acquainted with the growth pattern, as expressed in standard values, of the population to which the child belongs. The principal characteristics of growth and somatic differentiation are height, weight, and sexual maturation. By plotting a child's individual measures and degree of maturation in growth diagrams, its growth can be defined as relatively tall or small, advanced or retarded, underweight or overweight. Comparison with growth curves enables us to define growth velocity in terms like advanced, average, or delayed, although it has to be remembered that especially in adolescents individual growth curves as a rule differ from cross-sectional curves (see Chapter 5, Volume 1).

As a population's growth can be subject to considerable changes, established standard values can grow obsolete. Thus standard values should be determined periodically, and up-to-date values should be put at the disposal of centers of child and school health care and clinical pediatrics.

3. The proportions that man creates in his world are largely determined by the distance between the soles of the feet and the topmost point of the head (freely adapted from Garn, 1966).

The social impact of secular growth shift, taking into consideration its direction, duration, and extent, is considerable. A logical concomitant of a change in stature is, for example, that clothing, furniture, and tools have to be redesigned, and that the architecture of houses, schools, offices, and workshops has to be adapted. An earlier development of maturation characteristics, accompanied by an earlier onset of biological adulthood, will undoubtedly have an effect upon legislation, jurisdiction, and education.

Interest in human growth gave rise to nonscientific, emotional presuppositions, as well as to scientific, anthropometric research.

Several investigators of anthropometrics (Kiil, 1939; Oppers, 1963; Udjus, 1964) mention the fact that the Romans used to classify their soldiers according to height and that Roman historians showed an interest in the height of peoples with whom they were at war. However, exact figures for those days are not available, and Kiil (1939) assumes that ancient descriptions of the impressive heights of Germans and Gauls might well be explained by the fact that historians had a tendency to depict an enemy taller than he actually was, which would then constitute an excuse in case of defeat and an even greater triumph in case of victory. The magical conception that people in primitive communities are generally taller than those in "weakened, civilized" societies may also be an influence.

Villermé (1829) published *Mémoire sur la taille de l'homme en France,* a work that makes very readable matter even nowadays, but which apparently has been read by few investigators, for it is only quoted in Quételet (1835), Zeeman (1857), and Kiil (1939).

Especially those authors who so very diligently tried to classify mankind according to race, using height as a variable, ignored the quintessence of Villermé's studies:

> Mais avant, je dois prévenir qu'il ne s'agit point, dans ce travail, des différences que l'on peut observer entre des hommes de races ou d'origines différentes, mais bien des différences qui résultent à la longue, parmi les hommes d'une même origine ou race, des conditions différentes dans lesquelles ils vivent ou ont été élevés.

> [First of all, however, I must stipulate that this work will not deal with differences that can be established between persons of different races or ethnic origins. It will deal with differences that, in the long run, are developed between individuals of one and the same race or ethnic origin, as a result of differences in living conditions or in conditions in which they grew up.]

Villermé studied the official statistical data on conscripts of the 1800–1810 drafts as to height, age at which adult physical development was attained, percentages of rejections on account of ill health, and occupational status. The data for this study had been collected by the prefects of the French departments in 1812 and 1813. Villermé related these data to descriptions of living conditions and environments, likewise collected by the prefects. For all departments Villermé established a relationship between standards of life in the various districts, mean height, and percentage of rejected conscripts. Factors acting upon a department's standard of life were the following: trade and industry, child labor, and geographical situation (e.g., high mountains or marshy land, low land, and soil conditions in general). Villermé conclusively demonstrated that there was no direct correlation between linear growth and climate, latitude, or elevation above sea level. These factors may, however, have an effect upon *aisance ou bien la misère* (well-being or poverty).

In his article, numbering under 50 pages, Villermé further remarked that: when in 1808 the age of conscription was lowered, the increase in number of "undersized" conscripts in poor areas proved to be larger than that in areas having a high level of prosperity, thus indicating that growth in poor regions was retarded; there was a clear relationship between mean height and the percentage of rejections on account of ill health; when the standard for minimum height was lowered the percentage of rejections on account of ill health increased, thus indicating that small persons tended to be less healthy than tall persons; the majority of Frenchmen did not attain a mean height of 164.4 cm before they were 20 or 21 years of age, and in regions where living conditions were unfavorable this height was not attained before the ages of 22 or 23; in several departments linear growth continued until the age of 25; wars had a negative effect upon secular growth since they caused the preferential elimination of conscripted, genetically tall men; governments could promote a population's health and growth by improving general living conditions; urban populations tended to be taller than rural populations.

In another publication Villermé (1828) demonstrated that mortality and socioeconomic level were highly interrelated.

Only one of the above-mentioned observations do we consider to be incorrect, viz., the one saying that those rejected as conscripts have a greater opportunity of transmitting undesirable physical characteristics to their descendants, and that wars, therefore, lead to a decrease in height.

Villermé's publication induced the astronomer and statistician Quételet (1830, 1831) to study the median height of conscripts of the 1823–1827 drafts, coming from a number of towns and from the surrounding countryside in the Belgian province of Brabant. He found the median height for conscripts from the towns to lie in the range 164.0–165.4 cm, and for conscripts from the countryside in the range 162.3–163.1 cm. The difference was almost 2 cm. He demonstrated, on the basis of data obtained from 900 conscripts of the same draft in Brussels who were measured at the ages of 19, 25, and 30, that growth might go on till after the age of 25. Mean adult height in Belgium was found to be 167.5 cm at the age of 25, and 168.4 cm at the age of 30. After a study of records of the 19th century, Oppers (1963) confirmed that, in general, adult height in the Netherlands was not attained before the age of 25. Like Villermé, Quételet established differences in height according to socioeconomic level: most of the 17- to 20-year-old pupils of a Brussels grammar school proved to be taller than 170 cm.

Villermé and Quételet conducted surveys in subpopulations showing almost constant differences as to linear growth, morbidity, and mortality. These differences were considered to be the result of differences in living conditions. The very important interpretation of both authors is that changes in living conditions must be accompanied by changes in linear growth, morbidity, and mortality. Villermé brought up the point of governmental responsibility:

> Une conséquence importante et toute nouvelle découle de tout ce que j'ai rapporté: c'est que, non-seulement la santé des hommes, mais encore leur stature, dépendent en partie du degré de civilisation, de la prospérité ou du malheur public, et que, très-souvent, les gouvernements pourraient à leur gré, en travaillant de tout leur pouvoir au bonheur général, allonger la taille commune des hommes qui leur sont soumis.

> [From all that has been reported by me, the following important and completely new conclusion can be derived: not only man's health, also his body height is partly determined by degree of civilization, by general prosperity or misfortune, and very often governments are in a position to advance the average height of those that are subjected to them, by improving general well-being as much as they can.]

4. Secular Changes in Sexual Maturation

Pubic hair, breast development, and the onset of menstruation are the most prominent characteristics of sexual maturation in girls, whereas growth of genitalia and testes, and pubic hair are the yardsticks for sexual maturation in boys.

Cristescu *et al.* (1966) suggest that the importance of the first menstruation in the process of maturation may differ according to the age at menarche: for example, after an early menarche more anovulatory bleedings may occur than after a late menarche. Moreover a lot of girls show a shorter or longer period of amenorrhea shortly after menarche. By assessing the degree of karyopycnosis in urinary sediment smears Baanders-van Halewijn and de Waard (1968 *a,b*) found that both menarche and the subsequent onset of ovulatory cycles, occurred at a younger age in (taller) Dutch girls than in (smaller) Bantu girls.

Since ancient times the age at menarche was a matter of interest (Amundsen and Diers, 1969, 1973), but critical data are not available before the 19th century. Standard values for other sexual maturation characteristics have been established only recently in Great Britain (Marshall and Tanner, 1969, 1970), in the Netherlands (van Wieringen *et al.*, 1971), and, including the first ejaculation for boys, in Hungary (Dezsö, 1965). Concerning the secular trend of maturation, there are as yet only data available on the age at menarche.

Tanner (1962, 1966; Tanner and Eveleth, 1975) has made several reviews on international publications about menarche. The diagram in Figure 1 is from Berenberg (1975). The oldest published data suffer from disadvantages both of sampling and technique: most of the pre-1920 values concern hospital patients and the recollected age of menarche, that is the age at which menarche was remembered to have occurred many years after the event. Nevertheless, the data given in the figure

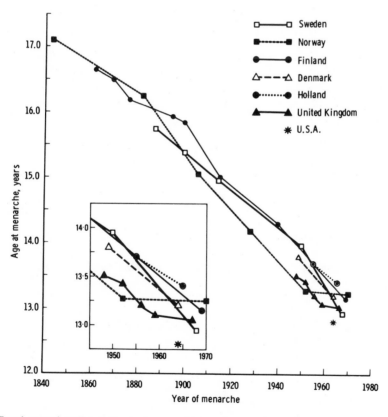

Fig. 1. Trend toward earlier menarche in some European countries, and compared with the U.S.A. (Reprinted from Berenberg, 1975, with permission.)

are convincing: menarche in many countries in Europe has started 2–4 years earlier during the last hundred years, which represents a rate of 3–4 months per decade.

Vlastovsky (1966) and Harper and Collins (1972) found evidence for fluctuations in the secular trend of maturation paralleling times of war and economic crises.

The problems concerning the secular trend of menarche as discussed in recent literature are: (1) What are the environmental factors that influence differences in age at menarche between populations, which may also be related to secular changes? (2) Will secular changes in growth (height and weight) always be combined with secular changes in maturation? (3) Is there an "age threshold," that is, a genetic minimum age at menarche? (4) If it is found for certain populations that the trend seems to have come to an end, is that "because naturally there has to come an end to it," or because the environmental factors do not improve further? (5) The tallest populations do not have the earliest age at menarche; is the cause of the relative independence between growth and maturation genetically based or do hitherto unknown environmental factors play a role? (6) There seems some contradiction in the fact that a secular increase in height is related to a secular trend of earlier age at menarche, while it is also reported for some subpopulations that the taller girls, with an earlier onset of menstruation, will have an adult stature equal to that of girls who have menarche later (Dreizen *et al.*, 1967).

Poorly-off girls have menarche later than better-off girls in nearly all populations studied. Figure 2 shows the results of some studies concerning socioeconomic status and menarche as discussed by Tanner and Eveleth (1975). The same relationship between socioeconomic status and mean age at menarche is found in: Naples, Italy (Carfagna *et al.*, 1972) in 1969–1970: high economic level, 12.24–12.31, and low economic level, 12.69–12.99 (ranges according to middle and low educational level, respectively); Tirupati, South India (Indira Bai and Vijayalakshmi, 1973): 13.11 and 13.35, respectively; Istanbul, Turkey (Neyzi *et al.*, 1975), where only 20% of the children are attending schools: 12.4 and 13.2, respectively.

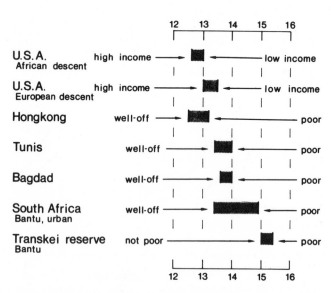

Fig. 2. Age at menarche in well-off and poor girls. (Reprinted from Berenberg, 1975, with permission.)

Rona (1975; Rona and Pereira, 1974), however, found the reverse in Santiago, Chile, in 1970: age at menarche in girls of lower socioeconomic status was 12.30 years and in those of high status 12.96. They ascribe these differences to differences in the ethnic origin: girls with grandparents born in Europe had menarche later than those with grandparents of South American origin.

It is obvious that in most countries the occupational status of the father is highly correlated with expenditure on food, housing, and medical care. As a rule differences in height and maturation according to the professional status of the father are predictable. For Great Britain, however, several investigators (Douglas, 1964; Dann and Roberts, 1969; Roberts *et al.,* 1971) found little or no association between menarcheal age and social class, whereas for height and weight there remained clear differences between children of different socioeconomic groups. For the British subpopulations investigated by Roberts *et al.* it became clear that although there were no social class differences, a strong difference in age at menarche remained according to family size: 0.18 years per additional sib in families with 1–4 children.

For girls in Constanza, Roumania, Štukovský *et al.* (1967) reported a significant linear relation between age at menarche (mean 13.47 years) and number of siblings: menarche was later by an average of 2.1 months for every additional sibling. They ascribe this relationship to general socioeconomic and nutritional conditions in the family.

The conclusion is that in general better socioeconomic conditions and smaller family size lead to an earlier onset of menstruation, but that growth and maturation are reacting in a somewhat independent way.

Dann and Roberts (1973) found by the recall method among students of the university college of Swansea a lowest age at menarche of 12.6 years for girls born in 1946. For later year-groups the menarcheal age was over 12.7 years. Tanner (1975) reported that the menarcheal age for London schoolgirls did not change from 1959 (13.05 years) till 1966–1967 (13.02 years). For the 1966–1967 survey the corresponding heights and weights have not been published yet.

There is some indication of independence between secular changes of height and the trend in menarche in the data of Oslo children: Brundtland *et al.* (1975) mentioned an ongoing secular increase in height from 1950 till 1970, while in an earlier publication Brundtland (1973), in comparing data of 1952 and 1970, found no change in menarcheal age. At the mean age at menarche in 1970, 13.24 years, the median height for girls can be estimated, by interpolation from the published Oslo data for attained height, as 159.4 cm.

The Swedish 1965–1971 survey (Ljung *et al.,* 1974) indicates for 13-year-old girls an increase in height of 3.6 cm since a 1938–1939 survey (Broman *et al.,* 1942) and a shift to an earlier menarcheal age of 0.73 years since a 1950 survey (Romanus, 1952). At the mean age of menarche in 1970 (12.94 years) the median height, as calculated by interpolation, was 157.2 cm. At age 13.0 Oslo girls measured 158.4 cm and Swedish girls 158.6 cm.

In Hungary (Eiben, 1969, 1972) the mean menarcheal age is 13.23 years, for which age a median height of about 150 cm can be estimated.

For Moscow girls in 1965, Vlastovsky (1966) reported an age at menarche of 13.0 years. Taking into account the height for 13.0-year-old boys, as that for girls is not given, it appears that in Russia menarche also takes place at a relatively early age.

In 1965 Dutch 13.0-year-old girls had a median height of 157.2 cm, the age at menarche was 13.4 years, for which age a median height of 158.7 cm can be interpolated. For Dutch girls later data are not available, but since 1965 Dutch 18-year-old conscripts increased 2.7 cm in height. The Norwegian, Swedish, and Dutch data were not analyzed according to socioeconomic status.

Analyzing an anthropometric survey in 1970 in New South Wales (Australia), Jones *et al.* (1973) found the mean age at menarche for daughters of migrants from South Europe to be 12 years, 5 months, and for daughters of migrants from other parts of Europe (mainly northwest Europe) to be 13 years, 1 month. At the age of 13½ years, the mean heights of these groups were 153 cm and 155–157 cm, respectively.

Thus there are populations which, after a long period of positive secular changes, have reached very tall median statures in childhood, adolescence, and young adulthood, while the menarcheal age has been lowered to about 13.0 years. Other populations in less well-to-do or rather poor living conditions still have a distinctly shorter stature, but the onset of menstruation has changed to an even lower age. It will be very interesting to see, in future, if the tall populations will continue their positive trend in height and if that will be associated with a halt in the trend toward earlier maturation (the Oslo situation), or with an ongoing lowering of the menarcheal age (the Swedish situation). The same question holds for populations with a relatively small stature and a rather early menarche: if the environmental factors improve one can expect a positive trend in height, but one wonders if the onset of menstruation can shift to a still earlier age.

It is expected, based on the well-known differences nowadays between relatively tall and relatively short populations with paradoxically relatively late and relatively early menarche, respectively, that in future there will be thresholds, in terms of minimum age for menarche, that will differ from population to population or from area to area. One feels inclined to assume genetic causes for these (expected) differences in minimum menarcheal age, as is generally done for the existing northwest–southeast difference in Europe (Tanner and Eveleth, 1975). But Johnston (1974), in a comprehensive publication, points out that there is a truly multivariate control of age at menarche. The most frequently reported associations (but not in terms of a cause-and-effect relationship) are the correlations with the other maturation aspects, with fatness (early menarche) and linearity in build (late maturation), and with height. The environmental aspects are very complex. Disease delays maturation, differences according to urbanization, to family size, and to socioeconomic status reflect mainly nutritional differences, but there may also be factors such as season and daylight. The conclusion of Johnston is that only research that adopts a multifactorial design can bring us farther.

5. Secular Changes in Height

Tanner (1966) reviewed more than a hundred publications concerning secular changes in growth and maturation in different countries. The following remarks are taken from his publication: (1) Except for data based on the medical examination of conscripts (see also Section 5.2), the earliest publications concern small, often positively or negatively selected, samples. (2) The start of the positive trend in the 19th century coincided with the moment that industrialization began to improve socioeconomic conditions. For the declining mortality there was the same association with socioeconomic development. (3) The trend started directly after birth. At the age of 5–7, height increased 1–2 cm/decade. (4) The gain in weight was approximately proportional to that in height; thus the change was in size and not in proportion. (5) War and hunger slowed the trend. (6) After World War II in several countries the positive trend was stronger than in the first part of the 20th century. (7) If adult stature had remained constant all the time, all the gain in children's height would be due to earlier maturation. But adult height increased also (about 10 cm) and was reached at an earlier age: for males this age lowered from about 26 years to 18 years. (8) In 1938 peak height velocity for Swedish boys was reached one year earlier than in 1883. According to the standard deviation of height, which is maximal during peak height velocity, in 1825–1837 Norwegian boys had their growth spurt at 17, and in the 1930s at 14.

Tanner stated that environmental changes were chiefly responsible for the change in menarcheal age and for that portion of the greater size in children which reflected earlier maturation. The trend in adult height was perhaps due to genetic as well as environmental factors.

The problems concerning secular changes in height and weight, as discussed or touched on in recent literature, are: (1) If for a certain population there is no difference in height after an interval of several years, is that because the genetically controlled maximum is reached or is it due to a leveling off of the environmental factors? (2) Will the range of height diminish further if the positive trend goes on? (3) Will adult height increase further if in childhood the positive trend continues? (4) What changes in postwar environmental conditions can be considered as causal factors for the still-increasing height in those countries where the positive trend was already very advanced? (5) Is the interpretation of increasing height only positive, or are there also negative aspects?

Nationwide representative surveys of secular growth have seldom been made and only in a very few countries (Jordan *et al.,* 1975). It is therefore impossible to follow closely the secular growth trends of all age groups over prolonged periods of time. Data on secular trends have to be gathered from papers that were published at prolonged intervals. Another disadvantage is that the sampled subpopulations show large differences as to size, socioeconomic conditions, ethnic origins, and representativeness in respect of the area or age group concerned.

Studying children of school age is comparatively easy. In the top classes of primary schools, however, a significant selection occurs as pupils having high standards of attainment may go on to postprimary schools, whereas the others will lag behind. There are clear differences in school achievements according to socioeconomic background.

It is extremely difficult to obtain nationally representative samples of children before and after the age at which attendance at school is compulsory. Infants and toddlers can usually be examined only in as far as they are seen at welfare centers. Children not getting medical guidance, or getting guidance at a pediatrician's private office, cannot be sampled. Adolescents can be biometrically examined almost exclusively insofar as they are attending school. These adolescents constitute a highly selected group as to socioeconomic status compared to the group of adolescents that is already employed. In the Netherlands, where secondary education is selective, there is a strong correlation between the level of education (elementary technical for boys and domestic science for girls, secondary modern and grammar-type, respectively, for both sexes) and social class, and therefore also between level of secondary education and body height (van Wieringen, 1972). Although university and college students represent a group that can easily be approached for anthropometrical research (Grant and Hitchens, 1953), they also represent a highly selected group as to socioeconomic status. As had been expected, students were found to be taller than their contemporaries (Durnin and Weir, 1952; Grant and Wadsworth, 1959; Aubenque, 1957, 1963; Bakwin and McLaughlin, 1964; van Wieringen, 1972).

In 1860 the mean heights of conscripts in Norway and Sweden were 169.0 and 166.4 cm, respectively (Udjus, 1964), and in 1865 in the Netherlands, 165.0 cm. Nowadays the mean heights of conscripts in these countries have increased to about 180 cm, whereas in a country like Greece in 1947 the mean height of conscripts was 166.7 cm and in 1965 still nearly the same: 167.2 cm (Valaoras, 1970). Recent literature indicates that Scandinavian and Dutch children and adolescents belong to the tallest populations.

In trying to answer the questions above, it seems reasonable to analyze the growth pattern of those populations where the positive secular growth shift is most advanced. The author is aware that in doing so a lot of publications are neglected. But for the purpose of this paper, it is less important to know in general which other populations show a positive or negative trend in the last decades. For that reason we will concentrate on the results of surveys in Sweden, Norway, and the Netherlands.

5.1. School Children and Adolescents

Ljung *et al.* (1974) reported data from a longitudinal study in 1965–1971 on Swedish school children aged 9–17 years. Comparison with Swedish data from 1883 and 1938–1939 confirmed that children had become larger at all ages, while adult

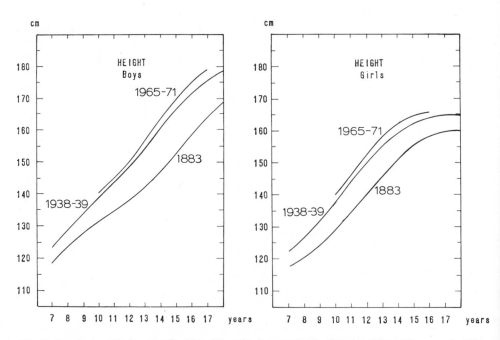

Fig. 3. Secular trend in growth of height of Swedish boys and girls. (Reprinted from Ljung *et al.*, 1974, with permission.)

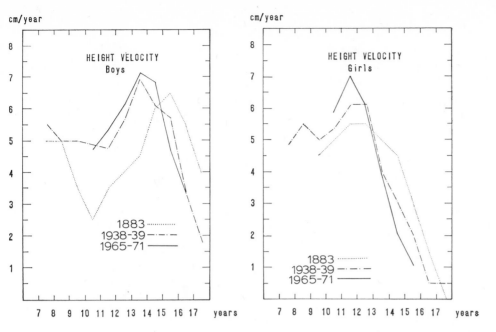

Fig. 4. Secular trend in age of adolescent spurt. Height velocity for Swedish boys and girls. (Reprinted from Ljung *et al.*, 1974, with permission.)

height had also increased considerably (Figure 3). Since 1938 the trend in Sweden seemed to be more pronounced for girls than for boys. For both sexes the gain in centimeters per decade was greater in the interval 1883 to 1938–1939 than in the interval 1938–1939 to 1965–1971. Height velocity, as determined by the increments in height of successive age groups, indicated that the adolescent growth spurt at the three survey dates was at 12.8, 12.2, and 11.6 years for girls, and at 15.2, 14.3, and 14.0 years for boys (Figure 4).

Brundtland *et al.* (1975) compared their 1970 survey of Oslo school children and adolescents with an investigation in the early and mid-50s in Bergen (Sundal, 1957). In the age range 7–16 years the Oslo children showed a gradually increasing lead in stature, for girls from 1.5 to 3 cm, and for boys from 1 to 4.5 cm. After subtracting geographical variations in height the conclusion was that the Oslo children must have increased in height over the last two decades.

The first nationwide surveys on height, weight, and maturation in the Netherlands were performed in 1955 (de Wijn and de Haas, 1960) and in 1965 (van Wieringen *et al.*, 1971). The mean heights as found in the following investigations are plotted in diagrams with the 1965 centiles P_3, P_{50}, and P_{97} for attained height (Figures 5 and 6): in 1865, orphans and school children separately (Commissie Kinderarbeid, 1869); in 1905, schoolboys of lower class at Utrecht (Moquette, 1907); in 1916, school children of lower class at Amsterdam (Gemeente Amsterdam, 1916); in 1935, school children of lower class at Rotterdam (Vermet, 1938); and in 1955, a select nationwide sample. In the data of 1865 the better-off group of school children was taller than the worse-off group of orphans. This is found in all anthropometric surveys all over the world. Still, in 1965 differences of 1–3 cm according to the occupational status of the father were found for Dutch children in the age range 1–19 years. In terms of the secular process the better-off groups were, and are, more advanced in the positive trend.

From the middle of the 19th century till 1965 the pattern of height increase in the Netherlands has roughly been as follows: in 5-year-olds the shift amounted to 14–15 cm, in 10-year-olds to 17–18 cm, in girls at the age of 13 to 22 cm, in boys at the age of 14 to 21 cm, in 18-year-old males to 18 cm, in 19-year-old males to 13 cm, in male adults to 10–11 cm, and in female adults of a lower socioeconomic status to over 13 cm.

De Wijn (1975) reported data of nationwide surveys on 8-year-old Dutch school children (mean age 8.5 years), which were performed every three years since 1959. The positive trend still continues: the median height for boys increased from 133.0 cm in 1965 to 133.9 cm in 1974, and for girls these heights were 131.6 and 133.0 cm, respectively. The heights of both sexes of the 1974 sample exceeded the heights of Oslo coevals in 1970 by about 1.5 cm.

In 1973–1974 the heights of 6–14-year-old boys and girls in the West German state of Bavaria were equal to those of the Dutch 1965 survey (Kunze, 1974).

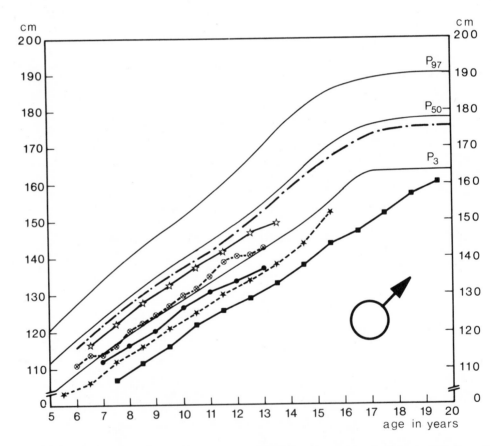

Fig. 5. Secular growth shift in the Netherlands, 1865–1965. Mean heights for boys in 1865 (■——■ orphans; ⋆---⋆ school boys), 1905 (●——● Utrecht school boys, lower class), 1916 (⊗——⊗ Amsterdam school boys, lower class), 1935 (☆——☆ Rotterdam school boys, lower class), 1955 (—.—.—. nationwide, unselected), and 1965 (P_3, P_{50}, P_{97} nationwide, unselected).

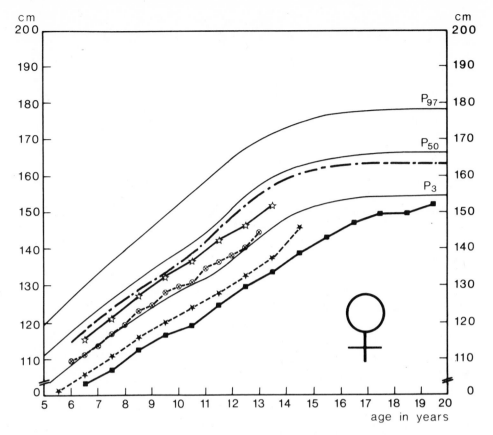

Fig. 6. Secular shift in the Netherlands, 1865–1965. Mean heights for girls in 1865 (■——■ orphans; ★---★ schoolgirls), 1916 (⊗----⊗ Amsterdam schoolgirls, lower class), 1935 (☆——☆ Rotterdam schoolgirls, lower class), 1955 (—·—·— nationwide, unselected), and 1965 (P₃, P₅₀, P₉₇ nationwide, unselected).

5.2. Dutch Conscripts, 1851–1975 Drafts

The only biometrical data covering prolonged periods of time concern young adult conscripts who were measured when they were medically examined. It is remarkable that these data have only occasionally been used for exhaustive studies of secular changes of growth.

Nationally representative biometrical data on the heights of Dutch conscripts date back to the draft of 1851. An analysis of the findings will be presented here, first, in order to underline the epidemiological significance of such material and, second, because some of the results yield an insight into universally applicable principles of secular change. Some of the details are, of course, specifically related to the social and socioeconomic development in the Netherlands. This particularly applies to velocity and extension of changes in the various parts of the "height spectrum."

For the 1851–1862 drafts only the percentages of "undersized" conscripts (those smaller than 157 cm) have been published. For the 1863–1940 drafts the only data available are frequency distributions with various class intervals. The 1941–

1947 drafts have not been systematically examined, and data on heights were not published. The data available for the 1948–1970 drafts were classified into groups with an interval not smaller than 5 cm, with the exception of the 1950,1955, 1960, 1965, and 1970 drafts, the data of which were classified into 1-cm groups. After 1970 yearly 1-cm classifications have been made.

Classification into groups differing 5 cm or more is too broad to allow calculation of centiles. The historical data have therefore been arranged and studied as shown in Figure 7: years of draft upon the ordinate, percentages upon the abscissa. The curves represent the percentages of conscripts smaller in stature than the indicated height. These heights will further be termed "class limits." The percentages of 10, 50, and 90 are accentuated by horizontal lines.

The crossing of a curve and a percent line is identical to the concept of centile: for instance, the point at which a curve crosses the 50% line accurately indicates the median. The angles at which the curves of the class limits cross the 10, 50, and 90% lines indicate the velocity of the shift for the corresponding part of the height spectrum.

The prolonged positive growth shift in the Netherlands, applying the percentages of undersized conscripts of the 1851–1862 drafts, began shortly after 1855—to be precise, at the 1858 draft.

For the 1865 draft median height was found to be 165 cm, for the 1917 draft it was 170 cm, for the 1952 draft median height was estimated at 175 cm, and for the 1975 draft it was 180 cm (180.1 cm, computed exactly). Conscripts of the 1865 and 1917 drafts were 19 years of age, those of the 1952 and 1975 drafts were 18 years of

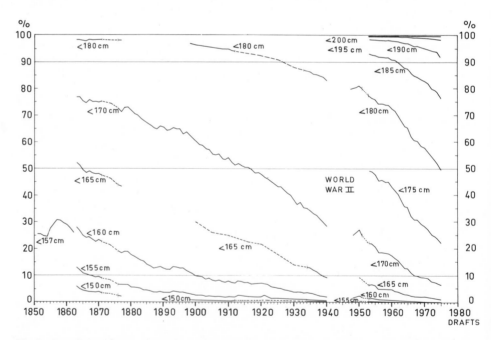

Fig. 7. Cumulative frequency, in percentages, of the height of Dutch conscripts, 1851–1975 drafts. Year of draft = year of birth plus 20 years; age of examination: 18 for 1851–1862 and 1948–1975 drafts, 19 for 1863–1891 and 1913–1940 drafts, and 20 for 1892–1912 drafts.

Table I. Percentiles (in cm) of the Heights of Specified Postwar Drafts in the Netherlands[a]

P_x	1950	1955	1960	1965	1970	1971	1972	1973	1974	1975
P_3	160.8	162.5	163.6	165.2	166.1	166.5	166.7	166.9	167.1	167.4
P_{10}	165.2	166.8	167.7	169.1	170.3	170.5	170.8	171.1	171.1	171.5
P_{50}	174.1	175.3	176.0	177.4	178.7	178.9	179.2	179.5	179.7	180.1
P_{90}	182.6	183.9	184.5	185.8	187.1	187.4	187.8	188.0	188.2	188.8
P_{97}	186.8	187.9	188.5	190.1	191.3	191.6	191.9	192.2	192.4	193.3

[a]Percentiles calculated from frequency distributions with 1-cm intervals.

age. Height shift among 18-year-olds is about 19 cm, because in 1865 the height increment for that age was approximately 4 cm/year, as was demonstrated by Oppers (1963).

Because postwar growth curves after the age of 18 scarcely show an increase, changes in adult height can be analyzed without corrections for age. In 110 years the secular shift of the median has been 15 cm. The first 5 cm took 52 years, the second 5 cm 35 years, and the third 5 cm 23 years. Mean secular shifts per decade therefore amounted to 1.0 cm, 1.4 cm, and 2.2 cm, respectively. In connection with this, it should be noted that in the last decade (the 1965–1975 drafts) median height increased by 2.7 cm.

At first (up to the 1877 draft), the positive secular trend was mainly apparent among groups of small or average stature (i.e., the groups smaller than 170 cm), whereas its effect upon the taller groups was negligible. It was not until after 1900 that the percentage of conscripts taller than 180 cm increased. In recent decades the height shift has been considerable for all components of the height spectrum.

In just over a century, from the 1867 draft until the 1975 draft, centiles P_{10} and P_{90} changed as follows: P_{10} from 155.0 cm to 171.5 cm, the increment being 16.5 cm, and P_{90} from, approximately, over 175 cm to 188.8 cm, the increment being over 13 cm. The total range of height has therefore narrowed. Recent data, however, suggest caution as to the obvious conclusion that, if positive secular growth continues, narrowing of the distribution will likewise continue because variation due to genetic causes will become increasingly noticeable. In Table I centiles P_3, P_{10}, P_{50}, P_{90}, and P_{97} are presented for the postwar drafts of which the frequency distributions for height were available in 1-cm classes. Shifts in percentile values were calculated in 5-year-intervals, and for the 1970–1975 drafts in 1-year intervals (Table II). From the draft of 1950 to the draft of 1960 the shift of the lower percentile values was more rapid than that of the higher percentile values; from 1960 to 1965 the shifts ran practically parallel to each other, and in the two periods from 1965 to 1970 and from 1970 to 1975 the higher percentile values seemed to change fastest. We may conclude that before the 1960 draft there was a narrowing in range, whereas after the 1960 draft the range at least remained constant.

Data on heights of conscripts, collected and processed by region in the first half of the 19th century, showed large fluctuations, strongly differing according to time and area. Zeeman (1861) directly related the annual fluctuations in percentage of undersized conscripts in various regions to the price of rye. Crop failures, bringing a rise in prices, implied starvation for a large part of the population that was already characterized by malnutrition, pauperization, complete social depression, slum-dwelling, illiteracy, child labor, and high morbidity and mortality rates with frequent

Table II. Shifts (in mm) of Percentiles of the Heights of Specified Postwar Drafts in the Netherlands

P_x	1950–1955	1955–1960	1960–1965	1965–1970	Interval between drafts 1970–1971	1971–1972	1972–1973	1973–1974	1974–1975	1970–1975
P_3	17	11	16	9	4	2	2	2	3	13
P_{10}	16	9	14	12	2	3	3	0	4	12
P_{50}	12	7	14	13	2	3	3	2	4	14
P_{90}	13	6	13	13	3	4	2	2	6	17
P_{97}	11	6	16	12	3	3	3	2	9	20

peaks as a result of epidemics, for instance of cholera and smallpox. As in those days adult height was not attained before the age of 20, a year of famine also affected linear growth of boys in the year before conscription.

The beginning of the prolonged positive secular growth shift coincided with the start of a period of gradual economic development, from 1850 until 1873. Afterwards, the positive trend has slowed down, come to a temporary stop, or turned in the opposite direction several times. This can be seen in Figure 7 from the paths of the class-limit curves that go down less steeply, run horizontal, or go up, respectively.

Short interruptions of the increase in level of prosperity have been reported to occur round about the years of 1857 and 1866. Moreover in 1866–1867 the Netherlands was afflicted by a cholera epidemic. A direct relationship between these factors and a period of retrogression noted in the positive secular trend for the 170-cm and 180-cm class limits at the 1867–1871 drafts could not be demonstrated. There existed a strong relationship, however, between the agrarian crisis that lasted from 1888 until 1896, and the flat paths of the curves for 155-, 160-, and 170-cm during this period.

The not very impressive flat paths of the curve for 155-cm (drafts 1918–1921), the curve for 160-cm (drafts 1918–1920), and the curve for 170-cm (draft 1919) may reflect the effects of World War I. The serious economic crisis of the 1930s is noticeable in the flat paths of the curves for 155- and 160-cm of the 1931–1934 drafts. The consequences of the crisis for the curves of the tallest height groups (170 cm and 180 cm) are scarcely noticeable.

The foregoing affords an illustration of the phenomenon, observed by Zeeman (1861), that unfavorable living conditions mostly affect groups of lower socioeconomic level, groups in which the number of persons of small stature is comparatively large.

The inhibiting effect of World War II upon secular growth shift in Holland was larger than that of World War I and the crisis of the 1930s. World War II affected the class limits of both tall and short groups. From the fact that the curves for 160-cm and 165-cm of the 1950 draft show peaks that are even higher than the corresponding levels of the 1940 draft, it can be deduced that in this period also the groups of small stature received a more severe setback than the other groups.

After the war secular changes for all components of the height spectrum were more rapid than ever before. In the years between 1950 and 1953 unemployment in the Netherlands was unusually high. Four years after the unemployment crisis, that is for the 1957–1959 drafts, the paths of all curves were horizontal. In order to relate the crisis to the arrest in the positive secular shift, the following two hypotheses have to be framed: (1) Growth retardation as a result of unfavorable living conditions is especially marked in the growth spurt during adolescence and catch-up is not complete during the next four years. (Growth spurt in boys occurs at an age of about 14, and the age at which boys are medically examined for military service is 18.) This hypothesis is supported by some observations: The war-induced famine in 1944–1945 particularly retarded the height growth of Amsterdam girls between the ages of 12 and 14 (the age at which growth spurt occurs in girls), and conscripts in whom war-induced growth retardation was largest, viz., conscripts of the 1950 draft, had an average age of 14½ during the famine winter of 1944–1945. (2) An economic recession, in terms of unemployment apparently affecting only a small part of the population, has to involve a group that is large enough to demonstrate the inhibiting effect upon longitudinal growth for an entire year-group. From income statistics and surveys on patterns of expenditure of families of different socioeconomic status, this appeared to be the case during the above-mentioned period of crisis. The recession coincided with a decrease in protein consumption in large parts of the population (van Wieringen, 1973). Strictly speaking, it has not yet been demonstrated that there is a close relationship between unemployment and growth retardation, but all the signs indicate that the relationship exists.

Analysis of historical data on height and weight of Dutch conscripts leads to, or endorses, the following conclusions: (1) Socioeconomic and sociohygienic changes are closely followed by changes in secular growth, although nowadays it takes a few years before the effect in 18-year-olds becomes manifest. The method that is to be preferred, if one wants to establish the effect of socioeconomic changes as soon and as clearly as possible, seems to be the frequent, periodic measurement of boys at the age of 14, and of girls at the age of 12, that is, at ages of the adolescent growth spurt. (2) As a result of changes in living conditions, secular changes (positive or negative) may occur at all ages. (3) Adult height is still increasing. (4) The negative influences of times of recession and war appear to inhibit the growth of groups with small stature more than those of tall stature. (5) Up to the 1877 draft, that is, at the beginning of the positive secular shift, the groups of small stature (in general also being the groups that were poorest) got the benefit of the gradually improving socioeconomic conditions. After the 1877 draft there was also an increase in the percentages of conscripts of heights between 170 cm and 180 cm. The period up to 1900 is characterized by a "catch-up phenomenon." During the second period, in the years after 1900, the shift is considerable for all components of the height spectrum, although up to the 1960 draft it was somewhat more marked for the group of smallest stature than for the tallest group. If the latter group reaches a "genetic maximum," a continuing general amelioration of living conditions would lead to the onset of a third period that again would be characterized by a "catch-up phenomenon," because, for the time being, the secular shift would go on only in the groups of small height. However, even in the Netherlands, having a highly advanced secular growth shift, this third period has not yet begun. On the contrary, during the last decade the distribution appears to have remained constant, as the tallest group also has grown taller.

6. Secular Changes in Weight–Height Relationship

Nutritional states should not be described simply in terms of weight-for-age (Waterlow, 1973). What has been said above makes clear that attained height is of primary importance for the assessment of a population's nutritional status. Weight-for-height represents a second important indicator.

In studies on secular growth shift special attention has been paid to changes in height-for-age and weight-for-age. The few authors that suggested that the weight–height relationship might possibly show a secular shift, suggested it mainly in passing. From a decrease in the W/H^3 index, Broman *et al.* (1942) inferred that, from 1883 until 1939, the increment in height in Sweden was larger than that in weight. As the weight–height relationships of populations having inhibited growth appear to be different from populations showing more rapid growth increments, de

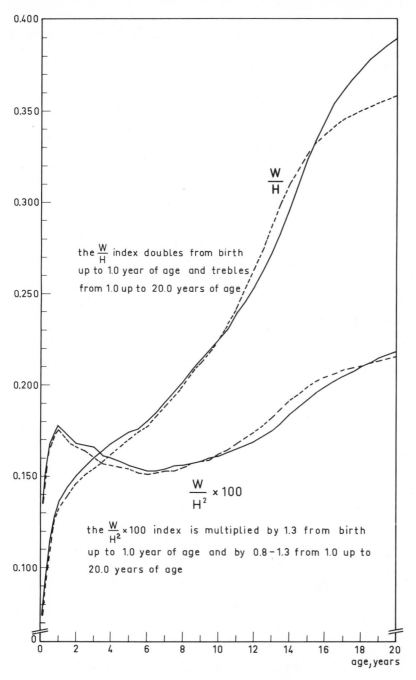

Fig. 8. Indices W/H and $W/H^2 \times 100$ for attained median height and weight, boys (—) and girls (---) 0.08–20.0 years of age, in the Netherlands in 1965.

Wijn (1952) has argued that it is incorrect to assess weight-for-height without taking into consideration the ages of the children examined. From the decrease in the W/H^3 index for Czechoslovakian children between the years of 1895 and 1951, it was inferred by Prokopec (1961) that body proportions had become more slender. In international comparisons he found that the value of this index is smaller for populations of tall height than for populations of short height. For London school children between the years of 1949 and 1959, Scott (1961) established an increase both in height and in the weight–height relationship. Kemsley *et al.* (1962) demonstrated that adult weight–height relationships were showing secular differences as well as regional differences. Wolanski (1967) mentioned that in cases of positive secular changes in growth, the weight–height ratio gradually diminished.

As is shown for the weight–height relationship in the Dutch 1965 survey (van Wieringen, 1972), the indices W/H and W/H^2 are not constant, but vary considerably with age (Figure 8). Therefore indices are better not used for analyzing changes in weight-for-height.

For an accurate assessment of the distribution of the weight–height relationship, a very large sample is needed. It is, therefore, quite understandable that, in most cross-sectional studies on height and weight, standard values for the weight–height relationship are lacking. As a result, weight-for-height distributions can seldom be used to ascertain whether there have been secular changes in the weight–height relationship.

If only standard values for age are available, plotting attained (median or mean) weight for attained (median or mean) height in one diagram appears to be the method of choice in order to compare the weight–height relationship of various populations (Hiernaux, 1964). This method is inadequate, however, if there are

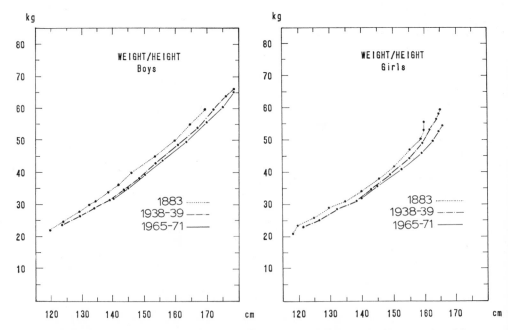

Fig. 9. Secular trend in weight for height of Swedish boys and girls, 1883–1971. (Reprinted from Ljung *et al.*, 1974, with permission.)

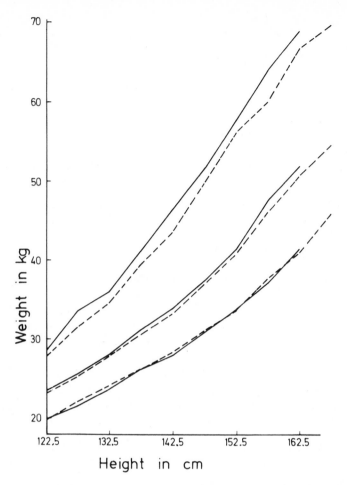

Fig. 10. Weight–height percentiles for girls: ---- Bergen 1956; —— Oslo 1970. (From Brundtland *et al.*, 1975. Reprinted with permission of *Acta Paediatr. Scand.*)

large differences in adult height of the populations, as comparisons at the end of the growth curves cannot be made in such a case. By this method Ljung *et al.* (1974) found that Swedish children have increased proportionally more in height than in weight (Figure 9), while Brundtland *et al.* (1975) reported a higher weight-for-height for the taller Oslo girls than found in the Bergen survey (Figure 10). By comparing the results of several surveys performed between 1893 and 1970, Maaser (1974) concluded that for the West German region Dortmund/Westfalia no change in the weight–height relationship had taken place.

Based on an analysis, by the same method, of international data from several periods, it has been concluded (van Wieringen, 1972) that the positive secular shift in longitudinal growth is not necessarily accompanied by a relative weight increment or decrease. Regardless of the age periods, populations of small as well as of large heights were found to be light or heavy in weight-for-height. This may be explained by the effect of differences in quality or quantity of nutrition, being of different importance in regard of linear growth and weight increment.

7. Discussion

Secular shifts in growth and maturation are inherent to plant and animal life, including that of humans.

7.1. Theories

The most important exogenous factors are nutrition and disease. Indirect factors, exerting an influence by way of nutrition and an improved state of health, are the attainment of improved socioeconomic living conditions: guaranteed minimum income, social care, transport, pure drinking water, sewerage, technologically and hygienically proper food preparation (both in manufacturing industries and in the kitchen at home), improved sanitary conditions, health care, national health insurance, medical care, and education.

It is very difficult to compare growth of populations having different economic systems, cultural patterns, educational systems, welfare facilities, nutritional habits, patterns of morbidity and mortality, as well as sanitary and medical facilities. Even in countries that have an advanced, and on the face of it comparatively homogeneous, socioeconomic level, there proves to be a correlation between still-prevailing differences in environmental factors, and old established differences in growth, maturation, morbidity, and mortality.

In the light of the potential of populations to undergo changes in the growth pattern, it is attractive to consider concurrent differences in growth and maturation among population groups as differences in the stage of the secular process. It is also attractive to invoke the influence of genetic factors for explaining concurrent growth differences. In view of the strong environmental influences on growth and maturation, it is only recently that clear genetically controlled differences between populations have been identified (Tanner and Eveleth, 1975).

The close relationship between changes in exogenous factors and in growth pattern is generally accepted as a suitable explanation for positive, as well as negative, secular shifts. Controversy exists, however, as to whether changes in the hereditary control of growth and maturation contribute to the positive secular trend of the last century. The most important genetic change is supposed to be the increase of the number of heterozygous combinations, which in the breeding of animals and plants may lead to bigger specimens and greater yield (heterosis). A definite explanation of hybrid vigor is not yet given, while the term heterosis remains ambiguous in spite of the many attempts to define it (Mangelsdorf, 1964; Shull, 1964). Even before World War II the concept of heterosis as a factor in the positive secular growth shift was discussed. One of the most direct indications that outbreeding in human beings may lead to larger stature was given by Hulse (1957). He found that grown-up sons of parents who came from different Swiss villages ("exogames") measured 2 cm taller than sons of parents who both came from the same village ("endogames"). But as Hulse had no measurements of the parents, there is no proof that the heights of the heterozygous sons differed more from the heights of their fathers than was the case in the other group. Besides Hulse indicated clearly that there existed cultural differences between the two groups; it is curious that he denied that the smaller family size in the heterozygous group was an important difference as far as growth is concerned. Heterosis can only be proven if all the above-mentioned environmental factors are excluded in comparing inbred and outbred populations. As inbreeding in human isolates is quite another situation than inbreeding in experimental studies on plants and animals, the hypothesis that heterosis is a contributing factor in the positive secular trend has no strong basis.

If one rejects heterosis and other genetic changes, as the present writer is inclined to do, the very prolonged positive shift in growth inevitably leads to the question whether environmental factors are still improving. This question is even more relevant where the years after World War II are concerned: in several industrialized countries the positive growth shift during the postwar period has been more striking than ever before (Cone, 1961; Udjus, 1964; van Wieringen, 1972).

In the Netherlands improvements in nutrition are observed even after 1950, but from a qualitative point of view nutrition is still suboptimal, especially in groups of lower socioeconomic status (de Wijn, 1968, 1970). Since World War II infant nutrition has been subject to considerable change: there has been a decrease in breast-feeding, whereas the use of formulae prepared and sterilized in dairy factories has increased. At infant welfare centers advice on feeding has also changed: eggs, fish, liver, meat, grated cheese, vegetables, and cereals are given at an earlier age than before. In the past few years special attention has been paid to the prevention of iron-deficiency anemia and extra iron is added to infant food. From a publication by Bakwin (1964) it can be seen that these changes in infant food are similar to those in the United States. According to Meyer (1967), resorption of toxins, occurring because of a disturbance of normal equilibrium in the intestinal flora, may retard an infant's growth. But in infants that are fed in accordance with modern standards serious intestinal infections hardly ever occur.

Median height of children decreases as family size increases (Grant, 1964; Goldstein, 1971). According to Tanner (1962) the decrease in family size, the average number of children being 6 in 1880 and 2–3 in 1940, accounts for a positive growth shift of 2.5 cm in itself. In the last decades fertility in the Netherlands decreased as follows: after a marriage of 14 years the mean number of deliveries per woman was 2.84 when married in 1937, 2.74 for 1942, 2.59 for 1947, and 2.50 for 1952 and 1957. According to the mean number of legitimate births after a marriage of 5 years, this trend is still continuing: for women married in 1958, 1963, and 1968 the averages were 1.61, 1.58, and 1.42, respectively (Centraal Bureau voor de Statistiek, 1974).

In developed countries, postwar mothers tend to be taller and better nourished than their own mothers from the years between 1920 and 1940 (Luyken, 1970). They also have been ill less often. The present generation again has a better prenatal start than the previous one.

In industrializing countries the improvements in living conditions manifested themselves in an impressive decline in child morbidity and mortality. In the beginning the better chance for survival occurred practically without direct (socio-) medical interference. In the postwar period the continuing decrease in morbidity and mortality of children has partly been the result of improved living conditions and partly the result of preventive medicine (child health care, mass vaccination programs, amelioration of sanitary conditions), and improved treatment facilities: children are ill less often, and their diseases do not last as long and are not as serious as before.

J. C. VAN WIERINGEN

Drawing a comparison between the processes of secular growth shifts in different countries is beyond the scope of the present paper. Such a comparison would certainly yield large differences as to moment of onset, direction, velocity, and extent of the changes. An explanation for these changes, however, requires a profound understanding of technological, socioeconomic, sociohygienic, and cultural developments in the countries concerned. Prokopec (1968) mentioned, for example, that in the period between 1895 and 1961 the secular shift for Czechoslovakian girls was larger than that for boys. This phenomenon may not be explained merely in terms of biologically determined differences between both sexes in regard to susceptibility to changes in living conditions, because changes in the cultural pattern itself may have been different for girls and boys. In the Netherlands the shift in linear growth between 1865 and 1965 has been identical for school children of both sexes. Whether culturally determined differences in living conditions have been of influence in this respect cannot be clearly ascertained, but this seems to be unlikely.

Investigators of secular changes in man should have an understanding of the relationship between living conditions and growth as it has been in the past. The two factors directly related to growth shifts are changes in nutrition and in morbidity. But there are many more environmental factors that are liable to changes. It would not be correct, however, to single out only some of them and present them as (contributory) causes of secular trends, independent of nutrition and morbidity. There also have been large shifts in growth without radical changes in external factors, such as pattern of activity (traffic, sport, rest, sleep, labor), microclimatic conditions (differing from only one fireplace in the house to central heating), or educational programs.

Secular changes in growth and development are to be regarded as indicators of socioeconomic and sociohygienic conditions and a population's state of health. It therefore seems safe to assume that regularly performed height measurements among populations that are technologically underdeveloped are to be preferred to vague concepts such as "income per capita" or "gross national product," when, for instance, it is wished to ascertain accurately and rapidly whether living conditions, and as a consequence health, have improved by means of development aid.

If one wishes to compare secular changes in different populations, one should take into account the following characteristics: (1) the particular part of the population that is showing secular changes, according to age group, social class, and region; (2) attained height, attained weight, weight–height relationship, and age at which sexual maturation characteristics are being developed should be combined in the same survey; (3) velocity and extent of changes; (4) changes in the distribution of the variables; (5) changes in extreme distribution centiles, for instance, P_3 and P_{97}; (6) age of adolescent growth spurt; (7) age at which adult height is being attained; and (8) connection with changes in socioeconomic conditions, nutritional pattern, morbidity, and mortality.

Because of the fact that the secular shift of the entire height spectrum of Dutch conscripts, being among the tallest conscripts in the world with a median height of 180 cm, is still continuing, it is hard to predict at what point genetically determined height growth will have reached its maximum. If in some countries the prolonged positive secular trends appear to have come to an end in samples of individuals from the same socioeconomic background over time (Bakwin and McLaughlin, 1964; Brundtland *et al.*, 1975), it seems more probable that the phenomenon has been brought about by socioeconomic and sociohygienic living conditions in those particular countries rather than by the fact that positive secular trends would have reached their biological maxima.

No predictions can be made as to the course of regional, national, or universal socioeconomic conditions, nor as to environmental factors connected with them. The only conclusion that is warranted so far is that in years to come there will be periods of adversity as well as of prosperity; therefore, inhibition of growth and development will be either more or less strong.

7.3. Implications

The question is whether continuation of growth acceleration, earlier occurring maturation, and increase in adult height have to be regarded as positive factors in respect of a population's health or as indicators of an approach to "optimal health." If environmental factors furthering growth velocity were also found to be the cause of pathological traits, or if it were found that the effect of tallness as such is pathogenic, secular shifts in growth as health indicators would of course have to be interpreted differently. Not taking into account unfounded, mostly emotional views in which rapid growth and early maturation are disapproved of, continuation of the present trend would have to be regarded as negative if the following relationships would be demonstrated: (1) Factors furthering growth velocity have a causative connection with those forms of pathology that are showing an increase when standards of living are high. From a theoretical point of view it is conceivable, for instance, that saturated fatty acids are conducive to a positive secular shift in growth, whereas it is also assumed that their occurrence in a rather large quantity in nutrition advances the onset of atherosclerosis. (2) Tallness is a negative factor as to optimal health; for example, if relative deficiencies in nutrients will be developed as a result of a too strong neutralization of growth-inhibiting factors. (3) Tallness upsets the body's stability and hemodynamics. (4) High birth weight causes difficulties in parturition. (5) Linked to more rapid growth and greater height, there is a tendency to obesity.

No support could be encountered in the literature for the first three suppositions.

It has been established by Selvin and Janerich (1971) that birth weights in New York state between the years of 1959 and 1967 were showing a downward tendency for all birth orders, although especially for the higher ones. According to the authors the phenomenon is accounted for by the decrease in number of children per family, as well as by the decrease in age at which women are giving birth. Positive secular trends are therefore not unalterably accompanied by increments in birth weight.

In Rantakallio's view (1968), a high birth weight involves a rather high risk of perinatal mortality. The percentage of children having IQs under 80 would be larger among children of high birth weight than among children of normal birth weight (Babson *et al.*, 1969). It has been inferred from a survey by McEwan and Murdoch (1966), however, that maternal mortality and perinatal infant mortality at birth weights higher than 4.3 kg are not necessarily increased.

In some European countries, taller height has been accompanied by a decrease in weight-for-height. At present, there is an increase in weight-for-height in the comparatively tall population of the United States (Statistical Bulletin, 1973).

The increase in the weight–height relationship at an advanced age, especially where it concerns persons having had a comparatively low weight in childhood, appears to be associated with the occurrence of certain diseases in adults (Abraham *et al.*, 1971).

Food studies, as well as our own analysis of the weight–height relationship, support the view that, provided nutrition contains an adequate supply of calories, height growth is primarily determined by the types of nutrients. As far as fatty tissues are concerned, weight–height relationships probably are mainly determined by, on the one hand, supply of calories, and on the other hand, expenditure of energy on growth processes and physical activity. Disturbance of this balance (as a result of unrestricted overfeeding and/or insufficient physical activity) may lead to obesity in populations of short as well as of tall stature. Such a disturbance is not inherent in a secular shift in height.

It can be concluded, therefore, that the positive secular shift in height is still to be regarded as a positive indicator of a population's health, and that, at present, it is impossible to indicate a particular height in a particular population as being "optimal."

With a view to future epidemiologic–biometric research, the following recommendations may be worth considering:

1. Every decade nationwide surveys on height, weight, and sexual maturation characteristics should be performed, special attention being given to certain socioeconomic groups.

2. "Height-for-age" has to be added to the indicators of the nutritional state of a population.

3. Surveys on the somatic and mental morbidity patterns in childhood (two components of one phenomenon) ought to be combined with biometric research into growth and maturation.

4. Surveys on individual and secular changes in the weight–height relationship during growth and adulthood should be linked with epidemiological studies of diseases in the elderly, e.g., coronary heart disease and diabetes mellitus.

5. Medical examination of conscripts and the collection of biometric data among this segment of the population should be given their proper place in epidemiological research.

6. The data of growth surveys in various countries should be set against the growth and maturation standards of the population with the most advanced positive secular growth shift.

8. References

Abraham, S., Collins, G., and Nordsieck, M., 1971, Relationship of childhood weight status to morbidity in adults, *HSMHA Health Rep.* **86**:273.

Amundsen, D. W., and Diers, C. J., 1969, The age of menarche in classical Greece and Rome, *Hum. Biol.* **41**:125.

Amundsen, D. W., and Diers, C. J., 1973, The age of menarche in medieval Europe, *Hum. Biol.* **45**:363.

Aubenque, M., 1957, Note documentaire sur la statistique des tailles des étudiants au cours de ces dernières années, *Biotypologie* **18**:202.

Aubenque, M., 1963, Note sur l'évolution de la taille des étudiantes, *Biotypologie* **24**:124.

Baanders-van Halewijn, E. A., and de Waard, F., 1968*a*, Menstrual cycles shortly after menarche in European and Bantu girls; Epidemiological aspects, *Hum. Biol.* **40**:314.

Baanders-van Halewijn, E. A., and de Waard, F., 1968*b*, Menstrual cycles shortly after menarche in European and Bantu girls; Endocrine and cytologic aspects, *Hum. Biol.* **40**:323.

Babson, S. G., Henderson, N., and Clark, W. M., 1969, The preschool intelligence of oversized newborns, *Pediatrics* **44**:536.

Bakwin, H., 1964, The secular change in growth and development, *Acta Paediatr.* **53**:79.

Bakwin, H., and McLaughlin, S. M., 1964, Secular increase in height; is the end in sight, *Lancet* **2**: 1195.

Berenberg, S. R., ed., 1975, *Puberty, Biologic and Psychosocial Components,* Stenfert Kroese, Leiden.

Broman, B., Dahlberg, G., and Lichtenstein, A., 1942/43, Height and weight during growth; review of the literature, *Acta Paediatr.* **30**:1.

Brundtland, G. H., 1973, Menarcheal age in Norway: Halt in the trend towards earlier maturation, *Nature* **241**:478.

Brundtland, G. H., Liestöl, K., and Walloë, L., 1975, Height and weight of school children and adolescent girls and boys in Oslo 1970, *Acta Paediatr. Scand.* **64**:565.

Carfagna, M., Figurelli, E., Matarese, G., and Matarese, S., 1972, Menarcheal age of schoolgirls in the district of Naples, Italy, in 1969–70, *Hum. Biol.* **44**:117.

Centraal Bureau voor de Statistiek, 1974, Huwelijksvruchtbaarheid, een cohortanalyse 1937–1971.

Chase, H. C., 1972, The position of the United States in international comparisons of health status, *Am. J. Public Health* **62**:581.

Commissie Kinderarbeid, 1869, Rapport der commissie belast met het onderzoek naar de toestand der kinderen in fabrieken, Min. Binnenl. Zaken, 's-Gravenhage.

Cone, T. E., 1961, Secular acceleration of height and biologic maturation in children during the past century, *J. Pediatr.* **59**:736.

Cristescu, M., Petrovici, O., and Onofrei, M., 1966, Sur l'accélération du développement des caractères sexuels secondaires, *Ann. Roum. Anthropol.* **3**:65.

Dann, T. C., and Roberts, D. F., 1969, Physique and family environment in girls attending a Welsh college, *Br. J. Prev. Soc. Med.* **23**:65.

Dann, T. C., and Roberts, D. F., 1973, End of the trend? A 12-year study of age at menarche, *Br. Med. J.* **3**:265.

Deszö, Gy., 1965, Budapestifiúk gonádérésének idöpontja, *Anthropol. Közl.* **9**:151.

de Wijn, J. F., 1952, Over een onjuiste methode van het beoordelen der voedingstoestand van schoolkinderen ener bevolking, *Voeding* **13**:102.

de Wijn, J. F., 1968, De veranderingen in het Nederlandse voedingspatroon; rapport van de commissie van de voedingsraad ter oriëntering omtrent de voeding en voedingstoestand in Nederland, *Voeding* **29**:490.

de Wijn, J. F., 1970, Geneeskundige beschouwingen over het zich veranderende Nederlandse voedings- en activiteitenpatroon in de laatste dertig jaren, *Huisarts Wet.* **13**:217.

de Wijn, J. F., 1975, Somatometrisch onderzoek, in: *Zesde oriënterend onderzoek omtrent de voeding en voedingstoestand van 8-jarige schoolkinderen in Nederland,* p. 1, Voedingsraad, Rijswijk.

de Wijn, J. F., and de Haas, J. H., 1960, Groeidiagrammen van 1-25-jarigen in Nederland, Ned. Inst. Praev. Geneeskd., Leiden.

de Wijn, J. F., and Rusbach, H. W., 1961, Het hemoglobinegehalte van Nederlandse schoolkinderen en adolescenten, *Ned. Tijdschr. Geneeskd.* **105**:1028.

Douglas, J. W. B., 1964, *The Home and the School,* Macgibbon and Kee, London.

Dreizen, S., Spirakis, C. M., and Stone, R. E., 1967, A comparison of skeletal growth and maturation in undernourished and well-nourished girls before and after menarche, *J. Pediatr.* **70**:256.

Durnin, J. V. G. A., and Weir, J. B. de V., 1952, Statures of a group of university students and of their parents, *Br. Med. J.* **1**:1006.

Eiben, O. G., 1969, Growth and development from the point of view of evolutionary trends, *Symp. Biol. Hung.* **9**:131.

Eiben, O. G., 1972, Genetische und demographische Faktoren und Menarchealter, *Anthropol. Anz.* **33**:205.

Garn, S. M., 1966, Body size and its implications, *Rev. Child Dev. Res.* **2**:529.

Garn, S. M., and Russell, A. L., 1971, The effect of nutritional extremes on dental development, *Am. J. Clin. Nutr.* **24**:285.

Gemeente Amsterdam, 1916, Verslag omtrent een onderzoek naar den voedingstoestand van een deel der schoolbevolking, ingesteld in Januari 1916 te Amsterdam, door het gemeentelijk geneeskundig schooltoezicht, Gemeenteblad afd. 1, bijlage E.

Goldstein, H., 1971, Factors influencing the height of seven year old children—results from the national child development study, *Hum. Biol.* **43**:92.

Grant, M. W., 1964, Rate of growth in relation to birth rank and family size, *Br. J. Prev. Soc. Med.* **18**:35.

Grant, G., and Hitchens, R. A. M., 1953, Some observations on the heights and weights of undergraduates with special reference to the university of Wales, *Br. J. Prev. Soc. Med.* **7**:60.

Grant, M. W., and Wadsworth, G. R., 1959, The height, weight, and physical maturity of Liverpool schoolgirls, *Med. Off.* **102**:303.

Habicht, J.-P., Yarbrough, C., Martorell, R., Malina, R. M., and Klein, R. E., 1974, Height and weight standards for preschool children. How relevant are ethnic differences in growth potential? *Lancet* **1**:611.

Hagen, W., 1962, The secular acceleration of growth and the individual, *Mod. Probl. Paediatr.* **7**:8.

Harper, J., and Collins, J. K., 1972, The secular trend in the age of menarche in Australian schoolgirls, *Aust. Paediatr. J.* **8**:44.

Hiernaux, J., 1964, Weight/height relationship during growth in Africans and Europeans, *Hum. Biol.* **36**:273.

Hulse, F. S., 1957, Exogamie et heterosis, *Arch. Suisses Anthropol. Gén.* **22**:103.

Indira Bai, K., and Vijayalakshmi, B., 1973, Sexual maturation of Indian girls in Andhra Pradesh (South India), *Hum. Biol.* **45**:695.

Johnston, F. E., 1974, Control of age at menarche, *Hum. Biol.* **46**:159.

Jones, D. L., Hemphill, W., and Meyers, E. S. A., 1973, Height, Weight and Other Physical Characteristics of New South Wales Children. 1. Children Aged 5 Years and Over, New South Wales Department of Health.

Jordan, J., Ruben, M., Hernandez, J., Bebelagua, A., Tanner, J. M., and Goldstein, H., 1975, The 1972 Cuban national child growth study as an example of population health monitoring: Design and methods, *Ann. Hum. Biol.* **2**:153.

Kemsley, W. F. F., Billewics, W. Z., and Thomson, A. M., 1962, A new weight-for-height standard based on British anthropometric data, *Br. J. Prev. Soc. Med.* **16**:189.

Kiil, V., 1939, Stature and growth of Norwegian men during the past two hundred years, *Skr. Nor. Vidensk. Akad.* **2**(6):1–175.

Koch, E. W., 1953, Die Akzeleration und Retardation des Wachstums und ihre Beziehungen zum erreichbaren Höchstalter des Menschen, *Dtsch. Gesundheitswes.* **8**:1492.

Kunze, D., 1974, Perzentilkurven von Körpergrösze und Körpergewicht 6- bis 14jähriger Kinder, *Klin. Pediatr.* **186**:505.

Ljung, B.-O., Bergsten-Brucefors, A., and Lindgren, G., 1974, The secular trend in physical growth in Sweden, *Ann. Hum. Biol.* **1**:245.

Luyken, R., 1970, De voeding van de zwangere en zogende, *Voeding* **31**:498.

Maaser, R., 1974, Eine Untersuchung gebräuchliger Längen/Gewichtstabellen-Zugleich ein Vorschlag für ein neues Somatogramm 0–14jähriger Kinder, *Monatschr. Kinderheilk.* **122**:146.

Mangelsdorf, P. C., 1964, Hybridization in the evolution of maize, in: *Heterosis* (J. W. Gowen, ed.), p. 175, Hafner Publishing Company, New York/London.

Marshall, W. A., and Tanner, J. M., 1969, Variation in the pattern of pubertal changes in girls, *Arch. Dis. Child.* **44**:291.

Marshall, W. A., and Tanner, J. M. 1970, Variation in the pattern of pubertal changes in boys, *Arch. Dis. Child.* **45**:13.

McEwan, H. P., and Murdoch, R., 1966, The oversized baby, a study of 169 cases, *J. Obstet. Gynaecol. Br. Commonw.* **73**:734.

Meyer, A., 1967, Das Problem des Akzeleration und des Wachstums, *Med. Ernaehr.* **8**:97; 135.

Moore, W. M., 1966, Human growth in secular perspective, *Postgrad. Med.* **40**:A-89.

Moquette, J. J. R., 1907, Onderzoekingen over volksvoeding in de gemeente Utrecht, Thesis, Utrecht.

Nellhaus, G., 1970, Head circumference growth in North American negro children (letter), *Pediatrics* **46**:817.

Neyzi, O., Alp, H., and Orhon, A., 1975, Sexual maturation in Turkish girls, *Ann. Hum. Biol.* **2**:49.

Oppers, V. M., 1963, Analyse van de acceleratie van de menselijke lengtegroei door bepaling van het tijdstip van de groeifasen, thesis, Amsterdam.

Owen, G. M., Nelsen, C. E., and Garry, P. J., 1970, Nutritional status of preschool children: hemoglobin, hematocrit and plasma iron values, *J. Pediatr.* **76**:761.

Prokopec, M., 1961, Některé závěry z antropometrických výzkumů m ládeže z let 1947 až 1957, *Acta Fac. Rerum Nat. Univ. Comenianae Anthropol.* **6**(1):81 (publication 4).

Prokopec, M., 1968, A survey of growth and development studies carried out in Czechoslovakia, with proposals for future research to the IBP and comments on methodology, *Mater. Pr. Anthropol.* **75**:97.

Quételet, A., 1830, Sur la taille moyenne de l'homme dans les villes et dans les campagnes, et sur l'âge òu la croissance est complètement achevée, *Ann. Hyg. Publique (Paris)* **3**:24.

Quételet, A., 1831, Recherches sur la loi de la croissance de l'homme, *Ann. Hyg. Publique (Paris)* **6**:89.

Quételet, A., 1835, *Sur l'homme et le développement de ses facultés, ou: Essai de physique sociale,* Bachelier, Paris.

Rantakallio, P., 1968, The optimum birth weight, *Ann. Paediatr. Fenn.* **14**:66.

Roberts, D. F., Rozner, L. M., and Swan, A. V., 1971, Age at menarche, physique and environment in industrial north east England, *Acta Paediatr. Scand.* **60**:158.

Romanus, T., 1952, Menarche in schoolgirls, *Acta Gen. Stat. Med.* **3**:50.

Rona, R., 1975, Secular trend of pubertal development in Chile, *J. Hum. Evol.* **4**:251.

Rona, R., and Pereira, G., 1974, Factors that influence age at menarche in girls in Santiago, Chile, *Hum. Biol.* **46**:33.

Scott, J. A., 1961, Report on the Heights and Weights (and Other Measurements) of School Pupils in the County of London in 1959, County Council, London.

Selvin, S., and Janerich, D. T., 1971, Four factors influencing birth weight, *Br. J. Prev. Soc. Med.* **25**:12.

Shull, G. H., 1964, Beginnings of the heterosis concept, in: *Heterosis* (J. W. Gowen, ed.), p. 14, Hafner Publishing, New York/London.

Statistical Bulletin, 1973, Recent growth trends in childhood and adolescence, *Stat. Bull. Metrop. Life Insur. Co.* **53**:8.

Štukovský, R., Valšík, J. A., and Bulai-Ştirbu, M., 1967, Family size and menarcheal age in Constanza, Roumania, *Hum. Biol.* **39**:277.

Sundal, A., 1957, The Norms for Height (Length) and Weight in Healthy Norwegian Children from Birth to 15 Years of Age, *Univ. Bergen Arbok Med. Rekke,* **10**:1–19.

Tanner, J. M., 1962, *Growth at Adolescence,* 2nd ed., Blackwell, Oxford.

Tanner, J. M. 1966, The secular trend towards earlier physical maturation, *T. Soc. Geneesk.* **44**:524.

Tanner, J. M. 1975, Trend towards earlier menarche in London, Oslo, Copenhagen, The Netherlands and Hungary, *Nature* **243**:95.

Tanner, J. M., and Eveleth, P. B., 1975, Variability between populations in growth and development at puberty, in: *Puberty; Biologic and Psychosocial Components* (S. R. Berenberg, ed.), p. 256, Stenfert Kroese Publishers, Leiden.

Udjus, L. G., 1964, *Anthropometrical Changes in Norwegian Men in the Twentieth Century,* Universitetsforlaget, Oslo.

Valaoras, V. G., 1970, Biometric studies of army conscripts in Greece, *Hum. Biol.* **42**:184.

van Wieringen, J. C., 1972, Secular Changes of Growth, 1964–1966 Height and Weight Surveys in the Netherlands in Historical Perspective, thesis, Leiden.

van Wieringen, J. C., 1973, Lichaamslengte en werkloosheid, *Mens Onderneming.* **27**:30.

van Wieringen, J. C., Wafelbakker, F., Verbrugge, H. P., and de Haas, J. H., 1971, *Growth Diagrams, 1965, Netherlands,* Wolters-Noordhoff, Groningen.

Vermet, P., 1938, *Onderzoek naar den voedingstoestand van Rotterdamse schoolkinderen in de jaren 1935, 1936 en 1937, vooral in verband met de werkloosheid,* G.G. en G.D., Rotterdam.

Villermé, L.-R., 1828, Mémoire sur la mortalité en France, dans la classe aisée et dans la classe indigente, *Mem. Acad. R. Med.* **1**:51.

Villermé, L.-R., 1829, Mémoire sur la taille de l'homme en France, *Ann. Hyg. Publique (Paris)* **1**:351.

Virey, 1816, Géant, in: *Dictionnaire des sciences médicales par une société de médecins et de chirurgiens,* p. 546, Panckoucke, Paris.

Vlastovsky, V. G., 1966, The secular trend in the growth and development of children and young persons in the Soviet Union, *Hum. Biol.* **38**:219.

Waterlow, J. C., 1973, Note on the assessment and classification of protein–energy malnutrition in children, *Lancet* **2**:87.

Wolanski, N., 1967, The secular trend: Microevolution, physiological adaptation and migration, and their causative factors, in: *Proceedings of the Seventh International Congress of Nutrition,* (J. Kühnau, ed.), Hamburg, 1966, Vol. 4: Problems of World Nutrition, p. 96, Pergamon, Oxford.

Zeeman, J., 1857, Verslag namens de commissie voor statistiek, voorgedragen op de algemeene vergadering te Zwolle, den 25sten Junij 1857, *Ned. Tijdschr. Geneeskd.* **1**:481.

Zeeman, J., 1861, Rapport van de commissie voor statistiek over de lotelingen van de provincie Groningen 1836–1861, *Ned. Tijdschr. Geneeskd.* **5**:691.

17

The Influence of Exercise, Physical Activity, and Athletic Performance on the Dynamics of Human Growth

DONALD A. BAILEY, ROBERT M. MALINA, and ROY L. RASMUSSEN

1. Introduction

This chapter deals with two important questions: What are the effects of exercise and athletic activity on growth? What is the present state of knowledge in regard to these effects?

A survey of the literature concerned with exercise and growth reveals a surprising lack of information. Early reviews on the subject by Rarick (1960) and Espenschade (1960) have recently been brought up to date by Malina (1969a), Elliott (1970), and Rarick (1974). These reviews reach the general conclusion that a certain minimum of physical activity is necessary to support normal growth. But what this minimum is or should be and what effect more intensive activity may have remain to be determined. Rather pessimistically, Cumming (1976) states, "We do not know the 'optimal' amount of physical activity for childhood, and likely never will."

There are certain things we do know, however. We know that the impulse to physical activity in children is strong and that such activity probably constitutes one

DONALD A. BAILEY ● College of Physical Education, University of Saskatchewan, Saskatoon, Canada. **ROBERT M. MALINA** ● Department of Anthropology, The University of Texas, Austin, Texas. **ROY L. RASMUSSEN** ● Department of Physical Education, St. Francis Xavier University, Antigonish, Nova Scotia, Canada.

of the great needs of life. We also know that considerable numbers of children are starting serious athletic training and intensive exercise programs at earlier ages than ever before. Further, we know that more and more adults are becoming interested in exercise programs as health problems associated with sedentary living become more pronounced. Thus, two specific questions have increasing relevance today: What (if any) are the effects of exercise or physical training on the growth of children? What are the effects on functional capacity when maturity is reached?

Intensive physical activity subjects the developing organism to a variety of stresses, which give rise to significant circulatory, respiratory, thermal, chemical, and biomechanical responses. The degree of response varies with the timing, duration, and intensity of the training stimulus. Adaptive responses generated through vigorous physical training during youth are believed by many to result in persistently favorable influences on the organism during growth and development and into adulthood (Skinner, 1968; Malina, 1969*a*, 1975*a;* Bailey, 1973; Rarick, 1974). It should be noted, however, that physical activity is only one of many factors that may affect the growing child, so that the precise role of properly graded training programs in influencing growth and development is not completely understood.

It should also be noted that a significant amount of our knowledge on the effects of physical activity on the developing organism is derived by extrapolation from studies of experimental animals, and the concept of species specificity must be recognized. Much of the research has involved only relatively brief periods of activity or inactivity. Training programs of varying intensity and duration are commonly used, and are usually described as being mild, moderate, and severe exercise stresses without more specific definition. Training, however, is a continuum, ranging from mild work to vigorous, severely stressing activity. It is not a single entity and varies considerably in type, duration, and intensity. In addition, variable and composite age groups and relatively small sample sizes are commonly used in training studies. Further, very few studies have been conducted in which the exercise factor has been regularly applied and changes monitored over a sufficiently long period of time during the growing years. Difficulties inherent in conducting longitudinal programs with humans are obvious. Studies on children are further confounded by problems inherent in any attempt to disentangle training-induced effects from changes accompanying normal growth and development, which are processes variable in expression and magnitude.

Traditionally, the evaluation of physical growth has relied on the application of anthropometric techniques. The comparison of these physical measures with known standards permits an appraisal of the anthropometric status of the individual relative to age and sex peers. Similarly, the measurement of size, shape, and body composition has added to the assessment of a child's growth status. Indices of developmental or biological age define an individual's attainment of a particular size and shape in terms of age as related to a standard population.

These more commonly used measurements have proven invaluable to the study of normal and abnormal growth. However, researchers have become aware of limitations in this traditional approach and have continued to search for more efficient techniques. The classical measurements are gross; it has recently become apparent that they are limited when examining the effects of specific stimuli on growth. Simply measuring increments in height, weight, or volume will not distinguish between the two basic types of cellular growth: increase in cell number (hyperplasia) and increase in the size of the individual cells (hypertrophy). Thus, many recent investigations into the growth phenomenon have been made by researchers utilizing biochemical techniques to determine cellular change.

Obviously, the complex problems associated with exercise and growth will continue to require sophisticated research solutions. The general aim of this chapter is to consolidate what is known in order that the problems may be viewed in perspective. Specifically, the sections which follow are attempts to select, describe, and evaluate the pertinent literature on exercise and growth from the earlier basic anthropometric approaches to the more recent biochemical studies.

2. Bone

Bone growth is in itself a complex process comprising a unique series of events; the effects of specific exercise programs upon this process have not been thoroughly studied. The basic shape of a long bone is determined by one's genes and is established embryologically in cartilage. The development of this cartilaginous model into an adult bone entails growth in length and width, changes in density, plus the maintenance of shape and integrity. When one speaks about growth of bone or about the effects of exercise on bone growth, care must therefore be taken to distinguish among the above processes.

2.1. Bone Length and Diameter

In an early study of growing dogs, Howell (1917) found that the stress and strain of use had little effect on long-bone growth early in life, but that later the effects of disuse retarded diametric growth by as much as 20–25%. This led Howell to conclude that the length of bone and the basic bony configuration are determined by inherent factors, whereas growth in diameter of bones is influenced by exercise.

In his classic review of the chronic effects of exercise, Steinhaus (1933) presented evidence on the effects of immobilization and denervation on the growth of bone in the extremities of young animals (dogs and rabbits). The bones of inactive limbs were lighter in weight, with less mineral matter and greater water content. These bones were also longer and more slender, with radiographic and histologic signs of more activity at the epiphyseal zone. The increased length was attributed to the absence of growth-retarding compressive forces and the slenderness to the removal of muscular tensions which serve as the "normal stimulus to lateral growth."

More recently, Lamb *et al.* (1969) reported on the effects of enforced swimming on the growth of rats. Sedentary animals had significantly longer tibias and femora than exercising rats. Thus, forced exercise in prepubertal rats resulted in reduced length of the long bones. Corresponding data in children are difficult to find. Observations by Kato and Ishiko (1966) are suggestive of detrimental effects of excess compressive forces on epiphyseal growth of the lower extremities of children. They reported 116 cases (out of 4000 observations) of reduced epiphyseal growth in the lower extremities of Japanese children exposed to strenuous physical labor. Epiphyses of the distal femur and proximal tibia showed earlier closure by several years; hence, stature was reduced. Kato and Ishiko attributed such premature closure to excessive loads carried on the shoulders. It should be carefully noted that the environment in which these children were reared and in which they worked was quite poor economically and substandard nutritionally. Suboptimal nutrition, however, has not been shown to cause earlier epiphyseal closure.

Studies utilizing experimental tensile and compressive stresses produce similar results in animals. Gelbke (cited by Evans, 1957) looked at the effects of such stresses on endochondral bone growth by subjecting the olecranon epiphysis to a tensile stress and the distal humeral epiphysis to a compressive stress in the same young dog. After 4–15 weeks of such stresses, the epiphyseal cartilage at both sites was replaced by osseous spongiosa, which is resistant to mechanical forces. Each bone had apparently sacrificed its growth potential in order to retain its configuration and stability in light of continuous mechanical stress.

On the other hand, the observations of Buskirk *et al.* (1956) suggest that persistent physical activity stimulates bone growth in length. Buskirk and colleagues found the dominant hands and forearms of seven nationally ranked tennis players to have longer bones than the nondominant extremity. Since the tennis players in this sample had participated extensively in this activity during their teenage years, the authors attributed the observed laterality differences in length to the effects of vigorous exercise on bone growth during the adolescent years. Prives (1960) reported similar results for subjects engaged in specialized sports and work occupations as did Helela (1969), who noted that Finnish men engaged in manual work had thicker bones.

Ivanitsky (cited by Rarick, 1974) reported that long participation in athletics effects a change in bone diameter and internal structure of the bone; the femur of long-time soccer players is frequently larger in diameter than that of the nonathlete, and that the marrow cavity of the tibia in runners active for over five years is often enlarged. The question of whether large bones resulted from participation as opposed to heredity was, however, unanswered.

One of the best examples that can be offered to demonstrate how bones grow and develop in reaction to stress placed on them has been documented by Houston (1978) in a case study of a boy who was born with only one instead of two bones in his lower leg (Figure 1). Orthopedic surgeons operated on this handicapped infant when he was 2 years old and moved the fibula, which is normally a thin strut bone, centrally beneath the femur so that it could bear weight (Figure 2). When this boy was 3½ years old, after 18 months of weight-bearing (Figure 3), the fibula which was doing the work of the tibia assumed essentially the shape, size, and strength of the missing tibia. This is a good example of the effect of stress on bone growth and development. In this case, the initial absence of the tibia was genetic but the later development of the fibula was due to stress.

While it is impossible to say how much stress or activity is necessary for optimal bone growth, research supports a statement by Trueta (1968) that intermittent energetic compression of the entire cartilage, aided by gravity, weight-bearing, and muscle contractions, is indispensable to keep children's bones growing at the required rate.

2.2. Bone Density

Exercise is known to increase bone density while inactivity decreases density. Inactivity—such as prolonged periods of recumbency, periods of immobilization in casts, and more recently prolonged space flights—leads to an increase in urinary excretion of calcium and nitrogen, and a decalcification of bones. Resumption of

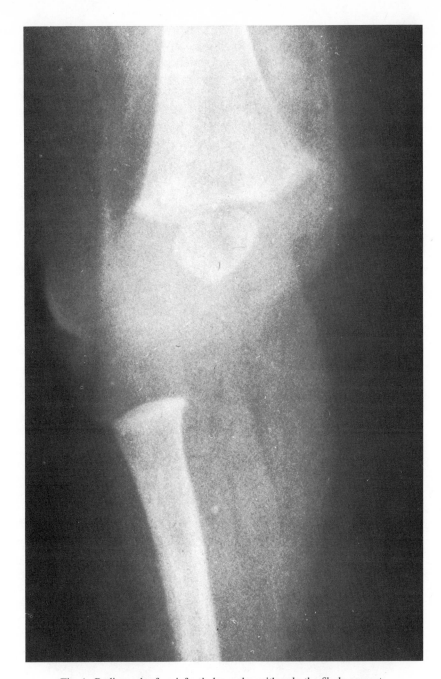

Fig. 1. Radiograph of an infant's lower leg with only the fibula present.

normal physical activity corrects this imbalance, although Kottke (1966) notes that several years may be needed to restore immobilization-induced loss of calcium from the skeleton.

Evidence obtained from humans who have become bedfast gives some insight into the role which physical activity plays in maintaining the normal structure of bone tissue. Asher (1947), in pointing out the dangers of prolonged bed rest, states that bones which are not subjected to normal use lose calcium and that the absence of weight-bearing may delay the complete union of fractured bones. According to Abramson (1948), a normal healthy person who is placed in bed in a plaster fixation increases his calcium loss for 5–6 weeks. An associated nitrogen loss during this period demonstrates that complete physical inactivity depletes the normal stores of both calcium and protein. Upon restoration of a program of physical activity there occurs a positive calcium and nitrogen balance.

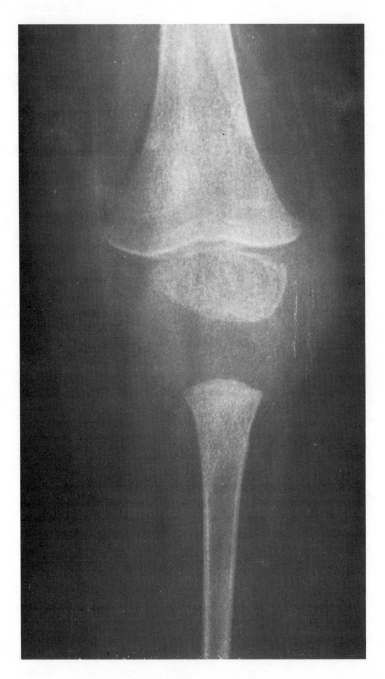

Fig. 2. Radiograph of the lower leg after the fibula was surgically moved centrally beneath the femur.

Donaldson *et al.* (1970) subjected three healthy young adult males to bedrest for 30 weeks. The measured calcium loss averaged 4.2% of the total body calcium, but each weight-bearing heel bone, the calcaneus, disproportionately lost 25–45% of its mineral. As activity was reinstituted at the end of the experiment, the calcaneus remineralized at about the same speed at which the calcium had been lost, even though the over-all body calcium balance remained negative during the 3-week period of reambulation. Bone loss is even more rapid in the weightlessness of

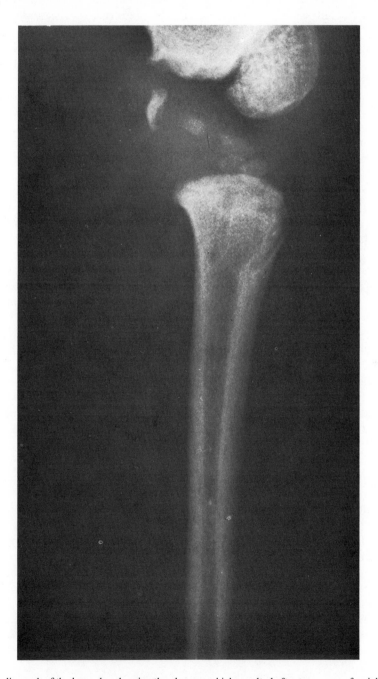

Fig. 3. Radiograph of the lower leg showing the changes which resulted after two years of weight bearing.

space flights with density loss of 3% from the calcaneus in 4- to 14-day Gemini missions (Vose, 1974).

Experimental evidence as to whether weight-bearing or muscle activity is more important in preventing disuse osteoporosis is conflicting. Tower (1939), in a denervation study, eliminated neurotrophic action as a growth stimulant and concluded that weight-bearing was the primary factor in the maintenance of bone density. In Donaldson's experiment, the osteoporosis could be prevented by quiet standing for 2 hr/day. Issekutz *et al.* (1966) also found that 3 hr of quiet standing reduced the calcium decline, whereas up to 4 hr of exercise in the supine position did not. Wyse and Pattee (1954), however, found that humans with paralyzed extremities did not benefit from weight-bearing on the tilt table; they concluded that muscular activity was more important. Gillespie (1954), in experiments with kittens, noted that bone loss in paralyzed limbs was due entirely to loss of muscle power and tone. Plum and Dunning (1958), who studied polio patients, and Abramson (1961), who studied paraplegics, emphasized that ambulation and muscular activity were necessary to prevent bone loss.

Experimental work by Kharmosh and Saville (1965) on animals with motor denervation shows a substantial relationship between loss in muscle mass and loss in bone ash. Other studies with rats indicate positive effects of activity programs on bone density and breaking force (Saville and Smith, 1966). At any given body weight, the exercised animals had more calcium in the axial (except skull) and appendicular skeletons than the control animals (Saville and Whyte, 1969).

The observations of Buskirk *et al.* (1956) on tennis players—whose dominant hand and forearm bones were wider and more robust than the nondominant member—suggest an exercise-mediated effect during the growing years. More recently, Nilsson and Westlin (1971) reported greater bone densities in young adult athletes 22–23 years of age, compared to nonathletes. A clear gradient from highest densities in "top" athletes through "ordinary" athletes and "exercising" controls to "nonexercising" controls was apparent. Among the athletes, those participating in sports involving heavy loads on the lower limbs generally had the higher densities; it was speculated that physical activity during the growing years may have had persistent beneficial effects.

Activity-related bone mineralization data for children are not as conclusive as those for adults. In a study of bone mineralization of the dominant and nondominant arms of amateur baseball players 8–19 years of age, Watson (1973) reported marked mineralization and width differences only for the dominant humerus compared to the nondominant member. However, inconsistent patterns of greater mineralization and width in the dominant radius and ulna were noted. Differences in the mineral content between the dominant and nondominant humeri increased with age, while those for the dominant and nondominant radii and ulnae did not. Although these observations suggest possible activity-mediated effects, mineral dominance was not related to indicators of physical activity stress (limb girths and grip strength) after body weight was controlled in the analysis. These indicators of physical stress are essentially estimates of muscularity, and radiographic studies of bone and muscle show little relationship between widths of the two tissues (Malina, 1969*b*).

The exact mechanisms by which physical activity promotes increased density are not clear. Saville and Smith (1966) suggest that it is the specific effect of muscular contraction acting on bone. Other mechanisms postulated are neural, circulatory, and, in the case of limb bones, compression or weight-bearing stress (Lamb, 1968).

In summary, a statement by Houston (cited by Bailey, 1976) is worth quoting.

> Bones are not solid unchanging structures as people once thought. Instead, a bone, like any other organ in the human body, is a *dynamic* structure. New cells are laid down and old cells are taken away. New calcium ions enter the scaffolding as old calcium ions return to the blood. The body has a very precise balancing mechanism to maintain normal levels of calcium and phosphate in our blood. For this reason, it takes extreme deficiency of calcium or vitamin D in our diet, or a marked abnormality of the parathyroid glands to cause demineralization of bone. Yet only a few weeks of inactivity often cause noticeable deossification—loss of half the calcium from a bone. So the amount of activity we get is much more important than the amount of milk we drink. If we are active, our bones will be well mineralized and both bones and muscles will be strong. This is true in children, true in adult life and true in old age. As a radiologist, I see almost every day, dramatic changes appearing in bones and muscles from disuse. It seems obvious that continued exercise is necessary for the maintenance of normal bone and muscle strength—and normal health. This is not new or original but it is important. Inactivity is very harmful.

3. Muscle

Increases in muscle size, strength, and endurance are probably among the best recognized effects of muscular exertion. "The fact that persistent use of muscles causes their enlargement and a correlated increase in their strength has been known ever since there were boys . . ." (Steinhaus, 1968). The increased size theoretically could be accounted for by an increase in the size of existing cells or an increase in the number of cells or both. In muscle tissue, these possibilities have been subjected to experimentation for nearly a century. The nature of investigation has varied from gross studies of muscle weight to biochemical determinations of protein and DNA concentrations. Nonetheless, controversy as to the mechanics of muscle growth still exists, and Morpurgo's (1897) statement might well apply today:

> As certain as is the fact that the mass of voluntary muscle increases in response to greater work, so uncertain is our knowledge concerning the mechanism that underlies the enlargement. There is no lack of assertions in the literature that deal with this subject in more or less decisive fashion and interpret the activity hypertrophy as either a true hypertrophy in the sense of Virchow or a combination of hyperplasia and hypertrophy, but exhaustive proof is everywhere lacking.

3.1. Histological Studies of Exercise and Muscle Growth

In 1895 Morpurgo suggested that an increase occurred in the number of radialis muscle fibers during the first two weeks of postnatal life in the male rat. He noted a definite increase in fiber number from birth to 16 days of age and postulated that following these first two weeks of life, growth in the cross-sectional area of the entire musculature was attributable to enlargement of individual fibers only. To test this hypothesis Morpurgo (1897) removed the sartorius from one side of each of two adult dogs and took a cross-sectional count at the exact midpoint of each muscle. After the incisions had healed the dogs were made to run on a treadmill over a 53-day period. Following the training period the contralateral sartorius was removed and compared. Although the exercised muscles were some 54% larger in cross-sectional area than the pre-exercise controls, Morpurgo attributed the increased muscle mass to a true hypertrophy of individual fibers and not to the appearance of new fibers. The increase in size was attained with no increase in the number of muscle fibers. No evidence could be found to suggest that exercise caused a lengthening of the muscle fibers, an increase in the number of nuclei fusing to form the muscle syncytium, or an increase in the number of fibrilli. This latter finding led Morpurgo to suggest that hypertrophy was the result of increased sarcoplasmic, not contractile, proteins.

Siebert (1928), feeling that Morpurgo's (1897) results may have been determined by the type of work his animals were made to perform (duration as opposed to effort) and knowing that in adult animals (as used by Morpurgo) the tendency of tissues to separate is generally less than it is in young animals, reinvestigated the nature of muscular response to exercise. Two dogs from the same litter, 3–4 months old when the experiment began, were trained in work of the effort type. The animals ran on an inclined treadmill and effort was slowly increased until it reached a maximum of 10 km in 1½ hr at an incline of 33%. In addition, each dog was made to carry a ballast equal to 20% of the body weight. At the end of six months of training, the sartorius of each dog was excised and compared with its contralateral which had been removed before the training period began. Even though the fibers comprising the active muscles were found to be more numerous (5 and 6% greater) and in each case favored the exercised muscle, Siebert felt the observed differences were within the limits of normal variation and did not provide proof of fiber increase. "The skeletal muscle itself hypertrophies only by simple enlargement of cells, as a result of excessive work" (Siebert, 1928). Siebert lent support to the findings of Morpurgo (1897) when he concluded:

> From certain embryonic months until the end of the extrauterine body development—a length of time that certainly represents a period of growth—the muscle mass increases only by enlargement and not by multiplication of cells.

Kleeberger (1932) electrically stimulated the ear muscles on one side of three rabbits and observed increases of as much as 40% in weight and 90% in cross-sectional area. The increases, according to Kleeberger, were attributable entirely to the enlargement of already existing cells. Holmes and Rasch (1958) found an increase in volume of the sartorius muscle in rats after running for seven weeks up to 840 m/day, but like Morpurgo (1897) did not encounter an increased number of fibers. Klebanova (1959) conducted an experiment similar to Morpurgo's (1897) and noted hypertrophy of individual fibers in dog sartorius muscle, but no hyperplasia.

Hettinger (1960) exposed a group of dogs to regular exercise on a treadmill, while a second group was injected once each week with testosterone. There were nearly identical increases in muscle weight and diameter of the muscle fibers in the two groups, but there was no indication of hyperplasia.

The preceding studies indicate that muscle growth due to exercise is the result of hypertrophy or simple enlargement of existing cells. Most of these investigations used adult animals as subjects, however. The effect of exercise on muscle tissue during the growing years is still unclear, and the possibility of hyperplasia in young animals cannot be discounted entirely.

Microscopic examinations of biopsy and autopsy specimens of human muscle by Durante and his associates (1953) indicated the possibility of an increased number of fibers in hypertrophied muscles. Denny-Brown (1955) reported a twofold increase in myofibrillar counts made on gastrocnemius muscle of a cat assumed to have undergone work of high intensity and forceful dynamic contraction over an activity period of three months. Goldspink (1964) reported in biceps brachii of young mice a three- to fourfold increase in myofibrils and 30% gain in fiber area without corresponding increase in muscle weight. The animals were made to lift progressively increasing weights in order to obtain food from a unique pulley-operated food-rewarding device.

Thus, although the majority of postnatal histological studies suggest that an increase in muscle size due to exercise is entirely due to hypertrophy, more recent findings suggest that the question is still open, especially during early years of growth.

3.2. Biochemical Studies of Exercise and Muscle Growth

Surprisingly few studies have used biochemical techniques to study the influence of exercise on muscle growth. Helander (1961) found in guinea pigs 4–6 weeks of age, after they had run 1000 m daily on a treadmill for three months, a 15% increase in myofilament protein concentration without any change in sarcoplasmic protein and muscle weight. This was the first biochemical investigation of work hypertrophy and also the first study to report increased myofibrillar proteins with running. Previous histological evidence in running animals denied any rise in contractile proteins and strongly supported the contention that hypertrophy was totally sarcoplasmic (Morpurgo, 1897; Holmes and Rasch, 1958). A tentative explanation for the apparent contradictions was offered by Gordon *et al.* (1966), who pointed out that Helander's (1961) animals were young and undergoing active growth, whereas subjects in the anatomical study were adult. Gordon and his associates (1966) outlined their views:

> Growth alone, however, cannot account for Helander's findings of increased concentration of contractile protein in his running guinea pigs, because control animals submitted to restricted and usual activity showed no rise in equal period of time. Nevertheless, it is possible that exercise stimulus may have an enhanced effect on actively growing animals as contrasted with adults.

A recent approach that offers a new perspective for examining hypertrophy and hyperplasia of muscle tissue during growth is based upon the recognition that for most tissues the amount of DNA per cell nucleus is constant. This finding provides a chemical unit which can be applied in studying the growth phenomenon. Using DNA content of skeletal muscle tissue as an index of nuclear proliferation, i.e., cell number, and the ratio of protein to DNA as an index of cell size, Cheek and colleagues (Cheek *et al.*, 1965, 1971; Cheek and Hill, 1970) have shown that hypertrophy of normally growing muscle fibers is accompanied by proliferation of their nuclei.

Christensen and Crampton (1965) examined gastrocnemius muscles of 1-year-old rats exposed to forceful exercise and found increased DNA, RNA, and protein values. No difference in protein/DNA ratios was observed between the exercised rats and a control group. The authors concluded that exercise stimulated muscle growth and increments in protein and DNA were corresponding in degree. Gordon *et al.* (1967a) trained rats in running or swimming (prolonged endurance-type exercise) and noted increased sarcoplasmic and decreased myofibrillar concentrations in the quadriceps and gastrocnemius muscles.

Buchanan and Pritchard (1970) investigated the DNA content of the tibialis anterior muscle of the rat from birth to 50 weeks of age. Between birth and puberty the total DNA of the muscle (including its tendon) increased 25-fold in females and 30-fold in males. Thereafter DNA content remained constant. More importantly they reported that strenuous daily running exercise in 4-week-old males for 2–7 weeks significantly increased DNA levels above normal controls. In other words, skeletal muscle nuclei population was facilitated by exercise.

Bailey *et al.* (1973) studied the effect of daily swimming exercise on prepubescent rats. DNA concentration of the gastrocnemius muscle was found to be significantly greater in animals that had exercised from 3 to 12 weeks compared to sedentary controls, lending support to the suggestion that exercise may stimulate nuclear proliferation in muscle tissue still in a dynamic state.

Hubbard *et al.* (1974) examined the effects of growth and endurance training on the protein and DNA content of rat gastrocnemius, soleus, and plantaris muscles between 39 and 123 days of age. The trained groups were forced to exercise on a motor-driven treadmill under a shock-avoidance contingency. The animals were exercised 30 min/day, 5 days a week with the treadmill set at a 6° incline for up to 12 weeks. The first week, the animals were exercised at a belt speed of 15 m/min. Each week the treadmill speed was increased by 5 m/min until the fourth week when the animals were running at 30 m/min. This speed was maintained throughout the remainder of the training period. Although the concentration of muscle DNA on a wet-weight basis fell markedly between 35 and 50 days of age, the total number of nuclei per muscle nearly tripled. Their data suggested that endurance exercise may slow the rate and prolong the phase of nuclei accumulation. In contrast to the findings of Gordon *et al.* (1967*b*), endurance training did not selectively stimulate the hypertrophy of muscle sarcoplasmic or myofibrillar protein.

Goss (1966) has suggested that the ultimate size of the muscle fiber is ordained by how many nuclei are included in it. Whether exercise imposed during the hyperplastic growth phase of skeletal muscle tissue is facilitative to the proliferation of nuclei is an open question. The studies of Buchanan and Pritchard (1970), Bailey *et al.* (1973), and Hubbard *et al.* (1974), while suggestive, are inconclusive; further investigations are warranted.

The functional implications of reported biochemical responses of muscle tissue to activity also need examination. In an extensive review of the problem, Jeffress and Peter (1970) indicate increased functioning at the molecular level in skeletal muscle tissue as a result of the biochemical responses to chronic exercise. Edgerton *et al.* (1969), for example, found an increase in the proportion of skeletal muscle fibers having high oxidative enzyme activity in exercised as opposed to sedentary rats. The authors suggest this as a possible mechanism through which muscle tissue adapts to the chronic overload of vigorous physical activity. Working with adult guinea pig skeletal muscle tissue, Barnard *et al.* (1970*a*) found an increase in mitochondrial concentration per gram of muscle with prolonged treadmill running. Histochemical examination of fiber types indicated a relative increase in the number of red fibers in trained muscles. On the other hand, there were no significant differences in the contractile properties of skeletal muscle in the trained and control animals (Barnard *et al.*, 1970*b*).

It would seem, in conclusion, that the biochemical investigations dealing with the effects of exercise on muscle growth and function have reaffirmed the complexity of the question, as noted by Haylor (1975) in a study that provided the basis for much of this discussion on skeletal muscle tissue. Problems of definition and interpretation need to be resolved and clearly further research is called for. Chapter 10 of this volume provides a more detailed discussion of biochemical research on muscle tissue and provides further insight into the problems of definition and interpretation which seem to characterize this area of research.

4.1. Body Composition: Leanness–Fatness

Physical activity is an important factor in the regulation and maintenance of body weight. Weight, however, is a heterogeneous mass, quite frequently partitioned into two components in growth and training studies: lean body mass and fat. Physical training generally produces an increase in lean body mass and a corresponding decrease in body fat, quite frequently without any appreciable change in body weight (Parizkova, 1963, 1968b; Dobeln and Eriksson, 1972). The magnitude of change, however, varies with the intensity and duration of the training regime. In growing children the difficulty in separating the effects of growth *per se* and training presents a further problem.

Wells *et al.* (1962, 1963), for example, studied the effects of a 5-month daily physical training program on 34 adolescent girls as compared to an equal number of control subjects. The findings showed a definite change in body composition, a significant increase in lean tissue and a corresponding reduction in fat in the trained group. No such changes were reported for the control subjects. On the other hand, Moody *et al.* (1972) did not observe changes in gross body composition after 15 weeks in an activity program of walking and jogging involving 12 adolescent girls, mean age 16.5 years. The girls did show, however, a small decrease in fatness as shown in smaller skinfold thicknesses.

In one of the more comprehensive studies, Parizkova (1968a, 1970, 1974) has reported the results of longitudinal observations on teenage boys, age 11–18 years, exposed to different degrees of sports participation and physical training over a 7-year period. Needless to say, subject attrition as the study progressed was considerable ($N = 143$ in 1961; $N = 39$ with complete data in 1968). Three levels of training were compared: regularly trained (intensive; 6 hr/week); trained, but not on a regular basis (went to sport schools; about 4 hr organized exercise per week); and untrained (about $2\frac{1}{2}$ hr/week including school physical education). Sample sizes for these three groups at the conclusion of the study were 8, 18, and 13, respectively. The sample was from a generally similar socioeconomic and educational background and was also of similar maturational status. The different activity groups did not differ markedly in selected anthropometric dimensions, although the group with the highest physical activity level was on the average taller and heavier. It should also be noted that the groups did not differ significantly in relative body composition at the beginning of the study. At the end of the first 4 years and then at the end of the 7 years, the most active boys had significantly more lean body mass and less fat than the least active boys. The relative body composition differences were most marked between the most and least active groups of adolescent boys. Interestingly, in the data for the small samples of boys who completed the study, the group classified as intermediate in physical activity tended to be shorter, lighter, and leaner than the least active group of boys. Indeed, Parizkova (1974) raises the question: "Is it possible that moderate exercise tends to be detrimental to growth whereas greater quantities or intensities of physical activity enhance growth?" More probably, the results represent sampling variations and the small number of subjects completing the study.

Concerning fatness itself, there is generally more concern with the correction of overweight children and adolescents than with its prevention. Data cited earlier clearly indicate that physical activity is important in regulating levels of fatness. A problem that merits consideration is the potential of physical activity programs, initiated early in life, in a preventive role. Experimental evidence suggests that training early in life effectively reduces the rate of fat-cell accumulation in growing rats, thus producing a significant reduction in body fat later in life (Oscai *et al.*, 1974). Adipose cell growth may result from increases in fat-cell number or size. Estimates of fat-cell development in humans indicate that cells increase in number up to early adolescence, after a period of rapid increase in the first year of life (Brook, 1972; Knittle, 1972). One can inquire, therefore, as to the influence of physical training on fat-cell number in humans and/or as to the age at which cell numbers can be influenced by activity programs. It should be noted, however, that rapid growth of fat-cell number during early infancy may be the critical factor in producing obesity, as some obese adolescents have 3–4 times the normal number of fat cells (Knittle, 1972).

4.2. Physique

Physique is a composite, referring to an individual's body form, i.e., the conformation of the entire body as opposed to emphasis on specific features (Tanner, 1953). Studies of training and physique, however, have generally considered height–weight relationships and specific features. Several studies of adolescents, for example, indicate beneficial effects of short-term training on body bulk, specifically muscular development (Malina and Rarick, 1973). Changes in aspects of physique are usually most apparent in body parts specifically trained. Godin (1920), in an early study, for example, considered the effects of gymnastics training on boys 14–18 years of age. The gymnasts were slightly taller and heavier than nongymnasts but were especially larger in thoracic and forearm measurements. Parizkova (1968a), in the longitudinal study of boys 11–15 years with different training levels (see previous section), noted that the most active boys developed, on the average, relatively narrower pelves in relation to stature and shoulder breadth at the end of four years, even though the study groups had similar values at the start of the study. There was, however, much overlap in the indices of the different activity groups, such that the significance of the change is not clear. It could reflect differential timing of the adolescent spurt in height, thus influencing the ratio of hip width to stature. Parizkova and Carter (1976) also considered the stability of anthropometric body types in this longitudinal sample over a 7-year period. The distribution of body types did not show differences among the three activity groups, i.e., high, moderate, and low training levels. These data would seem to suggest no marked effect of different physical training programs on anthropometric body type. Rather, individuals changed considerably in anthropometric body type between 11 and 18 years. These changes occurred in a random fashion and were not attributable to physical activity. The extent of variation was such that all boys changed in body-type ratings at least once, and 67% changed in component dominance. These observations thus suggest that individual variation in anthropometric body-type stability over adolescence confounds the evaluation of possible training-related changes.

One can inquire as to the relationship between age, physique, and response to training programs. In his classic review Steinhaus (1933), for example, reported increased muscle girths in children and youth with training. However, during periods of rapid growth in body length (presumably the adolescent height spurt), exercise had less effect on muscle girths than in periods when the tendency to grow in width predominated. In this regard it should be noted that the peak adolescent gain in weight and muscle tissue generally occurs after the peak height velocity in boys and girls (Stolz and Stolz, 1951; Tanner, 1962, 1965) and that the apex of the male adolescent strength spurt occurs after both the adolescent height and weight spurts (Carron and Bailey, 1974).

5. Physiological Function

A traditional approach that has been used to study the influence of exercise or physical activity on functional growth has been to compare athletes with nonathletes, or to compare subjects classified according to various levels of physical activity.

For instance, Astrand *et al.* (1963) conducted a detailed investigation of 30 ''top'' Swedish girl swimmers. The girls were 12–16 years old and had been training extensively for some years. Some of them trained 28 hr a week and covered, in that time, a distance of 65,000 m. The findings of the study indicated that the girls who experienced the rigorous swim training had a mean maximal oxygen uptake of 3.8 liters/min as compared to 2.6 liters/min in average girls of the same age. Not only was their oxygen uptake larger but the dimensions of some components of their oxygen-transporting system were also significantly larger (heart volume, lung volume, and total hemoglobin).

The increased dimensions correlated well with each other as well as with maximal oxygen uptake. Thus it was concluded that the hard physical training had given these girls an increased size of the organs involved in the oxygen-transport system, thereby enabling an increase of the maximal oxygen uptake. There was no indication that the hard physical training had caused any damage to these girls.

Ten years after the first investigation, the same girls were restudied (Eriksson *et al.*, 1971). During this time all the girls had stopped their regular training and most did not engage in any specific physical activity in their spare time. Their daily lives were filled with studying, working in their professions, doing domestic work, working with their children, and so on. Due to the very low grade of physical activity, the girls showed a pronounced decrease in their aerobic power. The mean value for the girls' maximal oxygen uptake had decreased by 29% over the 10 years. On the other hand, the dimensions of the lungs and heart were relatively unchanged. As a consequence, there was a change in the correlation between their dimensions and the maximal oxygen uptake. The girls appeared to have retained the organic capability or high function even though their actual uptake had decreased significantly, due to their markedly reduced activity levels.

Eriksson *et al.* (1971) concluded that hard physical training influences normal development of the child. No deleterious effects were demonstrated. Furthermore, it would seem that those organic dimensions which are related to the oxygen-transport system may have been influenced by the training regime. In commenting on this study, Astrand (1967) speculated,

> During adolescence there may be a second chance to improve those dimensions which are of importance for the oxygen transport system. This is an interesting problem, especially with regard to physical education in school. It may not be possible to repair later in life what is neglected during the adolescent years.

As part of this valuable study, it is unfortunate that a control group of nonswimmers had not been investigated over the same time period, as this may have provided further insight into the relative involvement of training and natural endowment.

Ekblom (1970) used a different approach to try to gain some insight into the influence of physical training on functional growth. He noted that if physical training is started in a grown man there is an upper limit in functional capacity over which one cannot go, even if the physical training is very hard and extended over many years. He investigated the effect of physical training on boys during adolescence to try to answer the question of when in life this upper limit is set. Is it strictly genetic or can it be influenced by physical training before and during puberty? He studied six 11-year-old boys in a 6-month training program using a group of seven untrained boys of the same age for controls. The aerobic working capacity of the trained group improved 15%; that of the untrained group remained unchanged. Five of the six boys in the training group remained in training for an additional 26 months with the results that their oxygen uptake increased a total of 55%, the heart volume increased by 45%, and the vital capacity was augmented by 54%. These changes were considerably more than would be expected from age-dependent increases in body dimensions. Apparently an extrinsic environmental factor like exercise was involved along with intrinsic genetic growth factors.

Daniels and Oldridge (1971) conducted a training study on 14 boys during a 22-month period. The age of the boys ranged from 10 to 15 years. The training was supervised by a professional track trainer, and the boys followed the middle-distance runners' training program. While the boys displayed no increase in their maximum oxygen uptake per kilogram of body weight, there was a 5% increase in their maximal oxygen uptake when expressed dimensionally per height squared.

In a longitudinal study, Sprynarova (1974) followed the development of functional capacity of boys 11–18 years of age, who were divided into three groups, depending on their previous physical activity. Annual determinations of maximal oxygen uptake on the treadmill revealed that the most physically active group were consistently higher than the other two groups throughout the age range studied. It was concluded that the pubertal period is especially susceptible to training effects. These findings are in agreement with a similar longitudinal study of boys aged 8–15 years conducted by Bailey (1973), who reported functional differences between boys classified according to activity patterns on the basis of a teacher response to a questionnaire.

In contrast to the findings of these studies, separate investigations by Bar-Or and Zwiren (1972) and Mocellin and Wasmund (1972) report no apparent change in functional capacity due to increased training or increased physical activity. Thus there is still some uncertainty about the effect physical training has on young individuals.

For instance, Klissouras and colleagues in a series of twin studies (Klissouras 1971, 1972; Weber *et al.*, 1976) have concluded that maximal oxygen uptake in a population exposed to common environmental forces may be entirely determined by heredity, but the relative contribution of heredity may be reduced to about 50% with the operation of extreme environmental conditions such as physical training.

A further unanswered question is this: Do functional changes as a result of training or activity during youth persist into adult years?

A study by Saltin and Grimby (1968) provides some evidence in answer to this question. They compared the adjustability to effort of three groups of subjects in the age group 50–59 years. One group consisted of men who were former athletes who had given up all sports 20 years earlier and followed a sedentary occupation. A second group comprised men who had achieved the same athletic performance in their youth as their former colleagues in cross-country running or long-distance skiing but had kept up regular training during the adult years. A third group consisted of men who were nonathletes during youth. The results confirm the hypothesis: maximal oxygen uptake was 30 ml/min/kg body weight in the men who were nonathletes during youth, 38 ml for the former youthful athletes who followed sedentary adult living patterns, and 53 ml for the former athletes who had kept up regular activity in the adult years. Generally speaking, functional capacity as an adult appeared to be at least partially a function of activity during the growing years.

Similarly, a study by Leaf (1973) on long-lived people is of interest. He evaluated the living patterns and lifestyle of very old people living in three geographic locations in the world where longevity of the people has been documented to be well above normal expectations. The one common link among the old people of the three communities was a high degree of physical activity.

> The old people of all three cultures share a great deal of physical activity. The traditional farming and household practices demand heavy work, and male and female are involved from early childhood to terminal days. Superimposed on the usual labor involved in farming is the mountainous terrain. Simply traversing the hills on foot during the day's activities sustains a high degree of cardiovascular fitness.

On the question of whether physical training as a youth can affect mature functional capacity, the evidence while not yet clear-cut is suggestive and indicates that an increase in physical activity for children is advisable not only for the immediate effect on the child, but also over the long term as it applies to adult health.

"Top" athletes, for instance, who continue with regular physical training after stopping sport competition, have been shown to have a lower incidence of coronary heart disease and a reduced death rate due to cardiac diseases (Schnorr, 1971).

6. Muscular Strength and Motor Performance

6.1. Muscular Strength

Strength increases with chronological age and is characterized by a growth curve very similar to the curves of growth in height and weight with respect to the growth spurt pattern (Jones, 1947; Howell *et al.*, 1965; Carron and Bailey, 1974).

Sex differences appear to exist in the growth and development of strength (Jones, 1947; Howell *et al.*, 1965; Espenschade and Eckert, 1967; Asmussen 1973). Differences are not great until approximately age 13 years (Jones, 1947; Asmussen, 1973), after which they increase in magnitude.

Evidence for the relationship between growth in strength and growth in body size parameters of height and weight was provided by the studies of Jones (1949), Howell *et al.* (1965), Rarick and Oyster (1964), Clarke (1971), and Carron and Bailey (1974).

For young males, the relationship of strength and body growth patterns has been established by investigating the timing of peak strength velocity with reference to the timing of peak height and peak weight velocity. Stolz and Stolz (1951) and Carron and Bailey (1974) have conducted similar investigations. Stolz and Stolz

(1951) indicated that the apex timing for strength and weight was closer than for strength and height. They reported that the peak in strength gains occurred approximately 14 months following the occurrence of peak height velocity and approximately 9 months after peak weight velocity.

Carron and Bailey (1974) found that the maximum increments for composite, upper, and lower body strength were obtained one year after the occurrence of peak height velocity and peak weight velocity.

Tanner (1962) stated that the weight peak occurs about six months following the height peak and that the muscles appear to have their spurt about three months after peak height velocity. (Thus peak muscle growth occurs previous to peak weight velocity.) This was judged from longitudinal data of limb muscle breadths measured by X ray. Tanner (1962) added that if peak velocity of muscular growth occurred only shortly after peak height velocity and shortly before peak weight velocity, the muscles must grow first in size and only later in strength.

In the case of young women, menarche provides a focal biological rhythm and an accurate indication of sexual maturity to which the growth and development of strength can be related. Jones (1947) found that regardless of the age at which menarche fell, the peak rate of growth in strength occurred shortly before this reference point. On the average, Jones (1947) claimed that girls attained their maximum body growth rate at the age of 12.5 years, and half a year later at age 13 had attained approximately 75% of their terminal pull strength and 90% of their thrust strength. Full adult attainment for thrusting strength was met as early as 13.5 or 14.0 years.

The literature indicates that individual differences in strength tend to be highly stable over successive years but decrease in stability as the time between measures is increased (Rarick and Smoll, 1967; Carron and Bailey, 1974).

Clarke (1971), reporting on the Medford Boys' Growth Study, drew the following basic conclusions regarding the strength characteristics of athletes: boys who make, and are successful on, athletic teams are definitely superior to their peers in muscular strength, and the gross strength of athletes participating in all interscholastic sports is consistently high at all school levels.

Finally, the "trainability" of muscle strength has been investigated by several authors. Rohmert (1968) studied the increase in isometric strength in several muscle groups in 8-year-old boys and girls who were given a series of standard training programs. He also conducted a parallel series of studies on adults for comparative purposes. The findings of Rohmert's work have been interpreted as showing that children are naturally adapted to a lower degree of strength utilization in daily life and therefore may be easier to train than adults. This interpretation is supported in experiments by Ikai (1967), who found that another parameter, muscular endurance, could be increased considerably more in children, especially in the age bracket 12–15 years, than in adults.

This evidence, although far from conclusive, does appear to indicate the need for physical activity during childhood and adolescence.

6.2. Motor Performance

Proficiency in motor activities is an important component of the child's behavioral repertoire and is essential for participation in active games and athletics. This section briefly considers the effects of age trends and sex differences on performance in selected fundamental motor tasks requiring power, speed, agility, and coordination. Although motor skills are many, those commonly used to assess a youngster's motor ability are running, jumping, and throwing tasks, and variants of these skills (Espenschade and Eckert, 1974).

Motor performance improves with age in boys to 17 or 18 years. Note that these ages are generally the last year of high school, after which the available data are not as extensive. Some data for college students indicate continued improvement in motor performance into the early 20s. Jumping and throwing show some improvement during male adolescence, suggestive of a performance spurt in these power events. Performance of girls improves through childhood, but reaches a plateau at approximately 14 years of age, with little improvement afterwards (Espenschade and Eckert, 1974).

Sex differences are evident during childhood and adolescence. Boys, on the average, perform better than girls in running, jumping, and throwing tasks. The sex differences in performance become progressively marked during adolescence. The slopes of the performance curves are steep for boys, while those for girls are rather flat at this time. In a longitudinal series of children (Espenschade, 1940), average performances of girls in all motor tasks except throwing fall within one standard deviation of the boys' means during early adolescence. From 14 years of age on, however, the average performances of girls are consistently outside the limits included by one standard deviation below the boys' averages. Few girls, if any, approximate the throwing performance of boys as adolescence progresses (Espenschade, 1940; Espenschade and Eckert, 1974; Malina, 1974).

A child's level of motor performance is in part related to age, size, physique, body composition, and biological maturity status. These biological correlates of performance have been treated elsewhere in depth (Malina, 1975b), and only an overview of selected relationships will be presented. Evidence suggests, for example, negative effects of excess body weight, fatness, and endomorphy on motor performance items involving movement of the entire body. The magnitude of correlations of size, physique, composition, and maturity status with motor performance, however, is generally low to moderate, and as such has limited predictability. Further, size, physique, and maturity effects on performance are generally more apparent at the extremes of comparisons, for example, extreme endomorphs relative to extreme ectomorphs, or early- compared to late-maturing children.

Maturity-associated variation in motor performance is marked in boys, with early-maturing boys generally superior to their late-maturing peers. On the other hand, motor performance of adolescent girls is poorly related to maturity status, with some suggestion of better performance levels in later-maturing girls (Espenschade, 1940; Clarke, 1971).

The complex interrelationships among biological characteristics and motor performance would seem to suggest a possible role for other factors in determining a youngster's performance. These include, for example, social class, ethnicity, birth order, rearing practices, ordinal position, and motivation, to mention several.

7. Childhood and Adolescent Athletics

There have been numerous discussions on the role of competitive sports in childhood and adolescence, and it is not within the scope of this chapter to consider the positive and negative aspects of the situation. This section offers an overview of selected physical growth and maturity characteristics of the young athlete, most commonly between 9 and 16 years of age. Participation in sport is concerned to a large extent with intensive physical activity in the form of training for a specific sport or event and with the competition *per se*. As such, there is concern about the effects of regular training and competition on the growing organism.

Young athletes are usually defined in terms of success on interscholastic or community teams (especially football, hockey, baseball, basketball, and track), in national and international competitions (especially swimming, track, and gymnastics), and in selected athletic club and age-group competitions (especially swimming, gymnastics, and tennis). Young athletes are also a highly selected group, usually on the basis of size and skill. Needless to say, larger size is an advantage in a number of sports and can be a limiting factor in others. Physique might also be a limiting factor, especially at the somatotype extremes. Endomorphy tends to be negatively related to performance on a variety of motor tasks, while ectomorphy tends to be negatively related to strength. Mesomorphy generally correlates well with strength and performance. Note, however, that correlations between somatotype components and performance are generally low to moderate and are limited in predictive meaning (Malina, 1975b). In addition, the complex interrelationships of maturity status, size, physique, and composition with strength and motor performance must be considered in an evaluation of the growth of young athletes (Tanner, 1962; Clarke, 1971; Malina and Rarick, 1973; Malina, 1974).

7.1. Young Male Athletes

Data considering maturity status and relationships to size, physique, and body composition in young athletes are not extensive, are limited to several sports, and are not entirely consistent within sports and across sports. Before looking at participants in specific sports, one should consider the early and frequently misquoted study of Rowe (1933). Comparing growth in height and weight of junior high school athletes and nonathletes (matched on age, height, and weight) over a 2-year period from 13.75 to 15.75 years, Rowe noted that the athletes were bigger, but grew at a slower rate over the 2-year period. Since biological maturity status of the sample was not controlled, Rowe (1933) stressed that the observed differences probably reflect differential timing of the adolescent spurts: " . . . since the athletic group is composed of boys who have matured earlier, age considered, than the group of non-athletic boys, the athletic boy is not going to grow as much as the non-athletic boy over the period studied." In a similar study comparing male junior high school athletes and nonathletes, Shuck (1962) reported greater heights, weights, and "speeds of growth" (as assessed by the Wetzel grid) among the athletes. The athletes showed no marked acceleration or retardation in height and weight growth that could be related to athletic participation. With only one exception, years of participation, total games played, and length of season apparently had no marked effect on growth in height and weight of the junior high school athletes.

7.1.1. Baseball

Data from the 1957 Little League World Series (Krogman, 1959) suggested that success of the young baseball players was related to their skeletal maturity status. Of the 55 finalists, 11 years to 13 years 11 months, 71% had skeletal ages in advance of their chronological ages, and 29% had skeletal ages delayed relative to their chronological ages. When the Little League participants were viewed as early, average, or late maturers, 5 boys had delayed skeletal ages, 25 had skeletal ages in the average or normal range, and 25 had advanced skeletal ages. In the words of Krogman (1959), "In general, the successful Little League ball player is old for his age, i.e., he is biologically advanced. This boy succeeds, it may be argued, because he is more mature, biologically more stable, and structurally and functionally more advanced."

Similar results were noted in a study of the pubic-hair development (Crampton's 3-point scale) of 112 participants in the 1955 Little League World Series (Hale, 1956). Most of the young athletes were pubescent (37.5%) or postpubescent (45.5%). Maturity also appeared related to position and batting order: starting pitchers, first basemen, and left fielders were generally postpubescent, and all boys who batted in the fourth position were postpubescent. In terms of height and weight, the postpubescent boys equaled the standards for 14-year-old boys. In contrast to these observations on Little League finalists, successful baseball players at the upper elementary and senior high school level in the Medford Boys' Growth Study (Clarke, 1968, 1971, 1973) did not show significant maturity, physique, and size differences relative to nonparticipants.

7.1.2. Basketball

Skeletal maturity, physique, and body size did not significantly differ between basketball athletes and nonparticipants from 9 through 12 years of age in the Medford study (Clarke, 1968, 1971, 1973). However, at the junior high level (12–15 years) height, weight, and skeletal age differed between the participants and nonparticipants. This emphasizes the variability characteristic of male adolescence and the size advantage of early-maturing boys at this time. At both the upper elementary and junior high levels (9–15 years), the basketball athletes were also significantly better in strength, the standing broad jump, and shuttle run. At the high school level (15–17 years), the morphological characteristic distinguishing basketball players was, as one might expect, height. Although not significant at all ages, the athletes tended to be mesomorphic. They were also consistently stronger in arm strength and faster in the shuttle run. The lack of significant skeletal maturity differences at the senior high school ages probably reflects the attainment of skeletal maturity by many youngsters as well as the catch up of late-maturing boys.

7.1.3. Football

Young football athletes in the Medford Boys' Growth Study (Clarke, 1968, 1971, 1973) were significantly advanced in skeletal age, strength, the standing broad jump, and the shuttle run at all ages from 10 through 15 years compared to nonparticipating boys. They were significantly taller and heavier between 12 and 15 years, suggesting a somewhat earlier adolescent spurt. At the senior high school level, the football players were most often characterized by a consistently greater strength index, which is, undoubtedly, in part a function of training procedure. Although not consistent from 15 through 17 years, the football athletes were significantly taller and more mesomorphic in some of the age group comparisons.

Novak (1965) compared a group of 18 boys who were active school athletes in football and basketball with a control group of 20 nonathletic boys. The adolescent boys who participated in athletics were leaner and more muscular than the nonparticipants. Total body fat as estimated by skinfold thickness and densitometry was significantly less for the athletic boys with a correspondingly greater muscle mass as measured by creatinine excretion. Novak attributed the lean body and fat differences between the athletes and nonathletes to the training effects.

7.1.4. Ice Hockey

Among 280 tournament "PeeWee" hockey players 10–13 years of age, Bouchard and Roy (1969) found skeletal age, on the average, slightly delayed relative to chronological age (11.7 ± 1.1 and 12.0 ± 0.4 years, respectively). When partitioned by position, boys playing defense were slightly advanced skeletally and taller and heavier than goalkeepers and forwards. Within specific age groups, however, correlations between skeletal age and game performance of forwards and defensemen (goals and assists) were low (−0.30 to +0.28).

7.1.5. Track

Track athletes in the Medford study (Clarke, 1968, 1971, 1973) did not differ from nonathletes in maturity, size, or physique at the upper elementary level (9–12 years). They were, however, significantly stronger (11–12 years) and performed better in the jump and shuttle run. At the junior high level (12–15 years), the track athletes were significantly advanced in skeletal maturity, height, strength, strength endurance, the standing broad jump, and the shuttle run compared to the nonparticipants. The track athletes were also significantly more mesomorphic and less endomorphic at 14 and 15 years of age. At the senior high school level, the track athletes and nonparticipants were not significantly different in height, but differed in physique (mesomorphy), perhaps representing selection for skill in track events and perhaps the catch up of late-maturing boys.

Among 101 boys (11–18 years) enrolled in a camp for track and field athletes, Cumming et al. (1972) reported a composite skeletal age slightly in advance of the group's chronological age. Although the difference is small, it is in the direction reported earlier for track athletes at the junior high school level (Clarke, 1971). Skeletal age was a better predictor of performance in track and field tests than chronological age, but the correlations were at best moderate (partial r's controlling for height and weight ranged from 0.27 to 0.50).

7.1.6. Swimming

Data for young male swimmers are not as extensive as for the preceding sports. A survey of the 28 "best young Hungarian swimmers" 8–15 years of age showed skeletal age to be slightly advanced, on the average, over chronological age (Bugyi and Kausz, 1970). Another study of 30 Hungarian swimmers (16 males and 14 females, but the results were not analyzed by sex) had a distribution of skeletal age that indicated more swimmers of normal and accelerated skeletal maturity than retarded (Szabo et al., 1972; Szabo-Wahlstab et al., 1973). Interestingly, the average chronological ages of the retarded and accelerated groups were 12.5 and 12.7 years, respectively, while that of the normal group was 13.4 years. The accelerated maturity group, as one would expect, was heavier than the normal group, which in turn was heavier than the retarded group. The same pattern was also apparent for maximal oxygen uptake per kilogram of body weight, vital capacity, and 100-m free-style swimming speed. Although sexes are combined in this study, the results illustrate the interrelatedness of maturity, size, functional capacity, and performance in a group of young swimmers who had a training history of five years.

Observations by Andrew *et al.* (1972) on young male competitive swimmers (8–18 years from swim clubs), over a 3-year period, indicated greater body size after 12 years of age. Similar results for 28 young male swimmers (10–16 years, representing swim teams) were reported by Cunningham and Eynon (1973). The young male swimmers did not differ in height relative to the standard, but were slightly heavier. In comparisons of eleven age 13-, twelve age 14-, and ten age 15-year-old swimmers to each other and to selected reference populations, Sobolova *et al.* (1971) found the athletes leaner, but not different in height and weight. Since the maturity status of the swimmers in these studies was not considered, it is difficult to draw firm conclusions from that data relative to the contribution of physical training to the growth of young swimmers.

In older swimmers (mean age 17.3 ± 1.4 years) Andersen and Magel (1970) reported observations similar to those noted for younger competitive swimmers. The ten champion swimmers were taller, heavier, and fatter (sum of three skinfolds) than nonathletic school boys of the same age. Maturity status of those older swimmers was not considered, but a study of 15 international-caliber male swimmers under 20 years of age showed no difference between chronological age and skeletal age (Bouchard and Malina, 1977). Even when the small sample of swimmers was divided into those above or below 19 years of age, the skeletal age of the younger athletes was only slightly in excess of their chronological age.

7.1.7. Cycling

Studying the effects of training for bicycle racing, Berg and Bjure (1974, see also Berg, 1972) followed a small group of 13 boys, 10–15 years old for 3 years. The heights and weights of the boys equaled the standard, but they had a low body fat content at the start of the study. It should be noted that the boys had been training 1–2 years prior to the study. After 3 years of training the boys showed only small increases in body cell mass and total body water relative to body weight and thus little fluctuation in relative body fatness. Since no control group was similarly followed, Berg and Bjure (1974) comment: " . . . it is not possible to state . . . if long-term changes observed here were actually due to training or to be expected at these different stages of puberty in all adolescent boys."

7.2. Young Female Athletes

Maturity and growth observations on young female athletes are not as extensive as those for young male athletes. In light of current trends, however, such data will become more available as young girls participate more in sports.

7.2.1. Swimming

Data for young female athletes are most available for competitive swimmers, beginning with the now classic study of girl swimmers by Astrand and his colleagues (1963). Of the 30 swimmers 39 had already attained menarche giving a mean age of 12.9 years (range 11.0 to 14.9 years). This average age was slightly earlier than the Swedish standards by about half a year. Similarly, Bugyi and Kausz (1970) found a small sample of the "best eight young Hungarian girl swimmers" significantly accelerated in skeletal age relative to chronological age. In a comparison of 34 young swimmers 10–12 years of age who reached the final of a national age-group competition and those who reached the semifinals, Bar-Or (1975) found that the finalists were somewhat more mature in the development of secondary sex characteristics (breasts and pubic hair) even though they were slightly younger chronologically. In contrast to the suggested earlier maturation in young female swimmers, data for 14 international-caliber swimmers under 20 years of age indicate a skeletal age delayed relative to chronological age (Bouchard and Malina, 1977).

Growth characteristics of the young female swimmers have been considered in detail by Astrand *et al.* (1963). As a group, the girls were taller than the Swedish standards and had normal weight for height. Examination of their school records indicated that the girls had been taller than average since 7 years of age. After age 12 their heights appeared to deviate (accelerate) more from the standards than before this age. Mean deviation in height for age expressed as a fraction of the standard deviation was +0.61 ± 0.81 at the time of the study, while the mean deviation at seven years of age was +0.37 ± 0.87. This acceleration, however, was not related to the intense swimming training. It is perhaps related to their somewhat earlier menarche: eight girls attained menarche between 11.0 and 11.9 years, seven girls in the 12th year, ten girls in the 13th year, and four girls in the 14th year. Menarche usually follows peak height velocity (Tanner, 1962), so that one might expect the girls to be somewhat taller.

Bar-Or (1975) reported greater height and lean body mass in young female swimming finalists compared to the semifinalists 10–12 years of age, while Andrew *et al.* (1972) observed greater body size after 12 years of age in competitive female swimmers, 8–18 years of age, over a 3-year period. On the other hand, the 19 competitive female swimmers (11–16 years of age), reported by Cunningham and Eynon (1973), did not differ from a height–weight reference standard. As in the case of the young boys in these studies, maturity status was not considered so it is difficult to draw firm conclusions relative to the effects of physical training on the growth of young swimmers.

7.2.2. Track

Observations on 158 young female track athletes are limited. Among girls 12–18 years of age enrolled in a camp for track and field athletes, Cumming *et al.* (1972) noted a composite skeletal age slightly delayed relative to the group's chronological age. Skeletal age did not predict performance in track and field in these girls as well as it did for boys. After controlling for body size, skeletal age was related to performance in only four of nine track and field events (partial r's ranged from 0.009 to 0.81). These results are generally similar to those on nonathlete adolescent girls reported by Espenschade (1940). Among these girls, performance, including several track and field-related items—50-yard dash and standing broad jump—was not related to skeletal age or the age at menarche. Later-maturing girls, however, tended to be the better performers.

7.2.3. Gymnastics

Focusing on body composition changes in 10 young female gymnasts (13–18 years) longitudinally over a 5-year period, Parizkova (1963, 1973) noted fluctuations in body fatness relative to the intensity of training. Body density also showed similar variation with intensity of training. Compared to seven girls who did not train regularly, the gymnasts had significantly less fat and more lean body mass after five years, even though the two groups did not differ in height and weight. These observations are thus similar to those observed in young males undergoing intensive physical training programs.

Maturity status of young female gymnasts was not considered in Parizkova's (1963, 1973) data. Other data for seven international-caliber female gymnasts under 20 years of age indicate a skeletal age delayed relative to chronological age (Bouchard and Malina, 1977).

7.3. Retrospective Studies of Menarche in Older Female Athletes

Maturity status of female athletes can also be inferred from retrospective studies of adult athletes regarding their reported ages at menarche. In such a study of American college-age track and field athletes, Malina *et al.* (1973) found a mean age of menarche of 13.58 ± 1.3 years, which was significantly later than that reported for nonathlete samples of American girls. Spirduso *et al.* (n.d.) also observed later reported ages at menarche among Olympic volleyball aspirants (14.2 ± 0.9 years) and college-level volleyball players (12.7 ± 1.11 years) compared to nonathlete college women. On the other hand, Erdelyi (1962) reported similar ages at menarche for Hungarian female athletes participating in a variety of events and the Hungarian population as a whole (13.6 years). Although not entirely conclusive, some of these data on reported ages at menarche by young adult female athletes and on skeletal maturity of international caliber athletes would seem to indicate an association between delayed maturation and successful athletic performance. This appears to be in contrast to somewhat earlier maturity observed in young swimmers (Astrand *et al.*, 1963; Bugyi and Kausz, 1970; Bar-Or, 1975), but in the same direction as that noted in Canadian girls at a track and field camp (Cumming *et al.*, 1972) and in Canadian figure skaters (Ross *et al.*, 1976). Note, however, that with the exception the series of girls of Astrand *et al.* (1963), which included several international record-holders and/or champions, the young female athletes in the other studies are primarily age-group and club competitors and not of national or international caliber.

7.4. Overview

Young athletes of both sexes grow as well as nonathletes, i.e., the experience of athletic training and competition does not have harmful effects on the physical growth and development of the youngster. The young trained athlete is also generally leaner, i.e., has a lesser percentage of body weight as fat. Maturity relationships are not consistent across sports. Young male athletes more often than not tend to be advanced maturationally compared to nonathletes. These differences seem more apparent in sports or positions within sports where size is a factor. Maturity data for young female athletes are variable. Young female swimmers appear to be maturationally slightly advanced, although not as markedly so as in the case of young male athletes in some sports. On the other hand, data derived from college-age female athletes (menarche) and from international-caliber female athletes (skeletal age) suggest that delayed maturity is related to successful performance in swimming, track and field, volleyball, and gymnastics. In light of such maturity differences within and between sports, it is difficult to draw firm conclusions about the effects of sports training and participation on the growth of the young athlete. Although there are numerous opinions suggesting that the larger size and/or optimal growth of the young athlete is due to training, few studies have really considered the complex interrelationships of chronological age, biological age (skeletal, menarche), and body size. Further, most of the data are derived from small cross-sectional samples, and the small amount of longitudinal data available is treated in a cross-sectional manner. This immediately emphasizes the need for longitudinal studies of young athletes and for appropriate controls.

There is also a need for follow-up studies of young athletes, centering on the question of the persistence of training-associated changes. Many changes in response to short-term training are generally not permanent and vary with the quantity of training. This is especially clear in the fluctuating levels of fatness in young gymnasts over five years (Parizkova, 1963, 1973). Fatness varied inversely with the quantity of training. The need for continued activity is strikingly evident in the follow-up study (Eriksson *et al.,* 1971) of the young female swimmers studied by Astrand *et al.* (1963) (see above).

Related to the persistence of activity-related changes are "clinical" reports on young athletes in certain sports, particularly baseball, football, and tennis. Epiphyseal injuries and epiphysitis in the adolescent athlete, although not extremely common, do represent a potential growth-influencing factor, and perhaps bring about an unevenness of growth. Fortunately, most of the epiphyseal injuries which occur in young athletes are amenable to medical treatment or correction. The risk of permanent damage is, nevertheless, always present. The data reported, however, are clinical cases, and growth of the skeletal element involved is usually not considered or followed after recuperation or repair. In spite of best medical care, epiphyseal injuries in the long bones of youngsters may give trouble with possible deformity. Further, "residual handicaps" (Larson, 1973) associated with childhood athletic injuries might influence the young athlete's subsequent activity habits in adulthood.

In summary, young athletes are comparable in growth status and progress to nonathletes. Maturity-associated variation in size and physique is a significant factor in comparing athletes and nonathletes, especially during the circumpubertal years. Data for a variety of sports are more readily available for young male athletes than they are for young female athletes. However, the data for young athletes are not really extensive, and some sports are notably omitted (e.g., wrestling and soccer).

8. Conclusion

This chapter has been an attempt to review and assess what is known about the effects of exercise and athletic endeavor on structural and functional growth. If some questions have been answered or are close to resolution, some others have received only partial answers—and new questions are arising constantly as investigators become increasingly interested in this complex area. However, data is being collected more rapidly than ever before, and long-overdue studies are now being undertaken. Hopefully, continuing research will provide answers to many of the questions posed in this review.

Abramson, A. S., 1948, Atrophy of disuse, *Arch. Phys. Med.* **29**:562–570.

Abramson, A. S., and Delagi, E., 1961, Influence of weight-bearing and muscle contraction on disuse osteoporosis, *Arch. Phys. Med. Rehab.* **42**:147–151.

Andersen, K. L., and Magel, J. R., 1970, Physiological adaptation to a high level of habitual physical activity during adolescence, *Int. Z. Angew. Physiol. Einsch. Arbeitsphysiol.* **28**:209–227.

Andrew, G. M., Becklake, M. R., Guleria, J. S., and Bates, D. V., 1972, Heart and lung functions in swimmers and nonathletes during growth, *J. Appl. Physiol.* **32**:245–251.

Asher, R. A. J., 1947, The dangers of going to bed, *Br. Med. J.* **2**:967–968.

Asmussen, E., 1973, Growth in muscular strength and power, in: *Physical Activity, Human Growth and Development* (G. L. Rarick, ed.), pp. 60–79, Academic Press, New York.

Astrand, P.-O., 1967, Commentary—Symposium on physical activity and cardiovascular health, *Can. Med. Assoc. J.* **96**:760.

Astrand, P.-O., Engstrom, L., Eriksson, B. O., Karlberg, P., Nylander, I., Saltin, B., and Thoren, C., 1963, Girl swimmers, *Acta Paediatr. Suppl.* 147.

Bailey, D. A., 1973, Exercise, fitness and physical education for the growing child—a concern, *Can. J. Public Health* **64**:421–430.

Bailey, D. A., 1976, The growing child and the need for physical activity, in: *Child in Sport and Physical Activity* (J. G. Albinson and G. M. Andrew, eds.), pp. 81–93, University Park Press, Baltimore.

Bailey, D. A., Bell, R. D., and Howarth, R. E., 1973, The effect of exercise on DNA and protein synthesis in skeletal muscle of growing rats, *Growth* **37**:323–331.

Barnard, R. J., Edgerton, V. R., and Peter, J. B., 1970a, Effect of exercise on skeletal muscle. I. Biochemical and histochemical properties, *J. Appl. Physiol.* **28**:762–766.

Barnard, R. J., Edgerton, V. R., and Peter, J. B., 1970b, Effect of exercise on skeletal muscle. II. Contractile properties, *J. Appl. Physiol.* **28**:767–770.

Bar-Or, O., 1975, Predicting athletic performance, *Physician Sport Med.* **3**:80–85.

Bar-Or, O., and Zwiren, L. D., 1972, Physiological effects of increased frequency of physical education classes and of endurance conditioning on 9–10 year old girls and boys, in: *Proc. 4th Int. Symp. Paediatr. Work Physiol.,* Wingate Institute, Israel.

Berg, K., 1972, Body composition and nutrition of adolescent boys training for bicycle racing, *Nutr. Metab.* **14**:172–180.

Berg, K., and Bjure, J., 1974, Preliminary results of long-term physical training of adolescent boys with respect to body composition, maximal oxygen uptake, and lung volume, *Acta Paediatr. Belg.* **28** (Suppl.):183–190.

Bouchard, C., and Malina, R. M., 1977, Skeletal maturity in a Pan American Canadian team, *Can. J. Appl. Sport Sci.* **2**:109–114.

Bouchard, C., and Roy, B., 1969, L'age osseux des jeunes participants due Tournoi International de Hockey Pee-Wee de Quebec, *Mouvement* **4**:225–232.

Brook, C. G. D., 1972, Evidence for a sensitive period in adipose-cell replication in man, *Lancet* **2**:624–627.

Buchanan, T. A. S., and Pritchard, J. J., 1970, DNA content of tibialis anterior of male and female white rats measured from birth to 50 weeks, *J. Anat.* **107**:185.

Bugyi, B., and Kausz, I., 1970, Radiographic determination of the skeletal age of young swimmers, *J. Sports Med. Phys. Fitness* **10**:269–270.

Buskirk, E. R., Andersen, K. L., and Brozek, J., 1956, Unilateral activity and bone and muscle development in the forearm, *Res. Q.* **27**:127–131.

Carron, A. V., and Bailey, D. A., 1974, Strength development in boys from 10 through 16 years, *Monogr. Soc. Res. Child Dev.* **39**(4), 37 pp.

Cheek, D. B., and Hill, D. E., 1970, Muscle and liver cell growth: Role of hormones and nutritional factors, *Fed. Proc.* **29**:1503–1509.

Cheek, D. B., Powell, G. K., and Scott, R. E., 1965, Growth of muscle cells (size and number) and liver DNA in rats and Snell Smith mice with insufficient pituitary, thyroid, or testicular function, *Bull. Johns Hopkins Hosp.* **117**:306–321.

Cheek, D. B., Holt, A. B., Hill, D. E., and Talbert, J. L., 1971, Skeletal muscle cell mass and growth: The concept of the deoxyribonucleic acid unit, *Pediatr. Res.* **5**:312–328.

Christensen, D. A., and Crampton, E. W., 1965, Effects of exercise and diet on nitrogenous constituents in several tissues of adult rats, *J. Nutr.* **86**:369.

Clarke, H. H., 1968, Characteristics of the young athlete: A longitudinal look, *Kinesiology Review, 1968,* pp. 33–42, American Association for Health, Physical Education and Recreation, Washington, D.C.

Clarke, H. H., 1971, *Physical and Motor Tests in the Medford Boys' Growth Study,* Prentice-Hall, Englewood Cliffs, New Jersey.

Clarke, H. H. (ed.), 1973, Characteristics of athletes, *Phys. Fitness Res. Digest* Series 3, No. 2 (April).

Cumming, G. R., 1976, The child in sport and physical activity—medical comment, in: *Child in Sport and Physical Activity* (J. G. Albinson and G. M. Andrew, eds.), pp. 67–77, University Park Press, Baltimore.

Cumming, G. R., Garand, T., and Borysyk, L., 1972, Correlation of performance in track and field events with bone age, *J. Pediatr.* **80:**970–973.

Cunningham, D. A., and Eynon, R. B., 1973, The working capacity of young competitive swimmers, 10–16 years of age, *Med. Sci. Sports* **5:**227–231.

Daniels, J., and Oldridge, N., 1971, Changes in oxygen consumption of young boys during growth and running training, *Med. Sci. Sports* **3:**161–165.

Denny-Brown, D., 1955, Degeneration, regeneration and growth of muscle. A symposium, *Am. J. Phys. Med.* **34:**210.

Dobeln, W. von, and Eriksson, B. O., 1972, Physical training, maximal oxygen uptake and dimensions of the oxygen transporting and metabolizing organs in boys 11–13 years of age, *Acta Paediatr. Scand.* **61:**653–660.

Donaldson, C. L., Hulley, S. B., Vogel, J. M., Hattner, R. S., Bayers, J. H., and McMillan, D. E., 1970, Effect of prolonged bed rest on bone mineral, *Metabolism* **19:**1071–1084.

Durante, G., cited by Adams, R., 1975, in: *Diseases of Muscle,* p. 215, Harper and Row, New York.

Edgerton, V. R., Gerchman, L., and Carrow, R., 1969, Histochemical changes in rat skeletal muscle after exercise, *Exp. Neurol.* **24:**110–123.

Ekblom, B., 1970, Physical training in normal boys in adolescence, *Acta. Paediatr. Scand.* **217:**60.

Elliott, G. M., 1970, The effects of exercise on structural growth. A review, *J. Can. Assoc. HPER* **36:**21.

Erdelyi, G. J., 1962, Gynecological survey of female athletes, *J. Sports Med. Phys. Fitness* **2:**174–179.

Eriksson, B. O., Engstrom, I., Karlberg, P., Saltin, B., and Thoren, C., 1971, A physiological analysis of former girl swimmers, *Acta Paediatr. Scand. Suppl.* **217:**68–72.

Espenschade, A., 1940, Motor performance in adolescence, *Monogr. Soc. Res. Child Dev.* **5**(24):1–126.

Espenschade, A., 1960, The contributions of physical activity in growth, *Res. Q.* **31:**351–364.

Espenschade, A. S., and Eckert, H. M., 1967, *Motor Development,* Charles E. Merrill, Columbus, Ohio.

Espenschade, A., and Eckert, H. M., 1974, Motor development, in: *Science and Medicine of Exercise and Sport,* 2nd ed. (W. R. Johnson and E. R. Buskirk, eds.), p. 322, Harper, New York.

Evans, F. G., 1957, *Stress and Strain in Bones,* Charles C Thomas, Springfield, Illinois.

Gillespie, J. A., 1954, The nature of the bone changes associated with nerve injuries and disuse, *J. Bone Joint Surg.* **36B:**464–472.

Godin, P., 1920, *Growth during School Age* (translated by S. L. Eby), Gorham Press, Boston.

Goldspink, G., 1964, The combined effects of exercise and reduced food intake on skeletal muscle fibers, *J. Cell Comp. Physiol.* **63:**209.

Gordon, E. E., Kowalski, K., and Fritts, M., 1966, Muscle proteins and DNA in rat quadriceps during growth, *Am. J. Physiol.* **210**(5):1033.

Gordon, E. E., Kowalski, K., and Fritts, M., 1967a, Adaptations of muscle to various exercise, *J. Am. Med. Assoc.* **199:**103.

Gordon, E. E., Kowalski, K., and Fritts, M., 1967b, Protein changes in quadriceps muscle of rat with repetitive exercise, *Arch. Phys. Med.* **48:**296.

Goss, R. J., 1966, Hypertrophy versus hyperplasia, *Science* **153:**1615.

Hale, C. J., 1956, Physiologic maturity of Little League baseball players, *Res. Q.* **27:**276–284.

Haylor, L. R., 1975, The influence of exercise on DNA and protein synthesis in the gastrocnemius muscle of growing rats, pp. 16–20, Unpublished M. Sc. thesis, University of Saskatchewan, Saskatoon, Canada.

Helander, E. A. S., 1961, Influence of exercise and restricted activity on the protein composition of skeletal muscle, *Biochem. J.* **78:**478–482.

Helela, T., 1969, Variations in thickness of cortical bones in two populations, *Ann. Clin. Res.* **1:**227–231.

Hettinger, T., 1960, Der einfluss des testosterons auf muskulatur und kreislaud, *Med. Mitt. Schering.* **21**(4):2.

Holmes, R., and Rasch, P. J., 1958, Effect of exercise on number of myofibrils per fiber in sartorius muscle of the rat, *Am. J. Physiol.* **195:**50.

Houston, C. S., 1976, cited in: *Child in Sport and Physical Activity* (J. G. Albinson and G. M. Andrew, eds.), p. 82, University Park Press, Baltimore.

Houston, C. S., 1978. The radiologist's opportunity to teach bone dynamics. *J. Can. Assoc. Radiol.* **29:** (in press).

Howell, J. A., 1917, An experimental study of the effect of stress and strain on bone development, *Anat. Rec.* **13**:233–253.

Howell, M. L., Loiselle, D. S., and Lucus, W. G., 1965, *Strength of Edmonton School Children,* University of Alberta, Edmonton, Canada.

Hubbard, R. W., Smoake, J. A., Matther, W. T., Linduska, J. D., and Bowers, W. S., 1974, The effects of growth and endurance training on the protein and DNA content of rat soleus, plantarius, and gastrocnemius muscles, *Growth* **38**:171.

Ikai, M., 1967, Trainability of muscular endurance as related to age, in: *Proceedings of ICHPER 10th Int. Congr.,* Vancouver, B.C., Canada, 1968.

Issekutz, B., Jr., Blizzard, J. J., Birkhead, N. C., and Rodahl, K., 1966, Effect of prolonged bed rest in urinary calcium output, *J. Appl. Physiol.* **21**:1013–1020.

Jeffress, R. N., and Peter, J. B., 1970, Adaptation of skeletal muscle to overloading—a review, *Bull. L. A. Neurol. Soc.* **35**:134–144.

Jones, H. E., 1947, Sex differences in physical abilities, *Hum. Biol.* **19**:12–25.

Jones, H. E., 1949, *Motor Performance and Growth,* University of California Press, Berkeley.

Kato, S., and Ishiko, T., 1966, Obstructed growth of children's bones due to excessive labor in remote corners, in: *Proceedings of International Congress of Sport Sciences* (K. Kato, ed.), p. 479, Japanese Union of Sports Sciences, Tokyo.

Kharmosh, O., and Saville, P. D., 1965, The effect of motor denervation on muscle and bone in the rabbit's hind limb, *Acta. Orthop. Scand.* **36**:361–370.

Klebanova, E. A., 1959, Concerning the enlargement of the fiber numbers of skeletal muscle during hypertrophy, cited in Chiakulus, J. J., and Pauly, J. E., 1965, A study of post natal growth of skeletal muscle in the rat, *Anat. Rec.* **152**:55.

Kleeberger, R., 1932, The effects of excessive activity on muscle tissue, cited in: *Toward an Understanding of Health and Physical Education* (A. H. Steinhaus), 1963, pp. 87–88, Wm. C. Brown, Dubuque, Iowa.

Klissouras, V., 1971, Heritability of adaptive variation, *J. Appl. Physiol.* **31**:338–344.

Klissouras, V., 1972, Genetic limit of functional adaptability, *Int. Z. Angew. Physiol. Einschl. Arbeitsphysiol.* **30**:85–94.

Klissouras, V., Pirnay, F., and Petit, J.-M., 1972, Adaptation to maximal effort: Genetics and age, *J. Appl. Physiol.* **35**:288–293.

Knittle, J. L., 1972, Obesity in childhood: A problem in adipose tissue cellular development, *J. Pediatr.* **81**:1048–1059.

Kottke, F. J., 1966, The effects of limitation of activity upon the human body, *J. Am. Med. Assoc.* **196**:825–830.

Krogman, W. M., 1959, Maturation age of 55 boys in the Little League World Series, 1957, *Res. Q.* **30**:54–56.

Lamb, D. R., 1968, Influence of exercise on bone growth and metabolism, *Kinesiology Review, 1968,* pp. 43–48, American Association for Health, Physical Education and Recreation, Washington, D.C.

Lamb, D. R., Van Huss, W. D., Carrow, R. E., Heusner, W. W., Weber, J. C., and Kertzer, R., 1969, Effects of prepubertal physical training on growth, voluntary exercise, cholesterol and basal metabolism in rats, *Res. Q.* **40**:123–133.

Larson, R. L., 1973, Physical activity and the growth and development of bone and joint structures, in: *Physical Activity: Human Growth and Development* (G. L. Rarick, ed.), pp. 32–59, Academic Press, New York.

Leaf, A., 1973, Getting old, *Sci. Am.* **3**:45, 229.

Malina, R. M., 1969a, Exercise as an influence upon growth, *Clin. Pediatr.* **8**:16–26.

Malina, R. M., 1969b, Quantification of fat, muscle and bone in man, *Clin. Orthop.* **65**:9–38.

Malina, R. M., 1974, Adolescent changes in size, build, composition and performance, *Hum. Biol.* **46**:117–131.

Malina, R. M., 1975a, Exercise and growth, *Na'Pao, A Saskatchewan Anthropol. J.* **5**:3–8.

Malina, R. M., 1975b, Anthropometric correlates of strength and motor performance, *Exercise Sport Sci. Rev.* **3**:249–274.

Malina, R. M., and Rarick, G. L., 1973, Growth, physique and motor performance, in: *Physical Activity: Human Growth and Development* (G. L. Rarick, ed.), pp. 125–153, Academic Press, New York.

Malina, R. M., Harper, A. B., Avent, H. H., and Campbell, D. E., 1973, Age at menarche in athletes and non-athletes, *Med. Sci. Sports* **5**:11–13.

Malina, R. M., Spirduso, W. W., Tate, C., and Baylor, A. M., 1978, Menarche in athletes at different competitive levels and in different sports, *Med. Sci. and Sports* (in press).

Mocellin, R., and Wasmund, U., 1972, Investigations on the influence of a running-training programme or the cardio-vascular and motor performance capacity in 8 to 9 year old boys and girls, in: *Proc. 4th Int. Symp. Paediatr. Work Physiol.,* Wingate Institute, Israel.

Moody, D. L., Wilmore, J. H., Girandola, R. N., and Royce, J. P., 1972, The effects of a jogging program on the body composition of normal and obese high school girls, *Med. Sci. Sports* **4:**210–213.

Morpurgo, B., 1895, On the nature of functional hypertrophy of voluntary muscle, *Arch. Sci. Med.* **19**(16)327.

Morpurgo, B., 1897, Veber activats hypergrophic den wilkurlichen muskeln, *Virchows Arch.* **150:**552.

Nilsson, B. E., and Westlin, N. E., 1971, Bone density in athletes, *Clin. Orthop.* **77:**179–182.

Novak, L. P., 1965, Body composition and clinical estimation of desirable body weight, Proceedings of the Seventh National Conference on the Medical Aspects of Sports, American Medical Association Philadelphia.

Oscai, L. B., Babirak, S. P., McGarr, J. A., and Spirakis, C. N., 1974, Effect of exercise on adipose tissue cellularity, *Fed. Proc.* **33:**1956–1958.

Parizkova, J., 1963, Impact of age, diet, and exercise on man's body composition, *Ann. N.Y. Acad. Sci.* **110:**661–674.

Parizkova, J., 1968*a,* Longitudinal study of the development of body composition and body build in boys of various physical activity, *Hum. Biol.* **40:**212–225.

Parizkova, J., 1968*b,* Body composition and physical fitness, *Curr. Anthropol.* **9:**273–287.

Parizkova, J., 1970, Longitudinal study of the relationship between body composition and anthropometric characteristics in boys during growth and development, *Glas. Antropol. Drus. Jugosl.* **7:**33–38.

Parizkova, J., 1973, Body composition and exercise during growth and development, in: *Physical Activity: Human Growth and Development,* (G. L. Rarick, ed.), pp. 97–124, Academic Press, New York.

Parizkova, J., 1974, Particularities of lean body mass and fat development in growing boys as related to their motor activity, *Acta Paediatr. Belg.* **28**(Suppl.):233–243.

Parizkova, J., and Carter, J. E. L., 1976, Influence of physical activity on stability of somatotypes in boys, *Am. J. Phys. Anthropol.* **44:**327.

Plum, F., and Dunning, M. F., 1958, The effect of therapeutic mobilization on the hypercalciuria following acute poliomyelitis, *Arch. Intern. Med.* **101:**528–536.

Prives, M. G., 1960, Influence of labor and sport upon skeletal structure in man, *Anat. Rec.* **136:**261.

Rarick, G. L., 1960, Exercise and growth in: *Science of Medicine of Exercise and Sports* (W. R. Johnson, ed.), pp. 440–465, Harpers, New York.

Rarick, G. L., 1973, Competitive sports in childhood and early adolescence, in: *Physical Activity: Human Growth and Development* (G. L. Rarick, ed.), pp. 364–386, Academic Press, New York.

Rarick, G. L., 1960, Exercise and growth, in: *Science of Medicine of Exercise and Sports* (W. R. Johnson, ed.), pp. 440–465, Harpers, New York.

Rarick, G. L., and Oyster, N., 1964, Physical maturity, muscular strength and motor performance of young school age boys, *Res. Q.* **35:**523–531.

Rarick, G. L., and Smoll, F. L., 1967, Stability of growth in strength and motor performance from childhood to adolescence, *Hum. Biol.* **B9:**295–306.

Rohmert, W., 1968, Rechts-Links- Verlich bei isometrischem Armmuskeltraining mit verschiedenem Trainingereiz bei achtzahrigen Kindern, *Int. Z. Angew. Physiol. Einschl. Arbeitsphysiol.* **26:**363.

Ross, W. D., Brown, S. R., Faulkner, R. A., Vajda, A., and Savage, M. V., 1976, Monitoring growth in young skaters, *Can. J. Appl. Sport Sci.* **1:**163–167.

Rowe, F. A., 1933, Growth comparisons of athletes and non-athletes, *Res. Q.* **4:**108–116.

Saltin, B., and Grimby, G., 1968, Physiological analysis of middle-age and old former athletes: Comparison with still active athletes of the same ages, *Circulation* **38:**1104.

Saville, P. D., and Smith, R., 1966, Bone density, breaking force and leg muscle mass as functions of weight in bipedal rats, *Am. J. Phys. Anthropol.* **25:**35–39.

Saville, P. D., and Whyte, M. P., 1969, Muscle and bone hypertrophy: Positive effect of running exercise in the rat, *Clin. Orthop.* **65:**81–88.

Schnorr, P., 1971, Longevity and causes of death in male athletic champions, *Lancet* **1:**1364.

Shuck, G. R., 1962, Effects of athletic competition on the growth and development of junior high school boys, *Res. Q.* **33:**288–298.

Siebert, W. W., 1928, Investigations on hypertrophy of the skeletal muscle, reprinted from *Z. Clin. Med.* **109:**350, in: *Classical Studies on Physical Activity,* 1968 (R. C. Brown and G. S. Kenyon, eds.) pp. 229–236, Prentice-Hall, Englewood Cliffs, New Jersey.

Skinner, J. S., 1968, Longevity, general health, and exercise, in: *Exercise Physiology* (H. B. Falls, ed.), pp. 219–238, Academic Press, New York.

Sobolova, V., Seliger, V., Grussova, D., Machovcoca, J., and Zelenka, V., 1971, The influence of age and sports training in swimming on physical fitness, *Acta Paediatr. Scand. Suppl.* **217:**63–67.

Sprynarova, S., 1974, Longitudinal study of the influence of different physical activity programs on functional capacity of the boys from 11 to 18 years, *Acta Paediatr. Belg.* **28**(Suppl.):204–213.

Steinhaus, A. H., 1933, Chronic effects of exercise, *Physiol. Rev.* **13:**103–147.

Steinhaus, A. H., 1955, Strength from Morpurgo to Muller—a half century of research, *J. Assoc. Phys. Ment. Rehab.* **9:**147–150.

Stolz, H. R., and Stolz, L. M., 1951, *Somatic Development of Adolescent Boys,* Macmillan, New York.

Szabo, S., Doka, J., Apor, P., and Somogyvari, K., 1972, Die Beziehung zwischen Knochenlebensalter, funktionellen anthropometrischen Daten und der aeroben Kapazitat, *Schweiz. Z. Sportmed.* **20:**109–115.

Szabo-Wahlstab, S., Doka, J., Apor, P., and Somogyvari, K., 1973, Metacarpal age, anthropometric and functional anthropometric measurements and aerobic capacity, in: *Physical Fitness* (V. Seliger, ed.), pp. 387–390, Charles University, Prague.

Tanner, J. M., 1953, Growth and constitution, in: *Anthropology Today* (A. L. Kroeber, ed.), pp. 750–770, University of Chicago Press, Chicago.

Tanner, J. M., 1962, *Growth at Adolescence,* 2nd ed., Blackwell, Oxford.

Tanner, J. M., 1965, Radiographic studies of body composition in children and adults, in: *Human Body Composition* (J. Brozek, ed.), pp. 211–236, Pergamon, Oxford.

Tower, S. S., 1939, Reaction to muscle denervation, *Physiol. Rev.* **19:**1–48.

Trueta, J., 1968, *Studies of the Development and Decay of the Human Frame,* W. B. Saunders, Philadelphia.

Vose, G. P., 1974, Review of roentgenographic bone demineralization studies of the Gemini space flights, *Am. J. Roentgenol.* **121:**1–4.

Watson, R. C., 1973, Bone Growth and Physical Activity in Young Males, Unpublished doctoral dissertation, University of Wisconsin, Madison.

Weber, G., Kartodihardjo, W., and Klissouras, V., 1976, Growth and physical training with reference to heredity, *J. Appl. Physiol.* **40:**211–215.

Wells, J. B., Parizkova, J., and Jokl, E., 1962, The Kentucky physical fitness experiment, *J. Assoc. Phys. Ment. Rehab.* **16:**69–72.

Wells, J. B., Parizkova, J., Bohanan, J., and Jokl, E., 1963, Growth, body composition and physical efficiency, *J. Assoc. Phys. Ment. Rehab.* **17:**37–40; 56–57.

Wyse, D., and Pattee, C. J., 1954, Effect of the oscillating bed and tilt table on calcium, phosphorus and nitrogen metabolism in paraplegia, *Am. J. Med.* **17:**645–661.

18

The Low-Birth-Weight Infant

FREDERICK C. BATTAGLIA and
MICHAEL A. SIMMONS

1. Definitions

1.1. Birth before the End of a Normal Gestation (Preterm) and after an Abnormal Rate of Intrauterine Growth (Small-for-Gestational Age)

Low-birth-weight infants (< 2500 g) may result from pregnancies terminating before the completion of a normal gestational period (preterm infant) or from pregnancies during which the rate of intrauterine growth is abnormally slow, regardless of the duration of the gestation (SGA or small-for-gestational-age infant). Ever since the World Health Organization (1950) defined prematurity on the basis of birth weight (< 2500 g) alone, there has been slow but increasing recognition that a significant percentage of low-birth-weight infants are due to a decreased intrauterine growth rate rather than preterm delivery (Gruenwald, 1963).

The importance of distinguishing between a preterm infant and an SGA infant is reflected not only in the biological implications for interpreting various phenomena of intrauterine growth and development but also in planning specific clinical management. The obstetrician confronted with an infant of low birth weight can anticipate neonatal survival rates which vary significantly with gestational age (Figure 1). An infant weighing 1000 g has a much lower mortality rate if he is a 34-week SGA than if he is 28 weeks and of appropriate size for gestational age (AGA). Recognizing that a given infant has intrauterine growth retardation (IUGR) may alter obstetrical planning for the route of delivery, since the incidence of asphyxia and intrapartum abnormalities is higher among SGA infants than their AGA peers, particularly at term (Low *et al.*, 1972).

FREDERICK C. BATTAGLIA • Department of Pediatrics, University of Colorado School of Medicine, Denver, Colorado. *MICHAEL A. SIMMONS* • Departments of Pediatrics and Obstetrics, The Johns Hopkins University School of Medicine, Baltimore, Maryland.

FREDERICK C. BATTAGLIA AND MICHAEL A. SIMMONS

Recent advances in obstetrical practice have occurred partly in response to the necessity for a more accurate estimate of both fetal weight and gestational age. The reliability of the estimate of gestational age made from the menstrual history is no longer dismissed out of hand when uterine size is repeatedly smaller than anticipated from the gestation age. In fact, it has been shown that maternal menstrual histories form rather reliable clinical data under most circumstances (Treloar *et al.*, 1967). A widely used neonatal mortality risk prediction scale relies exclusively on maternal dates for gestational age assignment (Lubchenco *et al.*, 1972). Recently, the most frequently used assessment of uterine size has been done using McDonald measurements (McDonald, 1910). Routine obstetrical visits should confirm dates of conception (using menstrual data, onset of quickening, and first auscultation of fetal heart tones) and record graphically gestational age vs. uterine size (by McDonald measurement). Additional information about fetal size and growth rate may be obtained utilizing ultrasound measurements of biparietal diameter (Campbell and

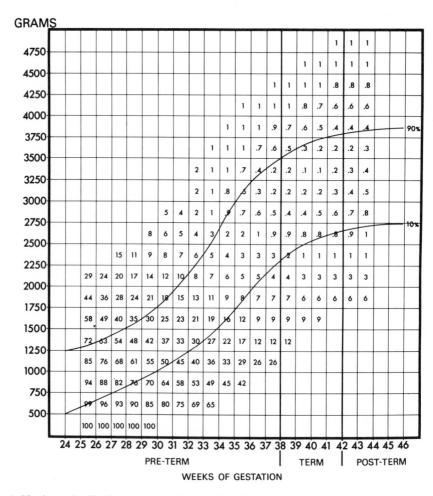

Fig. 1. Newborn classification and neonatal mortality risk by birth weight and gestational age. From Lubchenco *et al.* (1972). Interpolated data based on mathematical fit from original data; University of Colorado Medical Center newborns, July 1, 1958–July 1, 1969.

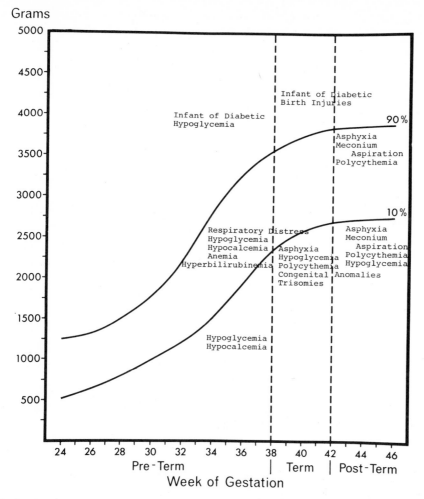

Fig. 2. Clustering of various clinical problems by birth weight and gestational age distribution.

Dewhurst, 1971). Various chemical and histological tests on amniotic fluid have been used to predict fetal maturity, but only amniotic fluid creatinine (Pitkin and Zwirek, 1967) and lecithin/sphingomyelin ratios (Gluck *et al.*, 1971) have received widespread acceptance.

It is equally important for the pediatrician to distinguish between a preterm infant and a SGA infant. Not only does mortality risk vary (Figure 1) but specific clinical problems can be anticipated by an infant's birth weight/gestational age distribution. Figure 2 outlines certain conditions that might be anticipated on this basis alone. Figure 3 demonstrates how different birth weight/gestational age groups have markedly different incidences of neonatal hypoglycemia, defined as blood glucose less than 30 mg/dl. An incidence of hypoglycemia of 67% among preterm, SGA infants demands the routine administration of parenteral glucose to such infants. Furthermore, a recent study has shown that some SGA infants, instead of developing hypoglycemia, may develop marked elevations in blood glucose producing nonketotic hyperosmolality of the plasma (Dweck and Cassady, 1974). Hypo-

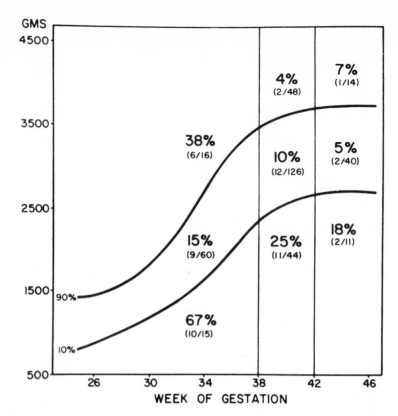

Fig. 3. Incidence of hypoglycemia in newborn infants, classified by birth weight and gestational age. Glucose levels < 30 mg/100 ml prior to first feeding. From Lubchenco and Bard (1971).

glycemia may be particularly severe in SGA infants who also have polycythemia and hyperviscosity. Hyperviscosity may lead to reduced cerebral perfusion, further complicating the effect of a reduced arterial glucose concentration.

1.2. Methods of Distinguishing Preterm from SGA

1.2.1. Physical

A definitive gestational age assessment is important in view of the occasional unreliability of obstetrical dates. Thus, various techniques for the clinical prediction of gestational age have been developed (Dubowitz *et al.*, 1970; Lubchenco, 1970; Usher, 1970; Robinson, 1966). We have used the methodology described by Lubchenco (1970) with some recent minor modifications. Tables I and II outline the various physical and neurological parameters to be expected at given gestational ages. A data sheet is prepared on each infant which is an exact facsimile of Tables I and II to assess each infant's gestational age. The data sheet is included in each patient's chart. Certain cautions must be observed in order to make this form of gestational age assessment reproducible and reliable. The physical variables (Table I) should be assessed within the first 4 hr of life, while the neurological assessment (Table II) is best delayed until 12–24 hr of life. The neurological variables are undependable in very ill infants since they rely on motor tone which may be

Table I. Clinical Estimation of Gestational Age, An Approximation Based on Published Data[a]

Examination first hours

Weeks gestation: 20 21 22 23 24 25 26 27 28 29 30 31 32 33 34 35 36 37 38 39 40 41 42 43 44 45 46 47 48

Physical findings		Description across weeks gestation
Vermix		Appears (21). Covers body, thick layer (24–37). On back, scalp, in creases (38). Scant, in creases (40–41). No vermix (44–48)
Breast tissue and areola		Areola & nipple barely visible, no palpable breast tissue (24–28). Areola raised (34). 1–2 mm nodule (36). 3–5 mm (38); 5–6 mm (39). 7–10 mm (41–42). ?12 mm (46)
Ear	Form	Beginning incurving superior (35). Incurving upper 2/3 Pinnae (37). Well-defined incurving to lobe (43–44)
	Cartilage	Pinna soft, stays folded (24–28). Cartilage scant returns slowly from folding (33–34). Thin cartilage springs back from folding (37–38). Pinna firm, remains erect from head (43)
Sole creases		Smooth soles & creases (24–27). 1–2 anterior creases (33). 2–3 anterior creases (35). Creases anterior 2/3 sole (36). Creases involving heel (39–40). Deeper creases over entire sole (43–47)
Skin	Thickness & appearance	Thin, translucent skin, plethoric, venules over abdomen, edema (24–28). Smooth, thicker, no edema (34). Pink (37). Few vessels (39). Some desquamation, pale pink (40–41). Thick, pale, desquamation over entire body (43–47)
	Nail plates	Appear (20). Nails to finger tips (37). Nails extend well beyond finger tips (46)
Hair		Appears on head (20). Eye brows & lashes (25–26). Fine, woolly, bunches out from head (31–33). Silky, single strands, lays flat (38). ?Receding hairline or loss of baby hair, short, fine underneath (43–46)
Lanugo		Appears (20). Covers entire body (25–27). Vanishes from face (34–35). Present on shoulders (40). No lanugo (45)
Genitalia tests	Scrotum	Testes palpable in inguinal canal (30–33). In upper scrotum (37–38). In lower scrotum (44). Few rugae (32–34). Rugae, anterior portion (37). Rugae cover (40–41). Pendulous (44)
	Labia & clitoris	Prominent clitoris, labia major small, widely separated (32–34). Labia majora larger, nearly cover clitoris (37). Labia minora & clitoris covered (43)
Skull firmness		Bones are soft (24–25). Soft to 1" from anterior fontanelle (30–31). Spongy at edges of fontanelle, center firm (36). Bones hard, sutures easily displaced (40). Bones hard, cannot be displaced (44)
Posture	Resting	Hypotonic, lateral decubitus (22–24). Hypotonic (27–28). Beginning flexion thigh (31). Stronger hip flexion (33). Frog-like (35). Flexion all limbs (37). Hypertonic (39–40). Very hypertonic (45)
Recoil–Leg		No recoil (26). Partial recoil (35). Prompt recoil (43)
Arm		No recoil (27–28). Begin flexion, no recoil (34). Prompt recoil may be inhibited (38–39). Prompt recoil after 30″ inhibition (45)
Horizontal positions		Hypotonic, arms & legs straight (30–32). Arms & legs flexed (36). Head & back even, flexed extremities (38–39). Head above back (44)

[a] From L.O. Lubchenco, in press, *The High-Risk Infant*, W. B. Saunders, Philadelphia, with permission.

FREDERICK C.
BATTAGLIA AND
MICHAEL A. SIMMONS

Table II. Clinical Estimation of Gestational Age, An Approximation Based on Published Data[a]

Confirmatory neurologic examination to be done after 24 hours

	Physical findings	Weeks gestation (20–48)
Tone	Heel to ear	No resistance (24–28); Some resistance (31); Impossible (35)
	Scarf sign	No resistance (24–25); Elbow passes midline (33); Elbow at midline (37); Elbow does not reach midline (44)
	Neck flexors (head lag)	Absent (27)
	Neck extensors	Head begins to right itself from flexed position (34); Good righting cannot hold it (36–37); Holds head few seconds (39); Head in plane of body (40); Keeps head in line c̄ trunk > 40' (41–42); Holds head (45); Turns head from side to side (46)
	Body extensors	Straightening of legs (34); Straightening of trunk (37–38); Straightening of head & trunk together (44–45)
	Vertical positions	When held under arms, body slips through hands (30); Arms hold baby legs extended? (35); Legs flexed good support c̄ arms (42)
Flexion angles	Popliteal	No resistance (24); 150° (30); 110° (33); 100° (36); 90° (39); 80° (45)
	Ankle	45° (34); 20° (38); 0 (41); A pre-term who has reached 40 weeks still has a 40° angle
	Wrist (square window)	90° (30); 60° (34); 45° (37); 30° (39); 0 (41)
Reflexes	Sucking	Weak not synchronized c̄ swallowing (27); Stronger synchronized (33); Perfect (36)
	Rooting	Long latency period flow, imperfect (28); Hand to mouth (32); Brisk, complete, durable (37); Perfect hand to mouth (40)
	Grasp	Finger grasp is good strength is poor (29); Stronger (35); Can lift baby off bed involves arms (40); Hands open (46)
	Moro	Barely apparent (23); Weak not elicited every time (27); Stronger (33); Complete c̄ arm extension open fingers, cry (35); Arm adduction added (40); Begins to lose moro (47)
	Crossed extension	Flexion & extension in a random, purposeless pattern (27); Extension but no adduction (32); Still incomplete (35); Complete (45)
	Automatic walk	Minimal (31); Begins tiptoeing good support on sole (33); Fast tiptoeing (37); Heel-toe progression whole sole of foot (40); A pre-term who has reached 40 weeks walks on toes (44); ? Begins to lose automatic walk (47)
	Pupillary reflex	Absent (25); Appears (30); Present
	Glabellar tap	Absent (24); Appears (33); Present
	Tonic neck reflex	Appears (36); Present after 37 weeks
	Neck-righting	Absent (28); Appears (34); Present

Week columns: 20 21 22 23 24 25 26 27 28 29 30 31 32 33 34 35 36 37 38 39 40 41 42 43 44 45 46 47 48

[a] From L.O. Lubchenco, in press, *The High-Risk Infant*, W.B. Saunders, Philadelphia, with permission.

abnormal in any sick infant. We have found no truly reliable signs to assess gestational age under 28 weeks. At any other gestational age, the reliability of clinical estimates is no greater than ±2 weeks. Therefore adjustment of obstetrical dates should be avoided if the discrepancy between the obstetrical dates and clinical dates is within this error range.

1.2.2. Neurological

The assessment of gestational age using both physical and neurological variables remains a subjective one, despite efforts to quantitate this process (Dubowitz *et al.*, 1970). Interobserver error can be significant even if a specific system is used regularly. However, the appearance of a few neurological responses are sufficiently reliable to be generally useful. The appearance of pupillary reaction to light at about 30 weeks, the glabellar tap reflex at about 33 weeks, and the traction response at about 34 weeks are all useful and reliable landmarks. Nevertheless, even with these few consistent responses, there may be a variability of ±2 weeks, and no reliable neurological patterns of development exist to assess gestational age less than 28 weeks.

In the context of these limitations, there has been great interest in developing objective measures of gestational age, particularly as they relate to neurological maturation. The developmental patterns on serial electroencephalographs and motor-nerve conduction velocities have both been used to assess gestational age.

EEG patterns, in particular comparison of sleep vs. arousal states, have been used to predict gestational age (A. H. Parmalee, Jr., *et al.*, 1970; A. H. Parmalee, 1967, 1968; Dreyfus-Brisac, 1965), but the difficulty in interpretation and the lack of precision have limited the usefulness of this technique. The evaluation of the electroencephalograms of immature infants is very difficult and requires unique expertise.

Although the measurement of motor-nerve conduction time in small infants may be technically difficult, these measurements are objective and easily quantified. Motor-nerve conduction velocity has been shown to be highly correlated with gestational age, with little influence of birth weight or growth retardation (Blom and Finnström, 1968; Schulte *et al.*, 1968).

Figure 4, taken from Schulte *et al.* (1968), demonstrates the close correlation of gestational age and motor-nerve conduction velocity. This technique is currently the most reliable objective method for documenting gestational age, even at ages less than 30 weeks, but precision is still within ±1–2 weeks.

The striking correlation of both EEG patterns and motor nerve conduction time with gestational age would support the observations of Saint-Anne Dargassies that the rate of neurological maturation is minimally affected by extrauterine influences (i.e., maternal diseases, nutrition) or by the rate of intrauterine somatic growth (Saint-Anne Dargassies, 1955). Recent reports that stressful pregnancies may accelerate neonatal functional maturity (Gluck and Kulovich, 1973) must be regarded as speculative until careful quantification of neurological maturity under such conditions is provided. At the present time, the rate of objective neurological maturation in infants appears to be a remarkable biological constant.

Finally, a word of caution should be introduced regarding *all* attempts to equate aspects of maturation with chronological age. Such attempts can only be crude approximations of the true chronologic age of a patient, regardless of whether one is assessing CNS or pulmonary function. Somehow, in considering fetal life,

FREDERICK C.
BATTAGLIA AND
MICHAEL A. SIMMON

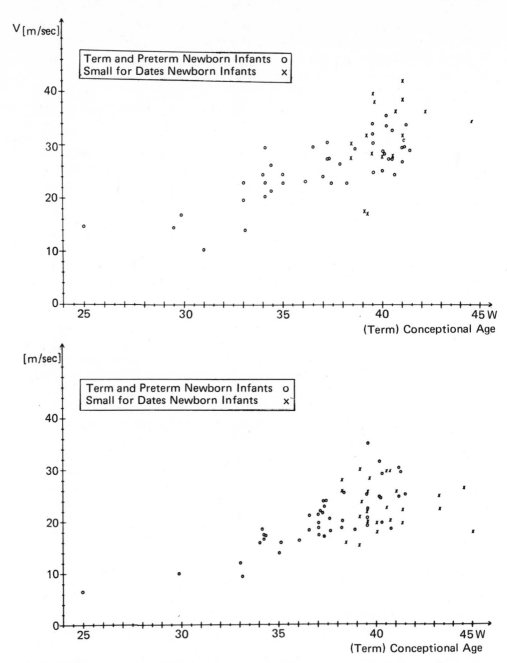

Fig. 4. Ulnar (top) and tibial (bottom) motor nerve conduction velocity increases with conceptional age. From Schulte *et al*. (1968).

this obvious fact is often overlooked although it is well accepted about later childhood. If one child reaches the onset of menarche at age 12 and another at age 15, we do not assume that their birth dates are in error but rather recognize the fact that maturation of the same organ system occurs at different chronological age in children. Too frequently tests are compared against the estimated gestational age of the fetus and the closer the agreement with the gestational age the more "precise" the test is assumed to be. Only when conception can be dated with precision will we be able to determine the chronological age of a fetus precisely and view that age against a varying spectrum of organ-system maturation.

1.2.3. Laboratory

Before beginning a discussion of some of the laboratory tests currently in use to assess fetal maturity, it is important to reemphasize that these evaluations also must be considered against a backdrop which is not the chronological age of the fetus but rather the maturational level of the particular organ system the test reflects.

As outlined previously, several laboratory measurements have been employed to assist in the estimation of fetal maturity. Amniotic fluid assayed for estrogen (Alleem *et al.,* 1969; Bolognese *et al.,* 1971), human lactogen (Teoh *et al.,* 1971), osmolality (Miles and Pearson, 1969; Mattison, 1970; O'Leary and Feldman, 1970), cytologic studies with Nile blue sulfate (Bishop and Corson, 1968; Sharma and Trussel, 1970), creatinine (Pitkin and Zwirek, 1967), lecithin/sphingomyelin ratio (Gluck *et al.,* 1971), protein and prealbumin (Touchstone *et al.,* 1972), and alpha-fetoprotein have been used to assess various aspects of fetal maturity (Seppala and Ruoslakti, 1972). For obvious reasons, few of these tests can be applied to newborn infants.

In the past radiographic methods were used for predicting fetal maturity, and these could also be used in the newborn infant. Besides the hazards of radiation, these methods are neither reliable nor sensitive measures of maturity. Distal femoral epiphyses have been used as an index of maturity (usually gestational age > 36 weeks) since the early 1950s (Christie, 1949). Utilizing the presence of both proximal tibial epiphyses and distal femoral epiphyses, Schreiber showed an excellent correlation between the presence of epiphyseal centers and the incidence of gestational age greater than 37 weeks. Still, several infants who lacked these centers were mature by other criteria (Schreiber *et al.,* 1962, 1963; Russell, 1969). Thus, the absence of these epiphyseal centers does not guarantee immaturity; but if both are present, there is a nearly 100% correlation with a gestational age of 37 weeks or greater.

A recent suggestion that the X-ray calcification of the first and second molars *in utero* can be used to predict maturity of 36 weeks or over has met with limited usefulness in obstetrics (Lemons *et al.,* 1972) because of the difficulty in obtaining satisfactory radiologic exams. Perhaps this tool could be satisfactorily applied in the neonatal period, but it would still not be helpful below a gestational age of 36 weeks, where the major problems in maturity assessment exist.

In summary, considering the potential hazards of radiation exposure, there seems little reason to use such techniques whose effectiveness is greatest in confirming a term pregnancy since that is the very period in gestation when many other tests are equally effective.

1.3. Relationship of Outcome in LBW (Preterm or SGA) to Parental Potential

One important aspect of the evaluation of fetal growth involves the careful assessment of growth during a particular pregnancy in the context of the previous reproductive performance of those parents. This aspect of evaluation is often overlooked by obstetricians and yet is instrumental in arriving at conclusions concerning the health of the infant as well as the health of the mother. A number of studies have shown that maternal size is one of the determinants of newborn size.

FREDERICK C.
BATTAGLIA AND
MICHAEL A. SIMMONS

This aspect has been stressed by Tanner and Thomson (1970), who have suggested that an "expected" fetal size should consider maternal height and weight. For obstetricians following a pregnancy, this factor is not likely to be of much clinical significance since the effect of maternal size on an infant's birth weight in the absence of disease would be well within the variance of the estimate of fetal size by any of the current clinical techniques. However, in multiparous patients the past reproductive performance should be carefully assessed. Certainly the loss of a previous infant as a spontaneous abortion, stillbirth, or neonatal death is associated with an increased risk in the current pregnancy and thus should lead to the same careful observation and laboratory examinations associated with any high-risk pregnancy. This is generally done in obstetrics. What is often not done, however, is to record graphically the previous liveborn infants by birth weight and gestational age. In this way, any striking change in previous pregnancy outcome which is unassociated with a fetal or neonatal death becomes more apparent. The value of considering an infant's intrauterine growth against that of his siblings was demonstrated in the study of Turner (1971) on infants with congenital rubella infection. That study demonstrated that when the infants with rubella infection were compared with a population as a whole, the incidence of intrauterine growth retardation appeared to be approximately 40%. In contrast to this there was an 80% incidence

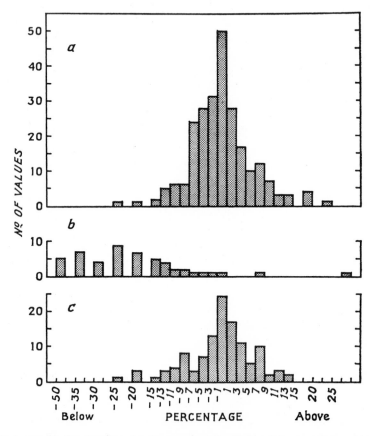

Fig. 5. Distribution (a) of each birth weight from mean of the sibship of the control group; (b) of the birth weight of the rubella baby from the mean of the sibship (rubella baby excluded); (c) of each birth weight from the mean of the sibship (rubella baby excluded) of the rubella group. From Turner (1971).

of IUGR when the infected infants were compared to their uninfected siblings born either before or after the target pregnancies (Figure 5). When following high-risk pregnancies, one finds frequently that the first one or two children born to a couple are children whose birth weights were above the 50th or 75th percentile, while the birth weight of an infant of a later pregnancy falls below the 25th or even the 10% percentile. Such infants often show many of the metabolic problems associated with intrauterine growth retardation. Such a change in pregnancy outcome should not only alert the pediatrician to the possibility of metabolic problems developing in the infant but should also alert the obstetrician to a reevaluation of the mother's health in order to determine whether chronic renal disease or chronic hypertensive disease have appeared subsequent to the change in pregnancy outcome. For these reasons, the past medical history should emphasize the past reproductive history and include graphic recording of the previous siblings' birth weight and gestational age distribution. Standards for comparing a newborn's weight with the weight expected from that of the preceding siblings have been published by Tanner *et al.* (1972).

2. Etiology of Low-Birth-Weight Infants

2.1. Factors Affecting Length of Gestation

2.1.1. Species

Placentation is one of the characteristics which distinguishes mammals from all other vertebrates. Yet, within mammals there is a wide diversity in the length of time this organ, the placenta, must survive and in the amount of fetal tissue a kilogram of placenta must supply with nutrients. In general the length of gestation is prolonged by the larger newborn size of the species. Primates as a group are distinguished by a longer gestation period for the same newborn size than other mammals. Primates, man included, can be said to build their babies slowly. Stated more precisely, Huggett and Widdas (1951) showed that within orders of mammals there was a linear relationship between the cube root of the birth weight and the reduced gestation time for the different species in an order. However, they also demonstrated that among orders of mammals there were striking differences in growth rates, of approximately 1000-fold.

Recently Sacher and Staffeldt (1974) have proposed a theory to account for the marked differences in length of gestation among eutherian mammals. Their hypothesis is that the rate of growth of the brain is the slowest of any organ in the mammal and that this rate of growth is the single common limiting factor. Figure 6, taken from their report, illustrates the linear relationship between the cube root of brain weight and the length of gestation among all the eutherian mammals. Thus, length of gestation increases with increasing brain size at birth and decreases with increasing litter size. Primates, producing one young with a large brain, have a correspondingly long gestation time.

2.1.2. Litter Size

As alluded to in the study by Sacher and Staffeldt, increasing litter size decreases the length of gestation. This observation holds true in man as well. McKeown and Record (1952) presented a clear demonstration of the effect of

FREDERICK C.
BATTAGLIA AND
MICHAEL A. SIMMONS

multiple pregnancy in man on both the duration of pregnancy and the fetal growth rate. Figure 7, from their study, shows how the duration of gestation is reduced in man as more and more fetuses are packed into the uterus. The reason for an earlier onset of labor in multiple pregnancy is unknown. It may be due in part to distention of the uterus acting as a mechanical trigger for the onset of labor. Currently there is considerable interest in the hypothesis that the fetal pituitary–adrenal axis plays an important role in determining the onset of labor. It may be the cumulative effect of this pathway in a number of fetuses which contributes to the shortened gestation. In any case it is important for both the obstetrician and pediatrician to recognize the fact that an early delivery should be anticipated in all multiple pregnancies. Thus, triplets have an expected date of confinement approximately 5 weeks earlier than singleton fetuses. Beginning at the time in gestation when fetal growth is first

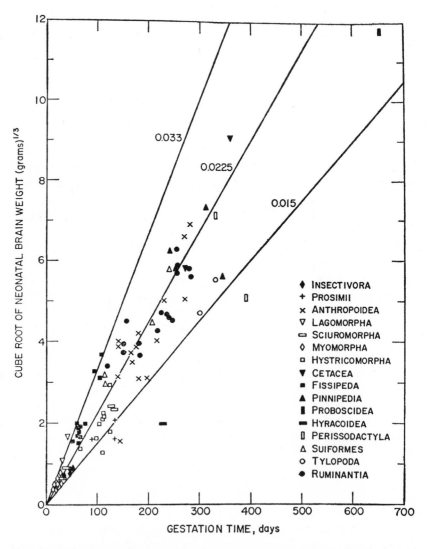

Fig. 6. Cube root of neonatal brain weight as ordinate, against gestation time on the abscissa. Lines are drawn for visual guidance to the range of growth rates observed. From Sacher and Staffeldt, 1974, *The American Naturalist* **108**:593. © 1974 University of Chicago.

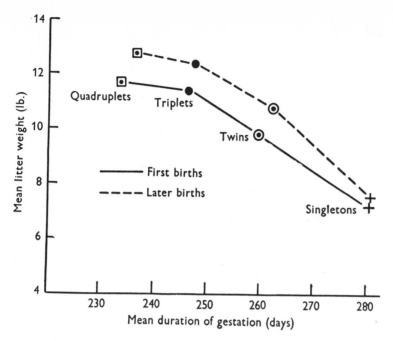

Fig. 7. Mean litter weight and mean duration of gestation of first and later births. From McKeown and Record (1952).

retarded, the increment in litter weight is approximately the same regardless of the number of fetuses. This observation suggests the possibility of a maximum rate at which the mother, through the uterine circulation and placenta, can supply nutrients for growth.

2.1.3. Maternal Age

The effect of maternal age on fetal growth rate and duration of gestation is one which increases the risks to the infant at either end of the reproductive age group. In general, reproduction around the time of menarche is associated with an increased incidence of preterm delivery, but a normal rate of intrauterine growth. Thus, prematurity rates are increased in teenage pregnancies. The incidence of premature deliveries may be decreased significantly by careful medical management of preeclampsia, a complication which also occurs with increased frequency among teenage pregnant women. In this regard the introduction in obstetrics of a number of tests that act as early-warning signals in selecting out women who are at high risk of developing preeclampsia later in pregnancy may be particularly useful in the management of teenage pregnancies.

Menarche has come at an earlier chronological age in recent years. This is demonstrated effectively in Figure 8, from the study of Frisch (1972). She holds the controversial view that menarche tends to occur at a critical body weight, and thus it is at least possible for very young children to become pregnant today. The repercussions of such a pregnancy pertain not only to that pregnancy but to subsequent pregnancies as well. More specifically, the question can be asked how does a pregnancy in a 12-year-old affect subsequent pregnancies when that child reaches a prime reproductive age group (such as the early 20s). Very little informa-

**FREDERICK C.
BATTAGLIA AND
MICHAEL A. SIMMONS**

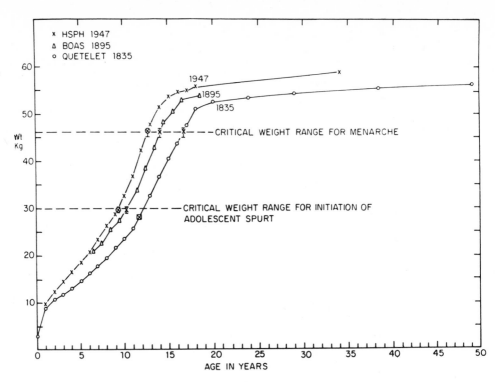

Fig. 8. Relationship of age of menarche to weight in kilograms over three time periods. From Frisch (1972).

tion is available on this point, but our own experience has suggested that a poor pregnancy outcome may occur in later pregnancies, regardless of the outcome of the initial pregnancy (abortion, stillbirth, etc.).

Advanced maternal age is associated with intrauterine growth retardation. Figure 9, taken from the study of Lobl *et al.* (1971), shows the striking reduction in infant birth weight occurring in the firstborn child when the mean maternal age is 38 years. If advanced maternal age and fetal growth retardation are coupled with a postterm pregnancy, intrapartum fetal distress is common and very often severe.

2.1.4. Maternal Factors

A shortened gestation often occurs secondary to what might be considered direct mechanical factors such as anomalies of the uterus (e.g., bicornuate or septate uteri) and "incompetent cervix." In general, both anatomic abnormalities of the uterus and the syndrome of "incompetent cervix" can lead to termination of the pregnancy at such an early date that the problem may present as a history of repeated spontaneous second-trimester abortions. A number of surgical procedures have been devised to enforce the cervical connective tissue in the therapy of "incompetent cervix" with considerable success.

2.1.5. Premature Rupture of Membrane

Recent studies in sheep have served to emphasize the role of the pituitary–adrenal axis of the fetus as one of the major pathways controlling the onset of

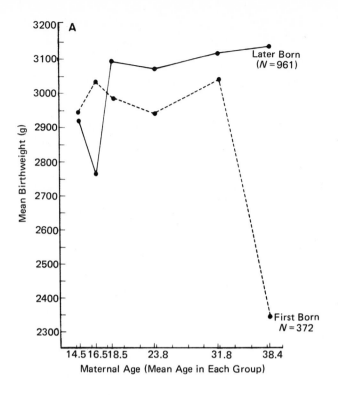

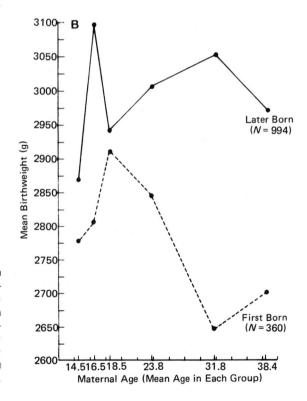

Fig. 9. (A) Mean birth weight at each maternal age comparing firstborn children to later-born black males ($N = 1333$). (B) Mean birth weight at each maternal age comparing firstborn children to later-born black females ($N = 1354$). From Lobl et al., 1971, *Johns Hopkins Medical Journal* **128:**347. © 1971 Johns Hopkins University Press.

FREDERICK C.
BATTAGLIA AND
MICHAEL A. SIMMONS

parturition (Liggens, 1969). In addition, the clinical observation that some anencephalic human fetuses with adrenal hypoplasia have prolonged gestations has tended to strengthen the hypothesis that this pathway may be equally important in man. Recently, there have been reports suggesting that a number of obstetrical complications, including premature rupture of membranes (PROM), may act as a stress triggering the release of glucocorticoids in the fetus, accelerating the maturation of a number of organ systems including the lung.

However, it is not yet clear whether the fetal pituitary–adrenal axis plays a major role in man for the control of parturition. The data are conflicting. Honnebier and Swaab (1973) pointed out that when the effect of anencephaly on gestation length is examined carefully, mean gestation length is not prolonged but rather there is a much greater variability in the length of gestation with more pregnancies terminating preterm and postterm than in normal pregnancy. Similarly, when the effect of premature ruptured membranes on the incidence of RDS was examined independent of gestational age, Jones *et al.* (1975) could not document any reduction in the incidence of RDS with PROM. Studies in subhuman primates had shown that if the fetus were removed from the uterus prematurely, the pregnancies could continue to term and the placenta could then be delivered at the expected time. Thus, at this point, the relationship between the fetal pituitary–adrenal axis and the onset of parturition in man must be considered a tenuous one deserving further study and clarification.

2.2. Factors Affecting Intrauterine Growth Rate

2.2.1. Nutrition

Nutrition has an important impact on growth and development at any age. Undernutrition of the mother affects the fetus and produces a reduced organ size in all organs. This effect occurs in man as well as in other mammals. Recently there has been considerable interest in defining the effect of undernutrition on brain growth and brain development. These efforts were encouraged by the hypothesis that there was a critical period in brain growth when undernutrition would have a maximum and lasting effect on brain development. However, some confusion was introduced by these studies when attempts were made to extrapolate from phenomena in small mammals to similar effects on man, without a sufficient appreciation of the marked differences in fetal growth rate and length of gestation among eutherian mammals which we have discussed earlier. Clinical studies have demonstrated quite clearly the lasting impact on brain development which occurs in man when undernutrition extends throughout intrauterine development and includes the first few years of postnatal life. It is also clear that undernutrition confined to fetal life produces growth retardation and thus a smaller birth weight of the infant at term. This has been shown in many mammals and was clearly described in man in the early studies by Smith (1947) and later by Stein and Susser (1975) on pregnancy outcome during the wartime famine imposed on Holland in World War II. More recently essentially similar findings were obtained in the INCAP studies in Central America (Lechtig *et al.,* 1975). In one of these studies on the relationship of acute starvation or severe chronic malnutrition to pregnancy outcome in man, several general characteristics emerged. Maternal weight gain is reduced; the percentage of reduction in maternal weight gain is the largest effect on undernutrition. Next in order of magnitude is the reduction in placental weight and finally a reduction in fetal body weight and length. In the Dutch famine studies the effect seemed limited to those fetuses exposed to undernutrition in the third trimester, the magnitude of effect roughly corresponding to the effect of a reduced fetal growth observed in multiple pregnancies which also is a third-trimester phenomenon.

The challenge to fetal survival presented by maternal starvation can be contrasted to that presented by maternal hemorrhage or other acute stress. In the latter example the stress to the mother stimulates the release of catecholamines which are potent vasoconstrictors of the uterine circulation (Barton *et al.*, 1974). Their effect is dose-related, and this can produce virtually total constriction of the uterine vascular bed. In such a situation, in which the mother's life is threatened acutely, there would be little point in adaptive mechanisms which aimed at fetal survival without maternal survival. Thus with hemorrhage, a rather acute stress, the fetus is sacrificed to preserve maternal survival. On the other hand, undernutrition is a chronic challenge to which the mother would be expected to adjust and survive, and the adaptations that are made can be interpreted as an attempt to maintain fetal growth and survival even in the face of increased metabolic and nutritional demands placed on the mother. Thus, fetal growth, although definitely retarded, is sustained while normal maternal weight gain in pregnancy may be sacrificed.

2.2.2. Placental Size and Function

The placenta is a fetal organ which is as essential as its heart to its survival. The role of the placenta as a determinant of fetal growth and ultimate fetal size has been thoroughly studied in animals where placental size at birth and placental function can be compared. The function of the placenta encompasses many areas that are the province of a number of different organ systems after birth. Thus, diversity of roles presents some obvious difficulties in attempting to evaluate placental function. In man tests of placental function are still rather crude and of doubtful reliability. Thus, in man the relationship between fetal growth and placental growth has been inferred from the determinations of placental weight, total DNA, and other measurements of total tissue mass. Animal studies have clearly demonstrated that placental size is a major determinant of fetal size (Alexander, 1964; Kulhanek *et al.*, 1974). Figure 10, taken from the study of Kulhanek *et al.*, shows that if one compares placentas of the same gestational age, the smaller placenta has decreased

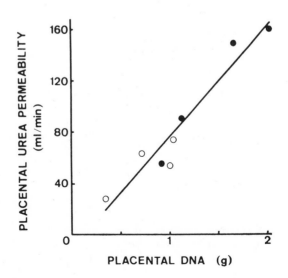

Fig. 10. Among fetuses of approximately the same age (140–150 days), placental urea permeability is proportional to placental DNA (first-order regression analysis); • = singletons; ○ = twins. From Kulhanek *et al.* (1974).

FREDERICK C.
BATTAGLIA AND
MICHAEL A. SIMMONS

functional capacity as well as decreased size. Since the small placenta in both man and animals invariably determines a small birth weight at term, the question can then be asked what the factors are which regulate placental development. While not all of the factors are known, Alexander's work has shown that to some extent placental development is determined at the time of implantation. In sheep, for example, the number of cotyledons developing in the sheep placenta ultimately affects total placental size at term. Thus, of approximately 90 potential implantation sites in the sheep uterus, the animal which uses all 90 sites will have a larger placenta at term than one utilizing only 20 or 30 sites.

One of the principal functions of the placenta is to ensure continuous supply of nutrients to the embryo and fetus for growth. The topic of fetal nutrition is dealt with in other sections of this treatise. At this point, however, it is worth reviewing some aspects of placental development which determine the supply of nutrients to the fetus. Three distinct stages of placental development can be recognized in both man and other mammals. The first stage is implantation. This phase plays a crucial role in determining the success of subsequent stages of placental development. There is considerable interest today regarding how implantation is affected in predictable ways when an insemination is mistimed with respect to ovulation and the zygote is the product of either an aging sperm or an aging ovum. Certainly zygote loss is increased under these conditions, presumably reflecting problems in the implantation stage of placental development. A second stage is that of rapid

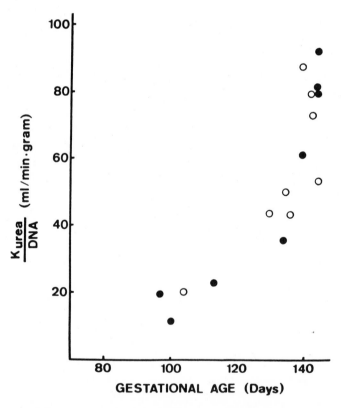

Fig. 11. Urea permeability per gram of placental DNA rises steeply in the last 2 months of gestation; • = singletons; ○ = twins. From Kulhanek *et al.* (1974).

growth which is completed when total DNA and the weight of the placenta are maximal. The last stage is that of maturation and relates to the striking increase in placental function which occurs in late gestation despite relatively little change in placental weight or DNA. Figure 11, also taken from the study of Kulhanek *et al.*, illustrates this phenomenon quite clearly. Several investigators, notably Gruenwald (1970), have encouraged the assumption that the fetus in late gestation may be growing slowly by some limitation imposed by a placenta unable to cope with its demands. This assumption seems invalidated by the physiologic data which establish the progressive increase in function and maturation of the placenta despite a constant organ size.

In both man and other mammals, when fetal IUGR occurs associated with the small placenta, the fetal/placental weight ratio is significantly increased. Thus, there is evidence of some compensatory growth of the fetus despite the limitation imposed by a small placenta, although the compensatory growth is not complete and the fetus is smaller than normal. The larger fetal-placental weight ratio implies a narrow margin of safety for the fetus which becomes more apparent in the added stress of parturition. This may, in part, account for the higher incidence of fetal distress during the intrapartum period in pregnancies complicated by fetal growth retardation. For this reason it is imperative that pediatricians and obstetricians be prepared for problems during labor and for the possibility of a difficult resuscitation in the delivery room in pregnancies characterized by intrauterine growth retardation recognized prior to delivery.

2.2.3. Maternal Disease

In general the ability of a woman to carry a pregnancy to term is an effective test of her general health. Almost all serious diseases in the mother are associated with some infertility, an increased incidence of premature onset of labor, and a high incidence of preterm delivery. In part, this is determined by the severity of the illness and in part, by the specific nature of the disease. For example, in the case of acute infections of the mother such as cholera, influenza, etc., fetal mortality and morbidity are related to the extent of the illness in the mother. If her circulation, including cardiac output, distribution of cardiac output, and perfusion pressures are maintained and if oxygenation and hydration are also well-maintained, the outcome for the fetus is satisfying. In the case of diseases which directly affect the development of the placenta, or affect the vascular development of the uterus, such as maternal collagen vascular disease, chronic vascular disease, etc., the severity of the illness in the mother may be an indirect reflection of the impact of that disease on the fetus. It is not the purpose to review here the many diseases that can present during pregnancy but rather to reemphasize the importance to the mother of a careful assessment of pregnancy outcome as a reflection of the mother's general health. Whenever there is an unexpected change in pregnancy outcome, the possibility that this reflects the presence of a chronic disease previously unrecognized in the mother should be considered and carefully evaluated. In this regard, it should perhaps be emphasized that one expects a lower perinatal mortality and morbidity with the second pregnancy compared to the first and a larger infant in the second and third pregnancy compared to the first. Thus the appearance of an SGA infant in subsequent pregnancies should stimulate a careful review and reassessment of the mother's health.

FREDERICK C.
BATTAGLIA AND
MICHAEL A. SIMMONS

2.2.4. High Altitude

High altitude is associated with a reduction in infant birth weight presumably through an effect on fetal oxygenation. Exposure to the lower oxygen tension of the inspired air at elevations exceeding 3000 m lowers both the maternal arterial and the maternal uterine venous Po_2. The latter Po_2 determines the umbilical venous Po_2, which is the most oxygenated blood of the fetus (Figure 12) (Makowski *et al.*, 1968). Infants born at high altitudes have lower birth weights and larger placentas (Lichty *et al.*, 1957; Kruger and Arias-Stella, 1970) than those born at sea level, suggesting some compensatory hypertrophy of the placenta. There have been numerous reports stressing the variety of cardiovascular effects on the infants and children after birth. This has included reports of primary pulmonary hypertension in children and, perhaps of most interest, a marked increase in the incidence of patent ductus arteriosus (Khoury and Hawes, 1963; Sime *et al.*, 1963). Peñaloza *et al.* (1964) reported the incidence of patent ductus arteriosus was 30 times greater in infants and children living above 4500 m (see Figure 13).

2.3. Clinical Conditions Associated with Low-Birth-Weight Infants

Many clinical problems in the mother have been associated with intrauterine growth retardation or premature delivery in the fetus. As a rule, the woman's ability to carry a pregnancy to term is a subtle test of her general health.

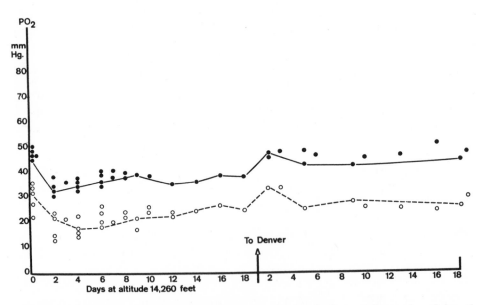

Fig. 12. Oxygen tension in uterine and umbilicalveins. On the abscissa are the number of days the animals were observed at high altitude followed by the number of days the animals were observed at Denver. On the ordinate is plotted the oxygen tension (mm Hg) in both the uterine and umbilical veins of all 6 animals. Closed circles = uterine vein. Open circles = umbilical vein. The solid line connects the oxygen tension in the uterine vein of animal 0005 and the broken line connects the oxygen tensions in the umbilical vein of that same animal. From Makowski *et al.* (1968).

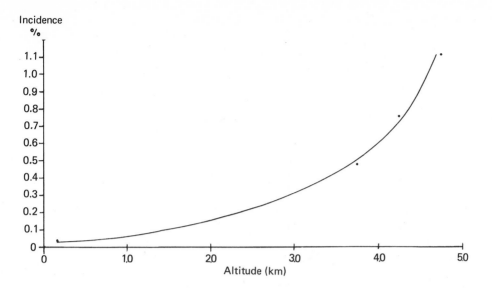

Fig. 13. The incidence of patent ductus arteriosus increases gradually as altitude above sea level increases, so that between 4500 and 5000 m the frequency is 30 times greater than at sea level. The relationship between the level of altitude and the incidence of ductus arteriosus approximately follows a parabolic curve. From Peñaloza *et al.* (1964).

2.3.1. Maternal Infection

Maternal infection with rubella or cytomegalovirus is often associated with fetal and placental infection and thus intrauterine growth retardation. Infants demonstrating the complete spectrum of the rubella syndrome are usually growth retarded; the detrimental effects of this condition on growth are readily apparent when the infected infants are compared with a cohort of their normal siblings (Turner, 1971). The growth retardation in rubella infants affects weight predominantly but may also lead to a decrease in body length and head circumference (Desmond *et al.,* 1969). The intrauterine growth retardation accompanying congenital toxoplasmosis infection can lead to severe reduction in weight, length, and head circumference. Infection with cytomegalovirus may be associated with intrauterine growth retardation but more often results in an infant who is normally grown but demonstrates the clinical stigmata of thrombocytopenia, hepatitis, and, infrequently, meningoencephalitis. Congenital infection with herpes is usually of acute onset and thus not associated with significant growth retardation.

Other infections in mothers have also been linked to the etiology of low birth weight. Kass has reported that the incidence of asymptomatic bacilluria among mothers delivering prematurely is significantly greater than among mothers who have term infants (Abramowicz and Kass, 1966). Other studies have been unable to document such a relationship. Significant infections in women often cause decreased fertility rates (i.e., parasitic infections). Other maternal infections (i.e., tuberculosis, syphilis, literis, vibrio fetus, and malaria) have not been associated with intrauterine growth retardation, but they have increased the incidence of prematurity.

The mechanisms by which maternal infection may lead to abnormalities of intrauterine growth and prematurity are unknown. Direct invasion of virus into fetal tissues may lead to impairment of cell division (Boué and Boué, 1969). Cheek (Hill *et al.*, 1970) has shown that infants infected congenitally with rubella virus may have both an impairment in numbers of cells and in cell size, presumably as a direct reflection of viral injury. One can only speculate whether other fetal pathogens may have similar effects.

2.3.2. Drug Use

Maternal drug abuse tends to cluster in certain socioeconomic strata. Thus it is difficult to isolate a particular drug usage as a cause of low-birth-weight infants. Naeye *et al.* (1973) have reported that infants born to heroin-addicted mothers have intrauterine growth retardation. There was no documentation of an increased incidence of preterm delivery among these infants. It has been suggested that another major effect of heroin usage on reproductive performance is a decrease in fertility and fecundity.

Chronic alcohol abuse in a mother may lead to a particular clinical syndrome. This syndrome includes microcephaly, short palpebral fissures, prominent epicanthal folds, micrognathia associated with abnormalities of joints, heart, and external genitalia. Intrauterine growth retardation was invariably present, and a majority of infants had mental retardation (Jones *et al.*, 1974; Jones and Smith, 1973). Maternal alcohol abuse appears to have a direct effect on fetal growth and development. An indirect effect may also play a role via various nutritional deficiencies which may be common among alcoholic patients. Animal studies suggest that heroin may lead directly to abnormalities of intrauterine growth in the absence of any confounding social or nutritional factors (Taeusch *et al.*, 1973).

Numerous studies have shown a reduction in mean birth weight in infants born to mothers who smoke. This reduction in infant birth weight is due to intrauterine growth retardation (Ounsted, 1971*a,b*; Frazier *et al.*, 1961).

2.3.3. Specific Obstetrical Problems

Maternal diseases which may affect placental function through vascular changes may be associated with intrauterine growth retardation as has been outlined in the previous section. Pregnancy-induced hypertension is one of the entities that is commonly associated with such growth anomalies. The ultimate growth of the fetus under such conditions is dependent on the duration of hypertensive disease in pregnancy (Järvinen *et al.*, 1958) (Figure 14). Late-onset toxemia such as that which occurs in pregnancies during adolescence has less direct effect on ultimate fetal growth than long-term underlying maternal hypertensive disease.

In a retrospective study, Ounsted demonstrated that over 30% of the mothers of growth-retarded infants had significant hypertension (Ounsted, 1971*a,b*). In reviewing infants with intrauterine growth retardation at the University of Colorado Medical Center, Lubchenco *et al.* (1968) found that one third were born to mothers with hypertension. Unfortunately, the precise nature of the maternal hypertension was not well defined in any of these studies. Long-standing essential hypertension in a mother had been shown to be associated with an increased rate of fetal death, which was even higher if the hypertension became exaggerated during pregnancy (Butler, 1969).

Other obstetrical problems occurring in the third trimester, such as abruptio placenta, placenta previa, and premature rupture of membranes, are not directly

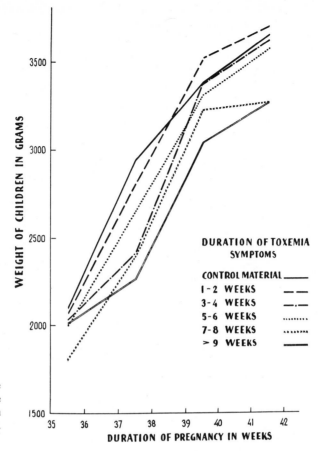

Fig. 14. This illustration shows the effect of toxemia symptoms on the birth weights of infants delivered in different weeks of pregnancy. From Järvinen *et al.* (1958).

associated with abnormalities of fetal growth, as would be anticipated due to the time of onset of the problem and the nature of the prompt delivery which usually occurs. Uterine anatomic anomalies (bicornate or septate uteri) may be associated with both preterm delivery and more infrequently intrauterine growth retardation. Significant growth retardation has been noted in conjunction with chorioangioma of the placenta (Battaglia and Woolever, 1968).

2.3.4. Multiple Pregnancy

Multiple pregnancies have long been associated with a higher incidence of preterm deliveries (McKeown and Record, 1953). In addition, a significant degree of intrauterine growth retardation may be associated with multiple pregnancy. Significant growth retardation does not occur in twin pregnancies prior to 34 weeks, but after this time there is a rapid decrease in the rate of intrauterine growth in twin pregnancies, resulting in small-for-gestational-age infants (Lubchenco, 1970).

2.3.5. Chromosomal and Congenital Anomalies

Specific chromosomal anomaly syndromes may be associated with intrauterine growth retardation. Trisomy 16–18, trisomy 13–15, and Down's syndrome have all been associated with intrauterine growth retardation. In trisomy 16–18 the occurrence of intrauterine growth retardation is an almost invariable accompaniment of the syndrome. Trisomy 16–18 infants may also be born postterm. The association of growth retardation with Down's syndrome and trisomy 13–15 is less frequent.

A wide variety of congenital malformations unassociated with chromosomal abnormalities are associated with deviations of intrauterine growth rate (Van den Berg and Yerushalmy, 1966). In a series of 9000 infants collected from the University of Colorado Medical Center by Lubchenco, the over-all incidence of congenital anomalies was 3.3%, with the highest incidence occurring in infants of less than 30 weeks gestation and the next highest incidence in term small-for-gestational-age infants.

2.3.6. Primary-Growth-Failure Syndromes

Certain syndromes associated with primary growth failure during infancy have also been associated with abnormality of intrauterine growth. The Silver-Russell syndrome and Seckel's bird-headed dwarfism are two such entities. Infants with primary-growth-failure syndromes are noted during infancy to have an abnormal rate of growth, not only of weight, but also of height and other measures of skeletal development. The precise mechanism whereby such growth failure occurs is not presently known. Over half the entities associated with congenital malformations and ultimate short stature give some evidence of the onset of growth retardation prior to birth (Smith, 1970).

3. Physiology of Abnormal Intrauterine Growth

3.1. Metabolic Rate

Oxygen consumption in animal species during late gestation averages 6–8 ml/kg/min. Although similar measurements are not available for the human, it seems likely that soon after birth oxygen consumption falls to a level below that in fetal life. Oxygen consumption in the first 12 hr of life averages between 4.5 and 5.5 ml/kg/min (Hey, 1969; Hill and Rahimtulla, 1965; Brück, 1961). These basal rates of oxygen consumption, which appear to be significantly lower than those rates in animal species studied during fetal life, increase fairly rapidly over the first 10 days of life so that mean oxygen consumption at 24 hr of age averages approximately 6.5 ml/min/kg and by 10 days of age averages 7 ml/min/kg (Hill and Rahimtulla, 1965).

There has been disagreement about the relation between gestational age and absolute rates of oxygen consumption. Sinclair (1970) has stated that the oxygen-consumption measurements among preterm infants show a rise with time similar to that among term infants, but the absolute values tend to be slightly lower per unit weight than in full-term infants of the same postnatal age. Sinclair has likewise stated that babies who are of a small-for-gestational age systematically have higher rates of oxygen consumption than their normally grown peers (Sinclair, 1970). Sinclair and Silverman (1966) initially proposed that the apparent increase in metabolic rate could be accounted for by the relatively larger brain, as a percentage of body weight, in small-for-gestational-age infants. The metabolic rate of brain tissue may be two times the minimal metabolic rate of body tissues as a whole, and thus an increased relative brain size could lead to an increase in total metabolic rate.

In this study, Sinclair and Silverman noted that although oxygen consumption was higher on a per kilogram of body weight basis, there was no increased oxygen consumption related to head circumference among small-for-dates infants.

In careful studies by Hill and Robinson (1968), these findings of variations in oxygen consumption between the small-for-gestational age premature and term infant could not be documented. These investigators could show no systematic variations in either the absolute values of oxygen-consumption measurements at a given postnatal age or in the trend to increased oxygen consumption over the first 10 days of life among these three groups of infants. They state that the values of oxygen consumption are directly proportional to birth weight in all babies regardless of their gestational age at any given postnatal age. They could demonstrate no relationship of oxygen consumption to gestational age *per se*.

The obvious discrepancy in these observations cannot be definitely explained. It has been suggested that among some small-for-gestational-age infants there is a substantial rise in the minimal rate of oxygen consumption noted on the fourth postnatal day, presumably in association with an increased caloric intake (Scopes and Ahmed, 1966). Sinclair has speculated that in the face of decreased substrate availability (in particular, glucose) among some small-for-gestational-age infants during the first few days of life, there may be a direct effect on the limitation of cerebral oxygen consumption. Although this limitation of basal oxygen consumption could only be secondary to severe substrate deficiency or depletion, it is possible that some small-for-gestational-age infants might have a reduced oxygen consumption during the first few days of life on the basis of decreased substrate availability. Thus, the above discrepancy between the interpretation of Sinclair and that of Hill might be explained. Such an explanation would not, however, account for the continuing identical oxygen consumptions noted throughout the first several days of life among all infants, regardless of gestational age, in the study by Hill.

The presumed increase in basal metabolic rate noted in some studies of small-for-gestational-age infants has been implicated in the increased caloric demands that such infants may demonstrate. However, the small differences demonstrated in the oxygen-consumption rates of small-for-gestational-age infants in the studies reporting such observations would not lead to a significantly increased caloric requirement, particularly in comparison to the increased caloric demands that may occur because of growth rate, thermal stress, and activity.

At the present time it would appear that there is at least no clinically significant difference in oxygen consumption per kilogram among infants at a given postnatal age regardless of their gestational age or rate of intrauterine growth.

3.2. Thermoregulation

Preterm and low-birth-weight infants are considered homeotherms, although their mechanisms for maintaining thermostability are severely limited. All low-birth-weight infants are at a disadvantage in terms of heat exchange because of their high surface-area to body-weight ratios, their postural characteristics, and their diminished surface-tissue insulation functions. No abnormality has been found in either vasoconstriction or vasodilatation in response to cold or heat stress among low-birth-weight infants.

Regardless of their rate of intrauterine growth, all infants show an appropriate rise in heat production in the face of cold stress (Hey, 1969; Brück, 1961), but the

FREDERICK C.
BATTAGLIA AND
MICHAEL A. SIMMONS

extent to which metabolic rate increase is limited in low-birth-weight infants as compared to term, appropriately grown infants is unclear. There is some disagreement about the differences in response to cold stress between low-birth-weight infants who are appropriately grown and those with intrauterine growth retardation. Hey demonstrated that, regardless of the etiology of the low birth weight (intrauterine growth retardation or prematurity), the degree of rise in metabolic rate in response to cold stress was similar in infants of the same weight. In contrast, Silverman and Agate (1964) have shown a gestational effect, with a greater response in metabolic rate among infants who were gestationally more mature as compared to gestationally immature infants of the same weight. These initial findings were confirmed in a later study of minor cold stress in low-birth-weight infants who were premature and had intrauterine growth retardation (Silverman *et al.*, 1966).

In general, the mechanism of this increase in heat production among low-birth-weight infants differs from that in the adult homeotherm. Shivering thermogenesis does not play a major role in response to cold, although voluntary muscular activity may contribute to heat production. It seems likely that nonshivering thermogenesis is responsible for most of the rise in heat production in response to cold stress. The site of this chemical thermogenesis is most likely brown fat tissue which is present in significant concentrations in newborn infants (Aherne and Hull, 1966). The uncoupling of oxidative phosphorylation which exists in brown fat may account for its high rate of heat production.

The responses to heat stress in the adult animal include peripheral vasodilatation, which favors increased blood flow to the surface and thus increased heat loss to the environment, and sweating, which increases evaporative heat losses. After some initial controversy it is now apparent that infants do sweat, depending on their degree of maturity. Prior to about 36 weeks of gestational age, little sweating occurs in response to significant heat stresses. Thus, preterm infants are at significant risk of heat stress, but small-for-gestational-age infants, even of very low birth weight, may have more normal sweating responses if they are born after 36 weeks of gestation.

Small-for-gestational-age infants tend to have fewer problems with temperature instability than premature infants of the same birth weight. Since the metabolic responses of these two groups of infants seem to be independent of gestational age, the most reasonable explanation for the increased resistance to cold stress among small-for-gestational-age infants as compared to preterm infants of the same birth weight is their capacity for altering their posture (that is, reducing their effective surface area by flexing their arms and legs onto their chest and abdomens) and their increased capacity for voluntary and spontaneous muscular activity which in itself may be a source of heat production.

3.3. Water Spaces

With the demonstration that body composition in infants with intrauterine growth retardation is similar to body composition in infants suffering postnatal malnutrition, it would seem likely that total body water and the partition of total body water might also be similar in the two conditions. Several studies have demonstrated that in postnatal malnutrition there is an expanded total body water which is primarily accounted for by an increase in extracellular fluid water (Hansen *et al.*, 1965; Kerpel-Fronius, 1960; Kerpel-Fronius and Kovach, 1948). These studies have demonstrated that the more severe the degree of malnutrition, the greater the increase in body water and in the expansion of the extracellular fluid space.

Cassady has shown that among intrauterine-growth-retarded infants a marked expansion in albumin space exists at birth, but these fluid spaces rapidly adjust within the first 12 hr of life; consequently, by the end of that period of time, the plasma volumes of intrauterine-growth-retarded infants are identical to those of mature infants. Intrauterine-growth-retarded infants have an increase in total body water, due principally to expansion of the extracellular fluid space (Cassady, 1970). The more severe the growth retardation, the more striking this observation.

These findings are not surprising given the pattern of changes in body water with growth. Growth is characterized by a decrease in total body water as a percentage of total body weight and a decrease in extracellular fluid water as a percentage of total body water. In infants with significant growth retardation it is possible that this process is interrupted, and at any given gestational age in a growth-retarded infant body water will represent a higher percentage of body weight, and extracellular fluid water, a higher percentage of total body water.

These observations of alterations in total body water and in the distribution of body water have clinical implications in planning fluid therapy in growth-retarded infants, particularly in the presence of any compromise secondary to asphyxia, etc. Frank edema, similar to that seen in postnatal protein–calorie malnutrition, may develop in the infant with intrauterine growth retardation; fluid therapy, including colloid administration, should be carefully planned. Fairly rapid spontaneous correction of the disorders in extracellular fluid expansion can be expected among all infants, including those with growth retardation.

4. Problems and Management of LBW Infants Related to Intrauterine Growth

4.1. Resuscitation: OB

All low-birth-weight infants, whether SGA or preterm, are at risk of perinatal asphyxia and its attendant neonatal complications, including central nervous system (CNS) depression. In reviewing the condition of infants with birth weights < 1200 g born at Colorado General Hospital in 1975, the observation was made that the one-minute Apgar score was usually depressed to < 3 even though umbilical venous and umbilical arterial pH and base excess most frequently were within the normal range. This discrepancy between frequently low Apgar scores in the very immature baby and normal umbilical venous and arterial blood gas studies raises the question of whether the Apgar scoring system is an adequate assessment in the very-low-birth-weight infant. It is our belief that Apgar scores are inappropriate as a means of estimating viability for these very immature infants. The point deserves emphasis since, often on the basis of appearance and initial Apgar score, a poor prognosis is assumed and prompt and effective resuscitation is omitted. With the understanding of the impossibility of predicting, on the basis of the initial Apgar score and general appearance, what the infant's ultimate outcome will be, it is necessary to institute prompt and effective resuscitation beginning at the moment of birth.

FREDERICK C.
BATTAGLIA AND
MICHAEL A. SIMMONS

Both the obstetrician and the pediatrician have a responsibility in the resuscitation of very-low-birth-weight infants. The obstetrician's responsibility includes the recognition of specialized care that a high-risk infant is likely to require, and thus where he chooses to deliver such an infant and with whom he chooses to work during this difficult period are of critical importance in the ultimate outcome of the infant. Similarly the pediatrician must be specially skilled both in the knowledge of the problems of the low-birth-weight infant in the immediate perinatal period and in the technical skills which are required for resuscitating such infants.

There has been too great an emphasis on a pharmacological approach to resuscitation, since the majority of the resuscitative needs required by infants in the immediate neonatal period are simple mechanical ones. Attention to temperature control by providing a heat source in the delivery room and by rapid drying of the infant is the first step in the proper support and resuscitation of any infant. The rapid provision of adequate oxygenation and ventilation, either by increasing the inspired oxygen concentration that the infant breathes or by mechanical ventilation using bag and mask or intubation, are essential first steps. Occasionally, either because of severe asphyxia or intrapartum volume loss, such infants will need external cardiac massage in order to assure cardiac output and thus tissue oxygenation. Delivery-room emergencies include other mechanical factors which may lead to the impossibility of resuscitation, for example, pneumothorax or an upper airway obstruction.

Few infants fail to respond to the above conservative mechanical measures. Thus, pharmacological intervention is rarely necessary. In the 1975 series at Colorado General Hospital only one infant among 94 with birth weights < 1500 g required additional steps beyond the mechanical ones. It should be emphasized that the lack of pharmacological intervention in the delivery room or stabilization environment occurred despite the fact that 50% of the infants had initial Apgar scores of 3 or less. We feel strongly that the routine use of the alkali therapy and various cardiac stimulants in the delivery room should be actively discouraged, and our results in resuscitating infants without these agents support that position.

Stabilization efforts begin in the delivery room and continue throughout the first few hours and then subsequently in the nursery environment. Supportive care includes attention to ventilation, oxygenation, cardiac output, tissue perfusion, glucose homeostasis, fluid and electrolyte homeostasis, and acid–base homeostasis.

4.2. First Few Hours of Life

4.2.1. Specific Problems Related to Preterm Delivery

The specific problems that develop over the first week of life in preterm infants are directly related to the developmental stage of various organ systems. These include the following: in the lung, hyaline membrane disease (HMD); in the liver, abnormalities of bilirubin conjugation and excretion; in the brain, the high incidence of intracranial hemorrhage; and in the gastrointestinal tract, the difficulties with nutrition.

Since the initial description of surfactant deficiency in infants dying of HMD (Avery and Mead, 1959), it has become generally accepted that deficiency of surfactant (which is necessary for end-expiratory alveolar stability) is a major factor leading to the development of HMD. Management consists of techniques designed to improve alveolar hypoventilation and right-to-left shunting which commonly occurs. General principles of supportive care, including attention to adequacy of perfusion, are critically important in the management of this disease. Traditionally, the provision of supplemental inspired oxygen has been the hallmark of therapy, with the use of mechanical ventilation contemplated only in the face of significant respiratory failure. Over the past several years there has been a more aggressive ventilatory approach to infants with HMD, largely precipitated by the description of Gregory *et al.* (1971) of the value of continuous positive airway pressure (CPAP) in the management of this disease. The use of positive end-expiratory pressure (PEEP) during mechanical ventilation or CPAP alone in a spontaneously ventilating infant has led to significant improvement in the respiratory management and survival of these infants. Marked improvement in the design of the respirator used in small infants has undoubtedly contributed to the over-all improvement in survival rates for HMD. In most cases infants with an inspired oxygen requirement of greater than 60% during the first 6 hr of life benefit from the use of moderate continuous positive airway pressure (4–6 cm H_2O).

Because of the implications of both cold and heat stress on metabolic and physiologic homeostasis in small infants, it is critical that a neutral thermal environment be provided in the nursing area. Thermal neutrality has been defined as that temperature zone in which an infant's basal oxygen consumption is at a minimum. Generally this correlates with skin temperatures of between 36 and 36.5°C. It is within this range that an infant's skin temperature should be maintained during his nursery stay.

Preterm infants (particularly if they are extremely preterm and of low birth weights) may be born with a significantly diminished total hemoglobin concentration. The existence of this anemia may further compromise clinical management of other common first-hour problems. In an infant with hypotension, the existence of significant anemia jeopardizes systemic oxygen delivery. In infants with respiratory distress from hyaline membrane disease or other etiologies, a significant part of the pathophysiology is a diminished oxygen content because of a low pO_2 in the arterial blood. If significant anemia is added to this arterial oxygen desaturation, further problems in oxygen transport and delivery will exist. When coexistent clinical problems do occur, we have attempted to maintain central hematocrits greater than 45% by frequent packed-red-cell or whole-blood transfusions.

Preterm infants are at high risk of developing systemic infections. Various deficiencies in their host–defense mechanisms have been described, including low IGM levels, low complement levels, poor chemotaxis, and decreased phagocytic killing. Equally important perhaps are environmental factors rather than host origin factors. Frequently, preterm infants are delivered after pregnancies complicated by early and prolonged rupture of the amniotic membranes. Such premature rupture of membranes predisposes an infant to amnionitis and to the intrauterine acquisition of bacterial infection. Many manipulative procedures are necessary in order to carry out current management practices for high-risk infants. Almost all of these techniques predispose to the breakdown of epidermal or endothelial barriers to bacterial invasion.

Multiple nutritional problems exist in preterm low-birth-weight infants. Many of these problems may begin in the first few hours of life. The primary source of calories during this period is the parenteral glucose received. The incidence of hypoglycemia in preterm appropriately grown infants is not as striking as that of their undergrown peers (both in the preterm and term age groups), but hypoglycemia nonetheless occurs in preterm infants and close monitoring of blood glucose concentration is required.

FREDERICK C.
BATTAGLIA AND
MICHAEL A. SIMMONS

The incidence of intrapartum asphyxia among small-for-gestational-age infants is significantly higher than among their normal-sized peers. This increased incidence of asphyxia is presumably secondary to a limitation in placental function in undersized infants. During the stress of the intrapartum period, an already compromised placenta may be unable to meet the oxygenation and nutritional requirements of the fetus. Problems with asphyxia should be anticipated in any infant who is small for gestational age even in the absence of overt intrapartum findings of fetal distress. Meconium staining of the amniotic fluid may be the only definitive sign of intrapartum fetal distress, and this management problem must be anticipated. Prevention of meconium aspiration is the most effective management technique because once meconium aspiration is established there is little to offer except ventilatory therapy. The increased incidence of meconium aspiration syndrome among small-for-gestational-age infants is explainable on the basis of a higher incidence of intrapartum asphyxia in such infants and by the fact that often the meconium that is passed is diluted in a much smaller volume of amniotic fluid than would be present in an AGA infant (normally grown). Thus, the stimulus for meconium passage *in utero* occurs more frequently in SGA infants, and meconium is present in a higher concentration.

Hypoglycemia occurs two to four times more frequently among small-for-gestational-age infants than among their peers of normal size. The explanation for the cause of this high incidence of hypoglycemia remains speculative. The high incidence of asphyxia in intrapartum distress noted in SGA infants has suggested to some that there may be an increased stimulus to *in utero* glucose utilization during this period. On the basis of decreased glycogen content in the livers of SGA infants, it has been speculated that decreased glycogen availability largely accounts for the high incidence of hypoglycemia in the first several hours of life. If metabolic rates are indeed increased in small-for-gestational-age infants, this would be an additional stress on limited stores of precursors.

The frequency and severity of hypoglycemia in such infants would suggest that there are explanations other than deficient glycogen stores. For instance, in a normal term infant where the incidence of hypoglycemia is minimal, glycogen stores average 2–3 g/kg. This amount of glycogen would support the blood glucose concentration in such an infant for 12–24 hr. However, even after this time period such infants rarely develop hypoglycemia; so it seems likely that among SGA infants mechanisms other than simple glycogen deficiency may account for the abnormalities of glucose homeostasis.

Recent studies have suggested that gluconeogenesis may also be impaired in SGA infants. Increased levels of gluconeogenic precursors in small-for-gestational-age infants have been found (Haymond *et al.,* 1974). Glucose levels in both control and SGA infants were shown to be inversely proportional to the level of gluconeogenic precursors. These precursors included lactate, pyruvate, and the following gluconeogenic amino acids: alanine, glycine, serine, aspartate, and methionine. There were marked differences in serum alanine during the first 24 hr. Alanine infusions into small-for-gestational-age infants result in minimal increases in blood glucose concentrations. Although it is impossible to interpret the kinetics of glucose formation and utilization from studies reporting only the absolute concentrations of various substrates, these studies would suggest that small-for-gestational-age infants may have abnormalities in the rate of *de novo* glucose synthesis from gluconeogenic precursors. On the other hand, certain infants who are small for gestational age have been shown to have plasma glucose clearance rates which are significantly faster than those of normal infants (Soltész *et al.*, 1972).

Regardless of its etiology, hypoglycemia must be prevented by the provision of glucose in adequate amounts. Even during fetal life, cerebral glucose utilization rates are equal to those of adult life (Makowski *et al.*, 1972). Accordingly, we have defined the lower level of acceptable glucose concentration at approximately 40–45 mg/dl and would provide a glucose intake to any infant sufficient to maintain that level. The incidence of hypoglycemia is so high among preterm SGA infants (~60%) that we routinely begin an intravenous glucose infusion in the SGA preterm infant. There may be justification for the routine administration of glucose to all SGA infants regardless of their gestational age since even in the term SGA group the incidence of hypoglycemia may exceed 25%.

Polycythemia may occur in any birth-weight/gestational-age category but is most common among infants who are small for gestational age (Humbert *et al.*, 1969). Venous hematocrits of over 60% were found in half of the infants who were term and small for gestational age. The etiology of polycythemia in SGA infants is not well understood. It has been assumed that SGA infants may have had chronic *in utero* hypoxia and that this was the stimulus to erythropoiesis. The extrauterine correlate of such a situation would be high-altitude, hypoxia-stimulated erythropoiesis. Infants with polycythemia may have neurological, pulmonary, or cardiac symptoms. Classical neurological symptomatology includes lethargy alternating with jitteriness and hyperirritability when disturbed, poor feeding, and occasionally frank seizures. Recently, more subtle neurological symptomatology has been reported using the Brazelton neonatal assessment scale (Wirth *et al.*, 1975). Other symptoms may include respiratory distress, peripheral cyanosis, radiographic findings of pulmonary venous congestion, and an enlarged heart. There is also an increased incidence of hyperbilirubinemia.

In symptomatic infants, partial plasma exchange to lower the venous hematocrit has been performed. This procedure is justified on the basis that the long-term prognosis for normal growth and development of children who are symptomatic from polycythemia is poor (Gross *et al.*, 1973). A report of a controlled study (Wirth *et al.*, 1975) suggested that early treatment with partial plasma exchange transfusion was effective in improving outcome at 2 and 6 weeks of age. At present, in any symptomatic infant who has polycythemia (often confirmed with viscosity measurements), we elect to perform a partial plasma exchange in an effort to lower the central hematocrit to 50% or the capillary hematocrit to 60%.

Congenital infection commonly occurs among small-for-gestational-age infants. Such infections may be completely asymptomatic except for their effect upon intrauterine growth. In other cases, congenital infection with rubella, CMV, or toxoplasmosis may present in an acute and fulminant clinical form. The signs and symptoms of infection of viral origin are indistinguishable from those of bacterial septicemia. In screening for the etiology of small-for-gestational-age infants, we have taken viral cultures from the nose, throat, and urine, as well as obtained a cord-blood IgM or an IgM level from the baby in the initial few hours of life. It remains difficult to interpret IgM levels because of the effect of birth weight, gestational age, and postnatal age on normal values. However, a cord-blood or infant blood sample obtained in the first few hours of life which has an IgM content of over 12 mg/dl is highly suggestive of an *in utero* infection. Contamination with maternal blood can be excluded by measuring simultaneously a cord IgA which should also be very low if it is a pure fetal sample.

Other problems peculiar to SGA infants may have their onset during the first few hours of life but generally present during the subsequent days. These problems include abnormalities of calcium homeostasis, difficulty in institution of adequate enteral nutrition, and a unique susceptibility to pulmonary hemorrhage. Continued neurological assessment, particularly in SGA infants, is essential as one attempts to predict prognosis and ultimate outcome.

4.2.3. Specific Problems Related to Perinatal Asphyxia

Perinatal asphyxia refers to the clinical syndrome which may result from hypoxia at any time during the intrapartum period or the immediate postpartum period in the infant. The clinical problems stem from a chain of physiologic events triggered by the hypoxic stimulus. These events include the release of various vasoactive compounds (e.g., catecholamines, bradykinin, prostaglandins) which affect, not only total cardiac output, but more significantly, its distribution to various organs. Some organs such as the brain, kidneys, and adrenals, besides having a high blood flow during fetal life, show marked vasodilation and an increase in organ blood flow in response to hypoxia. It is then not surprising that hemorrhage in these organs may occur in an infant following severe perinatal asphyxia. Other organs, such as the gastrointestinal tract, vasoconstrict in response to hypoxia. The association of severe perinatal asphyxia with many cases of necrotizing enterocolitis (NEC) may be due in part to this response. The changes in blood flow to various organs directly affect the nutrition and metabolism of that organ. Recently, a high incidence of both hypocalcemia and hypoglycemia has been reported in newborn infants who have had perinatal asphyxia. The significance of these observations arises from the fact that these are both easily correctable problems, and both may contribute to an increased incidence of seizures in newborn infants during the second and third days of postnatal life following perinatal hypoxia. Thus, the infant who has been subjected to a hypoxic stress tends to perfuse a core group of organs at the expense of increased tissue hypoxia for a more "peripheral" group of tissues. Many of the problems which develop in these infants during the first week of postnatal life following intrapartum or immediate postpartum asphyxia reflect the hypoxia sustained by underperfused tissues at the time of the hypoxic stress. Cardiac function may also be affected in part by inadequate cardiac nutrition. It is not unusual to find a somewhat enlarged heart and an increased heart rate in newborn infants recovering from fetal distress. This finding is often coupled with pulmonary congestion, representing both increased lymphatic and venous congestion. Together these problems produce a syndrome of respiratory distress characterized by oxygen dependency, tachypnea, and pulmonary hilar congestion typical of the "wet-lung" syndrome. Clotting problems may further compromise organ perfusion, including pulmonary perfusion.

Thus, clinical efforts should be directed in a two-pronged attack on these problems. The first aspect concentrates on early recognition of fetal distress. This area of obstetrics is developing rapidly, including improvements in monitoring techniques and the establishment of a better conceptual base for the interpretation of such data. An integral part of early recognition includes the forewarning of pediatric colleagues to prepare for the resuscitation and immediate support of an infant delivered after fetal distress. The second aspect includes prompt support of both metabolic and physiologic needs of the infant. This support includes blood volume expansion when necessary, early hydration, and maintenance of glucose and calcium concentrations. In general, one should attempt to provide support in all areas of physiology very early in the postnatal period and then reduce these various levels of support as they are no longer required. In this sense it is far easier to practice "preventive" pediatrics than to attempt to compensate for a clinical situation which is rapidly deteriorating.

4.3. First Week of Life

4.3.1. Water and Electrolyte Balance

The total body water content of an infant and the distribution of that water between the intracellular and extracellular space are dependent on gestational age, body weight, and postnatal age. Because of the expanded extracellular fluid volumes of newborn infants, it has been suggested that water administration to a high-risk infant was unnecessary during the first and second days of life. As parenteral glucose infusions were used to treat or prevent hypoglycemia, the improved condition of infants receiving increased water intakes in the first 48 hr of life became clear. However, in the early phase of parenteral fluid administration it was not recommended to add electrolytes to the intravenous infusates. This was based on the belief that infants did not require electrolyte administration for the first and second days of life because of their expanded extracellular fluid volumes. As more experience was gained with the use of parenteral fluids in high-risk infants, it became apparent that large insensible water losses were common in such infants and the renal function was definitely improved when adequate amounts of water were administered. The volumes of water that were required to replace insensible water losses, establish normal urine flow rates, and provide adequate glucose in the form of 10% dextrose frequently led to hyponatremia in infants. It is now recommended that electrolyte supplementation be introduced during the first day of life.

The volumes of water that are required during the first week of life are directly dependent on an infant's insensible water losses. The volume of such insensible losses varies considerably with the baby's size and surface area, and his thermal environment which can be complicated by therapy such as radiant heat warmers or phototherapy lamps. Insensible water losses can vary from 35 to 150 ml/kg. Thus, it is critical to monitor each infant so that appropriate volumes of water may be administered.

A common electrolyte disturbance occurring on most nursery services is hyponatremia. In almost every case hyponatremia is a reflection of inappropriate water administration and not a direct result of abnormal sodium losses. Since the pathophysiology of such hyponatremia is directly related to excessive water administration, the therapy is water restriction, not sodium administration. If hyponatremia exists in the immediate neonatal period, it may be secondary to an inappropriate rate of water administration to the mother during labor. Hyponatremia may also be associated with neonatal meningitis, asphyxia, and birth injuries involving head trauma. A mechanism commonly presumed to be present in these situations is inappropriate ADH excretion, and the therapy required is significant fluid restriction. Hyponatremia is rarely secondary to total body salt depletion. During rapid growth the maintenance sodium requirements are 3 meq/kg/day; if an infant is being maintained on a low-sodium formula, the sodium in his diet may not be sufficient to meet the sodium requirements of his growth rate. Occasionally, small preterm infants develop primary renal salt loss with a urine sodium concentration as high as 200 meq/liter.

Hypernatremia as a result of excessive sodium administration can occur in the neonatal age group. Frequently this is a complication of the use of sodium bicarbonate in the treatment of neonatal acidosis. Hypernatremia reflects not only the excessive amounts of sodium that may be administered in the treatment of a metabolic acidosis but also the inability of the infant to excrete a sodium load rapidly. The risks of hypernatremia and hypertonicity include expansion of the extracellular fluid space and its potential deleterious effects on cardiorespiratory function and an apparent correlation with hemorrhagic phenomenon.

Hyperkalemia may occur in high-risk neonates in association with renal failure and/or excessive endogenous potassium availability from necrotic tissue areas. In general, the most common cause of hyperkalemia in the newborn period is in association with necrotizing enterocolitis when massive amounts of tissue injury may occur leading to an increase in the amount of potassium and hydrogen which must be cleared and excreted by the infant's kidney. Since renal insufficiency may also occur in these conditions, profound hyperkalemia and acidosis may develop rapidly. The first indication of hyperkalemia may be an abnormality noted on the heart rate monitor and confirmed by electrocardiogram such as QRS complexes which appear similar to an idioventricular rhythm. The earlier changes of elevated T waves are often not detected.

Hypokalemia is less frequent but may occur after intensive diuretic therapy without appropriate potassium replacement. Any child on diuretic therapy should have frequent determinations of the serum chloride and potassium, and appropriate replacement prescribed.

4.3.2. Nutrition during the Neonatal Period

The problems of complete nutritional support continue to be among the most difficult medical problems encountered in an intensive-care nursery. The precise nutritional requirements of very-low-birth-weight infants are still ill-defined. A variety of feeding regimens have been proposed which are often mutually contradictory.

The total caloric requirement for sustained growth in infants has been defined as 120 cal/kg/day to achieve a growth rate of 1–2% a day. Although this may be an overestimation in the healthy full-term infant, caloric intakes much greater than this figure may be required in certain infants during the acute and recovery phase of serious illnesses and during catch-up growth.

The protein requirements of preterm babies are not precisely defined but appear to be between 3 and 4 g/kg/day. Davidson et al. (1967) demonstrated that protein intakes of 6 g/kg/day were associated with azotemia without significant improvement in terms of weight gain. In addition he demonstrated a decreased rate of growth on protein intakes of 2 g/kg/day. In a later study Kagan et al. (1972) demonstrated that there was no significant difference in true growth rate for infants on high-protein feedings. Much of the weight gain associated with high-protein intakes was accounted for by mineral and water retention rather than by true growth. A 3-year follow-up study by Goldman has suggested that high-protein feedings given early in the postnatal period to low-birth-weight infants may be associated with poor neurologic outcome at 3 years of age. Thus, it seems reasonable to provide between 2 and 4 g/kg/day of protein in routine feedings for preterm infants.

Carbohydrate in the diet provides an easily absorbed, readily utilized source of energy and carbon for growth. During fetal life approximately 45% of the total caloric requirements for metabolism and growth is met by carbohydrate (James *et al.*, 1972). In human breast milk, carbohydrate also represents approximately 45% of the total calories. The extent to which carbohydrate and fat are required as carbon sources for energy, thus sparing protein and ensuring growth, is not well established. Given sufficient calories, positive nitrogen balance can be achieved in patients maintained on carbohydrate-free diets (Hoover *et al.*, 1975; Blackburn *et al.*, 1973*a,b*). In general, one can rely on carbohydrate absorption and utilization in even the most immature preterm infants, although there have been some reports of marked hyperglycemia developing on moderate carbohydrate intake (Dweck and Cassady, 1974). Studies have demonstrated the efficient utilization of partially hydrolyzed starch in newborn infants, and this remains an effective method of increasing caloric concentration in a milk diet.

In fetal life there is minimal utilization of fat as a source of carbon for growth and as a metabolic fuel (James *et al.*, 1971). However, human breast milk has approximately 50% of its total calories supplied as fat. Newborn infants and particularly preterm infants have a lower proportion of the total fat ingested which is absorbed (Katz and Hamilton, 1974). The inefficiency of fat absorption reflects in part the gradual maturation of bile salt production and excretion by the liver which occurs during the later third of gestation (Watkins *et al.*, 1973, 1975). Long-chain saturated fatty acids are absorbed less efficiently than unsaturated fatty acids by infants. In addition, the specific structure of a triglyceride is important in determining its efficiency of absorption (Filer *et al.*, 1969). A recent study has demonstrated that the absorption of medium-chain triglycerides in low-birth-weight infants is greater than that of complex lipids (Roy *et al.*, 1975). Thus, medium-chain triglycerides may provide an additional source of calories under special circumstances.

Because of the inefficiency and incoordination of sucking prior to approximately the 36th week of gestation, feedings are usually provided by intermittent gavage. Recently in an effort to overcome the abnormalities of stomach volume capacity and gastrointestinal motility, the use of continuous slow nasojejunal feedings has been suggested (Cheek and Staub, 1973). This technique has proven to be a valuable adjunct in certain selected infants who have abnormalities of gastrointestinal function. It does not offer any advantage as a routine technique. The use of gastrostomy in preterm infants has been discarded.

The requirements for vitamins, minerals, and trace metals are even less well defined than the total caloric requirements. However, low-birth-weight infants do not have adequate vitamin intake from the formula which they receive even though that formula is always supplemented. Thus we have administered supplemental multiple vitamins including A, C, D, and the B-complex vitamins to all low-birth-weight infants.

4.3.3. Bilirubin Metabolism

Serum bilirubin concentrations in newborn infants are higher than at any other time in life. Serum bilirubin concentrations < 12 mg/dl are regarded as within the "physiological" range. The peak of physiological jaundice occurs between the third and fifth day in full-term infants and between the fifth and seventh day in preterm

FREDERICK C.
BATTAGLIA AND
MICHAEL A. SIMMONS

infants. Preterm infants have significantly higher bilirubin concentrations than either term AGA or SGA infants.

The concern with elevated bilirubin concentrations in preterm infants relates directly to the danger of bilirubin encephalopathy or kernicterus. Since the study of Mollison and Cutbush (1949), the association between severe unconjugated hyper-bilirubinemia and kernicterus has been firmly established (Mollison and Cutbush, 1954; Hsia *et al.*, 1952).

Preterm infants are at higher risk of developing bilirubin encephalopathy than full-term infants but the exact explanation for this susceptibility has not been established. A number of factors which would increase the risk for preterm infants have been described. Lower concentrations of serum albumin and reduced bilirubin binding capacity may play a major role. In addition, immature infants may have an increased permeability of the blood–brain barrier, although this has largely been discredited on the basis of a study demonstrating no difference between the adult and newborn animal in bilirubin penetration of the central nervous system (Diamond and Schmid, 1966) and since bilirubin encephalopathy has occurred in adults. The frequent occurrence of acidosis (which may reduce albumin binding capacity for bilirubin) and hypoxemia (which may increase the susceptibility of cells to the toxic effects of bilirubin) among preterm infants may be additional contributing factors. There is no definitive guideline for a safe serum bilirubin concentration in preterm infants. The cumulative risk of the bilirubin concentration, the degree of immaturity of the infant, serum protein concentration, severity of the infant's illness, and the existence of acidosis and hypoxemia are all taken into account in deciding when therapy should be instituted.

The reasons for the exaggerated hyperbilirubinemia among preterm infants are better understood. Preterm infants have a higher bilirubin production rate than normal adults (Maisels *et al.*, 1971). This increased bilirubin production rate is probably due to the higher circulating red cell volume, decreased red cell life-span, and increased incidence of ineffective erythropoiesis. In addition to the increased bilirubin production, there is animal evidence that newborn primates have defective uptake of bilirubin into the hepatocyte and defective conjugation and excretion of bilirubin into bile. The Y protein receptor which is important in hepatic uptake of bilirubin is very low in the immediate neonatal period and may not reach adult levels until the end of the first week of postnatal life (Levi *et al.*, 1970). Marked impairment in conjugation capacity of the newborn's liver has also been reported in monkeys (Gartner and Lane, 1971) and in the human (Lathe and Walker, 1958), although conflicting evidence in the human has been presented (DiToro *et al.*, 1968).

Physiological jaundice in the newborn may be associated with an exaggerated enterohepatic recirculation of bilirubin (Poland and Odell, 1971). Among preterm infants it is common to have episodes of poor gastrointestinal motility and thus to delay feedings during a period of acute illness. With decreased gut motility and emptying there could be an increase in the enterohepatic circulation of bilirubin with resultant hyperbilirubinemia. In addition, a preterm infant may have a persistent fetal hepatic circulatory pattern which diverts a large percentage of portal blood flow away from circulation through the liver to the ductus venosus into the right heart. If the ductus venosus were to remain patent postnatally, a significant portion of the bilirubin in the portal blood flow would be shunted from the circulation of the liver.

With the high incidence of hyperbilirubinemia among preterm infants and their unique susceptibility to the toxic effects of bilirubin, it is important to prevent dangerous levels of hyperbilirubinemia. This requires frequent monitoring of a preterm infant's bilirubin concentration and the prompt institution of therapy when appropriate. Basically two therapeutic modalities exist for the treatment of hyperbilirubinemia: exchange transfusion and phototherapy. The experience with exchange transfusion was developed for the management of erythroblastosis fetalis but gradually has been extended for the management of all forms of marked hyperbilirubinemia.

Exchange transfusion has continued as the primary mode of therapy for the management of severe hyperbilirubinemia. The indications for exchange transfusion are based on a cumulative risk assessment previously outlined. As newer methods are developed for the assessment of free bilirubin concentration or of reserve binding capacity of albumin for bilirubin, more specific guidelines for exchange transfusion and other forms of therapy will be developed.

Over the past 10 years phototherapy has been introduced for the management of hyperbilirubinemia in both preterm and term infants. Exposure to blue light leads to a fall in serum bilirubin concentration in preterm and term infants and can be used to reduce the incidence of hyperbilirubinemia among preterm infants.

The exact mechanism whereby light of 450–460 nm wavelenth acts to lower bilirubin is unclear. In infants with Crigler–Najjar syndrome, lipid-soluble bilirubin can be demonstrated to be broken down to water-soluble products which may be excreted in the bile and in the urine (Callahan *et al.*, 1970). In experimental animals treated with phototherapy, Ostrow (1971) has demonstrated that nearly half of a radioactive bilirubin load can be recovered in the intact, unconjugated form from bile and only a portion of the remaining radioactivity was recoverable in the form of dipyrrol compounds. The exact mechanism by which phototherapy leads to an increased bilirubin clearance awaits further study.

Usually the bilirubin level has stabilized by one week of postnatal age in preterm infants. The presence of prolonged hyperbilirubinemia or the existence of conjugated hyperbilirubinemia are each suggestive of underlying disease processes, in particular, unsuspected infection. Among infants with prolonged hyperbilirubinemia or conjugated hyperbilirubinemia, careful attention should be paid to rule out the possibility of urinary-tract infection or viral infections by appropriate cultures.

4.3.4. Calcium Metabolism

During fetal life, the total calcium concentration and calcium activity in fetal serum are higher than in maternal serum, suggesting active transport of calcium across the placenta. The role of parathormone of fetal origin vs. parathormone of maternal origin and of vitamin D in regulating placental transport of calcium is still unknown. The bulk of calcium accretion in the fetus occurs after the 30th week of gestational age; therefore, the premature infant tends to have smaller total body calcium stores available for calcium homeostasis.

Neonatal hypocalcemia tends to occur with highest frequency at two times in the neonatal period, one peak occurring in the first 48 hr of life and another after the seventh day of life. The latter peak occurs predominantly in full-term newborn infants who are being fed a modified cow's milk formula with a low calcium/phosphate ratio. The presumed etiology for this neonatal tetany of later onset is

FREDERICK C.
BATTAGLIA AND
MICHAEL A. SIMMONS

hyperphosphatemia which leads to hypocalcemia. The hyperphosphatemia has been attributed to the increased phosphate load from formulas derived from cow's milk, in concert with diminished phosphate excretion because of relative renal immaturity. However, maternal vitamin D metabolism also plays a role in determining both the calcium transfer to the fetus, particularly in the third trimester of pregnancy, and the incidence of tetany in the newborn. Studies from the University of Edinburgh have shown a higher incidence of late tetany of the newborn in those infants whose development during the latter third of gestation occurred in the winter months, with the least maternal exposure to sunlight. These children also showed defects in enamel formation of the deciduous teeth in those areas of the tooth in which enamel deposition occurs during the third trimester of gestation (Purvis *et al.*, 1973; Roberts *et al.*, 1973).

Among preterm infants, hypocalcemia most often occurs early in the postnatal period. It has been clearly shown that even preterm infants respond normally to an exogenous administration of parathormone. Presumably their hypocalcemia is secondary to decreased parathyroid responsiveness and consequent deficits of parathormone concentrations (Tsang and Oh, 1970). At present we have no adequate explanation for the development of transient primary hypoparathyroidism. In the preterm infant, several factors may be operating to increase the frequency of hypocalcemia. In a very immature infant, total calcium stores available for homeostasis may be deficient. The stress associated with difficult deliveries or asphyxia has been implicated in the etiology of hypocalcemia. In addition, low-birth-weight infants and preterm infants are prone to fetal distress in the intrapartum period.

Tsang and Oh have shown that parathyroid responsiveness is determined by gestational age, not birth weight, so that in the absence of other factors leading to hypocalcemia (such as stress, asphyxia, bicarbonate administration, etc.), a small-for-gestational-age infant should not be expected to have a higher incidence of hypocalcemia than his normally grown peer. Although there have been reports associating hypocalcemia with an increased incidence of apnea and with cardiomegaly and heart failure, there is no good evidence that hypocalcemia untreated in the immediate neonatal period leads to any long-term sequelae. Often hypocalcemia is detected only because of the high index of suspicion of low calcium levels in preterm infants. Such infants may be asymptomatic even in the face of significant hypocalcemia.

In general, those term infants who develop early-onset hypocalcemia have encountered significant stress of asphyxia during the intrapartum or immediate postpartum period. Any full-term infant who has developed hypocalcemia without perinatal stress should be evaluated carefully for the possibility of primary maternal hyperparathyroidism. Presumably, the elevated levels of maternal parathormone produce increased maternal serum calcium concentrations and thus increased fetal calcium concentrations. This depresses any fetal endogenous parathyroid activity and leads to functional hypoparathyroidism.

4.3.5. Continued Neurological Assessment

Much attention has been paid to the ultimate prognosis of premature infants in terms of physical growth and long-term neurologic sequelae. In contrast to this, the continued assessment of the neurological growth and development of such infants while in the neonatal unit is frequently neglected. It is important to continue the thorough neurological assessment made at birth on a routine and periodic basis

throughout the infant's stay in a nursery. Techniques for such continued neurological assessment can vary from classical neurological tests (motor tone, strength and
activity, deep tendon reflexes, testing of sensory modalities, etc.) to more recent
developments such as the Brazelton Neonatal Behavioral Assessment scale. The
limitations of classical neurological techniques in the nursery are well known. Since
the greatest interest is in evaluating cognitive functioning which is poorly tested by
all traditional neurological techniques, the Prechtl scale and, to a greater extent, the
Brazelton scale have allowed continued neurological assessment to focus more on
behavioral adaptations of the infant that may occur over time. Although the
Brazelton scale is a rather recent innovation, it seems likely that its sequential use
on a newborn service may prove to be a useful technique of continued neurological
assessment.

If the neurological examination technique described by Lubchenco and outlined in Table II is used repeatedly over the several weeks that a child is in the
hospital, one can detect the fact that an infant may not be progressing in achieving
new neurological landmarks. Should such deficits appear, we would recommend
using the Brazelton scale to detect any other behavioral abnormalities.

Recent studies have suggested that the cognitive function of a neonate may be
measurable by various techniques of assessing attention span and extinction
responses. The exact role that such techniques will play in the future in the
continued neurological assessment of high-risk infants is not clear, but these
techniques have already opened for investigation a whole new area of neurological
research in high-risk infants.

4.3.6. The Parental Factor

Attention should be given during the first few hours of life to the special needs
of parents of high-risk infants and of their influence on the growth and development
of their offspring. In the past, nursing regulations have limited the access of parents
and other nonmedical personnel to the wards because of the fear of communicable
illness. This philosophy presupposes that the risk to the infant of the acquisition of
infection is greater than the risk of social deprivation caused by parental separation.
The idea that the infant of low birth weight has physical problems of such an
overwhelming dimension that social interactions which all normal infants undergo
during the first week of life do not deserve consideration has been challenged
recently. In part, this challenge rests on the demonstration of the safety, with regard
to infection and potential disturbance of nursing and medical routine, in having
parents visit freely a neonatal intensive-care unit after careful instruction in "isolation" techniques. The small additional risk of a communicable illness is more than
justified by the advantages of having the parents participate in the care of their
infant.

Animal studies in nonprimates that are now over 25 years old should have
suggested to us a number of years ago the potential importance of early maternal–
infant attachment in man. Konrad Lorenz in studies on imprinting demonstrated the
importance of early imprinting on subsequent infant behavior, growth, and development in several animal species. A major factor in such imprinting seemed to be the
availability of a "mother" figure immediately after birth and constantly thereafter.

That attachment processes occur in man has been suspected by careful and
observant clinicians, but the data to support such attachment processes have only
recently been collected. Studies in man have focused primarily on the mechanisms

FREDERICK C.
BATTAGLIA AND
MICHAEL A. SIMMONS

and importance of maternal aspects of the attachment. Bibring (1959) has identified many behavioral changes that occur in a mother during her pregnancy and many of these changes can be viewed as the initial steps of the maternal–infant attachment. Kennell and Klaus (1971) have emphasized the importance of these changes in behavior during pregnancy in their outline of the steps of maternal–infant attachment. Klaus *et al.* (1972) have likewise emphasized the significance of the first few hours and days of the life of an infant to the ultimate attachment of a mother to her infant. Kempe, at the University of Colorado, has demonstrated that certain variables of the pathological attachment process may be accurately predicted by a mother's initial responses to her infant immediately after birth. This would suggest that definitive attachment of a mother to an infant has occurred prior to delivery and that various pathological processes can intervene even at this early stage.

In small preterm infants, because of their physical demands and the sophistication of the medical and nursing care that is required, it has been common to exclude the mother from participating in the care of her infant simply because she lacks the technical skills and expertise to be of obvious help. Only recently, as the results of maternal–infant separation studies have become available, has it become a recommended practice to involve mothers in early contact with their infant and indeed occasionally in the care of their premature infant. The importance of early contact and involvement on ultimate outcome of maternal attachment has been demonstrated in two controlled prospective studies (Klaus and Kennell, 1970; Leifer *et al.*, 1972). The parents' early contact and involvement with their infant has led to a better understanding of the nature and significance of their infant's illness as well as to an ease in communicating to them any changes in their infant's condition. We urge physicians taking care of such infants not to emphasize the high expected mortality rates in small infants because almost always this emphasis is unrealistically pessimistic. With a more positive emphasis on the infant's chance of survival, the parents' involvement in an attachment with the child may be more secure since their behavior can be devoted toward this attachment rather than toward pregrieving in anticipation of a loss.

The emphasis on helping the parents treat their premature infant as their own child rather than as a very ill third party has been helpful in promoting maternal–infant attachment. Because of its possible nutritional advantages and also because it is another way of involving the mother and father in the care of their infant, more emphasis has been placed on providing small premature infants with breast milk. Elaborate precautions about sterility, processing, and storage of breast milk are necessary before such a program is undertaken, but there can be no doubt that among those mothers who do routinely provide breast milk for their small infant's feedings, this contribution which the mother is allowed to make creates a special feeling of involvement in the care of her infant.

4.4. Growth during the First Few Months of Postnatal Life

4.4.1. Evaluation of Growth and "Catch-Up" Growth

By the end of the first week of life, nutrition of the LBW infant has improved to the point where total caloric intake is adequate for growth. The second peak incidence of neonatal tetany occurring at the end of the first week of life in infants fed a cow's milk formula is a reflection of the increased dietary intake and thus of the increasing importance of the composition of the diet to the infant's health.

Problems relating to the gastrointestinal tract occur more frequently at this time, again reflecting the fact that this organ system is now tested more thoroughly by an increased dietary intake. These problems include disorders of motility (e.g., chalasia) and of absorption (e.g., lactose intolerance).

With additional protein and carbohydrate intake, inborn errors of metabolism present with increasing severity of signs and symptoms. Galactosemia, maple-syrup urine disease, etc., may present with a metabolic acidosis which becomes worse and includes more obvious signs of illness. Despite the severity of the signs, some of the intolerances to protein or carbohydrate may prove to be transient and do not preclude a normal dietary intake at a later date. The recent report of Danks *et al.* (1975) of a 20-week-old female infant with metabolic acidosis and dehydration secondary to tyrosinemia is an example of such a clinical course. In spite of the initial severity of the illness which required a marked reduction in phenylalanine and tyrosine intake for a period of time, the child eventually tolerated a normal dietary intake of these and other amino acids without biochemical abnormality at the age of 17 months.

Similarly, the function of other organ systems reflects the continued adaptation required for extrauterine life. Oxygen consumption of the newborn infant, both term, preterm, and SGA infants, increases throughout the first several months of life (Figures 15 and 16) (Hill and Robinson, 1968). The increase in metabolic rate in preterm babies occurs at a time when many of the infants may be recovering from some form of respiratory distress in the immediate newborn period. With recovery, pulmonary vascular resistance decreases. If there is a patent ductus arteriosus

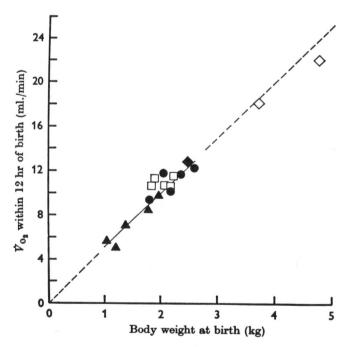

Fig. 15. Triangles: young preterm. Circles: preterm. Squares: small-for-dates. Diamonds: babies of diabetic mothers. The symbol has been filled in when the baby's weight is appropriate to its gestational age and has been left open when the weight is either lower or higher than usual for gestational age, i.e., open for small-for-dates and large-for-dates. The dashed line is defined V_{O_2} ml/min = 5 × body weight, kg. From Hill and Robinson (1968).

FREDERICK C.
BATTAGLIA AND
MICHAEL A. SIMMONS

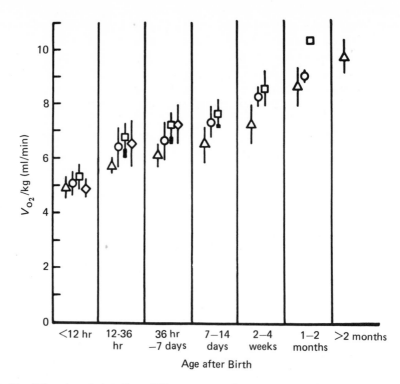

Fig. 16. The different symbols indicate different groups; the mean ±2 SE of mean is shown. Where a filled rectangle is attached to the small-for-dates group, it indicates that the mean differs from that of the young preterm group at P < 0.05. From Hill and Robinson (1968).

present, a large left-to-right shunt through the ductus will then occur. At the same time, increasing anemia may be present, in part physiologic and in part iatrogenic, from frequent chemical determinations in the first week of life. Thus, the demands for increased organ perfusion may occur at precisely the same time as an increasing shunt from a large patent ductus arteriosus and a decrease in oxygen capacity in the blood. Other organ-system function then becomes affected by this chain of physiologic events. For example, this may include diminished gastrointestinal function and thus affect the nutritional status of the infant. However, when adequate caloric intake is established, both SGA and preterm babies show comparable rates of postnatal growth. Numerous attempts to accelerate the growth rate of small infants by a variety of feeding techniques which include increased caloric intake in the early neonatal period have not significantly altered the postnatal growth during the first

Table III.   Weekly Increments in Body Weight (Mean ± SD)[a]

| Week | Control (g) | Photototherapy | |
		Continuous (g)	Intermittent (g)
1	97.3 ± 76.4	90.9 ± 74.5	105.3 ± 77.7
2	139.0 ± 69.8	184.5[b] ± 55.5	193.6[b] ± 82.3
3	162.4 ± 62.3	225.3[b] ± 62.5	195.5[b] ± 74.2
4	161.7 ± 82.7	168.1 ± 103.9	179.4 ± 63.1

[a]From Wu et al., 1974, J. Pediatr. 85:563, with permission.
[b]Indicates values significantly different (P < 0.05) from the control group.

few months of life, although such attempts have blunted or abolished the weight loss which occurs in the first week of postnatal life. An example of "catch-up" growth is found in the infant's response to phototherapy. Infants receiving phototherapy show a greater weight loss than a control group of comparable age and weight. However, their weight gain in the second and third weeks of life after phototherapy is discontinued is greater, and their final body weights at the end of the third week of life are no different from the control group (Table III) (Wu *et al.*, 1974).

4.4.2. Continued Assessment of Neurologic Development

The later neonatal period is particularly important for the careful reassessment of neurologic development. It is during this time when the infant has recovered from many of the acute illnesses of the first few days of life that the physician attempts to evaluate the residual impact of such illnesses on neurologic development. The techniques for this assessment are still rather crude. However, many recent reports have demonstrated the usefulness and potential clinical significance of "habituation" tests or tests of "boredom or fatigue" to repeated auditory or visual stimuli. Such tests requiring the recognition of a stimulus, the memory of the stimulus as alike or different from previous stimuli, and a change in a response to the stimulus provide a more direct assessment of cortical function. An example of such tests for the evaluation of neurologic development in the neonatal period is given in the study of Brackbill *et al.* (1974). For the same purpose, attempts are being made to evaluate the father's and the mother's capacity for healthy "parenting" and in this manner to provide early recognition of those families where increased emotional and other support will be required to avoid child-battering, physical or emotional. The later phase of neonatal development is also a period when decisions are made regarding child placement (i.e., foster care, adoption, etc.), thus increasing the need for a careful and thoughtful appraisal of neurologic development from a prognostic standpoint.

4.5. Long-Term Effects

4.5.1. Physical Growth

The effect on adult size of a premature birth or some degree of intrauterine growth retardation is not consistent. In some cases, such as infants of diabetic mothers, it would appear that the body proportions characteristic of these patients as infants persist at least into adolescence. Thus Farquhar (1969) reported that the height was somewhat reduced and the body-weight-to-height ratio increased in children of diabetic mothers, suggesting an increase in body fat to height proportions. In trisomy 21, the Silver–Russell syndrome, and many other syndromes associated with intrauterine growth retardation and congenital anomalies, there is significant shortness of stature persisting throughout childhood and into early adolescence. It is possible that some of the changes in body composition may be advantageous to the child. The decrease in fat cell number in SGA infants reported by Brook (1972) could be advantageous if it persists into adult life. The question of whether there is a relationship between the rate of maturation in early life and the rate of aging in adult life has not been adequately studied. Certainly, if one considers mammals as a whole, there is a striking relationship between length of gestation and life-span of the organism; mammals completing intrauterine development slowly also exhibit a slower aging process.

4.5.2. Neurologic Sequelae

FREDERICK C.
BATTAGLIA AND
MICHAEL A. SIMMONS

Neurologic sequelae of a low birth weight have been amply documented in the past. A number of studies in the 1950s quoted an incidence of $\sim 50\%$ for major CNS problems in the follow-up of babies with birth weights < 1500 g. Spastic diplegia was reported during this same decade in $\sim 20\%$ of preterm low-birth-weight babies. However, these problems have been markedly reduced in frequency. More recent follow-up studies on infants < 1500 g report an incidence of $\sim 10\%$ for major neurologic sequelae. In recent follow-up studies of preterm babies spastic diplegia has not been found at all. This latter observation has stimulated considerable speculation regarding the possible etiology of spastic diplegia.

4.5.3. Impact of Perinatal Events in Adulthood

The effect on adult health of events occurring in the perinatal period is being documented by an increasing number of studies. Such studies have suggested in animal experimentation that the administration of a drug in the perinatal period may alter the animal's response to that drug during adulthood. Other drug exposure in the perinatal period may initiate events in the tissues of the fetus which are recognized only in adulthood. The administration of stilbestrol to pregnant women in the late 1940s and 1950s provides a striking example of the delayed manifestations of the effects of a drug. In a few female infant fetuses exposed to the drug, a change in cell biology was initiated in the vagina which led to the development of a rare adenocarcinoma of the vagina in adult life (Ulfelder, 1973). It is important to emphasize that these fetuses were entirely normal at birth and normal throughout childhood as well. Only in adulthood did the problem present. Furthermore, the fact that this tumor was extremely rare even in elderly women and virtually unknown in young women was sufficient to stimulate a retrospective study designed to explore the possibility of some environmental trigger to explain its occurrence in young women. Otherwise, the relationship of time between exposure *in utero* to stilbestrol and a tumor in adulthood might still be undetermined. Similarly, in a prospective study of the effects of X rays of the abdomen in pregnant women on their offspring, an increased fertility rate was noted in the female fetuses exposed to radiation *in utero* (Meyer and Tonascia, 1973). Thus, an environmental effect on an organ in fetal life, in this case, the ovary, did not produce a recognizable effect until adult life, when the women reached the child-bearing age group.

Along similar lines, emphasizing the continuum of growth and development, aging and reproduction, one could consider the shifting age of menarche toward a younger age group in recent years. This trend may now have arrested in most Western cultures (see chapter 16). The trend seems to represent a cumulative effect of better nutrition (as well as other factors) which begins in intrauterine life.

5. References

Abramowicz, M., and Kass, E. H., 1966, Pathogenesis and prognosis of prematurity, *N. Engl. J. Med.* **275**:878, 938, 1001, 1053.

Aherne, W., and Hull, D., 1966, Brown adipose tissue and heat production in the newborn infant, *J. Pathol. Bacteriol.* **91**:223.

Alexander, G., 1964, Studies on the placenta of the sheep (*Ovis aries* L.). Placental size, *J. Reprod. Fertil.* **7**:289.

Alleem, F. A., Pinketon, J. H. M., and Neill, D. W., 1969, Clinical significance of the amniotic fluid oestrial level, *J. Obstet. Gynaecol. Br. Commonw.* **76:**200.

Avery, M. E., and Mead, J., 1959, Surface properties in relation to atelectasis and hyaline membrane disease, *Am. J. Dis. Child.* **97:**517.

Barton, M. D., Killman, A. P., and Meschia, G., 1974, Response of ovine uterine blood flow to epinephrine and norepinephrine, *Proc. Soc. Exp. Biol. Med.* **145:**996.

Battaglia, F. C., and Woolever, C. A., 1968, Fetal and neonatal complications associated with recurrent chorioangiomas, *Pediatrics* **41:**62.

Bibring, G. L., 1959, Some considerations of the psychological processes in pregnancy, *Psychoanal. Study Child* **14:**113.

Bishop, E. H., and Corson, S., 1968, Estimation of fetal maturity by cytologic examination of amniotic fluid, *Am. J. Obstet. Gynecol.* **102:**654.

Blackburn, G. L., Flatt, J. P., Clowes, G. H. A., and O'Donnell, T. E., 1973*a,* Peripheral intravenous feeding with isotonic amino acid solutions, *Am. J. Surg.* **125:**447.

Blackburn, G. L., Flatt, J. P., Clowes, G. H., O'Donnell, T. F., and Hensle, T. E., 1973*b,* Protein sparing therapy during periods of starvation with sepsis or trauma, *Ann. Surg.* **177:**588.

Blom, S., and Finnström, O., 1968, Motor conduction velocities in newborn infants of various gestational ages, *Acta Pediatr. Scand.* **57:**377.

Bolognese, R. J., Corson, S. L., Touchstone, J. C., and Lakoff, R. M., 1971, Correlation of amniotic fluid estriol with fetal age and well-being, *Obstet. Gynecol.* **37:**437.

Boué, A., and Boué, J. G., 1969, Effects of rubella virus infection on the division of human cells, *Am. J. Dis. Child.* **118:**45.

Brackbill, Y., Kane, J., Manniello, R. L., and Abrasom, D., 1974, Obstetric premedication and infant outcome, *Am. J. Obstet. Gynecol.* **118:**377.

Brook, C. G. D., 1972, Evidence for a sensitive period in adipose-cell replication in man, *Lancet* **2:**624.

Brück, K., 1961, Temperature regulation in the newborn infant, *Biol. Neonate* **3:**65.

Butler, N. R., ed., 1969, British Perinatal Mortality Survey 1958. Perinatal problems; the second report of the 1958 British Perinatal Mortality Survey, Livingstone, Edinburgh.

Callahan, E. W., Thaler, M. M., Karon, M., Bauer, K., and Schmid, R., 1970, Phototherapy of severe unconjugated hyperbilirubinemia: Formation and removal of labeled bilirubin derivatives, *Pediatrics* **46:**841.

Campbell, S., and Dewhurst, C. J., 1971, Diagnosis of the small-for-dates fetus by serial ultrasonic cephalometry, *Lancet* **2:**1002.

Cassady, G., 1970, Body composition in intrauterine growth retardation, *Pediatr. Clin. North Am.* **17:**79.

Cheek, J. A., and Staub, G. F., 1973, Naso-jejunal alimentation for premature and full-term newborn infants, *J. Pediatr.* **82:**955.

Christie, A., 1949, Prevalence and distribution of ossification centers in the newborn infant, *Am. J. Dis. Child.* **77:**355.

Danks, D. M., Tippett, P., and Rogers, J., 1975, A new form of prolonged transient tyrosinemia presenting with severe metabolic acidosis, *Acta Pediatr. Scand.* **64:**209.

Davidson, M., Levine, S. Z., Bauer, C. H., and Dann, M., 1967, Feeding studies in low-birth-weight infants. I. Relationships of dietary protein, fat, and electrolytes to rates of weight gain, clinical courses, and serum chemical concentrations, *J. Pediatr.* **70:**695.

Desmond, M. M., Montgomery, J. R., Melnick, J. L., Cochran, G. G., and Verniand, W., 1969, Congenital rubella encephalitis: Effects on growth and early development, *Am. J. Dis. Child.* **118:**30.

Diamond, I., and Schmid, R., 1966, Experimental bilirubin encephalopathy. The mode of entry of bilirubin 14-C into the central nervous system, *J. Clin. Invest.* **45:**678.

DiToro, R., Lupi, L., and Ansanelli, V., 1968, Glucuronation of the liver in premature babies, *Nature* **219:**265.

Dreyfus-Brisac, C., 1965, Sleep of premature and full-term neonates, *Proc. R. Soc. Med.* **58:**6.

Dubowitz, L., Dubowitz, V., and Goldberg, C., 1970, Clinical assessment of gestational age in the newborn infant, *J. Pediatr.* **77:**1.

Dweck, H. S., and Cassady, G. E., 1974, Glucose intolerance in infants of very low birth weight. I. Incidence of hyperglycemia in infants of birth weights 1,100 grams or less, *Pediatrics* **53:**189.

Farquhar, J. R., 1969, Prognosis for babies born to diabetic mothers in Edinburgh, *Arch. Diseases Childh.* **44:**36.

Filer, L. J., Mattson, F. H., and Fomon, S. J., 1969, Triglyceride configuration and fat absorption by the human infant, *J. Nutr.* **99:**293.

Frazier, J. M., Davis, G. H., Goldstein, H., and Goldberg, I. D., 1961, Cigarette smoking and prematurity: A prospective study, *Am. J. Obstet. Gynecol.* **81:**988.

Frisch, R. E., 1972, Weight at menarche: Similarity for well-nourished and undernourished girls at differing ages, and evidence for historical constancy, *Pediatrics* **50**:445.

Gartner, L. M., and Lane, D., 1971, Hepatic metabolism and transport of bilirubin during physiological jaundice in the newborn rhesus monkey, *Pediatr. Res.* **5**:413 (abstract).

Gluck, L., and Kulovich, M., 1973, Lecithin isphingomyelin ratios in amniotic fluid in normal and abnormal pregnancy, *Am. J. Obstet. Gynecol.* **115**:539.

Gluck, L., Kulovich, M. V., Borer, R. C., Brenner, P. H., Anderson, G. G., and Spellacy, W. N., 1971, Diagnosis of the respiratory distress syndrome by amniocentesis, *Am. J. Obstet. Gynecol.* **109**:440.

Gregory, G. A., Kitterman, J. A., Phibbs, R. H., Tooley, W. H., and Hamilton, W. K., 1971, Treatment of idiopathic respiratory-distress syndrome with continuous positive airway pressure, *N. Engl. J. Med.* **284**:1333.

Gross, G. P., Hathaway, W. E., and McGaughey, H. R., 1973, Hyperviscosity in the neonate, *J. Pediatr.* **82**:1004.

Gruenwald, P., 1963, Chronic fetal distress and placental insufficiency, *Biol. Neonate* **5**:215.

Gruenwald, P., 1970, Intrauterine growth, in: *Physiology of the Perinatal Period* (U. Stave, ed.), pp. 3–27, Appleton-Century-Crofts, New York.

Haymond, M. W., Karl, I. E., and Pagliara, A. S., 1974, Increased gluconeogenic substrates in the small-for-gestational age infant, *N. Engl. J. Med.* **291**:322.

Hansen, J. D. L., Brinkman, G. L., and Bowie, M. D., 1965, Body composition in protein–calorie malnutrition, *S. Afr. Med. J.* **39**:491.

Hey, E. N., 1969, The relation between environmental temperature and oxygen consumption in the newborn baby, *J. Physiol.* **200**:589.

Hill, J. R., and Rahimtulla, K. A., 1965, Heat balance and the metabolic rate of newborn babies in relation to environmental temperature and the effect of age and weight on basal metabolic rate, *J. Physiol.* **180**:239.

Hill, J. R., and Robinson, D. C., 1968, Oxygen consumption in normally grown small-for-dates and large-for-dates newborn infants, *J. Physiol.* **199**:685.

Hill, D. E., Arellano, C. P., Izakawa, T., Holt, A. B., and Cheek, D. B., 1970, Studies in infants and children with congenital rubella: Oxygen consumption, body water, cell mass, muscle and adipose tissue composition, *Johns Hopkins Med. J.* **127**:309.

Honnebier, W. J., and Swaab, D. F., 1973, The influence of anencephaly upon intrauterine growth of fetus and placenta and upon gestation length, *J. Obstet. Gynaecol. Br. Commonw.* **80**:577.

Hoover, H. C., Jr., Grant, J. P., Gorschboth, C., and Ketcham, A. S., 1975, Nitrogen-sparing intravenous fluids in postoperative patients, *N. Engl. J. Med.* **293**:172.

Hsia, D. Y., Allen, F. H., Gellis, S. S., and Diamond, L. K., 1952, Erythroblastosis fetalis: Studies of serum bilirubin in relation to kernicterus, *N. Engl. J. Med.* **247**:668.

Huggett, A. St. G., and Widdas, W. F., 1951, The relationship between mammalian foetal weight and conception age, *J. Physiol.* **114**:306.

Humbert, J. R., Abelson, H., Hathaway, W. E., and Battaglia, F. C., 1969, Polycythemia in small for gestational age infants, *J. Pediatr.* **75**:812.

James, E., Meschia, G., and Battaglia, F. C., 1971, A-V differences of free fatty acids and glycerol in the ovine umbilical circulation, *Proc. Soc. Exp. Biol. Med.* **138**:823.

James, E. J., Raye, J. R., Gresham, E. L., Makowski, E. L., Meschia, G., and Battaglia, F. C., 1972, Fetal oxygen consumption, carbon dioxide production, and glucose uptake in a chronic sheep preparation, *Pediatrics* **50**:361.

Järvinen, P. A., Panhamaa, P., and Kinnunen, O., 1958, The effect of the duration of toxemia on the weight of the fetus, *Ann. Chir. Gynaecol. Fenn.* **47**(Suppl. 81):76.

Jones, K. L., and Smith, D. W., 1973, Recognition of the fetal syndrome in early infancy, *Lancet* **2**:999.

Jones, K. L., Smith, D. W., Streissguth, A. P., and Myrianthopoulos, N. C., 1974, Outcome in offspring of chronic alcoholic women, *Lancet* **1**:1076.

Jones, M. D., Jr., Burd, L. I., Bowes, W. A., Jr., Battaglia, F. C., and Lubchenco, L. O., 1975, Failure of association of premature rupture of membranes with respiratory-distress syndrome, *N. Engl. J. Med.* **292**:1253.

Kagan, B. M., Stanincova, V., Felix, N. S., Hodgman, J., and Kalman, D., 1972, Body composition of premature infants: Relation to nutrition, *Am. J. Clin. Nutr.* **25**:1153.

Katz, L., and Hamilton, J. R., 1974, Fat absorption in infants of birthweight less than 1300 gm, *J. Pediatr.* **85**:608.

Kennell, J. H., and Klaus, M. H., 1971, Care of the mother and of the high-risk infants, *Clin. Obstet. Gynecol.* **14**:926.

Kerpel-Fronius, E., 1960, Volume and composition of body fluid compartments in severe infant malnutrition, *J. Pediatr.* **56**:826.

Kerpel-Fronius, E., and Kovach, S., 1948, The volume of extracellular body fluids in malnutrition, *Pediatrics* **2**:21.

Khoury, G. H., and Hawes, C. R., 1963, Primary pulmonary hypertension in children living at high altitude, *J. Pediatr.* **62**:177.

Klaus, M. H., and Kennell, J. H., 1970, Mothers separated from their newborn infants, *Pediatr. Clin. North Am.* **17**:1015.

Klaus, M. H., Jerauld, R., Kreger, N. C., McAlpine, W., Steffa, M., and Kennell, J. H., 1972, Maternal attachment. Importance of the first post-partum days, *N. Engl. J. Med.* **286**:460.

Kroushop, R. W., Brown, E. G., and Sweet, A. Y., 1975, The early use of continuous positive airway pressure in the treatment of idiopathic respiratory distress syndrome, *J. Pediatr.* **87**:263.

Kruger, H., and Arias-Stella, J., 1970, The placenta and the newborn infant at high altitudes, *Am. J. Obstet. Gynecol.* **106**:586.

Kulhanek, J. F., Meschia, G., Makowski, E. L., and Battaglia, F. C., 1974, Changes in DNA content and urea permeability of the sheep placenta, *Am. J. Physiol.* **226**:1257.

Lathe, G. H., and Walker, M., 1958, The synthesis of bilirubin glucuronide in animal and human liver, *Biochem. J.* **70**:705.

Lechtig, A., Delgado, H., Lasky, R., Yarbrough, C., Klein, R. E., Habicht, J., and Béhar, M., 1975, Maternal nutrition and fetal growth in developing countries, *Am. J. Dis. Child.* **129**:553.

Leifer, A. D., Lederman, P. H., Barnett, C. R., and Williams, J. A., 1972, Effects of mother–infant separation on maternal attachment behavior, *Child Dev.* **43**:1203.

Lemons, J. A., Huhns, L. R., and Poznanski, A. R., 1972, Calcification of fetal teeth as an index of fetal maturation, *Am. J. Obstet. Gynecol.* **114**:628.

Levi, A. J., Gatmaitan, Z., and Arias, I. M., 1970, Deficiency of hepatic organic anion-binding protein, impaired organic anion uptake by liver and "physiologic" jaundice in newborn monkeys, *N. Engl. J. Med.* **283**:1136.

Lichty, J. A., Ting, R. Y., Bruns, P. D., and Dyar, E., 1957, Studies of babies born at high altitude, *Am. J. Dis. Child.* **93**:666.

Liggens, G. C., 1969, The foetal role in the initiation of parturition in the ewe, in: *Foetal Autonomy* (G. E. W. Wolstenhome and M. O'Connor, eds.), pp. 218–231, J. S. Churchill, London.

Lobl, M., Welcher, D. W., and Mellits, E. D., 1971, Maternal age and intellectual functioning of offspring, *Johns Hopkins Med. J.* **128**:347.

Low, J. A., Boston, R. W., and Pancham, S. R., 1972, Fetal asphyxia during the intrapartum period in intrauterine growth-retarded infants, *Am. J. Obstet. Gynecol.* **113**:351.

Lubchenco, L. O., 1970, Assessment of gestational age and development at birth, *Pediatr. Clin. North Am.* **17**:125.

Lubchenco, L. O., and Bard, H., 1971, Incidence of hypoglycemia in newborn infants classified by birth weight and gestational age, *Pediatrics* **47**:831.

Lubchenco, L. O., Hansman, C., and Backstrom, L., 1968, Factors influencing fetal growth, in: *Aspects of Praematurity and Dysmaturity,* Second Nutricia Symposium, Groningen, May 10–12, 1967 (J. H. P. Jonxis, H. K. A. Visser, and J. A. Troelstra, eds.), pp. 149–166, H. E. Stenfert Kroese, N. V., Leiden, Holland, and Charles C Thomas, Springfield, Illinois.

Lubchenco, L. O., Searls, D. T., and Brazie, J. V., 1972, Neonatal mortality rate: Relationship to birth weight and gestational age, *J. Pediatr.* **81**:814.

Maisels, M. J., Pathak, A., Nelson, N. M., Nathan, D. G., and Smith, C. A., 1971, Endogenous production of carbon monoxide in normal and erythroblastotic newborn infants, *J. Clin. Invest.* **50**:1.

Makowski, E. L., Battaglia, F. C., Meschia, G., Behrman, R. E., Schruefer, J., Seeds, A. E., and Bruns, P. D., 1968, Effect of maternal exposure to high altitude upon fetal oxygenation, *Am. J. Obstet. Gynecol.* **100**:852.

Makowski, E. L., Schneider, J. M., Tsoulos, N. G., Colwill, J. R., Battaglia, F. C., and Meschia, G., 1972, Cerebral blood flow, oxygen consumption, and glucose utilization of fetal lambs *in utero, Am. J. Obstet. Gynecol.* **114**:292.

Mattison, D. R., 1970, Amniotic fluid osmolality, *Obstet. Gynecol.* **36**:420.

McDonald, E., 1910, The duration of pregnancy with a new rule for its estimation, *Am. J. Med. Sci.* **140**:349.

McKeown, T., and Record, R. G., 1952, Observations on foetal growth in multiple pregnancy in man, *J. Endocrinol.* **8**:386.

McKeown, T., and Record, R. G., 1953, The influence of placenta size on foetal growth in man with special reference to multiple pregnancies, *J. Endocrinol.* **9**:418.

Meyer, M. B., and Tonascia, J. A., 1973, Possible effects of x-ray exposure during fetal life on the subsequent reproductive performance of human females, *Am. J. Epidemiol.* **98**:151.

Miles, P. A., and Pearson, J. E., 1969, Amniotic fluid osmolality in assessing fetal maturity, *Obstet. Gynecol.* **34:**701.

Mollison, P. L., and Cutbush, M., 1949, Haemalytic disease of the newborn: Criteria of severity, *Br. Med. J.* **1:**123.

Mollison, P. L., and Cutbush, M., 1954, Haemalytic disease of the newborn, in: *Recent Advances in Paediatrics* (D. M. T. Gairdner, ed.), pp. 110–132, The Blakiston Co., New York.

Naeye, R. L., Blanc, W., Leblanc, W., and Khatamee, M. A., 1973, Fetal complications of maternal heroin addiction: Abnormal growth, infections, and episodes of stress, *J. Pediatr.* **83:**1055.

O'Leary, J. A., and Feldman, M., 1970, Amniotic fluid osmolality in the determination of fetal age and welfare, *Obstet. Gynecol.* **36:**525.

Ostrow, J. D., 1971, Photocatabolism of labeled bilirubin in the congenitally jaundiced (Gunn) rat, *J. Clin. Invest.* **50:**707.

Ounsted, M., 1971*a*, Biological factors and fetal growth, *Dev. Med. Child Neurol.* **13:**524.

Ounsted, M., 1971*b*, Fetal growth, in: *Recent Advances in Paediatrics* (D. Gairdner and D. Hull, eds.), pp. 23–62, J. and A. Churchill, London.

Parmalee, A. H., 1967, Electroencephalography and brain maturation, in: *Regional Development of the Brain in Early Life* (A. Minkowski, ed.), pp. 459–470, Blackwell, Oxford.

Parmalee, A. H., 1968, Maturation of EEG activity during sleep in premature infants, *Electroencephalogr. Clin. Neurophysiol.* **24:**319.

Parmalee, A. H., Jr., Akiyama, Y., Schultz, M. A., Wenner, W. H., Schulte, F. J., and Stern, E., 1970, Electroencephalogram in active and quiet sleep in infants, in: *Clinical Electroencephalography of Children* (P. Kelleway and I. Petersén, eds.), pp. 77–88, Grune and Stratton, New York.

Peñaloza, D., Arias-Stella, J., Sime, F., Recavarren, S., and Marticorena, E., 1964, The heart and pulmonary circulation in children at high altitudes, *Pediatrics* **34:**568.

Pitkin, R. M., and Zwirek, S. J., 1967, Amniotic fluid creatinine, *Am. J. Obstet. Gynecol.* **98:**1135.

Poland, R. L., and Odell, G. B., 1971, Physiologic jaundice: The enterohepatic circulation of bilirubin, *N. Engl. J. Med.* **284:**1.

Purvis, R. J., Barrie, W. J. McK., MacKay, G. S., Wilkinson, E. M., Cockburn, F., Belton, N. R., and Forfar, J. O., 1973, Enamel hypoplasia of the teeth associated with neonatal tetany: A manifestation of maternal vitamin-D deficiency, *Lancet* **2:**811.

Roberts, S. A., Cohen, M. D., and Forfar, J. O., 1973, Antenatal factors associated with neonatal hypocalcemic convulsions, *Lancet* **2:**809.

Robinson, R. J., 1966, Assessment of gestational age by neurological examination, *Arch. Dis. Child.* **41:**437.

Roy, C. C., Ste-Marie, M., Chartrand, L., Weber, A., Bard, H., and Doray, B., 1975, Corrections of malabsorption of the preterm infant with a medium chain triglyceride formula, *J. Pediatr.* **86:**446.

Russell, J. G. B., 1969, Radiologic assessment of fetal maturity, *J. Obstet. Gynaecol. Br. Commonw.* **75:**208.

Sacher, G. A., and Staffeldt, E. F., 1974, Relation of gestation time to brain weight for placental mammals: Implications for the theory of vertebrate growth, *Am. Nat.* **108:**593.

Saint-Anne Dargassies, S., 1955, La maturation neurologique du prématuré, *Etud. Neo-natales* **4:**71.

Schreiber, M. H., Menachof, L., Gunn, W. G., and Biehuesen, F. C., 1962, Reliability of visualization of the distal femoral epiphysis as a measure of maturity, *Am. J. Obstet. Gynecol.* **83:**1249.

Schreiber, M. H., Nichols, M. M., and McGanity, W. J., 1963, Epiphyseal ossification center visualization, *J. Am. Med. Assoc.* **184:**504.

Schulte, F. J., Michaelis, R., Linke, I., and Nolte, R., 1968, Motor nerve conduction velocity in term, preterm, and small-for-dates newborn infants, *Pediatrics* **42:**17.

Scopes, J. W., and Ahmed, I., 1966, Minimal rates of oxygen consumption in sick and premature newborn infants, *Arch. Dis. Child.* **41:**407.

Seppala, M., and Ruoslakti, E., 1972, Alpha-fetoprotein in amniotic fluid: An index of gestational age, *Am. J. Obstet. Gynecol.* **114:**505.

Sharma, S. D., and Trussel, R. R., 1970, Value of amniotic fluid examination in the assessment of fetal maturity, *J. Obstet. Gynaecol. Br. Commonw.* **77:**215.

Silverman, W. A., and Agate, F. J., Jr., 1964, Variation in cold resistance among small newborn infants, *Biol. Neonate* **6:**113.

Silverman, W. A., Sinclair, J. C., and Agate, F. C., Jr., 1966, Oxygen cost of minor changes in heat balance of small newborn infants, *Acta Pediatr. Scand.* **55:**294.

Sime, F., Banchero, N., Peñaloza, D., Gamboa, R., Cruz, J., and Marticorena, E., 1963, Pulmonary hypertension in children born and living at high altitudes, *Am. J. Cardiol.* **11:**143.

Sinclair, J. C., 1970, Heat production and thermoregulation in the small-for-date infant, *Pediatr. Clin. North Am.* **17:**147.

Sinclair, J. C., and Silverman, W. A., 1966, Intrauterine growth in active tissue mass of the human fetus with particular reference to the undergrown baby, *Pediatrics* **38**:48.

Smith, C. A., 1947, Effects of maternal undernutrition upon the newborn infant in Holland, *J. Pediatr.* **30**:229.

Smith, D. W., 1970, *Recognizable Patterns of Human Malformation: Genetic, Embryonic and Clinical Aspects,* W. B. Saunders, Philadelphia.

Soltész, Gy., Mestyán, J., Schultz, K., and Rubecz, I., 1972, Glucose disappearance rate and changes in plasma nutrients after intravenously injected glucose in normoglycaemic and hypoglycaemic under- weight newborns, *Biol. Neonate* **21**:184.

Stein, Z., and Susser, M., 1975, The Dutch famine, 1944–1945, and the reproductive process. I. Effects on six indices at birth, *Pediatr. Res.* **4**:70.

Taeusch, H. W., Carson, S. H., Wang, N. S., and Avery, M. E., 1973, Heroin induction of lung maturation and growth retardation in fetal rabbits, *J. Pediatr.* **82**:869.

Tanner, J. M., and Thomson, A. M., 1970, Standards for birthweight at gestation periods from 32 to 42 weeks, allowing for maternal height and weight, *Arch. Dis. Child.* **45**:566.

Tanner, J. M., Legarraga, H. A., and Tanner, A., 1972, Within-family standards for birthweight, *Lancet* **2**:193, 1314.

Teoh, E. S., Spellacy, W. N., and Buhi, W. C., 1971, Human chorionic somatomammotropin (HCS): A new index of placental function, *J. Obstet. Gynaecol. Br. Commonw.* **78**:673.

Touchstone, J. C., Glazer, L. G., Bolognese, R. J., and Corson, S. L., 1972, Gestational age and amniotic fluid protein patterns, *Am. J. Obstet. Gynecol.* **114**:58.

Treloar, A. E., Behn, B. G., and Cowan, D. W., 1967, Analysis of gestational interval, *Am. J. Obstet. Gynecol.* **99**:34.

Tsang, R., and Oh, W., 1970, Neonatal hypocalcemia in low birth weight infants, *Pediatrics* **45**:773.

Turner, G., 1971, Recognition of intrauterine growth retardation by considering comparative birth- weights, *Lancet* **2**:1123.

Ulfelder, H., 1973, Stilbestrol, adenosis, and adenocarcinoma, *Am. J. Obstet. Gynecol.* **117**:794.

Usher, R. H., 1970, Clinical and therapeutic aspects of fetal malnutrition, *Pediatr. Clin. North Am.* **17**:169.

Van den Berg, B. J., and Yerushalmy, J., 1966, The relationship of the rate of intrauterine growth of infants of low birthweight to mortality, morbidity, and congenital anomalies, *J. Pediatr.* **69**:531.

Watkins, J. B., Ingall, D., Szczepanik, P., Klein, P. D., and Lester, R., 1973, Bile salt metabolism in the newborn. Measurement of pool size and synthesis by stable isotope technic, *N. Engl. J. Med.* **288**:431.

Watkins, J. B., Szczepanik, P., Gould, J. B., Klein, P., and Lester, R., 1975, Bile salt metabolism in the human premature infant. Preliminary observations of pool size and synthesis rate following prenatal administration of dexamethasone and phenobarbital, *Gastroenterology* **69**:706.

Wirth, F. H., Goldberg, K. E., and Lubchenco, L. O., 1975, Neonatal hyperviscosity: Incidence and effect of partial plasma exchange transfusion, *Pediatr. Res.* **9**(695):372 (abstract).

World Health Organization, Expert Group on Pre-Maturity: Final Report, 1950, Technical Report Series No. 27, WHO, Geneva.

Wu, P. Y. K., Lim, R. C., Hodgman, J. E., Kokosky, M. J., and Teberg, A. J., 1974, Effect of phototherapy in preterm infants on growth in the neonatal period, *J. Pediatr.* **85**:563.

19

Growth Dynamics of Low-Birth-Weight Infants with Emphasis on the Perinatal Period

INGEBORG BRANDT

1. Introduction

1.1. Short Review of Studies to Date and Their Problems

In the past there have been contrary findings on the dynamics of postnatal growth of preterm infants. One group of authors has reported that preterm infants remain below the standards of full-term infants (Ylppö, 1919; Glaser *et al.*, 1950; Blegen, 1952; Alm, 1953; Douglas and Mogford 1953; Seckel and Rolfes, 1959; Woolley and Valdecanas, 1960; Falkner *et al.*, 1962; Lubchenco *et al.*, 1963*a;* Drillien, 1961, 1964; Robinson and Robinson, 1965; Davies and Russel, 1968; Wedgwood and Holt, 1968; Benedikt and Csáki, 1969; Babson, 1970). A second group has observed no growth differences between preterm and full-term infants (Levine and Gordon, 1942; Hirschl *et al.*, 1948; Kahl, 1950; Hess, 1953; Schwinn, 1954; Rossier *et al.*, 1958; Dann *et al.*, 1964; Van den Berg, 1968; Walsh, 1969; Dweck *et al.*, 1973; Stewart and Reynolds, 1974; Beck and Van den Berg, 1975; Bourlon *et al.*, 1975; Davies, 1975; Fitzhardinge, 1975).

Only some authors in each group allowed for prematurity in their calculation of age, that is, they calculated age as postmenstrual or corrected (rather than postnatal) age throughout. Some of the discrepancies in the reports may be due to not making this allowance in the early years (as well as to differences in social class, origin, postnatal care, and nutrition of the subjects). Some reports failed to differen-

---

INGEBORG BRANDT • University Children's Hospital Bonn, Bonn, Germany

tiate preterm infants into those who were appropriate in weight or size for gestational age (AGA) and those who were small for gestational age (SGA); thus the results have been confused since growth and development of SGA infants—who represent about 30–40 percent of infants in developed countries with a birth weight of 2500 g and below—are different from AGA preterm infants (Scott and Usher, 1966; Yerushalmy, 1970; Cruise, 1973; Brandt, 1975).

Little is known about the dynamics of postnatal (and especially perinatal) growth of preterm infants because velocity curves in relation to postmenstrual age—which are more instructive than distance curves—are seldom seen in the present literature. For assessment of growth according to postmenstrual and to corrected age, and for early diagnosis of abnormal growth, velocity standards are necessary. They are obtainable only from longitudinal studies. In the past, findings from longitudinal studies often have been reported only cross-sectionally.

1.2. Definitions

1.2.1. Critical Evaluation of the Different "Intrauterine Growth Curves"

Intrauterine growth standards are necessary to quantitate normal fetal growth and to diagnose fetal growth retardation. These "intrauterine" curves are based on postnatal measurements of infants born at different postmenstrual ages (cross-sectional). Figure 1 gives a synopsis of intrauterine weight curves from different authors. Between 33 and 37 postmenstrual weeks the 10th percentiles of Lubchenco *et al.* (1963*b*), Hosemann (1949*a*), Gruenwald (1966), and Hohenauer (1973) agree quite well, whereas the curves of Thomson *et al.* (1968) and Babson (1970) run at a higher level, as do Sterky's curves (1970), not plotted here. From 28 to 32 postmenstrual weeks the 10th percentile of the U.S. standards of Brenner *et al.* (1976), not listed in Figure 1, agrees with the corresponding Lubchenco curve, and subsequently until 40 weeks it runs at a slightly higher level. This synopsis shows that before 38 weeks the Lubchenco curves are appropriate for the classification of preterm infants with known postmenstrual age according to their intrauterine growth pattern. Beyond the 38th postmenstrual week the Lubchenco curves flatten, possibly because of intrauterine growth retardation caused by the high altitude of Denver (1600 m), where the infants were measured.

1.2.2. Postmenstrual Age, Conceptional Age

Postmenstrual, fetal, or gestational age is the time from the first day of the mother's last menstrual period until birth; in the preterm infant, postmenstrual age should be used at all examinations up to 40 postmenstrual weeks, irrespective of birth. The definitions postmenstrual age and postconceptional age (time from conception to birth) have to be clearly separated. The confusing use of "conceptional age" for the time from the first day of the mother's last menstrual period until birth plus postnatal time of life (Schulte, 1974; Prechtl and Beintema, 1976) is to be avoided. Conceptional age is an established term for the obstetrician, anatomist, and anthropologist, but it should be used as the time from conception to birth.

For differentiation between a newly born infant of 35 postmenstrual weeks and an infant of the same age, born at 28 postmenstrual weeks and with a postnatal age of 7 weeks it is suggested that two ages be given, the postmenstrual age at birth and the postnatal age.

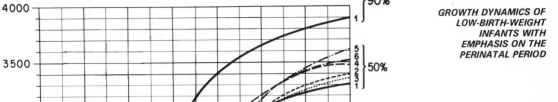

Fig. 1. "Intrauterine" weight curves for the third trimester of gestation, based on birth weight measurements of infants born at different postmenstrual ages (cross-sectional).

1.2.3. Appropriate for Gestational Age (AGA) and Small-for-Gestational Age (SGA) Preterm Infants, Synonyms

Preterm infants are defined as those born with a postmenstrual age of less than 38 weeks (i.e., 37 weeks + 6 days or less), irrespective of birth weight (McKeown and Gibson, 1951; Battaglia and Lubchenco, 1967). The other definition (following an official WHO recommendation from 1950) of preterm infants, those with a postmenstrual age of less than 259 days (American Academy of Pediatrics—Committee on Fetus and Newborn, 1967; Working Party to Discuss Nomenclature based on Gestational Age and Birth-weight, 1970) is unacceptable because infants who are born 21 days before regular term (280 days) cannot in any way be considered, and treated, as "mature."

Preterm infants with normal intrauterine development, here taken as birth

Table I. Synonyms for Small-for-Gestational-Age Infants

Small-for-dates (SFD)	Butler and Bonham (1963)
Chronic fetal distress	Gruenwald (1963)
Small-for-gestational-age (SGA)	Lubchenco et al. (1966)
Intrauterine malnutrition	Scott and Usher (1966)
Fetal growth retardation, intrauterine growth retardation (IUGR)	Wigglesworth (1966)
Small-for-date infants	Drillien (1970)
Light-for-dates	Neligan (1970)
Dysmature	Dewhurst et al. (1972)
Fetal malnutrition, undergrown in utero	Miller and Hassanein (1973)
Fetal deprivation of supply (FDS)	Hagberg et al. (1976)

weight between the 10th and 90th percentiles of Lubchenco et al. (1963b), are defined as appropriate for gestational age (AGA).

Small-for-gestational-age (SGA) infants are defined as those with a birth weight below the 10th percentile of Lubchenco et al. (1963b) or the 10th percentile of Hosemann (1949a).

In the literature there are many synonyms for newborn infants with a birth weight below the 10th percentile of intrauterine growth charts. Of the synonyms in Table I, the term SGA is most commonly used.

1.2.4. Perinatal Period

The perinatal period is defined as the time from 28 postmenstrual weeks to the seventh day after term. In the preterm infant, the perinatal period covers the third trimester of pregnancy of the full-term infant.

1.3. Relevance of Growth Studies in the Perinatal Period

1.3.1. Cross-Sectional and Longitudinal Studies; Establishment of Standards

Older reports of anatomists about fetal growth in the third trimester of pregnancy are mostly based upon the relations of head circumference, weight, and length among one another, or to other external dimensions of the body (Jackson, 1909; Scammon and Calkins, 1929). Such growth changes could not be evaluated in relation to time since few specimens were accompanied with clinical histories from which the postmenstrual age could be estimated with any accuracy.

The first "intrauterine growth standards" for the perinatal period were established by Hosemann (1949a) and by Lubchenco et al. (1963b). Since these growth standards were based on cross-sectional data (i.e., single postnatal measurements of infants born at different postmenstrual ages), they cannot provide information on growth velocity. Moreover, the currently available "intrauterine growth curves" (cross-sectional) indicate a decline in the rate of growth near the expected date of delivery (term) which could be due simply to a reduction of nutrient supply late in gestation. Such a decline does not occur in the postnatal growth of preterm infants of the same age (Babson, 1970; Brandt, 1976b). The few longitudinal studies of the perinatal period have either not considered postmenstrual age (O'Neill, 1961), or the examination intervals have been too long (Cruise, 1973; Fitzhardinge, 1975) for calculation of effective velocities.

With the increasing use of ultrasound techniques for intrauterine growth evaluation and the establishment of intrauterine growth standards, appropriate growth has become more and more important as an indicator of fetal health, since fetal growth retardation has been identified as one major risk factor in perinatal mortality and morbidity (Yerushalmy, 1970; Usher and McLean, 1974). For the preterm infant, postnatal growth also is an indicator of his well-being. Therefore, standards for immediate postnatal growth are essential for the evaluation of preterm infants.

Since it has been shown that severe maternal undernutrition has its greatest effect on fetal growth in late gestation, that is, in the third trimester (Naeye *et al.*, 1973; Stein and Susser, 1975; Lechtig *et al.*, 1976), and may affect growth in all dimensions, adequate nutrition of preterm infants in the perinatal period for optimal growth and development has gained increasing significance.

1.3.2. Head Circumference Growth; Association with Brain Development

The close relationship between head growth and brain development in the normally growing infant in the first year of life has been shown in many studies (Jackson, 1909; Ylppö, 1919; Bray *et al.*, 1969; Winick and Rosso, 1969a; Dobbing, 1970). Only recently Buda *et al.* (1975) have reported a "strong linear relation between occipito-frontal head circumference and calculated skull volume in infants with normally shaped skulls" with a correlation coefficient of 0.97. It is to be expected that this relationship also holds for the fetus in the third trimester of gestation and for the preterm infant in the perinatal period. Postnatal regular measurements of head circumference are of prognostic value because there exists a relationship between small head circumference and developmental retardation, for instance, in the lower IQ found in the smaller member of a twin pair (Babson *et al.*, 1964; Babson and Phillips, 1973; Bazso *et al.*, 1968; Hohenauer, 1971; Brandt, 1975). Winick (1973) reports that head circumference reflects the severity of functional deficits in the survivors of marasmus. In children who died with marasmus during the first year of life, the reduced head circumference accurately reflects the reduced brain weight. Concerning the exact extent of the period of the human brain growth spurt, there exists some controversy (Davison and Dobbing, 1966; Dobbing, 1970, 1974a,b; Dobbing and Sands, 1973; Winick, 1968, 1974; Winick and Rosso, 1969b), but regardless of these different views, there is agreement that around term brain growth is rapid and involves glial multiplication (DNA synthesis), myelination, and establishment of synaptic connections (Dobbing, 1974b; Herschkowitz, 1974).

1.4. Methods

1.4.1. Assessment of Postmenstrual Age

Knowledge of the postmenstrual age at birth is a prerequisite to growth evaluation of preterm infants. In the Bonn longitudinal study, on which the growth data presented here are based, postmenstrual age at birth has been determined with particular care from (1) obstetric history; (2) clinical external characteristics (Farr *et al.*, 1966); and (3) especially repeated neurological examinations according to a schedule developed on the model of Joppich and Schulte (1968), Prechtl and Beintema (1968), Robinson (1966), and Saint-Anne Dargassies (1955, 1966, 1970).

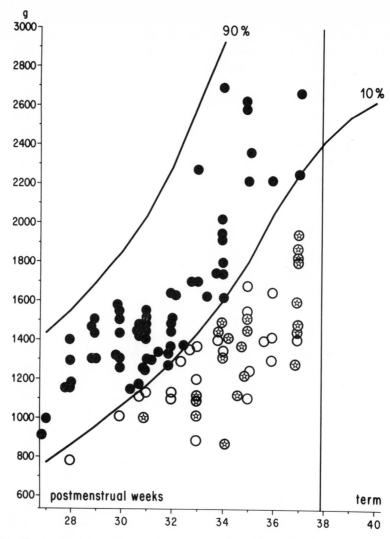

Fig. 2. Classification by birth weight according to fetal age of the Bonn preterm infants (percentiles from Lubchenco): within the 10th and 90th percentiles 64 AGA (points); below the 10th percentile 43 SGA (circles); 21 of these, the circles with an asterisk, mark the SGA infants with catch-up growth.

Some of the mothers have come from the infertility unit of the Bonn Universitäts-Frauenklinik, and the date of conception is known by measurements of basal temperature. In addition the intrauterine growth of some infants is known by repeated ultrasound measurements (Hansmann, 1974, 1976).

1.4.2. Clinical Material

The growth standards presented here are based on a longitudinal study of growth and development of preterm infants from birth to the sixth year. Between 1967 and 1975 all preterm infants with a birth weight of 1500 g and below admitted to the Bonn Universitäts-Kinderklinik,* and some of higher birth weight chosen at random, have been included. By the criteria of postmenstrual age and birth weight, the 107 preterm infants of the Bonn study (Figure 2) are classified as follows:

1. Sixty-four infants, AGA, 40 of them with a postmenstrual age of 32 weeks and below, and a birth weight of 1500 g or less, all falling between the 10th and 90th percentiles of Lubchenco—the solid circles.

2. Forty-three infants, SGA, 35 of them with a birth weight of 1500 g or less, all falling below the 10th percentile of Lubchenco—the open circles. Twenty-one of these (the circles with an asterisk) belong to a recent special group exhibiting catch-up growth.

In this sample of 43 SGA infants, maternal undernutrition has not been found, and this is in accord with the experience of Ounsted and Taylor (1971).

A further 80 full-term infants were included in the study as controls.

All the anthropometric measurements—approximately 4000—have been made by the author, at weekly to fortnightly intervals before term, at monthly intervals during the first year, at three-month intervals in the second year, and at six-month intervals thereafter. Since the growth velocity in the last 10 postmenstrual weeks is very rapid and changes quickly, frequent examinations at short intervals are essential. The infants were measured supine to 2½ years of age (supine length), and thereafter in the erect position (height). Head circumference was measured in the horizontal plane including the maximum occipitofrontal circumference, using a nonstretch glass-fiber tape and employing light pressure.

From these data, percentiles, means, and standard deviations of growth in weight, length, and head circumference were calculated. The final curves, presented here, required no smoothing and were plotted from the raw data. The growth charts cover the age spans of perinatal period, birth to 18 months, and birth to 5 or 6 years, respectively. This plan seemed best in order to obtain sufficient separation between the percentile curves in the perinatal period and in the first six months of life when growth is very rapid.

The infants were born predominantly to middle- and upper-middle-class parents. The growth curves are representative not only of Bonn, but also of the population from other parts of Germany, since after World War II many people from all parts of Germany moved to the Bonn region.

1.4.3. Correction for Postmenstrual Age

For calculation of the growth standards for preterm infants, a correction for postmenstrual age has been made; that is, discounting the time of prematurity from chronological age. Although this is also done in the standards of Gairdner and Pearson (1971) and of Tanner and Whitehouse (1973), in many other studies preterm infants are plotted without allowing for their true age.

In full-term infants the date of birth (independent of duration of pregnancy) is the point of departure for calculation of postnatal growth standards, while for the preterm infant the expected date of delivery (EDD) is the point of departure. Forty weeks are taken as the base for a mean normal duration of gestation (Hosemann, 1949a; Bjerkedahl et al., 1973).

---

*In Bonn, the small preterm infants of the surrounding obstetric departments are cared for at the University Children's Hospital.

For data analysis, weeks and months have been converted into decimals of a year (decimal age).

Growth velocity values are small amounts compared with distance values; therefore, the measurements on which the standards are based have to be done meticulously. Measuring errors are potentially doubled when calculating velocity from two measurements. Further, it is essential to measure all children at the same ages or to correct individual measurements to such set ages. With weekly measurements in the perinatal period, when growth velocity is very rapid, variations of 2–3 days will distort or flatten the curve. For that reason, the actual examination ages have been corrected to the exact target ages by linear interpolation. When the examination results are taken together in time classes (for instance of one month), the velocity curves would flatten seriously since longer and shorter intervals are "averaged."

1.4.4. Consideration of Midparent Height and Midparent Head Circumference

The hereditary background of a child has a great influence on his growth pattern and final size. In the Bonn study, therefore, height and head circumference of all parents of the preterm infants and the full-term control infants have been measured to give midparent heights (i.e., the averages of the father and the mother). Allowing for parental stature is important not only for assessment of the preterm infants' later growth status, but also for comparison with other growth studies (see p. 575) when considering the representativeness of the data.

2. General Growth Patterns of AGA Preterm Infants; Distance and Velocity

2.1. Weight

2.1.1. Distance

Perinatal Period. Before "term," the extrauterine weight gain (longitudinal) of the preterm infants (Figure 3) is small; from 32 to 40 postmenstrual weeks the weight curve follows approximately the 10th percentile of the "intrauterine curves" from different sources (cross-sectional).

Before and After Term until 18 Months. At term and at one month the weight of the preterm infants is still low (Figure 4) and significantly less than that of the full-term control infants ($P < 0.01$). Two months after term the weight curve of the preterm infants has reached that of the full-term control infants, and from then on there is no significant difference between the groups. Therefore the percentiles from 2 to 18 months in Figure 4 are based on preterm and full-term infants combined.

Two to Five Years. From 2 to 5 years, weight growth in the preterm infants is similar to that of the full-term control infants, although there is a (statistically insignificant) tendency of the preterm infants to have slightly lower values. The percentiles from 2 months to 5 years in Figure 5 are based on preterm and full-term infants combined. All the curves represent the unsmoothed data, which appear to be Normally distributed.

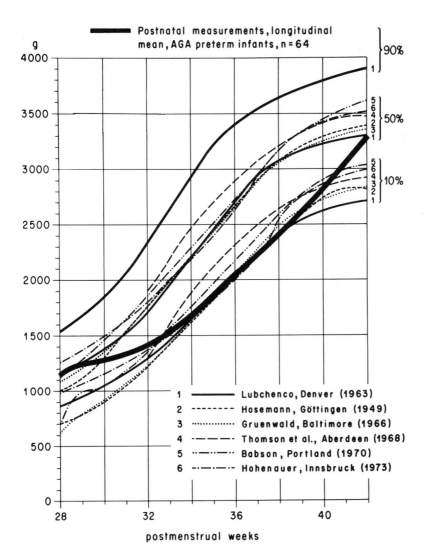

Fig. 3. Mean weight curve of the Bonn AGA preterm infants, $N = 64$, postnatal measurements, longitudinal, compared with "intrauterine" standards (10th and 50th percentiles) from different authors.

Sex Differences. From 30 to 40 postmenstrual weeks, the Bonn Study preterm boys are heavier than girls except at 34 weeks where the mean weights are the same. At 31 to 32 weeks, the weight difference of 100 g is significant ($P < 0.05$); between 33 and 37 weeks the difference, still 100 g (except at 34 weeks), has become nonsignificant ($P > 0.05$). But shortly before the expected date of delivery (EDD), the boys gain more weight than the girls, and at 38–40 weeks become 200–300 g heavier ($P < 0.025$).

This is in reasonable conformity with the findings of Thomson *et al.* (1968) that the sex difference does not appear before about 30 weeks. These authors found that birth weights by sex were practically identical at 32–33 postmenstrual weeks and then gradually diverged, males being about 150 g heavier at term than females. The charts of Tanner and Thomson (1970) incorporate these figures and are given for

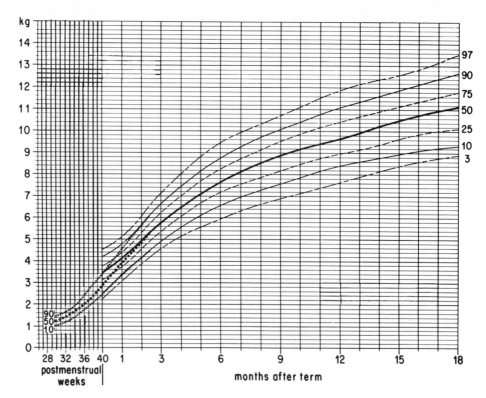

Fig. 4. Percentile curves for postnatal weight development in AGA preterm and full term infants, longitudinal; · · · · from 30 postmenstrual weeks to 2 months after term AGA preterm infants, $N = 64$; —— from 0 to 1 month full term infants, $N = 80$; ——from 2 to 18 months AGA preterm and full-term infants combined, $N = 144$.

sexes separately. Babson (1970) presents percentiles of fetal growth by sex, but gives no figures for the sex difference: ''from 37 weeks on the separation in weight becomes more marked and by 40 weeks of gestation the average discrepancy in weight is about 150 g.''

The Bonn results agree also with the study of Hohenauer (1973), in which from 35 weeks onwards boys were heavier than girls, this difference being significant between 37 and 42 postmenstrual weeks. Before 34 weeks the sex differences in weight were inconsistent. Besides a combined chart for boys and girls, he also presented curves for boys and girls separately from 33 postmenstrual weeks.

Hosemann (1949a) gives no figures for sex differences in weight before term, and his intrauterine curves are for boys and girls combined. Lubchenco et al. (1963b) report differences of about 100 g between the weights of boys and girls, and presented her intrauterine curve for sexes combined. Gruenwald (1966) gives his

→

Fig. 5. Percentile curves for weight of AGA preterm and full-term infants from birth to 5 years, longitudinal; . . . . . . from 30 to 39 postmenstrual weeks AGA preterm infants, $N = 64$, 10th, 50th and 90th percentiles; —— from 0 to 1 month full-term infants, $N = 80$; —— from 2 to 60 months AGA preterm and full-term infants combined, $N = 144$, (a) boys. $N = 72$; (b) girls. $N = 72$.

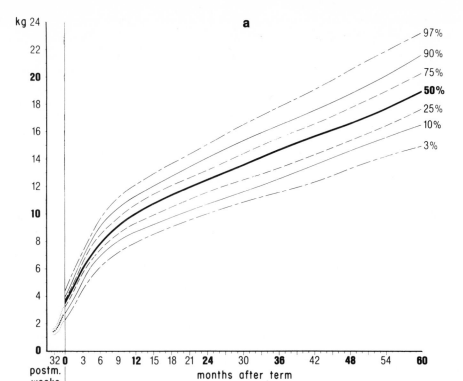

a

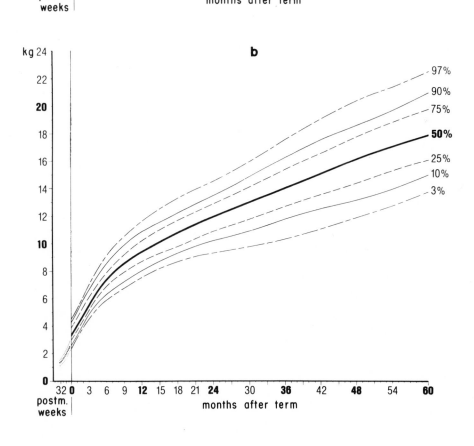

b

birth-weight results during the third trimester for boys and girls combined. According to Usher and McLean (1969), male infants have consistently been found to be about 5% heavier than female infants. Since the numbers in his study were insufficient to give an accurate indication of sex differences at each postmenstrual age, the growth standards are presented for both sexes together.

After term until 5 years, the Bonn Study preterm boys continue to be heavier than girls, the mean sex difference of 200–500 g being not significant except at 6 months, with a significant weight difference of 400 g ($P < 0.025$). A similar tendency has been observed in the full-term control infants, where from birth to 5 years boys are heavier than girls.

Babson (1970) did not study sex differences; Cruise (1973) did not calculate them, although the weight differences were inconsistent with a tendency of the boys to be heavier. From longitudinal studies of full-term infants, relatively constant mean sex differences in weight from birth to five years are reported, with boys heavier.

Necessary Period for Age Adjustment for Prematurity. From birth to 21 months, there are significant differences in weight amounting to 1.3–0.4 kg between corrected and uncorrected age. From 24 months onwards this difference between weight at corrected and uncorrected age decreases and becomes nonsignificant. After 24 months it becomes unnecessary to make the age correction.

Comparison with Other Studies. Earlier longitudinal studies in which the postmenstrual age at birth has not been taken into account, and in which AGA have not been separated from SGA, have not been used for comparison.

In the study of Babson (1970), based on repeated measurements from birth to one year of 12 infants born between 27 and 29 weeks, and of 12 infants with a postmenstrual age of 31 to 33 weeks, the weight curve paralleled those from normal U.S.A. standards, but at a lower level throughout the first year of life. At 12 months the mean weight of 8.8 and 9.2 kg in these two groups corresponded to the 25th percentile of the Bonn standards.

The mean weight at 3 years of the 39 preterm infants, born at 28–32 postmenstrual weeks, in the longitudinal study of Cruise (1973) is below that of the Bonn standards, by 0.2 kg for the girls and 0.7 kg for the boys. Compared with local standards, the preterm infants' weight is markedly low by more than 1.0 kg.

Stewart and Reynolds (1974) report that the weight distribution of 95 children with a birth weight of 1500 g and below in their follow-up study, at a mean age of 4 years, 9 months, compared with the London growth standards of Tanner *et al.* (1966) was as follows: 11% were on or above the 75th percentile; 33% lay between the 25th and 74th percentiles; 31% were between the 10th and 24th percentiles; and 25% were below the 10th percentile.

The preterm infants (birth weight from 1501 to 2500 g) in the study of Beck and Van den Berg (1975) "approach closely the weight of the normal-birth-weight infants by the age of two years."

Fitzhardinge (1975), who has examined 65 preterm infants with a mean birth weight of 1440 g at 3, 6, 9, and 12 months, reports that the mean weight of the boys agrees well with the percentiles of the Harvard Growth Study, whereas the girls tend to be lighter (no figures given).

In the French study of Bourlon *et al.* (1975) on 119 preterm infants from birth to 18 months, based on the records of the hospital and on records of later consultations, the mean weight curve lies in the area between 0 and -1 SD of national standards (Sempé *et al.,* 1964).

569

GROWTH DYNAMICS OF
LOW-BIRTH-WEIGHT
INFANTS WITH
EMPHASIS ON THE
PERINATAL PERIOD

Table II. Weight (in kg) from 1 to 36 Months (International Comparison)[a]

| Age (mo.): | 1 | | 3 | | 6 | | 9 | | 12 | | 18 | | 24 | | 36 | |
Sex:	B	G	B	G	B	G	B	G	B	G	B	G	B	G	B	G
AGA Preterm Infants																
Bonn[b]																
Mean	4.0	3.8	5.9	5.7	7.8	7.4	8.9	8.6	9.6	9.4	10.9	10.7	12.1	11.8	14.2	13.8
50th Percentile	4.0	3.7	5.8	5.5	7.6	7.3	8.9	8.4	9.6	9.3	11.1	10.6	12.1	11.9	13.9	13.8
Full-Term Infants																
Basel[c]																
50th Percentile	4.0	3.9	6.0	5.5	8.1	7.5	9.4	8.6	10.1	9.5	11.3	10.9	12.5	12.1	14.6	14.3
Holland[d]																
50th Percentile	4.2	4.0	5.8	5.5	7.9	7.4	9.3	8.9	10.5	10.0	12.1	11.5	13.3	12.8	15.6	14.9
London[e]																
Mean	4.1	3.7	5.9	5.4	7.9	7.4	9.3	8.7	10.2	9.6	11.5	10.9	12.5	11.8	14.6	14.4
London and Oxford[f]																
50th Percentile	4.3	3.9	5.9	5.6	7.9	7.4	9.2	8.7	10.2	9.7	11.6	11.1	12.7	12.2	14.7	14.3
Paris[g]																
Mean	3.9	3.9	5.8	6.0	7.9	7.5	8.9	8.3	9.9	9.3	11.2	10.5	12.2	11.6	14.2	13.6
Stockholm[h]																
Mean	4.0	3.8	5.9	5.6	7.8	7.5	9.1	8.9	10.2	9.9	11.7	11.3	12.9	12.4	14.8	14.5
50th Percentile	4.0	3.8	5.8	5.6	7.7	7.4	9.1	8.8	10.2	9.8	11.6	11.2	12.9	12.3	14.7	14.4
U.S.A.[i]																
50th Percentile	4.3	4.0	6.0	5.4	7.9	7.2	9.2	8.6	10.2	9.5	11.5	10.8	12.6	11.9	14.7	13.9
Zürich[j]																
Mean	3.9	3.8	5.8	5.4	7.7	7.2	9.0	8.5	10.1	9.5	11.6	10.9	12.7	12.0	15.0	14.2
Bonn[b]																
Mean	4.2	4.1	6.0	5.6	8.0	7.4	9.2	8.5	10.2	9.4	11.5	10.9	12.6	12.0	14.9	14.5
50th Percentile	4.2	4.0	6.1	5.7	8.0	7.3	9.3	8.5	10.2	9.4	11.5	11.0	12.5	12.0	15.0	14.6

[a]B = boys; G = girls.
[b]Brandt (1977).
[c]Heimendinger (1964a,b).
[d]van Wieringen et al. (1971).
[e]Falkner (1958).
[f]Tanner et al. (1966).
[g]Sempé et al. (1964).
[h]Karlberg et al. (1976).
[i]National Center for Health Statistics (1976).
[j]Prader and Budliger (1977).

International comparisons from 1 to 36 months in Table II show the Bonn Study preterm infants' weight to be similar to that of full-term infants of other European studies and similar to the U.S.A. National Center for Health Statistics (1976) (NCHS) charts, with a tendency to be less heavy. For instance, at the age of 6 months the preterm girls' weight mean (7.4 kg) or 50th percentile (7.3 kg) agrees within 100 g with the results from the Netherlands (van Wieringen *et al.,* 1971), from London (Falkner, 1958; Tanner *et al.,* 1966), Stockholm (Karlberg *et al.,* 1976), Zurich (Prader and Budliger, 1977), and the U.S.A. (National Center for Health Statistics, 1976). At the age of 3 months the preterm boys' weight mean (5.9 kg) or 50th percentile (5.8 kg) agrees within 100 g with the results from Basel (Heimendinger, 1964*a,b*), from the Netherlands, from London, Paris (Sempé *et al.,* 1964), Stockholm, the U.S.A. (NCHS), and Zurich.

2.1.2. Velocity

Perinatal Period to 18 Months. Velocity curves of intrauterine or extrauterine growth in weight in the perinatal period, derived from longitudinal data, have not been available before. The "intrauterine" mean weight velocity in the third trimester, as derived from cross-sectional distance data of "intrauterine curves" from different authors calculated in grams per month are given in Table III. The maximum velocity of about 1050 g/month is reached at about 32–33 weeks, after which the velocity falls.

Table III. *"Intrauterine" Mean Weight Velocity (in g/month), Derived from Distance Data (Cross-Sectional) of Different Sources;* [a] *Sexes Combined*

Postmenstrual weeks (age center)	Mean weight velocity (g)
28.5	596 g
29.5	706 g
30.5	625 g
31.5	742 g
32.5	1050 g
33.5	1014 g
34.5	1008 g
35.5	896 g
36.5	918 g
37.5	771 g
38.5	588 g
39.5	557 g

[a]Hosemann, 1949*a:* from the 50th percentile; McKeown and Record, 1952: from the mean; Lubchenco *et al.,* 1963*b:* from the 50th percentile; Gruenwald, 1966: from the 50th percentile; Thomson *et al.,* 1968: from the 50th percentile; Kloosterman, 1969: from the 50th percentile; Usher and McLean, 1969: from the mean; Babson *et al.,* 1970: from the 50th percentile; Hansmann and Voigt, 1973: from the mean; Hohenauer, 1973: from the mean; Nickl, 1972: from the mean; Milner and Richards, 1974: from the mean; Hinckers and Hansmann, 1975: from the mean; and Iffy *et al.,* 1975: from the mean.

571

GROWTH DYNAMICS OF
LOW-BIRTH-WEIGHT
INFANTS WITH
EMPHASIS ON THE
PERINATAL PERIOD

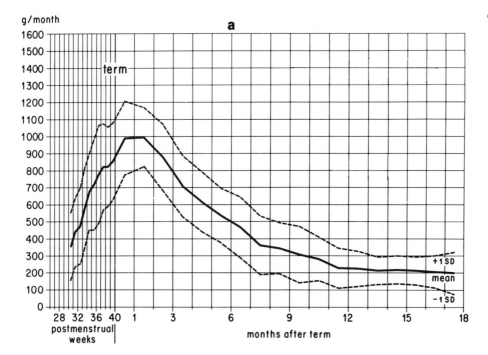

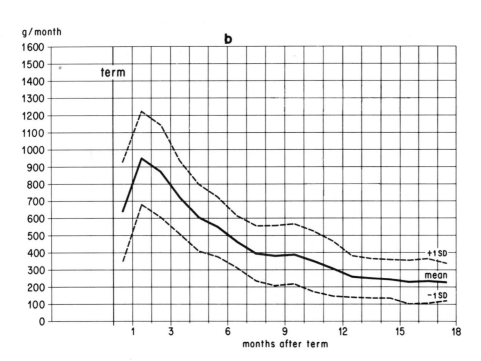

Fig. 6. Velocity curves of weight from 30.5 postmenstrual weeks to 17.5 months after term, mean and one standard deviation, unsmoothed values. (A) Preterm infants (boys and girls), appropriate for gestational age, $N = 64$; (b) full-term infants (boys and girls) $N = 80$.

The extrauterine mean weight velocity of the preterm infants, calculated in grams per month, before term, ranges only from 380 g to 890 g/month and is significantly less ($P < 0.005$) than in the first and second month after term, when it is 1000 g/month (Figure 6). The initial postnatal retardation is caught up by the preterm infants in the first month after term, when they put on significantly more weight than the full-term control infants (1000 g vs. 640 g). In the second month, weight velocity of the preterm infants is a little, but not significantly, higher than that of full-term infants (1000 g vs. 950 g); thereafter, the curves coincide without a significant difference (Figure 6).

As is to be expected, weight velocity has proved to be the growth indicator that is most dependent on caloric intake. The introduction of early and high-caloric feeding of preterm infants according to the Tizard scheme (1972, 1973) in July 1972 (Weber *et al.,* 1976) has led to an extrauterine weight velocity similar to that suggested for the fetus in the third trimester (Table III).

Since most of the preterm infants in the Bonn Study were born before the quantitative change in immediate postnatal nutrition in 1972, the altered growth dynamics are shown here in an individual curve. Figure 7 shows the weight velocity of an AGA preterm girl, born in the 32nd postmenstrual week after normal intra-uterine development with a birth weight of 1250 g (line of solid circles) plotted against the Bonn standards (mean and 1 SD). With an energy intake of 150–160 cal/kg/day in the 33rd postmenstrual week, a velocity peak of 1200 g/month is reached; this is in accordance with the "intrauterine" mean weight velocity between 1000 and 1050 g, calculated from different sources, as shown in Table III. The psychomotor development of this girl is above average.

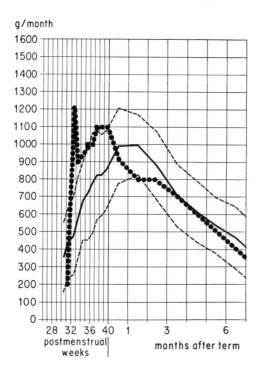

Fig. 7. Postnatal weight velocity from 31.5 postmenstrual weeks to 6 months after term of an AGA preterm girl, born in the 32nd week with a birth weight of 1250 g, line of solid circles: plotted against the Bonn standards (mean and one standard deviation).

KG/YEAR

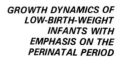

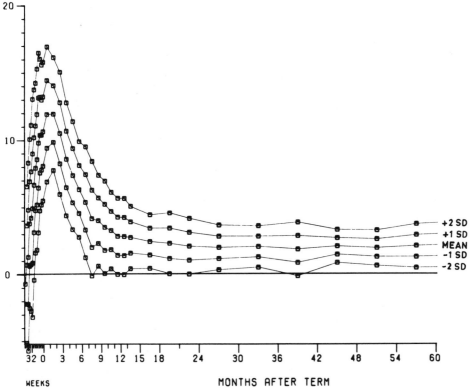

Fig. 8. Computer plot of weight velocity, from 30 postmenstrual weeks to 60 months after term, calculated in kg per year, mean with one and two standard deviations; AGA preterm infants, $N = 64$.

The high caloric nutrition leads to a better weight gain and thus to a shorter hospital stay of the preterm infants. Falkner *et al.* (1962) showed that a high caloric nutrition of infants of very low birth weight reduces significantly ($P < 0.01$) the number of days in the hospital.

Two to Five Years. From 2 to 5 years weight velocity of the preterm infants corresponds to that of the full-term control infants without significant difference, but with a slight tendency for the preterm infants to gain less weight. In general, weight velocity calculated in grams per month slows down from its peak of 1000 g at 0.5 and 1.5 months, to 200 g at 16.5 months. In the following months until 5 years, weight velocity remains relatively constant between 170 and 200 g/month, with little or no sex difference.

When weight velocity is calculated in kilograms per year, as shown in a computer plot, from 30 postmenstrual weeks to 60 months after term (Figure 8), the results of the preterm infants agree well with those of the full-term control infants and with infants from other studies. For example, at 2.75 years the weight velocity of the preterm infants amounts to 2.1 kg/year, compared to 1.96 in the London study (Tanner *et al.*, 1966), to 2.0 kg in Stockholm (Karlberg *et al.*, 1976), and to 2.2 kg/year in Zurich (Prader and Budliger, 1977).

2.2. Supine Length/Height

INGEBORG BRANDT

2.2.1. Distance

Perinatal Period. Before term, postnatal growth in length of the preterm infants is, like weight, clearly small (Figure 9); from 31 to 40 postmenstrual weeks the 50th percentile of supine length (longitudinal) runs markedly below the 50th percentiles or means of "intrauterine curves" from different sources (cross-sectional). In Figure 9, one has to make allowance for the fact that the Lubchenco curve flattens beyond 37 weeks, for reasons mentioned previously.

Before and After Term until 18 Months. Between term and 18 months the mean length difference between the preterm infants and the full-term control infants decreases from 2.5 cm (term) to 1.2 cm (18 months); from 0 to 18 months the difference remains significant ($P < 0.01$). Only after 21 months is the difference of 0.9 cm no longer significant. In Figure 10, the 3rd, 50th, and 97th percentiles of the preterm infants from 32 postmenstrual weeks to 4 months after term (line of dots, shaded area) are compared with the percentiles of the full-term control infants. The

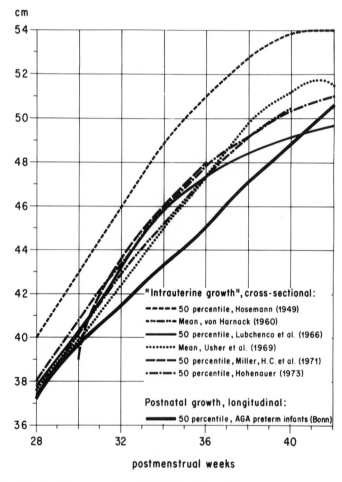

Fig. 9. Supine length, 50th percentile, of the Bonn AGA preterm infants, $N = 64$, postnatal measurements, longitudinal, compared with "intrauterine" standards (mean or 50th percentile) from different authors.

50th percentile of the preterm infants corresponds at term to the 10th, and at 4 months to the 25th percentile of the full-term infants.

Two to Five Years. From two to five years the curves of preterm and full-term infants show no statistically significant difference, although the preterm infants tend to be 0.3–1.2 cm smaller on average. Since the midparent height of the preterm infants of 168.3 cm is 1.5 cm less than of the full-term infants (169.8 cm), the small difference in height between the groups may be attributed to genetic factors and not to preterm birth.

A synopsis of midparent height from other European studies (Tanner *et al.*, 1970) in Table IV demonstrates that the midparent height of the Bonn preterm infants of 168.3 cm is similar to that of the Zurich study of Prader and Budliger (1977; cited by Tanner *et al.*, 1970) with 168.7 cm. The midparent height of the Bonn full-term control infants of 169.8 cm agrees with that of the parents in the study from Stockholm (Karlberg *et al.*, 1976) of 170.2 cm.

In the growth chart from birth to 60 months in Figure 11, the data of the preterm infants from 21 to 60 months are combined with those of the full-term control infants, as there is no significant difference between the groups. Since the mean sex difference of 1.0–2.0 cm between 3 and 60 months is relatively constant, boys and girls are represented in a combined chart. However, since significant sex

575

*GROWTH DYNAMICS OF
LOW-BIRTH-WEIGHT
INFANTS WITH
EMPHASIS ON THE
PERINATAL PERIOD*

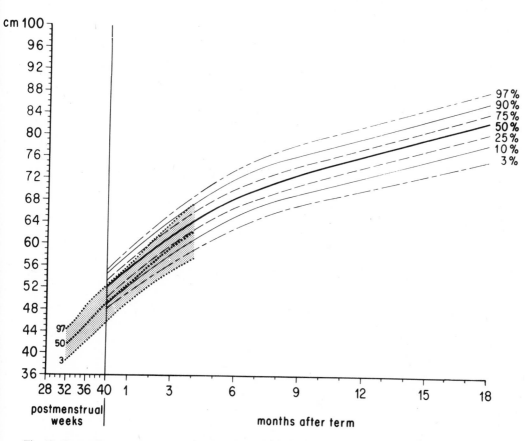

Fig. 10. Percentile curves (3rd, 50th, and 97th) for supine length of AGA preterm infants (boys and girls) from 32 postmenstrual weeks to 4 months after term (line of dots, shaded area) compared with percentiles (3rd to 97th) of the full-term controls from 0 to 18 months (continuous and broken lines).

differences do occur, a table of standards for each is also given (Table V) so separate standards may be constructed if desired. The standards are from unsmoothed Normally distributed data.

Sex Differences. From 30 to 40 postmenstrual weeks (longitudinal) there are—according to the Bonn results—significant sex differences in supine length of 0.6–1.4 cm in favor of the boys. Except in the Tanner–Thomson standards (1970), most of the "intrauterine" growth-curve (cross-sectional) sex differences in length have not been given (Hosemann, 1949b; Lubchenco et al., 1966; Usher and McLean, 1969; Miller and Hassanein, 1971; Nickl, 1972). Hohenauer (1973, 1975), who also combined boys and girls in his "intrauterine curves," mentions that on average boys are 0.2–1.15 cm longer than girls.

After term until 5 years, preterm boys continue to be bigger than girls, the mean difference of 0.6–2.0 cm being significant ($P < 0.05$ to $P < 0.01$), except at 1 and 2 months, and 5 years. Also from other studies of full-term infants there have been reported similar and relatively constant mean sex differences from 0 to 5 years, with boys being larger.

Necessary Period for Age Adjustment for Prematurity. When a comparison is made between the data of preterm infants at corrected and uncorrected ages, there is a significant mean difference in supine length, decreasing from 7.7 cm at birth to 1.4 cm at three years.

From 3.5 to 5 years the difference decreases to 0.9–1.0 cm and becomes nonsignificant. From 3.5 years onwards, for the assessment of height growth in preterm infants, it is not necessary to correct for age, but there is in fact a mean difference of about 1 cm.

Comparison with Other Studies. Earlier studies are not used for comparison where postmenstrual age has not been considered and where AGA and SGA have not been separated.

In the study of Babson (1970), the preterm infants' length from birth to 12 months remains below normal U.S.A. standards; the growth curve runs parallel at a lower level. At 12 months the infants of this Portland study are on average 0.9 cm smaller than in Bonn. The growth graphs for infants of varying gestational age of Babson and Benda (1976), derived from different sources, agree well with the Bonn results.

In the longitudinal study in Buffalo (Cruise, 1973), from birth to 3 years the preterm infants remain on average 4–5 cm smaller than the full-term control infants.

Table IV. Synopsis of Midparent Heights from European Studies

	Mothers and fathers number of pairs measured	Midparent height (mean in cm)
Brussels[a]	41	167.0
London[a]	55	166.9
Stockholm[a]	114	170.2
Zurich[a]	19	168.7
Bonn[b]		
Preterm infants	61	168.3
Full-term infants	79	169.8

[a]Tanner et al. (1970).
[b]Brandt (1977).

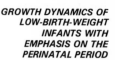

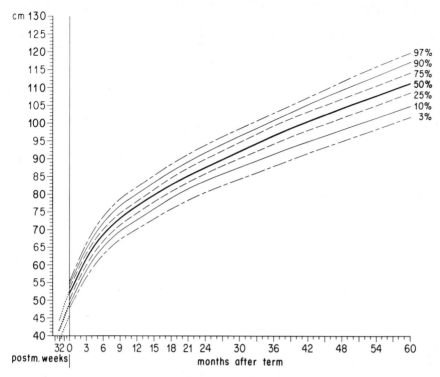

Fig. 11. Percentile curves for supine length/height of AGA preterm and full-term infants (boys and girls) from birth to 5 years, longitudinal; . . . . from 32 to 40 postmenstrual weeks AGA preterm infants, N = 64; 3rd, 50th, and 97th percentiles; —— from 0 to 18 months full term infants, N = 80; —— from 21 to 60 months AGA preterm and full-term infants combined, boys (N = 72) and girls (N = 72).

Stewart and Reynolds (1974) compared the height of 89 preterm infants with a birth weight of 1500 g and below at a mean age of 4 years, 9 months with the London standards (Tanner *et al.*, 1966). Fifteen percent of the children lay on or above the 75th percentile, 42% were between the 25th and 74th percentiles, 23% between the 10th and 24th percentiles, and 20% below the 10th percentile.

In the Oakland study from 1 to 10 years (Beck and Van den Berg, 1975), only low-birth-weight infants with a birth weight of 1501 g and above were included, and the preterm infants were only 0.7 cm less at the age of 5 years.

Bourlon *et al.* (1975) reported that from birth to 18 months the length of preterm infants was similar to full-term infants in the Paris longitudinal study (Sempé *et al.*, 1964).

The growth of the preterm infants from Montreal in the study of Fitzhardinge (1975) from birth to 12 months corresponds approximately with the standards of the Harvard Growth Study, i.e., is at a lower level than the Bonn standards.

Table VI shows, in an international comparison, from 1 to 36 months that the Bonn preterm infants' length after the age of 18 months is similar to that of full-term infants of other European studies and of the U.S.A. (NCHS Charts, 1976). At 36 months the preterm boys and girls are larger than full-term infants from Hamburg, London, Paris, and the U.S.A.; they agree well with the results from Stockholm; the preterm girls are only 0.3 cm smaller than the Dutch girls, who are among the tallest in the world.

Table V. Supine Length/Height (Distance) in cm for Boys and Girls—AGA Preterm and Full-Term Infants[a]

Age	N	Boys M	SD	SE	Percentiles 10	50	90	N	Girls M	SD	SE	Percentiles 10	50	90	Difference[b] between boys and girls
Months															
0	P 27	49.4	1.7	0.3	46.9	49.3	51.8	36	48.5	1.6	0.3	46.6	48.4	51.1	0.9 S.
	F 41	51.8	1.9	0.3	49.0	52.0	54.4	36	51.6	2.2	0.4	48.7	51.0	54.7	0.2 N.S.
1	P 26	53.0	1.9	0.4	50.6	52.9	55.6	36	52.4	1.6	0.3	50.7	52.0	53.5	0.6 N.S.
	F 42	55.4	1.9	0.3	52.5	55.7	57.8	31	54.9	2.3	0.4	51.8	55.1	58.2	0.5 N.S.
2	P 25	56.8	2.1	0.4	54.6	56.7	59.3	36	56.0	1.9	0.3	53.3	55.9	58.5	0.8 N.S.
	F 41	58.8	2.1	0.3	55.5	58.8	61.4	31	58.0	2.4	0.4	54.4	58.2	61.4	0.8 N.S.
3	P 25	60.1	2.3	0.5	57.6	59.6	63.5	36	59.1	2.1	0.4	55.8	59.2	62.1	1.0 S.
	F 43	62.0	2.1	0.3	59.2	62.1	64.9	31	60.8	2.5	0.5	57.6	60.9	64.7	1.2 S.
4	P 25	63.2	2.3	0.5	60.8	62.3	67.3	35	62.0	2.3	0.4	58.1	61.8	65.4	1.2 S.
	F 43	64.8	2.5	0.4	61.1	64.5	68.5	31	63.3	2.6	0.5	60.0	63.5	67.4	1.5 H.S.
5	P 25	65.8	2.6	0.5	62.8	65.2	70.3	35	64.3	2.3	0.4	60.7	64.5	67.9	1.5 H.S.
	F 42	67.3	2.3	0.4	64.3	67.3	70.5	31	65.6	2.9	0.5	62.1	65.8	70.0	1.7 H.S.
6	P 25	68.0	2.5	0.5	65.2	68.1	72.1	35	66.4	2.2	0.4	63.1	66.5	69.4	1.6 H.S.
	F 42	69.4	2.3	0.4	67.1	69.0	72.9	31	67.9	2.9	0.5	64.4	67.5	72.3	1.5 S.
7	P 25	69.7	2.5	0.5	66.8	69.8	73.8	35	68.1	2.4	0.4	64.8	68.1	71.8	1.6 H.S.
	F 42	71.0	2.4	0.4	68.4	70.5	74.8	31	69.5	3.0	0.5	65.8	69.2	74.4	1.5 H.S.
8	P 25	71.1	2.6	0.5	68.3	70.5	75.5	35	69.5	2.5	0.4	66.5	69.8	73.0	1.6 H.S.
	F 42	72.2	2.5	0.4	69.2	71.9	75.8	30	70.9	2.9	0.5	67.1	70.8	75.2	1.3 S.
9	P 26	72.6	2.6	0.5	69.8	72.0	76.7	35	71.0	2.6	0.4	67.6	71.3	74.1	1.6 H.S.
	F 42	73.5	2.5	0.4	70.3	73.4	77.0	30	72.3	2.6	0.5	68.6	72.9	76.0	1.2 S.
10	P 26	73.7	2.5	0.5	70.9	73.1	77.7	35	72.3	2.6	0.4	69.0	73.0	75.2	1.4 S.
	F 42	74.6	2.4	0.4	71.1	74.7	77.6	30	73.6	2.8	0.5	69.4	73.9	77.5	1.0 N.S.

579

GROWTH DYNAMICS OF
LOW-BIRTH-WEIGHT
INFANTS WITH
EMPHASIS ON THE
PERINATAL PERIOD

Age	Group	N	M	SD	SE				N	M	SD	SE				Diff[b]	Sig
11	P	26	74.7	2.5	0.5	71.7	73.8	78.7	35	73.5	2.6	0.4	70.4	73.8	76.6	1.2	S.
	F	42	75.9	2.5	0.4	73.0	76.1	78.9	30	74.6	2.8	0.5	70.7	75.0	78.7	1.3	S.
12	P	26	75.9	2.7	0.5	72.3	75.0	79.8	35	74.6	2.6	0.4	71.7	74.9	78.0	1.3	S.
	F	42	77.2	2.7	0.4	73.7	77.5	80.7	29	75.7	3.0	0.6	72.0	75.6	79.8	1.5	S.
15	P	26	79.3	2.8	0.6	75.5	79.0	83.1	33	77.7	2.7	0.5	74.8	77.4	81.5	1.6	S.
	F	42	80.4	3.0	0.5	77.1	80.4	84.0	30	78.9	3.0	0.6	75.1	79.2	82.6	1.5	N.S.
18	P	25	82.1	2.5	0.5	78.6	81.8	85.6	32	80.6	3.0	0.5	77.0	80.9	84.5	1.5	S.
	F	42	83.3	3.0	0.5	79.4	83.3	86.8	29	82.1	3.1	0.6	78.5	82.6	86.1	1.2	N.S.
21	P/F	66	85.7	2.9	0.4	82.1	85.7	89.1	58	84.5	2.9	0.4	80.5	84.7	88.4	1.2	S.
Years																	
2	P/F	65	88.3	3.1	0.4	84.4	88.4	91.8	56	87.1	2.9	0.4	82.9	87.0	91.0	1.2	S.
2.5	P/F	67	93.0	3.4	0.4	88.8	93.2	97.2	57	91.6	3.6	0.5	86.6	92.0	95.9	1.4	S.
3	P/F	64	97.2	3.6	0.5	92.5	97.6	101.4	57	96.0	3.8	0.5	90.7	96.3	100.5	1.2	S.
3.5	P/F	61	101.2	3.8	0.5	96.6	101.1	106.0	57	99.7	4.0	0.5	94.1	100.2	104.7	1.5	S.
4	P/F	56	104.9	4.1	0.6	100.0	104.8	109.8	51	103.5	4.2	0.6	97.3	104.0	108.3	1.4	S.
4.5	P/F	53	108.8	4.2	0.6	103.2	108.8	113.8	50	107.0	4.3	0.6	100.5	107.3	112.0	1.8	S.
5	P/F	52	112.3	4.4	0.6	106.5	111.7	117.8	46	110.3	4.3	0.6	104.1	110.3	115.9	2.0	S.

[a] P = pretem infants; F = full-term infants; N = number of infants; M = mean; SD = standard deviation; SE = standard error of the mean; N.S. = P > 0.05; S. = P < 0.0125 to P < 0.05; H.S. = P < 0.01.

[b] Difference between boys and girls = mean of the boys minus mean of the girls.

Table VI. Supine Length/Height (in cm) from 1 to 36 Months (International Comparison)[a]

Age (mo.):	1		3		6		9		12		18		24		36	
Sex:	B	G	B	G	B	G	B	G	B	G	B	G	B	G	B	G
AGA Preterm Infants																
Bonn[b]																
Mean	53.0	52.4	60.1	59.1	68.0	66.4	72.6	71.0	75.9	74.6	82.1	80.6	87.9	86.7	96.9	95.7
50th Percentile	52.9	52.0	59.6	59.2	68.1	66.5	72.0	71.3	75.0	74.9	81.8	80.9	87.8	86.8	97.1	96.2
Full-Term Infants																
Hamburg/Kiel[c]																
Mean	53.0	52.5	61.4	59.1	67.2	66.4	72.8	70.9	76.3	74.8	81.4	79.1	87.6	86.2	96.2	94.9
Holland[d]																
50th Percentile	55.5	54.5	61.6	60.5	68.5	67.2	73.3	71.9	77.0	75.6	83.8	82.5	88.9	87.7	97.5	96.5
London[e]																
Mean	53.9	52.5	60.2	58.2	66.7	64.8	71.3	69.3	75.0	73.3	81.2	79.7	86.5	84.5	94.1	94.1
London and Oxford[f]																
50th Percentile	54.0	53.0	60.7	59.0	68.2	65.5	72.7	70.2	76.3	74.2	82.1	80.5	86.9	85.6	94.2	93.0
Paris[g]																
Mean	53.2	52.4	59.8	58.7	66.5	64.8	70.7	69.0	74.5	72.8	80.3	78.8	85.6	84.2	94.2	92.8
Stockholm[h]																
Mean	54.5	53.7	61.3	59.9	68.0	66.4	72.4	71.1	76.4	75.2	83.4	82.2	88.9	87.8	97.5	96.4
50th Percentile	54.6	53.8	61.3	60.0	67.9	66.3	72.3	70.7	76.4	75.1	83.1	82.1	88.6	87.5	97.1	96.1
U.S.A.[i]																
50th Percentile	54.6	53.5	61.1	59.5	67.8	65.9	72.3	70.4	76.1	74.3	82.4	80.9	87.6	86.5	96.5	95.6
Zürich[j]																
Mean	53.7	53.0	60.9	59.5	67.9	66.0	72.6	70.9	76.2	74.5	82.6	80.8	88.3	86.6	97.2	95.4
Bonn[b]																
Mean	55.4	54.9	62.0	60.8	69.4	67.9	73.5	72.3	77.2	75.7	83.3	82.1	88.5	87.6	97.4	96.8
50th Percentile	55.7	55.1	62.1	60.9	69.0	67.5	73.4	72.9	77.5	75.6	83.3	82.6	89.1	88.5	97.7	97.6

[a]B = boys; G = girls.
[b]Brandt (1977).
[c]Spranger et al. (1968).
[d]van Wieringen et al. (1971).
[e]Falkner (1958).
[f]Tanner et al. (1966).
[g]Sempé et al. (1964).
[h]Karlberg et al. (1976).
[i]National Center for Health Statistics (1976).
[j]Prader and Budliger (1977).

2.2.2. Velocity

581

*GROWTH DYNAMICS OF
LOW-BIRTH-WEIGHT
INFANTS WITH
EMPHASIS ON THE
PERINATAL PERIOD*

Perinatal Period to 18 Months. The extrauterine mean growth velocity in supine length, from 29.5 to 39.5 postmenstrual weeks, calculated in centimeters per month, is relatively constant (Figure 12a) and differs significantly from the velocity in the first month after term, which is itself equal in AGA preterm and full-term infants (Figure 12b) (Brandt, 1976c).

From 1.5 to 7.5 months, preterm infants grow significantly faster than full-term infants ($P < 0.05$). Then to 18 months the growth velocity of preterm and full-term infants is either identical or the preterm infants grow a little faster (but without significant difference).

Two to Five Years. From 2 to 5 years supine length/height velocity of the preterm infants is either identical to that of the full-term control infants or the preterm infants grow a little faster, but again without significant difference. The data suggest a protracted but complete "catch-up" of the initial postnatal retardation.

In general, length velocity, calculated in centimeters per month, slows down from a peak of 4.0 cm/month at 35.5 and 36.5 postmenstrual weeks (Figure 12a) to about 0.5 cm/month at 5.75 years.

When length velocity is calculated in centimeters per year, as shown in a computer plot from 30 postmenstrual weeks to 60 months after term (Figure 13), the preterm infants grow at 44.6 cm/yr in the 2nd month, 29.7 cm/yr in the 5th, and 17.5 cm/yr in the 8th month—always significantly faster than the full-term control infants. From 9.5 months to 5.5 years, growth velocity of preterm infants is on average 0.1–0.8 cm/yr faster than of the full-term control infants. The growth velocity of the preterm infants corresponds also to the results of other longitudinal studies of full-term infants.

Sex differences in growth velocity from birth to 5 years are inconsistent and small.

Velocity curves for growth of preterm infants in the perinatal period and until 5 years are not available in the present literature. In the longitudinal study of Cruise (1973), where the preterm infants were measured every 13 weeks in the first year, mean increments in the first 13 weeks of 11.7 cm for girls and of 12.4 cm for the boys correspond to velocities of 3.9 and 4.1 cm/month, respectively; this agrees with the Bonn results.

2.3. Head Circumference

2.3.1. Distance

Perinatal Period. Before term from 28 to 39 postmenstrual weeks, head circumference growth of the Bonn preterm infants (Figure 14) (postnatal measurements, longitudinal) is similar to that of the "intrauterine curves" (cross-sectional). The 50th percentile of Lubchenco *et al.* (1966) is crossed in the 37th postmenstrual week; at term, the mean growth curve of the AGA preterm infants corresponds well with the 50th percentile of Hosemann (1949c), and the means of von Harnack (1960), Usher and McLean (1969), Miller and Hassanein (1971), and Nickl (1972), not plotted here. The slight initial growth retardation of the preterm infants may be due to adaptation to extrauterine life and to delayed feeding in some infants.

Before and After Term until 18 Months. At term and thereafter until 18 months, head circumference of preterm infants is identical with that of the full-term

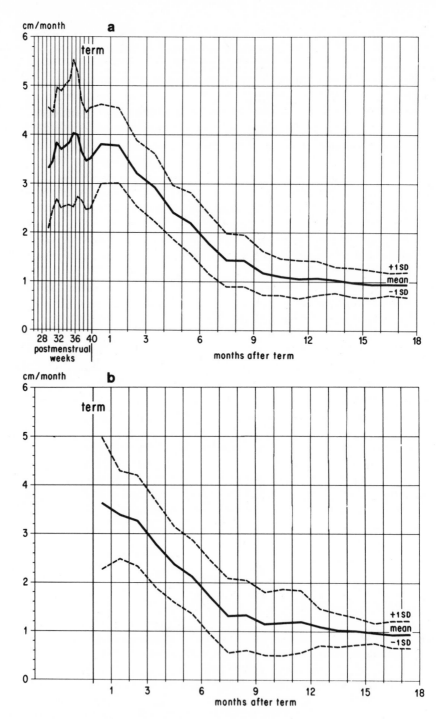

Fig. 12. Velocity curves of supine length from 29.5 postmenstrual weeks to 17.5 months after term, mean and one standard deviation, unsmoothed values. (a) Preterm infants (boys and girls), appropriate for gestational age, $N = 64$; (b) full-term infants (boys and girls), $N = 80$.

CM/YEAR

583

GROWTH DYNAMICS OF
LOW-BIRTH-WEIGHT
INFANTS WITH
EMPHASIS ON THE
PERINATAL PERIOD

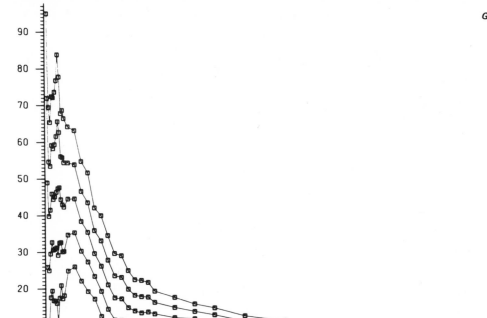

Fig. 13. Computer plot of supine length/height velocity from 30 postmenstrual weeks to 60 months after term, calculated in cm per year, mean with one and two standard deviations; AGA preterm infants, $N = 64$.

control infants (Brandt, 1976*a,b*). Therefore, the percentile curves in Figure 15 are based on preterm and full-term infants combined. The standards are from unsmoothed Normally distributed data.

Contrary to weight and length, head circumference growth is not diminished.

Two to Six Years. From 2 to 6 years the head circumference curves of preterm and full-term infants continue to be identical. In both groups also the mean head circumference of the parents agrees well (with no significant difference; $P > 0.20$).

	Mean head circumference (cm)	
	Fathers	Mothers
Bonn AGA preterm	57.3 ($N = 59$)	54.7 ($N = 60$)
Bonn full-term	57.6 ($N = 78$)	54.8 ($N = 80$)

There was a close relationship between the head circumference at 3 years of the daughters, and head circumference of the mothers for preterm infants ($r = 0.63$), and for the full-term girls ($r = 0.74$). For the fathers and the sons the coefficient was less than 0.44.

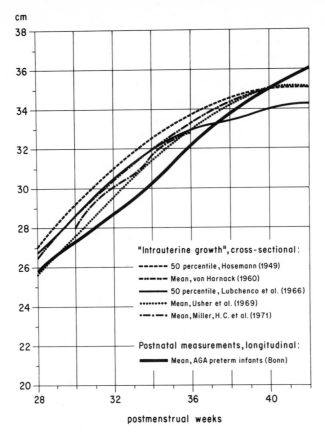

Fig. 14. Mean head circumference curve of the Bonn AGA preterm infants, $N = 64$, postnatal measurements, longitudinal, compared with "intrauterine" standards (mean or 50th percentile) from different authors.

Sex Differences. From 30 to 40 postmenstrual weeks (longitudinal) there are significant sex differences in head circumference of 0.6–1.4 cm with boys larger (Figure 16). In their "intrauterine growth" chart, Lubchenco *et al.* (1966) found no significant difference between sexes at any age. Usher and McLean (1969) did not mention sex differences because the numbers involved in their series were insufficient to give acceptable indications. Miller and Hassanein (1971) found no sex differences. According to the intrauterine curves of Hohenauer (1974), between 35 and 40 weeks there is a mean difference between boys and girls of 0.1–0.8 cm.

After term until 6 years, preterm and full-term boys continue to have a larger head circumference than girls; the mean difference amounts to 0.8–1.5 cm and remains significant.

From other studies of full-term infants, a relatively constant mean sex difference of 0.6–1.4 cm on average is reported between birth and 6 years; in longitudinal studies, the sex difference fluctuates only between 0.6 and 0.8 cm.

Necessary Period for Age Adjustment for Prematurity. When a comparison is made between the results of the preterm infants at corrected and at uncorrected age, there is a significant mean difference in head circumference from birth to 17 months which decreases during this period from 6.3 to 0.5 cm. With further decreasing growth velocity this difference diminishes and becomes nonsignificant at 18 months, being then 0.4 cm. At 6 years it is 0.04 cm. This "assimilation" of the non-age-corrected curve to the corrected curve which is caused only by the rapid decrease of growth velocity (see p. 591), has led, in some earlier publications (Sobel and

Falkner, 1969; Szénásy, 1969), to the erroneous assumption of catch-up growth in AGA preterm infants. For practical use of the standards of head circumference growth in preterm infants, it is unnecessary to correct for age from 18 months onwards.

Comparison with Other Studies. Few longitudinal data are available before term on head circumference growth where the age has been corrected for prematurity. According to Reiche (1916), who does not give values, the growth of preterm infants is dependent on the time from conception, independent of chronological age, and in the last postmenstrual weeks head circumference growth is "considerably more rapid" than in the first 6 months after term. Ylppö (1919), in his Berlin longitudinal study on about 350 preterm infants, did not give standards for head circumference growth, but he reports a "megacephalus" of the preterm infants which, however, is not big in relation to the postmenstrual age, but is due to the growth dissociation of other body dimensions; "the brain and therewith the head are following their own laws of growth, relatively independent of growth disorders of the other parts of the body." Lang (1962) reports in his preterm infants at term a greater head circumference than in full-term infants, but subsequently, at 12 weeks after term, there is a coincidence of the curves. In the study of Babson (1970), based on repeated measurements of 24 AGA preterm infants from 27 to 33 postmenstrual weeks, head circumference growth, after initially lower values at two weeks, is similar to that of national standards. The head circumference values for the Kansas preterm infants, reported by Miller and Hassanein (1971), derived from single measurements of infants at 3 days postnatal age, and with postmenstrual ages of 30–35 weeks, are slightly greater than the Bonn longitudinal values for preterm infants.

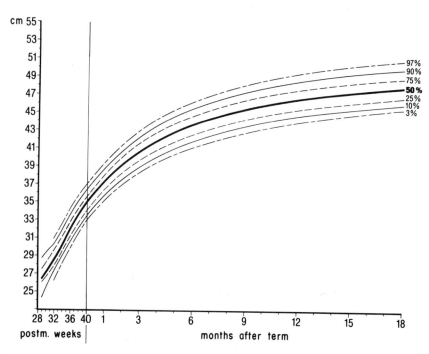

Fig. 15. Percentile curves of head circumference growth (longitudinal) from 29 postmenstrual weeks to 18 months after term of AGA preterm infants ($N = 64$), and of full-term infants ($N = 80$) combined, unsmoothed values.

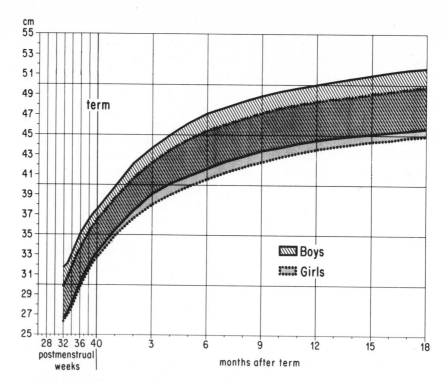

Fig. 16. Standard chart for growth in head circumference from 32 postmenstrual weeks to 18 months after term, comparison of boys and girls, mean and two standard deviations, unsmoothed values. Continuous line with striped area: AGA preterm and full-term boys, $N = 72$; line of dots with shaded area: AGA preterm and full-term girls, $N = 72$.

At a postmenstrual age of 36 weeks, the mean head circumference of 32.8 cm (SD = 1.08) of the Kansas preterm infants ($N = 76$, cross-sectional) corresponds with the mean head circumference of the Bonn preterm infants of 32.9 cm (SD = 1.1) at 37 postmenstrual weeks ($N = 64$, longitudinal). The data of Finnström (1971), based on single measurements on the first or second day of life, are not suitable for comparison because there are only 6 preterm infants with a postmenstrual age of less than 225 days (32.14 weeks), and the other infants are grouped in intervals of two weeks.

Cruise (1973) reports mean measurements of head circumference of low-birth-weight infants from birth to 3 years who are arranged in postmenstrual age groups of 28–32 weeks and 33–36 weeks. The infants were measured in the first year every 13 weeks so that comparison with the Bonn standards before term is not feasible. At three years, the preterm infants in her study are smaller in head circumference than local full-term control infants by 1.2 cm for the girls and 0.9 cm for the boys.

In the study of Davies (1975) head circumference growth has been compared in two groups of preterm infants with a birth weight of 1500 g and below, born from 1961 to 1964 ($N = 38$) and from 1965 to 1970 ($N = 73$), and with different postnatal nutrition and different environmental temperature control. Mean head circumference of the earlier born (1961–1964) preterm infants, given significantly fewer calories after birth, was smaller and similar to that of SGA preterm infants of the same period. From this it is concluded that these preterm infants have become growth retarded after birth. Mean head circumference of the later born (1965–1970) preterm infants, given significantly more calories after birth, was not diminished.

587

*GROWTH DYNAMICS OF
LOW-BIRTH-WEIGHT
INFANTS WITH
EMPHASIS ON THE
PERINATAL PERIOD*

Head circumference of the preterm infants from Montreal, with a mean birth weight of 1440 g, in the study of Fitzhardinge (1975), agrees with the international standards of Nellhaus (1968). Further details in the form of tables or values are not given. The infants were not measured in the perinatal period.

In the French study of Bourlon *et al.* (1975) head circumference growth of the preterm infants from 31 to 37 postmenstrual weeks agrees with the Bonn results. Subsequently, up to 40 weeks the French results are 0.4–0.5 cm below the Bonn standards. From 1 to 18 months the mean curve of the French preterm infants is 0.6–0.9 cm below that of the Bonn study, but agrees with the French standards for full-term infants (Paris longitudinal study, Sempé *et al.*, 1964).

In these last three studies, where postmenstrual age has been considered and the infants have been grouped into AGA and SGA, growth of head circumference agrees with corresponding national standards.

Follow-up results of 28 preterm infants (Grassy *et al.*, 1976) with a birth weight of 1000 g and below showed regular head circumference growth in the girls, whereas most of the boys remained below the mean of the Nellhaus (1968) curve.

In some studies the conclusion still prevails that preterm infants exhibit catch-up in head circumference during the first year of life. Sobel and Falkner (1969) report much greater increments of head circumference in preterm infants than in those of full-term infants, and that at 3 months of age (uncorrected) preterm infants, on average, have a similar head circumference to full-term infants. Others have recorded a smaller head circumference in preterm infants. According to Justesen (1962), the head circumference of preterm infants at 12 years remains on average 0.5–1.5 cm smaller than that of full-term infants. Baum and Searls (1971*b*) report that 18-year-olds, who were preterm infants, had a smaller head circumference than those who were full term.

A correction for the dolichocephaly of preterm infants by the formula presented by Baum and Searls (1971*a,b*) did not essentially change the present Bonn results, because making the correction decreased the head circumference of our full-term infants to nearly the same extent as it did the normal infants of Bayley and Davis (1935).

Table VII shows, in an international comparison from 1 to 36 months, that the Bonn preterm infants' head circumference is similar or bigger than that of full-term infants of other European studies and of the U.S.A. (NCHS Charts, 1976). At twelve months the Bonn preterm boys' mean head circumference of 47.1 is similar to those from the U.S.A. and Zurich. At 36 months the preterm girls' mean head circumference of 49.5 cm agrees well with the data of Hamburg (Spranger *et al.*, 1968), London (Falkner, 1958), Oxford (Westropp and Barber, 1956), and Zurich (Prader and Budliger, 1977).

2.3.2. Velocity

Perinatal Period to 18 Months. The close relationship between head circumference growth and brain development in the normally growing infant during the first year of life has been demonstrated in many studies (Jackson, 1909; Bray *et al.*, 1969; Winick and Rosso, 1969*a;* Dobbing, 1970; Buda *et al.*, 1975). It may be concluded, therefore, that rapid growth velocity in head circumference indicates a period of rapid brain development.

From 30.5 to 39.5 postmenstrual weeks the extrauterine growth velocity of head circumference, calculated in centimeters per month, is significantly higher than after term (Figure 17). It amounts to 3.3 cm/month in the 31st postmenstrual

INGEBORG BRANDT

Table VII. Head Circumference (in cm) from 1 to 36 Months (International Comparison)[a]

Age (mo.): Sex:	1 B	1 G	3 B	3 G	6 B	6 G	9 B	9 G	12 B	12 G	18 B	18 G	24 B	24 G	36 B	36 G
AGA Preterm Infants																
Bonn[b]																
Mean	38.0	37.0	41.4	40.3	44.5	43.2	46.1	45.0	47.1	46.1	48.6	47.4	49.5	48.3	50.5	49.5
50th Percentile	38.1	37.2	41.6	40.5	44.6	43.5	46.1	45.3	46.9	46.2	48.5	47.2	49.5	48.2	50.3	49.7
Full-Term Infants																
Hamburg/Kiel[c]																
Mean	36.2	35.6	40.2	39.3	43.0	42.0	44.8	43.5	46.4	44.9	48.3	47.2	49.5	48.7	50.7	49.5
London[d]																
Mean	37.4	36.5	40.7	39.7	43.8	42.7	45.7	44.6	46.8	45.7	48.3	47.0	49.3	47.9	50.4	49.5
Oxford[e]																
Mean	37.3	36.5	40.7	39.8	43.6	42.5	45.7	44.6	46.8	45.6	47.9	47.0	49.1	48.0	50.4	49.5
Paris[f]																
Mean	37.0	36.1	40.3	39.0	43.3	42.0	45.4	43.9	46.7	45.3	48.1	46.7	49.0	47.5	50.2	49.0
Stockholm[g]																
Mean	36.9	36.1	40.2	39.2	43.3	42.1	45.1	44.0	46.4	45.2	48.0	46.6	48.9	47.6	49.9	48.7
50th Percentile	37.0	36.1	40.2	39.2	43.2	42.1	45.0	44.0	46.4	45.2	48.0	46.7	49.0	47.7	50.0	48.7
U.S.A.[h]																
Mean	37.2	36.4	40.6	39.5	43.8	42.4	45.8	44.3	47.0	45.6	48.4	47.1	49.2	48.1	50.5	49.3
Zurich[i]																
Mean	37.0	36.4	40.5	39.7	43.8	42.6	45.8	44.6	47.0	45.8	48.6	47.4	49.5	48.3	50.8	49.5
Bonn[b]																
Mean	37.6	36.5	41.3	39.9	44.2	42.7	46.1	44.4	47.3	45.7	48.6	47.1	49.5	48.2	50.7	49.4
50th Percentile	37.6	36.4	41.3	39.8	44.5	42.5	46.2	44.3	47.5	45.8	48.7	47.1	49.3	48.0	50.7	49.4

[a]B = boys; G = girls.
[b]Brandt (1977).
[c]Spranger et al. (1968).
[d]Falkner (1958).
[e]Westropp and Barber (1956).
[f]Sempé et al. (1964).
[g]Karlberg et al. (1976).
[h]National Center for Health Statistics (1976).
[i]Prader and Budliger (1977).

589

GROWTH DYNAMICS OF
LOW-BIRTH-WEIGHT
INFANTS WITH
EMPHASIS ON THE
PERINATAL PERIOD

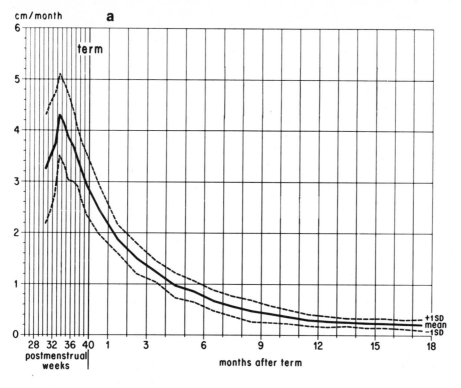

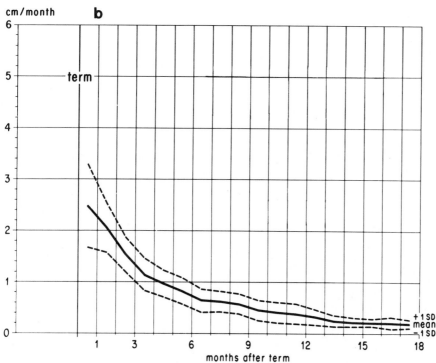

Fig. 17. Velocity curves of head circumference growth from 30.5 postmenstrual weeks to 17.5 months after term, mean and one standard deviation, unsmoothed values. (a) Preterm infants (boys and girls), appropriate for gestational age, $N = 64$; (b) full-term infants (boys and girls), $N = 80$.

week, to 3.9 cm/month in the 32nd, reaches a peak of 4.3 cm/month in the 34th week, and decreases to 3.2 cm/month in the 39th week (Brandt, 1976*a,b*).

After term, velocity decreases rapidly. It amounts to 2.5 cm/month in the first month, similar to that of the full-term control infants with 2.4 cm/month. Until 18 months, the velocity curve of the AGA preterm infants also agrees with that of the full-term control infants (Figure 17); the differences between the groups from the first to the 18th month are not significant except in the 4th month, where preterm infants grow faster, and in the 9th and 12th month where full-term infants grow somewhat faster. Thus, a period of rapid head circumference growth extends from the 31st postmenstrual week until the 6th month after term. The curve in Figure 17 clearly shows the considerable decrease of velocity from 4.3 cm/month at its peak in the 34th postmenstrual week to 0.8 cm/month (one fifth) in the 6th month after term. This high velocity of head circumference in the perinatal period corresponds to a phase of rapid brain development (Winick, 1968; Dobbing, 1974*b*) and suggests a great energy requirement of preterm infants.

Beyond 6 months the curve further flattens markedly. At 16.5 months, head circumference velocity has diminished to 0.2 cm/month on average. This is only $\frac{1}{20}$ of the mean peak velocity in the 34th postmenstrual week.

The individual development of three AGA preterm boys, born between 29 and 32 postmenstrual weeks, in comparison with the mean velocity curve of the Bonn AGA preterm infants is shown in Figure 18. The timing of peak growth velocity in these three infants agrees well with that of the mean curve. The caloric intake in cal/kg/day, called the energy quotient (EQ), at peak velocity of infant number one (line of dots) was 140; the second infant (broken line) had 139 cal/kg/day, and the third (broken line with dots) was fed 160 cal/kg/day. The standard formula given consisted of cow's milk, adjusted to breast milk constitution with 75 cal/100 ml. The content per 100 ml was: fat, 3.3 g; protein, 2.3 g with a casein/lactalbumin ratio of 40/60; and lactose, 8.6 g. In some cases 3–5% glucose was added to the formula. In a few other cases, the peak has been shifted a little to the right or left, giving a phase-difference effect similar to that for the growth spurt in height at adolescence described by Tanner *et al.* (1966). This effect alters the mean velocity curve and makes it broader and flatter, compared with that of individual infants. Because of the lack of awareness about such a period of very high growth velocity in head circumference at a certain postmenstrual age, sometimes incipient hydrocephalus has been, or is, erroneously diagnosed, or its onset is feared (Falkner *et al.*, 1962).

Although head circumference has been shown to have a growth pattern least affected by premature birth, there still exists an increased vulnerability during the spurt. In some infants with delayed feeding, low caloric intake, or illness in the early days after birth, growth velocity of head circumference was decreased or was even zero. In some cases an interruption of the spurt by illness was observed, leading to a two-peaked velocity curve.

During the growth spurt, preterm infants need high-calorie diets because their nutritional requirements before term are higher then than later (Smith, 1962; Tizard, 1973; Bergmann, 1976; Fomon *et al.*, 1976). When preterm infants have been reported in former studies to have a smaller head circumference than full-term infants, this could be attributed to delayed and low-calorie feeding in the first postnatal weeks. Unfortunately, there is still much controversy about the actual energy requirements and optimal nutrition of very small preterm infants.

The individual velocity data further demonstrate that, after term, no AGA preterm infant has a significantly higher growth velocity than full-term control infants. Only in SGA preterm infants with catch-up growth (see p. 599) may a significantly increased velocity occur for 1–2 months after term. Accelerated growth in head circumference beyond this time in preterm infants is pathological.

If age is not corrected for prematurity, the peak of the velocity curve becomes flat and spread out and does not demonstrate the individual growth rate (Figure 19). From 0.5 to 17.5 months, mean growth velocity of head circumference at uncorrected age is significantly higher than at corrected age ($P < 0.0005$ to $P < 0.05$). Calculated in centimeters per month it amounts to 3.2 cm in the first, 3.4 cm in the second, and to 2.5 cm in the third month. Subsequently the velocity decreases, and at 17.5 months the curve is at 0.22 cm/month compared with 0.19 cm/month for the corrected curve; only at 19.5 months do both curves agree without a significant difference. A similar observation can be made for height velocity at puberty; here also the peak broadens and flattens when the different maturity stages of children of the same chronological age are not considered in calculating the mean growth spurt and when the data are treated only cross-sectionally (Tanner *et al.*, 1966).

Two to Five Years. From 2 to 5 years, head circumference velocity of the preterm infants agrees with that of the full-term control infants without significant difference.

Thus, head circumference velocity, calculated in centimeters per month, slows down from its peak of 4.3 cm/month at 33.5 postmenstrual weeks (Figure 17) to 0.04 cm at 4.25 years (i.e., by more than $\frac{1}{100}$). In the following months until 6 years, the velocity remains at the level of about 0.04 cm/month.

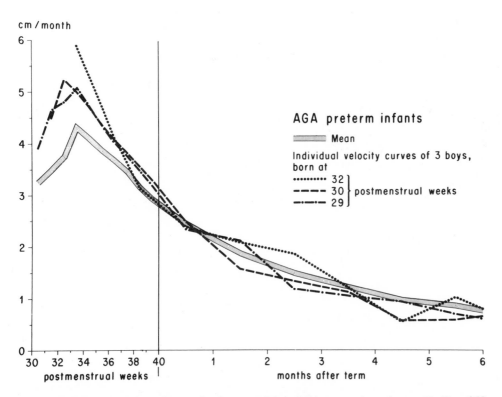

Fig. 18. Individual head circumference development of three AGA preterm boys, born at 29, 30, and 32 postmenstrual weeks, plotted against the mean growth velocity curve of the Bonn AGA preterm infants.

This enormous decrease in head circumference velocity from the perinatal period to 4 years is demonstrated in a computer plot from 30 postmenstrual weeks to 60 months after term (Figure 20). Here velocity is calculated in centimeters per year, giving the mean with one and two standard deviations; it amounts to 22.7 cm/year in the 2nd month, to 11.7 cm/year in the 5th, and to 6.9 cm/year in the 8th month. At 2.25 years the head grows at 1.29 cm/year, and at 4.25 years only at 0.45 cm/year, remaining at about this level until 6 years. The velocity agrees with that of the full-term control infants and with full-term infants from other studies.

Sex differences in head circumference velocity from birth to 5 years are inconsistent and small. From 0.5 to 5.5 months, the boy's head grows faster, but the difference only reaches significance at 1.5 months. From 6.5 to 12.5 months, the girl's head grows faster, but only at 9.5 months is the difference significant. From 13.5 months to 5 years, sometimes the girl's, and sometimes the boy's head grows (not significantly) faster, or the difference is zero. None of the authors dealing with head circumference velocity reports any consistent sex difference (Nellhaus, 1968; Kantero et al., 1971; Tanner, 1973; Cruise, 1973; Grantham-McGregor and Desai, 1973).

Velocity curves of head circumference growth in the perinatal period and up to 5 years are not available in the present literature. Davies and Davis (1970) calcu-

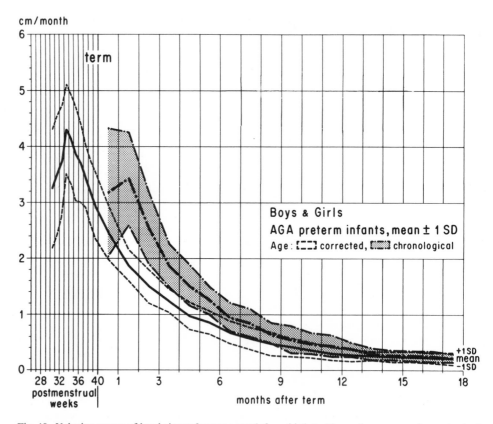

Fig. 19. Velocity curves of head circumference growth from birth to 18 months, mean and one standard deviation. Comparison of corrected age and uncorrected (= chronological) age; from 0.5 to 17.5 months decreasing significant difference of 6.53 to 0.32 cm. Continuous line: corrected age; broken line and dots, shaded area: uncorrected age.

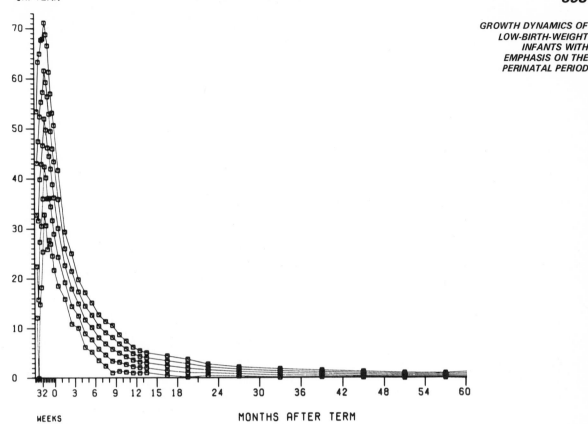

Fig. 20. Computer plot of head circumference growth velocity from 30 postmenstrual weeks to 60 months after term, calculated in cm per year, mean with one and two standard deviations; AGA preterm infants, $N = 64$, unsmoothed results.

lated a mean daily head growth rate of 0.1 cm before term for AGA preterm infants who were born between 1965 and 1968; that is, 3.0 cm/month. Cruise (1973) reported a mean increment in head circumference of 9.3 cm "from birth to 13 weeks" in 22 girls and 17 boys, all born at 28–32 postmenstrual weeks who were measured every 13 weeks. This corresponds to a mean growth velocity of 3.1 cm/ month for both sexes in the first 3 months. In that study, velocity is not represented in relation to postmenstrual age or to corrected age of the preterm infants, but only the increment in intervals of 13 weeks and one year. Miller and Hassanein (1971) did not calculate velocity from their cross-sectional head circumference measurements of preterm infants, but the data demonstrate a difference of 1.0 cm from 33 to 34 postmenstrual weeks, which corresponds to a mean velocity of 4.2 cm/month and agrees with the peak of 4.3 cm in the 34th postmenstrual week of the Bonn preterm infants. Subsequently, the mean weekly head circumference increments in the preterm infants of Miller decrease, just as in the Bonn study. For calculation of velocity standards in the perinatal period, head circumference measurements at short intervals are necessary, since growth velocity in this period is very rapid and changes quickly (Figure 17).

Table VIII. Head Circumference Increments (in cm), Boys and Girls Combined, Mean and Standard Deviation (International Comparison)[a]

| Increment period in months | Full-term infants | | | | | | | | Preterm infants, corrected age Bonn[e] | |
| | Oxford[b] | | Berkeley[c] | | Stockholm[d] | | Bonn[e] | | | |
	M	SD	M	SD	M	SD	M	SD	M	SD
1– 3	3.4	1.21	3.0	0.62	3.2	0.65	3.5	0.41	3.4	0.30
3– 6	3.2	1.04	3.3	0.68	3.0	0.60	2.9	0.26	3.1	0.22
6–12	3.2	1.15	3.4	0.75	3.2	0.40	3.1	0.20	2.9	0.17
12–18	1.5	0.81	1.4	0.60	1.5	0.50	1.4	0.10	1.4	0.28

[a]M = mean; SD = standard deviation. [c]Eichhorn and Bayley (1962). [e]Brandt (1976a).
[b]Westropp and Barber (1956). [d]Karlberg et al. (1968).

After term, the deceleration of head growth velocity observed beyond the sixth month corresponds well with the examination of Diemer (1968) of the development of the capillaries in the brain. He showed a rapid increase of capillary density until the end of the sixth month and a deceleration thereafter; the density of capillaries in the grey matter of the frontal lobe reached at the end of the sixth month already corresponded to that of the adult brain. According to Winick (1968), increase in DNA content of the brain begins to level off at about the time of birth and the absolute content reaches a maximum by five months after term, indicating that little cell division occurs in the human brain after that time. Most recent data (Winick, 1974) suggest that there is a peak in cell division around 26 postmenstrual weeks and again around term and that cell division continues to about 18 months. Dobbing (1970) has reported the human brain "growth spurt" as extending from about 30 weeks of gestation to 18 months or 2 years of age. Recently Dobbing (1974a; Dobbing and Sands, 1973) extended the human brain growth spurt "from midpregnancy well into the third or fourth postnatal year" with a peak around term, whereas the peak in head circumference growth velocity occurs shortly before term. According to Dobbing (1974c), "it is to be expected that curves of circumferential growth show different timing from curves of volume growth (brain weight)."

In Table VIII is shown an international comparison of head circumference increments between 1 and 18 months of the Bonn AGA preterm infants and of full-term infants from Oxford, Berkeley, Stockholm, and Bonn. There is good agreement between the four groups of full-term infants and the preterm infants. For example, between 3 and 6 months, the increment of 3.2 cm in the study from Oxford (Westropp and Barber, 1956) and of 3.0 cm in the study from Stockholm (Karlberg et al., 1968) is almost identical with the Bonn preterm infants' increment of 3.1 cm. Consequently, the opinion cannot be held any longer that the normal growth pattern of preterm infants is that of "catch-up" growth.

3. Growth Patterns of SGA Preterm Infants: Catch-Up

Although the causes of intrauterine growth retardation are heterogeneous, three major groups can be differentiated:

1. Fetal factors: Genetic defects, chromosomal abnormalities, malformations, infections, radiation, or hypoxemia. The outcome for these infants is dependent on the quality of postnatal care, as well as the primary causative factors.

2. Placental factors: Extremely small placentas, unfavorable site of implantation, anomalies of form and of cord, disturbances of differentiation, severe maternal disease affecting the placenta.

3. Maternal factors [which cannot always be separated from placental factors (Gruenwald, 1975)]: "Vascular insufficiency," toxemia, multiple pregnancy, genetically small mother, smoking, undernutrition, low social class; those factors having after birth no further direct influence upon infant growth.

The fate of the SGA infants from groups 2 and 3 is dependent on *duration, timing,* and *severity* of adverse intrauterine factors and on postnatal care and nutrition. For example, not only neonatal hypoglycemia, but also lack of energy supply during the brain growth spurt in the perinatal period, are risk factors for such infants who, according to Burmeister and Romahn (1974), lack adipose tissue and are deficient in reserve stores of glycogen.

In most of the infants with intrauterine growth retardation in the Bonn study, maternal and placental factors have been causative; in some cases the condition seemed to be familial (Scott and Usher, 1966; Tanner *et al.,* 1972; Brandt, 1975).

Since the classic publication of Prader *et al.* (1963) on catch-up growth following illness or starvation, this concept has become accepted in pediatrics and has attained increasing relevance. The subject of this publication was catch-up growth in height, weight, and bone age during the 2nd to the 13th year of life. However, there is still little literature on catch-up growth of head circumference after intrauterine growth retardation, and there is no information about the timing of such a catch-up (Falkner, 1966; Babson and Henderson, 1974).

3.1. Classification of SGA Preterm Infants into Two Groups

Among the 107 preterm infants (AGA and SGA) of the Bonn study there were 43 SGA (infants with malformations or prenatal infection have been excluded for this report). Figure 21 shows their birth weight. All but four were 1500 g or below. Head circumference and supine length at birth according to postmenstrual age are compared with the percentiles of Lubchenco *et al.* (1963b, 1966).

Analysis of outcome enables us to classify those infants by (1) whether or not catch-up in head growth occurred (defined as below), and (2) whether or not the neuromotor and psychological outcome was favorable, that is, whether their DQ and IQ was similar to those of the AGA preterm and the full-term infants and whether they were free of cerebral palsy, cerebellar dysfunction, or disturbance in coordination (p. 601). This gave rise to only two groups: group A, with catch-up growth of head circumference and with a favorable development ($N = 21$, 14 girls and 7 boys, circles with an asterisk in Figures 2 and 21); and group B, without catch-up growth of head circumference and with a less favorable development ($N = 22$, 17 girls and 5 boys, circles in Figures 2 and 21). There were no cases in the other two possible groups.

As a definition one might say that catch-up occurs when SGA preterm infants, significantly below the mean at birth, reach the value of the full-term control infants; this is achieved by a period of significantly more rapid growth velocity than occurs in the control infants.

Of the 21 infants in group A, 15 were born after 1971, and the mothers of four of the six earlier-born infants have participated in a prospective study on pregnancy progress and child development of the Deutsche Forschungsgemeinschaft and have taken advantage of intensive prenatal care. In 14 of the 15 infants born since 1971 intrauterine growth retardation has been demonstrated by repeated ultrasound measurements of the biparietal and thorax diameters (Hansmann and Voigt, 1973; Hansmann, 1974, 1976; Brandt, 1976c). In these cases, the pregnant woman was intensively monitored throughout pregnancy. Thirteen infants in Group A were delivered by elective cesarean section. Four infants were the first infants alive after abortion, stillbirth, or neonatal death.

Most of the infants in group B (12 of 22) were born before 1971, three of 22 being delivered by cesarean section.

At birth the infants in both groups (Figure 21: A, circles with an asterisk; B, circles) were mostly low not only in weight, but also in head circumference and supine length. Figure 21 further shows that between group A with catch-up growth and group B without catch-up, there is no difference in the degree of growth retardation. In those infants whose intrauterine development has been followed with repeated ultrasound measurements (Brandt, 1976c; Brandt and Hansmann,

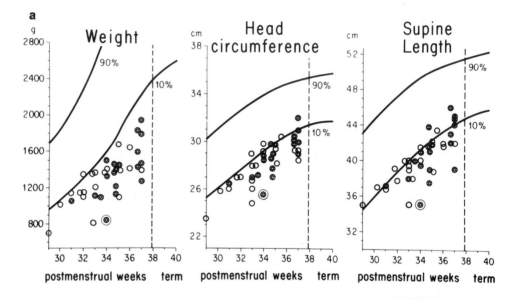

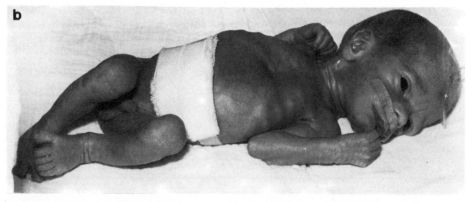

Fig. 21. (a) Weight, head circumference, and supine length after birth according to fetal age, SGA preterm infants of group A (circles with an asterisk), $N = 21$, and SGA of group B (circles), $N = 22$, compared with the percentiles from Lubchenco *et al.* (1963b, 1966). (b) SGA boy of group A, born by elective cesarean section in the 35th postmenstrual week with a birth weight of 850 g, at the age of 8 days. The birth measurements (encircled) are reduced in all three parameters, with the head least affected.

597

*GROWTH DYNAMICS OF
LOW-BIRTH-WEIGHT
INFANTS WITH
EMPHASIS ON THE
PERINATAL PERIOD*

1977; Hansmann, 1976), the growth retardation has been shown to take place in the third trimester of pregnancy. In most of the cases the trunk [measured by the thorax diameter (Hansmann, 1976)] is retarded more distinctly and earlier than the head. Theoretically, the proof of actual catch-up is possible only if the genetic potential of the individual is known or the deviation from a growth channel has been observed. For specific designation of the criteria involved, the intrauterine growth pattern and the timing of retardation are important.

As an example of relatively proportional retardation, Figure 21 shows a SGA boy of group A, born by elective cesarean section in the 35th postmenstrual week, with a birth weight of 850 g at the age of 8 days. The birth measurements (circled in Figure 21) are reduced in all three parameters, with the head least affected. The placenta showed multiple infarction. The pregnancy was uneventful, the mother smoked about five cigarettes per day, and there was a family history of multiple abortions. The mother previously aborted and had had a SGA preterm infant of 700 g birth weight who died immediately after birth.

It has been possible to examine all the 43 SGA preterm infants (except one handicapped infant in group B who left the study after 7 months) at regular intervals (see page 563).

In the next section of this chapter we will discuss in detail only the group of 21 SGA preterm infants with catch-up growth of head circumference (circles with an asterisk in Figures 2 and 21) with analysis of the background of this catch-up.

Head circumference catch-up growth has been chosen here as a criterion for separation of the SGA preterm infants into groups A and B. Because of the rapidly decreasing velocity of head circumference after 6 months (see p. 590), the possibility for catch-up after intrauterine or early postnatal growth retardation seems to exist only within about the first 6 months after birth. After that age any considerable degree of catch-up becomes more and more unlikely. According to Stoch and Smythe (1976) "timing is important, as catch-up growth may be possible with good nutrition if it occurs within the period of time when growth of that particular structure normally continues." In all the SGA infants of group A, the catch-up in head circumference was accomplished before the corrected age of 6 months.

As far as the catch-up in length and/or weight after intrauterine growth retardation is concerned, it may be possible also at later ages, and that *independently* of whether catch-up has or has not occurred in head circumference in the first months of life. In line with this, Varloteaux *et al.* (1976) have reported their follow-up study of 128 low-birth-weight (LBW) infants. At the age of 13–14 years, the SGA infants still were significantly retarded in head circumference compared with AGA controls, whereas in height and weight there was no significant difference. In a follow-up of children who suffered severe undernutrition in infancy in the South Africa study of Stoch and Smythe (1976), an improvement of nutrition at later ages, i.e., after 2 years, led to a catch-up in weight and length although the retardation in head circumference persisted.

3.2. Head Circumference

3.2.1. Distance

The postnatal head circumference measurements of the 21 SGA preterm infants in group A from 34 postmenstrual weeks to one month after term (Figure 22) are significantly below the Bonn standards (Brandt, 1975, 1976b). By two months after term they approach the value of the AGA preterm infants; the difference of 0.5

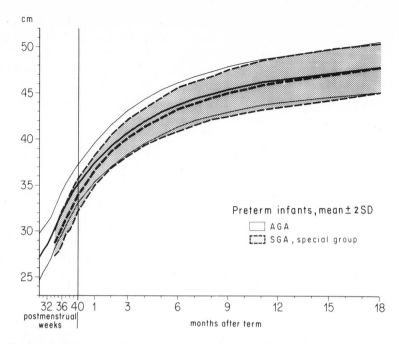

Fig. 22. Head circumference growth (mean and two standard deviations) from 34 postmenstrual weeks to 18 months after term of SGA preterm infants, special group (= A), N = 21, shaded area with broken lines; plotted against the Bonn standards for AGA preterm infants, N = 64, continuous lines.

cm is not significant, $P > 0.05$. At the age of 3 months the difference is 0.4 cm (nonsignificant), and by 12 months the curve of group A has reached to within 0.2 cm of the value of the control infants, where they remain until 3 years. This signifies that the SGA infants have exhibited almost complete catch-up after their prenatal retardation. The slight persistent difference of 0.2–0.4 cm in the graph (Figure 22) between the groups can be ascribed to the sex difference in head circumference since in both groups of SGA infants there is a preponderance of girls (see p. 584).

The head circumference growth curve of the 22 SGA infants in group B starts at 34 postmenstrual weeks—similar to group A—significantly below the Bonn standards (Brandt, 1976*b*). Contrary to group A, the infants in group B do not exhibit catch-up, but remain at 6 years significantly below the Bonn standards. Their mean curve lies at about −2 SD (Brandt, 1975).

The mean head circumference of the mothers in group A, 54.47 cm (SD = 1.15 cm), agrees well with that of the mothers of group B, 54.56 cm (SD = 1.40 cm). Between the fathers there is an insignificant difference of 0.65 cm. Therefore the growth differences between the groups cannot be ascribed to genetic influences.

Until recently, catch-up growth of head circumference, and thus presumably of the brain, following intrauterine growth retardation has not been reported except by Falkner (1966) and Babson and Henderson (1974). Stoch and Smythe (1963) stated that in infants exposed to malnutrition during the first year of life, there was a significantly smaller head circumference and lower IQ at the age of 7 years than in a group of normally nourished infants. Davies and Davis (1970) and Davies (1975) have demonstrated that SGA infants have smaller heads on long-term follow-up than AGA infants. Babson *et al.* (1964, Babson and Phillips, 1973) reported in their study of nine pairs of monozygotic twins, with a birth-weight difference of more than 25%, that there still remained, at the age of 19 years, a significantly smaller head circumference in the smaller twin; six of the smaller twins had been subjected to a delay in postnatal feeding in the first week of life. Hohenauer (1971) showed in 20 sets of twins of the same sex a significantly smaller head circumference of the smaller twins at the mean age of 8½ years. In 426 "small-for-dates" in a British study of 16,955 infants born in 1970, there was a significantly smaller head circumference ($P < 0.001$) at the age of 22–23 months than that of a control group (Chamberlain and Davey, 1975).

Falkner (1966) reported one pair of full-term monozygotic male twins with a birth weight difference of 48% (1.35 kg), in whom the smaller twin under good nutrition had greatly reduced a head circumference difference of 4.0 cm at birth by the second examination at the age of 9 months. Recently, Babson and Henderson (1974) have reported normal one-year head circumference measurements in 5 of 10 full-term girls with intrauterine growth retardation but gave no information on the timing of the catch-up. The other 5 girls of the Babson study without catch-up of head circumference (i.e., at or below the 10th percentile), had an IQ of 10 points less, on average, than the girls with catch-up. Also, Dobbing (1974a,b), contrary to his earlier publication (1970), discussed the possibility of catch-up if birth liberates the baby from some growth restriction.

3.2.2. Velocity

Catch-up growth of head circumference of the SGA preterm infants in group A occurred in a period of significantly more rapid growth velocity than in the AGA preterm infants from the 36th postmenstrual week ($P < 0.01$) to the second month

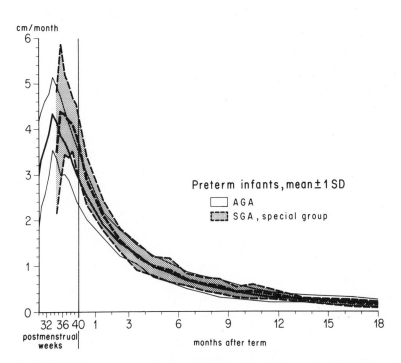

Fig. 23. Head circumference growth velocity (calculated in cm per month, mean and one standard deviation) from 34.5 postmenstrual weeks to 18 months after term of SGA preterm infants, special group (= A), $N = 21$, shaded area with broken lipnes; plotted against the Bonn standards for AGA preterm infants, $N = 64$, continuous lines.

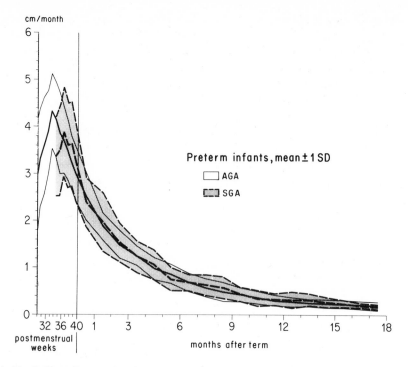

Fig. 24. Head circumference growth velocity (calculated in cm per month, mean and one standard deviation) from 34.5 postmenstrual weeks to 17.5 months after term of SGA preterm infants without catch-up growth, group B, $N = 22$, shaded area with broken lines; plotted against the Bonn standards for AGA preterm infants, $N = 64$, continuous lines.

after term ($P < 0.005$) (Figure 23). Subsequently, growth velocity until 3 years (here plotted until 18 months) is similar in both groups, with no significant differences.

Contrary to this, head circumference velocity of the SGA in group B is similar to that of the Bonn standards for AGA preterm infants, except at 39.5 weeks, where the SGA grow significantly faster. However, there is no demonstrable catch-up in head circumference (Figure 24).

The velocity standards of head circumference growth of healthy AGA preterm and full-term infants have made it possible to evaluate the growth of a special group (A) of SGA preterm infants. Although this group is small and heterogeneous, the timing of the catch-up occurred in all of the infants during a period of very rapid growth of head circumference, and it was accomplished before the third to fourth month after term. The exact time-span in which catch-up in head circumference growth after intrauterine growth retardation is possible has still to be defined. Both the extent and the age at which such catch-up growth in head circumference will be possible under good nutritional conditions after early growth retardation needs further research.

3.2.3. Growth Patterns of Individual SGA Infants

The timing, extent, and duration of intrauterine retardation and of subsequent postnatal catch-up growth of head circumference are demonstrated in a combined chart of the individual development of a girl (Figure 25) born in the 34th postmenstrual week by elective cesarean section, with a birth weight of 1130 g. Prenatal ultrasound measurements of the biparietal and thorax diameter are plotted against

601

*GROWTH DYNAMICS OF
LOW-BIRTH-WEIGHT
INFANTS WITH
EMPHASIS ON THE
PERINATAL PERIOD*

the Bonn standards of Hansmann (1974, 1976). The growth retardation in the third trimester is evident. After birth (Figure 25) the growth curve of head circumference starts more than 2 SD below the Bonn standards and reaches the normal range at one month after term. The catch-up velocity is evident, with a peak of 5.8 cm/month at 36.5 weeks, which is more than 2 SD above the normal mean of 3.7 cm/month (SD = 0.7 cm). The psychomotor development is above average. The causes of intrauterine growth retardation of this firstborn child of a 23-year-old mother are unknown. The pregnancy was uneventful, although the mother carried on working until two weeks before delivery, which may be relevant.

The next combined chart (Figure 26) shows the intrauterine retardation and the postnatal catch-up growth of another SGA preterm girl, born in the 37th postmenstrual week with a birth weight of 1280 g. The curve of the biparietal diameter runs in the range of 2 SD below the mean, whereas the curve of the thorax—more affected by intrauterine restriction—is markedly below 2 SD. Underneath in the graph are plotted the postnatal measurements of head circumference. As is to be expected, at birth the growth curve starts markedly below 2 SD and, after a period of very rapid growth, reaches the normal range in the third month after term. The child's psychological development is above average in spite of unfavorable environmental conditions (Brandt, 1975). This child was the firstborn of an 18-year-old mother, a heavy smoker, who carried on working until 2½ months before delivery. In the third trimester of gestation she suffered from toxemia with pyelonephritis and hyperemesis.

Further evidence for catch-up comes from the head circumference growth of some monozygotic twins (Table IX). In three pairs, who differed in birth weight by 14%, 25%, and 31%, the smaller SGA twin makes up a difference in head circumference at birth of 1.5 cm, 2.0 cm, and 2.5 cm, respectively, before the sixth month.

The SGA infants with catch-up growth (group A) also show in their mental development, to date, better results than the SGA in group B without catch-up of head circumference. Their developmental and intelligence quotients (Griffiths, 1964; Cattell, 1966; Stanford-Binet of Terman and Merrill, 1965) are similar to those of the AGA preterm infants and of the full-term control infants (Brandt and Schröder, 1976). From 2 to 48 months there is no significant difference between the groups ($P > 0.40$).

Contrary to the favorable psychological development of the SGA preterm infants with catch-up growth (group A), the developmental and intelligence quotients of the SGA infants without catch-up growth, group B, are highly significantly below the full-term control infants from 6 to 72 months, although in the first 5 months the Griffiths DQ scores are not significantly different (Brandt, 1975; Brandt and Schröder, 1976). This is in accordance with reports from other studies that the development of SGA preterm infants has been, in general, less favorable than that of AGA infants (Collins and Turner, 1971; Francis-Williams and Davies, 1974; Brandt, 1975; Hagberg *et al.*, 1976; Varloteaux *et al.*, 1976).

3.3. Analysis of Factors Favoring Catch-Up Growth

3.3.1. Early Diagnosis of Intrauterine Growth Retardation by Ultrasound Follow-Up Studies and Prenatal Care

In pregnancies with a risk of intrauterine growth retardation, an accurate assessment of fetal growth and development can be made by repeated ultrasound measurements of the biparietal and thorax diameters (Hansmann, 1974, 1976). This

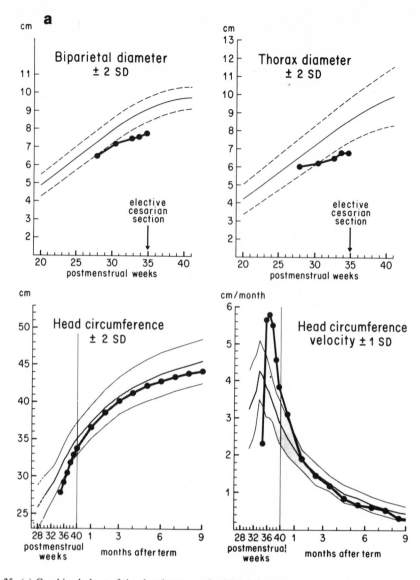

Fig. 25. (a) Combined chart of the development of a SGA infant, born in the 34th postmenstrual week with a birth weight of 1130 g. Above ultrasound measurements of intrauterine growth of biparietal and thorax diameter, plotted against the Bonn standards (Hansmann, 1974, 1976). Below postnatal measurements of head circumference (distance and velocity), plotted against the Bonn standards for AGA preterm and full-term infants (Brandt, 1976a). (b and c) The girl at the chronological age of 2 weeks and of 18 months.

enables an early diagnosis of fetal growth retardation to be made. An essential prerequisite in obtaining valid results in the third trimester of pregnancy is the timely estimation of gestational age of the fetus. This can be assessed between 7 and 14 weeks of gestation by crown–rump measurements with an accuracy of ±5 days (Hackelöer and Hansmann, 1976). In the second trimester of pregnancy, ultrasound measurements of the biparietal diameter give an estimate of fetal age with an accuracy of ±6 days (Hansmann et al., 1977).

603

GROWTH DYNAMICS OF
LOW-BIRTH-WEIGHT
INFANTS WITH
EMPHASIS ON THE
PERINATAL PERIOD

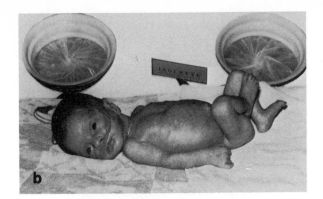

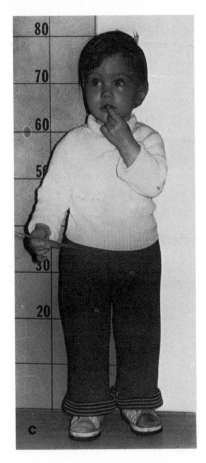

Fig. 25. (*continued*)

After an early diagnosis of fetal growth retardation by ultrasound, an intensive program of monitoring fetal well-being was instigated in most of the cases in the Bonn study. This included hospitalization, daily estimations of urinary estrogen excretion, serum human placental lactogen (HPL), and fetal heart-rate pattern in the cardiotocogram (CTG). With regard to the choice of the optimal delivery date, pathological findings in the CTG have been found to be most important. In addition, estimation of fetal lung maturity by determination of the lecithin–sphingomyelin

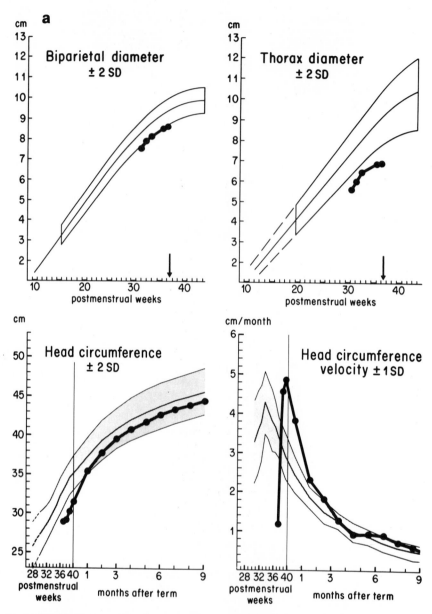

Fig. 26. (a) Combined chart of the development of a SGA infant, born in the 37th postmenstrual week with a birth weight of 1280 g. Above, ultrasound measurements of intrauterine growth of biparietal and thorax diameter, plotted against the Bonn standards (Hansmann, 1974, 1976). Below, postnatal measurements of head circumference (distance and velocity), plotted against the Bonn standards for AGA preterm and full-term infants (Brandt, 1976a). (b and c) The girl at the chronological age of 8 days and of 13 months.

605

GROWTH DYNAMICS OF
LOW-BIRTH-WEIGHT
INFANTS WITH
EMPHASIS ON THE
PERINATAL PERIOD

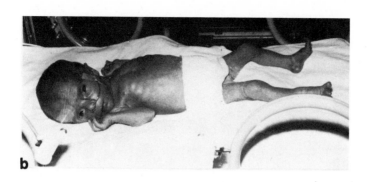

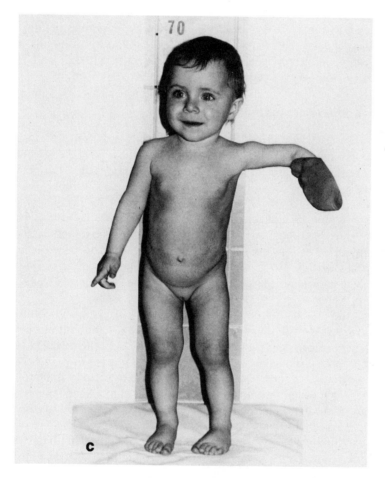

Fig. 26. (*continued*)

Table IX. Catch-Up Growth of Head Circumference in Monozygotic Twins

	Gestation (weeks)	Birth weight difference (%)	Difference of head circumference SGA to AGA twin (cm)		
			At birth	At 6 months	At 12 months
Pair 1					
Twin I SGA	34	14	−1.5	+0.1	0.0
Twin II AGA					
Pair 2					
Twin I AGA	37	25	−2.0	+0.1	+0.1
Twin II SGA					
Pair 3					
Twin I AGA	35	31	−2.5	+0.2	+0.5
Twin II SGA					

ratio is taken into consideration. The risk of progressive intrauterine dystrophy has to be weighed against the risk of immaturity. The majority of infants in group A (13 of 21) were delivered by elective cesarean section (Brandt and Hansmann, 1977).

3.3.2. Nutrition and Postnatal Care

As already described (p. 572), postnatal weight gain of small preterm infants is dependent, essentially, on caloric intake. Within the 8-year period of the Bonn study, which was started in 1967, there have been quantitative changes in immediate postnatal feeding practices. The quality of the standard formula given remained unchanged: cow's milk, modified to breast milk, with 75 cal/100 ml: fat, 3.3 g; protein, 2.3 g with a casein/lactalbumin ratio of 40/60; and lactose, 8.6 g. In some cases 3–5% glucose was added to the formula. Most of the infants in group A with catch-up growth (13 of 21) were born after the introduction of early feeding (i.e., 3–6 hrs after birth) in July 1972, and eight infants in addition had the advantage of immediate high-calorie feeding according to the Tizard scheme (1972, 1973) (group I according to Weber et al., 1976).

Most of the infants in group B (15 of 22) were born before the introduction of early feeding in July 1972. Seven infants were fed early, but did not belong to the group with high-calorie feeding according to the Tizard scheme (1973).

Figure 27 above shows the weight velocity in grams per day, and also the mean caloric intake per kilogram body weight per day and the energy quotient in the first 10 days after birth of both groups of SGA preterm infants. The infants with higher caloric intake and catch-up growth (group A, line of dots) started to gain weight, on average, at the second postnatal day in contrast to the infants without catch-up (group B, continuous line), who started to gain weight at the ninth day on average. The mean time required for group A infants to regain their birth weight was 3 days, although some infants had no loss in weight after birth; group B required 10 days, on average, to regain their birth weight.

From group B, Figure 28 shows as an example the postnatal growth of head circumference of a SGA boy, born in 1967 in the 35th postmenstrual week with a birth weight of 1410 g. At the second postnatal day feeding was started with an EQ of 12, at the ninth postnatal day he reached an EQ of 82. He slowly started to gain weight 9 days after birth and regained his birth weight at the 15th postnatal day. His postnatal growth of head circumference runs in the area of −2 SD, and his growth velocity of head circumference (right on the photo) did not exceed the range of 1 SD. This indicates that this boy has not caught up after the prenatal retardation. At 6 years his intelligence quotient (Stanford-Binet) was 88, compared with an IQ of 102 of his AGA twin sister.

From group A, for example, the SGA girl in Figure 25 who exhibited the striking catch-up has started to gain weight by the second postnatal day, with an intake of 94 cal/kg/day (EQ). Her birth weight is surpassed at 3 days. She reached her full caloric intake at the 5th postnatal day with an energy quotient of 160. This infant belongs to the group who had early and high-calorie feeding according to Tizard (1973; Weber *et al.*, 1976).

Some time ago a relationship between delayed feeding and neurological defects was suspected by Drillien (1964). For example, the preterm infants of her study, born between 1953 and 1954 in Edinburgh, lost up to 20% of their weight after birth, and did not regain their birth weight until the 33rd postnatal day. These infants, thus, became SGA infants after birth.

Churchill (1973) has reported, in a group of preterm infants with spastic diplegia, a significantly greater postnatal weight loss and a significantly longer time for regaining birth weight than in normal infants of similar birth weight and risk

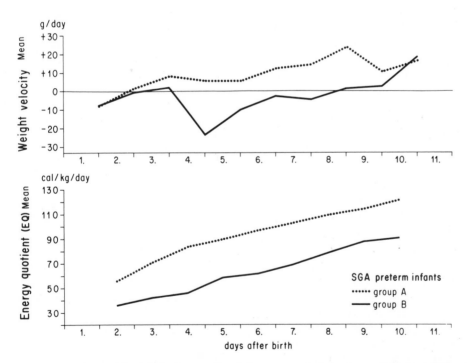

Fig. 27. Above Mean weight velocity in g per day; below: mean caloric intake per kg body weight and day, the energy quotient (EQ) from the 2nd to the 11th day of life. Line of dots = SGA preterm infants with catch-up growth (group A), $N = 21$; continuous line = SGA preterm infants without catch-up growth (group B), $N = 22$.

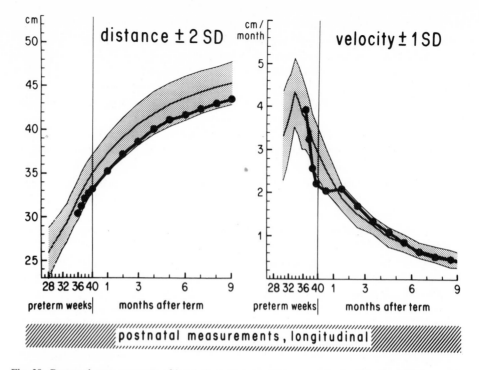

Fig. 28. Postnatal measurements of head circumference (distance and velocity) of a SGA boy without catch-up growth (group B) in Fig. 27, plotted against the Bonn standards for AGA preterm and full-term infants (Brandt, 1976*b*).

factors ($P < 0.01$). The proportion of SGA infants among the preterm infants is not mentioned, but it is a possibility that in these infants severe lack of energy and/or hypoglycemia led to serious impairment of brain development.

Lubchenco *et al.* (1972) compared mean postnatal caloric intake, in the first week, in 113 preterm infants, born between 1947 and 1953, with a birth weight of 1500 g and below. Later intelligence quotients (IQ) were determined. They concluded that the low mean energy quotient (EQ) of 25–45 in the first week was a major factor in the unfavorable development of these infants. The proportion of SGA infants is not mentioned, but it is possible about one third were growth retarded at birth, and postnatal food restriction during a period of very rapid brain development may have added to the impairment.

Generally, in infants who have had some prenatal restriction, a postnatal state of hunger is a further disadvantage. Because of lack of energy reserves, they are especially susceptible to immediate postnatal hypoglycemia and to postnatal malnutrition in the sense of Winick's (1971) ''double deprivation,'' with possible lasting adverse effects on mental development.

The capacity for catch-up depends on the duration, timing, and severity of restriction. Further, the developmental stage of the tissue involved appears to be a critical determinant of the effects of refeeding on catch-up growth (e.g., of head circumference). After a restriction in the third trimester, when neuronal multiplication is already completed, and the growth ''spurt'' of the brain mainly consists of

609

*GROWTH DYNAMICS OF
LOW-BIRTH-WEIGHT
INFANTS WITH
EMPHASIS ON THE
PERINATAL PERIOD*

glial multiplication, myelination, and dendritic growth (Dobbing, 1974b), catch-up might be possible as long as the brain remains in this developmental stage.

The different growth patterns, demonstrated in the computer plots of velocity in weight (p. 573), length (p. 583) and head circumference (p. 593) of AGA preterm infants from birth to 5½ years, suggest that the capacity for catch-up depends also on the velocity. If growth has slowed down significantly, as for head circumference in the tenth month to one tenth of its mean peak velocity in the perinatal period, and in the sixth year to one hundredth, one may assume that beyond this stage there is but little chance for catch-up after an intrauterine or early postnatal retardation. Length and weight velocity, on the other hand, decrease in the tenth month only to one third of their mean peak velocity in the perinatal period, and in the fifth year to about one sixth. These data might suggest a continuous capacity for a catch-up if nutrition is adequate.

There is still some controversy about the optimal nutrition for very small SGA preterm infants. Early and high-calorie feeding was tolerated very well by the SGA infants in the Bonn study. Catch-up growth, of course, requires energy (Forbes, 1974); since growth velocity must exceed the norm throughout the entire recovery period, food must be provided in excess of normal requirements until recovery is complete. The present results suggest that the amount of calories given immediately after birth seems to play an important role in further growth and development, especially of very small SGA infants with a birth weight around 1000 g. The small SGA infants are in danger of a period of immediate postnatal deprivation, when the caloric intake lies below the basic metabolic rate and below the requirements for heat production. In the SGA infant there is a relative hypermetabolism during the first day, and even more so beyond 2–3 days of postnatal age (Sinclair, 1970). The SGA infant with a supposed rapid rate of brain growth shows a substantially increased rate of oxygen consumption compared with an AGA of similar weight, probably due mainly to the much larger brain (Sinclair, 1970). According to Sinclair (1970), the metabolic rate of an SGA infant with a birth weight of 860 g amounts to 54. But it has to be remembered that "minimal" metabolic rates *underestimate* the true maintenance requirements, because the contributions of muscular activity, thermal stress, and fecal loss must also be allowed for. In addition, the caloric requirements for growth itself (that may be very high) have to be considered.

The question of the extent to which intrauterine deprivation can be tolerated without resultant damage, and of the conditions under which prenatal growth retardation of head circumference can be overcome by catch-up after birth, needs further research. Even if several factors are involved in intrauterine growth retardation, there exists—at least in cases of malnutrition—the chance to overcome resultant deficits postnatally by early and adequate nutrition.

4. Conclusions

For a proper analysis of growth patterns of preterm infants, both postmenstrual age at birth and the assessment of intrauterine development (appropriate for gestational age or small for gestational age) have to be considered. Further, the genetic background, social environment, and nutrition—immediately after birth as well as in the first year of life—play important roles in the evaluation of growth.

In the perinatal period, growth—especially in head circumference—is more rapid than thereafter and emphasizes the high caloric requirements of small preterm infants. This is particularly so for those infants who are born small for gestational age. It is suggested that one main reason for the less favorable outcome for growth in some former studies might have been low-energy and delayed nutrition immediately after birth.

If nutrition and care are adequate, the growth pattern of AGA preterm infants depends largely on postmenstrual age, irrespective of postnatal (i.e., chronological) age.

Head growth is least affected after a short period of undernutrition and is the first to regain the original growth pattern when nutrition has become adequate. An initial retardation of weight is compensated for significantly earlier than a retardation in length which needs 18 months, on average, to compensate, according to the present results.

The period necessary for age adjustment for prematurity depends mainly on the postmenstrual age at birth, on growth velocity of the different variables (which decreases with increasing age), and on increasing absolute values. For weight, there are significant differences between corrected and uncorrected age from birth to 21 months. From 24 months onward this difference becomes nonsignificant, so that for the assessment of weight growth, age need not be corrected in preterm infants.

For supine length/height the difference between corrected and uncorrected age remains significant until 3 years; from 3.5 years it becomes nonsignificant. For the assessment of height growth, age need not be corrected in preterm infants, but until five years there remains a mean difference of about 1 cm, i.e., an increment of 2 months, which should be considered.

For head circumference there are significant differences between corrected and uncorrected age from birth to 17 months; thereafter they become nonsignificant. Hence it follows that from 18 months, on average, to evaluate head circumference growth in preterm infants, age need not be corrected any further; the remaining difference at 6 years amounts to only 0.04 cm. This "assimilation" of the non-age-corrected curve to the corrected curve, caused by the very rapid decrease of growth velocity from its peak of 4.3 cm/month at 33.5 postmenstrual weeks to 0.04 cm/month at 4.25 years (a drop of more than one hundredth), has led, in some earlier reports, to the erroneous assumption of a "catch-up" growth pattern in normal AGA preterm infants.

The present results are in accordance with the findings of various other studies and show that growth and development of even very small AGA preterm infants is similar to that of full-term control infants, provided that the environmental conditions are favorable, and the age is corrected for prematurity. This has been further proved by international comparisons with full-term infants.

In a group of 21 SGA preterm infants, a catch-up growth of head circumference—as an accepted indicator of brain development—has been shown during early and high-calorie feeding. Head circumference catch-up growth has been chosen here as an important criterion since the possibility of catch-up after intrauterine retardation, according to the Bonn findings, seems to exist only in the first 6–8 postnatal months; thereafter growth velocity decreases markedly. Analysis of the main relationships of this catch-up reveals, besides the quality of postnatal care and immediate postnatal feeding, the importance of prenatal care and of early diagnosis of intrauterine growth retardation.

611

*GROWTH DYNAMICS OF
LOW-BIRTH-WEIGHT
INFANTS WITH
EMPHASIS ON THE
PERINATAL PERIOD*

For growth in weight and length, a catch-up may occur at later ages independent of whether there is catch-up or not in head circumference in the first months of life.

ACKNOWLEDGMENTS

This research was carried out with grants from the Bundesminister für Jugend, Familie und Gesundheit, from the Minister für Wissenschaft und Forschung des Landes Nordrhein-Westfalen, and from the Fritz Thyssen Stiftung which are gratefully acknowledged.

For support of this study at the Universitäts-Kinderklinik Bonn (Director: Professor Dr. Heinz Hungerland until 1974, and Professor Dr. Walther Burmeister from 1975) my thanks are due to my colleagues and assistants.

I particularly wish to thank all the mothers and fathers in the study for their faithful cooperation, without which the frequent examinations at regular intervals would not have been possible.

My thanks are due to Dr. rer. nat. Herbert Kurz of the Gesellschaft für Mathematik und Datenverarbeitung Bonn, and to Dr. med. Ulf Voigt of the Institut für Medizinische Statistik, Dokumentation und Datenverarbeitung Bonn for statistical advice and for their help in writing the computer programs. The computations were performed on the IBM/370-168 of the Regionales Hochschul-Rechenzentrum der Universität Bonn.

I wish to thank Dr. Manfred Hansmann of the Bonn Universitäts-Frauenklinik, (Director, Professor Dr. H. J. Plotz) for his generosity in allowing me to use his prenatal ultrasound growth data.

5. References

Alm, I., 1953, The long-term prognosis for prematurely born children. A follow-up study of 999 premature boys born in wedlock and of 1002 controls, *Acta Paediatr. Scand.* **94:**1–116.

American Academy of Pediatrics—Committee on Fetus and Newborn, 1967, Nomenclature for duration of gestation, birth weight and intrauterine growth, *Pediatrics* **39:**935–939.

Babson, S. G., 1970, Growth of low-birth-weight infants, *J. Pediatr.* **77:**11–18.

Babson, S. G., and Benda, G. I., 1976, Growth graphs for the clinical assessment of infants of varying gestational age, *J. Pediatr.* **89:**814–820.

Babson, S. G., and Henderson, N. B., 1974, Fetal undergrowth: Relation of head growth to later intellectual performance, *Pediatrics* **53:**890–893.

Babson, S. G., and Phillips, D. S., 1973, Growth and development of twins dissimilar in size at birth, *N. Engl. J. Med.* **289:**937–940.

Babson, S. G., Behrman, R. E., and Lessel, R., 1970, Fetal growth: Liveborn birth weights for gestational age of white middle class infants, *Pediatrics* **45:**937–944.

Babson, S. G., Kangas, J., Young, N., and Bramhall, J. L., 1964, Growth and development of twins of dissimilar size at birth, *Pediatrics* **33:**327–333.

Battaglia, F., and Lubchenco, L. O., 1967, A practical classification of newborn infants by weight and gestational age, *J. Pediatr.* **71:**159–163.

Baum, J. D., and Searls, D., 1971*a,* Head shape and size of newborn infants, *Dev. Med. Child Neurol.* **13:**572–575.

Baum, J. D., and Searls, D., 1971*b,* Head shape and size of pre-term low-birthweight infants, *Dev. Med. Child Neurol.* **13:**576–581.

Bayley, N., and Davis, F. C., 1935, Growth changes in bodily size and proportions during the first three years: A developmental study of sixty-one children by repeated measurements, *Biometrika* **27**:26–87.

Bazso, J., Lipak, A., Soos, A., Malan, M., and Zsadanyi, O., 1968, The later effects of pre-natal malnutrition on the physical and mental development of twins, Proceedings of the 1st Congress of the International Association for the Scientific Study of Mental Deficiency, Montpellier, France, pp. 404–413.

Beck, G. J., and Van den Berg, B. J., 1975, The relationship of the rate of intrauterine growth of low-birth-weight infants to later growth, *J. Pediatr.* **86**:504–511.

Benedikt, A., and Csáki, P., 1969, Zur Frage der Entwicklung von Kindern mit niedrigem Geburtsgewicht, *Akad. Kladó* **7**:167.

Bergmann, K. E., 1976, Personal communication.

Bjerkedal, T., Bakketeig, L., and Lehmann, E. H., 1973, Percentiles of birth weights of single, live births at different gestation periods, *Acta Paediatr. Scand.* **62**:449–457.

Blegen, S. D., 1952, The premature child, the incidence, aetiology, mortality and the fate of the survivors, *Acta Paediatr. Scand.* **41**(Suppl. 88):3–71.

Bourlon, J.-M., Sempé, M., Genoud, J., and Bethenod, M., 1975, Développement somatique de 119 prématurés nés avant la 31ᵉ semaine et suivis jusqu'à 18 mois, *Pédiatrie* **30**:679–694.

Brandt, I., 1975, Postnatale Entwicklung von Früh-Mangelgeborenen, *Gynaekologe* **8**:219–233.

Brandt, I., 1976a, Normwerte für den Kopfumfang vor und nach dem regulären Geburtstermin bis zum Alter von 18 Monaten—Absolutes Wachstum und Wachstumsgeschwindigkeit, *Monatschr. Kinderheilkd.* **124**:141–150.

Brandt, I., 1976b, Dynamics of head circumference growth before and after term, in: *The Biology of Human Fetal Growth* (D. F. Roberts and A. M. Thomson, eds.), pp. 109–136, Taylor & Francis, London.

Brandt, I., 1976c, Growth dynamics before and after term, including catch-up growth: Nutritional influences on the immediate postnatal growth in weight, length and head circumference of preterm infants, Proceedings 5th European Congress of Perinatal Medicine, Uppsala 1976, pp. 221–229.

Brandt, I., 1977, Growth and development of AGA and SGA preterm infants from birth to 6 years. Results from a longitudinal study (in press).

Brandt, I., and Hansmann, M., 1977, Ultrasound diagnosis of intrauterine growth retardation and postnatal catch-up growth in head circumference—A combined longitudinal study, *Acta Univ. Carol. Med. Anthropol. Matern.* (in press).

Brandt, I., and Schröder, R., 1976, Postnatales Aufholwachstum des Kopfumfanges nach intrauteriner Mangelernährung, *Monatschr. Kinderheilkd.* **124**:475–477.

Bray, P. F., Shields, W. D., Wolcott, G. J., and Madsen, J. A., 1969, Occipitofrontal head circumference—an accurate measure of intracranial volume, *J. Pediatr.* **75**:303–305.

Brenner, W. E., Edelman, D. A., and Hendricks, C. H., 1976, A standard of fetal growth for the United States of America, *Am. J. Obstet. Gynecol.* **126**:555–564.

Buda, F. B., Reed, J. C., and Rabe, E. F., 1975, Skull volume in infants—methodology, normal values, and application, *Am. J. Dis. Child.* **129**:1171–1174.

Burmeister, W., and Romahn, A., 1974, Über die Entwicklung des Depotfetts und der Körperzellmasse von Neugeborenen bis zum Ende des Wachstums, *Monatschr. Kinderheilkd.* **122**:558–559.

Butler, N. R., and Bonham, D. G., 1963, *Perinatal Mortality,* Livingstone, Edinburgh.

Cattell, P., 1966, The Measurement of Intelligence of Infants and Young Children. First Reprinting of Revised Form from 1960, The Psychological Corporation, New York.

Chamberlain, R., and Davey, A., 1975, Physical growth in twins, postmature and small-for-dates children, *Arch. Dis. Child.* **50**:437–442.

Churchill, J. A., 1963, Weight loss in premature infants developing spastic diplegia, *Obstet. Gynecol.* **22**:601–605.

Cruise, M. O., 1973, A longitudinal study of the growth of low birth weight infants: I. Velocity and distance growth, birth to 3 years, *Pediatrics* **51**:620–628.

Collins, E., and Turner, G., 1971, The importance of the "small-for-dates" baby to the problem of mental retardation, *Med. J. Aust.* **2**:313–315.

Dann, M., Levine, S. Z., and New, E. V., 1964, A long-term follow-up study of small premature infants, *Pediatrics* **33**:945–955.

Davies, P. A., 1975, Perinatal nutrition of infants of very low birth weight and their later progress, *Mod. Probl. Paediatr.* **14**:119–133.

Davies, P. A., and Davis, J. P., 1970, Very low birth-weight and subsequent head growth, *Lancet* **2**:1216–1219.

Davies, P. A., and Russel, H., 1968, Later progress of 100 infants weighing 1000 to 2000 g at birth fed immediately with breast milk, *Dev. Med. Child Neurol.* **10**:725–735.

Davison, A. N., and Dobbing, J., 1966, Myelination as a vulnerable period in brain development, *Br. Med. Bull.* **22**:40–44.

Dewhurst, C. J., Beazley, J. M., and Campbell, S., 1972, Assessment of fetal maturity and dysmaturity, *Am. J. Obstet. Gynecol.* **113**:141–149.

Diemer, K., 1968, Grundzüge der postnatalen Hirnentwicklung, in: *Fortschritte der Pädologie* (F. Linneweh, ed.), Vol. 2, pp. 1–255, Springer, New York.

Dobbing, J., 1970, Undernutrition and the developing brain. The relevance of animal models to the human problem, *Am. J. Dis. Child.* **120**:411–415.

Dobbing, J., 1974a, The later growth of the brain and its vulnerability, *Pediatrics* **53**:2–6.

Dobbing, J., 1974b, The later development of the brain and its vulnerability, in *Scientific Foundations of Paediatrics* (J. A. Davis and J. Dobbing, eds.), pp. 565–577, Heinemann Medical Books, London.

Dobbing, J., 1974c, Personal communication.

Dobbing, J., and Sands, J., 1973, Quantitative growth and development of human brain, *Arch. Dis. Child.* **48**:757–767.

Douglas, J. W. B., and Mogford, C., 1953, The results of a national enquiry into the growth of premature children from birth to 4 years, *Arch. Dis. Child.* **28**:436–445.

Drillien, C. D., 1961, A longitudinal study of the growth and development of prematurely and maturely born children. VI. Physical development in age 2–4 years, *Arch. Dis. Child.* **36**:1–10.

Drillien, C. M., 1964, *The Growth and Development of the Prematurely Born Infant,* Livingstone, Edinburgh.

Drillien, C. M., 1970, The small-for-date infant: Etiology and prognosis, *Pediatr. Clin. North Am.* **17**:9–24.

Dweck, H. S., Saxon, S. A., Benton, J. W., and Cassady, G., 1973, Early development of the tiny premature infant, *Am. J. Dis. Child.* **126**:28–34.

Eichhorn, D. H., and Bayley, N., 1962, Growth in head circumference from birth through young adulthood, *Child Dev.* **33**:257–271.

Falkner, F., 1958, Some physical measurements in the first three years of life, *Arch. Dis. Child.* **33**:1–9.

Falkner, F., 1962, The physical development of children. A guide to interpretation of growth-charts and development assessments; and a commentary on contemporary and future problems, *Pediatrics* **29**:448–466.

Falkner, F., 1966, General considerations in human development, in: *Human Development* (F. Falkner, Ed.), pp. 10–39, W. B. Saunders, Philadelphia.

Falkner, F., Steigman, A. J., and Cruise, M. O., 1962, The physical development of the premature infant. I. Some standards and certain relationships to caloric intake, *J. Pediatr.* **60**:895–906.

Farr, V., Mitchell, R. G., Neligan, G. A., and Parkin, J. M., 1966, The definition of some external characteristics used in the assessment of gestational age in the newborn infant, *Dev. Med. Child Neurol* **8**:507–511.

Finnström, O., 1971, Studies on maturity in newborn infants. I. Birth weight, crown–heel length, head circumference and skull diameters in relation to gestational age, *Acta Paediatr. Scand.* **60**:685–694.

Fitzhardinge, P. M., 1975, Early growth and development in low-birthweight infants following treatment in an intensive care nursery, *Pediatrics* **56**:162–172.

Fomon, S. J., Ziegler, E. E., and Vázquez, H. D., 1977, Human milk and the small premature infant, *Am. J. Dis. Child.* **131**:463–467.

Forbes, G. B., 1974, A note on mathematics of catch-up growth, *Pediatr. Res.* **8**:929–931.

Francis-Williams, J., and Davies, P. A., 1974, Very low birthweight and later intelligence, *Dev. Med. Child Neurol.* **16**:709–728.

Gairdner, D., and Pearson, J., 1971, A growth chart for premature and other infants, *Arch. Dis. Childh.* **46**: 783–787.

Glaser, K., Parmelee, A. H., and Plattner, E. B., 1950, Growth pattern of prematurely born infants, *Pediatrics* **5**:130–143.

Grantham-McGregor, S. M., and Desai, P., 1973, Head circumference of Jamaican infants, *Dev. Med. Child Neurol.* **15**:441–446.

Grassy, R. G., Hubbard, C., Graven, St. N., and Zachmann, R. D., 1976, The growth and development of low birth weight infants receiving intensive neonatal care, *Clin. Pediatr.* **15**:549–553.

Griffiths, R., 1964, 3rd printing, *The Abilities of Babies. A Study in Mental Measurement,* University of London Press, London.

Gruenwald, P., 1963, Chronic fetal distress and placental insufficiency, *Biol Neonate* **5**:215–265.

Gruenwald, P., 1966, Growth of the human fetus. I. Normal growth and its variation, *Am. J. Obstet, Gynecol.* **94:**1112–1119.

Gruenwald, P., 1975, *The Placenta and Its Maternal Supply Line: Effects of Insufficiency on the Fetus,* Medical and Technical Publishing, Lancaster.

Hackelöer, B.-J., and Hansmann, M., 1976, Ultraschalldiagnostik in der Frühschwangerschaft, *Gynaekologe* **9:**108–122.

Hagberg, G., Hagberg, B., and Olow, I., 1976, The changing panorama of cerebral palsy in Sweden 1954–1970. III. The importance of foetal deprivation of supply, *Acta Paediatr. Scand.* **65:**403–408.

Hansmann, M., 1974, Kritische Bewertung der Leistungsfähigkeit der Ultraschalldiagnostik in der Geburtshilfe heute, *Gynaekologe* **7:**26–35.

Hansmann, M., 1976, Ultraschall-Biometrie im II. und III. Trimester der Schwangerschaft, *Gynaekologe* **9:**133–155.

Hansmann, M., and Voigt, U., 1973, Ultraschall-Biometrie des Feten unter besonderer Berücksichtigung der Gewichtsschätzung bei intrauteriner Malnutrition, *Perinatale Med.* **4:**330–335.

Hansmann, M., Voigt, U., and Lang, N., 1977, The assessment of fetal growth retardation by ultrasonic cephalo-thoracometry, *Acta Univ. Carol. Med. Anthropol. Matern.* (in press).

Harnack, G.-A. von, 1960, Das übertragene, untergewichtige Neugeborene, *Monatschr. Kinderheilkd.* **108:**412–415.

Heimendinger, J., 1964a, Die Ergebnisse von Körpermessungen an 5000 Basler Kinder von 2-18 Jahren, *Helv. Paediatr. Acta* **19** (Suppl 13):1–231.

Heimendinger, J., 1964b, Gemischt longitudinale Messungen von Körperlänge, Gewicht, oberem Segment, Thoraxumfang und Kopfumfang bei 1-24 Monate alten Säuglingen, *Helv. Paediatr. Acta* **19:**406–436.

Herschkowitz, N., 1974, Effekt der Unterernährung auf die Gehirnentwicklung, *Monatschr. Kinderheilkd.* **122:**240–244.

Hess, J. H., 1953, Experiences gained in a 30 years' study of prematurely born infants, *Pediatrics* **11:**425–434.

Hinckers, H. J., and Hansmann, M., 1975, Mütterlicher Glukosestoffwechsel, *Arch. Gynaekol.* **219:**283–284.

Hirschl, D., Levy, H., and Litvak, A. M., 1948, Körperliche und geistige Entwicklung der Frühgeborenen. Eine statistische Übersicht im Verlauf von 5 Jahren, *Arch. Pediatr.* **65:**648–653.

Hohenauer, L., 1971, Studien zur intrauterinen Dystrophie. II. Folgen intrauteriner Mangelernährung beim Menschen, *Paediatr. Paedol.* **6:**17–30.

Hohenauer, L., 1973, Intrauterines Längen- und Gewichtswachstum, *Paediatr. Paedol.* **8:**195–205.

Hohenauer, L., 1974, Bestimmung des intrauterinen Wachstums und der Reife des Neugeborenen, *Paediatr. Prax.* **14:**549–554.

Hohenauer, L., 1975, Personal communication.

Hosemann, H., 1949a, Schwangerschaftsdauer und Neugeborenengewicht, *Arch. Gynaekol.* **176:**109–123.

Hosemann, H., 1949b, Schwangerschaftsdauer und Neugeborenen-Grösse, *Arch. Gynaekol.* **176:**124–134.

Hosemann, H., 1949c, Schwangerschaftsdauer und Kopfumfang des Neugeborenen, *Arch. Gynaekol.* **176:**443–452.

Iffy, L., Jakobovits, A., Westlake, W., Wingate, M., Catenni, H., Kanofsky, P., and Menduke, H., 1975, Early intrauterine development: 1. The rate of growth of Caucasian embryos and fetuses between the 6th and 20th weeks of gestation, *Pediatrics* **56:**173–186.

Jackson, C. M., 1909, On the prenatal growth of the human body and the relative growth of the various organs and parts, *Am. J. Anat.* **9:**119–165.

Joppich, G., and Schulte, F. J., 1968, *Neurologie des Neugeborenen,* Springer, Heidelberg.

Justesen, K., 1962, Cranial circumference of premature children, *Acta Paediatr. Scand.* **51:**13–16.

Kahl, M., 1950, Über die körperliche und geistige Entwicklung Untermassiggeborener, *Arch. Kinderheilkd.* **138:**138–150.

Kantero, R.-L., and Tiisala, R., 1971, V. Growth of head circumference from birth to 10 years, *Acta Paediat. Scand., Suppl.* **220:**27–32.

Karlberg, P., Engström, I., Lichtenstein, H., Klackenberg, G., Klackenberg-Larsson, I., Stensson, J., and Svennberg, I., 1968, The development of children in a Swedish urban community. A prospective longitudinal study. I-VI, *Acta Paediatr. Scand., Suppl.* **187:**48–66.

Karlberg, P., Taranger, J., Engström, I., Lichtenstein, H., and Svennberg-Redegren, I., 1976, The somatic development of children in a Swedish urban community. A prospective longitudinal study, *Acta Paediat. Scand., Suppl.* **258:**1–148.

615

*GROWTH DYNAMICS OF
LOW-BIRTH-WEIGHT
INFANTS WITH
EMPHASIS ON THE
PERINATAL PERIOD*

Kloosterman, G. J., 1969, Over intra-uterine groei en de intra-uterine groei-curve, *Maandschr. Kinderge-neeskd.* **37**:209–225.

Lang, K., 1962, Die Entwicklung des Kopfumfanges bei Frühgeborenen. Eine Längsschnittstudie, *Monatschr. Kinderheilkd.* **110**:490–494.

Lechtig, A., Delgado, H., Martorell, R., Yarbrough, C., and Klein, R. E., 1976, Effect of maternal nutrition on infant growth and mortality in a developing country, Proceedings 5th European Congress of Perinatal Medicine, Uppsala, pp. 208–220.

Levine, S. Z., and Gordon, H. H., 1942, Physiologic handicap of the premature infant, *Am. J. Dis. Child.* **64**:274–312.

Lubchenco, L. O., Horner, F. A., Reed, L. H., Hix, I. E., Jr., Metcalf, D., Cohig, R., Elliot, H. C., and Bourg, M., 1963*a*, Sequelae of premature birth. Evaluation of premature infants of low-weights at ten years of age, *Am. J. Dis. Child.* **106**:101–115.

Lubchenco, L. O., Hansman, C., Dressler, M., and Boyd, E., 1963*b*, Intrauterine growth as estimated from live born birth-weight data at 24 to 42 weeks of gestation, *Pediatrics* **32**:793–800.

Lubchenco, L. O., Hansman, C., and Boyd, E., 1966, Intrauterine growth in length and head circumference as estimated from live births at gestational ages from 26 to 42 weeks, *Pediatrics* **37**:403–408.

Lubchenco, L. O., Delivoria-Papadopoulos, M., Butterfield, L. J., French, J. H., Metcalf, D., Hix, I. E., Danick, J., Dodds, J., Downs, M., and Freeland, M. A., 1972, Long-term follow-up studies of prematurely born infants. I. Relationships of handicaps to nursery routines, *J. Pediat.* **80**:501–508.

McKeown, T., and Gibson, J. R., 1951, Observations on all births (23,970) in Birmingham 1947. IV. Premature birth, *Br. Med. J.* **2**:513–517.

McKeown, T., and Record, R. G., 1952, Observations on foetal growth in multiple pregnancy in man, *J. Endocrinol.* **8**:386–401.

Miller, H. C., and Hassanein, K., 1971, Diagnosis of impaired fetal growth in newborn infants, *Pediatrics* **48**:511–522.

Miller, H. C., and Hassanein, K., 1973, Fetal malnutrition in white newborn infants: Maternal factors, *Pediatrics* **52**:504–512.

Milner, R. D. G., and Richards, B., 1974, An analysis of birth weight by gestational age of infants born in England and Wales, 1967 to 1971, *J. Obstet. Gynaecol. Br. Commonw.* **81**:956–967.

Naeye, R. L., Blanc, W., and Paul, C., 1973, Effects of maternal nutrition on the human fetus, *Pediatrics* **52**:494–503.

National Center for Health Statistics (1976), *NCHS Growth Charts, 1976,* Vital and Health Statistics, Series 11. Health Resources Administration, DHEW, Rockville, Maryland.

Neligan, G. A., 1970, Working party to discuss nomenclature based on gestational age and birthweight, *Arch. Dis. Child.* **45**:730.

Nellhaus, G., 1968, Head circumference from birth to eighteen years: Practical composite international and interracial graphs, *Pediatrics* **41**:106–114.

Nickl, R., 1972, Standardkurven der intrauterinen Entwicklung von Gewicht, Länge und Kopfumfang, pp. 1–43, Inaug.-Diss., Munich.

O'Neill, E. M., 1961, Normal headgrowth and the prediction of headsize in infantile hydrocephalus, *Arch. Dis. Child.* **36**:241–252.

Ounsted, M., and Taylor, M. E., 1971, The postnatal growth of children who were small-for-dates or large-for-dates at birth, *Dev. Med. Child Neurol.* **13**:421–434.

Prader, A., and Budliger, H., 1977, Körpermasse, Wachstumsgeschwindigkeit und Knochenalter gesunder Kinder in den ersten 12 Jahren (longitudinale Wachstumsstudie Zürich), *Helv. Paediatr. Acta* (Suppl. 37), 1–44.

Prader, A., Tanner, J. M., and Harnack, G. A. von, 1963, Catch-up growth following illness or starvation. An example of developmental canalization in man, *J. Pediatr.* **62**:646–659.

Prechtl, H. F. R., and Beintema, D. J., 1968, *Die neurologische Untersuchung des reifen Neugeborenen,* Georg Thieme Verlag, Stuttgart.

Prechtl, H. F. R., and Beintema, D. J., 1976, 2nd printing, *Die neurologische Untersuchung des reifen Neugeborenen,* Georg Thieme Verlag, Stuttgart.

Reiche, A., 1916, Das Wachstum der Frühgeburten in den ersten Lebensmonaten. II. Mitteilung: Das Wachstum des Brust- und Kopfumfanges, *Z. Kinderheilkd.* **13**:332–348.

Robinson, N. M., and Robinson, H. B., 1965, A follow-up study of children of low birth weight and control children at school age, *Pediatrics* **35**:425–433.

Robinson, R. J., 1966, Assessment of gestational age by neurological examination, *Arch. Dis. Child.* **41**:437–447.

Rossier, A., Michelin, J., and Camaret, 1958, Bilan éloigné de 156 prématurés de poids de naissance inférieur à 1500 g, *Arch. Fr. Pédiatr.* **15**:251–261.

Saint-Anne Dargassies, S., 1955, La maturation neurologique des prematurés, *Etud. Néo-Natales* **4:**71–116.

Saint-Anne Dargassies, S., 1966, Neurological maturation of the premature infant of 28 to 41 weeks' gestational age, in: *Human Development* (F. Falkner, ed.), pp. 306–325, W. B. Saunders, Philadelphia.

Saint-Anne Dargassies, S., 1970, Détermination neurologique de l'age foetal néonatal, in: *Journées Parisiennes de Pédiatrie,* pp. 311–326, Editions Médicales Flammarion, Paris.

Scammon, R. E., and Calkins, L. A. 1929, *The Development and Growth of the External Dimensions of the Human Body in the Fetal Period,* University of Minnesota Press, Minneapolis.

Schulte, F. J., 1974, The neurological development of the neonate, in: *Scientific Foundations of Paediatrics* (J. A. Davis and J. Dobbing, eds.), pp. 587–615, Heinemann Medical Books, London.

Schwinn, G., 1954, Untermassig-Geborene des 2. Weltkrieges. Ihre körperliche und geistige Entwicklung nach klinischer Aufzucht, *Z. Kinderheilkd.* **74:**507–518.

Scott, K. E., and Usher, R., 1966, Fetal malnutrition: Its incidence, causes, and effects, *Am. J. Obstet, Gynecol.* **94:**951–963.

Seckel, L. F., and Rolfes, L. J., 1959, Long-range studies of statural growth of the smallest surviving prematures, *Ann. Paediatr. Fenn.* **5:**75–93.

Sempé, M., Tutin, C., and Masse, N. P., 1964, La croissance de l'enfant de O à 7 ans, *Arch. Fr. Pédiatr.* **21:**111–134.

Sinclair, J. C., 1970, Heat production and thermoregulation in the small-for-date infant, *Pediatr. Clin. North Am.* **17:**147–158.

Smith, C. A., 1962, Prenatal and neonatal nutrition, *Pediatrics* **30:**145–156.

Sobel, E., and Falkner, F., 1969, Endocrine and genetic diseases of childhood. Normal and abnormal growth patterns of the newly-born and the preadolescent, in: *Endocrine and Genetic Diseases of Childhood* (L. I. Gardner, ed.), pp. 6–19, W. B. Saunders, Philadelphia.

Spranger, J., Ochsenfarth, A., Kock, H. P., and Henke, J., 1968, Anthropometrische Normdaten im Kindesalter, *Z. Kinderheilkd.* **103:**1–12.

Stein, Z., and Susser, M., 1975, The Dutch famine, 1944–1945, and the reproductive process. I. Effects on six indices at birth. II. Interrelations of caloric rations and six indices at birth, *Pediatr. Res.* **9:**70–76, 76–83.

Sterky, G., 1970, Swedish standard curves for intra-uterine growth, *Pediatrics* **46:**7–8.

Stewart, A. L., and Reynolds, E. O. R., 1974, Improved prognosis for infants of very low birthweight, *Pediatrics* **54:**724–735.

Stoch, M. B., and Smythe, P. M., 1963, Does undernutrition during infancy inhibit brain growth and subsequent intellectual development?, *Arch. Dis. Child.* **38:**546–552.

Stoch, M. B., and Smythe, P. M., 1976, 15-year developmental study on effects of severe undernutrition during infancy on subsequent physical growth and intellectual functioning, *Arch. Dis. Child.* **51:**327–336.

Szénásy, J., 1969, Über die diagnostische und prognostische Bedeutung der Gestaltung des Kopfumfanges bei Säuglingen, *Acta Paediatr. Acad. Sci. Hung.* **10:**345–358.

Tanner, J. M., 1970, 3. Physical growth, in *Carmichael's Manual of Child Psychology,* Vol. I (P. H. Mussen, ed.), pp. 77–155, Wiley & Sons, New York.

Tanner, J. M., 1973, Physical growth and development, in: *Textbook of Paediatrics* (J. O. Forfar and G. C. Arneill, eds.), Livingstone, Edinburgh.

Tanner, J. M., and Thomson, A. M., 1970, Standards for birthweight at gestation periods from 32 to 42 weeks, allowing for maternal height and weight, *Arch. Dis. Child.* **45:**566–569.

Tanner, J. M., and Whitehouse, R. H., 1973, Height and weight charts from birth to five years allowing for length of gestation for use in infant welfare clinics, *Arch. Dis. Child.* **48:**786–789.

Tanner, J. M., Whitehouse, R. H., and Takaishi, M., 1966, Standards from birth to maturity for height, weight, height velocity, and weight velocity: British children, 1965-I and -II, *Arch. Dis. Child.* **41:**454–471, 613–635.

Tanner, J. M., Goldstein, H., and Whitehouse, R. H., 1970, Standards for children's height at ages 2–9 years allowing for height of parents, *Arch. Dis. Child.* **45:**755–762.

Tanner, J. M., Lejarraga, H., and Turner, G., 1972, Within-family standards for birth-weight, *Lancet* **2:**193–197.

Terman, M. L., and Merrill, M. A., 1965, *Stanford–Binet Intelligenz-Test. Deutsche Bearbeitung von H. R. Lückert,* Verlag für Psychologie, Dr. C. J. Hogrefe, Göttingen.

Thomson, A. M., Billewicz, W. Z., and Hytten, F. E., 1968, The assessment of fetal growth, *J. Obstet. Gynaecol. Br. Commonw.* **75:**903–916.

Tizard, J. P. M., 1972, Ernährung in den ersten Lebenstagen, Jahresversammlung der Schweizerischen Gesellschaft für Pädiatrie in Biel.

617

GROWTH DYNAMICS OF
LOW-BIRTH-WEIGHT
INFANTS WITH
EMPHASIS ON THE
PERINATAL PERIOD

Tizard, J. P. M., 1973, Ernährung in den ersten Lebenstagen, *Paediatr. Fortbildk. Prax.* **37:**26–33.

Usher, R. H., and McLean, F. H., 1969, Intrauterine growth of live-born Caucasian infants at sea level: Standards obtained from measurements in 7 dimensions of infants born between 25 and 44 weeks of gestation, *J. Pediatr.* **74:**901–910.

Usher, R. H., and McLean, F. H., 1974, Normal fetal growth and the significance of fetal growth retardation, in: *Scientific Foundations of Paediatrics* (J. A. Davis and J. Dobbing, eds.), pp. 69–80, Heinemann Medical Books, London.

Varloteaux, C. -H., Gilbert, Y., Beaudoing, A., Roget, J., and Rambaud, P., 1976, Avenir lointain de 128 enfants agés de 13 à 14 ans, *Arch. Fr. Pédiatr.* **33:**233–250.

Van den Berg, B. J., Morbidity of low birth weight and/or preterm children compared to that of the "mature." I. Methodological considerations and findings for the first 2 years of life, *Pediatrics* **42:**590–597.

Van Wieringen, J. C., Wafelbakker, F., Verbrugge, H. P., and Haas, J. H. de, 1971, *Growth Diagrams, 1965 Netherlands Second National Survey on 0–24-Year-Olds,* Wolters-Noordhoff Publishing, Groningen.

Walsh, H., 1969, The development of children born prematurely with birth weights of three pounds or less, *Med. J. Aust.* **1:**108–115.

Weber, H. P., Kowalewski, S., Gilje, A., Möllering, M., Schnaufer, I., and Fink, H., 1976, Unterschiedliche Calorienzufuhr bei 75 "low birth weights": Einfluss auf Gewichtszunahme, Serumeiweiss, Blutzucker und Serumbilirubin, *Eur. J. Pediatr.* **122:**207–216.

Wedgwood, M., and Holt, K. S., 1968, A longitudinal study of the dental and physical development of 2–3 year-old children who were underweight at birth, *Biol. Neonate* **12:**214–232.

Westropp, C. K., and Barber, C. R., 1956, Growth of the skull in young children. Part I: Standards of head circumference, *J. Neurol. Neurosurg. Psychiatry* **19:**52–54.

Wigglesworth, J. S., 1966, Foetal growth retardation, *Br. Med. Bull.* **22:**13–15.

Winick, M., 1968, Changes in nucleic acid and protein content of the human brain during growth, *Pediatr. Res.* **2:**352–355.

Winick, M., 1971, Cellular growth during early malnutrition, *Pediatrics* **47:**969–978.

Winick, M., 1973, Relation of nutrition to physical and mental development, *Nutr. Dieta* **18:**114–122.

Winick, M., 1974, Nutrition and nucleic acid synthesis in the developing brain, *Proceedings XIV International Congress of Pediatrics, Buenos Aires,* Vol. 5, pp. 244–251.

Winick, M., and Rosso, P., 1969*a,* Head circumference and cellular growth of the brain in normal and marasmic children, *J. Pediatr.* **74:**774–778.

Winick, M., and Rosso, P., 1969*b,* The effect of severe early malnutrition on cellular growth of human brain, *Pediatr. Res.* **3:**181–184.

Woolley, P. V., Jr., and Valdecanas, L. Q., 1960, Growth of premature infants, *Am. J. Dis. Child.* **99:**642–647.

Yerushalmy, J., 1970, Relation of birth weight, gestational age, and the rate of intrauterine growth to perinatal mortality, *Clin. Obstet. Gynecol.* **13:**107–129.

Ylppö, A., 1919, Das Wachstum der Frühgeborenen von der Geburt bis zum Schulalter. Untersuchungen über Massen-, Längen-, Thorax- und Schädelwachstum bei 700 Frühgeborenen, *Z. Kinderheilkd.* **24:**111–178.

Index